International Scientific-Technical Conference on Innovative Engineering Technologies, Equipment and Materials (ISTC-IETEM-2018)

IOP Conference Series: Materials Science and Engineering
Volume 570

Kazan, Russia
5 – 7 December 2018

ISBN: 978-1-5108-9447-1
ISSN: 1757-8981

Printed from e-media with permission by:

Curran Associates, Inc.
57 Morehouse Lane
Red Hook, NY 12571

Some format issues inherent in the e-media version may also appear in this print version.

Institute of Physics
Dirac House, Temple Back
Bristol BS1 6BE UK

Phone: 44 1 17 929 7481
Fax: 44 1 17 920 0979

techtracking@iop.org

Additional copies of this publication are available from:

Curran Associates, Inc.
57 Morehouse Lane
Red Hook, NY 12571 USA
Phone: 845-758-0400
Fax: 845-758-2633
Email: curran@proceedings.com
Web: www.proceedings.com

Table of contents

Volume 570

International Scientific-Technical Conference on Innovative Engineering Technologies, Equipment and Materials 2018 (ISTC-IETEM-2018)

5–7 December 2018, Kazan, Russia

Accepted papers received: 18 June 2019
Published online: 15 August 2019

Preface

International Scientific-Technical Conference on Innovative Engineering Technologies, Equipment and Materials 2018 (ISTC-IETEM-2018)

Peer review statement

Papers

About waste disposal problem in Russian Federation

T M Abdullin, R G Sabirzyanov, I R Gilmanshin, N F Kashapov, S I Gilmanshina, A I Galeeva and E M Gadirova.....1

Comparative assessment of extraction methods of 3,4- benzopyrene from soils, bottom sediments and sewage sludge

G M Ahmadiev.....6

Additive technology for filter membranes

G M Akhmadiev.....9

Study of process of intensification of dehydration of secondary activated sludge for its utilization as a secondary energy source

A V Akhmerov, A L Osipov, A N Dolgova, G F Faizullina and L S Sabitov.....13

Ultrasonic method for measuring of gas bubbles in liquids

I H A Al-Umari, N F Kashapov, V G Saitkulov and M Fazlyyyakhmatov.....17

Study of the real work of constructions of translucent facade systems

A S Antonov, G N Shmelev, L S Sabitov, N F Kashapov and I A Galimullin.....22

A comparative analysis of machining of parts from 38H2MYuA by cutting tool of different materials

V I Astashchenko and I N Gilazov.....31

Efficiency increase of gas turbine work in the summer period

D S Balzamov, E Yu Balzamova, S R Ibatullin and L S Sabitov.....34

Analysis of the technique of software products for the selection of heat-insulating materials for heat networks pipelines

D S Balzamov, E Yu Balzamova, A I Khaibullina and L S Sabitov.....37

Studying the Effects of Mechanical Loads and Environmental Conditions on the Drive-Axle Performance

A Yu Barykin, M M Mukhametdinov, R Kh Takhaviyev and A D Samigullin.....40

Control of the process of pulsed laser thermal hardening based on emerging potential

D A Bashmakov.....43

Biodegradable fixator for sternum closure after median sternotomy

I Bebnev, R N Kashapov, N F Kashapov and L N Kashapov.....46

Structure formation in materials based on Fe-C-Si and Fe-C-Al systems (cast irons) prepared from dispersed mechanical engineering waste

R Bikulov.....51

The use of technogenic dispersed waste engineering to obtain compact forms of graphite in the materials of the system Fe-C-Al

R Bikulov.....59

Disadvantages of the selective laser sintering technology in the manufacture models for investment casting

J N Dementyeva, R N Kashapov, N F Kashapov and L N Kashapov.....65

Estimation of the state of a technical system with substantially non-linear characteristics

D N Demyanov.....69

Rebuilding of heavy machine tools

Z R Faskhieva and R M Khusainov.....73

Mathematical modeling of heat treatment of three-layered material

V L Fedyaev, V I Khaliulin, A V Belyaev and M Kh Faskhutdinov.....76

Research of a laser welding of thin-shit construction steels by continuous laser radiation

Az T Gabdrakhmanov, AT Galiakbarov, L N Shafigullin and T F Gabdrakhmanova.....82

Study of laser cutting of copper and copper alloys

Az T Gabdrakhmanov, I H Israphilov, L N Shafigullin and T F Gabdrakhmanova.....86

Rotor film reactor

L R Gainullina, V P Tutubalina and L S Sabitov.....90

Technologies for producing heat-conducting carbon foams by method of pitches carbonization under pressure

E R Galimov, E E Sharafutdinova, N Ya Galimova, F I Murataev, V M Samoylov and E A Danilov.....93

Development of technologies for producing heat-conducting syntactic carbon foams with specified operational properties

E R Galimov, E E Sharafutdinova, N Ya Galimova, V M Samoylov and E A Danilov.....104

Improving the efficiency of shaving through the use of wear-resistant coatings

V N Gilman, I P Balabanov and A I Faskhutdinov.....112

Methods of training specialists in the field of sustainable resource management and waste management

I R Gilmanshin, S I Gilmanshina, A I Galeeva, E M Gadirova and I A Galeev.....117

Formation of competencies of the 21st century by means of project activities

S I Gilmanshina, I R Gilmanshin and E V Dyatlova.....120

Digital learning technologies in the University education system

S I Gilmanshina, I R Gilmanshin and K G Bendjukevich.....124

Methods of identifying talent in the field of engineering and natural sciences

S I Gilmanshina, K G Bendjukevich, I R Gilmanshin, V A Glushkova and E M Gadirova.....128

Effects of moisture content on thermal conductivity of thermal insulation materials

A M Gusyachkin, L S Sabitov, A M Khakimova and A R Hayrullin.....131

RANS simulation for the prediction of heat transfer for staggered tube bundle in cross-flow

A I Haibullina, K V Chirukhin, L S Sabitov and A R Hayrullin.....136

Economy of energy conservation

S R Hajieva, E M Gadirova, N Sh Sharifova, S I Gilmanshina and I R Gilmanshin.....141

Research of the impact of low-frequency pulse force of lorentz on the process of motion of conducting liquid

V K Ilyin, D A Koppen, A I Khaibullina, O V Ilyin and L S Sabitov.....145

Simulation as a tool for supporting assembly systems

M Iskandarov and R Khisamutdinov.....152

Technological Support of Plasma Coating with the Plasmatron Power Modulation

A M Kadyrmetov, S N Sharifullin, G A Suhochev and E V Snyatkov.....155

Use of modified molecular sieves in mechanical filters

O R Karataev, V F Novikov, E S Karataeva and A A Kartashova.....159

Research of efficiency of work of heat-mechanical equipment by the thermography method

O R Karataev and E S Karataeva.....162

Development of turning-filtering mechanismof centrifuges

E S Karataeva, V A Lashkov, A A Brodt and M N Akhlyamov.....165

Automation of process of manufacturing products from a mixture of fluids fast polymerized in air using the jet centrifugal method

F G Karikh and L N Shafigullin.....170

Automation of metal and alloy melting processes using spectral analysis data on the composition of exit gases

F G Karikh, L N Shafigullin and G F Muhametzjanova.....174

Description of a complex technical system of urban electric transport from the standpoint of synergistic methodology

N F Kashapov, L S Sabitov, A E Auhadeev, R S Litvienko and I Z Gatiyatov.....178

New methodological approach to the development of the theory of electric traction of urban electric transport

N F Kashapov, L S Sabitov, A E Auhadeev, R S Litvienko and P P Pavlov.....184

Consideration of the influence of the reliability of elements of the urban electric transport system on its bandwidth

N F Kashapov, L S Sabitov, R S Litvinenko, A E Auhadeev and P P Pavlov.....191

The approach to the study of the reliability of the electric transport system of the city as a complex technical system

N F Kashapov, L S Sabitov, R S Litvinenko, A E Auhadeev and I Z Gatiyatov.....198

Modern problems of digitalization of agricultural production

N F Kashapov, M M Nafikov, M KH Gazetdinov, SH M Gazetdinov and A R Nigmatzyanov.....204

Development of digital and robotic agricultural production in the Russian Federation

N F Kashapov, M M Nafikov, Mansur M Nafikov and A R Nigmatzyanov.....211

The effect of tillage with various tools on its agrophysical and agrochemical properties

N F Kashapov, M M Nafikov, A R Nigmatzyanov and V V Medvedev.....214

The state of the technical level of domestic agricultural machinery

N F Kashapov, M M Nafikov, A R Nigmatzyanov and I G Gajnutdinov.....217

Influence of roughness on operational properties of details

I I Khafizov and I G Nurullin.....221

Influence of application of robotic systems on prospects and directions of development of electroerosive processings

I I Khafizov, I G Nurullin and N F Kashapov.....224

Application of the method of electroerosion treatment to improve the quality of the treated surface

I I Khafizov, I G Nurullin and Z B Sadykov.....227

The design of automotive alternator with high efficiency

A A Khafizov, R I Valiev, R S Akhmetshin and Yu I Shakirov.....231

Electric motor with electronic control unit for changing the polarity of the winding to improve reliability and energy efficiency

A A Khafizov, R S Akhmetshin and Yu I Shakirov.....234

Operation of brushes with increased service life on traction machines of electric vehicles

A N Khusnutdinov, E M Khusnutdinova and B M Butakov.....237

Issues related to energy saving in petrochemical sector and methods for their solution

E M Khusnutdinova, I A Konakhina and G R Khamidullina.....242

Current issues of resource and energy savings in the domestic oil and gas complex

I K Kiyamov, L S Sabitov, G I Kabirova, L Sh Akhtyamova and L Sh Iskhakova.....246

Actual issues of digital transformation of the Russian economy in modern conditions

I K Kiyamov, L S Sabitov, G I Kabirova, L Sh Akhtyamova and L Sh Iskhakova.....252

To a question from About modern state and prospects of development ah Russian oil and gas complex

I K Kiyamov, L S Sabitov, G I Kabirova, L Sh Akhtyamova and L Sh Iskhakova.....258

The study of the insulating properties of a new material based on straw and isoprene rubber C-nitrous vulcanization

O R Klyuchnikov, N A Martianov, I K Kiyamov, L S Sabitov, D A Morozov and I R Gilmanshin.....263

Investigation of Properties of Electro-metallized Coatings Deposited by Supersonic Air Stream with Aerosol Fluxing

A V Kolomeichenko, S N Sharifullin, V N Logachev and A A Izmalkov.....269

Flow resistance of ducting with triple-threaded auger inserts in laminar flow regimes

I A Konakhina, E M Khusnutdinova, A I Kadyirov and E K Vachagina.....275

Comparative analysis of friction losses in long pipelines with profiled channel flow area

I A Konakhina, E M Khusnutdinova, G R Khamidullina and E K Vachagina.....279

Engineering education in the system of modern innovation

I A Konakhina, G R Khamidullina, E M Khusnutdinova and N F Kashapov.....282

Features of the formation of interdisciplinary competencies in engineering education

S S Kosmodemyanskaya, I R Gilmanshin and Yu I Azimov.....285

Machine elements evaluation diagnostic devices development using compressed air on crankshaft bearings example

A T Kulakov, E P Barylnikova and A O Kulakov.....289

Designing of diffusion coatings on bimetallic parts (steel, bronze)

V K llyin, O V Ilyin and L S Sabitov.....293

Modifying surface properties of polyamide powders for selective laser sintering

A M Lyakhovich, R N Kashapov, N F Kashapov and L N Kashapov.....298

Features of very heavy crude oil transportation

A O Mamonova, O A Plugatyreva, E M Khusnutdinova and A N Khusnutdinov.....305

Modified gypsum-cement-pozzolanic composites reinforced with polypropylene fibers

R Kh Mukhametrakhimov, A R Galautdinov and I R Gilmanshin.....308

Development of high-strength cast iron for back-up layer of bimetallic products

G F Mukhametzyanova, I R Mukhametzyanov and M S Kolesnikov.....316

Ranking materials technologies by limiting characteristics of heat-resistant alloys and their longevity in the problems of import substitution

F I Murataev, E R Galimov and N Ya Galimova.....320

Substantiation of domestic material and welding technology for improving properties and competitiveness of pyrolysis furnace coils

F I Murataev, E R Galimov and N Ya Galimova.....327

Improving the quality of castings made of non-ferrous metal alloys when casting in metal molds

D L Pankratov and R V Gavariev.....334

The concept of quality management systems and monitoring of their effectiveness using "Lean production" concept

O A Plugatyreva and E M Khusnutdinova.....338

Investigation and explanation of mathematical tooling for accounting non-economic characteristics during the investment project effectiveness' assessing process

A S Puryaev, Zh A Puryaeva, A R Kharisova and A A Puryaev.....342

To determination of dependence of parameters of surface cracks in concrete and reinforced concrete volume elements

O V Radaikin, L S Sabitov, I K KIyamov and N F Kashapov.....346

To estimate the rigidity of bending reinforced concrete elements based on dependence of length macro trays from load

O V Radaikin, L A Sharafutdinov, L S Sabitov, Yu M Strelkov and N F Kashapov.....354

To determination of dependence of parameters of surface cracks in concrete and low-armed wall panels

O V Radaikin, L S Sabitov, I K Kiyamov and N F Kashapov.....360

Selection of thetoothdepthof the ball end mill based on the analysis of the working conditions of the tool

E A Ryabov, R M Khisamutdinov, O I Yurasova and S Y Yurasov.....367

Management of elastic displacements of a milling technological system based on the study of the main deformation axes

A R Sabirov and R M Khusainov.....371

Improving the efficiency of energy output from landfill gas by controlling the morphological composition of MSW

R G Sabirzyanov, N A Nurmatova, A I Galeeva, I R Gilmanshin and S I Gilmanshina.....375

Intelligent management of supply chains and outsourcing in mechanical engineering under "Industry 4.0" conditions

R A Sabitov, G S Smirnova, N U Elizarova and Sh R Sabitov.....381

Improving the process efficiency of helical gears' toothed rims at the stage of pre-production

D T Safarov, A G Kondrashov and I I Khafizov.....387

Bevel gears monitoring methods developmentin the total contact patch terms

D T Safarov and A G Kondrashov.....391

Four-point technique development for measuring spherical surfaces of machine parts

D T Safarov, A G Kondrashov and I I Khafizov.....395

Promising methods for enhancing the performance of boiler plants of centralized, decentralized and individual type of heat supply

T R Safin, I A Konakhina and G R Khamidullina.....399

Project activity as a stage of continuous engineering and technical education

R N Sagitova, I R Gilmanshin and A R Akchurina.....403

Changes in the CoCr alloys surface relief during plasma electrolytic treatment

L R Saleeva, R N Kashapov, L N Kashapov and N F Kashapov.....407

Development of a polymeric fuel tank with high barrier properties

L N Shafigullin, A M Sotnikov, N V Romanova, E S Shabaeva and D R Sarimov.....413

Promising processing technologies of glass-filled polyurethane materials

L N Shafigullin, N V Romanova, A T Gabdrakhmanov, E S Shabaeva and D R Sarimov.....417

The technological stresses of carbon-filled plastics arising during the pressing of prepregs

L N Shafigullin, N V Romanova, D I Israfilov and G D Shakirova.....421

Quantum chemical calculations of the vibrational spectrum of aliphatic nitro compounds. Cluster analysis of torsion oscillations

R M Shaikhullina, G M Hrapkovsky and M M Shaikhullina.....425

Assessment of the impact of technological parameters of hot stamping on the cost of the product

V G Shibakov, D L Pankratov, R S Nizamov and L M Nizamova.....429

Justification of requirements to quality and accuracy of blankfor precision stamping

V G Shibakov, D L Pankratov, R S Nizamov, L M Nizamova and R A Khairullin.....433

Revisiting the calculations techniques of the optimal planning methods for the local authority activities (on the Republic of Tatarstan example)

A M Shikhalev, I A Akhmetova, D P Vorontsov and A F Khananova.....437

The multicriteria estimation of the Russian Federation region districts promising for the logistic objects placement (on the example of the Republic of Tatarstan)

A M Shikhalev, I A Akhmetova, D P Vorontsov, G R Hamidullina, N K Kashapov and O N Rozhko.....440

Experimental study of the work of a temporary structure of a dismountable bar system

G N Shmelev, L I Khaidarov, L S Sabitov, L R Gimranov and N F Kashapov.....443

Numerical study and calculation of the structures of the stands with canopies on the wind impact, taking into account the terrain

G N Shmelev, L S Sabitov, I A Galimullin, L I Khaidarov and A S Antonov.....451

Characteristic features of the cooling capacity of aqueous polymer solutions

T V Shveyova, G F Muhametzyanova, V I Astashchenko, N N Zapadnova and E A Zapadnova.....458

The criterion of non-stationary cyclic fatigue of bodies, taking into account the presence of an incubation period of destruction

Latif H. Talibly.....464

Predicting the time of corrosion damage to a plate with a deep double-sided external undercut under stretching

L H Talibly, F B Imranov and A M Jafarova.....469

Determination of rational parameters of the assembly process of one-piece products which are based on modeling

A M Valiev and D L Pankratov.....475

The prospects of modern quality management approaches integration (on the example of the Republic of Tatarstan)

D P Vorontsov, A S Pirogova, A M Shikhalev, G R Khamidullina, N F Kashapov, A F Khananova and O N Rozhko.....479

Highly effective composite materials based on waste of technogenic sulfur

A A Yusupova, L R Barayeva, R T Akhmetova, A I Khatsrinov and A A Bobryshev.....485

About Modeling Kinematic Parameters of Deposition from Composite Electrochemical Coatings for Restoration of Details for Agricultural Machines

S Yu Zhachkin, S N Sharifullin, N A Penkov and A A Zavrazhnov.....489

Particular properties of operation of stationary lead-acid batteries at power distribution facilities

A D Ziganshin, L S Safiullin, A V Bogatkin, P P Pavlov, I Z Gatiyatov and A E Auhadeev.....492

Minimum density of irrigation of the rotor of a film centrifugal heat and mass-exchange apparatus

N H Zinnatullin, G N Zinnatullina, A I Haibullina, L S Sabitov and N.F. Kashapov.....496

Application of neural networks in object recognition tasks for ADAS systems

R R Ziyatdinov and R A Biktimirov.....505

Performance-based seismic evaluation methods for the estimation of inelastic deformation demands

M A Zubritskiy, O Y Ushakov and L S Sabitov.....508

Account for the contribution of higher modes under system seismic resistance estimation by nonlinear static method

M A Zubritskiy, O Y Ushakov and L S Sabitov.....515

Technological complexes of laser welding and hardening tool

V V Zvezdin, R M Hisamutdinov, R R Saubanov, R R Rakhimov and S S Sirkin.....522

The process control for improving the durability of the cutting edge of the teeth of the cutter at their recovery

V V Zvezdin, R M Hisamutdinov, R R Rakhimov, R R Saubanov and I H Israfilov.....526

Physical and mechanical properties of fiber reinforced gypsum-based composite

R Kh Mukhametrakhimov, L V Lukmanova and I R Gilmanshin.....530

Investigation of mechanical properties of fiber-cement board reinforced with cellulosic fibers

L V Lukmanova, R Kh Mukhametrakhimov and I R Gilmanshin.....536

IOP Conf. Series: Materials Science and Engineering **570** (2019) 011001 doi:10.1088/1757-899X/570/1/011001

IX International Scientific Conference "Innovative engineering technologies, equipment and materials - 2018" has been hosted by the exhibition center "Kazan Fair" within the framework of the 18th international specialized exhibition "Mechanical engineering. Metalworking. Kazan" and 12th specialized exhibition "TechnoWelding" on the 5th and 7th of December, 2018.

The conference program was including 6 sections of work:
Section 1 - Innovative materials, technologies and equipment in mechanical engineering
Section 2 - Innovative developments and economics in engineering
Section 3 - Chemical Engineering
Section 4 - Mathematical and physical modeling of information, technical, technological and management systems and processes
Section 5 - Innovative Welding Technologies in Industry

Participants and guests of the conference discussed the scientific and technical achievements in the field of development, design and manufacture of engineering products. See the results of studies of the country's leading universities in the field of robotics, new materials (including on the basis of production waste), waste management and recycling, mechanical engineering, energy, fuel cells, manufacturing economy.

Presidium of the session "Innovative development of mechanical engineering and economics" headed by Director of the Institute of Engineering Prof. Nail Faikovich Kashapov. The Presidium also included: Deputy. Director of the Institute for educational activities Ildar Hafizov Ilsurovich, Deputy. Director of the Institute for Science Gilmanshin Iskander Rafailevich and Linar Salihzanovich Sabitov.

Research and Innovation have been presented by the staff of the Engineering Institute of KFU, Institute of Naberezhnye Chelny, Institute of Chemistry named after A.M. Butlerov, Kazan State Power Engineering University and Kazan National Research Technical University named after A. N. Tupolev – KAI, Kazan State University of Architecture and Engineering.

Active participation in the conference were graduate students and students of the Engineering Institute. The experience of live communication with scientists, practitioners, representatives of the real economy will allow buduyuschim experts to determine its scientific direction for a successful solution of important problems of import substitution and enhancing competitiveness assigned to the engineering industry of our country.

IOP Conf. Series: Materials Science and Engineering **570** (2019) 011002 doi:10.1088/1757-899X/570/1/011002

Peer review statement

All papers published in this volume of *IOP Conference Series: Materials Science and Engineering* have been peer reviewed through processes administered by the proceedings Editors. Reviews were conducted by expert referees to the professional and scientific standards expected of a proceedings journal published by IOP Publishing.

IOP Conf. Series: Materials Science and Engineering **570** (2019) 012001 doi:10.1088/1757-899X/570/1/012001

About waste disposal problem in Russian Federation

T M Abdullin[1], R G Sabirzyanov[1], I R Gilmanshin[1,2], N F Kashapov[1], S I Gilmanshina[1], A I Galeeva[3] and E M Gadirova[4]

[1]Kazan (Volga Region) Federal University, 18 Kremlyovskaya str., Kazan 420008, Russian Federation
[2] Kazan National Research Technical University named after A N Tupolev, Kazan, Russia
[3]Kazan State Power Engineering University, Kazan, 420066, Russian Federation
[4]Baku State University, AZ 1148, Baku, Azerbaijan

Abstract. The article discusses the problems of waste disposal methods that do not cope with the task of environmentally friendly disposal of municipal solid waste (MSW). The comparison of destruction and thermal destruction methods (plasma gasification in particular) are given.

The collection of MSW without sorting and further disposal (> 90%) leads to land degradation (over 50 000 ha of land are occupied by landfills in Russia). In consequence, an increase in waste disposal to the environment (the amount of solid waste exceeds the capabilities of natural ecosystems). Incineration was achieving the target for a while, but now due to an increase in the amount of waste, its diversity and large accumulation utilization by the methods described above, inevitably carries with it a serious threat to the environment.

Purpose:

- Identification of optimal waste disposal methods in the Russian Federation.

Tasks:

- Analysis of waste management in the Russian Federation
- Comparison of destruction methods

Table 1 and 2 show more detailed statistics.

	Waste availability at the beginning of the reporting year	Waste generation for the reporting year	Receipt of waste from other organizations	Processed waste	Recycled waste	
					Total	of them for reuse (recycling)
Russian Federatio	36,275,388,506.639	6 220 638 673,598	183 466440.526	17,864,225.181	3 248 855	2 053 938 132,872

n					244,48 8	

Table 2

	Wastedisposal	Transfer of waste to other organizations					Availability in the organization at the end of the reporting year
	Total	for processing	Forrecycling	Forneutralization	for storage	for burial	
RussianFederation	Sixteen780168.018	12,908,930,117	148,490669,849	9,932,511,544	73,718,661.092	30,585,229,509	38 073 356 129,573

How much of all the waste is stored in specialized facilities for a period of more than eleven months for utilization , neutralization, disposal as well as the methods used in waste management in the Russian Federation can be found in Figure 1.

Figure 1. Information about waste management in E production and consumption in the Russian Federation for 2017 (in thousands of tons)

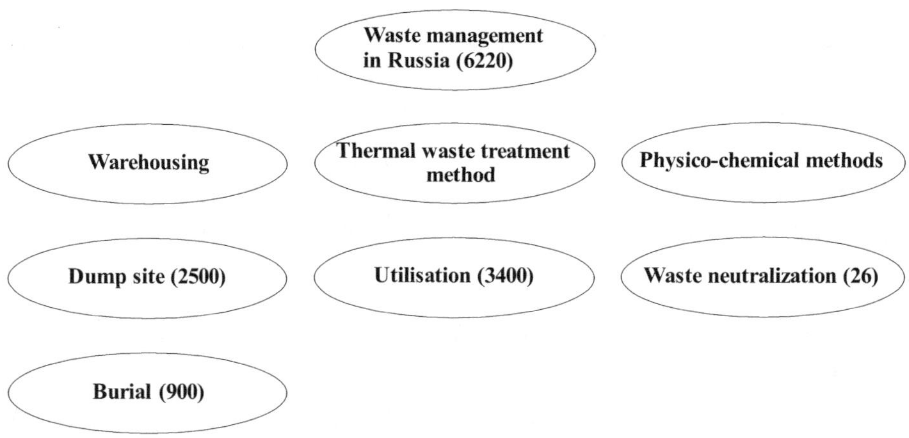

Waste disposal - isolation of waste that is not subject to further disposal in special storage facilities in order to prevent the ingress of harmful substances into the environment. (as amended by Federal Law of December 30, 2008 No. 309-FZ, of December 29, 2014 No. 458-FZ)

Waste disposal - the use of waste for the production of goods (products), works, services, including the reuse of waste, including the reuse of waste for its intended purpose (recycling), their return to the production cycle after appropriate preparation (regeneration), and extraction of useful components for their re-use (recovery). (as amended by the Federal Law dated 12.29.2014 N 458-FZ)

Disposal of waste - reducing the mass of waste, changing its composition, physical and chemical properties (including incineration and (or) disinfection at specialized facilities) in order to reduce the negative impact of waste on human health and the environment. (as amended by the Federal Law dated 12.29.2014 N 458-FZ)

Waste management looks clearly ineffective.

Consider the problems of basic methods of waste disposal and disposal methods in general.

Problems of used methods of waste disposal in the Russian Federation:

• Combustion in special kilns or in open dumps, leading to air pollution from combustion products.

• Burial in burial grounds, which leads to subsidence of the soil and pollution of groundwater, as well as the intensive formation of methane that can explode at the slightest spark

• Storage of garbage in specially designated landfills, which occupy vast territories, creates a threat to air traffic, attracting flocks of birds

• Recycling to secondary raw materials (the problem with this method is the lack of separate garbage collection)

With such problematic methods, waste stocks (availability in the organization at the end of the reporting year 38 073 129 tons.) It is necessary to introduce new methods and technologies.

Figure 2. Waste disposal methods

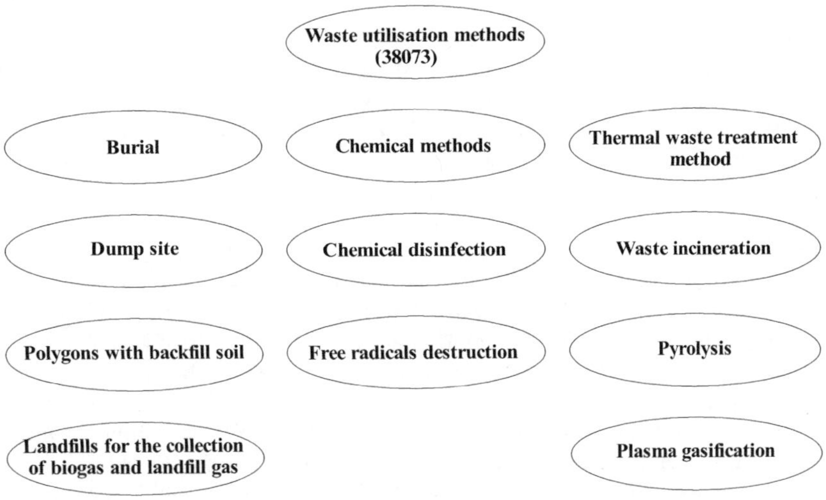

Russia already uses unsuccessfully the burial and storage - it takes and poisons the land for a long time.
Chemical methods are used effectively only to separate fractions and in the presence of mostly undivided garbage - this is a big problem.
Thermal methods are used effectively with a wide morphological spectrum, but not environmentally friendly or costly. The solution for thermal methods and consider them.

The main technological approaches in the implementation of the thermal method are:

- Pyrolysis
- Gasification

IOP Conf. Series: Materials Science and Engineering **570** (2019) 012001 doi:10.1088/1757-899X/570/1/012001

- Direct combution

Direct combustion is an inefficient process of obtaining thermal energy. Higher efficiency is ensured by the use of: some types of wood, solid waste from agro-industrial complex and solid domestic waste, peat. Direct combustion in a steam boiler with the further use of steam in a steam turbine is the traditional way of generating electricity.

The disadvantages of this technology include the high cost of equipment for small power plants with a capacity of less than 1 MW, large dimensions, significant fuel consumption, etc.

Pyrolysis is the process of thermal processing (450-1000°C) of raw materials under conditions of oxygen deficiency, with the possibility of further separation of various fractions of fuels.

The raw material base can be brown coal, shale, peat, coal oil sludge, solid household and industrial waste, wood waste, sludge, hydrolyzed lignin, waste from agro-industrial enterprises (manure, dung, husk, straw), etc.

The decomposition of the organic components of the waste in an oxygen-free atmosphere, after which the resulting concentrated gas mixture (HS) is sent to the afterburning chamber, where the transfer of toxic substances into less harmful ones.

Combustion products contain harmful substances CO and NO2. Which are removed with flue gases and have a harmful effect on the atmosphere and the environment.

Gaseous substances: nitrogen monoxide NO and nitrogen dioxide NO2 are formed in all combustion processes, but more often in the form of oxide.

Sulfur dioxide is formed by the oxidation of sulfur dioxide. The final product of the reaction is an aerosol or a solution of sulfuric acid in rainwater, which acidifies the soil.

Gasification - the transformation of the organic part of solid or liquid fuels into combustible gases during high-temperature (800-120°C) heating with an oxidizer (oxygen, air, water vapor, CO2, or, more often, a mixture of them). The resulting synthesis gas can be used directly to produce thermal or electrical energy or for their simultaneous production by cogeneration using gas generators combined into one energy complex with water boilers or diesel electric generators.

This filter technology is complex without plasma and has a large number of filters.

Plasma gasification - for solving a wide range of tasks, one of which is the conversion of any type of waste, including bio- waste, hazardous waste, into electricity / synthetic fuel (diesel fuel, ethanol) and other useful materials (a ton of waste is 1–1.3 MW / hr of electric power), which provides complete (more than 99.99%) degradation of all components of MSW in inert atmosphere at high temperatures (up to 5000^{0}C) transfer of the organic part of the waste into a gaseous state, free from toxic components and nonorganic part - in the glass-like slag compound.

For example, from 1 ton of solid waste consisting of 73% of MSW, 7% of rubber waste (mainly car tires) and 20% of coal get 40 kg of resin used in the boiler room, and 1500-2000 m^3 of wet gas. The use of plasma gasification is inherently associated with the Kyoto agreement to reduce the impact on the human atmosphere. The impact on nature and man is 10–15 times lower than world standards for MPCs.

Table 3

Comparison of thermal destruction methods

	Plasma gasification	Conventional gasification	Burning
Destruction	99%	90%	70%
Temperature	More than 1200C	Less than 1200C	Less than 1000C
The presence of resins and furans	Not	thereis	Lot

at the exit			
Ash content	1 %	10 %	30% toxic
Waste type	Any	In addition to certain inorganic materials	In addition to certain inorganic materials
Sorting	Not required	Is required	Is required
Flue gas emissions	Low	Average	High

Conclusion.

Plasma gasification is the most suitable method among the investigated ones. Because the hazardous cause to environment of it is much less due to decrease of the dioxins and complex compounds formation and ash residue comparing to the combustion of MSW.

Literature

[1] Shashkov IV 2015*Utilization and recycling of solid household waste* Publishing house FGBOU VPO "TSTU"

[2] Gilmanshin IR, Abdullin TM *Of medical equipment / medical systems /*

[3] Federal Law of December 30, 2009 No. 384*Technical Regulations on the Safety of Buildings and Structures*

[4] Sabirzyanov RG, Gilmanshin IR, Kashapov NF, Gilmanshina SI, Galeeva AI, Galeev IA, Krainova DR 2018 *Material and technical base of the laboratory for the thermal destruction of production and consumption waste for the training of specialists in the field of waste management* IOP Conf. Ser .: Mater. Sci. Eng. 412 **012061**

[5] Federal Law of July 22, 2008 No. 123-FZ *Technical Regulations on Fire Safety Requirements*

[6] Ershov A G, Shubnikov VL 2014 *Thermal waste disposal: theory and practice, myths and legends* Journal of MSWNo 5 pp47 – 52

[7] *On Environmental Protection* 2002 Federal Law No 7

[8] Kashapov NF, Gilmanshina SI, Galeeva AI, Galeev IA, Krainova DR 2018 IOP Conf. Ser .: Mater. Sci . Eng . 412 012001

[9] On Production and Consumption Wastes 1998 Federal Law No 89

[10] Bilitowski B 2007 Waste incineration: the German experience Municipal solid wasteNo 1 Pp 47 - 49

[11] On the sanitary and epidemiological well-being of the population1999 No 52

[12] Federal Law of November 23, 2009 No. 261-FZ "On Energy Saving and on Increasing Energy Efficiency and Amending Certain Legislative Acts of the Russian Federation"

[13] Gilmanshin IR, Kashapov NF, Gilmanshina SI, Kraynova D R,Sabirzyanov R G 2018 Thermal destruction of production and consumption wastes Proceedings of the IXth International Scientific and Technical Conference "Innovative machine-building technologies, equipment and materials - 2018 "(IMTOM-2018) pp 242-246

[14] Law of the Russian Federation 1992 No 2395-I "On Subsoil"

[15] Water Code of the Russian Federation 2006Federal Law No 74-FZ

[16] Gilmanshin I R, Kashapov N F, Gilmanshin S I, Kraynova D R,Abdullin TM 2018 Methods of research of plasma gasification technology of medical waste and production and consumption waste Proceedings of the IX International Scientific and TechnicalConference "Innovative machine-building technologies, equipment and materials - 2018 "(IMTOM-2018) pp 237-241

[17] Bernadiner M N, Bernadiner IM 2011 High-Temperature Processing and Disposal of Liquid, Pasty and Solid Industrial and Medical Wastes Ecology and industry of Russia pp 19 - 21

IOP Conf. Series: Materials Science and Engineering **570** (2019) 012002 doi:10.1088/1757-899X/570/1/012002

Comparative assessment of extraction methods of 3,4-benzopyrene from soils, bottom sediments and sewage sludge

G M Ahmadiev

Branch of Kazan Federal University in Naberezhnye Chelny, Kazan, Russia, Kremlevskaya Str., 18

GMAhmadiev@kpfu.ru, ahmadievgm@mail.ru

Annotation. The article is devoted to the scientific basis and principles of extraction methods of extraction of 3,4- benzopyrene (B(a)P) from soils, bottom sediments and sewage sludge, which can be used in monitoring studies, particularly in assessing the impact on the environment; in assessing the degradation changes in the soil; in monitoring the state of soils, as well as in assessing the natural and disturbed ecosystems from human activity in general; in the environmental regulation of soil pollution and other degradation processes. The data obtained may be useful in the development of environmental maps of priority pollutants. The results of studies of 3,4- B(a)P from soils, B(a)P sediments and sewage sludge can be used for zoning, actual and forecast parameters; in forecasting the environmental consequences of economic activity in the area; in assessing the risk of disasters; in assessing the feasibility and effectiveness of soil remediation contaminated with oil and petroleum products, during the environmental assessment, certification, certification of the territory or economic object of industrial and construction, industrial and transport, agro-industrial and industrial and commercial complexes.

Introduction

Nowadays a large-scale technospherization takes place in many countries all over the world, transforming them into urban areas. Accumulated waste is the cause of soil contamination with contaminants of biological and chemical origin [1].

Scientific substantiation, search of ecologically safe, economically justified, acceptable express methods and their application in practice as an innovative method of extraction of 3,4-B(a)P from sewage sludge is not a fully solved problem in applied science. Polyaromatic compounds (PAHs) occupy leading places among the common harmful factors in the habitat of plants, animals and human substances-ecotoxicants, in the complex socio-economic damage caused to the environment [2]. Most of them are formed in the enterprises engaged in oil and gas processing, industrial and energy complex, as well as the reason for their formation is road transport, chemical and oil refining industry. 3,4- B(a)P, due to its high toxicity, contaminant, carcinogenic and other unexplored and unknown properties, comes out on top in the priority pollution of the environment and in the negative impact on plants, animals and human health indicators.

3,4-B(a)P has a general structural chemical formula and is a representative of the class of polycyclic aromatic hydrocarbons (PAHs) – organic compounds, which are characterized by the presence of three or more condensed benzene rings in the chemical structure. The melting and boiling point of 3,4-B(a)P is 177°C and 456°C, respectively [2]. It is established that the studied organic chemical substance and the extracted compound has toxic, carcinogenic and mutagenic properties. Its

IOP Conf. Series: Materials Science and Engineering **570** (2019) 012002 doi:10.1088/1757-899X/570/1/012002

relative content is about 80% of the total amount of PAHs in polluted soils of urbanized and technosphere territories of Russian regions. Domestic and foreign scientific researchers have experimentally established that 3,4- B(a)P is a universal indicator for assessing the presence of the entire list of polycyclic aromatic hydrocarbons (PAHs) [2], so its presence is the main and determining criterion and indicator of the degree of well-being or environmental pollution of PAHs. Predisposing environmental factors and external sources of 3,4- B(a)P and they can be caused by natural, man-made, chemical and biochemical, microbiological and metabolic processes in soils, while its amount does not exceed the maximum permissible concentration (MPC), which is subject to mandatory control and is for the soil-20 ng/g [4].

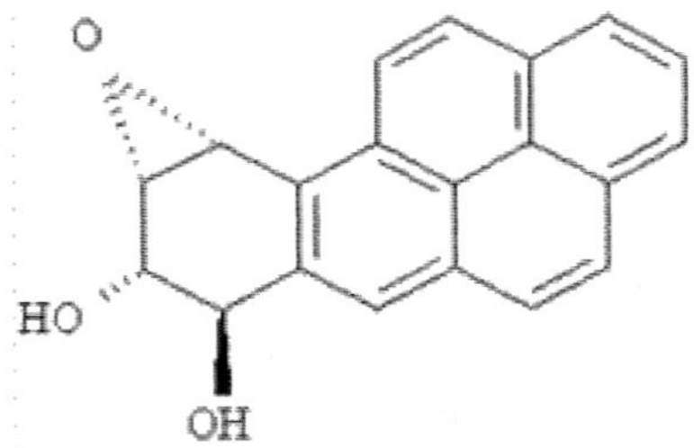

Figure 1. 3,4-B(a)P chemical formula

The aim of the paper is to provide a scientific justification and assessment of methods for extracting 3,4-B(a)P from soils, sediments and sewage sludge, and a comparative assessment and methodological analysis of the use of various well-known 3,4- B(a)P extraction methods from soils, sediments and sewage sludge.

Materials and Methods of Research.

Experiments and test results on the extraction of PAHs from soils with various organic solvents (methylene, acetone, acetone-hexane mixture) has shown that the most effective extraction of PAHs is with acetone for 20 minutes at a temperature of 120°C [5]. Therefore, one methylene chloride [6], 2-methoxyethanol [7], or a mixture of hexane, methylene chloride and methanol can be used for these purposes [8]. In another known method, alcohol was used as an extractant for PAHs (Polycyclic aromatic hydrocarbons) [8], while 25 ml of 2-methoxyethanol was added to 25 mg of soil. And then the mixture was placed on a rotary converter (shaker) and stirred for 15 minutes. The resulting extract after filtration was parted on a separating funnel with silica gel and eluted with pentane for subsequent determination by high-performance liquid chromatography (HPLC).

A method for the extraction of polycyclic aromatic hydrocarbons from the soil using acetone in the Soxlet apparatus is well known [10]. The share of extraction of the desired substance in the laboratory is insignificant and equals to 40% of the total PAH content in the soil obtained by hexane extraction. According to a similar method [11], PAHs were extracted with acetone and purified on a stainless steel column with a chromosphere.

Research results and discussion.

Nowadays a well- known method of extraction of polycyclic aromatic hydrocarbons from solid samples (RU 2018110 IPC [9] G01N 21/25; 1994-08-15) is environmentally safe and economically justified and effective in scientific and practical terms [12].

Currently, the method of saponification, consisting in the hydrolysis of organic compounds to extract 3,4- B(a)P from soil samples, bottom sediments and sewage sludge, using an ultrasonic bath, hexane extraction, purification by solid-phase extraction and HPLC analysis, is most often used. Yaroschuk A.V., Maksimenko E.V., Borisenko N. I. in the article "The Development of Extraction Methods of Benz (a) pyrene from Soil" published in the journal, cite their own research results. // Izvestiya vuzov. Severo-Kavkazskiy region. Estestvennyie nauki. Prilozhenie. - 2003. - №9 - P. 44-47.

IOP Conf. Series: Materials Science and Engineering **570** (2019) 012002 doi:10.1088/1757-899X/570/1/012002

The authors in a positively protected patent for inventions №2018110 (RU) described "A Method for Extracting Polycyclic Aromatic Hydrocarbons from Solid Samples" more detailed [12].

The device developed and proposed by the authors [16], is prepared by the state standard procedure (GOST 17.4.4.02-84 Methods of sampling and preparation of samples for chemical, bacteriological, helminthological analysis; Introduced 19.12.84.- 9 P. - Group 019) [14]. The content of Benz(a)pyrene in the analyzed samples was calculated by the method of external standard (absolute calibration) for standard solutions of different concentrations [16,17].

Conclusion.

Thus, this paper sets out the scientific basis and principles for assessing methods for extracting 3,4-Benz(a)pyrene from soils, bottom sediments and sewage sludge.

Reference

[1] Ilnitskiy A P 1975 *Carcinogenic hydrocarbons in soil, water and vegetation* Carcinogens in the environment pp 53-71

[2] GOST 17.4.4.02-84 *Methods of selection and samples preparation for chemical, bacteriological, helminthological analysis*; Introduced 19.12.84. - 9 S. - Group 019

[3] Andraves E F 1984 *Organic chemical reactions in supercritical water* J.Chromatogr v 284 No 2 pp 487-90

[4] Maseda C e.a. *Alkylation Reactions in Near-Critical Water in the Absence of Acid Catalysts*J. Chromatogr v 490 No 2 pp 313-27

[5] Aner V et.al. 1990 *Innovations in chemical reaction processes using supercritical water: an environmental application to the PAH extraction* Anal Chim. Acta V 237 No 2 pp 451-57

[6] Warzecha L 1993 *The analysis of PAHs nitrogen derivatives in the sewage sludges of upper Silesia, Poland* Chem. Anal. No 3 v 38 pp 303-13, 571-73

[7] RU №2018110 *Method of extraction of polycyclic aromatic hydrocarbons from solid samples MOD 9 G01N 21/25* Date of publication 1994-08-15.

[8] Barnabas I J e.a. 1995 *Extraction of polycyclic aromatic hydrocarbons from highly contaminated soils using microwave energ*y Analyst V 120 No 7 pp 1897-1904

[9] Hartmann R 1996 *Subcritical Water Extractor for Mars Analog Soil Analysis Astrobiology* Inst. J.Environ. Anal.Chem V 62 No 2 pp 161-73

[10] Gennadiev A N 1996 *Geochemistry of polycyclic aromatic hydrocarbons in rocks and soils* M.: Izd-vo MGU 196 p

[11] Lancar L.T. e.a. 1999 *Extraction PAH from soils* J.Chim. Phys. Et phys.-chim. Boil V 96 No 3 pp 352-63

[12] Bartulevich Ya, Yagov G V 2001 *Analytical methods for the determination of PAHs in environmental objects* Drinking water No 6 pp 11-14

[13] Yaroschuk A V, Maksimenko E V and Borisenko N I 2003 *Development of methods for extracting Benz(a)pyrene from soil* Izvestiya Vuzov. North Caucasus region. Natural science. Application No 9 pp 44-47

[14] A A Galkin, V V Lunin 2005 *Water in sub- and supercritical states is a universal medium for chemical reactions* Advances in chemistry T 74 No1 pp 24-40

[15] GN.2.1.7.2041-06 *Maximum permissible concentrations (MPC) of chemicals in soil 2006* Hygienic standards. M.: Federal center of hygiene and epidemiology of Rospotrebnadzor 11 p

[16] Kolesnikov S I, Sushkova S N, Minkina T M and Mandzhieva S S 2013 *Method of extraction of 3,4-Benz(a)pyrene from soils, bottom sediments and sewage sludge.* Patent for invention, No 2 485 109

[17] *Strategy of development of the industry on processing, utilization and neutralization of waste of production and consumption for the period till 2030. Order Of The Government Of The Russian Federation on January 25, 2018.* No 84-r

IOP Conf. Series: Materials Science and Engineering **570** (2019) 012003 doi:10.1088/1757-899X/570/1/012003

Additive technology for filter membranes

G M Akhmadiev
Kazan Federal University, Naberezhnye Chelny Institute (branch), 423812, Russia, Naberezhnye Chelny, 10 u Syuyumbike Ave.
GMAhmadiev@kpfu.ru, ahmadievgm@mail.ru

Abstract. Separation methods using membranes are used in many industries, in particular, in the field of environmental protection for the production of drinking water and the treatment of industrial liquid waste, in the chemical industry, petrochemistry, pharmaceuticals, food industry and in the field of biotechnology. In the framework of this research, new methods for membranes manufacturing are proposed, which in comparison with the known technical solutions from the point of view of reliability and performance, make it possible to obtain more results and also provide greater variability in the choice of substrate shapes and channels inside the substrate. The most effective technology is a method of manufacturing a membrane for tangential filtration of a fluid, and this membrane contains: a substrate having a three-dimensional structure and formed by a monolithic ceramic porous body in which paths are made to circulate the fluid being filtered, and at least one separating filter layer deposited on the wall of the circulation paths.

Introduction

At present, a more detailed analysis of the latest scientific research shows that in the field of additive technology, an ordinary technical and technological process will take place in the future, however, the science needs to cope with a lot of challenges before it reaches these boundaries. Therefore, state and departmental scientific organizations ought to take urgent and concrete measures. The problems of additive technologies of the present time are described by the following economic factors: the high cost of polymeric materials, the complexity of the process at all stages of technical and technological solutions [1,2,3]. The purpose of this work is a more detailed analysis of the known technical and technological solutions using additive technology to obtain filtration membranes. The work presented relates to a number of filter elements and the essence consists in the preparation of a membrane and includes a modified tangential filtration of a fluid. At the same time, the indicated membrane can and must also contain in the technical solution: a substrate having a three-dimensional structure and formed by a monolithic ceramic porous body, in which the main paths are made for the circulation of the filtered fluid and the separation filter layer deposited on the wall of the circulation paths, in which the three-dimensional structure of the substrate is obtained by means of additive technology.

Material and research methods

According to the well-known technology, the proposed three-dimensional structure of the substrate is cut into sections using a computer-aided design program, while these sections are created alternately in the form of elementary layers located one above the other and sequentially functionally connecting them between themselves.

Research results and discussion

IOP Conf. Series: Materials Science and Engineering **570** (2019) 012003 doi:10.1088/1757-899X/570/1/012003

The known technology for obtaining filtering membranes demands the strict observance of the following two steps. At the first stage, a uniform continuous layer of powder of constant thickness is intended to form a ceramic porous body over an area exceeding the cross-sectional pattern of the specified formed porous body at the reservoir level. In the second stage, in accordance with the pattern and determined for each layer, a part of the applied material is locally compacted to create an elementary layer. At the same time, these two stages are repeated in order to simultaneously associate the elementary layer formed in this way with the previous layer, gradually increasing the required three-dimensional shape. Separation methods using membranes are used in many industries, in particular, in the field of environmental protection for drinking water and processing industrial liquid waste, in the chemical industry, petrochemistry, pharmaceuticals, food industry and in the field of biotechnology [4; 5; 6]. The developed filtering membrane forms a selective barrier and, under the action of a transfer force, ensures the passage or retention of certain components of the medium or liquid being processed. The passage or retention of the components may depend on their size relative to the pore size of the membrane, which behaves like a filter. Depending on the pore size of the filtration membrane, these technologies are called microfiltration, ultrafiltration or nanofiltration. There are membranes with different structures and textures. As a rule, the membranes contain a porous substrate, which provides the mechanical strength of the membrane and gives it shape, that is, determines the filtering area of the membrane. One or several layers with a thickness of several microns are applied to this substrate, which provide separation and are called separating filter layers, separating layers or active layers. During the separation, the filtered fluid flows through the separation layer, then this flowable fluid spreads through the porous texture of the substrate, after which it moves to the outer surface of the porous substrate. This part of the treated fluid that has passed through the separation layer and the porous substrate is called the filtrate (permeate) and is collected in a collecting chamber enclosing the membrane. The other part is called a concentrate (retentate) and is most often sent back to the treated fluid at the inlet of the membrane using a closed circulation loop. Usually, a substrate of the required shape is made first by extrusion, then it is sintered at a certain temperature and for a sufficient time to provide the required hardness, while maintaining the necessary open and interconnected porous texture in the resulting ceramics. This method forces to perform straight channels, inside which a separating filter layer or separating filter layers are then applied and sintered [1]. Thus, using this technology, the filtering membrane is subjected to at least two sintering operations. Organic binders added to the paste prior to sintering, burn completely during sintering of the substrate. This method is satisfactory, but it is still necessary to achieve an increase in the profitability of the method and the maximum limitation of conceptual flaws. In the framework of this research, new methods for the membranes manufacturing are proposed, which make it possible to obtain a gain in terms of reliability and performance in comparison with the known technical solutions, and also provide greater variability in the choice of substrate shapes and channels inside the substrate.

Conclusion

The most efficient, economically and laboriously, technology is a method of manufacturing a filtering membrane for tangential filtration of a fluid, while this membrane contains:

- a substrate having a three-dimensional structure and formed by a monolithic ceramic porous body in which the paths for the circulation of the filtered fluid are made, and
- at least one separation filter layer deposited on the wall of the circulation paths, in which the three-dimensional structure of the substrate is obtained by additive technology. Next, the three-dimensional structure of the substrate is dissected into sections using a computer-aided design program, while these sections are created alternately in the form of elementary layers located one above the other and successively interconnected by repeating the following two stages, in which:

1) a uniform continuous layer of powder of constant thickness is applied, intended for the formation of a ceramic porous body on an area exceeding the cross-sectional pattern of the specified formed porous body at the reservoir level;

IOP Conf. Series: Materials Science and Engineering **570** (2019) 012003 doi:10.1088/1757-899X/570/1/012003

2) in accordance with the pattern and determined for each layer, a part of the applied material is locally compacted to create an elementary layer, while these two steps are repeated in order to simultaneously associate the elementary layer formed in this way with the previous layer, gradually increasing the required three-dimensional shape.

References

[1] Gogins Mark A, Uik Tomas M 2006 *Uzel vozdushnogo fil'tra i sposob fil'tracii* [Air filter assembly and filtering method] Patent na izobretenie 2274482

[2] Behrris Marti A, Gogins Mark A and Uik Tomas M 2006 *Fil'truyushchij ehlement (varianty), fil'truyushchaya sistema (varianty) i sposob fil'tracii (varianty)* [Filter element (options), filtering system (options) and filtering method (options)] Patent na izobretenie 2278722

[3] Gillingehm Gehri R, Gogins Mark A and Uik Tomas M 2007 *Konstrukciya fil'tra i sposob fil'tracii (varianty)* [Filter design and filtering method (options)] Patent na izobretenie 2301105

[4] Dojen Villi, Molenbergs Bart 2017 *Fil'truyushchij ehlement, vypolnennyj s vozmozhnost'yu promyvki obratnym potokom* [Filter element adapted for backwashing] Patent na izobretenie 2610075

[5] Lekosh Filip, AnketijZHerom 2018 *Sposob izgotovleniya fil'truyushchih membran pri pomoshchi additivnojtekhnologii i poluchennye membrany* [A method of manufacturing a filtering membrane using additive technology and the resulting membrane] Patent na izobretenie 2663768

[6] http://kak-bog.ru/additivnye-tehnologii-chto-eto-takoe-i-gde-primenyayutsya

IOP Conf. Series: Materials Science and Engineering **570** (2019) 012004 doi:10.1088/1757-899X/570/1/012004

Study of process of intensification of dehydration of secondary activated sludge for its utilization as a secondary energy source

A V Akhmerov[1], A L Osipov[1], A N Dolgova[1,2], G F Faizullina[1], L S Sabitov[1,3]

[1]Kazan State Power Engineering University, 51 Krasnoselskaya street, Kazan, 420066, Russian Federation
[2]Rostov State Transport University" 344038 Rostov-on-Don, pl. Rostov Rifle Regiment of the People's Militia, 2
[3]Kazan Federal University, 18 Kremlyovskaya street, Kazan, 420008, Russian Federation

akhm@mail.ru

Abstract. The processes of intensification of dehydration of secondary activated sludge for its utilization as a secondary energy source afterword are considering in this article. Experiments have shown the effectiveness of using asymmetric period pulse action method to accelerate the necessary processesand hydro transport of condensed sludge.

1. Introduction

Modern water treatment plants have multi-stage water purification technology, including the latest biotechnologies, during operation of which a large amount of secondary sludge is formed. Secondary sludge is advisable to consider as a valuable secondary energy resource [1-3], which can be used to produce pyrolysis gas for the subsequent generation of thermal energy. However, for burning in pyrolysis furnaces, it is necessary to remove excess moisture from the activated sludge.

Active sludge is an unstable heterogeneous biological system, so the issue of removing excess moisture using traditional technologies as centrifugation, and various filtration methods, flotation, etc. do not give satisfactory results. The most common traditional method of removing moisture is settling of waste sludge in large-diameter tanks, but settling is a very slow process and there is difficulty in evacuating the settled sludge. This method allows to achieve only a slight dehydration to 98.5%. Such high humidity is due to the fact that the process takes place in a practically stationary environment, and a large amount of "bound" water in the sludge itself counteracts further dehydration.

The new approach involves mathematical modeling of the three main processes in the apparatus: sewage settling, sewage filtration, and evacuation of the polydisperse phase of condenced sludge [5-10].

The purpose of this work is to study the fast dehydration of activated sludge to find technological solution. That is possible only after a series of laboratory tests, such as determining the rheological

IOP Conf. Series: Materials Science and Engineering **570** (2019) 012004 doi:10.1088/1757-899X/570/1/012004

properties of sludge at different humidity, selecting the optimal pulsation modes (frequency, amplitude, duty cycle), etc.

2. Instruments and methods

The object of the study was the return sludge from the water treatment plant in Naberezhnye Chelny (the time between sampling and laboratory test - 4 hours) with a flow rate of 6000 m3 / day and the mass content of dry matter is 6-10 g / l. The measurement of the dry matter content of sludge was carried out by filtering on a blue ribbon paper filter and drying of the precipitate at a temperature of 105 ° C. The result was 8.6 g / l. Next, to study the properties of the return sludge, the installation of a vibrating screen with a No. 1 sieve (cell size 0.3 mm) was used, the filtration time was 6 minutes, the colloidal phase with a mass of 31.2 g / l was selected. According to visual observations, most of the sludge left with water. When using sieve No. 2 (cell size 0.056 mm), the filtration time was 6 minutes, the colloidal phase with a mass of 98.6 g / l and a water content of 94% is isolated. When visually observed, the filtered water contained almost no suspension. Thus, it was established that the studied heterogeneous biological system is well filtered, and the apparatus should choose the largest cell in the range of 0.056 mm-0.300 mm while maintaining satisfactory characteristics of the filtered water (suspended particles, transparency, color, etc.).

The static sedimentation of sludge in a glass vessel was investigated; the sample volume was 1 liter. As a result of studies conducted after 2 h. 25 min. settling silt surfaced i. nitrification of activated sludge occurred. It can be concluded that the results of static settling are unsatisfactory, therefore, methods should be used that increase the efficiency of settling, for example, pulsating technologies.

The study of pulsating sedimentation was carried out on a laboratory bench, with a U-specimen, with a cross section of both knees of the apparatus 80x80 mm.

3. Results

The content of dry residue in condensed sludge from the bottom from the bottom of the U-sample is 18.5 g / l, from the bottom of the control tank -17 g / l, the volume of the original sludge is V = 4 l. Pulse frequency 15 pulse / min. The duration of the experiment was 3 hours and 37 minutes, the results are presented in figure.1.

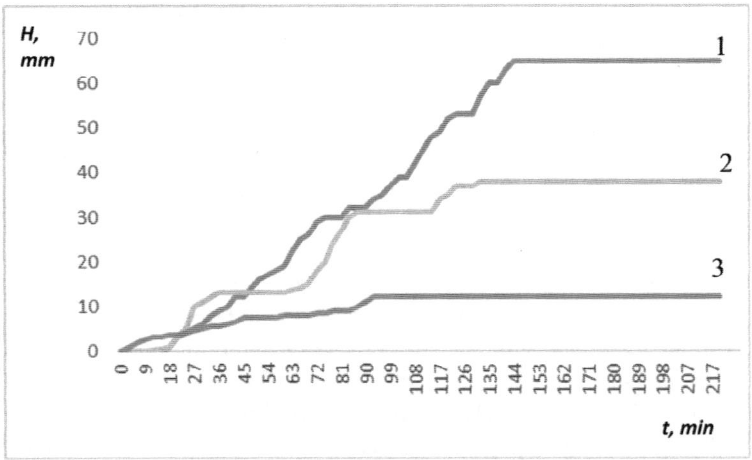

Figure 1. Dependence of the height of the column of clarified liquid on the time of exposure to biomass when pulsation is applied: 1 is the height of the column of clarified liquid in the pipe of the load , mm; 2- height of the clarified liquid column in the discharge pipe, mm; 3- height of the clarified liquid column in the measuring bowl, mm.

A study on the rotational effects (imposition of centrifugal forces, vibration) on the spent activated sludge. The imposition of centrifugal forces increases the deposition rate by 43%. The results are presented in Fig. 2

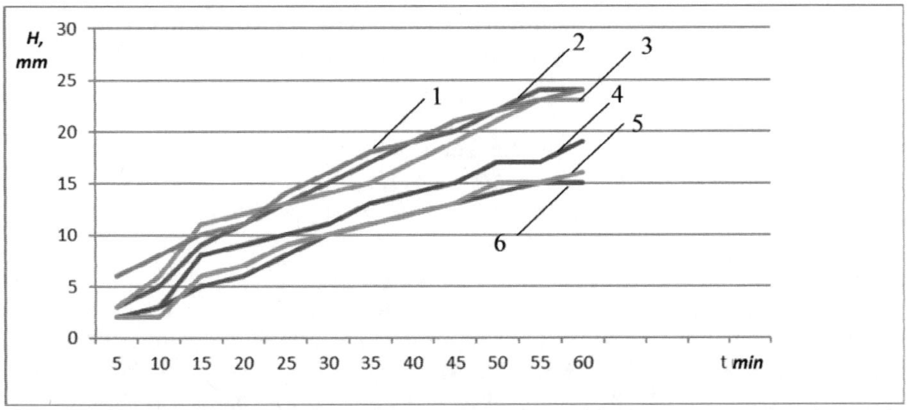

Figure 2. Dependence of the height of the column of clarified liquid on the time of exposure to biomass when applying centrifugal forces: 1- height of the column of clarified liquid at a rotation speed of 33 rpm, mm (1st experiment); 2 - the height of the column of the clarified liquid at a rotation speed of 78 rpm. mm (3rd experience); 3 - the height of the column of clarified liquid at a rotation speed of 45 rpm, mm (2nd experiment); 4,5,6 - height of the column of clarified liquid with static sedimentation, mm in the 1st, 2nd, 3rd experiments, respectively.

A study of vibration effects on spent activated sludge has been carried out. Impacts on the biomass of sludge were carried out using pulsed-oscillatory movements of the vibrating apparatus.

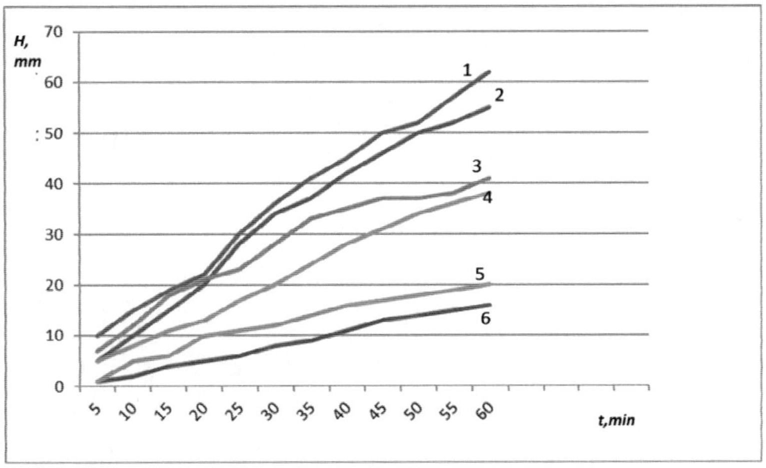

Figure 3. Dependence of the height of the column of clarified liquid on the time of exposure to biomass: when vibration 1 is the height of the column of clarified liquid under vibration exposure, mm (1st experiment); 3 - height of the clarified liquid column at a rotation speed of 78 rpm, mm (3rd experiment) 4- height of the clarified liquid column at a rotation speed of 33 rpm, mm (2nd experiment); 2,6,5 - height of the column of clarified liquid with static sedimentation, mm in the 1st, 2nd, 3rd experiments, respectively.

4. Conclusion

When pulsation is applied to the deposition zone of activated sludge, the process of water separation accelerates significantly (about twice) (in an hour - the water level is 110-120 mm with pulsation, 45-50 mm - without). Rotation (imposition of centrifugal forces) and vibration significantly accelerate (twice) the process of water separation (an hour later - the water level is 20 mm during pulsation, 10 mm - without). Low-frequency vibrations also contribute to sludge settling.

IOP Conf. Series: Materials Science and Engineering **570** (2019) 012004 doi:10.1088/1757-899X/570/1/012004

According to the results of the above experiments, it is advisable to combine rotational and pulsation stimulation of sludge sedimentation at frequencies close to the intrinsic resonant frequencies of the settler.

References

[1] Williams FA 1973 *Theory of Combustion* Moscow Science 29
[2] Pomerantsev In 1973 *Fundamentals of the practical theory of combustion* St. Petersburg Energy 264
[3] Kolyuzhny C 2006 *Wastewater treatment* Moscow Science 479
[4] Federal program “Integrated program for the development of biotechnology in the Russian Federation for the period up to 2020”: approved. Decree of the Government of Ros. The Federation of April 24, 2012. No. 1853p-P8. M .: 2012
[5] Guryanov AI, Marchenko GN 2014 *Industrial instrumental and technological systems* VNSTU Bulletin 17 13 269-273
[6] Guryanov AI, Rosenzweig AK 2005 *Conjugate modeling and design of pulsation apparatus* Kazan KGEU 200
[7] Guryanov AI, Egorov NI, Kagan SZ 1980 *Development and research of a pulsation coaxial extractor* Chemical industry 4 pp 232-34
[8] Guryanov AI, Kosterin AV, Sigal P A 1995 *Method of GASPAKA carrying out a continuous mass exchange process* Pat. 2050917 of the Russian Federation, BI number 36
[9] Ivanov B N, Guryanov A I, Gumerov A M 2009 *Wave Processes and Technologies for Oil* Kazan: Publishing House "FAN" 168
[10] Guryanov A I, Moryashov A A 1987 *Investigation of the marginal productivity and crushing of the dispersed phase in an extractor with asymmetric pulsation of flows*

IOP Conf. Series: Materials Science and Engineering **570** (2019) 012005 doi:10.1088/1757-899X/570/1/012005

Ultrasonic method for measuring of gas bubbles in liquids

I H A Al-Umari[1], N F Kashapov[1], V G Saitkulov[2] and M Fazlyyyakhmatov[1]
[1] Kazan (Volga Region) Federal University, 18 Kremlyovskaya str., Kazan 420008, Russian Federation
[2] Kazan National Research Technical University named after A.N. Tupolev, 31 K. Marksa str., Kazan 420111, Russian Federation

alumari1986@mail.ru

Abstract. The paper proposes to use an ultrasonic method to measure the content of gas bubbles in liquids. The task is relevant in various applications. Industrial ultrasonic analyzer STOK-101 was used to implement the method.

1. Introduction

Currently, a large number of methods and apparatus for the control of various impurities in liquids have been developed [1-4]. Oil products, wastewater and other liquids can act as the object of control. As pollutants and impurities can be solid particles, drops of other liquids. The analyzers are imperfect and have limited parameters.

Existing methods for determining the content of liquid droplets and suspended particles in water can be divided into the following categories:

- Laboratory experiments (colorimetric, gravitational, etc.);
– Methods to analyzing both laboratory and field conditions (UV, fluorescence);
- Acoustic measurement methods.

All of the above methods do not allow automatic real-time monitoring. In addition, the errors of most of these methods are quite large.

Of all these methods, acoustic or more precisely ultrasonic are the most promising.

2. Materials and methods

The method is based on reflection from air bubbles or other ultrasound defects.

The distance between the receiver and the defect, the type of defect (sphere, hole having a spherical bottom, a hole with a flat bottom, a plane, a side hole of a cylindrical shape), the filling of the defect (air or other gas) affect the amplitude of the reflection of the ultrasound signal from the defect. They have a greater impact on the reflection strength of the signal amplitude. The calculation of echo amplitudes for reflection from a sphere or a hole with a spherical bottom of diameter d is carried out as follows:

The attenuation of the signal amplitude when reflected from a sphere or a hole with a spherical bottom with a diameter *d* can be represented as [5]:

$$\frac{A_1}{A_0} = \frac{Sd}{4\lambda r^2}\exp(-2\delta r), \tag{1}$$

where A_0 is the emitted signal amplitude, A_1 is the received signal amplitude, S is the piezoelectric transducer area, d is the diameter of sphere, r is the distance from transducer to defect, λ is wavelength, δ is the attenuation coefficient.

The signal attenuation described by the first part of the formula is called diffraction. The second part of the formula (exp $(-2\delta r)$ describes the scattering of ultrasound in the medium.

The formula is valid if the diameter of the sphere or cylinder is greater than half the wavelength. If the defect diameter is less than half wavelength, then the reflection amplitude with a decrease in diameter during diffraction rounding of the defect in waves will decrease faster. Therefore, defects that are smaller than the wavelength are difficult to detect. For reflection from holes with a flat or spherical bottom, the formulas remain relevant, even if the diameters of the holes are smaller than the wavelength, since the side walls are an obstacle to bending around.

It is assumed that each point in the sphere is considered a secondary ultrasound emitter. The ultrasonic reflection can be represented as:

$$\left|\frac{P_1}{P_0}\right| = \frac{\lambda^2}{S}|I|^2 A_c, \tag{2}$$

where P_0 is the amplitude of the pressure tensor component emitted by the transducer, P_1 is the amplitude of the pressure tensor component received by the transducer, $I = \frac{S_a}{\lambda x}$ is function characterizing the delay in the far zone, $A_c = \frac{d}{4\lambda}$ dimensionless coefficient characterizing the reflectivity of the sphere.

In fact, there is reflection from a defect even if its diameter is smaller than the wavelength. In the region $d \ll \lambda$, with decreasing d, the amplitude decreases much faster ($A_c \approx 4.3d^3/\lambda^3$) than for $d > \lambda$.

In the case of a sphere with a diameter of significantly shorter wavelength:

$$\left|\frac{P_1}{P_0}\right| = \frac{\lambda^2}{S}\left|\frac{S}{\lambda x}\right|^2 \frac{4.3d^3}{\lambda^3} = \frac{4.3Sd^3}{\lambda^3 r^2}. \tag{3}$$

The dependence of the signal amplitude on the bubble volume is:

$$\left|\frac{P_1}{P_0}\right| = \frac{4.3Sd^3}{\lambda^3 r^2} = \frac{25.8S}{\pi\lambda^3 r^2}V, \tag{4}$$

where V is the bubble volume.

Multiplying the numerator and denominator of expression (4) by the density of the substance we get:

$$\left|\frac{P_1}{P_0}\right| == \frac{25.8S}{\pi\lambda^3 r^2 \rho}m, \tag{5}$$

where m is the bubble mass.

The amplitude of the ultrasound signal reflected from the bubble in the electrolyte is directly proportional to its mass at $r \ll \lambda$.

The real reflected signal also depends on the ratio of the acoustic resistances of water and gas. The reflection coefficient is determined by the formula:

IOP Conf. Series: Materials Science and Engineering **570** (2019) 012005 doi:10.1088/1757-899X/570/1/012005

$$R = \frac{Z_B - Z_E}{Z_B + Z_E}, \tag{6}$$

where Z_E is the acoustic impedance of the electrolyte, Z_B is the acoustic impedance of the gas bubble.

3. Experimental results and discussion

When emitting an ultrasonic signal in the form of continuous wave and receiving reflected signals, it is necessary to eliminate the induced signal to the receiving transducer. Practically, this cannot be done. One solution to this problem is a method of exciting a transmitting transducer with a pulsed signal, and receiving reflected pulses after some time from such a calculation so that at the time of their arrival, the amplitude of the induced signal from the sounding pulse is negligible. It was proposed to implement such a signal as a product of a single function, sinusoid and exponent:

$$f(t) = \gamma(t) \cdot \exp(-\frac{t}{\tau}) \cdot \sin(\omega \cdot t). \tag{7}$$

Moreover, the attenuation coefficient of the exponential component τ is chosen so that the noise level from the probe signal at the time of arrival of the pulses reflected from oil droplets and solid particles is minimal.

The study of the attenuation of ultrasound in the liquid allowed us to choose the frequency range of ultrasonic vibrations ~ 4 MHz, taking into account the fact that 90% of the suspended particles have a size not exceeding 5 μm in diameter.

The ultrasonic signal coming to the receiving transducer is complex and consists of a set of single pulses having an exponential shape. These pulses are reflected from gas bubbles, suspended liquid and solid particles and carry information about their mass content in wastewater. However, it is impossible to determine their content by simply measuring the amplitude of the signal. To solve this problem, a method based on comparing the reflected signal with the reference voltage and counting the number of pulses exceeding each level is developed.

Figure 1 shows the schematic diagram of the device by which the reflected signals from the air bubbles in the liquid are measured. The main elements of the device are: 1 is generator connected by a high-frequency output with a transmitting piezoelectric transducer, 2 is converter separately-combined type, placed in the body 3. The receiving piezoelectric transducer 4, connected to the input of the amplifier 5, the output of which is directed to the input of the controlled comparator 6. The output of the latter is connected to the logic input of the microcontroller 7, which is connected to the synchronizing output of the generator 1. The logic outputs of the microcontroller 7 are connected to the inputs of the controlled comparator 6. Data analysis is performed from a computer connected to the microcontroller via a USB port.

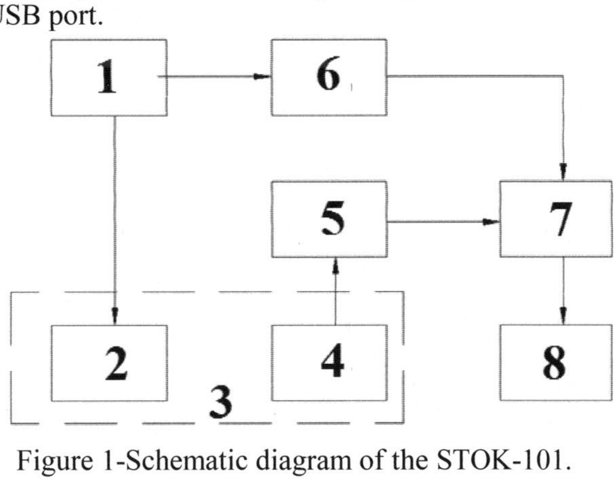

Figure 1-Schematic diagram of the STOK-101.

The main difficulty of the measurements is that with a large number of air bubbles (more than 100) there is an overlay of reflected signals. To solve the problem, an algorithm is used, which consists in summation of signals depending on the digitized reference voltage levels in the range of 8-120 mV. The number of reflected pulses was given by the formula:

$$Q = a + bM + cM^2 + dM^3, \tag{8}$$

and the coefficients A, B, C, D were determined by the least squares method based on the number n - by the total number of measurements:

$$Q_i = a + bM_i + cM_i^2 + dM_i^3, i = (1; n) \tag{9}$$

where Q_i are values of the total content of suspended gas bubbles and solid particles determined from the samples of wastewater by certified method (laboratory analysis) at the time of measurements by ultrasonic method. Taking into account the measured value M and previously defined coefficients A, B, C, D determine the content of suspended impurities in the liquid.

When conducting metrological tests, the device was calibrated for two samples, and then control measurements were carried out. As a result, the relationship between the values measured by ultrasound and the results of laboratory analysis of the samples was obtained. The correlation coefficient is 0.998. The measurement range is from 20 to 300 mg/l. The lower limit of the range is caused by noise from the probing signal.

When choosing the design of the ultrasonic oscillator, various options were considered and a generator with a shock excitation circuit was selected .The sensor is designed in such a way that it can be removed and installed back without stopping the movement of water (water pressure can reach 3 ATM.). The maximum distance of the sensor from the operator up to 700 m. In the receiver of ultrasonic oscillations installed single-stage amplifier with emitter follower input and output. To align the receiving piezoelectric element with the input of the amplifier, a transformer with a transmission coefficient 4 is installed, the amplifier gain is 160. To ensure stable amplification, the signal on the transformer is inverted. The receiver is equipped with a microcontroller that performs primary mathematical processing and transmits information to a computer for further processing, display and storage of measurement results. For microcontroller and computer developed algorithms and written programs. Information on telemetric communication channels can be transmitted to the Central control panel. The main parameters of the analyzer STOK-101: measurement Mode-flow, the frequency of emitted ultrasonic vibrations-3.4 MHz, measurement Error no more than 15 %.

The measuring unit of the analyzer is assembled in a cast sealed case of impact-resistant polystyrene, inside which there are two cast aluminum blocks. There is a generator in one block and a receiver in the other. All the cables are connected through cable glands. Thus, the dust-moisture protection of the analyzer is provided. The electronic components on the circuit Board receiver and generator, encapsulated and organized galvanic isolation based on optical couplers. This achieved the intrinsic safety of the electrical circuit of the device. Analyzer STOK-101 refers to the associated intrinsically safe electrical equipment group II according to GOST R 51330.13.

4. Conclusions

The article was devoted to the development of the ultrasonic method for measuring of the electrolyte condition in the discharge burning process. In this article, the mathematical model for describing gas bubbles in the electrolyte is proposed. The experimental setup of the measuring apparatus is proposed.

References

[1] Fazlyyyakhmatov M 2017 *J. Phys.: Conf. Series* **789** 012005
[2] Sagdiev R K, Denisov E S, Evdokimov Yu K, Fazlyyyakhmatov M G and Kashapov N F 2014 *IOP: Conf. Ser. Mater. Sci. Eng.* **69** 012012
[3] Denisov E S, Temyanov B K, Sagdiev R K and Fazlyyyakhmatov M G 2014 *IOP Conf. Series: Mater. Sci. Eng.* **69** 012014

IOP Conf. Series: Materials Science and Engineering **570** (2019) 012005 doi:10.1088/1757-899X/570/1/012005

[4] Evdokimov Yu K, Temyanov B K and Fazlyyyakhmatov M G 2014 *IOP Conf. Series: Mater. Sci. Eng.* **69** 012013
[5] Ermolov I N 1991 *Nerazrushayuschij kontrol'. Akusticheskie metody kontrolya* [Non-destructive control. Acoustical methods of control] (Moscow: High School) p 283

Study of the real work of constructions of translucent facade systems

A S Antonov[1], G N Shmelev[1*], L S Sabitov[2,3] , N F Kashapov[2] and I A Galimullin[1]

[1] Kazan State University of Architecture and Engineering, 1 Zelenaya street, Kazan, 420043, Russian Federation

[2] Kazan (Volga region) Federal University, 18 Kremlevskaya street, Kazan, 420008, Russian Federation

[3]Kazan State Power Engineering University, 51 Krasnoselskaya street, Kazan, 420066, Russian Federation

gn.smelev@mail.ru

Abstract. The issues of actual operation of structures, the most common stay-brace-and-cross-beam translucent facade systems used in the fencing of buildings and structures are studied, taking into account the spatial rigidity of the elements of the system. The existing building codes of the Russian Federation, in the absence of specialized methods for calculating such systems, offer static calculation schemes for the supporting frame without taking into account the other elements of the system, including the translucent modules themselves, which do not take into account the behavior of the actual structure and often leads to an increase in the cross section of the bearing elements and their value. Thus, in order to study the actual work and bearing capacity of the structures of translucent facade systems, numerical and experimental tests were carried out with modeling of the actual operating conditions, studies were performed for a fragment of the facade system, the stress-strain state of the system structures was analyzed taking into account their spatial work. Results were obtained on the actual operation of the structures, and behavioral features were revealed under the action of critical forces, which must be taken into account when further calculating and designing such systems.

1. Introduction

In recent years, facade systems, the structures of which include translucent elements, have become widespread in the Russian Federation as enclosing structures of buildings and structures. The most common of these are standard stay-brace-and-cross-beam systems. These systems are a frame in the form of, usually, aluminum cross-sections, on which translucent fences are installed, with fastening to bearing brackets, which in turn are anchored to the building foundation. General view and examples of the use of such facade systems are presented in figures 1 and 2.

IOP Conf. Series: Materials Science and Engineering **570** (2019) 012006 doi:10.1088/1757-899X/570/1/012006

Figure 1. Fragment of the facade using the stay-and-cross-beam system. YAWAL FA 50N HI (Poland)

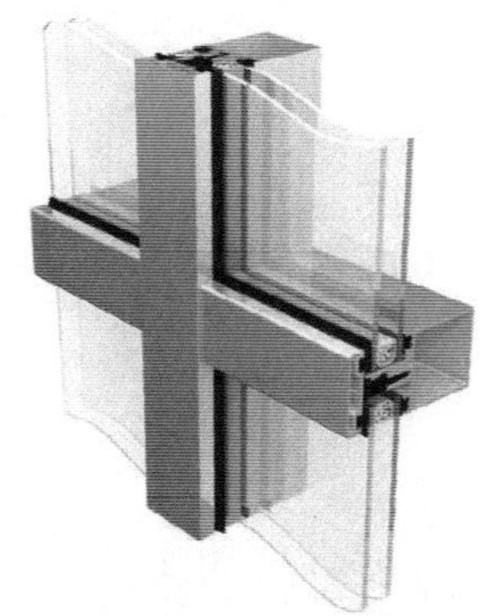

Figure 2. Fragment of the stay-and-cross-beam system

The main part of research in the field of translucent façade structures is aimed at studying the energy efficiency of these systems, however, when designing and performing static calculations at the design documentation development stage, there is a problem of correctness of using calculation methods with standardized calculation schemes in albums of technical solutions of manufacturers that do not take into account the specifics of work real design. Thus, in order to study the real work and determine the stress-strain state of the structural elements of translucent facade systems, it is necessary to carry out numerical and experimental tests with simulation of the actual operating conditions and taking into account the spatial work of the system elements to obtain results on the actual operation of the structures, to identify behavioral features under the action of critical forces that need to be considered in the further calculation and design of such systems.

2. Numerical studies

2.1. Numerical studies using existing calculation methods

Static calculations of the stay-brace-and-cross-beam of facade systems according to the existing building codes and methods of calculation consist of the separation of the supporting frame from the enclosing structures themselves. In order to determine the stress-strain state of the supporting structures of the stay-brace-and-cross-beam facade system, a fragment of the facade of the building with the most common dimensions and cross-sections of the supporting frame elements was selected. The fragment represents stay-braces - vertical load-bearing elements of the system from aluminum alloys, cross-beams - horizontal load-bearing elements of the system from aluminum alloys, these elements of the system perceive loads, from translucent modules installed on them, loads from wind and icy influences, efforts from loads from the supporting frame are transmitted to the construction base, usually, the ends of the monolithic reinforced concrete floor slabs of the building, through the supporting brackets made of aluminum alloys or steel.

The design scheme of a fragment of the facade system is modeled in the SCAD-Office, the skeleton scheme is set as auxiliary bars with hinged fastening at the edges, both for vertical stay-braces (detachment by bearing brackets mounted on a building foundation) and for horizontal auxiliary bars cross-beams (auxiliary bars cross-beams are hinged to the continuous rod of the bars). A simplified design of the auxiliary bar stay-brace is shown in Figure 3, the design diagram of the investigated fragment of the facade system is shown in Figure 4.

IOP Conf. Series: Materials Science and Engineering **570** (2019) 012006 doi:10.1088/1757-899X/570/1/012006

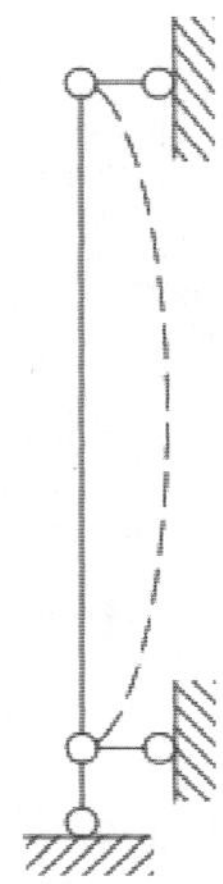

Figure 3. Simplified design of auxiliary bar-stay-brace

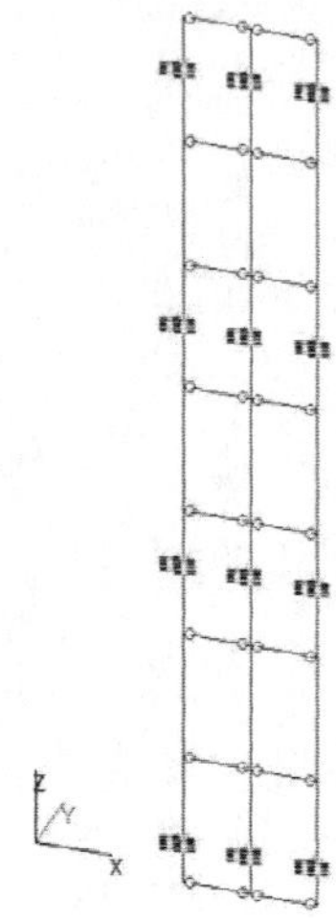

Figure 4. The design scheme of the studied fragment of the facade system

The elements of the auxiliary bar of the system are given appropriate stiffness and fixation. To the design scheme, constant loads from the own weight of the supporting frame were taken into account, which are taken into account by the software package, from the weight of translucent modules and temporary loads from wind and ice effects. The calculation was made on the first and second groups of limit states, as well as on the design requirements for these structures. According to the results of the calculation, a picture of the stress-strain state of the adopted sections of the supporting frame elements was obtained, so the stress values in the auxiliary bar elements — stay-braces, the auxiliary bar — cross-beams and bearing brackets do not exceed the design resistances of the materials of the products, the maximum deflections do not exceed the maximum allowable. However, the values of the adopted sections of the auxiliary bar - stay-brace do not meet the design requirements of the flexibility conditions in the glazing plane of the facade, which is unacceptable for further safe operation. To fulfill the requirements of the conditions of flexibility, measures are needed to detach the elements of the auxiliary bar - stay-brace, which is unacceptable in panoramic glazing, since the aesthetic qualities of the facade of the building are violated, or an increase in the cross-section of the auxiliary bar - stay-brace is necessary, which is also not the best solution, since this decision will lead to a significant increase in the cost of the entire facade of the building. The solution to this problem may be taking into account the remaining elements of the facade system in the design model, including the translucent modules themselves and the nodes from the junction with the supporting frame elements. Consideration of this issue is presented in subsection 2.2 of this article.

2.2. Numerical studies taking into account the spatial work of elements of the facade system

In order to determine the bearing capacity and deformability of the structures of the system of translucent facades, taking into account the spatial work of all elements of the system, numerical studies were carried out in the LIRA-SAPR software package. A finite element model of a facade system fragment has been created. The joints between the elements are modeled according to their actual stiffness. Nods of the base of the bracket are also modeled - parts of the monolithic reinforced concrete structures of the building frame with the corresponding characteristics, anchors and other structures that affect the operation of the system. Characteristics of materials are set according to the design and actual solutions of the investigated structures. Element materials are assigned in nonlinear formulation. The design scheme of the fragment under study, taking into account the remaining elements of the facade system, is presented in figure 5 (a and b).

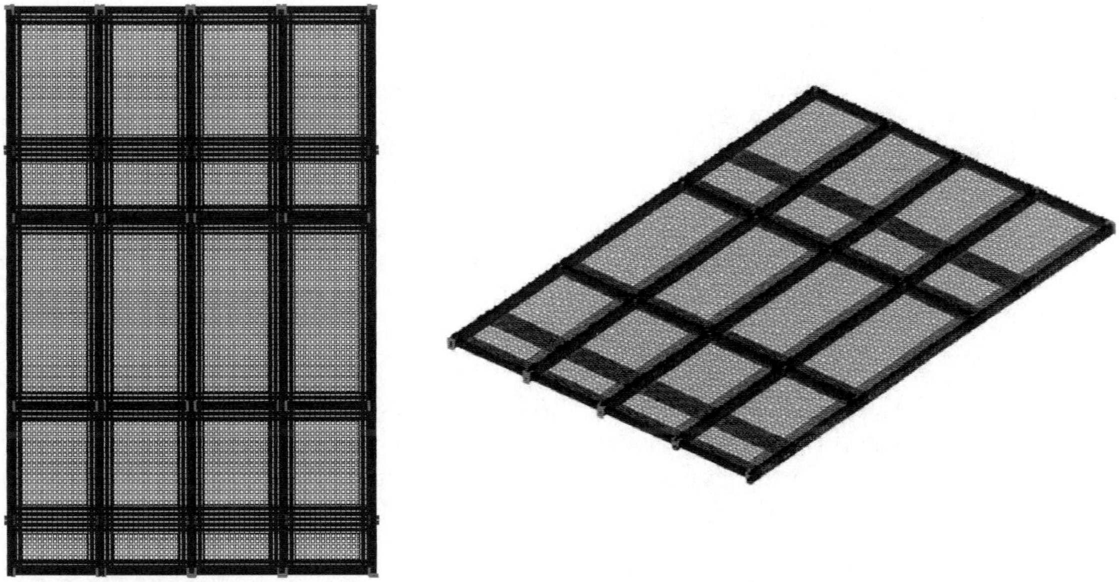
(a) (b)

Figure 5. General view of the calculated scheme of the fragment under study with the entered elements that are not taken into account when calculating according to the existing calculation methods.
(a) – top view, (b) – isometry

After the assignment of appropriate materials, stiffness characteristics, hinges and connections, the grid of three-dimensional finite elements of the model is divided. The contact surfaces and zones of greatest probable stresses are additionally detailed by a grid of finite elements of a smaller size. To the design scheme, constant loads from the own weight of the fragment structures were taken into account by the program complex, as well as temporary loads from wind and ice effects.

According to the results of calculations, a picture of the stress-strain state of the system structures was obtained, the values of stresses and strains did not exceed the maximum permissible values. In order to establish disk stiffness and determine areas of buckling, taking into account the work of the other elements of the system, additional calculations were made with application of single loads to obtain single movements reflecting the spatial work of facade systems of this type, as well as calculations with application of loads whose values are equal to the critical force for constructions of racks taking into account the spatial work of the fragment. The values of the critical force for each of the uprights were calculated on the basis of calculations of total and local stability, taken into account in the LIRA-SAPR software package. According to the results of the calculations, the zone of the most probable loss of stability of the fragment is determined - from the fastening by the bracket to the upper free end of the extreme stay-brace. Loss of stability stay-brace without the inclusion of the remaining elements of the system (single stay-brace) has a zone of total loss of stability in the middle of the span between the brackets in the glazing plane due to insufficient flexibility of the cross-section of the stay-brace. The values of the critical force for the last stay-brace, taking into account the inclusion of the remaining elements of the system, are 1500 kg, which is more than three times higher than the values of the critical force when calculating without including the other elements, i.e. single stay-brace. The results of determining the critical force, zones of instability and the overall stress-strain state are presented below in figures 6-7.

IOP Conf. Series: Materials Science and Engineering **570** (2019) 012006 doi:10.1088/1757-899X/570/1/012006

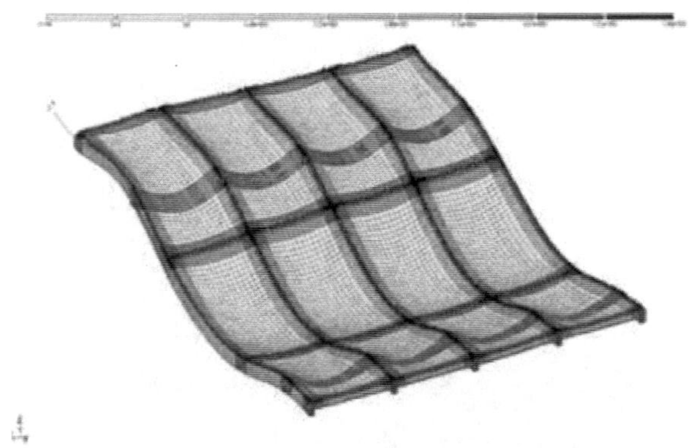

Figure 6. Mosaic of equivalent stresses under the action of the critical force on the last auxiliary bar - stay-brace. General view with glazing

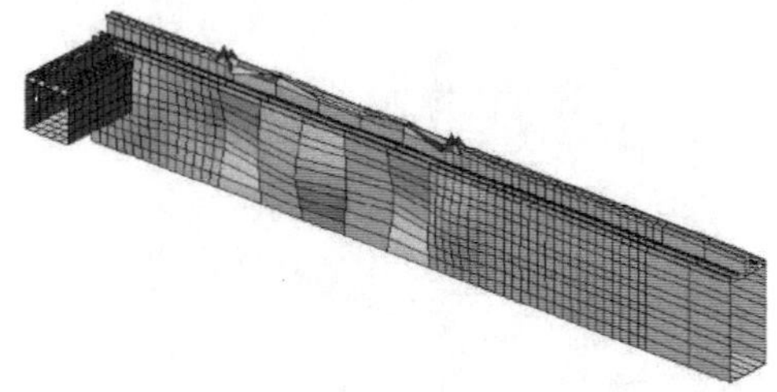

Figure 7. The form of loss of stability of the last auxiliary bar area - stay-brace under the action of critical force

According to the results of numerical studies of the structure of a fragment of a translucent facade system, taking into account the spatial work of all elements of the system, a picture of the stress-strain state of the system elements was obtained, the values of stresses and strains do not exceed the maximum allowable. According to the results of additional studies, the zone of the most probable loss of stability of the fragment was determined - from fastening with a bracket to the upper free end of the last stay-brace. Loss of stability of the stay-brace without the inclusion of the rest of the system elements (single stay-brace) has a zone of total loss of stability in the middle of the span between the brackets in the glazing plane due to insufficient flexibility of the cross-section of the stay-brace. The value of the critical force for last stay-brace is determined taking into account the inclusion of the rest of the system's elements, which is 1500 kg, which is more than three times higher than the values of the critical force when calculating without including the other elements, i.e. single stay-brace.

Thus, when other elements of the system are included in the work, the form of loss of stability of the elements of stay-brace changes, to determine the actual form of loss of stability (in the plane or from the glazing plane) it is necessary to conduct experimental studies in real conditions of operation of this system. With the help of the conducted numerical studies, the picture of the stress-strain state of the system elements was determined for their further control during field tests.

3. Experimental studies

3.1. Methods of testing a fragment of the facade system

To determine the actual bearing capacity, deformability and determine the form of the loss of stability of the stay-brace fragment of the system of translucent facades, experimental studies were conducted. Tests of a fragment of the facade system were carried out in laboratory conditions with the observance of all the parameters of the same design and actually installed at construction sites. A technique has been developed for testing a facade fragment for a critical force, determined from the results of numerical studies, applied as a vertical component of the load of its own weight of modules of a translucent facade to the extreme stay-brace fragment.

The test fragment 1 is mounted in the base 2 through the carrier brackets 3 using anchor fastening. The critical force is transmitted to the extreme stay-brace 4 through a specially designed load transfer system 5, which uniformly transfers the load on the stay-brace to the center of gravity, taking into account the remaining elements of the system fragment. The magnitude of the load is fixed by an electronic dynamometer 6 installed in the circuit. The direction of the force generated by the power winch 7, rigidly fixed to the reinforced concrete base with an anchor 8, is shown in the diagram. The load on a fragment of the facade system was supplied by steps constituting $N \approx 1/10–1/20$ of the

IOP Conf. Series: Materials Science and Engineering **570** (2019) 012006 doi:10.1088/1757-899X/570/1/012006

calculated critical force. When testing loading was carried out step by step with measurement at each step of:
- tensioning with strain gauges 9, installed on the stay-brace and the double-glazed window in the places of the highest rated voltages with the loss of stability, through an automated block of hardware and software measuring instruments;
- movements using linear displacement measuring instruments (three hour-type indicators 10 installed in the areas of calculated buckling and in the area of maximum calculated movements). The scheme of the test bench of the facade system fragment is presented in figure 8 (a and b).

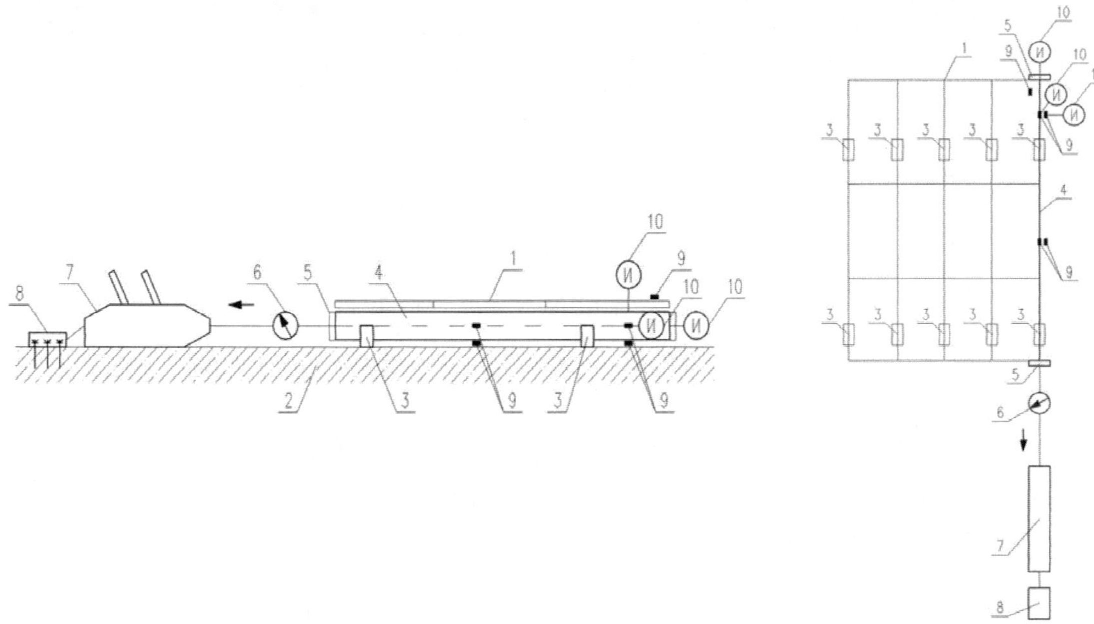

(a) (b)

Figure 8. General view of the test bench of a fragment of the facade system.
(a) – side view, (b) – top view

General view of the test bench of the system fragment is presented in figures 9-12.

Figure 9. General view of the tested fragment of the facade system

Figure 10. The area of calculated buckling of a fragment of the facade system. Installed strain gauges and dial gauges

IOP Conf. Series: Materials Science and Engineering **570** (2019) 012006 doi:10.1088/1757-899X/570/1/012006

Figure 11. D4 and D5 strain gauges installed on the auxiliary bar stay-brace in the area of the calculated buckling under the action of the critical force without taking into account the work of the other elements

Figure 12. Automated block of hardware and software measurement

3.2. Test results of a facade system fragment by stresses and strains

According to the test results, the values of stresses from strain gauges D1-D5 installed on a fragment of the facade system, and the values of deflections and movements of the stay-brace fragment of the facade system according to the readings of hour-type indicators under the effect of the calculated critical force were obtained. The readings of the D2 and D3 sensors installed in the zone of calculated loss of stability of the auxiliary bar - stay-brace begin to differ significantly when approaching the critical force, which is a sign of the beginning of buckling, similar to indications of indicators No. 1 and No. 2 installed in the zone of calculated loss of stability auxiliary bar - stay-brace. The values for sensor D3 and indicator No. 1 installed on the lower and upper surfaces of the stay-brace, respectively, are higher, therefore, stability is lost from the glazing plane. Graphs of stress and deflection versus load are presented in figures 13 and 14.

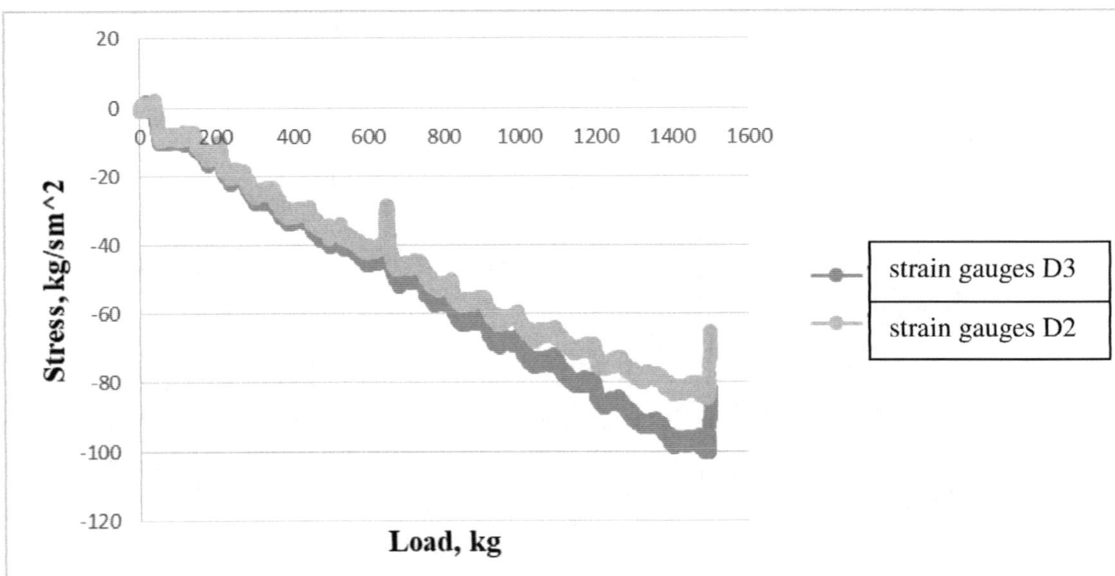

Figure 13. Stress measurement results for strain gauges D2 and D3 in the area of the calculated buckling of the auxiliary bar - stay-brace fragment of the facade system. Graph of stress versus load.

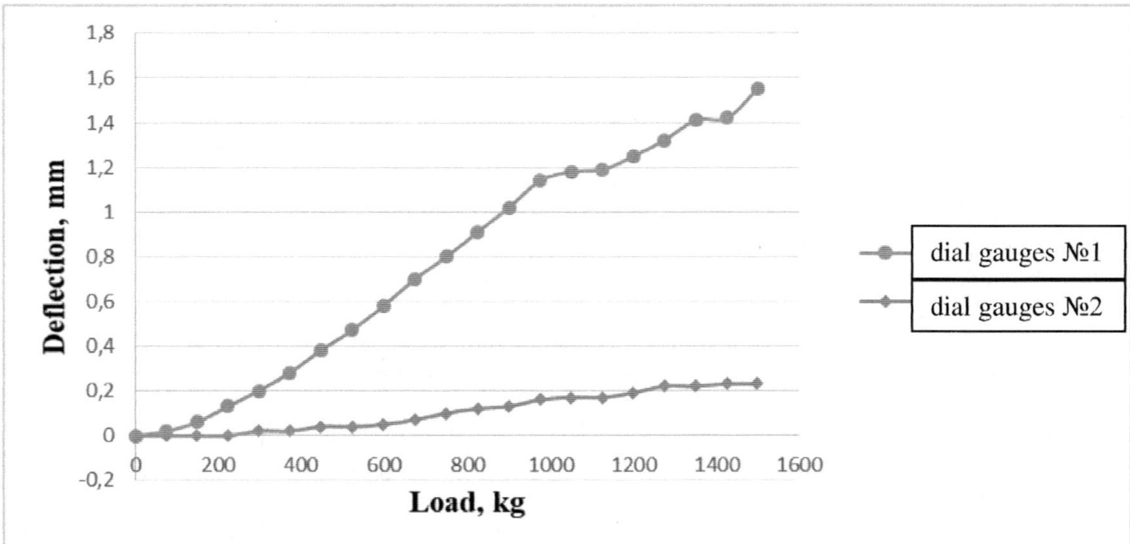

Figure 14. Measurements of auxiliary bar deflection - stay-brace of a facade system fragment using dial gauges No. 1 and No. 2, installed on the upper and side surfaces of the area of calculated buckling, respectively. Deflection versus load graph

4. Conclusion

According to the results of numerical and experimental studies of the structure of a stay-brace-and-cross-beam facade system with translucent modules, with and without spatial work of all elements of the system, a picture of the stress-strain state was obtained, the values of stresses, deflections and displacements were recorded. The actual picture of the stress-strain state was obtained and the actual shape of the stability loss of the stand of a fragment of the facade system under the action of the calculated critical force was established. Thus, it was experimentally established that the actual zone of buckling coincides with the calculated one and ranges from fastening with a bracket to the upper free end of the last auxiliary bar - stay-brace. Loss of stability of the stay-brace, assuming the absence of disk stiffness of a fragment of the facade system, i.e. without the remaining elements of the system (single stay-brace), assumed in the mid-span zone between the brackets in the glazing plane, were not observed due to insufficient flexibility of the cross-section of the stay-brace. The actual loss of stability according to the indications of strain gauges and dial gauges comes from the plane of glazing of the fragment. Therefore, the condition of insufficient flexibility in the glazing plane of the stay-braces of this facade system, provided that the remaining elements of the system are included in the work, is not decisive when performing static calculations of such systems. This requirement can be neglected, if additional numerical and experimental studies for each individual technical solution are made, since the existing methods for calculating such facade systems in the Russian Federation do not have clear requirements and standards.

References

[1] Fatemeh Pariafsai 2016 A review of design considerations in glass buildings *Frontiers of Architectural Research 5* pp 171-193

[2] M Cwyl, R Michalczyk, N Grzegorzewska and A Garbacz 2018 Predicting Performance of Aluminum - Glass Composite Facade Systems Based on Mechanical Properties of the Connection *Periodica Polytechnica Civil Engineering 62(1)* (Budapest: Budapest University of Technology and Economics) pp 259-266. DOI.org/10.3311/PPci.9988

[3] C Baniotopoulos, T Nikolaidis and G Moutsanidis 2016 Optimal structural design of glass curtain-wall systems *Proceedings of the Institution of Civil Engineers - Structures and Buildings, vol. 169, no. 6* (Birmingham: University of Birmingham) pp 450-457. DOI:10.1680/jstbu.13.00088

[4] M Santarsiero, Ch Louter and A Nussbaumer 2017 Optimal Laminated connections for structural glass components: a full-scale experimental study *Glass Struct. Eng. 2* (Switzerland: Springer International Publishing Switzerland) pp 79-101. DOI 10.1007/s40940-016-0033-2
[5] V A Silvestru, O Englhardt and J Schneider 2018 Linear adhesive connections at the edge of laminated glass panes: an experimental study under tensile, compressive and shear loading *Glass Struct. Eng.* (Springer). DOI.org/10.1007/s40940-018-0080-y
[6] A V Radkevych and K M Netesa 2015 Aspects definition of reliability evaluation facade systems from the view point of eurocode *Science and transport progress No. 4 (58)* (Ukraine: Dnipropetrovsk National University of Railway Transport). pp 205-212. DOI10.15802/STP2015/49287
[7] P A López-Jiméneza, M Mora-Péreza, G López-Patiñoa and M A B Escribano 2010 Model of ventilated façade in buildings by using CFD techniques *International Environmental Modelling and Software Society (iEMSs) International Congress on Environmental Modelling and Software Modelling for Environment's Sake, Fifth Biennial Meeting* (Ottawa, Canada)
[8] M Overend and K Zammit 2006 Wind loading on cladding and glazed façades *International Symposium on the Application of Architectural Glass, ISAAG*
[9] K Nore, B Blocken and J V Thue 2010 On CFD simulation of wind-induced airflow in narrow ventilated facade cavities: coupled and decoupled simulations and modelling limitations *Building and Environment*
[10] A M Fathy, J Planas and J M Sancho 2009 A numerical study of confined and freestanding masonry façades under wind load *Anales de Mecánica de la Fractura 26 Vol 1* pp 92-97

IOP Conf. Series: Materials Science and Engineering **570** (2019) 012007 doi:10.1088/1757-899X/570/1/012007

A comparative analysis of machining of parts from 38H2MYuA by cutting tool of different materials

V I Astashchenko, I N Gilazov

Kazan Federal University, Naberezhnye Chelny Institute, 423812, Russia, Nabereznye Chelny, Syuyumbike Ave., 10A
astvi-52@mail.ru

Abstract. The variant of mechanical treatment of gear parts made of steel 38H2MYuA is described Tool life of different materials - from R6M5K5 and ASP2052 is analyzed. It was found that for treatment of gear-wheels of steel 38H2MYUA by method of skiving, the most preferred variant is the use of tool made of powder steel than of high-speed steel.

The efficient choice of the cutting tool which possesses high physical and mechanical and operational characteristics, at blade processing of various hardware, is an actual task for any machine-building production, and especially at use of the modern equipment. The development of metalworking production is associated with both the creation of new and improvement of existing machine tools, and also cutting and auxiliary tools.

The cutting tool is a special object of metalworking technology. A variety of operating conditions and, consequently, the using of its cutting part cause various types of damage and failures of the technological system, and the tool wear rate is much higher than the wear rate of parts and components of the machine. Therefore, the performance of the technological system as a whole depends primarily on the quality of the cutting tool used [1-3].

Currently, in the engineering industry, tools from high-speed steel as well as from powder steel with various types of wear-resistant coatings are widely used. [2-6]

The article discusses the option of processing blanks parts made of steel 38H2 MYuA with internal gear. In the study the tool (see figure), made of steel R6M5K5 and powder alloy ASP2052, was used.

Figure 1. The detail and the tool for research.

IOP Conf. Series: Materials Science and Engineering **570** (2019) 012007 doi:10.1088/1757-899X/570/1/012007

To assess the tool life, a Power Skiving method was chosen.

The skiving technology is a high-performance processing of details like "disk" and gears with the processing of internal gears, and it is an alternative to pulling and notching. This method is characterized by high efficiency and very high cutting speed.

When machining gears wheels using the method Power Skiving, the issue of tool life is paramount. Therefore, when processing the details, two types of material for the tool were investigated:

1. High-speed steel R6M5K5;
2. Powder steel ASP2052.

Before operation, the tools were covered with wear-resistant coatings. As a base coating nitride alumotitan coating (AlTiN) was used.

On a small amount of the instrument coatings of alumino-titanium carbonitride (AlTiCN) and alumino-titanium-chromium nitride (AlTiCrN) were also used.

Table 1. Coating characteristics of aluminum-titanium nitride (AlTiN)

Microhardness HV0,06	3500±500
Friction coefficient for steel	0,7
Coating thickness, micron	2-4
Thermal threshold	900°C
Coating color	light gray anthracite
Key features	perfect oxidation resistance and high temperature resistance

Processing of blanks was carried out on a CNC machine for cutting teeth by the method of skiving Gleason 300 PS. The number of tools in the test was 10 pieces from each material. The hardness of the blanks was 241-287 HB.

The results of the study and tool life are shown in table 2. For the indicator of tool life the number of machined parts to the reference value of wear (20-30 microns) was taken or to the occurrence of a chip on the working surface of the tool.

Table 2. Tool life with different coating

Tool number	Tool material	Coating	Wear resistance of the part	Note
1	ASP2052	AlTiN	80	
2	ASP2052	AlTiN	80	
3	ASP2052	AlTiN	80	
4	ASP2052	AlTiN	80	
5	ASP2052	AlTiN	80	
6	ASP2052	AlTiN	80	
7	ASP2052	AlTiN	80	
8	ASP2052	AlTiN	76	Chip
9	ASP2052	AlTiN	100	
10	ASP2052	AlTiN	100	
1	P6M5K5	AlTiN	50	
2	P6M5K5	AlTiN	50	
3	P6M5K5	AlTiN	46	Chip
4	P6M5K5	AlTiN	44	Chip
5	P6M5K5	AlTiN	36	
6	P6M5K5	AlTiN	44	Chip

7	P6M5K5	AlTiN	49	Chip
8	P6M5K5	AlTiN	44	Chip
9	P6M5K5	AlTiN	56	Chip
10	P6M5K5	AlTiN	32	Chip
1X	ASP2052	AlTiCrN	73	
2X	ASP2052	AlTiCrN	71	Chip
1C	P6M5K5	AlTiCN	37	Chip

The results of the study showed that the average tool life of high-speed steel P6M5K5 is ≈45 parts, and of powder steel ASP2052 - ≈83 parts. In addition, it should be noted that when using a tool from R6M5K5, a chip from 0.4 to 1.2 mm was observed on it.

Thus, based on the results of the study, it can be noted that for machining gears of parts made of steel 38Kh2 MYuA using the Skiving method, it is most preferable to use a tool from powder steel rather than high-speed steel.

References

[1] Voynov B A 2010 *Wear-resistant alloys and coatings* Moscow: Mechanical engineering p 120

[2] Kondrashov A G, Safarov D T, Faskhutdinov A I and Davletshina G K 2017 Single-turn worm mills for conical round-tooth gears *Russian Engineering Research* vol 37 No 9 pp 812-13

[3] Astashchenko V I, Emelyanov D V and Gilazov I N 2018 The application of plasma-electrolysis boronizing of parts *IOP Conf. Series: Materials Science and Engineering* 412 **012004**

[4] Mukhametzyanova G F, Kolesnikov M S, Mukhametzyanov I R and Astashchenko V I 2017 Process research into metallic pipe wear of hot chamber die casting machines and methods ofin-creasing wear resistance *IOP Conf. Series: Materials Science and Engineering* vol 240 Is 1 **012053**

[5] Leshchenko G I 2003 *Plasma hardening and spraying* Kiev: Ecotechnology pp 26-36

[6] Tabakov V P 2008 *Formation of wear-resistant ion-plasma coatings of cutting tools* Moscow: Mechanical engineering p 311

IOP Conf. Series: Materials Science and Engineering **570** (2019) 012008 doi:10.1088/1757-899X/570/1/012008

Efficiency increase of gas turbine work in the summer period

D S Balzamov, E Yu Balzamova, S R Ibatullin, L S Sabitov
Federal State Budgetary Institution of Higher Education "Kazan State Power Engineering University ", Kazan, Krasnoselskaya Street, 51, 420066, Russian Federation

l.sabitov@bk.ru

Abstract. The operation of thermal power plants on the basis of steam-gas and gas-turbine installations in the summer period of the year is accompanied by a drop in the generation of electric energy, which leads to a decrease in the profitability of the station as a whole. The article discusses one of the options for improving the efficiency of gas turbines through the use of absorption chillers.

During operation of the gas turbine in normal conditions, the air content in the gas-air mixture reaches 97-98%. In the warm season, as the temperature rises, the air density decreases, which has a significant impact on the operation of the gas turbine unit (GTU). There is an increase in the energy expended on the process of air compression in a GTU compressor, which is quite clearly seen from formula (1).

$$l_{к} = c_p \cdot T_0^{к}\left(\pi_{к}^{\frac{k-1}{k}} - 1\right)\cdot\frac{1}{\eta_{oi}^{к}},\ \text{kJ / kg.} \qquad (1)$$

where c_p – isobaric heat capacity of air, kJ / kg • ° C;
T_0^k – air temperature before the compressor, ° C;
k – air adiabat index;
π – compressor compression ratio;
$\eta_{oiк}$ – compressor efficiency.

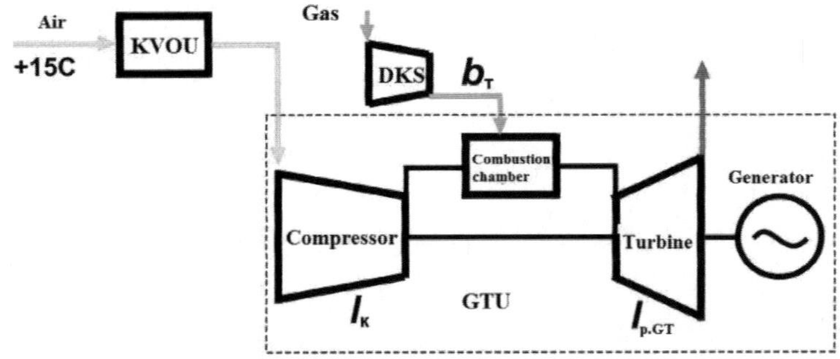

Fig. 1 Scheme of operation of GTU

IOP Conf. Series: Materials Science and Engineering **570** (2019) 012008 doi:10.1088/1757-899X/570/1/012008

The cost of compressing air in a compressor caused by an increase in outside air temperature consists of a part of the useful work of $l_{p.GT}$ gas turbine and part of the fuel burned in the combustion chamber $b_т$, the specific consumption of which increases with deviation of the nominal GTU operating mode. Thus, the cost of working on compressing air in a compressor is mathematically written in the form of expression (2).

$$l_к^{t_i} = \Delta l_{p.GT} + \Delta b_т \qquad (2)$$

where $\Delta l_{p.GT}$ – part of the useful work, perfect gas turbine, which is spent on the process of compressing air in the compressor;

$\Delta b_т$ – the increase in consumption of combustible fuel in the combustion chamber associated with the deviation from the nominal mode of operation of the GTU;

$l_к^{t_i}$ - work spent on the process of compressing air in the compressor.

In order to reduce the effect of high temperature of air compressed in the compressor on the generation of electrical energy from the combined-cycle plant (CCP), which includes the considered gas turbine unit (GTU), it is proposed to apply pre-cooling of the air before the compressor in the heat exchanger of the integrated air-cleaning unit (IACU). As a source of cold, it is proposed to use a lithium bromide absorption chiller (LBAC) [1].

In order to divert low-potential thermal energy from LBAC, a cooling tower installation is needed, and therefore an analysis of the cooling water flow at the existing thermal power plant (TPP) was performed and the possibility of connecting LBAC absorber cooling system and condenser to the existing TPP circulating water supply system was identified, thus eliminating , the cost of building an additional cooling tower for LBAC. Circulation water consumption for one installation of LBAC is 540 m3 / h (in total it is supposed to install two LBAC).

The analysis of underproduction of electricity in the period from April to October 2017 was carried out, the results of which are shown in Table 1.

Table 1. Electric power underproduction of a combined-cycle plant from April to October.

Month	**Underproduction electricity, thousand kW • h**
April	330,0745
May	1593,0244
June	2239,7154
July	3468,4574
August	4169,1656
September	391,9728
October	14,4805
Total	**12 206,89**

The analysis of electric power generation was carried out relative to the work schedule, which takes into account the dependence of the power of GTU on the outside temperature (Fig. 2) [2].

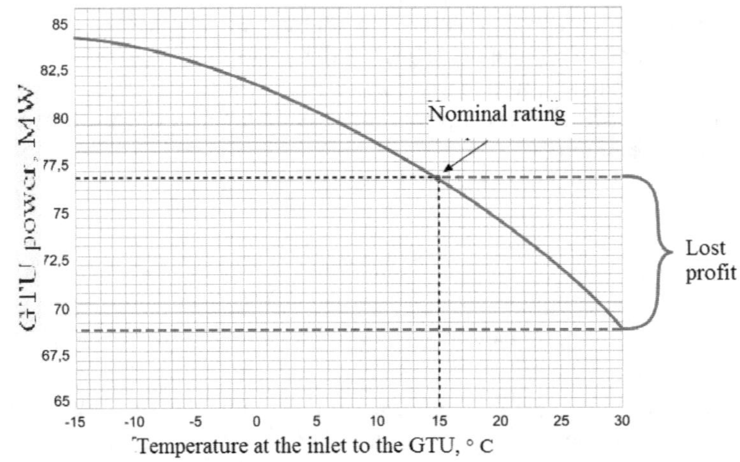

Fig.2. Dependence of GTU power on the outside air temperature

It can be seen from the graph that an increase in the ambient air temperature already to 30 ° C leads to a drop in the power of GTU by almost 9 MW, which leads to a decrease in the supply of electrical energy. All this, in turn, directly affects the payback period of the combined-cycle plant and the profit from the sale of electricity by an enterprise in the wholesale electricity and capacity market (WECM) [3].

The installation of LBAC, as one of the options for cooling the air in front of the compressor, allows you to eliminate the above problem and thereby ensure the production of electricity in the warm period of the year in the mode close to the nominal.

References

[1] Balzamov D S, Timershin B F 2017 *Overview of advanced technologies for generating companies* ModernScience No 2 pp 26-29

[2] Balsamov D S, Timershin B F 2017 *Promising technologies for enterprises of the energy industry* News in electric power industry No 5 (91) pp 38-40

[3] Kashapov N F, Gil'Manshin I R, Konahina I A 2014 *System analysis of the energy complex of engineering enterprise as a basic tool of effective energy management* IOP Conference Series: Materials Science and Engineering

[4] Dogadin D L, Anyukhin A B, Latypov G G, Krykin I N 2014 *The use of absorption lithium bromide refrigerating machines in the production cycle stations* Electric stations No 10 pp 40-46

[5] Radchenko A N, Kantor S A 2015 *Efficiency of methods for cooling air at the inlet of GTU compressor stations, depending on climatic conditions* Aerospace Engineering and Technology, No 1 (118) pp 95-98

[6] Baranenko A V, Popov A V, Timofeevsky L S 2001 *Energy saving absorption lithium bromide water cooling and water heating converters of heat* Engineering systems No 4

[7] Popov A V, Galimova L V, Slavin R B 2007 *Energy-saving system based on steam and gas turbine installation and absorption lithium bromide chiller of a new generation* Refrigeration equipment No 2

[8] Popov A V, Gorshkov V G, Levanov O S, Lystsov S O 2010 *Analysis of the effectiveness of various types of water-cooling machines in nuclear power plants* Heavy machinery No 4

[9] Kuzmina T G, Teslya E S 2013 *On increasing the power and efficiency of gas turbine engines in the warm season due to cooling of the cycle air* Gas turbine technologies No 1 pp 15-18

[10] Karpov V V, Mitin A A, Gyndenov S A 2010 *Improving the efficiency of GTU-TES in the warm season* Turbines and diesel engines No 1 pp 14-16

IOP Conf. Series: Materials Science and Engineering **570** (2019) 012009 doi:10.1088/1757-899X/570/1/012009

Analysis of the technique of software products for the selection of heat-insulating materials for heat networks pipelines

D S Balzamov, E Yu Balzamova, A I Khaibullina, L S Sabitov

Kazan State Power Engineering University, Kazan, Krasnoselskaya Street, 51, 420066, Russian Federation

l.sabitov@bk.ru

Abstract. This article provides an analysis of modern techniques and software systems for the selection of the optimal thermal insulation material for heat pipelines with various methods of installation.

The modern market of heat-insulating materials for heat networks is represented by numerous manufacturers producing a huge assortment of products with different cost, performance, physical and other characteristics. On the other hand, the heat supply system of Russia is characterized by a high length of pipelines of more than 250 thousand km, various conditions of their operation, installation methods, percentage of wear, parameters of heat carrier, soil, climatic conditions, etc. It is also worth considering the large variation in the value of the tariff for thermal energy in the regions of Russia, which is associated with numerous factors.

Every year, the circumstances described above complicate the decision making regarding the choice of the optimal heat-insulating material in heating networks, the use of which would correspond to the operating conditions typical for a given region with its minimum payback period.

To make it easier to choose the type of insulation product and its optimum thickness for a particular pipeline size, many manufacturers have developed software products. This is mainly characteristic of foreign companies that are striving to create comfortable conditions for the consumer and thereby increase sales.

Such programs are presented in Table 1 [1].

Table 1.

Software products for the selection and calculation of thermal insulation materials

№	The name of the software product
1	The program of JSC "Teploproject" for the calculation of thermal insulation using mineral wool cylinders ROCKWOOL (ROCKWOOL)
2	The program of JSC "Teploproekt" for the calculation of thermal insulation using mats ROCKWOOL TehMat

3	ROCKWOOL. The program of calculation of economic efficiency. Version 1.0
4	ROCKTEC 2.6.4. The program for calculating the parameters of thermal insulation materials in accordance with DS / EN ISO 12241: 1998
5	ArmWin 3.2. The program for calculating the parameters of thermal insulation ARMAFLEX
6	ThermaflexInsulation 3.1. The program for calculating the parameters of thermal insulation THERMAFLEX
7	FLEXALEN quick pipeline calculation program
8	K-PROJECT 1.0. The program of calculation (K-FLEX)
9	KaiCalc v6.0. Program for the selection of thermal insulation materials KAIFLEX

Wider functionality is implemented in the software product "Isolation 2.0", developed by STE "Truboprovod". The program has a library of the most frequently used heat-insulating materials, intended both for pipelines and for various containers and heat exchangers. The program takes into account the requirements of SNiP 41-03-2003 (currently replaced by SP 61.13330.2012), and also allows you to update the database with new materials. The program contains albums of technical solutions for thermal insulation materials and allows you to generate a list of installation work when isolating not only pipelines, but also valves, turns, plugs and other elements. When mats and cylinders are suitable for the object to be insulated, the program makes a choice in favor of cylinder insulation [2].

Analysis of the market for thermal insulation materials shows that the cost of mineral wool cylinders (both cut-out and rolled-up) significantly exceeds the cost of mats. Thus, analyzing [2], we can conclude that the "Insulation 2.0" program does not take into account such a factor as the recoupment of the insulating material, which is not very convenient when replacing the insulation in worn areas. It is also clear whether the program takes into account the change in the thermal conductivity coefficient when the thermal insulation is wetted on the flooded sections of the heating main.

The effect of thermal insulation moisture and the determination of heat loss is taken into account in such software as "TEPLOROTR-2" [3], and also considered in [4]. But these mathematical models are not focused on the choice of the optimal type of insulating material.

One of the most significant factors when choosing a thermal insulation material is corrosion, which can occur due to the effects of aggressive components contained in an insulating product. Thus, the materials used in thermal insulation structures should not emit harmful and foul-smelling substances, flammable and explosive substances in quantities exceeding the maximum allowable concentrations during operation.

as well as pathogenic bacteria, viruses and fungi.

The structure of thermal insulation for surfaces with a positive temperature must be included as mandatory elements.:

- • heat insulation layer;
- • coating layer;
- • fasteners.

Thus, the design of thermal insulation of pipelines and equipment must meet the following requirements [5]:

- energy efficiency - to have an optimal ratio between the cost of thermal insulation design and the cost of heat loss through insulation during the estimated service life;
- operational reliability and durability - to withstand operating, temperature, mechanical, chemical and other impacts during the estimated service life without reducing the heat-shielding properties and destruction;
- safety for the environment and operating personnel.

After completing the analysis of software systems and techniques, it can be stated that they are designed to solve "local" tasks related to one degree or another to the choice of thermal insulation material.

The urgency of developing a universal technique that allows you to make the choice of the optimal heat-insulating product, taking into account all factors of operation, is caused by the increasing requirements for the efficiency of heat supply and heat grid organizations [6], ranging from the formation of procurement documentation for the purchase of thermal insulation to installation work and commissioning of the site.

Thus, in this article, the task of the research is set, which is to create and test a mathematical model for choosing the optimal type of thermal insulation structure, taking into account the influence of external factors of operation.

References

[1] *The official website of the Altai thermal insulation company* [electronic resource]. Access mode is free.http://altaitk.ru.

[2] Kuznetsova T, Lisin S, Korelshtein L 2012 *Isolation: New Opportunities for a Unique Program* CADMASTER No 3 pp 82-88

[3] *Electronic journal Rosteplo.ru* [electronic resource]. Access mode is free.http://www.rosteplo.ru/soft/4/177

[4] Polovnikov V Yu 2006 *Mathematical modeling of thermal conditions of heat pipelines in conditions of insulation moisture*: Dis. Candidate of Technical Sciences: 05.14.04. - Tomsk 122 p

[5] Balsam D S, Balsamova E Yu, Akhmetova I G 2018 *The main requirements when choosing insulation materials for pipelines of heat networks* Scientific and technical progress: current and future directions of the future Vol II pp 194-97

[6] Konakhina I A, Khamidullina G R, Kuzmina M A, Khusnutdinova E M 2017 *Targeted energy monitoring method for analyzing industrial enterprises and housing and public utilities* Innovative machine-building technologies, equipment and materials pp 276-79

[7] Gurov A V 2013 *Experimental setup for measuring the thermophysical properties of heat-insulating materials by the method of a flat "instantaneous" source of heat* Metrology No 4 pp 16-24

[8] Ponomarev S V, Egorov M V, Lyubimova D A 2014 *Mathematical modeling of measurement errors of thermophysical properties of heat-insulating materials by the method of a flat "instantaneous" heat source* Metrology No 9 pp 23-25

[9] Shoikhet B V, Ovcharenko E G, Melekh A S 2001 *Regional standards for thermal insulation of industrial equipment and pipelines* Energy Saving No 6 pp 65-67

[10] Tsvetkov N A, Krivoshein Yu O, Khutornoy A N, Kolesnikova A V, Savrasov F V 2015 *Automated heterogeneous dispatching system and energy consumption management-* Bulletin of the Tomsk State University of Architecture and Civil Engineering No 11 pp 50-54

IOP Conf. Series: Materials Science and Engineering **570** (2019) 012010 doi:10.1088/1757-899X/570/1/012010

Studying the Effects of Mechanical Loads and Environmental Conditions on the Drive-Axle Performance

A Yu Barykin, M M Mukhametdinov, R Kh Takhaviyev, A D Samigullin

Kazan Federal University, 423800, 68/19 Mira avenue, Naberezhnye Chelny, Russia

samigullin86@mail.ru

Abstract: The paper discusses the problem of achieving the required output parameters of a truck drive under various operating conditions. The authors enumerate the final-drive and differential mechanical efficiency, the pre-load torque in the final-drive bearing assembly, the friction coefficient, and the differential locking factor as the most significant technical parameters of the drive axle and its components. The main factors that affect these parameters during operation are: (i) mechanical loads that transmission is exposed to when transmitting the torque from the engine to the drive wheels; (ii) the effects of external thermal and dynamic loads when driving on an uneven surface or in cold / hot climates. The paper describes the main dependences between the final-drive processes and the inter-wheel differential processes; it also demonstrates how the aforementioned factors affect the output parameters of the component. The assumptions hereof are based on the results of bench and field truck tests, including adverse-climate tests. We herein propose a method for estimating the drive-axle load, which is based on the weighted values of the aforementioned factors. We prove it feasible to use auto-adjusted heating of the drive-axle components. The degree of thermal exposure shall be determined with due account of environmental conditions, surface conditions, engine and transmission parameters. The control algorithm is supposed to automatically take into account any change in one or more factors. This approach to controlling the output parameters helps improve the efficiency and the longevity of the drive axle and truck as a whole.

1. Introduction

In running vehicles, transmission parts and components can be in a largely varying condition while still staying within acceptable ranges of control parameters. It is extremely important to evaluate the current condition and the residual life of such components by means of on-board controls and continuous monitoring of operating parameters [1].

The authors' analysis of the technical condition of truck-transmission components allows proposing the following classification of the power transfer process factors in an engine—transmission—drive-wheel system.

Temporary parametric deviation of a component for various reasons (including operation-induced change in temperature conditions). This factor expires as soon as the process ends or is de-intensified (e.g. when reducing the torque being transmitted).

Malfunction or consistent and prolonged parametric deviation of a component, which may be caused by e.g. superficial wear. Such deviations are detected before an adjustment is applied (clearance- or angular-backlash adjustments, etc.) regardless of whether such adjustments are automated or not.

Component failure, e.g. shock-induced gear breakage or shaft jamming. This factor apparently results in a full-stop of the process and necessitates a repair.

IOP Conf. Series: Materials Science and Engineering **570** (2019) 012010 doi:10.1088/1757-899X/570/1/012010

Operating a vehicle implies using direct and indirect diagnostic parameters for evaluating the health and longevity of components [2, 3]. In this case, on-board control options do not include a number of direct-reading methods due to the disassembly and assembly operations being not an option, or due to using incompatible equipment. The authors believe the most efficient method consists in on-board status monitoring based on the following easy-to-control parameters:

Constant or situational noise during component operation. This parameter can be perceived as a certain type of sound waves (e.g. hum, howl, or knock sounds). Therefore, registering such readings necessitates allocation of amplitude and frequency ranges specific to various power-transfer factors and different components/parts.

Constant or situational vibration of a component. This parameter is also associated with specific frequency and amplitude ranges; as such, its registration is similar to that of noise.

Component/part temperature changes during operation. In this case, one has to monitor either the temperature of a single part that is crucial to the operation of its associated component, or the volumetric-average temperature of the whole component. The first option results in greater precision, but it is also more difficult to implement.

Oil level in the drive-axle crankcase adjusted for rotation-induced spattering. The oil volume also has to be adjusted for heating or cooling against the baseline.

Aside from these main parameters, it is imperative to take into account the location of noise/vibration/heat sources in the adopted coordinates.

2. Methods

As found in this research effort, on-board monitoring requires tuning the evaluation parameters according to the following criteria.

First, the parameter readings must be within the qualifying range with due adjustment for various factors, e.g. temperature increase or ingress of sand and dust in the transmission oil.

Second, shall multiple controlled parametric anomalies be detected (e.g. noise and vibration at the same time), the dominant affecting factor must be identified.

Further on, it is necessary to estimate the need for applying any adjustments so as to restore the process to its normal parameters.

If parameters remain deviant, such deviation must be evaluated against the maximum permissible deviation values. The most efficient way consists in monitoring the closeness to maximum permissible wear of the connection-to-check (the frictional couple).

If necessary, one can consider limiting the mechanical (and therefore, the thermal) impact shall the controlled parameters be approaching their respective maximum permissible values; to that end, one has to use parametrically-triggered disabling devices that shut down the engine and the transmission while also preventing component overheating (if a heating device is present).

For specific vehicle operations, causative factors behind such deviations must be identified. On-board monitoring of the most important evaluation parameters is a must. Those include: angular position of the steering wheel and the steered wheels; gear shift frequency and sequence; speed; braking frequency and intensity.

3. Results and discussion

Performance indicators fall into two groups: external (i.e. road bumps), and driver- or computer-controlled (speed, etc.). Therefore, we can specify possible ways to limit mechanical and thermal loads, as mentioned above.

Friction-induced losses in transmission components determine the mechanical efficiency of the final drive and the differential [4]. To ensure the rigidity of the final drive, it is important that the pre-load of the bearing assembly remains stable, which greatly depends on the temperature and the transmitted torque, experiments have shown [5]. Friction coefficient and differential locking factor also depend on mechanical and thermal loads.

The authors have found it reasonable to monitor the intensity of friction in the drive-axle components so as to reliable estimate the residual life of the truck transmission. We have identified the load parameters in heavy-duty applications, e.g. in moderate-climate winter.

4. Summary

The method for estimating the drive-axle load is based on the weighted values of the aforementioned factors.

$U = \Sigma (\alpha i \; Ri)$,

where αi are the factor weights; Ri is the factor significance for the given operating conditions.

In particular, paper [6] states that when preparing a vehicle for driving after being long-parked in an open lot, the final-drive thermal balance is found as follows:

$Wn = 3{,}6 \, (Q1 + Q2 + Q3 + Q4) \, / \, ttp$,

where Wn is the power consumption of the on-board heater; Q1 is the heat required for the drive-axle parts; Q3 is the heat required for the transmission oil; Q3is the heat energy to be spent due to heat transfer to the environment when heating the final drive; Q4is the heat energy to be spent in unaccounted heat transfer processes; ttp is the time to heat.

Apparently, reliable evaluation of the adjustments made by the automatic transmission-process control system requires objective data on the intensity of heat transfer in the drive-axle — environment system.

5. Conclusions

The assumptions hereof are based on bench and road test data on large-capacity trucks. Environmental effects were taken into account under low-temperature moderate-climate conditions as well as in simulated effects of impurities on the transmission oil in the drive-axle crankcase.

Equipment used for tests was compliant with test specifications and regulations. The identified errors and discrepancies of theoretical and experimental data are within acceptable range.

References

[1] Barykin A Yu., Basyrov R R, Mukhametdinov M M 2014 On the System Analysis of Operating Conditions for KAMAZ Wheel Drives (K voprosu sistemnogo analiza uslovy ekspluatatsii privoda kolyos avtomobiley KAMAZ) *Scientific and Technical Volga Region Bulletin* Is 6 pp 74-76

[2] Kulakov A T, Gafiyatullin A A, Barylnikova E P 2014 Providing normal conditions of lubricating of diesel engine during its operation *IOP Conference Series: Materials Science and Engineering*- V 69 Is 1**012027**

[3] V V Lyandenbursky, M V Nefyodov, A S Ivlev et al 2015 Diagnosing the Final Drive in Trucks (Diagnostirovaniye glavnoy peredachi gruzovykh avtomobiley) *Problems of Vehicle Quality and Operation: Use and Development of Vehicles [Problemy kachestva i ekspluatatsii avtotransportnykh sredstv: Ekspluatatsiya i razvitiye avtomobilnogo transporta]* pp 199–205

[4] K Lorenz, Ch Dietrich, E Donges 1986 The Influence of a Limited Slip Differential on the Traction and Handling of Rear Axle Driven Cars – Part 2 *ATZ* No 3 pp 149-153

[5] Mukhametdinov M M 2012 Studying the Intensity of Pre-Load Drops in KamAZ Final-Drive Anti-Friction Bearings *In: Mechanical Engineering: Design, Engineering, Calculation, and Manufacturing and Repair Technology [Mashinostroyeniye: proyektirovaniye, konstruirovaniye, raschyot i tekhnologii remonta i proizvodstva]* pp 113–14

[6] A Yu Barykin, R Kh Takhaviyev 2017 Evaluating Energy Spent for Winter Operations of Truck Drive Axles [Otsenka zatrat energii v protsesse zimney ekspluatatsii vedushchego mosta gruzovogo avtomobilya] *Energy Saving. Science and Education (Energosberezheniye: Nauka i obrazovaniye):* pp 52–57.

[7] Barykin A Y, Takhaviev R K, Samigullin A D 2018 The research of thermal processes of the automobile chassis *International Journal of Mechanical and Production Engineering Research and Development*. Vol 8 Special Issue 8 pp 458-64

IOP Conf. Series: Materials Science and Engineering **570** (2019) 012011 doi:10.1088/1757-899X/570/1/012011

Control of the process of pulsed laser thermal hardening based on emerging potential

D A Bashmakov

Electrical Energy Industry and Electrotechnics Department, Naberezhnye Chelny Institute, Kazan Federal University, 68/19 Mira Ave., Naberezhnye Chelny, 423812, Russian Federation

bashmakovda@yandex.ru

Abstract. The article studies the relationship between the time power distribution of pulsed laser radiation and the emerging electric potential in metals. It is shown that the emerging potential characterizes the shape of the laser pulse during processing and characterizes the penetration depth of the thermal influence zone. It is revealed if the energy of laser action pulse does not exceed 12 J and duration is less than 1 μs the form of the emerging potential is the same for all types of time distribution of laser action power.

1. Introduction

In engineering laser thermal hardening is applied to improve the wear resistance of the lay shafts, crankshafts, gear wheels, working surfaces of the valves, valve rockers, valve seats etc. The quality of the hardened layer depends on the parameters of the laser technological complex (the energy of the radiation pulse, focal length, pulse duration, gaseous atmosphere, external fields, etc.). Laser hardening of the working surface of the metal product at capacities close to critical, which do not allow melting, cannot give stable surface quality indicators which are required in the production process. Therefore, it is required to monitor the depth of the thermal influence zone in real time.

2. Results

In the course of this work a method for determining the parameters of the generated laser radiation and the depth of the thermal influence zone was developed. This method consists in removing the oscillograms of the emerging potential in the metal plate under the influence of a laser pulse in real time and their further processing to determine the required adjustments in the generation parameters.

Analyzing the data of the emerging potential graphs, the differences in the time distributions of energy in the pulse are traced. The graphs of regular and Gaussian distribution are similar and are characterized by a smoother increase and decrease of potential in time (Fig. 1, 2), in contrast to the increasing and decreasing distribution (Fig. 3, 4). The similarity of the graphs of regular and Gaussian distribution is explained by the absence in the distribution of sharp power jumps and regularity. However, it should be noted that at the same energy in the laser pulse with a regular time distribution of energy in the pulse, the emerging potential is greater as well as the depth of the thermal influence zone (Fig. 5).

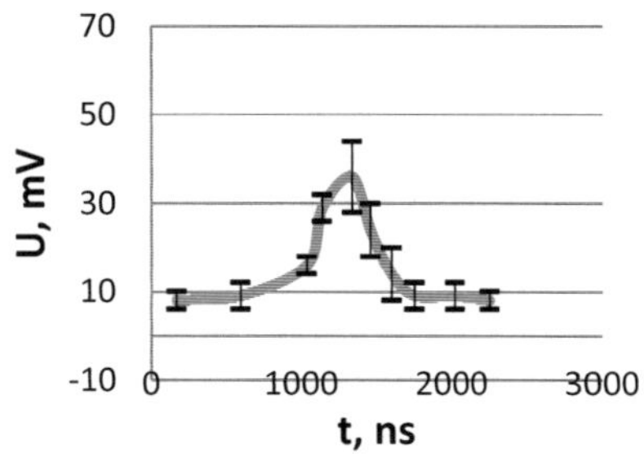

Fig 1. Graph of the dependence of the emerging potential on the time distribution of energy in the pulse: regular distribution, U = 300 V

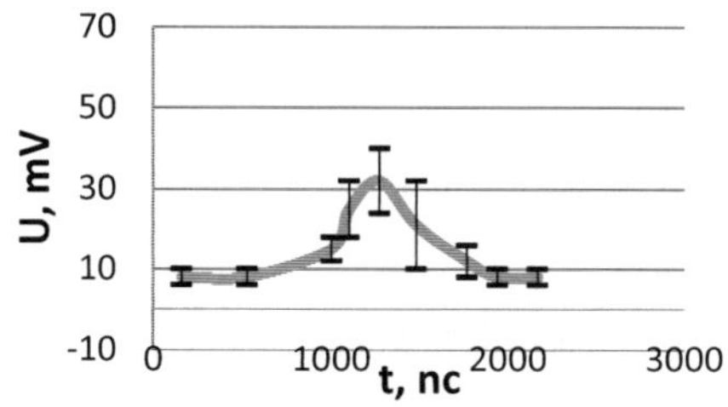

Fig. 2. Graph of the dependence of the emerging potential on the time distribution of energy in the pulse: Gaussian distribution, U = 300 V

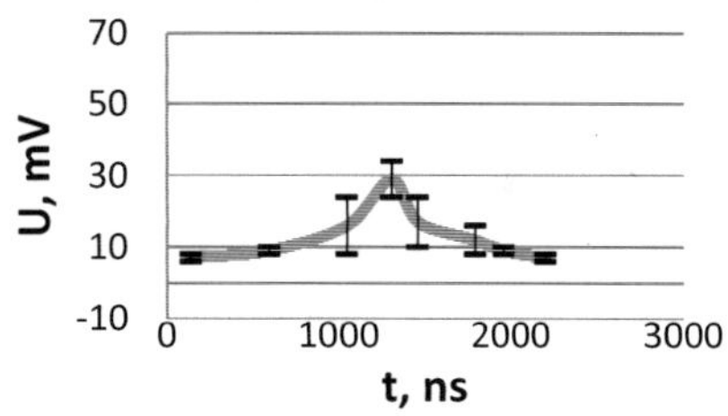

Fig. 3. Graph of the dependence of the emerging potential on the time distribution of energy in the pulse: increasing distribution, U = 300 V

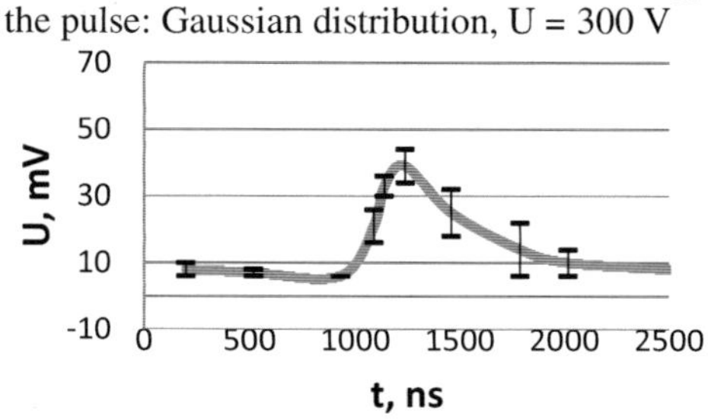

Fig. 4. Graph of the dependence of the emerging potential on the time distribution of energy in the pulse: decreasing distribution, U = 300 V

Increasing and decreasing distributions have graphical data which are characteristic of their time distribution of power in the pulse. Thus increasing distribution (Fig. 3) is characterized by a more gradual increase in potential, in contrast to decreasing distribution(Fig. 4), and a sharp decline to the average values with a further gradual decrease in potential. The decreasing distribution is characterized by a sharp increase in potential, followed by a very smooth decline.

We also carried out metallographic investigations with the aim of identifying the emerging features of the thermal influence zone influenced by of time power distribution and identify the relationship of emerging potential and depth of thermal influence zone.

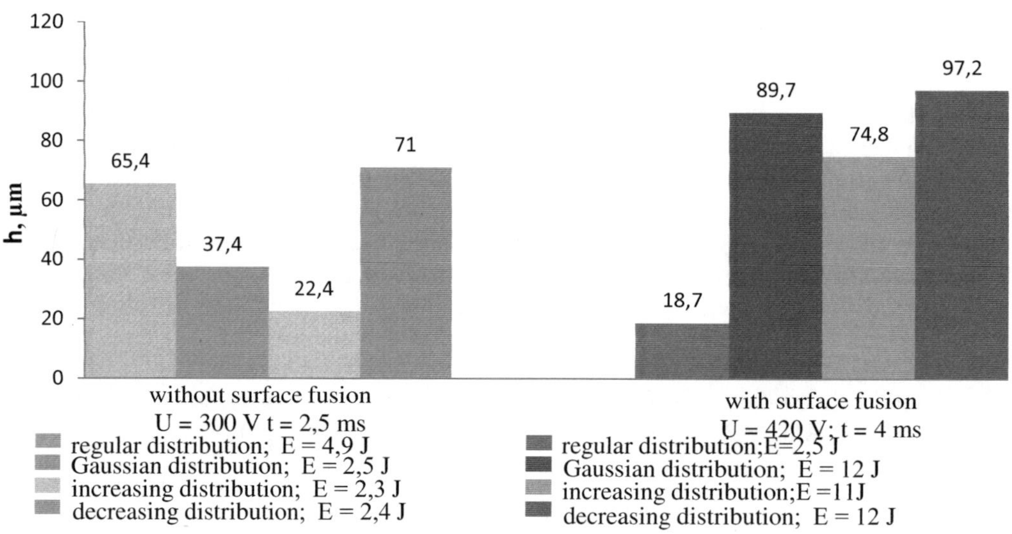

Fig. 5. Graph of the dependence of the depth of the thermal influence zone on the time distribution of energy in the pulse

The obtained values of the depth of thermal influence zone under the influence of a laser pulse with the surface fusion, except for the depth values under the influence of a regular time distribution of energy, have a smaller difference in values, while they are similar to the results of the influence without surface fusion. We can say that when the increases, the difference between the depth values, when exposed to different time distributions of energy, decreases.

In should be noted that on the graphs (Fig. 5) without surface fusion and with it, the greatest depth of thermal influence zone has pulses with decreasing time distribution of energy. The pulses with increasing time distribution have the lowest depth of thermal influence zone.

3. Conclusions

The identified relationship can be used to create a system of automatic control in real time. Such a system during processing will be able to accurately predict the depth of thermal influence zone as well as, if necessary, to adjust the power of laser radiation.

References

[1] Grigoryants A G, Shiganov I N, Misyurov A I 2006 *Tehnologicheskie processy lazernoj obrabotki* [Technological processes of laser processing]: Textbook for universities 664 p

[2] Gureev D M, Yamshchikov S V 2001 *Osnovy fiziki lazerov i lazernoj obrabotki materialov* [Fundamentals of laser physics and laser processing of materials]: Textbook. - Samara: Publishing house of Samara. state University 312 p

[3] Kayukov S, Gusev A A, Nesterov I G Zaychikov E G, Petrov A L 1996 *Vlijanie formy impul'sov lazernogo izluchenija na geometriju vanny rasplava* [The influence of the shape of pulses of laser radiation on the geometry of the molten bath] Physical and chemical materials processing, No 4 p 366

[4] Israfilov I H, Pesoshin V A, Zvezdin V V, Bashmakov D A 2010 *Upravlenie jelektrostaticheskim polem glubinoj lazernoj zakalki v metallah* [Control of electrostatic field of depth of laser hardening in metals] Bulletin of Kazan state technical University No 2 pp 47-49

[5] Bashmakov D A, Israfilov I H 2016 *Sposob opredelenija prostranstvenno-vremennogo raspredelenija jenergii v lazernom puchke* [Method of determining the space-time energy distribution in a laser beam] BEAM TECHNOLOGIES & LASER APPLICATION Proceedings of the International scientific and technical Conference pp 33-35

[6] Bashmakov D A, Zvezdin V V 2018 *Control of laser-field technological complexes for tool hardening* ARPN Journal of Engineering and Applied Sciences 13(8) pp 2979-82

IOP Conf. Series: Materials Science and Engineering **570** (2019) 012012 doi:10.1088/1757-899X/570/1/012012

Biodegradable fixator for sternum closure after median sternotomy

I Bebnev[1], R N Kashapov[1,2], N F Kashapov[1], L N Kashapov[1,3]
[1] Engineering Institute, Kazan Federal University, Kazan, 420008, Russia
[2] Laboratory of Radiation Physics, Zavoisky Physical-Technical Institute, FRC Kazan Scientific Center, Russian Academy of Sciences, Kazan, 420029, Russia
[3] Laboratory of Continuum Mechanics, Institute of Mechanics and Engineering, FRC Kazan Scientific Center, Russian Academy of Sciences, Kazan, 420111, Russia

kashramiln@gmail.com

Annotation. This article discusses the solutions to manufacture fixator for the osteosynthesis of the sternum from biocompatible and biodegradable material to prevent reopening.

Introduction

Currently, more than one million cardiac surgeries are performed worldwide and their number is growing due to an increase in the incidence of cardiovascular diseases in the population. Most operations are performed using a sternum dissection or sternotomy.

Standard closure after median sternotomy repeats approximation of the sternum using steel wires and simple suture technique. However, this closure technique was associated with several serious complications that prompted surgeons to look for new closure methods. These complications include instability of the sternum, dissection, and mediastinitis, as a result of bone compaction or destruction caused by steel wire. These conditions represent the main causes of morbidity and mortality after heart surgery.

The search for a convenient and optimal method for fixing the sternum, the main feature of which is universality, reliability and affordability, was carried out with the beginning of the development of cardiovascular surgery. The most common method of osteosynthesis of the sternum in our days for more than 50 years has been the fixation of fragments of the sternum with steel wire with many variations of suturing. Despite the entire duration of use, traditional methods do not exclude the possibility of complications in the postoperative wound.

Various methods of closure of the sternum, known to our days:

1. Steel wire closure (standard technique). The most common way to close the sternum is to use alternating trans and peristernal techniques.
2. Staples. Clips, which consist of the same material as the wire, prevent cutting the wire into the bone because of their larger contact area thus reduce the risk of deep wound infections.
3. Nitinol brackets with shape memory effect. Centofanti introduced a new technique for closing the sternum clamps. The material in the clips is called NITINOL (NITINOL - Nickel Titanium Naval Ordinance Laboratory) and consists of nickel and titanium. The clamps are implanted in the sternum.

4. Zipfix. Closing technology is called Zipfix, consisting of an implant similar to cable ties. This implant is made from PEEK (polyetheretherketone), which is biocompatible.
5. Sternal Talon. Sternal Talon is a new technique used to close the sternum after a sternotomy.
6. Kryptonite glue - bone glue for improved biocompatible closure of the sternum.
7. Posthorax Support Vest. Posthorax Support Vest is a vest designed for use around the chest after a heart operation.

One of the relatively new and reliable ways to close the sternum is the use of polymeric clamps of the "collar" type, the usefulness of which has been appreciated by cardiac surgeons.

The main disadvantage of all existing methods of connecting the sternum is that the materials with which the osteosynthesis was performed remain in the human body even after successful fusion of the sternum and no need to further support the bones. The use of clamps for osteosynthesis of the "collar" type allows to reduce the level of postoperative pain in patients due to reliable stable fixation of the sternum, reduce the clicking of the sternum in patients and its crepitus, and thus improve their quality of life and reduce the period of rehabilitation after surgery. A multicenter study with a large sample and long observation periods will further confirm the effectiveness and safety of this technique.

Therefore, the improvement of closure methods and the development of methods for osteosynthesis of the sternum after median sternotomy are promising tasks of modern cardiothoracic surgery.

The aim of the work is to develop a technology for manufacturing individual medical sternal fixers of the sternum from a biodegradable material.

Main part

Based on all methods of closure of the sternum, taking into account the experience of cardiac surgeons "University Hospital (RCH 2)" and taking into account their preferences, the design of the fixer of the "clamp" type with the most reliable geometry was modeled.

As a prototype of the offered fixer Zipfix sternum closure technology was chosen, which has an implant similar to cable ties, fig. 1. This implant is made from PEEK (polyetheretherketone), which is biocompatible. The closure technique uses cable ties wrapped around the open sternum with a needle. The number of cable ties varies, but usually these are five cable ties that are located around the sternum. Then the needle is removed, the cable ties are tightened and all excess valves are cut off. The reduced risk of bone eruption is due to the fact that implants have a relatively large contact width of 4.2 mm. Cable ties are wider than those used in steel wire closures. The contact area between the sternum and cable ties is then larger than the contact area between the sternum and steel wire, which results in lower pressure on the sternum. Implants are wrapped around the sternum in the intercostal space, which makes the risk of bleeding less than the standard technique in which wire can cut through the sternum in the transverse direction. Zipfix provides a lower offset than standard technology, which implies that the number of repetitive operations can also be reduced if Zipfix is used. It was actually shown at the Karolinska University Hospital, Sweden, that repeated operations were reduced when using Zipfix instead of standard equipment. The change was studied over a six-month period, and the total number of repeated operations changed from ten repeated operations for steel wires to two repeated operations for Zipfix implants.

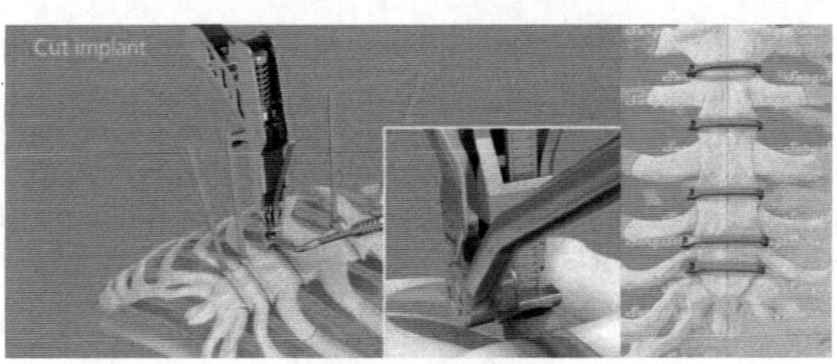

Figure 1 – ZipFix

IOP Conf. Series: Materials Science and Engineering **570** (2019) 012012 doi:10.1088/1757-899X/570/1/012012

After examining all the advantages and disadvantages of ZipFix, the following changes were included in the developed model:

1. The fastener-lock shown in Fig. 2. The reliability of fixing the fastener is due to its change, namely, the second clove is added to the lock to form a double point of contact with the teeth located on the main surface of the fixture, which will increase the adhesive strength of the lock and prevent the likelihood of the lock opening.

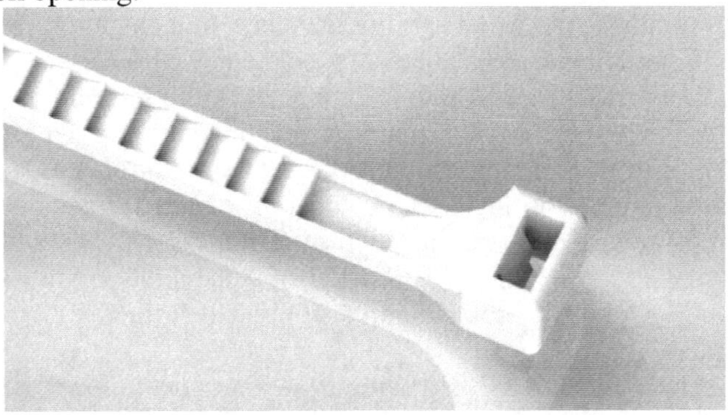

Figure 2 – Fastener-lock

2. Cogs are shown in figure 2. Cogs are enlarged over the entire surface of the fixer to fix it more firmly in the chest, similar teeth are made in the clasp-lock.
3. Needle in figure 3. To facilitate the technological production of the fixer model, it was decided to make a needle from the same material as the fixer web. The needle cross-section is made round for smooth expansion of the opening of soft tissues, which will ensure free passage of the retainer web. Needle diameter increased, starting from the tip (3 mm) to the base (5 mm).

The new design of the model clamp type "clamp" shown in Fig. 4, provides a rigid fixation for use in the primary closure of the sternum due to the new lock, which allows increasing fatigue strength. In addition, due to the large contact area of the implant with the bone, which leads to an optimized stress distribution, the effort to achieve bone dissection should be much higher. The fixer is easy to handle when the surgeon works, and it does not require additional equipment for its installation.

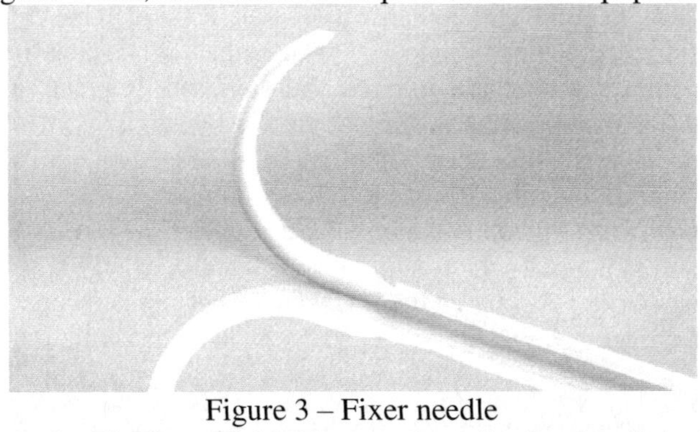

Figure 3 – Fixer needle

The main advantage over the ZipFix model is the creation of a fixative from biodegradable material. Polymers are used as biodegradable materials. The following table 1 shows the three types of material that are suitable for the manufacture of biodegradable fixatives.

The optimal material for the developed fixative was chosen polycaprolactone. This material is used for a long time in surgical practice. The following characteristics of polycaprolactone for biomedical application are known:

- does not cause allergic reactions;
- long resolved (completely resorption occurs within 12-24 months);

IOP Conf. Series: Materials Science and Engineering **570** (2019) 012012 doi:10.1088/1757-899X/570/1/012012

- loss of physical properties by 40-50% occurs in 2-3 months, which will allow the thorax to coalesce in time after the median sternotomy;
- leaves behind a skeleton of newly formed young connective tissue.

In fig. 5 drawing of the designed clamp with indicated dimensions is presented.

Table 1. Biodegradable materials

Properties	Material		
	PCL polycaprolactone [-(CH2)5-CO-O-]n	Polylactide $(C_3H_4O_2)_n$	Polyoxyalkanoates (ПОА)
Type	thermoplastic	thermoplastic	thermoplastic
Biodegradability, dissolving	yes	yes	yes
Melting temperature,°C	63	190-195	177
Flexural strength (23 °C), MPa	29	55,3	38
Tensile strength (23 °C), MPa	18	57,8	40
Density, g / cm^3	1,021	1,25	1,26
Stretching at break (23 °C), %	11	3,8	43
Young's modulus, GPa	7,4	0,384-0,481	1,4-3,5
Glass transition temperature, °C	-60	-60	(-50) - 4

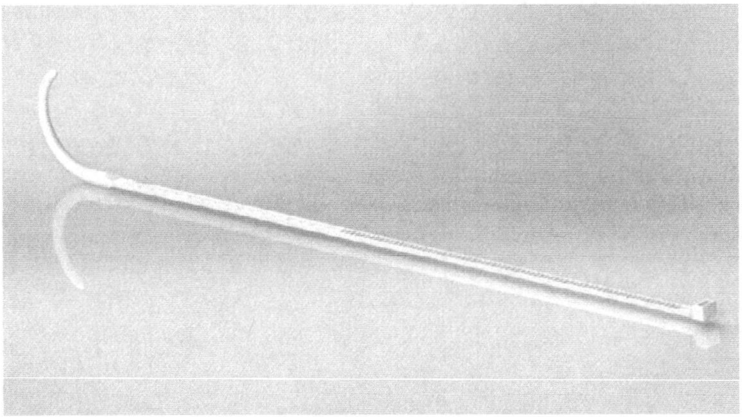

Figure 4 – Polycaprolactone fixator

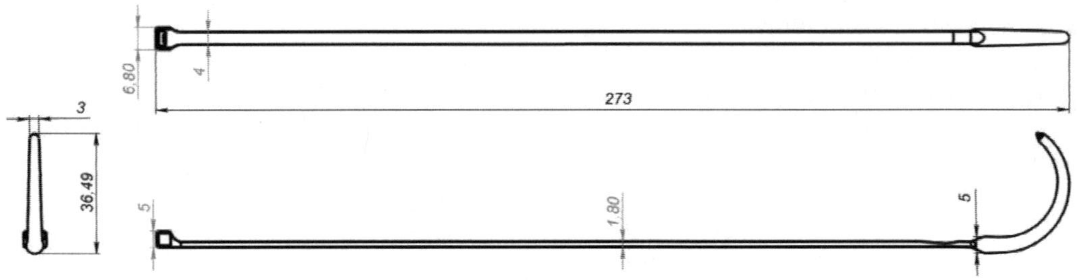

Figure 5 - Drawing developed fixator

Conclusion

The results of our study indicate that the fixers currently used require refinement and changes. The design of the fixer provides a rigid fixation for use in the primary closure of the sternum due to the new lock, which allows increasing fatigue strength. In addition, due to the large implant contact area with the bone, which leads to an optimized stress distribution, the effort to achieve bone dissection should be much higher. The fixer is easy to handle when the surgeon works, and it does not require additional equipment for its installation. The known biodegradable materials are analyzed and the most suitable polycaprolactone is selected.

Acknowledges

The work was carried out in the framework of the Base part of the state task of Ministry of Science and Higher Education of the Russian Federation (3.9399.2017/8.9)

References

[1] Kashapov L N, Kashapov N F, Kashapov R N 2016 *The application of additive technologies in creation a medical simulator-trainer of the human head operating field* Vol 134 Is 1 **012011**

[2] Kashapov RN, Korobkina AI, Platonov EV, Saleeva GT 2014 *The method of manufacture of nylon dental partially removable prosthesis using additive technologies* Vol 69 Is 1 **012026**

[3] *DePuy Synthes Companies* 2019 Available at: https://emea.depuysynthes.com/hcp/cmf/products/qs/sternal-zipfix-system (accessed 15 June 2019)

[4] Krutko E T, Prokopchuk N R, Globa A I 2014 *Technology biodegradable polymeric materials* Minsk: BGTU 105 p

[5] Peethes A A 2010 *Connection methods of the breast with different materials after middle sternotomy* Moscow

IOP Conf. Series: Materials Science and Engineering **570** (2019) 012013 doi:10.1088/1757-899X/570/1/012013

Structure formation in materials based on Fe-C-Si and Fe-C-Al systems (cast irons) prepared from dispersed mechanical engineering waste

R Bikulov[1]
[1]Kazan Federal University,NaberezhnyeChelny Institute, Peace Avenue, 68/19, NaberezhnyeChelne, 423812, Russian Federation

bikulov_uvr@mail.ru

Abstract.The paper presents fn alternative technological process for producing compact forms of graphite inclusions in the materials based on the Fe-C-Si and Fe-C-Al (cast iron) systems. It differs from traditional methods of obtaining compact forms of graphite inclusions, which consist in modifying the molten iron or annealing white iron: the proposed technological process is based on a synergistic complex of foundry-metallurgical factors affecting the cast iron in the stages of its melt preparation and crystallization in a mold.

1 Introduction

Production activities of large machine-building enterprises are associated with the formation in sufficiently large quantities and a diverse range of waste, including dispersed types, which are a big problem in involving them in the process both because of their dispersion, and the presence of non-metallic inclusions and metallic elements. The release of 1 ton of suitable blanks is accompanied by the emission of up to 30 kg of dust. To assess the scale of formation of iron-containing dispersed wastes of machine-building production, it suffices to give examples of the operation of electric arc furnaces in which melting dust is captured in an amount of 10...20 kg/t in the production of cast iron, as well as forging and thermal workshops, in which scale is formed in an amount of 20...50 kg / tonofproducts.

When other divisions of the machine-building complex work, dispersed iron-containing wastes are also formed: sludge from electrochemical and electroerosive treatment, electroplating, from wastewater treatment, shot blasting, spraying, grinding, grinding operations, exposure and other types of metal cutting. All this is taken to dumps, in which over 500 thousand tons of iron-containing sludge are buried. At the same time, society suffers significant environmental damage, and machine-building enterprises are forced to bear the burden of costs associated with the organization of burials and the payment of fines. However, the elemental composition of dispersed waste gives reason to consider them as a promising raw material in the form of substitutes for polymetallic ores and limestone. In industries related to the production of primary metal, technological solutions for the disposal of these wastes are known. First of all, this refers to the aglodomen production and alternative options for the production of primary metal: “Midrex”, “Corex”, “Romelt” [1...4]. However, the discussed processes are intended for the smelting of primary metals in the conditions of a metallurgical plant and are unacceptable for machine-building enterprises producing structural casting materials. Moreover, as the unfolding discussion shows [5...7], there is no complete clarity in terms of manufacturability, cost effectiveness and environmental friendliness of alternative processes for obtaining a primary metal, as well as utilization of dispersed production wastes in them.

Therefore, there is an urgent question about the maximum possible degree of waste disposal of the production of the machine-building complex inside it, the solution of which will make it possible to reduce the cost of manufactured products and improve the ecological situation in industrialized regions.

An effective solution to these problems can be found in the technological processes of procurement (foundry) engineering production.

2 Technological process for the production of materials based on Fe-C-Si and Fe-C-Al systems from iron-containing technogenic waste from mechanical engineering

The initial charge components for the implementation of this process are iron-containing dispersed waste engineering (ICWME), in which iron and other components are present both in the oxidized and in the reduced state. This utilization scheme ultimately provides for the production of materials based on Fe-C-Si and Fe-C-Al systems by synthesis from elements present in iron-containing machine-building waste. At the first stage of the technological process, a moistened mixture consisting of ICWME is obtained in the ratio of components, which makes it possible to obtain a melt with characteristics in terms of composition and properties required by the technological process. A solution of sodium chloride ($NaCl$) and iron trichloride ($FeCl_3$) together with etching waste from copper plates form the basis for the formation of the mixture. The last waste is saturated with copper ions, the presence of which in materials based on the Fe-C-Si and / or Fe-C-Al systems only improves its properties [8...10]. Pellets are made from the obtained composite molding material, which are dried in order to acquire sufficient strength necessary for transportation to the smelting unit.

Further granular composition is sent to a specialized melting unit. In this unit, the following physicochemical transformations proceed successively: heating the pellets to temperatures of 1000... 1300 K, metallizing the oxides, melting the granules with the appearance of the liquid slag phase and the metal phase and their overheating. As a result of physicochemical transformations, a melt of materials is synthesized based on the systems Fe-C-Si and / or Fe-C-Al. The proposed technology makes it possible to produce both cast irons and steel [11,12], in view of the fact that the synthesis of a cast alloy is carried out from elements that are restored in the smelting unit both in the liquid and solid phases. Metallization in the solid phase is carried out in a countercurrent granular composition and reducing gas. During the solid-phase metallization metals such as iron, molybdenum, copper, nickel are restored.

The main impurity in synthesized materials based on Fe-C-Si and / or Fe-C-Al systems is carbon, which has a reduction potential with respect to iron oxides and a carbon potential relative to reduced iron. Carbon saturation of the metal melt to a predetermined level is carried out as a result of cementing metallization.

The pellet composition, which underwent physicochemical heat treatment in a countercurrent reactor, enters the liquid bath of the melting unit as a result of the screw rotation. In the bath of the melting unit, under the action of the heat of an electric arc of direct current, the granular composition melts. Flag and metal phase are formed. In the process of liquid-phase metallization, such elements as silicon, manganese, tungsten, vanadium, aluminum, and magnesium are restored. In the melting unit, the anode is a non-consumable top electrode, and the cathode is a liquid metal bath with a bottom electrode connected to a DC source. Such a connection to a DC source implies an effective liquid-phase metallization in a melting furnace using electrolysis processes. The above proposed mode of melting of the granular composition creates conditions for obtaining high-quality alloy due to the refining effect of slag, possessing a high basicity and temperature, the presence of a reducing atmosphere that has a protective character in all phases of the smelter, as well as the effect of an electric current passing through the liquid metal melt [13].

3 The influence of technological features of the preparation of melt from dispersed waste engineering to the process of structure formation in materials based on systems Fe-C-Si and / or Fe-C-Al

The developed specialized melting unit, along with the fact that it allows to process ICWME, has another positive feature, which consists in obtaining high-quality materials based on Fe-C-Si and / or Fe-C-Al systems. This is achieved by the fact that dispersed wastes used as charge materials contain a lower content of sulfur and phosphorus (less than 0.02%). This circumstance allows the smelting of materials based on Fe-C-Si and / or Fe-C-Al systems without the use of highly basic slags and their processing of a liquid alloy. As a result of direct electric current passing through the liquid slag and metal, a change occurs in the structure of the metal matrix and graphite inclusions [13]. Such a liquid material based on Fe–C–Si and / or Fe–C–Al systems is prone to compact graphitization according to a synergistic model of this process [14,15], as well as experimental data [16,17].

One of the important factors contributing to compact graphitization of materials based on Fe-C-Si and / or Fe-C-Al systems is melt overheating. The higher the overheat of the melt, the higher its susceptibility to metastable crystallization. In this case, there are conditions for the occurrence of carbide deviations (fluctuations) in which the carbon activity determines a sufficiently large thermodynamic potential of graphitization. The value of the thermodynamic potential of graphitization is determined by the interaction of a number of factors and conditions, such as the temperature of overheating of the melt; the degree of melt refining from surface active substances; the presence of elements that form quasichemical ordering in the melt with iron and carbon.

The creation and activation of the thermodynamic potential of compact graphitization is possible by two mechanisms. The first mechanism of compact graphitization involves overheating of the melt to a supercritical state, in which the potential of compact graphitization is created on the basis of the formation of ordered macro-zones with pseudocarbide stoichiometry. The destruction of these zones, leading to the formation of point graphite - a building material for the formation of compact graphite inclusions, is carried out using graphitizing modifiers, either slow cooling or optimal supercooling and crystallization of highly superheated and deeply refined melts. In the second case, the potential of compact graphitization is created by entering the melt in the subcritical state, but at a temperature higher than the ionization temperature, a modifier (magnesium, yttrium, cerium containing additives) capable of forming metastable carbides in the melt with graphite clusters and iron. Thus, the thermodynamic mechanism of compact graphitization assigns an important role to the state of the melt of materials based on Fe-C-Si and / or Fe-C-Al systems, which has a strong influence on the structure formation in the workpiece.

The following critical threshold temperature effects were established by experiments [18... 20]: 1723...1753 K (first threshold), 1833...1853 K (second threshold). There are grounds [21] to consider the third threshold as 1883...1893 K, the fourth - 1973...1983 K.The appearance of different thresholds is caused by the presence of different factors with the initiating influence of some one. All factors influencing the effect of overheating of the melt of materials based on the systems Fe-C-Si and / or Fe-C-Al are classified by mechanisms of action at the macro-micro and submicroscopic levels. At the macro level, the effect of the following factors is noted: the disappearance of free germinal centers of graphite and, as a result, an increase in the propensity of materials based on Fe-C-Si and / or Fe-C-Al systems to metastable crystallization; refining the melt from non-metallic inclusions (silicates, sulfides, phosphides, oxides of refractory metals, nitrides, etc.); elimination of the "negative" heredity of the charge. Factors acting on the micro level include: restructuring of materials based on Fe-C-Si and / or Fe-C-Al systems with increasing temperature; polymorphic transformations in some elements of the melt (primarily iron, silicon, etc.); redistribution of chemical bonds; transition of the melt from the colloidal suspension state to the true solution state. The factors acting on the submicroscopic level are associated with donor – acceptor transitions in the melt elements; increasing the chemical activity of the elements of the melt; the occurrence of concentration deviations (fluctuations); their development into homogeneous nucleation, spinodal melt stratification.

The use of the listed mechanisms operating at different levels led to the development of technological measures for the intra-furnace melt processing: high overheating, isothermal holding at high temperatures of overheating, slag and thermo-slag electric arc treatment. The effectiveness of these technological measures of intrahep treatment is so high that it results in a compact form of graphite in materials based on the Fe-C-Si and / or Fe-C-Al systems without the use of traditional spheroidizing modification [21,22]. High-temperature and thermo-time processing of materials based on Fe-C-Si and / or Fe-C-Al systems contributes to its compact graphitization [23...27].

Analysis of the existing and obtained results in the issue of manufacturing high-quality materials based on Fe-C-Si and / or Fe-C-Al systems using the intra-furnace processing of the alloy and the theoretical rationale given above can be sufficiently attributed to the conditions of formation of materials for on the basis of Fe-C-Si and / or Fe-C-Al systems in a developed specialized melting unit. The formation of liquid materials based on the systems Fe – C – Si and / or Fe – C – Al is carried out in a highly superheated slag phase from elements that are in an oxidized state as a result of their metallization. In this case, difficulties with the negative influence of the heredity of charge materials disappear. The creation of a liquid metal phase is carried out by passing charge granules through a liquid slag phase. Passing through the liquid slag phase a droplet-shaped metal is subjected to deep refining due to the developed surface. Deep refining of metal is also enhanced by the electrocapillary effect under the action of direct current. At the same time non-metallic components are transferred to the slag, and the metal is deposited on the bottom. Overheating of the liquid phases in the melting bath is such that it overlaps the first and second thresholds. The effect of slag treatment is enhanced by the action of an electric arc, especially when used as provided in the proposed DC unit [21].

Thus, in the smelting unit under discussion, conditions are created for the manifestation of a synergistic effect of temperature, thermal, slag, electric arc and electric current treatments affecting the melt inclination of materials based on Fe-C-Si and / or Fe-C-Al systems to the formation of compact forms of graphite during crystallization with a high level of supercooling.

4 Metallographic examinations

The study analyzed the microstructure and shape of graphite inclusions of materials based on Fe-C-Si and / or Fe-C-Al systems synthesized by direct route from ICWME in a specialized electrothermal melting unit. Natural objects in the form of blanks of the piston group of the KAMAZ automobile, poured into sandy-clay forms [28, 29] served as samples. The structure of the rings is checked on the cross section, for which the section opposite the castle part of the ring was cut. The structure of the metal matrix, as well as the form of graphite inclusions, was evaluated according to State Standard 3443-87 "Castings from cast iron with various forms of graphite". The structure analysis was carried out on a Neofot 32 microscope. The chemical composition of materials based on the Fe-C-Al system of piston rings was characterized by the following elemental contents,% (wt.): C - 3.6 ... 4.0; Al - 2.4 ... 3.0; Mn - 0.5 ... 0.8; Cr - $\leq$0.3; Cu - $\leq$0,3; S - <0.02; P - <0.03. The chemical composition of materials based on the Fe – C – Si system of piston rings was characterized by the following elemental contents,% (wt.): C — 3.5 ... 4.0; Si - 3.2 ... 3.6; Mn - 0.4 ... 0.7; Mo - 0.014 ... 0.04; Cu - 0.75 ... 1.0; S - <0.02; P - <0.15.

Figure 1 shows the typical structures of materials based on the Fe-C-Si system, obtained in a specialized ICWME melting unit at various superheating temperatures without the use of modifiers.

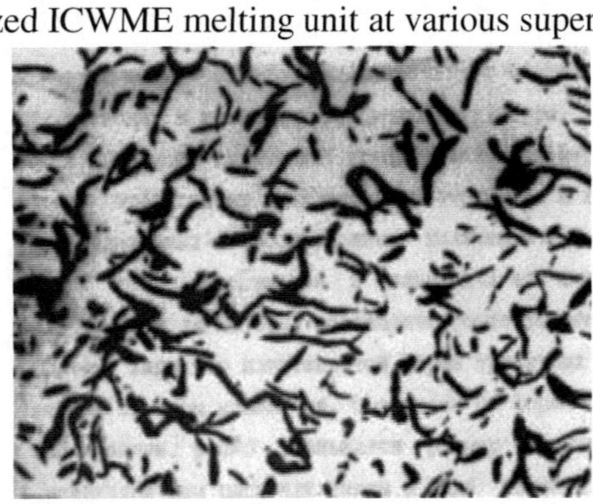
a

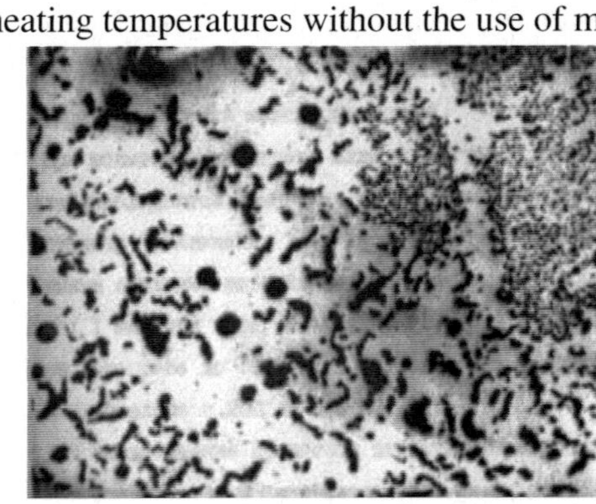
b

Figure 1 - Microstructure of materials based on the Fe-C-Si system, obtained in a specialized melting unit from ICWME with overheating up to 1773K (a) and 1873 K (b): a – x100; b - x100

The formation of various types of graphite in materials based on Fe-C-Si and / or Fe-C-Al systems is the result of direct melt carbon extraction during eutectic crystallization and further supersaturation of austenite with carbon. To obtain compact forms of graphite, high overheating of materials based on Fe-C-Si and / or Fe-C-Al systems is required, which in this case goes from a colloidal suspension state to a true solution due to deep refining from non-metallic inclusions, as well as from oxygen, sulfur and phosphorus. Analysis of changes in the nature of the formation of graphite according to Figure 1 confirms this.

So materials based on Fe-C-Si and / or Fe-C-Al systems, melted with overheating up to 1773K, have a lamellar rectilinear and vorticated form of graphite with a length varying from 25 to 90 microns. The size of graphite inclusions is 45 ... 50 microns. The distribution of inclusions in the structure of materials based on Fe-C-Si and / or Fe-C-Al systems is mostly uniform with an interdendritic shade. The share of graphite inclusions is about 6... 7%.

Increased overheating of liquid materials based on Fe–C–Si and / or Fe–C–Al systems in the melting process (1873 K) leads to a change in the graphite structure. The main form of graphite inclusions (≈60%) is a vermicular, tortuous shape.This graphite is represented in the structure as uniformly distributed and isolated inclusions. In the structure of materials, there is also observed a compact and lamellar form of graphite inclusions, the volume of which is, respectively, ≈ 15 and 25%.The size of graphite inclusions was significantly reduced to <15 μm with an interdendritic distribution in the form of colonies and points. Compact graphite has a spherical shape with dimensions of inclusions 15 ... 30 microns.In addition, there are in the structure and clusters of small inclusions. Further overheating of the melt of materials based on Fe-C-Si and / or Fe-C-Al systems leads to an increase in the dispersion of the pearlite base of the metal matrix and a change in the distance between the cementite plates when the liquid material is overheated based on Fe-C-Si and / or Fe systems -C-Al up to 1773 K up to 1 micron and with overheating up to 1873 K up to 0.7 micron.At the same time, the hardness of materials based on Fe–C–Si and / or Fe–C–Al systems increased from 200 HB to 215 HB.

Figures 2…5 show the effect of superheat temperature, carbon and part thickness on the microstructure of materials based on the Fe-C-Al system containing 1.7% silicon.

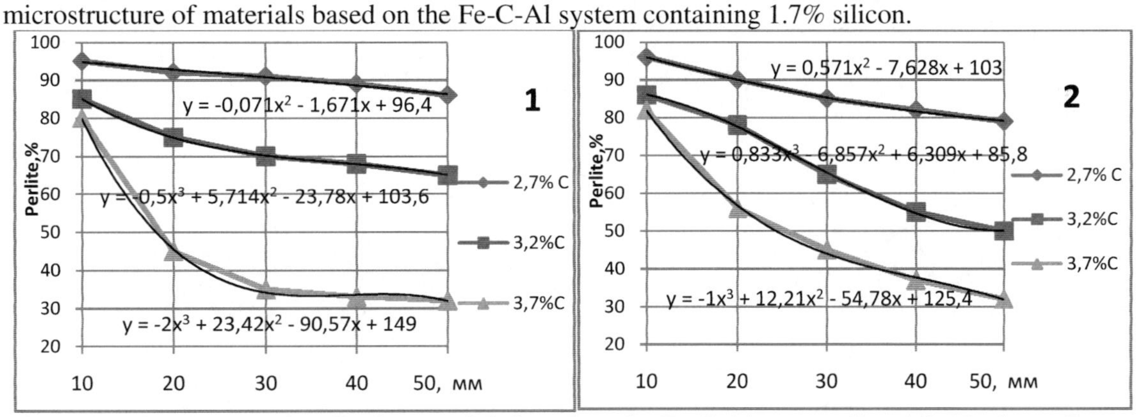

Figure 2 - The amount of perlite in materials based on the Fe-C-Al system (3 ... 4% Al), depending on carbon content, section thickness and superheat temperature: 1 - T = 1773 K; 2 - T = 1873 K; Δ- 3.7% C; □ - 3.2% C; ◊- 2.7% C

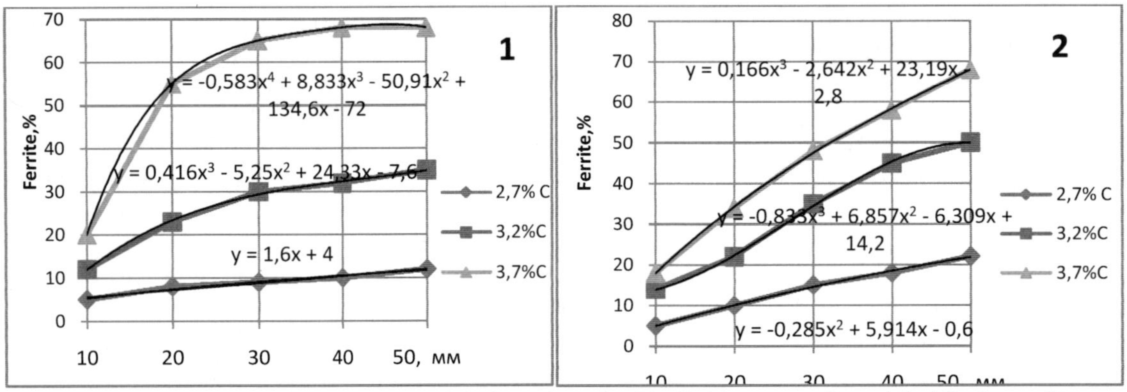

Figure 3 - Amount of ferrite in materials based on the Fe-C-Al system (3 ... 4% Al), depending on carbon content, section thickness and superheat temperature: 1 - T = 1773 K; 2 - T = 1873 K; Δ- 3.7% C; □ - 3.2% C; ◊- 2.7% C

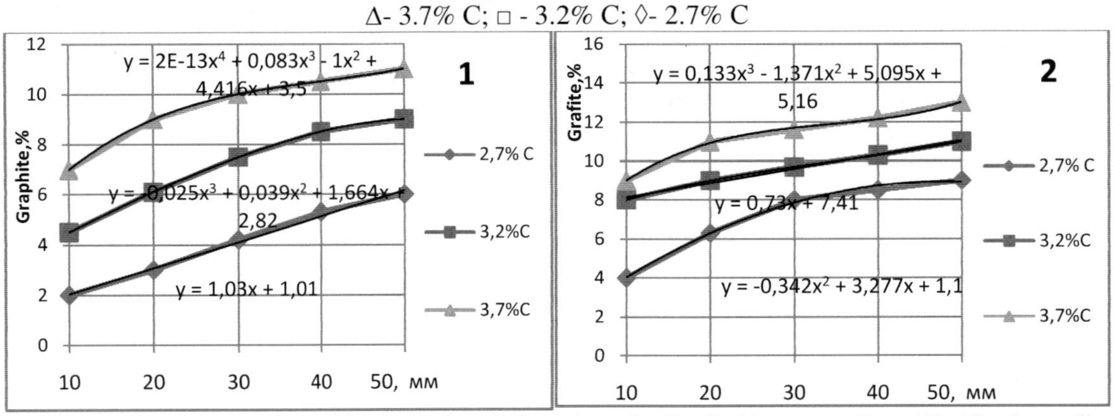

Figure 4 - The amount of graphite in materials based on the Fe-C-Al system (3 ... 4% Al), depending on carbon content, section thickness and superheat temperature: 1 - T = 1773 K; 2 - T = 1873 K; Δ- 3.7% C; □ - 3.2% C; ◊- 2.7% C

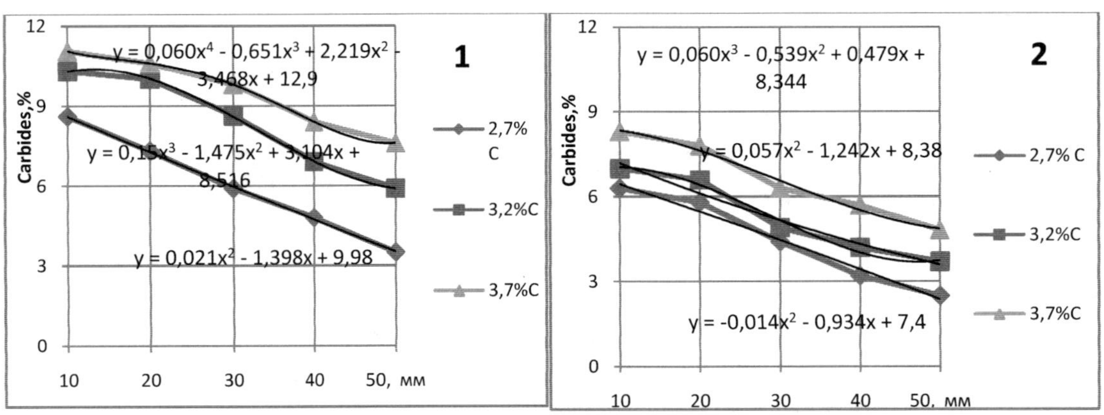

Figure 5 - The number of carbides in materials based on the Fe-C-Al system (3...4% Al), depending on carbon content, section thickness and superheat temperature: 1 - T = 1773 K; 2 - T = 1873 K; Δ- 3.7% C; □ - 3.2% C; ◊- 2.7% C

In low carbon (2.7% C) materials based on the Fe – C – Al system, small amounts of interdendritic carbides are found in the thinnest sections (10 mm). Suppression of these carbides is possible by increasing the thickness to 50 mm (Figure 5).

For blanks with small thicknesses (10 mm), overheating is effective from the point of view of suppressing carbides in low-carbon materials based on the Fe-C-Al system. For medium thicknesses

(25 mm) in materials based on the Fe-C-Al system containing 2.7% carbon, overheating results in the appearance of a large number of graphite globules distributed in the perlite-ferritic matrix. For large thicknesses (50 mm) in materials based on the Fe-C-Al system containing 3.7% carbon, overheating leads to the formation of significant quantities of crushed graphite.

The structure of the matrix and the morphology of graphite are very sensitive to the content of graphite. For a carbon content of 2.7%, globules of graphite of two different sizes are observed in the pearlite matrix. As the carbon content increases to 3.7%, the proportion of ferrite in the matrix increases significantly (Figure 3) and degenerate forms of graphite appear.

The interdendritic clusters of crushed graphite ("chunky") can be observed in thick sections (50 mm), especially with an increased carbon content in materials based on the Fe-C-Al system.

5 Conclusions and perspectives

The scientific fundamentals of the physicochemical processes of forming materials based on Fe-C-Si and / or Fe-C-Al systems with compact graphite inclusions without modifying the melt synthesized using dispersed industrial wastes containing elements in an oxidized state have been developed. The factors synergistically causing compact graphitization are established:

- direct synthesis of materials based on the systems Fe-C-Si and / or Fe-C-Al, excluding any heredity, including negative ones;
- formation of materials based on Fe-C-Si and / or Fe-C-Al systems from a melt of a drop-like state in a highly superheated slag phase, creating conditions for deep refining of the metal and its overheating into a supercritical state (T> 1853 K), forming carbide-forming fluctuations;
- electric current melt processing of materials based on systems Fe-C-Si and / or Fe-C-Al.

6 Acknowledgements

The work was supported by the Ministry of education and science of the Russian Federation, within the framework of the State task No 9.3236.2017/4.6, state Contract under the 220th resolution No14.z50.31.0023.

7 References

[1]YusfinYuS, TovarovskyIGandChernousovPI, Shatlov VA 1995Blast furnace - unit of the XXI century *Steel*No8 pp1-8

[2] Afonin SZ, ShneerovYaA 1995The Third Congress of Steelmakers *Steel*No 10 pp 26-33

[3] Romenets VA, Usachev AB andBalasovov AV, Churgel VO, Verein VG, Panov EM 1995Possibilities of using the Romelt process in the modernization of a factory of a complete metallurgical cycle *Steel*No 11 pp 64-67

[4] Protopopov EV, Shakirov KMandAizatulov RS, Fokin KS 1997Thermodynamic fundamentals of metal conversion with elements of direct reduction *IzvestiyaVuzov. Ferrous metallurgy*No 8 pp 13-17

[5] GavrilinIV 1998Remelting steel scale in induction furnaces *Foundry*No 8 pp 20-27

[6] Wegman EF 1992About the minimum theoretically possible fuel consumption in liquid-phase iron reduction furnaces *IzvestiyaVuzov. Ferrous metallurgy*No 5 pp 14-16

[8] Bobro YuG, Platonova LA 1981Features of the structure of copper-alloyed cast irons *New in metallography of cast iron*(Kyiv:Casting problems institute) pp 94-98

[9] Bobro Yu G, Dmitryuk N V andGusachuk D A 1997 High-copper cast iron with spherical graphite *Foundry* No 7 pp 9-11

[10] Bikulov RA, Karikh FGandSafronov NN 2012Cast insert of hot volume deformation from dispersed waste of mechanical engineering *Scientific and Technical Bulletin of the Volga region*No2 pp 112-115

[11]Bikulov RA, Safronov NN 2012Cast forging die tooling from dispersed waste of mechanical engineering *Foundry of Russia*No7 pp 12-14

[12] Minenko GN 1998Crystallization of gray iron in the treatment of current *Foundry*No 6 p 15

[13] Kuznetsov BL, Sazonov OA andBikulov RA1988 Synergetic model of compact graphitization and its application in the technology of producing high-quality cast irons *Ways to improve the*

quality and efficiency of casting processes: Materials of the Republican Scientific and Technical Conference(Odessa: OPI) pp 78-79

[14] KuznetsovB L 1991Foundry metallurgy of cast iron in the light of new scientific concepts *Problems of foundry metallurgy of cast iron: Interuniversity collection*(NaberezhnyeChelny: KamPI) pp 3-11

[15] Kumanin I B 1971*Foundry* (Moscow: Mechanical Engineering) p 320

[16] KhudokormovD N 1968*The role of impurities in the process of graphitization of cast iron* (Minsk: Science and Technology) p 156

[17] Khakimov OP, Tyagunov GV andChurkin BC, Novikov VF 1994The influence of the state of liquid iron on the formation of the structure and properties of castings *Foundry*No1 pp 10-11

[18] Baum B A 1988On the relationship of liquid and solid metallic states *Melt*No 2pp 15-18

[19] Tyagunov GV 1983The study of the properties of metal melts - the basis of a new direction in technology *Proc. Experimental studies of liquid and amorphous metals*(Sverdlovsk: UC, USSR Academy of Sciences) Part 2 pp 43-48

[20] Kuznetsov B L 1985 Production of globular and vermicular graphite cast iron without the use of spheroidizing modifiers *Foundry* No 8 pp18-19

[21] Kuznetsov BL, Jacobson AMandBikulov RA, Aniskovich II 1989TShED-processing is an important reserve for the intensification of high-quality cast iron smelting *Foundry*No 5 pp 12 - 13

[22] Sidorenko RA, Kharchuk MD andChermensky VI 1991Investigation of the joint action of the formation of graphite in alloys of metals of the iron subgroup with carbon *Problems of cast iron foundry metallurgy: Interuniversity collection*(NaberezhnyeChelny: KamPI) pp 16-21

[23] Sidorenko R A, Chermensky V I andKharchuk M D 1972 Growth conditions for graphite spherulites during cast iron hardening *News of the Academy of Sciences of the USSR. Metals*No 2 pp 159-165

[24] Savina L G, Muravyova E L, andBaryshev E E 1997 The effect of high-temperature melt processing on the particle size distribution and structure of the powder of the cast iron of grade SCH 18 *News of higher education institutions. Ferrous metallurgy*No 8 pp12-14

[25] Lubyany DA, Selyanin IF andBrylova TB, Galochkina SA 1997About the modes of smelting iron in an induction furnace *News of higher education institutions. Ferrous metallurgy*No 8 pp 75-77

[26] Aniskovich II, Sozonov OA 1991Thermal slag electric arc (TSHED) melt processing as a factor in the structure formation of cast iron in a casting *Problems of foundry metallurgy of cast iron: Interuniversity collection*(NaberezhnyeChelny: KamPI) pp 38-42

[27] Safronov NN, Bikulov RA 1996Recycling of foundry and forge production in the smelting of aluminum cast irons *Problems of the integrated use of ores: Materials of the 2nd International Symposium*(St. Petersburg: St. Petersburg State Technical University) pp 324-325

[28] SafronovNN,Bikulov RA 1996 Direct synthesis of cast aluminum cast irons *Young Science to the New Millennium: Materials of the International Scientific and Technical Conference. –* (NaberezhnyeChelny: KamPI) pp84

IOP Conf. Series: Materials Science and Engineering **570** (2019) 012014 doi:10.1088/1757-899X/570/1/012014

The use of technogenic dispersed waste engineering to obtain compact forms of graphite in the materials of the system Fe-C-Al

R Bikulov[1]
[1]Kazan Federal University, Naberezhnye Chelny Institute, Peace Avenue, 68/19, Naberezhnye Chelne, 423812, Russian Federation

bikulov_uvr@mail.ru

Abstract. Currently, the significant problem of domestic engineering is to improve the quality level of products based on efficient resource-saving technologies. In the paper the main emphasis has been placed on using in such a significant procurement industry of engineering, which is foundry, dispersed iron-containing with a different degree of oxidation of the designated waste element. Moreover, this use of waste is proposed to carry out both at the stage of preparation of the melt for blanks, and in the manufacture of a mold. We are talking about blanks of materials based on the system Fe-C-Al, the quality of which is largely determined by the shape of graphite inclusions.

1 Introduction

The proposed stable and economical technology of obtaining blanks from materials based on the Fe-C-Al system with a compact form of graphite inclusions is leant on the features of the structure formation processes during crystallization and cooling of alloys of materials based on the Fe-C-Al system.

The technical literature presents materials based on Fe-C-Si and / or Fe-C-Al systems with compact forms of graphite inclusions - these are high-strength materials based on Fe-C-Si and Fe-C-Al systems with spherical and "vermicular" graphite and malleable materials based on Fe-C-Si and Fe-C-Al systems with flaky graphite. The production of materials based on Fe-C-Si and Fe-C-Al systems with compact forms of graphite inclusions has a number of significant drawbacks. The technology for producing spherical shapes of graphite inclusions in cast irons is characterized by high process instability [1, 2] and requires careful adherence to the technological process. The task is achieved by graphitizing and spheroidizing modification of specially prepared liquid melt materials based on Fe-C-Si and / or Fe-C-Al systems with a given temperature and composition while limiting the content of surface-active substances. The process of obtaining materials based on Fe-C-Si and / or Fe-C-Al systems with compact forms of graphite is unstable due to the multi-factorial effect of a sufficient number of parameters (chemical composition of materials based on Fe-C-Si and / or Fe-C systems -Al, temperature of overheating and modification, particle size of the modifier, etc.). It causes instability of the modification process and, as a consequence, instability of the structure and properties of the resulting materials based on Fe-C-Si and / or Fe-C-Al systems.

IOP Conf. Series: Materials Science and Engineering **570** (2019) 012014 doi:10.1088/1757-899X/570/1/012014

The cost of the technological process of obtaining materials based on Fe-C-Si and / or Fe-C-Al systems with compact forms when modifying it is rather high. In case of intraform modification it is necessary to apply a modifier with a fraction of 1 ... 4 mm. The amount of modifier with a fraction of 1 ... 4 mm is not more than 40%, the rest amount must be recycled, which increases the high cost of the modifier.

The technology of annealing of cast white cast iron has a more stable technology in terms of obtaining compact forms of graphite inclusions. When implementing this technology, the requirements for the chemical composition of the material being melted based on the Fe-C-Si and / or Fe-C-Al systems are less stringent than on the production of high-strength materials based on the Fe-C-Si and / or Fe-C-Al systems. Basically, they mean to limiting the content of graphitizing elements (carbon and silicon). The long duration of annealing (30 ... 60 h) leads to an increase in the cost of the process due to substantial energy consumption, which implies the replacement of the production of ductile iron blanks with high-strength cast iron [3, 4].

2 Technological process of making compact forms of graphite in the materials of the system Fe-C-Al

The proposed technological process of making compact forms of graphite inclusions in materials based on the Fe-C-Al system differs from the above described by high stability and cost-effectiveness in the absence of strict requirements to the composition of the melted material based on the Fe-C-Al system [5...7]. The gist of the technological process is shown in Figure 1. It is based on the rapid cooling of a liquid material based on the Fe-C-Al system in a casting mold, which corresponds in composition to gray cast iron with a sufficiently high silicon content, so that white cast iron is obtained. This is achieved due to the fact that the liquid material based on the Fe-C-Al system is poured into a thermally active metallized form that is in the quenching medium in the form of a melt of a mixture of nitrate and alkali. Thus, the pouring of materials based on the Fe – C – Al system into the casting mold is combined with its quenching from the liquid state, which favorably affects the growth of the quality of the cast metal, in particular, by reducing the grain size [8].

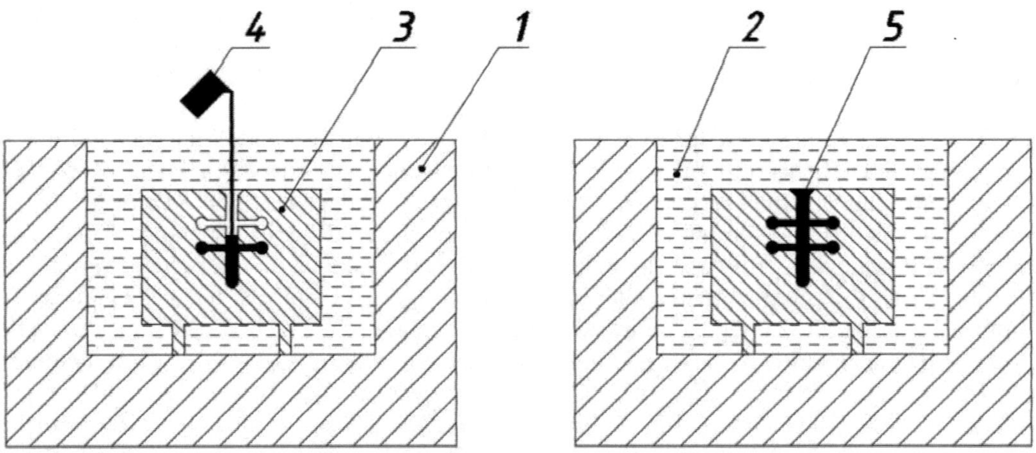

Figure 1 - Scheme of the technological process of obtaining blanks from materials based on the Fe-C-Al system with compact graphite: 1- metal pallet; 2 - melt quenching medium T = 1223-1273K; 3 - metallized form; 4 - liquid cast iron T = 1773K; 5 - casting

In the structure of materials based on the Fe-C-Al system, it is necessary to obtain white cast iron, as well as in the production of ductile iron. In the material based on the system Fe-C-Al at a concentration of> 9% Al, it becomes bleached. Sulfur in materials based on the Fe-C-Al system in the production of white iron is a desirable element. Based on the work [9, 10], guaranteed production of white cast iron is also achieved when beryllium and tellurium are introduced into the melt.

As can be seen from Figure 1, the temperature of the quenching medium in the left bath is maintained at 1223 ... 1273 K. This type of heat treatment involves the exposure of materials based on the Fe-C-Al system in the quenching medium to conduct the 1st graphitization step in it. Experiments have shown that the duration of this stage of graphitization is much lower than in the technological process of manufacturing ductile iron.

It takes two hours to complete making of the structure of austenite + graphite. The increased Al content in materials based on the Fe-C-Al system and its hardened state causes a high process rate.

It should be noted that sulfur has a positive effect in the acceleration of the 1st stage of graphitization. At this stage, compact forms of graphite in the form of small inclusions are formed. This process has no metal loss with scale.

3 Metallographic studies

Figure 2(a) and 2(b) show the microstructures of materials based on the Fe-C-Al system containing C - 3.6...4.0; Al - 2.4...3.0; Mn - 0.5...0.8; Cr - ≤0.3; Cu - ≤0.3; S - <0.02; P - <0.03 and 3.6 % C; 9.6 % Al; 0.4 % Mn; 0.017 % S; 0.019 P., respectively, obtained from iron-containing dispersed waste engineering in a specialized melting unit. This material based on the Fe-C-Al system was smelted with an overheat of the liquid phase equal to 1773 K. From the melt of the material based on the Fe-C-Al system, samples with a reduced thickness of 5 mm were obtained by pouring into a sandy and metallized form which was placed in a bath with a molten quenching medium at a temperature of 1223 K.

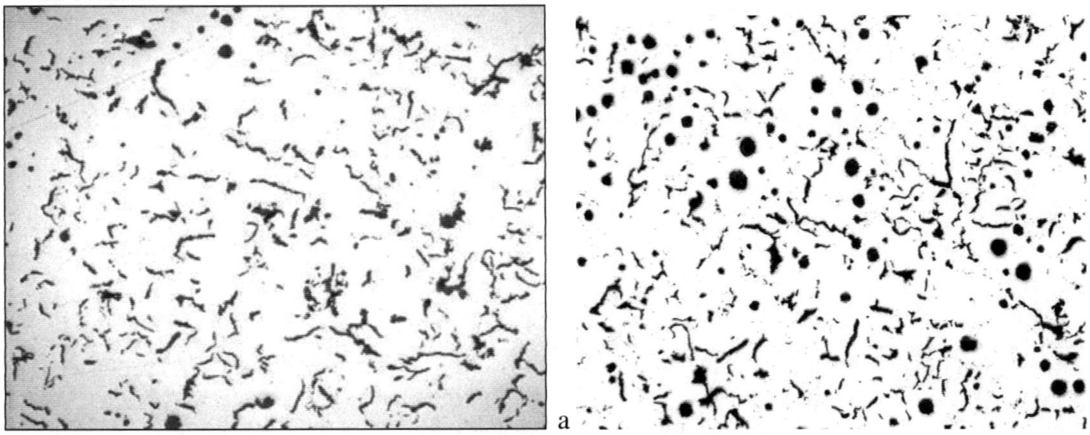

a b

Figure 2 - Microstructure of the material based on the Fe-C-Al system: a – in the cast state during solidification in the sand form; b - metallized in the melt of the quenching medium, respectively (a, b – x 100)

Metallographic studies of structural States of iron-carbon alloys were done on microscopes "Neofot-32" and "Apatin-2" (Germany) at magnification of 100 to 1000*. As an Etchant, a 4% solution of HNO_3 in alcohol was used. Assessment of the graphite inclusions in the materials on the basis of Fe-C-Al was carried out according to GOST 3443-87.

Figures 3...6 show the phase composition of materials based on the Fe-C-Al system containing 9.5% aluminum, 3.7% carbon depending on the silicon content, section thickness and exposure time in the metallized form: 1-1 hour; 2 - 2 hours ◊- 0.8% Si; □ - 1.3% Si; Δ- 1.8% Si.

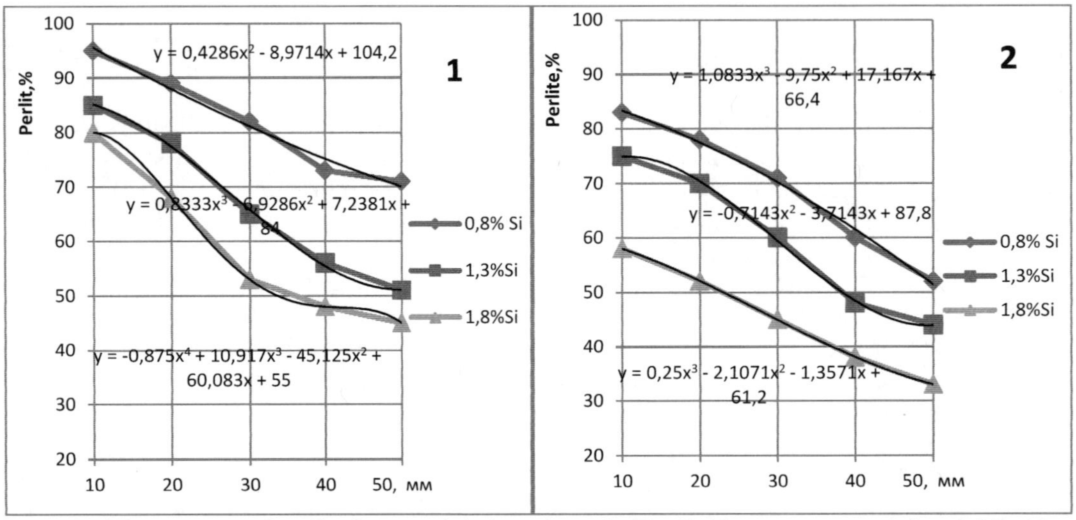

Figure 3 - The amount of perlite in materials based on the Fe-C-Al system, containing 9.5% Al and 3.7% C, depending on the silicon content, section thickness and holding time: 1-1 hour; 2 - 2 hours ◊- 0.8% Si; □ - 1.3% Si; Δ- 1.8% Si

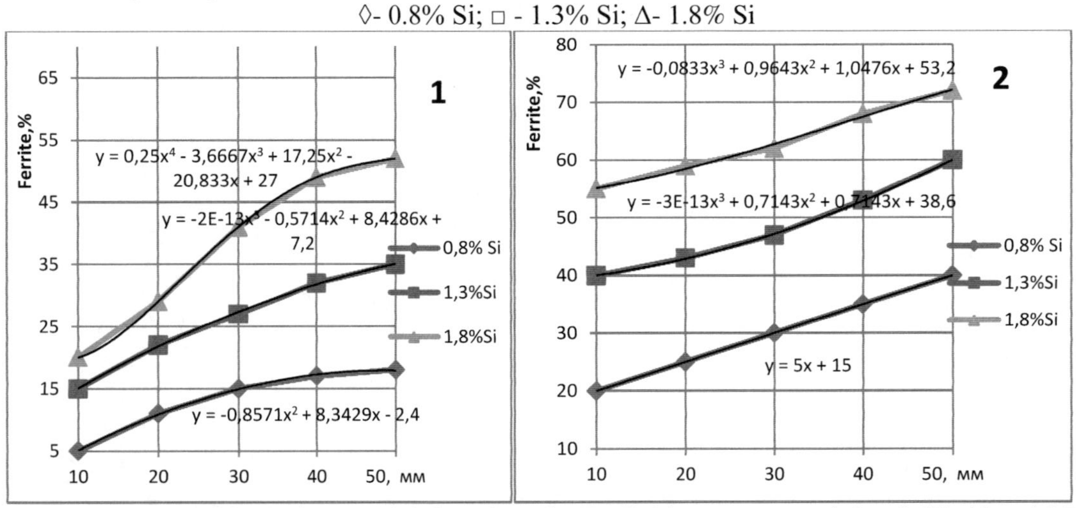

Figure 4 - Amount of ferrite in materials based on the Fe-C-Al system, containing 9.5% Al and 3.7% C, depending on silicon content, section thickness and holding time: 1-1 hour; 2 - 2 hours; ◊ - 0.8% Si; □ - 1.3% Si; Δ- 1.8% Si

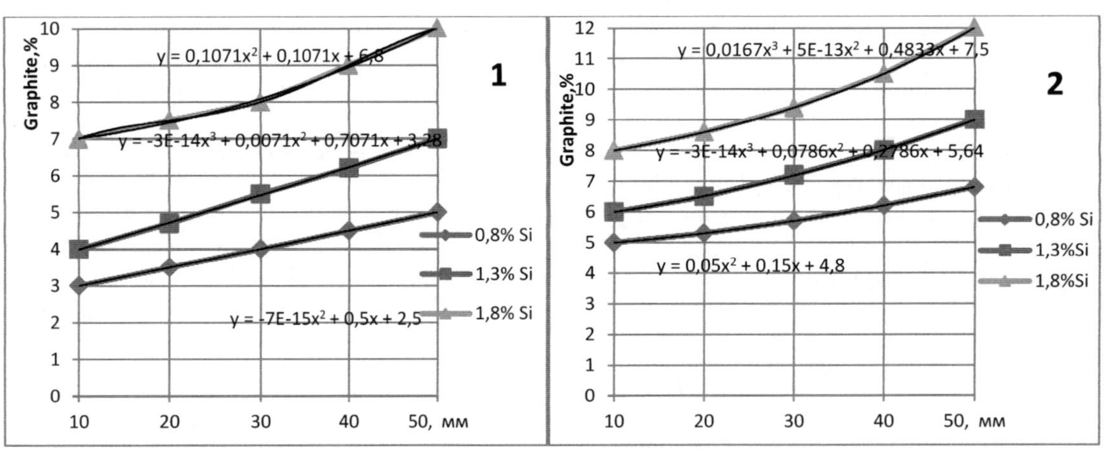

Figure 5 - The amount of graphite in materials based on the Fe-C-Al system, containing 9.5% Al and 3.7% C, depending on the silicon content, section thickness and holding time: 1-1 hour; 2 - 2 hours; ◊ - 0.8% Si; □ - 1.3% Si; Δ- 1.8% Si

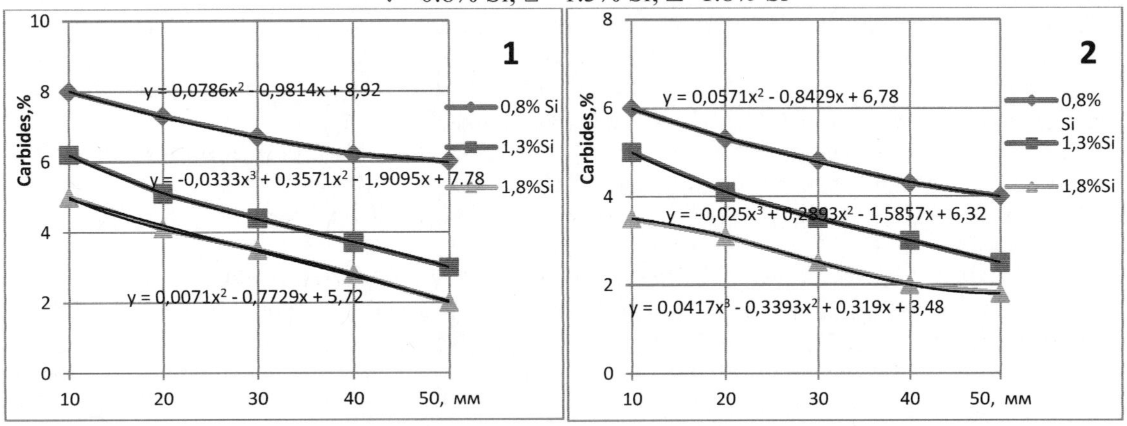

Figure 6 - The amount of carbides in materials based on the Fe-C-Al system, containing 9.5% Al and 3.7% C, depending on the silicon content, section thickness and holding time: 1-1 hour; 2 - 2 hours; ◊ - 0.8% Si; □ - 1.3% Si; Δ- 1.8% Si

The structure of the sample material on the basis of Fe-C-Al system cast in sand form is characterized by a plate straight shape of graphite with a small amount of swirling graphite form (10 %). The size of graphite inclusions varies in a wide range from 15 to 120 μm, but the preferred length is estimated at 50 μm. Graphite occupies about 6...7 % of the area.

The structure of a metal sample poured into a metallized form that is in a molten quenching medium immediately after crystallization acquires a finely dispersed austenitic-cementite structure of white iron. Heat treatment of samples in the form of an isothermal extract at a temperature of 1223 K for 2 hours, quenched from the liquid state, leads to accelerated graphitization of cementite. The released graphite is like a small inclusion of compact form with a diameter of 15 ... 60 microns or less.

An increase in the exposure time in the metallized form leads to an increase in the share of graphite due to a more complete decomposition of carbides, a decrease in perlite, an increase in ferrite, a decrease in carbides.

4 Conclusions and perspectives

This technology, which assumes the production of compact forms of graphite inclusions, is distinguished by stability with respect to the quality of graphite inclusions and metal matrix from other technological processes for producing materials based on the Fe-C-Al system currently used in the foundry industry. The stability of obtaining a compact form of graphite is provided by the specific features of the pouring of materials based on the Fe-C-Al system and the formation of the preform in a protective quenching medium. At the same time, there is no modification of materials based on the Fe-C-Al system and, as a consequence, the use of modifiers, which makes the proposed technological process economically effective. The problems associated with the provision of stringent requirements for the chemical composition of materials based on the Fe-C-Al system, in particular, the content of surfactants, are also eliminated.

5 Acknowledgements

The work was supported by the Ministry of education and science of the Russian Federation, within the framework of the State task № 9.3236.2017/4.6, state Contract under the 220th resolution №14.z50.31.0023.

IOP Conf. Series: Materials Science and Engineering **570** (2019) 012014 doi:10.1088/1757-899X/570/1/012014

References

[1] Alexandrov H H, Milman B S and Andreev V V, Illicheva L V 1977 Modern production methods and properties of high-strength cast irons *Organization of production and management* (Moscow: NIIEinformenergomash) №3 pp 64-66

[2] Zakharchenko E V, Levchenko Yu N and Gorenko V G 1986 *Castings of cast iron with spherical and vermicular graphite* (Kyiv: Naukova Dumka) pp 17-19

[3] Sabatinov B L, Abramenko Y E and Bekh N I 1988 *High-strength cast iron in the automotive industry* (Moscow: Mechanical Engineering) p 216

[4] Kuznetsov B L 1984 *Application of cast iron with spherical and vermicular graphite in the automotive industry* (Moscow: IPK) p 184

[5] Bikulov R A, Safronov N N 1998 *Innovation process of guaranteed compact graphitization in cast iron* Economic synergetics and innovation process: Materials of interuniversity scientific and practical conference (Naberezhnye Chelny: KamPI) pp 86-88

[6] Bikulov R A, Safronov N N 2010 *Technological process of guaranteed compact graphitization of cast irons* Foundry of Russia No 1 pp 34-36

[7] Bikulov R A, Safronov N N 2013 *Method for producing aluminum cast iron with compact graphite inclusions* Patent № 2487950 Russian Federation, IPC C21C 1/10

[8] Herman D E 1980 *Hardening from a liquid state Textbook for universities* (Moscow: Mechanical Engineering) p 230

[9] Zhukov A A, Savulyak V I and Pakhnushchiy I O 1998 *High-sulfur and sulfur-copper antifriction cast irons with improved machinability* Metallurgical science and heat treatment of metals No 3 pp 28 –30

[10] Zhukov A A 1996 *High-strength cast irons and cast steels with finely-divided compact graphite* Foundry No 10 pp 11–13

IOP Conf. Series: Materials Science and Engineering **570** (2019) 012015 doi:10.1088/1757-899X/570/1/012015

Disadvantages of the selective laser sintering technology in the manufacture models for investment casting

J N Dementyeva[1], R N Kashapov[1,2], N F Kashapov[1], L N Kashapov[1,3]
[1] Engineering Institute, Kazan Federal University, Kazan, 420008, Russia
[2]Laboratory of Radiation Physics, Zavoisky Physical-Technical Institute, FRC Kazan Scientific Center, Russian Academy of Sciences, Kazan, 420029, Russia
[3]Laboratory of Continuum Mechanics, Institute of Mechanics and Engineering, FRC Kazan Scientific Center, Russian Academy of Sciences, Kazan, 420111, Russia

kashramiln@gmail.com

Annotation. Preparation of production and development of new science-intensive products is a complex and time-consuming process, including the creation of product models for carrying out a series of experimental studies to refine computational models, and then manufacturing the necessary technological equipment. The use of digital production technologies can reduce the laboriousness of this process. The technology of selective laser sintering eliminates the need for the design and manufacture of casting equipment. This paper presents an analysis of the difficulties encountered when using the SLS process for the manufacture of polystyrene smelted models.

Introduction

Casting and smelting models are common methods of obtaining accurate, thin-walled, complex in the configuration of castings from various metals and alloys. Extensive industrial use of these methods began in the 60s of the last century. And it is associated with the development of equipment for investment casting and burning models of polystyrene foam. Styrofoam has good properties for this method: the burning temperature is within 560 °C, the burning time is from 10 to 20 s, the mass of the non-gasified residue reaches 0.015% of the mass of the model. In addition to polystyrene, there are other materials. For example, PMMA (Doro Chemical USA) [1] has a mass of non-gasified residue 20 times less than that of foundry polystyrene.

For the manufacture of polystyrene models used equipment, manufactured both by traditional methods, and using additive technologies. This can significantly reduce the time of pre-production. The technology of selective laser sintering is one of the most common technologies for growing polystyrene models.

During casting, melt pouring is carried out into the working cavity of an integral mold, obtained after preliminary burning or melting of the model from the mold. The ceramic shell is ignited and heated before melting.

Initially, polystyrene foam models (Replicast-CS [1]) traditionally produced in molds were used for the manufacture of shell molds, and with the advent of additive technologies at present, it became possible to use “grown” models. With all the advantages of selective laser sintering technology, there is no information in the literature about the existing shortcomings and problems that the manufacturer may encounter when using this technology [2]. Therefore, the purpose of this work

IOP Conf. Series: Materials Science and Engineering **570** (2019) 012015 doi:10.1088/1757-899X/570/1/012015

was to identify the difficulties encountered when using the SLS process for the manufacture of polystyrene smelted models

Main part

In experiments on selective laser sintering, we used the sPro 60 HD (3D Systems), USA. sPro 60 HD is an industrial selective laser sintering plant that creates physical models from powder compositions, mostly polymer ones. This installation is actively used in aircraft building, machine building, automotive industry, military-industrial complex, medicine for creating prototype models. Features: the construction area is 381x330x437 mm, the thickness of the layer is 0,08-0,15 mm, the printing speed is 1,0 l / h, the scanning system is ProScan CX (digital), the scanning speed is 6 m / s, the laser power is - 70 W / CO2 laser. The installation uses commercial materials - polyamide, polystyrene, polypropylene, glass-filled nylon 12.

It is known that the glass-transition temperature of the material is Tg = 89 ° C. Do not exceed this temperature on the surface of the layer to avoid sticking the material to the screed roller. The layer thickness is 150 μm. A laser power of 28 W is required for a layer of 150 μm. Laser scanning speed - 6 m/s.

CastForm PS powder (3D System) was used in the studies. It generates accurate low density samples suitable for investment casting [3]. These samples are coated with wax, then they can be polished to a smooth surface [4].

The mechanical properties of structural materials play a key role in their application. The basis for understanding their properties is information on how the material reacts to the operating external mechanical load, which are determined by such parameters as tensile strength, elongation at break, modulus of elasticity, plasticity coefficient, etc. For the evaluation of these properties, first of all, universal testing machines (UIM) during the testing of materials for tensile, compression, bending, shear, torsion. Universal testing machine GT-7000S with heat chamber UGT7001-HC6. Floor two-column machine with maximum effort 500 kgf. and 2000 kgf. Modern mechanics, high-precision force and displacement sensors, an external extensometer for elastomers, a user-friendly interface allow you to solve various research tasks.

Impact toughness was determined by the method of Charpy without a notch (GOST 4647), the energy of the hammer is 7.5 J.

The results of the mechanical tests of the obtained samples are presented in Figure 1 and in Table 1. Based on this information, it is possible to conclude that the low mechanical properties of the obtained samples. This will cause great difficulties in extracting samples from the construction chamber and cleaning it from "green" powder. The high brittleness of the samples will lead to their destruction when the powder is removed, especially those having a thickness of less than 1 mm. The test results for impact strength showed that it is equal to 1.498, 1.206, 1.353 kJ / m^2.

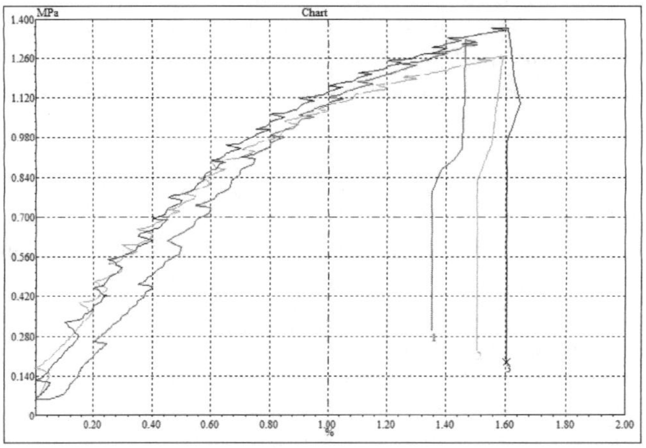

Figure 1 The results of the mechanical tests of polystyrene sls samples

Table 1. The results of mechanical testing of polystyrene sls samples

Test No	Elastic modulus kN/m²	Max. Load MPa	Max. Load kgf/cm²	Elongation Gauge %	Speed mm/min	Thickness mm
1	145800.609	1.3	13.5	1.4	50.0	2.98
2	113237.086	1.3	12.9	1.5	50.0	3.05
3	130114.602	1.4	13.9	1.6	50.0	2.91

Tests were conducted on the cultivation of complex elements of aviation products. Photographs of the samples are presented in Figure 2.

a б

Figure 2 Prototypes of polystyrene models, damaged during the removal of powder residues.

As can be seen, the samples are destroyed in the process of extracting powder from the construction and cleaning chamber. This causes certain difficulties at production plants, as it requires performers who are able to gently clean the products for this work.

To increase the strength of these models need to be impregnated with wax. Without wax impregnation, they cannot be covered with a ceramic coating and create a refractory shell form. At this stage, it may also be difficult to impregnate thin-walled or small-sized products. Figure 3a shows a sample of a monowheel after wax impregnation with observance of technological conditions. In Figure 3b, the temperature of the autoclave was not observed, which led to a distortion of the geometry of the blades of the monowheel.

a б

Figure 3 The results of waxing polystyrene models.

Conclusion

IOP Conf. Series: Materials Science and Engineering **570** (2019) 012015 doi:10.1088/1757-899X/570/1/012015

As a result of the research it was found that, despite all the advantages of the technology of selective laser sintering, it has several disadvantages in the manufacture of polystyrene models. The first is the low mechanical characteristics of the samples. The second is the need for waxing to increase strength. The third is the possibility of distortion of geometry in case of non-observance of technological procedures for impregnation with wax. And of course, there are new technologies for additive manufacturing, which work on PMMA material and do not use laser radiation. All this suggests the need to create new materials for the SLS process eliminating all the above disadvantages.

Acknowledges

The work was carried out in the framework of the Base part of the state task of Ministry of Science and Higher Education of the Russian Federation (3.9399.2017/8.9)

References

[1] Shuljak V S 2007 *Casting on gasified models* 408 p

[2] Shcmidt J and etc. 2015 *A novel process chain for the production of spherical SLS polymer powder with good flowability* The 7th World Congress on particle technology

[3] Kruth J.-P. and etc. Consolidation of polymer powder by selective laser sintering

[4] J Shcmidt, M Plata and etc. 2012 *Production of polymer particles below 5μm by wet grinding* Powder technology 228 p

Estimation of the state of a technical system with substantially non-linear characteristics

D N Demyanov

Naberezhnye Chelny Institute of Kazan (Volga region) Federal University, Suumbike 10 A, Naberezhnye Chelny city, the Russian Federation

demyanovdn@mail.ru

Abstract. The article considers the problem of estimating the state of the technical system described by nonlinear differential equations. An approach is proposed to design reduced-order asymptotic observers for systems where dynamic equations contain products of state variables. The solvability conditions for the synthesis problem in the form of linear matrix equations are determined. The possibility of reducing the original problem to the classical modal control problem is shown. Two modifications of the algorithm for calculating the coefficients of the state observer, which ensures that the estimation error is asymptotically zero, are proposed.

1. Introduction

Asymptotic state observers are often used to estimate the state of various technical systems [1, 2]. At the same time most design methods and algorithms imply the linearity of the model of the object, which is not always followed in practice [3]. Well-known results in designing observers of nonlinear systems state do not cover all known types of nonlinearities. Therefore, the problem of developing new methods of estimating, allowing to restore the state of nonlinear systems can be considered quite actual.

2. Problem specification

Consider the mathematical model of the technical system in the form of a set of equations in the state space. Let us introduce the following notation: X is the state vector of considering system, which includes two components X_1 и X_2; x_1 is the first element of the state vector; U is an input signal; Y is an output signal. We assume that the first part of the state vector forms the output signal of the system and the second part of it cannot be measured directly (respectively, $Y \equiv X_1$; $y_1 \equiv x_1$).

Consider the situation when equations of system dynamics include terms proportional to x_1 and other state variables:

$$\begin{cases} \dot{X}_1 = A_{11}X_1 + A_{12}X_2 + x_1F_{11}X_1 + x_1F_{12}X_2 + B_1U; \\ \dot{X}_2 = A_{21}X_1 + A_{22}X_2 + x_1F_{21}X_1 + x_1F_{22}X_2 + B_2U; \\ Y = X_1. \end{cases} \quad (1)$$

It is obvious that the model of the system can be represented in the form (1) only in the case when the input signal of the system is not transmitted directly to its output, and the state variable determining nonlinear terms in the dynamic equation arrives at the output signal of the system. It is required to synthesize the observer of the dynamic system state (1), which allows using the known Y and U to

IOP Conf. Series: Materials Science and Engineering **570** (2019) 012016 doi:10.1088/1757-899X/570/1/012016

form the vector of assessment Z in the following way:

$$\lim_{t\to\infty}\varepsilon(t)=\lim_{t\to\infty}[Z(t)-X_2(t)]=0\,. \tag{2}$$

3. Synthesis of the observer's equation

Let us compose the equation of the observer on the basis of the second equation of system (1), adding it to the term proportional to the difference of assessment and the true value of the vector X_2:

$$\dot{Z}=A_{21}Y+A_{22}Z+y_1F_{21}Y+y_1F_{22}Z+B_2U+L(A_{12}Z+y_1F_{12}Z-A_{12}X_2-y_1F_{12}X_2). \tag{3}$$

Substituting true values of terms $A_{12}X_2$ and $y_1F_{12}X_2$ from the first equation system (1) in equation (3), we get:

$$\dot{Z}=(A_{22}+LA_{12})Z+y_1(F_{22}+LF_{12})Z+(A_{21}+LA_{11})Y+y_1(F_{21}+LF_{11})Y+(B_2+LB_1)U-L\dot{Y}. \tag{4}$$

Let us form the equation of dynamics of the estimation error of the vector X_2 by subtracting from the equation (4) the second equation of the system (1):

$$\dot{\varepsilon}=(A_{22}+LA_{12})\varepsilon+y_1(F_{22}+LF_{12})\varepsilon\,. \tag{5}$$

In the general case the solution of equation (5) depends both on matrix coefficients on the right-hand side and on the form of the function y_1. It is obvious that not for all values of y_1 the equation (5) will provide the required dynamics of estimation error. To simplify the problem we choose the matrix L:

$$F_{22}+LF_{12}=0\,. \tag{6}$$

Whenever the condition (6) is fulfilled, the equation (5) will be the following:

$$\dot{\varepsilon}=(A_{22}+LA_{12})\varepsilon. \tag{7}$$

The asymptotic stability of the trivial solution of equation (7) can be ensured by the appropriate choice of the matrix L on the condition of visibility or identity of the pair (A_{22}, A_{12}). Obviously, this will satisfy the condition (2).

Let's write equation (4) with a constraint (6):

$$\dot{Z}=(A_{22}+LA_{12})Z+(A_{21}+LA_{11})Y+y_1(F_{21}+LF_{11})Y+(B_2+LB_1)U-L\dot{Y}\,. \tag{8}$$

The equation (8) determines the asymptotic state observer of reduced order for system (1). At the same time in its right part there is a term proportional to the derivative of the output signal. It is known [1] that in practice the introduction of the operation of numerical differentiation can lead to occurrence of significant errors, so it is recommended to avoid it.

To eliminate the operation of differentiation from the observer equation (8), we modify the calculation scheme. We introduce the auxiliary variable $w(t)$ as follows:

$$w(t)=Z(t)+LY(t). \tag{9}$$

Taking into account the relation (9), the equation (8) can be written as follows:

$$\dot{w}=(A_{22}+LA_{12})w+(A_{21}+LA_{11}-A_{22}L-LA_{12}L)Y+y_1(F_{21}+LF_{11})Y+(B_2+LB_1)U\,. \tag{10}$$

In this case the equation for assessing the vector X_2 will be the following:

$$Z(t)=w(t)-LY(t). \tag{11}$$

Thus, in the case of visibility or identity of the pair (A_{22}, A_{12}) by fulfilling the condition (6), one can form observer equations that ensure that assessment error of unmeasured component of the state vector of a technical system (1) is asymptotically zero. In this case two variants of the design scheme are possible – with differentiation of the output signal (8) and without it (10) - (11).

4. The analysis of solution conditions

Let us define conditions for the existence of an observer described by equations (8) or (10) - (11). Relator (6) is a linear matrix equation. The condition of its solution and expression for full set of solutions are formulated in terms of matrix canonization technology [4].

The solution of equation (6) exists if the condition [4] is satisfied:

$$F_{12}\overline{F}_{22}^{R}=0. \tag{12}$$

In this case a full set of solutions is determined by the expression:

$$\{L\}_\eta = -F_{22}\tilde{F}_{12} + \eta\overline{F}_{12}^{L}. \quad (13)$$

Here η is an arbitrary numerical matrix of the corresponding dimension. For an arbitrary matrix M symbols like $\overline{M}^R$ и $\overline{M}^L$ show nonzero matrices of maximal rank such as $M\overline{M}^R = 0$ и $\overline{M}^L M = 0$, respectively; and $\tilde{M}^R$ и $\tilde{M}^L$ matrices of maximum rank $\tilde{M}^L M\tilde{M}^R = I$; the matrix $\tilde{M} = \tilde{M}^R\tilde{M}^L$ is also introduced to simplify recording.

Taking into account the expression (13), the matrix of coefficients from the right side of equation (7) will look like:

$$A_{22} + LA_{12} = A_{22} - F_{22}\tilde{F}_{12}A_{12} + \eta\overline{F}_{12}^{L}A_{12}.$$

By analogy with the well-known result [5], we introduce new notations:

$$A^* = \left(A_{22} - F_{22}\tilde{F}_{12}A_{12}\right)^T; \; B^* = \left(\overline{F}_{12}^{L}A_{12}\right)^T. \quad (14)$$

To satisfy the condition (2), it is necessary and sufficient all the eigenvalues of the matrix to be located in the left semi-plane of the complex plane. If the pair (A^*, B^*) is fully Kalman controlled or stabilized, then the solution to the original problem can be obtained using modal control methods [6].

Thus, the solution of the problem is possible if condition (12) is fulfilled, and the pair (A^*, B^*) given by expression (14) is Kalman controlled or stabilized.

5. Algorithm of synthesis

Based on obtained results we formulate an algorithm for calculating coefficients of the state observer of a nonlinear dynamical system, the model of which can be represented in the form (1).

1. Set the matrix of coefficients of the model (1), the required law of distribution of poles of the observer, information on feasibility or non-feasibility of differentiation of the output signal.
2. Check the fulfillment of the condition (12). If it is fulfilled, then you should proceed to the next step, otherwise the proposed method is not applicable.
3. Calculate the matrix of coefficients by formulas (14) and estimate the controllability / stabilizability of the pair (A^*, B^*). If it is fully controllable or uncontrollable eigenvalues are located in the required area of the complex plane, then you should proceed to the next step, otherwise it is recommended to change requirements for the dynamics of the estimation process or use a different calculation algorithm.
4. Solve the problem of modal control by providing the choice of the matrix η to the desired location of eigenvalues of the matrix $A^* + B^*\eta^T$ ..
5. Calculate the matrix of observations using the formula (13).
6. Form equations of the observer. If differentiation of the output signal is allowed, then equation (8) is used. Otherwise, the equation of the observer's dynamics is given by formula (10) and the expression for estimating the vector X2 is given by formula (11).

The end of the algorithm.

6. Conclusions

A new method for estimating the state of the technical system is proposed, the model of which includes the product of state variables. The solvability conditions for the problem are formulated, as well as the step-by-step algorithm for calculating the coefficients of the observer, which ensures that the estimation error is asymptotically zero. The obtained results can be used in practice when building diagnostic systems for technical objects with non-linear characteristics.

7. Gratitude

This work was supported by the Russian Foundation for Basic Research (Grant No. 17-08-00516).

References

[1] Kuzovkov N T 1976 Modal Control and Monitoring Devices M 184 p

IOP Conf. Series: Materials Science and Engineering **570** (2019) 012016 doi:10.1088/1757-899X/570/1/012016

[2] Volkov V G, Demyanov D N 2017 *Designing Physically Realizable State Observer for Estimating the Kinematic Parameters of the Road Train* IOP Conference Series: Science and Engineering Vol 240 No **012067**

[3] Shafai B, Saif M 2015 *Proportional-integral Observer in Robust Control, Fault Detection, and Decentralized Control of Dynamic Systems* Studies in Systems, Decision and Control Vol 27 pp 13–43

[4] Bukov V N 2006 *Embedding Systems. Analytical Approach to the Analysis and Synthesis of Matrix Systems* Kaluga: Publishing House of N.F. Bochkareva 720 p

[5] Asanov A Z, Demyanov D N 2013 *Analytical Synthesis of Invariant Reduced-order State Observers* Optoelectronics, Instrumentation and Data Processing Vol 49 Issue 6 pp 546–52

[6] Ogata K 2010 *Modern Control Engineering* Fifth Edition New Jersey: Prentice Hall 905 p

IOP Conf. Series: Materials Science and Engineering **570** (2019) 012017 doi:10.1088/1757-899X/570/1/012017

Rebuilding of heavy machine tools

Z R Faskhieva and R M Khusainov

Branch of Kazan Federal University in Naberezhnye Chelny, 68/19 Mira Street, Naberezhnye Chelny, 423812, Russia

rmh@inbox.ru

Abstract. The article studies the questions of modernization reasonability, important tasks and activities of heavy cutting machines overhaul.

1. Introduction

Nowadays it is possible to find the use of heavy metal-cutting machines in production which have worked for more than 30 years. The restoration of their physical and moral deterioration requires an individual approach, depending on the conditions of production. One example of such machines is the horizontal boring machine with NC Skoda W200, which is used for drilling, boring, countersinking and milling operations in the production of heavy large parts, the weight of which reaches about 20 tons.

2. Description of the Problem

There is a problem of unsatisfactory accuracy of geometric shapes and the location of the machined surfaces, low positioning accuracy, presence of CNC system failures.

According to the test report on the geometric accuracy, the actual error exceeds the permissible values in several times, and some indicators - in a factor of ten [1,2].

In this regard, the decision to upgrade to restore geometric accuracy and improve the technical characteristics of the machine was made.

An important task of the overhaul and modernization of the equipment is to study all possible options for changing the design, as a result of which it will be more accurate, productive, reliable and safe in operation [3,4].

In modern conditions, when it comes to the reasonability of modernization of technological equipment, conflicting opinions can appear. Some believe that if there is a financial opportunity, it is better to buy new equipment. Others consider that modernization is expedient; firstly, for the economic reasons since the machine park is not enough updated at the enterprises. If we talk about the repair of small serial Chinese machines, then, perhaps, it is not necessary to invest in their modernization, because after three warranty periods parts wear begins, and the purchase of spare parts can take from 6 to 8 months. However, if we are talking about the repair of large and very heavy machines, as horizontal boring machine Skoda W200, worked for more than 30 years, such a new machine costs from 6 to 8 million euros. Technological progress is not as fast in the line of very large machines as in high-speed ones. There is no need for fast feed and high speed in heavy machines. Therefore, in this case, modernization is more appropriate than the purchase of a new machine.

3. Development of Activities for Modernization

According to the nature of the machine defects, identified in the tests for geometric accuracy, the following activities should be performed with major repairs:

IOP Conf. Series: Materials Science and Engineering **570** (2019) 012017 doi:10.1088/1757-899X/570/1/012017

1. Bed and saddle slideways repair.

The bed slideways and racks of a horizontal boring machine are efficiently repaired by grinding. When repairing heavy equipment, the most effective choice is mobile metal-cutting machines, which make it possible to provide high quality machining and save time and money.

The restoration of the saddle slideways should be performed using polymeric materials. The use of such polymers allows forming a response guide surface directly assembled with the saddle, reducing friction and wearing slideways on the machine bed.

2. Spindle unit repair with manufacturing new bushings and replacement of bearings and tool clamping mechanism.

3. The equipping of the machine with a new NC system, modern frequency converters complete with asynchronous electric motors and a DC Converter of the main drive, modern reference system for coordinate movements.

After equipping the machine with a modern NC system, it is possible to ensure the necessary processing accuracy using software control functions [5,6]. Previously, it was possible to influence the factors that affect the deviation from the geometric accuracy only by adjusting and repairing individual components of the machine. At present, modern NC systems are equipped with intelligent functions to reduce the impact of machine mechanics errors on the trajectory of the cutting edge of the tool. There are the following types of program compensations:

1. Backlash compensation. In the feed drive, the backlash leads to a delay in the movement of the movable body. To compensate for the backlash the movement with each change in direction is adjusted by the amount of correction entered for this axis in the corresponding machine data. The backlash value can be measured by testing the positioning accuracy.

2. Interpolation compensation. It allows to fix positioning errors, errors of misalignment of the movement, unevenness between the axes and the angular error. Correction values, equal to the error values, are determined by testing and recorded in the compensation table of the NC system memory with reference to the position along the axis. The compensated portion of the movement is divided into several sections of the path. The actual positions that limit these sections of the path are designated as "anchor points". The corresponding axis is corrected with linear interpolation between the reference points.

3. Dynamic preliminary control (compensation of error latency). With the help of preliminary control, the axial delay error can be reduced to zero. The delay results in a velocity-dependent error of the contour, especially in acceleration processes at the curves of the contour, such as circles and angles.

4. Friction compensation. It most strongly affects the error in the processing of rest friction in the slideways. The errors of the circuit caused by the rest friction can be well compensated by setting an additional motion pulse with the corresponding sign and amplitude.

5. Drift compensation. Temperature-dependent drift in the analog feed drives means that the analog speed control circuits must be controlled with a small set speed value other than zero to achieve a quiescent state. To prevent this error, a small additional set value of the engine speed is connected.

6. Electronic weighing balance. A mobile node of vertical flow actuators without weight trim inadvertently omitted after release and the error occurs on the axis. Electronic weighing balance prevents the axis with weight loads drawdown when the control is turned on. After releasing the brake, the available constant moment of weight balancing holds the position of the vertical axis.

Compensations are valid in all operating and control modes as soon as the necessary data is entered [7].

4. Conclusion

Implementation of the modernization of the machine with these measures should provide:

- efficiency compared to buying a new machine;
- improved processing performance;
- ability to handle complex surfaces;

IOP Conf. Series: Materials Science and Engineering **570** (2019) 012017 doi:10.1088/1757-899X/570/1/012017

- improved processing accuracy.

Reference

[1] GOST 8-82 1995 *Metal-Cutting Machines. General Requirements for Accuracy Test* (Moscow: IPK Publishing house of standards) pp 6-8

[2] GOST R ISO 230-1-2010 2011 Tes*ting of Machines. Part 1. Methods of measurement of geometrical parameters* (Moscow: Standartinform) p 166

[3] Labuig S 2017 Repair and Modernization of Machine Tools *Engineering Rhythm* (Moscow: Engineering Rhythm) **No 10** p 42-3

[4] Safarov D T, Kondrashov A G, Safarova L R, Glinina G F 2017 Energy Planning in Production Shops with Numerically Controlled Machine Tools *Russian Engineering Research.* (New York: Allerton Press) **Vol 37 No 9 pp** 827-34

[5] Khisamutdinov R M, Khusainov R M, Vedernikov Yu A, Petrov S M and Stupko V B 2013 *Geometric Accuracy of Machine Tools* (Naberezhnye Chelny: PC "Raduga") p 153

[6] Anikeeva O V 2016 Construction of Accuracy Estimate System of the Processed Surfaces *Scientific and Technical Bulletin of the Volga region* **No 6** pp 33-37

[7] Balabanov I P, Simonova L A and Balabanova O N 2015 Systematization of accuracy indices variance when modelling the forming external cylindrical turning process *IOP Conference Series: Materials Science and Engineering* (Bristol: IOP Publishing Ltd) **vol 86** 012010

IOP Conf. Series: Materials Science and Engineering **570** (2019) 012018 doi:10.1088/1757-899X/570/1/012018

Mathematical modeling of heat treatment of three-layered material

V L Fedyaev[1], V I Khaliulin[2], A V Belyaev[2] and Faskhutdinov M Kh[2]

[1]Institute of Mechanics and Engineering - Subdivision of the Federal State Budgetary Institution of Science «Kazan Scientific Center of the Russian Academy of Sciences», Lobachevsky str. 2/31, Kazan, 420111, Republic of Tatarstan, Russia
[2]Kazan national research technical university named after A.N. Tupolev, K. Marx street 10, Kazan, 420111, Tatarstan, Russian Federation

V.L. Fedyaev@yandex.ru, pla.kai@mail.ru, alexey-beljaev@mail.ru

Abstract. The paper presents mathematical models of the thermal behavior of the three-layered material in the presence of a thin air gap between the two of them. Both convective and radiative-convective heat transfers with the environment are considered. As a result of an acceptance of a set of assumptions, the relationships to estimate the change of the material temperature in the layers over time were found for each of these cases.

1. Introduction

The material under consideration is a three-layer plane wall (strip) infinitely extended in the longitudinal direction. The first layer has a thickness $h_т$, its material properties are characterized by density $\rho_т$, thermal conductivity $\lambda_т$, specific heat capacity $c_т$. The second layer is a relatively thin fabric with a thickness h_m freely lying on the first layer. In the general case, its material is porous, and you should distinguish between macroporosity characterizing the cavities between the threads of the fabric and the microporosity of the threads material.

2. Body text

Both the total porosity ε_m and the average (equivalent) thermophysical parameters – density ρ_m, thermal conductivity coefficient $\lambda_{\rm m}$, and specific heat capacity $c_{\rm m}$ – will possibly be anisotropic. The last layer is a layer of polymeric particles discretely located on the surface of the fabric. It is of great difficulty to take into account the effect of these discrete particles on the strip's thermal behavior. To avoid this, we replace the particles with an equivalent continuous layer with thickness h_p. Let us suppose that the parameters of this layer are close to the parameters of the material of the particles: density ρ_p, thermal conductivity λ_p, and specific heat capacity c_p.

The following important circumstance characterizing the structure of the strip under consideration is the conditions thermal contact of its layers. It is believed that the particles are contiguous to the fabric surface rather tightly there is ideal thermal contact between the materials of the introduced continuous layer and the reinforcing fabric. Since the fabric lies freely on the first layer its surface has a certain relief, an interlayer of gas (air) is formed in the interfacial region of the fabric layer and the first layer. If the radiative (radiant) heat transfer between the surfaces separated by a gas interlayer is not taken into account then the thermal conductivity of the contact is $1/R_k = 1/R_f + 1/R_g$, where $1/R_f$ is the thermal conductivity of the actual contact, $1/R_g$ is the conductivity of the gas interlayer.

Let us suppose that the material of this interlayer (conditional transition layer) with thickness h_c, is a mixture of the first layer, fabric, and air their volume concentration is $\beta_т$, β_m and β_a ($\beta_т + \beta_m + \beta_a = 1$). The averaged thermophysical parameters of this material are $\rho_c = \beta_т\rho_т + \beta_m\rho_m + \beta_a\rho_a$, $\lambda_c = 1/(\beta_т/\lambda_т + \beta_m/\lambda_m + \beta_a/\lambda_a)$, $c_c = \beta_т c_т + \beta_m c_m + \beta_a c_a$, the index "a" is used to denote the characteristics of air. Hence the thermal resistance of the considered layer is $R_c = h_c/\lambda_c$ its value is close to R_K.

Next, we single out the following heat transfer variants of the strip surfaces with the environment: double-sided convective and radiative-convective from the side of polymeric particles, convective on the outer surface of the first layer.

Convective heat transfer of the strip. Provided that the surface temperature of the strip that is in contact with the gaseous medium is equal to some average value T_w the thermophysical parameters of the gas weakly depend on the temperature at the surface of the strip's test section of length l, we accept [1, 2] that the average value of the convective heat transfer coefficient is $\bar{\alpha}_k = 0.66k_л(\lambda_g/l)\mathrm{Re}^{0.5}\,\mathrm{Pr}^{1/3}$ for laminar gas flow and $\bar{\alpha}_k = 0.04k_т(\lambda_g/l)\mathrm{Re}^{0.8}\,\mathrm{Pr}^{0.43}(\mathrm{Pr}/\mathrm{Pr}_w)^{0.25}$ for turbulent one. Here, k_l, $k_т$ – correction coefficients, λ_g – thermal conductivity coefficient of the surrounding gas environment, $\mathrm{Re} = wl/v_g$, $\mathrm{Pr} = v_g/a_g$, $a_g = \lambda_g/(\rho_g c_g)$; ρ_g, c_g, v_g – density, specific heat capacity and kinematic viscosity of the gas; $w = |u_т \pm u_g|$, $u_т$, u_g – motion speed of the strip and gaseous environment.

Because the strip is very long in the longitudinal direction thermal resistance in this direction is substantially higher than the resistance across the strip, therefore, the heat transfer in the transverse direction will be predominant in the strip. Considering the above circumstance, assuming that internal sources (sinks) of heat are absent the thermophysical parameters of the material of the layers change only slightly with temperature the corresponding thermo-adjoint problem is written. The main difficulties in solving this problem are connected with the mathematical description of heat distribution at the initial stage of heating (cooling) of the strip also called the inertial period. This period is characterized by the gradual heating up or cooling of the material of the layers with the boundary conditions playing a large role In our opinion, since the strip thickness is, in general, comparatively small the thermal conductivity of the material of the first layer, fabric and polymeric particles (excluding the intermediate layer) is quite large this period is not of great interest. The regular period follows the inertia period. During this period the entire strip is involved in the hearting (cooling) process the heating (cooling) rate in different layers is assumed to be the same and equal to a certain average speed.

Focusing on the regular period, we consider two mechanisms of the thermal behavior of the material of the layers. *In the first case,* we assume that in all strip layers the average temperature of the material is almost the same, characterized by the value $\bar{T}$. Accordingly, the average density of the strip material is

$$\bar{\rho} = \gamma_т\rho_т + \gamma_c\rho_c + \gamma_m\rho_m + \gamma_p\rho_p,$$

the average specific heat capacity $\bar{c} = \gamma_т c_т + \gamma_c c_c + \gamma_m c_m + \gamma_p c_p$, where $\gamma_т = h_т/h_0$, $\gamma_c = h_c/h_0$, $\gamma_m = h_m/h_0$, $\gamma_p = h_p/h_0$, $h_0 = h_т + h_c + h_m + h_p$.

From the condition of the heat balance of a strip section with the length l_0 over time $d\tau$ it follows that the thermal energy change in the volume of the strip $V_0 = l_0 h_0 \cdot 1$ of a unit width due to the heat transfer with the surrounding gaseous medium will be:

$$\bar{\rho}\cdot\bar{c}V_0 d\bar{T}=\left[\bar{\alpha}_{k1}\left(T_{g1}-\bar{T}\right)-\bar{\alpha}_{k2}\left(T_{g2}-\bar{T}\right)\right]S_0 d\tau\,, \tag{1}$$

where $S_0=l_0\cdot 1$ – a unit width strip surface area; T_{g1}, T_{g2} – the average temperature of the gas medium in contact with the first and the third layers of the strip, respectively; $\bar{\alpha}_{k1}$, $\bar{\alpha}_{k2}$ - the average convective heat transfer coefficients on opposite sides of the strip.

Solving equation (1) we find

$$\bar{T}=\bar{T}(\tilde{\tau})=T_0\exp(-\tilde{\tau})+\left(\frac{\bar{\alpha}_{k1}T_{g1}+\bar{\alpha}_{k2}T_{g2}}{\bar{\alpha}_k}\right)\left(1-\exp(-\tilde{\tau})\right). \tag{2}$$

Here $\tilde{\tau}=\tilde{\alpha}_k/\left(\bar{\rho}\cdot\bar{c}\cdot h_o\right)$, $\tilde{\alpha}_k=\bar{\alpha}_{k1}+\bar{\alpha}_{k2}$; T_0 - the average initial temperature of the strip layers.

In particular, for small values of the argument $\tilde{\tau}$ ($\tilde{\tau}\ll 1$) the formula (2) is simplified as follows:

$$\bar{T}=\bar{T}(\tilde{\tau})\approx T_0+\left(\bar{\alpha}_{k1}\Delta T_{g1}+\bar{\alpha}_{k2}\Delta T_{g2}\right)\tilde{\tau}/\tilde{\alpha}_k\,, \tag{3}$$

where $\Delta T_{g1}=T_{g1}-T_0$, $\Delta T_{g2}=T_{g2}-T_0$.

On the contrary, when the value $\tilde{\tau}$ is high ($\tilde{\tau}\gg 1$) (long time τ) the strip thickness h_0, average density, and specific heat capacity of its material are small; the influence of the multiplier $\exp(-\tilde{\tau})$ on the temperature $\bar{T}$ is insignificant.

As a consequence, approximately

$$\bar{T}\approx\left(\bar{\alpha}_{k1}T_{g1}+\bar{\alpha}_{k2}T_{g2}\right)/\tilde{\alpha}_k\,, \tag{4}$$

the average temperature of the strip is determined by mainly the ambient temperature and the intensity of the convective heat transfer on its surfaces.

In the second case, we assume that the heat transfer in the layers is carried out as a transfer in the "mixture" consisting of materials of the first layer, intermediate layer, fabric and polymeric particles in the absence of sources, sinks of heat. The thermal behavior of this "mixture" is described by the equation:

$$\partial T/\partial\tau=\bar{a}\,\partial^2 T/\partial y^2\,, \tag{5}$$

where thermal diffusivity $\bar{a}=\bar{\lambda}/(\bar{\rho}\bar{c})$, $\bar{\lambda}=1/\left(\gamma_т/\lambda_т+\gamma_c/\lambda_c+\gamma_m/\lambda_m+\gamma_p/\lambda_p\right)$ – the average coefficient of molecular thermal conductivity of the material of the mixture under consideration, $T=T(y,\tau)$ ($0\le y\le h_0$) – the strip temperature.

At the strip borders when

$$y=0:\ q_1=-\bar{\lambda}\left(\frac{\partial T}{\partial y}\right)=\bar{\alpha}_{k1}\left(T_{g1}-T\right), \tag{6}$$

$$y=h_0:\ q_2=-\bar{\lambda}\left(\frac{\partial T}{\partial y}\right)=\bar{\alpha}_{k2}\left(T-T_{g2}\right); \tag{7}$$

at the initial time $\tau=0$

$$T=T_0\,. \tag{8}$$

The problem specified in this way (5) – (8) is solved by the variable separation method. Omitting the details of this solution, we present only the final dependencies for estimating the strip temperature for small values of the parameter $F_0=\bar{a}\tau/h_0^2$:

$$T\approx A_1+A_2\bar{y}+\left(B_1\sin(\beta\bar{y})+B_2\cos(\beta\bar{y})\right)\left(1-\frac{\tilde{\alpha}_k\tau}{h_0\bar{\rho}\cdot\bar{c}}\right), \tag{9}$$

and for high F_0:

$$T=A_1+A_2\bar{y}\approx\left(\bar{\alpha}_{k1}T_{g1}+\bar{\alpha}_{k2}T_{g2}\right)/\tilde{\alpha}_k\,. \tag{10}$$

Here $\bar{y} = y/h_0$, coefficients

$$A_1 \approx \left(\alpha_{k1}^* T_{g1} + \alpha_{k2}^* T_{g2}\right)/\alpha_k^*, \quad A_2 \approx \alpha_{k1}^* \alpha_{k2}^* \left(T_{g2} - T_{g1}\right)/\alpha_k^*,$$
$$B_1 = B_1(\beta) = \alpha_{k1}^* (T_0 - A_1)/(\beta \alpha_{k0}^*), \quad B_2 = (T_0 - A_1)/\alpha_{k0}^*; \tag{11}$$

parameters $\beta \approx \sqrt{\alpha_k^*}$, $\alpha_k^* = \alpha_{k1}^* + \alpha_{k2}^*$, $\alpha_{k0}^* = 1 + 0{,}5\alpha_{k1}^*$, $\alpha_{k1}^* = \bar{\alpha}_{k1} h_0/\bar{\lambda}$, $\alpha_{k2}^* = \bar{\alpha}_{k2} h_0/\bar{\lambda}$.

We note that dependence (10) coincides with (4).

Radiative-convective heat transfer is realized from the side of polymeric particles located on the fabric. It is assumed that the working surface of the heating element is flat, located parallel to the flat surface of the fabric its longitudinal (relative to the fabric) size is $a_н$, and transverse is $b_н$, respectively, the area is $a_н b_н$. This element is fixed the fabric moves with the speed $u_т$ relative to it. The gaseous medium between the element and the fabric is considered optically transparent (does not absorb radiation) it moves at a speed $w = u_т \pm u_g$ with respect to the fabric, its average temperature is T_g. The surfaces of the heating element and the fabric are grey the absorption coefficient does not depend on the wavelength of the incident radiation, accordingly, on temperature and other characteristics of the radiating body. The emissivity factor (coefficient of radiation) of the heating element surface is denoted by ε_1, the surfaces of the fabric with polymeric particles – ε_2.

Next, we single out two variants of the thermal contact of the heating element with the surface temperature T_1 and area $S_1 = S_н$, and the fabric section with the temperature T_2 and area S_2. In the first case, we assume that the areas are commensurable ($S_1 \square S_2$) and reasonably large. In the second case, the heating element area is smaller than the area of the fabric section ($S_1 < S_2$).

According to the solutions provided in the literature in both cases the intensity of the resulting radiative heat flux is described by the equation:

$$q_r = \sigma_0 \varepsilon_{\text{reduced}} \left(T_1^4 - T_2^4\right), \tag{12}$$

where $\sigma_0 = 5.67 \cdot 10^{-8}\ \text{W}/\left(\text{m}^2 \cdot \text{K}^4\right)$ – Stefan-Boltzmann constant, in the first case, the reduced emissivity factor of the system under consideration $\varepsilon_{\text{reduced}} = 1/(1/\varepsilon_1 + 1/\varepsilon_2 - 1)$, in the second case, the reduced emissivity factor is approximately equal to

$$\varepsilon_{\text{reduced}} = 1\Big/\left(1/\varepsilon_1 + \frac{S_1}{S_2}(1/\varepsilon_2 - 1)\right).$$

It should be noted that for estimating ε_2 you can use the relationship:

$$\varepsilon_2 = \varepsilon_{2p}\zeta_p + \varepsilon_{2m}\left(1 - \zeta_p\right).$$

Here $\zeta_p = \pi d_p^2/\left(4S_{p0}^2\right)$, ε_{2p}, ε_{2m} – the emissivity factor of the particles and fabric surfaces, respectively.

At radiative-convective heat transfer which is often observed in engineering the total heat flow density is

$$q = q_r + q_k,$$

where q_k – specific heat flux which characterizes the convective heat transfer of the fabric surface with particles and gaseous medium ($T_{g2} = T_g$),

$$q_k = \bar{\alpha}_{k2}\left(T_2 - T_{g2}\right).$$

In engineering practice it is convenient to determine the radiation heat transfer coefficient in the same way as the convection coefficient is determined, assuming that [3, 4]

$$q_r = \alpha_r \left(T_1 - T_2\right). \tag{13}$$

From a comparison of relationships (12) and (13) it follows that

$$\alpha_r = \sigma_0 \varepsilon_{\text{reducced}} \left(T_1^4 - T_2^4\right) / \left(T_1 - T_2\right) = \sigma_0 \varepsilon_{\text{reducced}} \left(T_1^3 + T_1^2 T_2 + T_1 T_2^2 + T_2^3\right).$$

Since the temperature $T_1 > T_2$, approximately

$$\alpha_r \approx \sigma_0 \varepsilon_{\text{reduced}} T_1^3 \left(1 + T_2 / T_1\right). \quad (14)$$

Replacing in (14) the current temperature T_2 with the average temperature over the heating time of particles $\tilde{T}_2 = 0.5\left(T_{21} + T_{22}\right)$ (T_{21}, T_{22} – temperature T_2 before and after the heating end (given by technology)) we obtain the following estimate:

$$\alpha_r \approx \sigma_0 \varepsilon_{\text{reduced}} \tilde{T}_{12},$$

where multiplier is $\tilde{T}_{12} = \left(1 + \tilde{T}_2 / T_1\right) T_1^3$.

Taking into account that the radiative heat flux enters the strip material from the side of the particles layer, we write down:

$$q = \alpha_{rk} \left(T_2 - T_{g2}\right) - \alpha_r \Delta\tilde{T}. \quad (15)$$

Here $\alpha_{rk} = \alpha_r + \bar{\alpha}_{k2}$, $\Delta\tilde{T} = T_1 - T_{g2}$ – temperature difference which in this problem is considered to be a known constant value.

The representation of the total specific heat flux acting on the strip under consideration in the form (15) significantly simplifies the integration of the heat problem [5-8] since you can use the already available solutions for the average temperature $\bar{T}$ (2) and the temperature $T = T\left(\bar{y}, F_0\right)$ (9), (10) where the integration constants and parameter β are determined by the relationships (11).

The value $\tilde{\alpha}_k$ in (2) should be substituted by

$$\tilde{\tilde{\alpha}}_k = \bar{\alpha}_{k1} + \alpha_{rk},$$

the integration constant A_1,and parameter β (11) written in the form:

$$A_1 = \left(\alpha_{k1}^* \left(1 + \alpha_{rk}^*\right) T_{g1} + \alpha_{rk}^* T_{g2} + \alpha_r^* \Delta\tilde{T}\right) / \left(\alpha_{k1}^* + \alpha_{k1}^* \alpha_{rk}^* + \alpha_{rk}^*\right), \ \beta = \sqrt{\alpha_{k1}^* + \alpha_{rk}^*},$$

where $\alpha_{rk}^* = \alpha_{rk} h_0 / \bar{\lambda}$, $\alpha_r^* = \alpha_r h_0 / \bar{\lambda}$.

Accordingly, when determining A_2, B_1 these values of A_1 and β are used.

3. Conclusions

Under certain assumptions, quite simple relationships that allow estimating the thermal behavior dynamics of the three-layered strip taking into account the thermal resistance of the air interlayer and other parameters at its convective and radiative-convective heat transfer with the environment were obtained. These relationships can be useful when determining the rational technological regimes of heat treatment of the material under consideration.

References

[1] Isachenko V P, Osipova V A and Sukomel A S 1988 *Heat transfer* (Moscow: publishing house vysshaya shkola) 479 p

[2] Kutateladze S S 1990 *Heat transfer and hydrodynamics* (Moscow: publishing house Energoatomizdat) 367 p

[3] Bennett K O and Mayers D E 1966 *Hydrodynamics, heat and mass transfer* (Moscow: publishing house Nedra) 727 p

[4] Lukanin V N 2000 *Heat engineering* (Moscow: publishing house vysshaya shkola) 671 p

[5] Lunev N A, Shapkin A V, Kashapov R N, Kurylev D V, Kashapov L N and Kashapov N F 2018 Application of additive technologies in the process of technological preparation of aviation products *IOP Conference Series: Materials Science and Engineering* vol 412 number 012094

[6] Galimov E R, Shveyov A I, Shveyov I A, Lysanov D M and Galimova N Ya 2016 Development of combined technologies and multifunctional equipment for application of polymer powder

coatings *International Journal of Applied Engineering Research* vol 11, number 3 pp 1583-89

[7] Galimov E R, Sharafutdinov R F and Snigerev B A 2015 Mathematical model of the flow of polymer melt in the extruder forming die IOP *Conf. Series: Materials Science and Engineering* vol 86 number 012030

[8] Galimov E R, Gabdrakhmanov A T, Shafigullin L N and Ibragimov A R 2017 Surface thermohardening by the fast-moving electric arch *IOP Conf. Series: Journal of Physics: Conf. Series* vol 789 number 012010

IOP Conf. Series: Materials Science and Engineering **570** (2019) 012019 doi:10.1088/1757-899X/570/1/012019

Research of a laser welding of thin-shit construction steels by continuous laser radiation

Az T Gabdrakhmanov, AT Galiakbarov, L N Shafigullin, T F Gabdrakhmanova

Kazan Federal University, 423800, 68/19 Mira avenue, Naberezhnye Chelny, Russia

ATGabdrahmanov@kpfu.ru, veyron000@mail.ru

Abstract. The results of the laser welding by continuous radiation of thin-sheet low-carbon steels on the example of steel 08U presented in this paper. Metallographic studies of the laser weld were conducted. Tensile tests were carried out as a result of which it was determined that the nature of the fracture is adhesive, and after cohesive failure of the specimens, it is show the high quality of the welded joints at the beginning of the weld.

1. Introduction

Of the modern methods of welding is considered more effective the laser welding. The laser welding is an innovative method of metal processing [1-5]. This technology has a very wide scope of application, as it has many advantages. It can be used when working with the same and different metals, is actively used in electronic engineering and radio electronics. The convenience of the method is that such welding is carried out by partial or complete melting, in any position, at any angle. It is suitable for working with thin sheets of metal, and for large parts [6-10].

2. Experimental studies

The welding was performed using a LS-2 serial laser unit with a power of 2 kW manufactured by IRE-Polus with an IPG d38 f100W-150-200 optical head with a focal length of 150mm. The speed of movement of the instrument varied from 5 to 30 mm/s.

The object of the research was a cold-rolled steel 08U, structural carbonaceous quality with the addition of aluminum to give a steel elasticity. The steel 08U is used in the manufacture of car body parts, for example, in the production of KAMAZ trucks. The chemical composition of steel 08U was determined on a spark-meter spectrometer 100. The results are presented in Table 1.

Table 1. Chemical composition of steel 08U

C	Si	Mn	Ni	S	P	Cr	Al	Cu
0.052	0.011	0.139	0.006	0.007	0.007	0.024	0.031	0.022

The obtained results of the chemical composition of steel 08U correspond to GOST 9045.
The modes of laser welding are presented in table2.

IOP Conf. Series: Materials Science and Engineering **570** (2019) 012019 doi:10.1088/1757-899X/570/1/012019

Table 2. Modes of laser complexes for laser welding of steel 08U.

Mode	Laser power, W	Speed, mm / s	Focus, [mm]
1	800	30	-30мм
2	1200	30	-30мм
3	1500	30	-30мм
4	1500	30	-20мм
5	2000	30	-20мм
6	1200	10	-20мм

The appearance of the samples is shown in Figures 1-2:

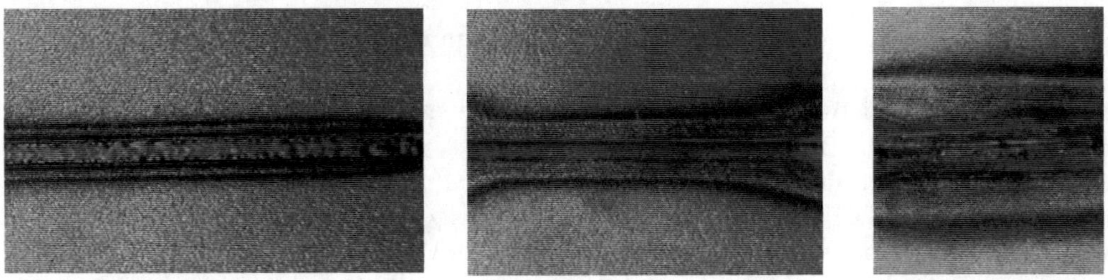

Sample No. 1 x20 Sample No. 3 x20 Sample No. 5 x20

Figure 1. Appearance of samples with laser welding

The quality of the laser weld is evaluated by geometrical parameters, microhardness, tensile tests.
The measurement of the geometric parameter of the laser weld is carried out over the cross section of the sample. Using the measuring scale of the microscope Altami MET 3M, the geometrical parameters of the laser welds were obtained. Figure 2.

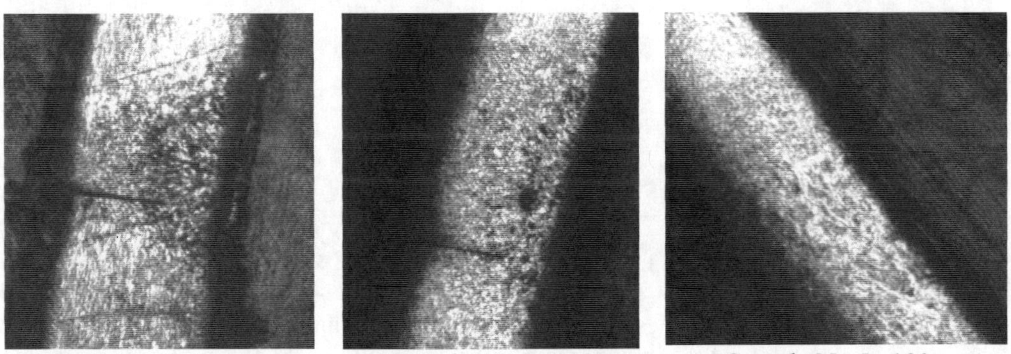

Sample No 1 x100 Sample No 3 x100 Sample No 5 x100

Figure 2. Transverse microsections of samples with pulsed laser welding

When determining the geometric parameters of the laser seam, it was found that the samples were welded on an automated laser system-2 №1, №2, №3 there is no through seam. A microhardness determination is carried out on a PMT-3 microhever meter. The principle of operation of the device is based on pushing the diamond pyramid into the material under test under a certain load and measuring the linear magnitude of the diagonal of the print obtained.
The results of changing the microhardness across the width of the weld of sample No. 5 are presented in the form of a graph for greater clarity Figure 3.

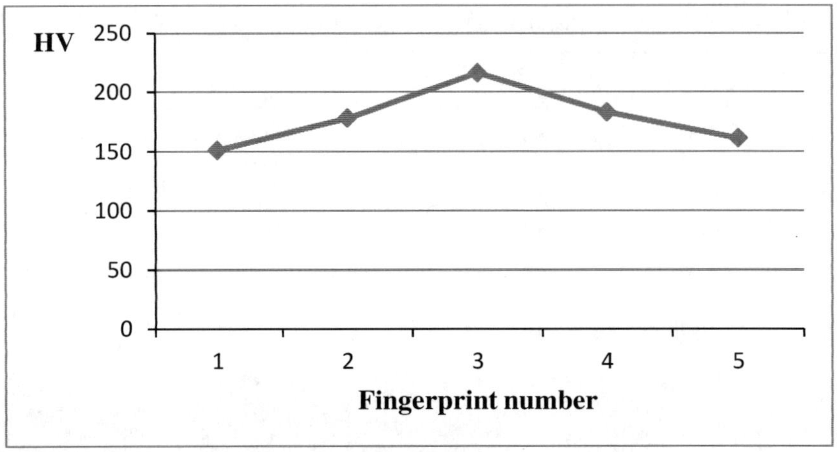

Figure 3. The change in microhardness across the width of the weld of sample No. 5

The determination of tensile strength was carried out according to the method "tensile test methods for thin sheets and tapes GOST 11701" and "tensile test methods GOST 1497-84". Cutting blanks for samples and manufacture of samples carried out according to GOST 1497-84.
The tensile strength is determined by special equipment - Universal tensile testing machine MIM-7LR-010. According to tensile tests, charts were constructed (Figure 4).

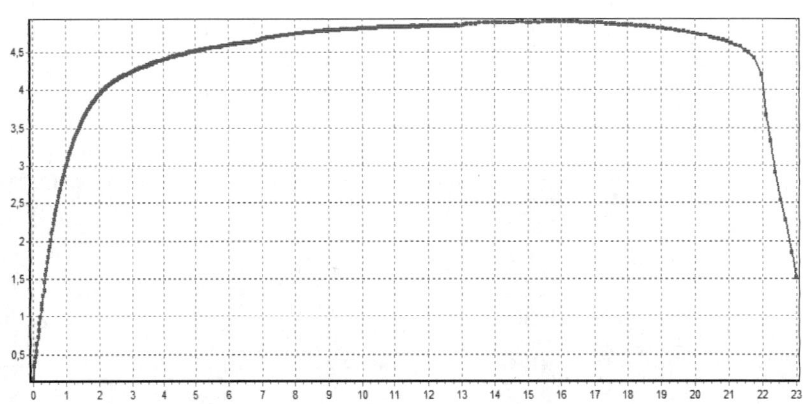

Figure 4. Graph - results on the gap sample number 5

In the process of tensile testing, adhesive and after cohesive failure of the specimens are observed, which shows the high quality of the welded joints at the beginning of the weld.

3. The conclusion

Experimental research on the effects of pulsed laser radiation on steel 08U shows the possibility of obtaining a weld with the required quality indicators. The localization of the welding zone ensures the rational design of the parts to be welded and minimizes permanent deformations.

References

[1] Grigoryants A G Shiganov I N Misyurov A I 2006 *Processes of laser processing (*Moscow: - Izd-vo MGTU im. N. Uh. Bauman) p 664

[2] Fedyaev V L Galimov E R Belyaev A V Dolgov A V and Samoylov V M 2018 *Evaluation of interphase boundaries dynamics at the contact of melt with solid material* (IOP Conf. Series: Materials Science and Engineering) vol 412 number 012015

[3] Kashapov N F Dautov I G Tkachenko L A Fadeev S A and Shaidullin L R 2018 *Calculation of*

IOP Conf. Series: Materials Science and Engineering **570** (2019) 012019 doi:10.1088/1757-899X/570/1/012019

the distribution of temperature in the form based on magnesium oxide for the casting of titanium products (Journal of Physics: Conference Series) vol 1058 issue 1 number 0120629

[4] Astashchenko V I Zapadnova N N Mukhametzianova G F and Shafigullina A N 2017 *Key concepts for production of high-quality parts* (IOP Conference Series: Materials Science and Engineering) vol 240 issue 1 number 012007

[5] Teregulov N G Sokolov B K Matveeva VS 2007 *Quality of the processed surface during laser cutting and its control* (Kumertau aviation industrial enterprise Institute of Metal Physics, Ural Branch of the Russian Academy of Sciences) Number 2 pp 62-72

[6] Kashapov N F Dementyeva J N Kashapov R.N Kashapov L N 2018 *Use of polystyrene production wastes in selective laser sintering processes* (IOP Conference Series: Materials Science and Engineering) Volume 412 Issue 1 number 012093

[7] Galimov E R Shveyov A I Shveyova T V Sharafutdinova E E Belyaev A V Sirotkina L V 2016 *Method of predicting the hardness of welded joints* (International Journal of Applied Engineering Research) Volume 11 Number 3 pp 1603-08

[8] A T Gabdrakhmanov Israphilov I H Galiakbarov AT Samigullin A D Gabdrakhmanov Al T 2016 *Improving the efficiency of plasma heat treatment of metals* (Journal of Physics: Conference Series) Volume 669 Article number 012014

[9] Khisamutdinov R M Zvezdin V V Saubanov Ruz R Israfilov I H Rakhimov R R Spirin A A 2016 *Study of processes of steels surfaces modification with highly concentrated energy flows* (IOP Conference Series: Materials Science and Engineering) Volume 669 Issue 1 Article number 012024

[10] Gabdrakhmanov A T Israphilov I H Galiakbarov A T 2017 *Study of the combined laser-plasma effect on metals* (IOP Conference Series: Materials Science and Engineering) Volume 240 Issue 1 Article number 012015

IOP Conf. Series: Materials Science and Engineering **570** (2019) 012020 doi:10.1088/1757-899X/570/1/012020

Study of laser cutting of copper and copper alloys

Az T Gabdrakhmanov, I H Israphilov, L N Shafigullin and T F Gabdrakhmanova

Kazan Federal University, 423800, 68/19 Mira avenue, Naberezhnye Chelny, Russia

ATGabdrahmanov@kpfu.ru, veyron000@mail.ru

Abstract. The processes of interaction of laser radiation with a surface layer of copper under laser irradiation, as well as the main factors affecting the absorption capacity of copper during surface treatment with a laser beam, are considered.

1. Introduction

Among laser technologies used at the present time in machine-building industries, a gas laser cutting of metallic and non-metallic materials become widely used [1-6]. The using of laser cutting in the blank production appears to be most efficient [7-10]. A wide range of thicknesses and grades of cut materials, almost any parameters of cut parts allow to manufacture parts of different sizes and geometric complexity.

It is possible to cut sheet material along a complex contour with an error from 100 to several micrometers using laser cutting technology.

Modern laser cutting systems allow to cut of thin-sheet materials at a speed of up to 120 m / min with an error of no more than 100 microns. It is necessary to increase the laser radiation power for maximize increase the thickness of the sheets to be cut and the cutting speed, so in recent years, the laser cutting technology has been commercialized at the power level of 5-6 kW.

Laser cutting of carbon and stainless steels is well studied. Cutting metals of the copper group has its own characteristics due to the high thermal conductivity of the material. Copper also has a large coefficient of heat capacity. It is imposes certain requirements on the equipment. In preparation for the process, it should be borne in mind that the laser cutting of copper the more difficult the thicker the plate. It is necessary to choose the right parameters of power and speed of the beam. The general rules are: the size of the laser spot should be as small as possible, and the power - high. It is possible to achieve a flat line of cut when the conditions of technology are observing. As a result of high-quality cutting, the edges of the product are not deformed. The copper sheet absorbs the radiation poorly. For this reason, the copper rolled cut at minimum speeds.

The research material was copper. The blanks were in the form of strip blanks with a thickness of 1 mm.

2. Experimental studies

The cutting was performed using a LS-2 serial laser unit with a power of 2 kW manufactured by IRE-Polus with an IPG d38 f100W-150-200 optical head with a focal length of 150mm. The cutting of metal samples was carried out with the installation of the focal plane on the workpiece surface. The speed of movement of the instrument varied from 5 to 30 mm / s. The laser head was not installed at a right angle to protect the laser from reflected radiation.

IOP Conf. Series: Materials Science and Engineering **570** (2019) 012020 doi:10.1088/1757-899X/570/1/012020

After cutting, the structure of the cutting surface, its geometry were studied and determined the roughness in accordance with GOST 2309-73; 2789-73. The height parameter Ra is selected as a quantitative indicator of the degree of roughness. The cut surfaces were photographed in studies using a digital camera and a microscope with magnifications from 50 to 200. A characteristic feature of the heat-affected zone is the presence of barb - residues of molten metal crystallized on the bottom (base) surface of the workpiece.

The barb deteriorates the quality of the laser cut and makes it necessary to remove the operation.

The specific features of the power tool - a huge energy density, concentrated on a small area of the material – it leads to high rates of heating and cooling of this area (up to 106 K/s). At the same time, the possibility of non-diffusion mechanisms of transformations is increased, there is a sharp structural and chemical heterogeneity and the effect of structural heredity is manifested clearly.

The laser cutting modes of copper are presented in table 1.

Table 1. Laser cutting modes.

Mode	Working gas	Gas pressure p, atm	Laser radiation speed υ, mm/s	Laser power N, W
1	Air	7	18	1000
2	Air	7	30	1200
3	Air	7	25	1000
4	Oxygen	13	18	1000
5	Oxygen	13	30	1000
6	Oxygen	13	30	800

The quality assessment criteria are: b - the width of the cut, h - the heat-affected zone, l - the size of the barb. Table 2 shows the quality parameters of the cutting surface dimensions.

Table 2. The cutting quality parameters.

Sample no.	Roughness Ra, µm	Barb l, mm	Heat affected zone N, mm	Cutting Width b, mm
1	1,528	0,430	1,402	0,196
2	0,608	0,171	1,273	0,159
3	-	-	-	-
4	-	0,257	0,837	0,267
5	0,891	0,289	0,618	0,235
6	1,106	0,225	1,040	0,253

The surface of the laser cut is characterized by a certain microgeometry, which is represented by a set of periodically recurring irregularities (waviness, striation, roughness) described by a set of parameters characterizing the average and maximum height of the irregularities, their width, and the distance between them. The view of the cut surfaces under various laser cutting conditions is shown in Fig. 1.

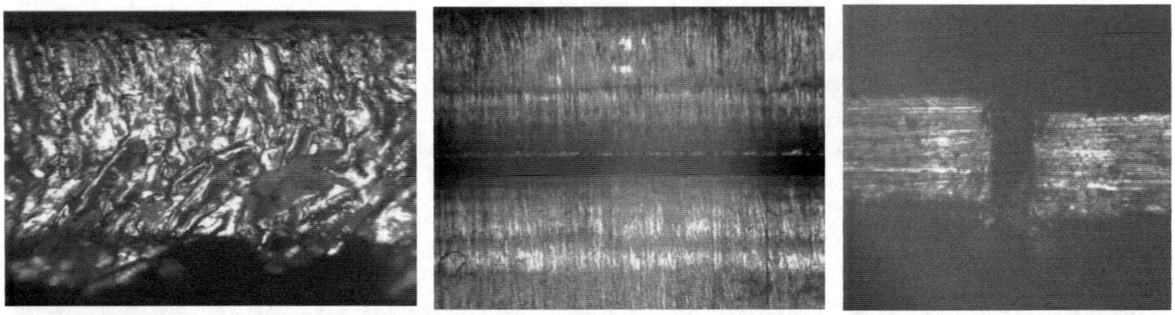

a)

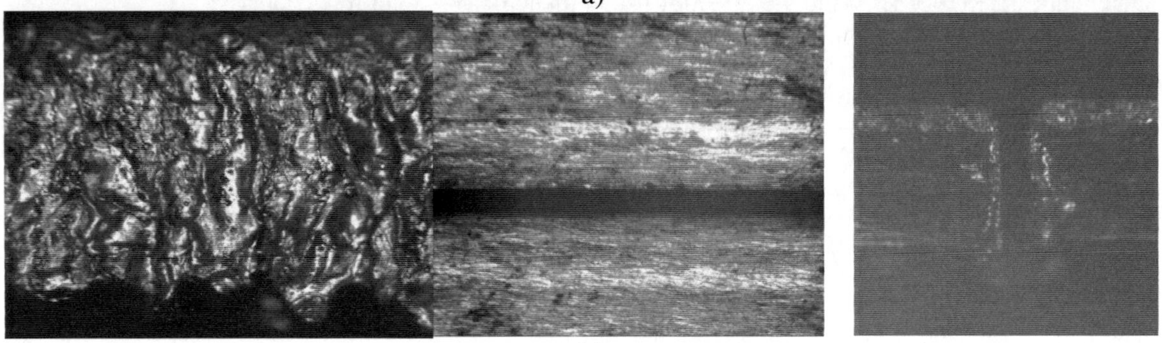

b)

Figure1. The surface of the laser cut copper 1 mm thick (left to right): cut edge × 200, top view × 50, in the section × 50; a) mode 1; b) mode 5;

3. The conclusion

For laser cutting of copper and its alloys, an oxygen-containing medium (air, oxygen) is necessary. When copper is exposed to laser radiation, a thin oxide layer is formed, which has a good absorption capacity, thus starting most of the radiation is absorbed, which leads to heating and removal of the material. This is confirmed by experiments using nitrogen and carbon dioxide as the working gas, without heating the copper billets, even at maximum power, all laser radiation is reflected from the surface. The use of oxygen as a working gas can reduce the power and increase the speed of laser radiation.

References

[1] Grigoryants A G Shiganov I N Misyurov A I 2006 *Processes of laser processing (*Moscow: - Izd-vo MGTU im. N. Uh. Bauman) p 664

[2] Fedyaev V L Galimov E R Belyaev A V Dolgov A V and Samoylov V M 2018 *Evaluation of interphase boundaries dynamics at the contact of melt with solid material* IOP Conf. Series: Materials Science and Engineerin v 412 No 012015

[3] Kashapov N F Dautov I G Tkachenko L A Fadeev S A and Shaidullin L R 2018 *Calculation of the distribution of temperature in the form based on magnesium oxide for the casting of titanium products* (Journal of Physics: Conference Series) v 1058 No 0120629

[4] Astashchenko V I Zapadnova N N Mukhametzianova G F and Shafigullina A N 2017 *Key concepts for production of high-quality parts* (IOP Conference Series: Materials Science and Engineering) v 240 No 012007

[5] Teregulov N G Sokolov B K Matveeva VS 2007 *Quality of the processed surface during laser cutting and its control* (Kumertau aviation industrial enterprise Institute of Metal Physics, Ural Branch of the Russian Academy of Sciences) No 2 pp 62-72

[6] Kashapov N F Dementyeva J N Kashapov R.N Kashapov L N 2018 *Use of polystyrene production wastes in selective laser sintering processes* (IOP Conference Series: Materials

Science and Engineering) v 412 No 012093

[7] Galimov E R Shveyov A I Shveyova T V Sharafutdinova E E Belyaev A V Sirotkina L V 2016 *Method of predicting the hardness of welded joints* (International Journal of Applied Engineering Research) v 11 No 3 pp 1603-08

[8] A T Gabdrakhmanov Israphilov I H Galiakbarov AT Samigullin A D Gabdrakhmanov Al T 2016 *Improving the efficiency of plasma heat treatment of metals* (Journal of Physics: Conference Series) v 669 **012014**

[9] Khisamutdinov R M Zvezdin V V Saubanov Ruz R Israfilov I H Rakhimov R R Spirin A A 2016 *Study of processes of steels surfaces modification with highly concentrated energy flows* (IOP Conference Series: Materials Science and Engineering) v 669 **012024**

[10] Gabdrakhmanov A T Israphilov I H Galiakbarov A T 2017 *Study of the combined laser-plasma effect on metals* (IOP Conference Series: Materials Science and Engineering) v 240 **012015**

IOP Conf. Series: Materials Science and Engineering **570** (2019) 012021 doi:10.1088/1757-899X/570/1/012021

Rotor film reactor

L R Gainullina[1], V P Tutubalina [1], L S Sabitov[1,2]

[1]Kazan State Power Engineering University, 51 Krasnoselskaya street, Kazan, 420066, Russian Federation
[2]Kazan Federal University, 18 Kremlyovskaya street, Kazan, 420008, Russian Federation

gainullina7819@mail.ru

Abstract. In this work the description of a design of the rotating film reactor and its features. The rotating film reactor is intended for processes of homogenization of various immiscible liquids, implementation of the heterogeneous processes proceeding in system liquid-liquid in a thin film including oxidation by hydrogen peroxide of organosulfur connections of oil fractions for the purpose of receiving sulfoxides and sulfones. The design of the multisection reactor allows to change temperature, concentration of reagents on zones and to regulate time of stay of reagents in a reaction zone.

The aim of the work was to study the possibility of using a rotor-film reactor for the desulfurization of oil fractions with hydrogen peroxide with simultaneous production of sulfoxides and sulfones - valuable compounds of interest to industry and agriculture [1-5].

As the object of study used oil fractions with $T_{кип}$ = 300-400 °C Romashkinskoye oil field with a total sulfur content of 1.41%, sulfide sulfur - 0.973% and residual sulfur - 0.437%.

The oxidation of organic sulfur compounds of the oil fraction in a rotary film reactor was controlled by analyzing samples of the reaction mass according to standard methods:

- determination of total sulfur according to GOST 19121-73, by sulfide and sulfoxide by the method of iodometric potentiometric titration and sulfonic sulfur by the difference of sulfide and sulfoxide sulfur [6];
- determination of the concentration of hydrogen peroxide in the sample (GOST 32460-2013).

The reactor (figure) consists of a body formed by the shells 1 and jackets 2. Inside the body, coaxially with it, is placed the shaft 3 with feeders and liquid distributors. Each feeder includes cups 4 and 5 with holes 6 attached to the shaft detachably by means of a bolted connection 7 and located inside stationary cylinders 8 and 9 having jagged edges 10. Stationary cylinders 8 and 9 are interconnected by means of ribs 11 and to a plate 12 installed in the case of the apparatus between two adjacent shells 1. The liquid distributor is made in the form of a hollow corrugated drum 13. At the upper end of this drum there is a disk 14 and a ring 15 in which there are vertical slots 17 and 16 that are offset each other and the corrugation of the drum located opposite the external protrusions, and at its lower end - a disk 18 connected to the shaft detachably by means of bolt fastening 19.

IOP Conf. Series: Materials Science and Engineering **570** (2019) 012021 doi:10.1088/1757-899X/570/1/012021

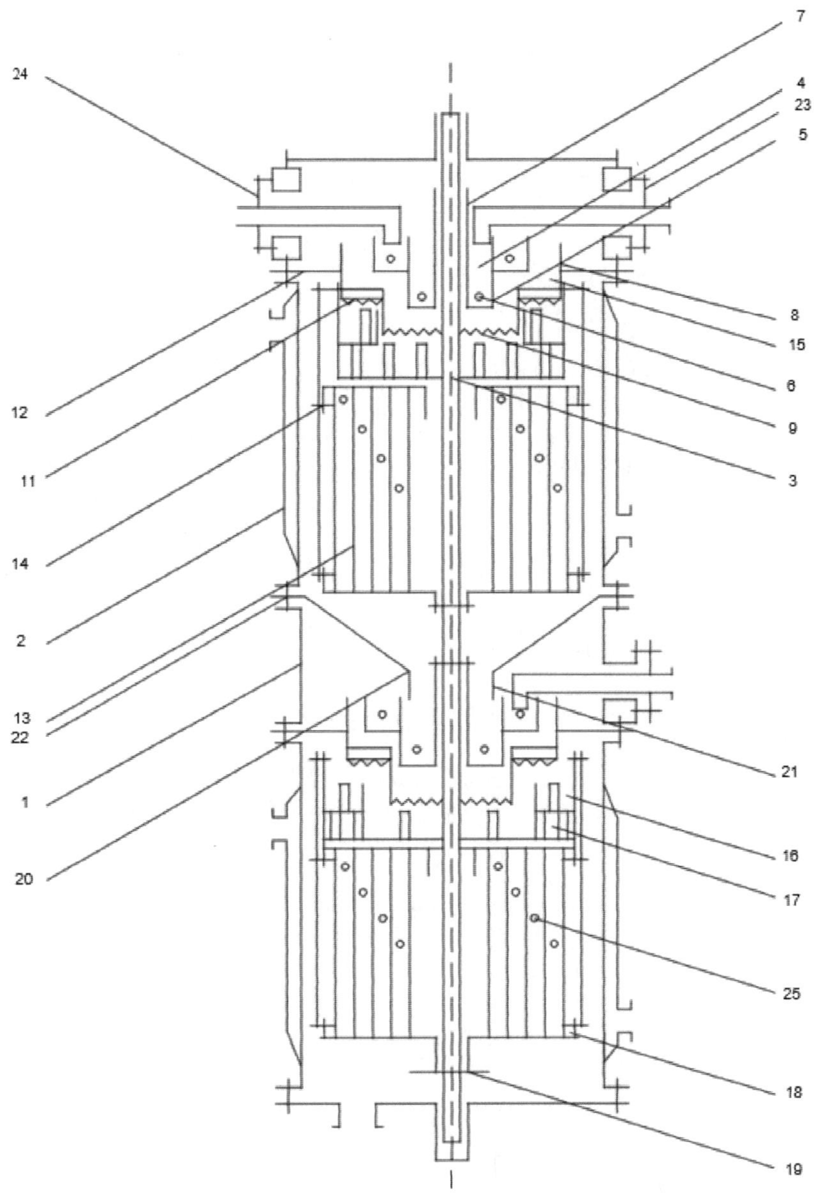

Figure 1. Rotating film reactor

Under each corrugated drum 13 a cone 20 is installed, to the lower base of which is attached a pipe 21 entering the cup 4, and to the upper a ring 22 mounted in a cone between two adjacent shells 1.
Rotary film multistage reactor operates as follows. The reagents through the nozzles 23 and 24 enter the cups 4 and 5, from where they are ejected through the openings 6 onto the inner surface of the fixed cylinders 8 and 9. As a result, one of the components falls on the ring 15 and the other into the pocket formed by the walls of the ring 15 and the disk 14 .
Since the slots 16 on the ring 15 and the slots 17 on the disk 14 are connected to the corrugations of the perforated drum 13 in fluid form, various reagents get into each two adjacent corrugations, and they are mixed directly on the heat exchange surface, where they are thrown through the holes 25 in the wall of the corrugated drum.
Next, the reaction mixture in a cone 20 and the pipe 21 is sent to the glass 4 of the underlying stage.

To determine the nature of the distribution of hydrogen peroxide [7] and organosulfur compounds that are in oil fractions at the height of a rotor-film reactor in a dynamic mode, experiments were carried out under the following conditions:
- for sulfoxides: acid catalyst consumption (acetic acid 1 g-mol / 1 g-atom of sulfide sulfur in the feedstock), hydrogen peroxide consumption 0.6 l / h, section temperature I, II, III – 80 °C;
- for production of sulfones: acid catalyst consumption (acetic acid 1 g-mol / 1 g-atom of total sulfur of the raw material, consumption of hydrogen peroxide 1.2 l / h, temperature in sections I, II, III – 100 °C;
- continuous installation time is 6 hours;
- consumption of raw materials (oil fraction) for the installation is 24 l / h.

The design of a multisection reactor allows changing not only the temperature, but also the concentration of reagents in zones, introducing various additives into the reaction mass, changing the concentration of catalyst and oxidizer at any stages of the process. In addition, by changing the height of individual sections, you can adjust the residence time of the reactants in the reaction zone.

References

[1] Zarifyanova MZ, Khusnutdinov I Sh, Gryaznov P I, Aristov I V, Vafina S D, Konstantinova A B 2013 *Oil sulfoxides. Report 2. Optimization of the choice of raw materials for the production of petroleum sulfoxides with the use of quantum-chemical calculations* News of higher educational institutions Series: Chemistry and Chemical Technology T 56 4 pp 8-11
[2] http://base.safework.ru/iloenc?d&nd=857300066&prevDoc=857300066 (date of the application 04.10.2018)
[3] Xiao- Feng Wu, Kishore Natte *The Applications of Dimethyl Sulfoxide as Reagent in Organic Synthesis* Advanced Synthesis & Catalysis 2016 (https://doi.org/10.1002/adsc.201501007).
[4] Fedorovtseva I A, Ramazanova AR, Bairashevsky I B, Kayralieva A I 2018 *Extraction purification of middle distillate fraction of gas condensate* Gas industry 6 (769) pp 132-36
[5] Shainsky J, Derry N-L, Leichtmann-Bardoogo Y, Wood T K, Fishman A 2009 *Rapid Methods for High-Throughput Detection of Sulfoxides* Appl Environ Microbiol **75(14)** pp 4711–19.
[6] Karaulova EN 1969 *Methods of analysis of organic compounds of oil, their mixtures and derivatives* Moscow Nauka pp 95-101
[7] Tutubalina VP 1995 *Organic sulfur compounds of oil fractions and methods for their isolation* Moscow Central Research Institute pp 92-112

IOP Conf. Series: Materials Science and Engineering **570** (2019) 012022 doi:10.1088/1757-899X/570/1/012022

Technologies for producing heat-conducting carbon foams by method of pitches carbonization under pressure

E R Galimov[1], E E Sharafutdinova[1], N Ya Galimova[1], F I Murataev[1], V M Samoylov[2], E A Danilov[2]

[1]Kazan national research technical university named after A N Tupolev, K. Marx St. 10, Kazan, 420111 Tatarstan, Russia
[2]JSC NIIgrafit, 2 Elektrodnaya str., Moscow, 111524, Russia

kstu-material@mail.ru

Abstract. Technology for producing syntactic carbon foams by the method of pitches carbonization under pressure is proposed. The components and optimal regime parameters for producing the foams with given thermophysical and strength characteristics are determined. The foams structure was studied using scanning electron microscopy.

1. Introduction
The technology for producing heat-conducting syntactic carbon foams obtained by the method of pitch carbonization under pressure is based on using the phenomenon of pore formation upon the release of volatile substances from coal tar or oil pitches their release is complicated by the application of external pressure.

2. Body text
The technological process consists of several stages. At the initial stage, the components are mixed. If it is necessary to work with pitches that have a mesophase yield of less than 25 wt. % the mesophase microspheres of PMUG brand or analogues obtained by heat treatment of the original pitch up to a temperature of the mesophase formation (400-500°C) with subsequent extraction of the soluble part in toluene can be added to increase the thermal conductivity of the resulting product. The mesophase microspheres should be added to increase the mesophase yield from the pitch up to 40-50 wt. %. Carbon micro- and nanoparticles (fine-grained soot, carbon nanotubes, natural graphite, graphene nanoparticles) can be added into the original pitch in an amount of 0.1-5.0 wt. % to increase product strength. When the loading is more than 1 kg, the components are mixed in a mixing machine with the Z-shaped blades at a temperature of 15-20°C higher than the softening temperature of the original pitch. When the loading is less than 1 kg, the mixing can be performed in a toluene solution.

The initial pitch and solvent (toluene) are loaded into a three-necked round-bottomed flask in an amount of 100 ml of solvent per 15 g of pitch the mixture is stirred by an anchor stirrer and boiled using a backflow condenser for 60 minutes. Then the mixture is cooled to a temperature of 40°C, and the solvent is removed by distillation under vacuum using a rotary evaporator followed by drying under vacuum using a rotary evaporator at a residual pressure of not more than 10 mm Hg and a temperature of 120°C for 60 minutes. After mixing, the mass is cooled to a temperature of 25°C and crushed to a particle size of less than 250 μm in a vibrating mill or hammer crusher.

In the second stage, the low-temperature carbonization under pressure is carried out. The moulding material of the crushed pitch with additives is placed in a cylindrical form made of carbon steel with a 0.5 mm thickness with one weld seam along the generatrix and one weld seam along the base. The form is filled not more than for 60% of the height. The form is placed in an autoclave with a limiting heating temperature of at least 600°C and a maximum attainable pressure of at least 16 MPa. Heating and pressure increase is carried out according to the mode shown in Figure 1. The pressure and temperature in the autoclave are controlled and regulated in close proximity to the container with samples.

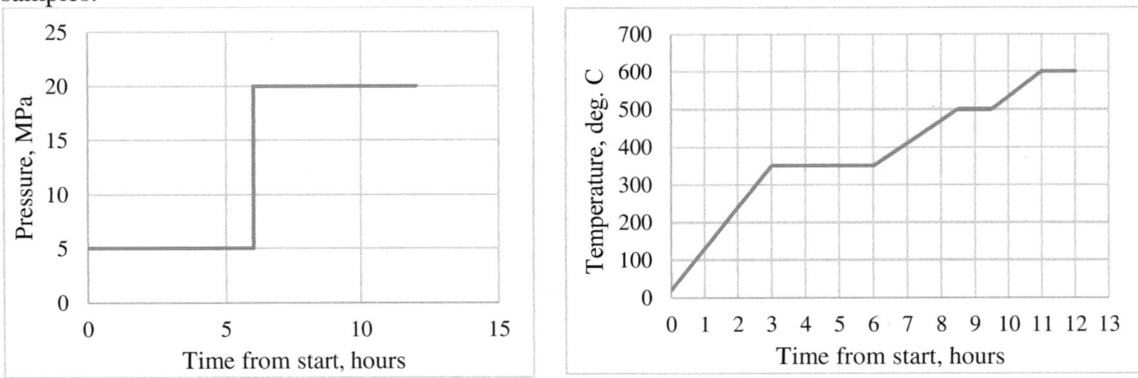

a) *b)*

Figure 1. Diagram of pressure increase (*a*) and temperature (*b*) for low-temperature carbonization of workpieces of foam under pressure (pressure relieve – after 12 hours, cooling - natural)

After cooling of the autoclave working area to a temperature of 450°C, the pressure can be reduced to atmospheric, for this, capture of toxic impurities is provided in the autoclave system (cold trap and trap with benzene). After cooling of the autoclave working area to a temperature of 150°C it is unloaded, containers with samples are taken out and cut.

At the third stage, carbonization is carried out. The obtained samples of the "green" foam were wrapped in a double layer of paper, placed in a steel container, covered by a 5-10 cm layer of broken graphite with a diameter of about 250 μm on top of which a 0.5-1.5 cm layer of high-temperature coal tar pitch with a diameter of 2.5 μm was poured it was covered with a steel lid and placed in a muffle furnace, equipped with a thermoregulator with the ability to control the speed and temperature of heating, as well as the exposure time. Samples in the container are heated at a rate of 2.5±0.5°C/min to a temperature of 900°C with exposure at the final temperature for 120 min after which the furnace is turned off and cooled to a temperature of 60±40°C for 12 hours. Samples are extracted, and overall dimensions, shrinkage, apparent density, ultimate compressive strength, and thermal conductivity coefficient are controlled. The obtained samples are placed in graphite crucibles with lids the filling (a scrap of artificial graphite production crushed to a particle size of less than 0.25 mm) is filled in. The crucibles are placed in an EVP-2500 (ЭВП-2500) vacuum furnace the working medium is pumped out to a residual pressure level of less than 1 mm Hg (it is desired to fill the furnace with argon to displace air and then pump it out to a residual argon pressure of less than 1 mm Hg). The furnace is heated for 8 hours to a temperature of 2100°C the samples are exposed for 2 hours then naturally cooled.

At the fourth stage, the graphitization process is carried out. The resulting samples are placed in graphite crucibles with lids the filling (a scrap of artificial graphite production crushed to a particle size of less than 0.25 mm) is filled in. The crucibles are placed in a graphitization furnace (such as Acheson or Tamman type) and heated to a temperature of 2700°C with exposure at the final temperature of at least 60 minutes.

In the next stage, the pyrocompaction is carried out. The obtained samples of graphitized foam are loaded into the pyrocompaction furnace chamber the furnace is sealed, blown through by argon, and

IOP Conf. Series: Materials Science and Engineering **570** (2019) 012022 doi:10.1088/1757-899X/570/1/012022

after that, the samples are heated and kept at a temperature of at least 1100°C and the methane pressure of 8-12 mm Hg for 20-120 hours.

The final stage is the mechanical treatment of the obtained workpieces (cutting, drilling, milling, and grinding at low feed rates by a tool with a Mohs hardness of more than 6).

Carbon foams were obtained in an industrial autoclave by pitch impregnation of carbon-carbon composite materials at a pressure of 8.0 MPa and a temperature of 500°C, followed by carbonization up to a temperature of 2200°C. The resulting foam had a non-uniform macrostructure: the outer layer was a material with a density of about 1.2 g/cm^3 and porosity observable with the naked eye, the inner layer had a density of about 0.7 g/cm^3 and distinct porosity.

Figure 2 shows the SEM images of the outer (dense) layer. It can be seen that the porosity structure of the foam is non-uniform – there are small channel pores and large cracks in the material structure. In terms of the structure of the pore walls, the material is quite homogeneous and has a graphite-like structure of a spherulitic type with a spherulites size of about 20 μm. The sections show that spherulites are graphite-like carbon material at the stage of pre-graphitization.

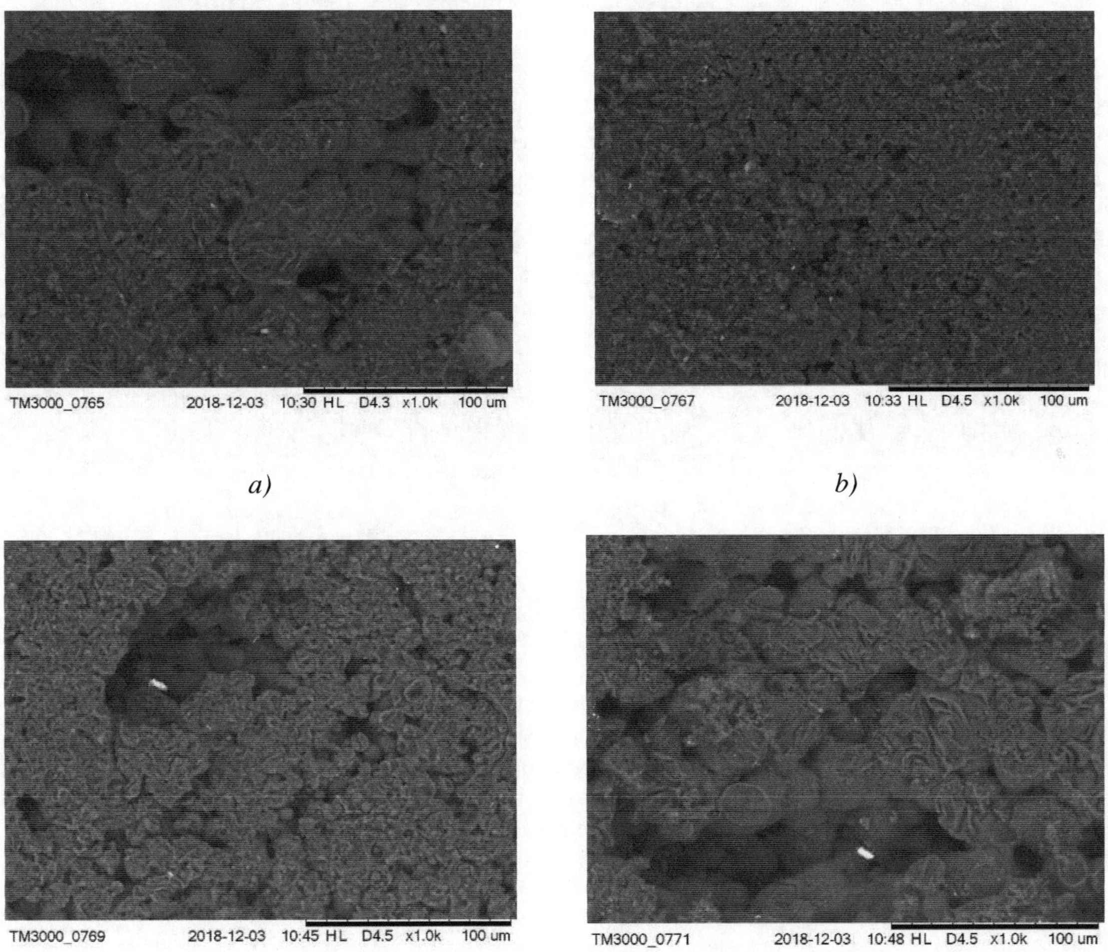

c) d)

e)

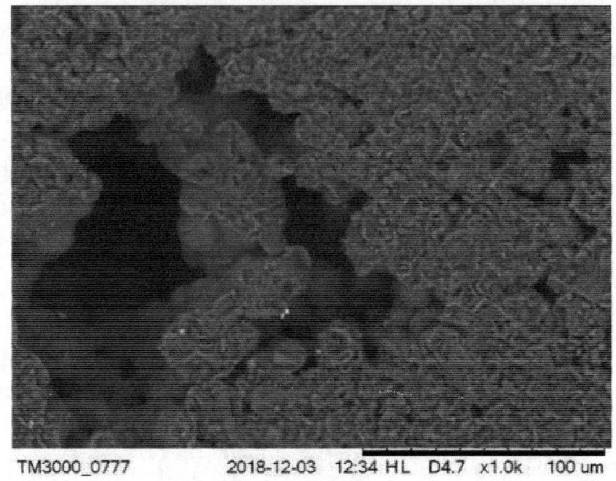

f)

g)

h)

Figure 2. SEM images of carbon foams obtained by carbonization under pressure (outer surface of the sample)

Based on the analysis of 156 fields, a macropore size distribution was plotted (figure 3). The data on the figure shows that the pores are mostly small (up to 17.6 μm) this, apparently, is the reason for the main porosity of the material. The average pore size is 12.24 μm at a high standard deviation for the sample (12.68 μm) which further confirms the polymodal nature of the distribution.

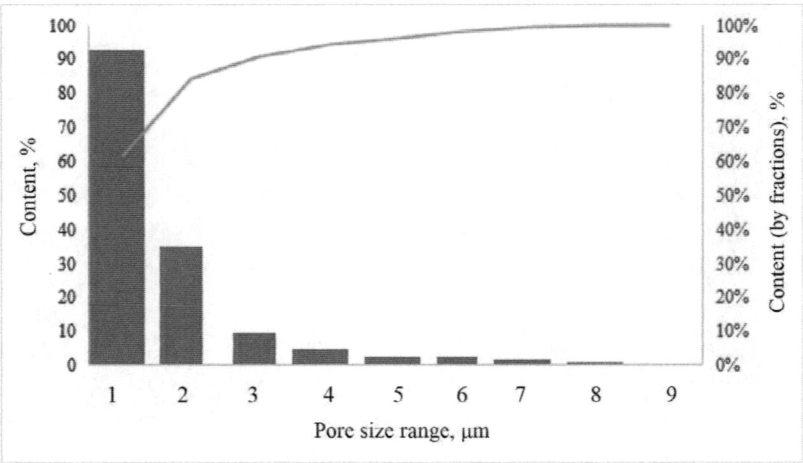

Figure 3. Macropores size distribution for carbon foams obtained by carbonization under pressure (outer surface of the sample): 1; 2; 3; 4; 5; 6, 7, 8, 9 – variation ranges of pore sizes respectively 1.02 – 9.32; 9.32 – 17.62; 17.62 – 25.92; 25.92 – 34.22; 50.82 – 59.12; 59.12 – 67.42; 34.42 - 42.52; 67.42 – 75.72; 42.52 - 50.82 µm

The internal porous part of the material is a foamy structure visible to the naked eye. Studies by the SEM method showed that the cuts and chips of the cell surface additionally contain canal-like macropores (figure 4) with a circular or elliptical section. In the interpore space, graphite has a striated structure and contains a large number of cracks with a width in the range of tens of nanometers. The secondary deposits of isometric particles are on the cell surface which indicates the carbon formation through the gas phase including due to volatile substances of the pitch.

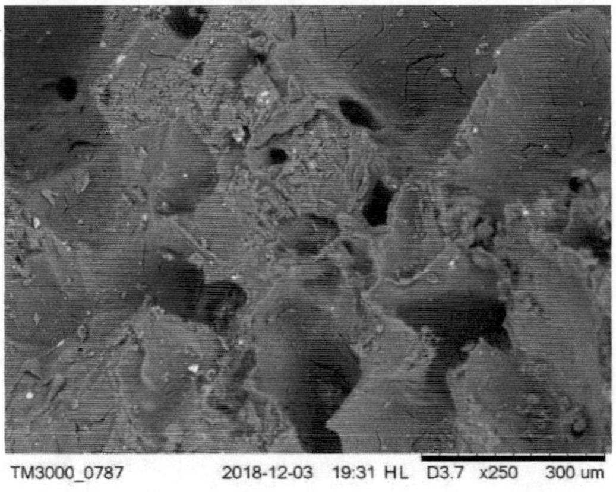

a)

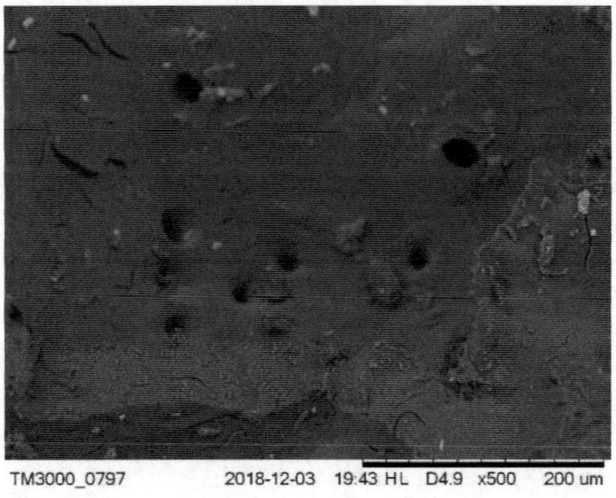

b)

IOP Conf. Series: Materials Science and Engineering **570** (2019) 012022 doi:10.1088/1757-899X/570/1/012022

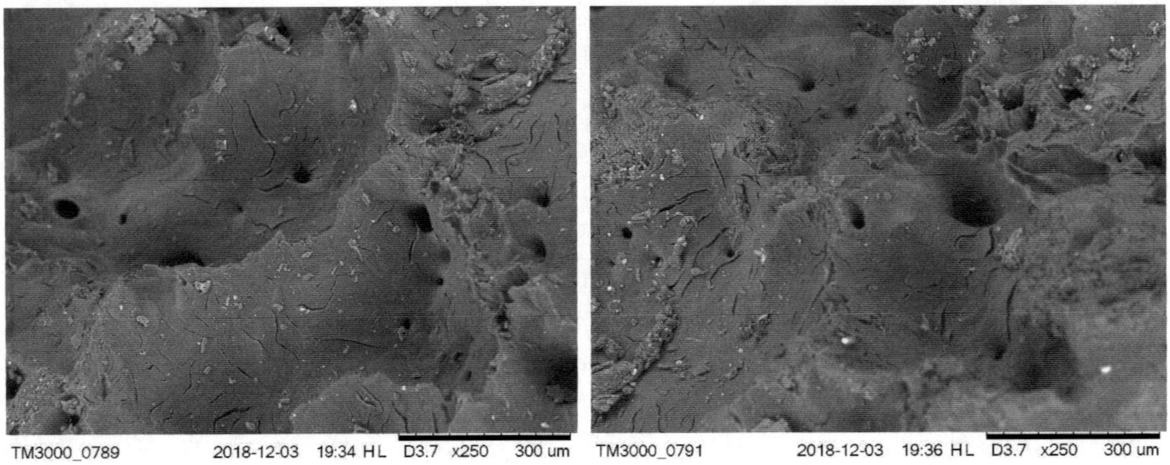

c) *d)*

Figure 4. SEM images of carbon foams obtained by carbonization under pressure (inner part of the sample)

Based on the analysis of 128 fields, the macropores size distribution was plotted (figure 5). The data on the figure shows that the pores in this part of the sample are larger (the main part is in the rage of 10-70 μm) this is the reason for the main porosity of the material. The pore diameter distribution is noticeable already at an average pore size of 41.79 μm at a standard deviation for a sample of 31.53 μm, which confirms the homogeneous nature of the material porosity.

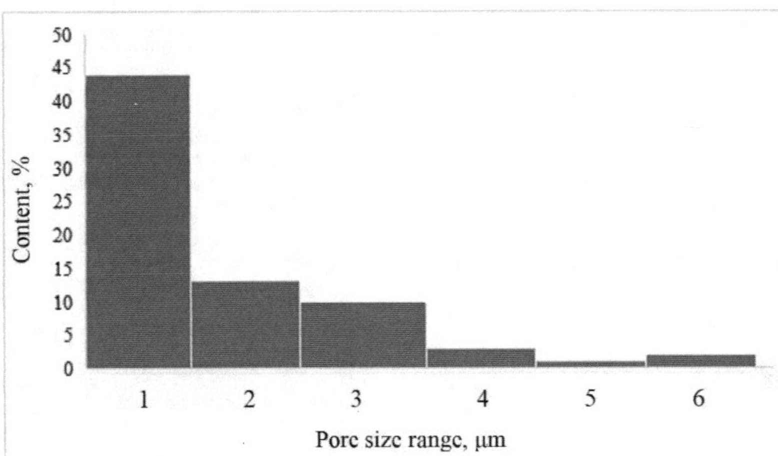

Figure 5. Macropores size distribution for carbon foams obtained by carbonization under pressure (inner part of the sample): 1; 2; 3; 4; 5; 6 – variation ranges of pore sizes respectively 10.80 – 35.84; 38.84 – 60.88; 60.88 – 85.92; 85.92 – 110.96; 110.96 – 136.00; > 136.00 μm

It should be noted that the wide pore size distribution does not allow classifying this class of carbon foams as syntactic.

Further, the experiments on carbonization under pressure of oil pyrolysis pitch with the softening temperature of 140°C at a pressure of 20 MPa and a temperature of 600°C were carried out, then the

IOP Conf. Series: Materials Science and Engineering **570** (2019) 012022 doi:10.1088/1757-899X/570/1/012022

samples were carbonized up to a temperature of 900°C processed in an electrovacuum furnace at 2100°C and graphitized in an Acheson type furnace up to a temperature of 2800°C.

Figure 6 shows the structure of the samples obtained by the SEM method. It can be seen that samples have pores of the approximately spherical cross-section with a narrow size distribution in the range of 250-300 μm, there are also inclusions of small spherulitic agglomerates of particles inside the pores which indicates the occurrence of secondary pyrolytic processes in the gas phase (soot formation and etc.). Moreover, the formed agglomerates have an isotropic structure and unexpressed texture despite the high final temperature of heat treatment.

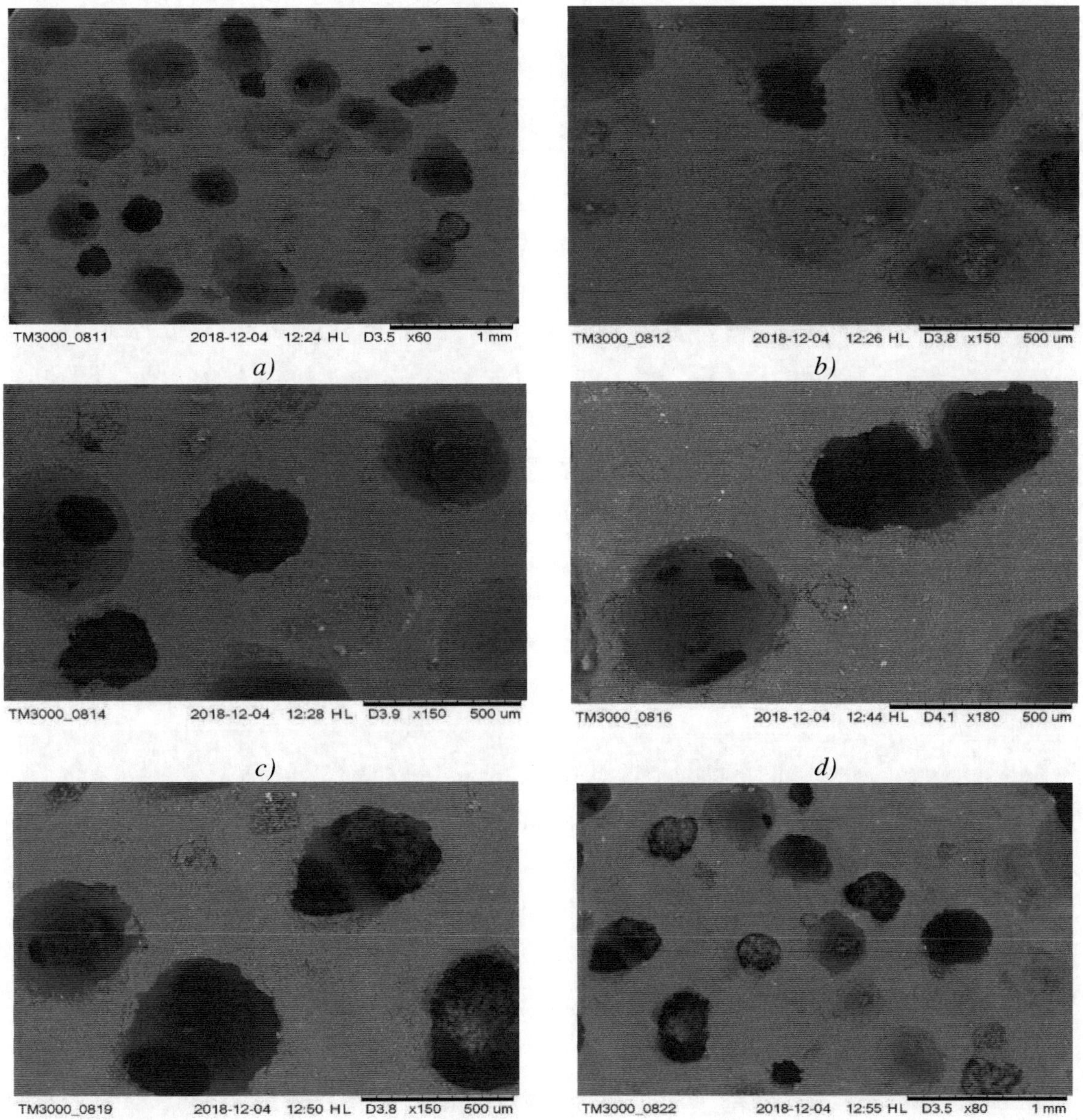

e) *f)*

Figure 6. SEM images of carbon foams obtained by carbonization under pressure of oil pyrolysis pitch

IOP Conf. Series: Materials Science and Engineering **570** (2019) 012022 doi:10.1088/1757-899X/570/1/012022

The graphitized part of the sample has an isotropic structure, which indicates an insufficient content of the mesophase in the material. The characteristic graphite texture is expressed explicitly. The 10 wt. % of mesophase microspheres were added to the initial pitch and the material was processed in similar conditions to increase the content of the mesophase in the material.

Figure 7 shows the microstructure images of the obtained graphitized foam. It can be seen that pores have approximately the same character, but the structure of the interpore walls differs sharply from the previous material – walls have the high degree of graphitization with the striated structure of graphite (figure 7, *c* and *d*) there are also secondary deposits (figure 7 *a* and *b*), however, mainly walls have more pronounced graphite nature. In some places, cracks and pores of the elongated shape are observed (figure 7 *e* and *f*), which is possibly related to the thermal stresses arising at the low-temperature stage of the process that is conditioned by the filler presence. The presence of cracks can lead to a noticeable decrease in material properties.

The samples were also obtained on the basis of a special oil pitch with a mesophase content of more than 80 wt. % to decrease the resulting thermal stresses. The samples were obtained according to a similar scheme.

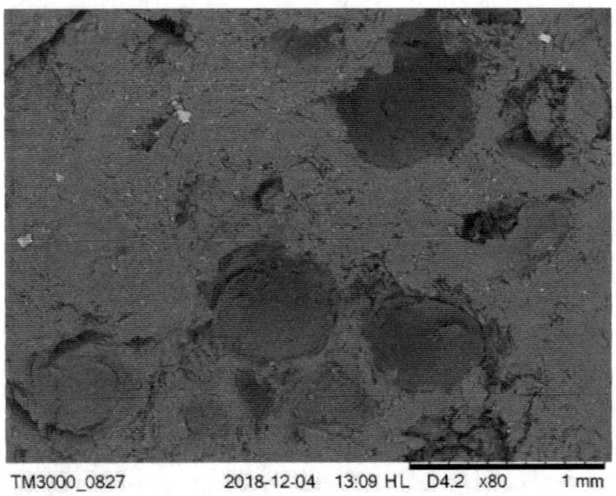

a)

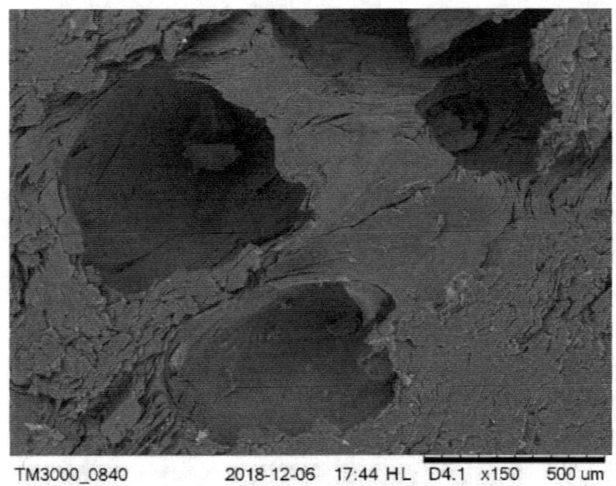

b)

c)

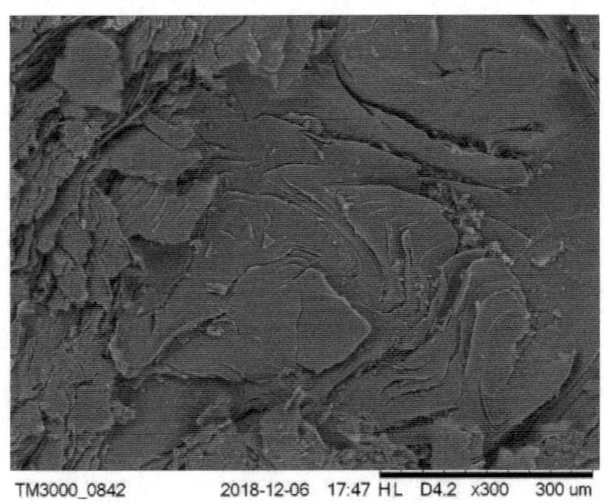

d)

e)

f)

Figure 7. SEM images of carbon foams obtained by carbonization under pressure of oil pyrolysis pitch with the addition of 10 wt. % of mesophase microspheres

Figure 8 shows the images of the material microstructure. It can be seen that material has a structure with the high degree of anisotropy of the graphite component similar to the previous one. The presence of cracks in the structure of the interpore walls which are visually comparable with those for the samples with mesophase microspheres can be seen.

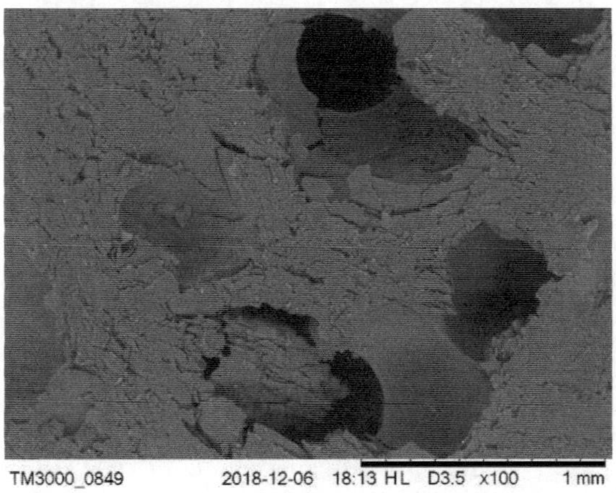

a)

b)

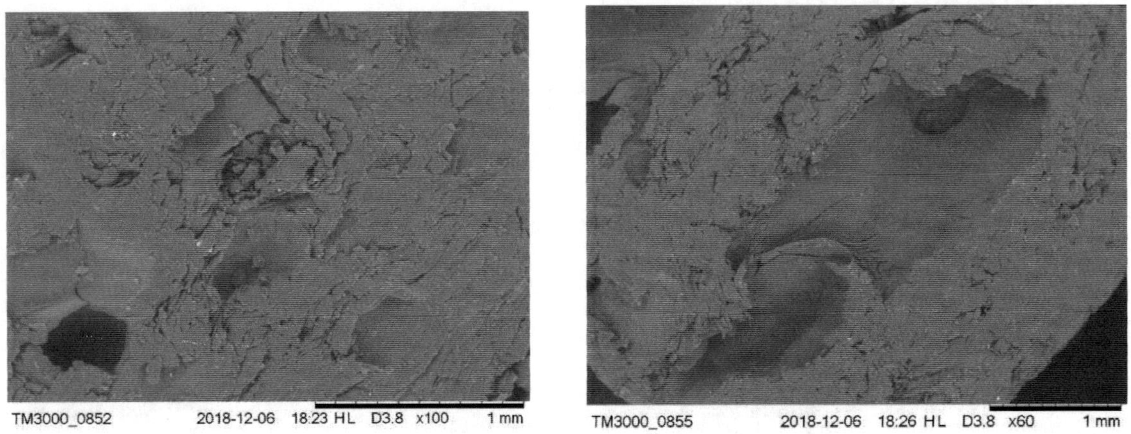

c) *d)*

Figure 8. SEM images of carbon foams obtained by carbonization under pressure of special mesophase pitch

At the same time, studies of mechanical and thermophysical properties have shown that the properties of the given group of samples have a higher level which indicates the slightly higher degree of perfection of the graphite obtained using this technology.

3. Conclusions

Thus, the studies showed that the technology for producing foams by carbonization under pressure provides a fairly high level of thermophysical (heat-conducting) and physical and mechanical properties of products.

References

[1] Tikhomirov V K 1975 *Foams. Theory and practice of their production and destruction* (Moscow: Khimiya Publ) pp 11-15

[2] Galimov E R, Fedyaev V L, Galimova N Ya, Takhaviev M S and Siraev A R 2017 Mathematical modeling of processes occurring during deposition of sprayed particles of polymeric powder *Journal of Physics: Conf. Series* vol 789 **012007**

[3] Salosina M O 2016 *The thermal conditions of the solar probe taking into account the impact of high-speed dust particles* (Moscow: Proceedings of MAI) p 86

[4] Shutov F A Syntactic polymer foams 1986 *Advances in Polymer Science* pp 63–123

[5] Inagaki M, Qiu J and Guo Q 2015 Review Carbon foam: Preparation and application *Carbon* Vol 87 pp128-52

[6] Anghelescu M S 2009 Integrated Engineering Thermal and Mechanical Analysis of Carbon Foam *Ph.D.* 122 p

[7] Chen Y, Chen B, Shi X, Xu H, Shang W, Yuan Y and Xiao L 2008 Preparation and electrochemical properties of pitch-based carbon foam as current collectors for lead acid batteries *Electrochimica Acta* vol 53 pp 2245-49

[8] Kashapov N F, Luchkin G S and Luchkin A G 2010 Vacuum technologies for the application of functional coatings *"Vestnik" of Kazan State Technological University* No 2 pp 340-45

[9] Galimov E R, Shveyov A I, Shveyov I A, Lysanov D M and Galimova N Ya 2016 Development of combined technologies and multifunctional equipment for application of polymer powder coatings *International Journal of Applied Engineering Research* vol 11 No 3 pp 1583-89

[10] Samoylov V M, Danilov E A, Galimov E R, Fedyaev V L, Galimova N Ya and Orlov M A 2017 Production of thermally conductive carbon foams and their application in automobile transport *IOP Conf. Series: Materials Science and Engineering* vol 240 012062
[11] Galimov E R, Fedyaev V L, Galimova N Ya, Gimranov I R and Marat M M 2017 Formation of polymeric powder coatings *IOP Conf. Series: Materials Science and Engineering* vol 240 **012014**
[12] Galimov E R, Fedyaev V L, Galimova N Ya, Gimranov I R and Siraev A R 2017 Dynamics of coalescence and spreading of liquid polymeric particles during coating formation *Journal of Physics: Conf. Series* vol 789 **012006**
[13] Shafigullin L N, Shafigullina G R, Sarimov D R and Romanova N V 2017 Analysis of carbon plastics destruction causes based on epoxy binding *Turkish online journal of design art and communication* vol 7 pp 1872-78
[14] Sokolova Y A, Shafigullin L N, Sokolova A G and Romanova N V 2018 Carbon fiber reinforced plastic technology *IOP Conference Series: Materials Science and Engineering* vol 365 issue 3 **032020**
[15] Shafigullin L N, Gumerov I F, Ibragimov A R, Shafugullina G R, Nizamova A I and Romanova N V 2018 Aspects of using accelerated weather testing methods for polymeric materials *IOP Conference Series: Materials Science and Engineering* vol 412 issue 1 **012069**

IOP Conf. Series: Materials Science and Engineering **570** (2019) 012023 doi:10.1088/1757-899X/570/1/012023

Development of technologies for producing heat-conducting syntactic carbon foams with specified operational properties

E R Galimov[1], E E Sharafutdinova[1], N Ya Galimova[1], V M Samoylov[2] and E A Danilov[2]

[1]Kazan national research technical university named after A.N. Tupolev, K. Marx street 10, Kazan, 420111, Tatarstan, Russian Federation
[2]JSC NIIgrafit, 2 Elektrodnaya str., Moscow, 111524, Russia

kstu-material@mail.ru

Abstract. Technology for producing the articles based on the syntactic carbon foams is proposed. Components and regime parameters for producing the foams of open-porous cellular structure with specified thermophysical and strength characteristics are determined.

1. Introduction

Syntactic carbon foams (SCF) are one of the modern and promising classes of carbon composite materials their distinctive feature is the presence of a regular porous structure which provides high properties compared to materials of an irregular porous structure at the same density level [1-3]. Carbon materials are notable for the dependence of thermophysical and other physical-chemical properties on the structure parameters which makes it possible to obtain heat-insulating or heat-conducting materials with a wide range of thermal conductivity coefficient – from less than 0.1 W/m·K (for heat-insulating materials) up to 180 W/m·K (for heat-conducting materials) – when varying the raw material base and technological methods of their processing.

Carbon foams can be used in the design of heat sinks and heat insulation, the creation of porous electrodes, and for many other applications. Therefore, research aimed at developing the SCFs with given thermophysical and physical and mechanical properties is relevant.

2. Body text

The objective of this research is to develop technologies for the production of heat-conducting SCFs with given service properties.

Research techniques. Research of physical and mechanical (apparent density, ultimate compressive strength, hardness) and thermophysical (thermal conductivity coefficient) properties was carried out to determine the influence of composition and ratio of components, production technology of SCFs on their structure and operational characteristics. Density, ultimate compressive strength, and hardness were determined by standard techniques. Evaluation of the thermal conductivity coefficient of the samples was performed by the method of a stationary flow at a temperature of 20°C. The study of the SCF microstructure was conducted by the scanning electron microscopy (SEM) on a Hitachi TM3000. The samples were directly attached to a conductive carbon tape. Shooting conditions: accelerating voltage 15 kV, vacuum 10^{-6} mm Hg.

The initial components for producing the foams. There are different approaches to the production of SCFs with a controllable set of properties [4-10]. This paper presents a technological process for

IOP Conf. Series: Materials Science and Engineering **570** (2019) 012023 doi:10.1088/1757-899X/570/1/012023

producing foams with an average value of thermal conductivity coefficient – 10-50 W/m·K. Foams were obtained by the technology involving the use of filler and binder. Carbon microspheres obtained by carbonizing hollow microspheres from the phenol-formaldehyde resin at a temperature of 900°C were used as the filler; high-temperature coal tar pitch with a softening temperature of 120°C was used as the binder. It should be noted that the properties of foams can be varied in a very wide range because of the changes in porosity, proportions between the bonding carbon material and microspheres, their size, the gaseous medium composition, and etc.

Experimental part. The technological process of obtaining the foam includes the sequential carrying out of the following stages: the mixing of components, the formation of "green" foam, low-temperature carbonization, carbonization, high-temperature heat treatment, pyrocompaction, mechanical processing.

In the first stage, weighted carbon microspheres (25 g), coal tar or petroleum pitch (25 g) in the form of plates or powder were placed in the three-necked round-bottomed flask with s volume of 500 ml. A backflow condenser with calcium chloride tube was attached to the flask, and a mixing element of a mechanical stirrer with the fluoroplastic working part and a hermetic seal was placed into it. Under mixing at a speed of 60 rpm, 250 ml of toluene were added into the mixture, heated in an oil bath, and boiled using a backflow condenser (IKA C-MAG HS7) under mixing at 100 rpm for 90 minutes. After that, the content of the flask was cooled to a temperature of 30±5°C and transferred in portions of 150±50 ml to a 500 or 1000 ml round-bottomed or pear-shaped flask. The mixture was evaporated using a rotary evaporator IR-2M (ИР-2М) (vacuum pump LVS 105T with the integrated vacuum gauge and residual pressure regulator) at 40°C and residual pressure of 250±50 mm Hg until complete removal of toluene. Next, the resulting mixture was dried under vacuum with mixing using a rotary evaporator IR-2M at 120°C and residual pressure of 12 mm Hg for 60 minutes. The obtained granulate was removed from the flask and crushed in a mill until obtaining a powder with a particle size of less than 250 microns.

Coal tar or petroleum pitch with a mesophase yield of at least 35 wt. % can be used as the main initial component. If it is necessary to work with pitches that have a mesophase yield of less than 25 wt. % the mesophase microspheres of PMUG (ПМУГ) brand or analogues obtained by heat treatment of the original pitch to a temperature of the mesophase formation (400-500°C) with subsequent extraction of the soluble part in toluene can be added to increase the thermal conductivity of the resulting product. The mesophase microspheres are added to increase the mesophase yield from the pitch up to 40-50 wt. %. Carbon micro- and nanoparticles (fine-grained soot, carbon nanotubes, natural graphite, graphene nanoparticles) can be added into the original pitch in an amount of 0.1-5.0 wt. % to increase product strength.

In the second stage, the formation of "green" foam was carried out. The powder was filled into a blind die with the inner diameter of 35.0±0.5 mm and molded using a PS-10 (ПС-10) press with a loading rate of 0.25-0.35 tf/min at a pressing pressure of 1.2 tf for 3 minutes. The molded workpiece was extracted, and overall dimensions, mass, and apparent density were controlled.

In the third stage, the low-temperature carbonization was carried out. The obtained samples of the "green" foam were wrapped in a double layer of paper, placed in a steel container, covered by a 5-10 cm layer of broken graphite with a diameter of about 250 μm on top of which a 0.5-1.5 cm layer of high-temperature coal tar pitch with a diameter of 2.5 μm was poured it was covered with a steel lid and placed in a muffle furnace, equipped with a thermoregulator with the ability to control the speed and temperature of heating, as well as the exposure time. Samples in a steel container were heated at a rate of 2.5±0.5°C/min to pitch mesophase formation temperature (450°C) with an exposure at a final temperature for 150 min after which the furnace was turned off, cooled for 12 h to 60±40°C and samples were extracted.

In the fourth stage, the carbonization was carried out. The obtained samples of the "green" foam were wrapped in a double layer of paper, placed in a steel container, covered by a 5-10 cm layer of broken graphite with a diameter of about 250 μm on top of which a 0.5-1.5 cm layer of high-temperature coal tar pitch with a diameter of 2.5 μm was poured it was covered with a steel lid and

IOP Conf. Series: Materials Science and Engineering **570** (2019) 012023 doi:10.1088/1757-899X/570/1/012023

placed in a muffle furnace. Samples in a steel container were heated at a rate of 2.5±0.5°C/min to a temperature of 900°C with an exposure at a final temperature for 120 min after which the furnace was turned off, cooled for 12 h to 60±40°C and samples were extracted.

In the fifth stage, high-temperature heat treatment was carried out. The obtained samples were placed in graphite crucibles with lids backfilling (a scrap of artificial graphite production crushed to a particle size of less than 0.25 mm) was filled in. The crucibles were placed in an EVP-2500 (ЭВП-2500) vacuum furnace the working medium was pumped out to a residual pressure of less than 1 mm Hg (it is desired to fill the furnace with argon to displace air and then pump it out to a residual argon pressure of less than 1 mm Hg). The furnace was heated for 8 hours to 2100°C the samples were exposed for 2 hours then cooled naturally and extracted. The resulting samples were placed in graphite crucibles with lids the backfilling (a scrap of artificial graphite production) was filled in. The crucibles were placed in a graphitization furnace with the resistive principle of heating (such as Acheson or Tamman type) and heated to a temperature of at least 2700°C with exposure at a final temperature of at least 60 minutes. Alternatively, graphitization to a temperature of 2700-3000°C can be carried out in a tubular or chamber furnace of a suitable design that provides an inert atmosphere (oxygen content not more than 10^2 vol. %).

In the next stage, the pyrocompaction was carried out. The obtained samples of graphitized foam were loaded into the pyrocompaction furnace chamber the furnace was sealed, blown through by argon, and after that, the samples were heated and kept at a temperature of 800-1100°C and a pressure of 8-12 mm Hg for 20-120 hours.

In the last stage, if necessary, the obtained workpieces of foam were subjected to mechanical treatment by cutting, drilling, milling, grinding at low feed rates by a tool with a Mohs hardness of more than 7.

Comprehensive analysis of the apparent density, strength and thermophysical properties of SCFs obtained on the basis of carbon microspheres and coal tar or petroleum pitches were carried out. It is shown that the apparent density of the samples changes linearly depending on the content of microspheres in the initial mixture (figure 1) which allows adjusting the material density changing the ratio of filler and binder.

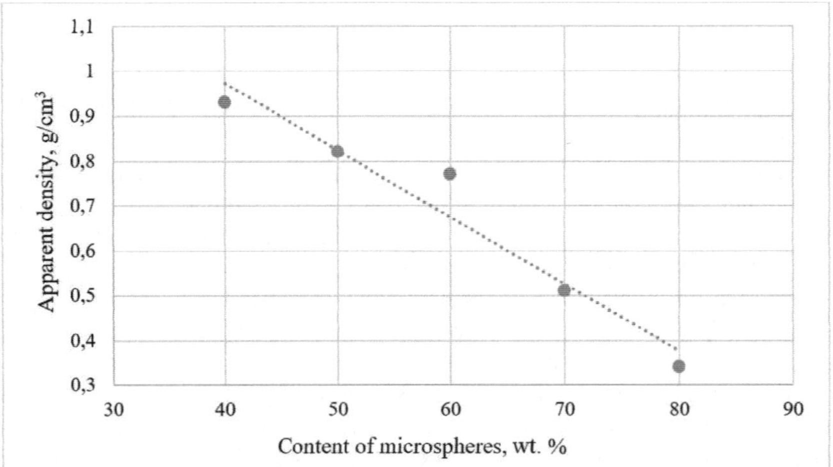

Figure 1. Change in apparent density of foam samples depending on the content of carbon microspheres

Strength is a fundamental requirement applicable to the structural materials. As a rule, the level of 20 MPa makes the materials acceptable for mechanical treatment using available methods (milling, cutting, and drilling). Figure 2 shows that SCFs produced according to the proposed technology satisfy this requirement when the microspheres content is 40 wt. % and less which allows their use for manufacturing of products of complex geometric shapes for various functional purposes. In general, the strength expectedly decreases with increasing content of microspheres which corresponds to an increase of the material porosity.

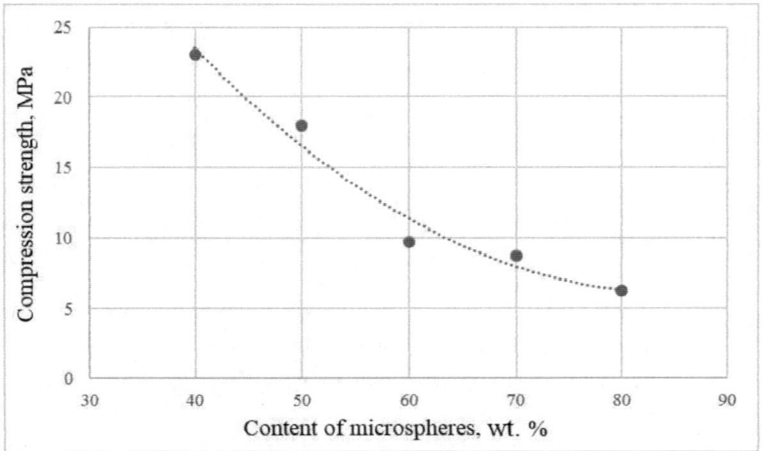

Figure 2. Change in compression strength of foam samples depending on the content of carbon microspheres

Hardness is also an important property that determines the possibility of mechanical treatment of materials. The data in figure 3 shows that the hardness of the studied foams samples allows their processing without using a special tool. With an increase of the microsphere content, the material hardness decreases which suggests that out of two trends – an increase of the average hardness because of the higher hardness of the coke of phenol-formaldehyde resin (microspheres material) compared to graphite from pitch, and decrease in it because of increase of the material porosity – the second one is predominant.

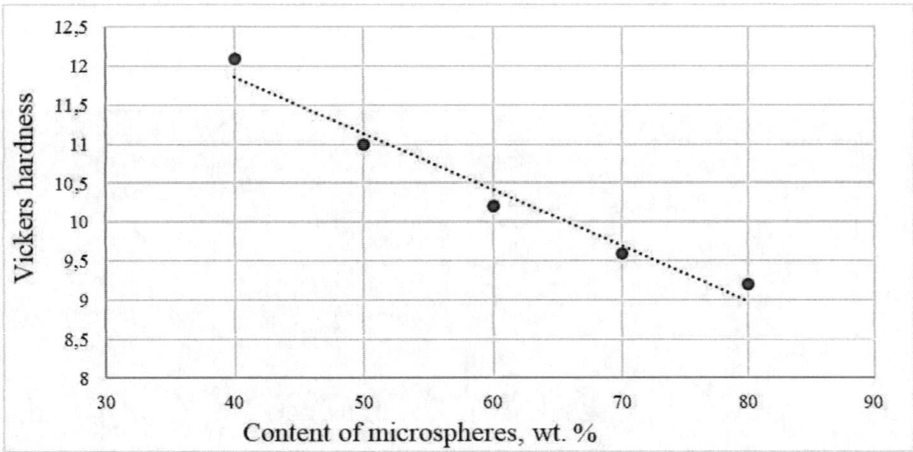

Figure 3. Change in Vickers hardness of foam samples depending on the content of carbon microspheres

Thermophysical properties determine the suitability of materials for use as products of high-temperature heat sink. The data in figure 4 shows that the thermal conductivity coefficient depends linearly on the content of microspheres and its values change approximately in the range from 10 to 50 W/m·K, that is, these foams fall in between thermal insulation materials with low thermal conductivity and isotropic heat-conducting graphite (90-110 W/m·K).

IOP Conf. Series: Materials Science and Engineering **570** (2019) 012023 doi:10.1088/1757-899X/570/1/012023

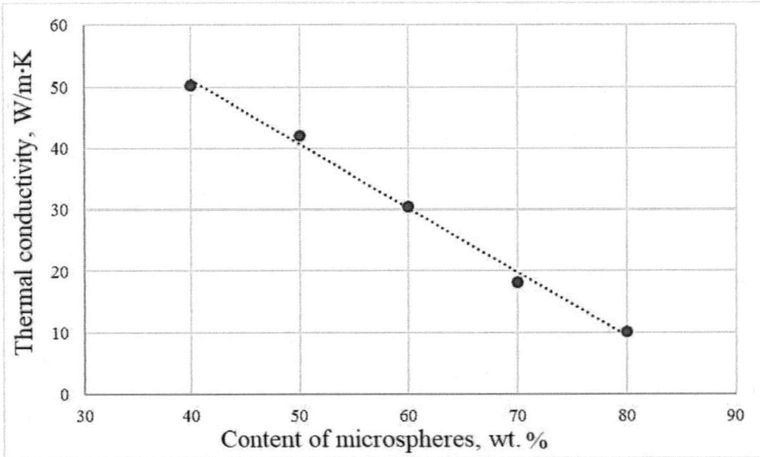

Figure 4. Change in thermal conductivity coefficient of foam samples depending on the content of carbon microspheres

The material microstructure was studied by scanning electron microscopy (figure 5). It can be seen that microspheres with holes or partially destroyed ones, probably because of the release of volatile substances during carbonization, form the basis of the obtained SCFs. The size of the obtained microspheres is less than 90 μm (the average is about 60 μm) which are connected by dense layers of binder (a product of the pitch carbonization). The binder interlayers are markedly textured which indicates the binder graphitization with the formation of graphite a striated structure (figure 5, *c-e*). Formation of small (less than 10 μm) isometric particles and chains of particles should also be noted (figure 5, *g-j*) which indicates the secondary processes occurrence in the gas phase (in the sort of "liquid-vapor-crystal") with the formation of soot-like particles from light fractions of pitch or toluene residues that were not removed. The peculiarity of this class of materials is the formation of a layered structure of the microspheres walls which indicates both the process of coke formation and subsequent binder graphitization on their surface and the change of the graphitizability degree of the wall material (figure 5 *f, h, j*).

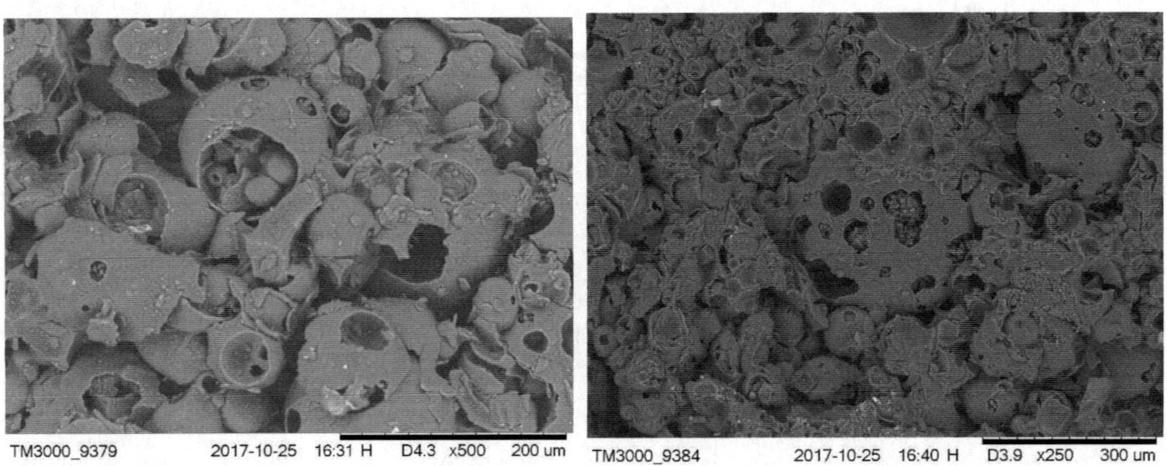

a) *b)*

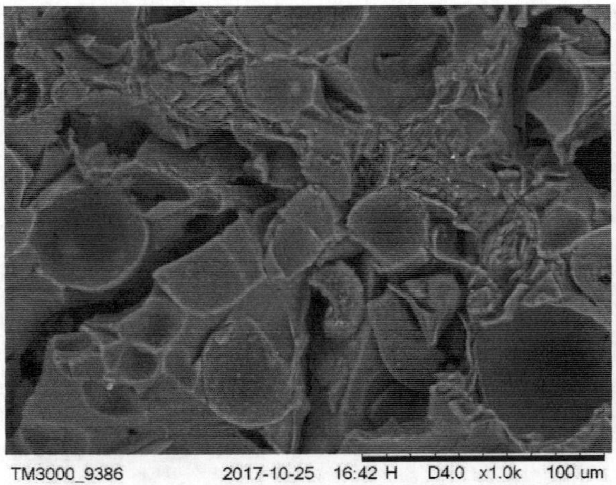
TM3000_9386 2017-10-25 16:42 H D4.0 x1.0k 100 um

c)

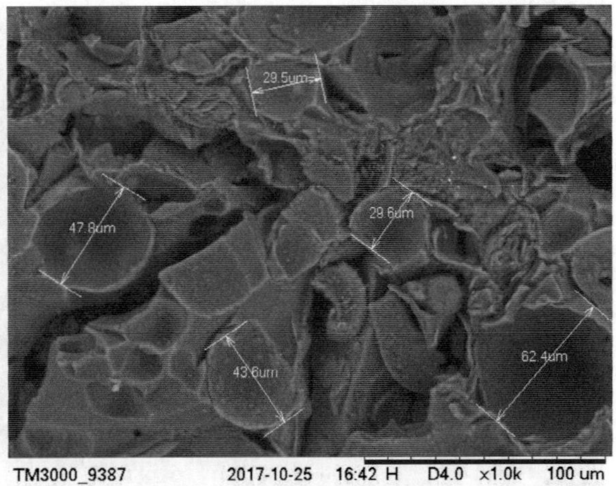
29.5um
47.8um
29.6um
43.6um
62.4um
TM3000_9387 2017-10-25 16:42 H D4.0 x1.0k 100 um

d)

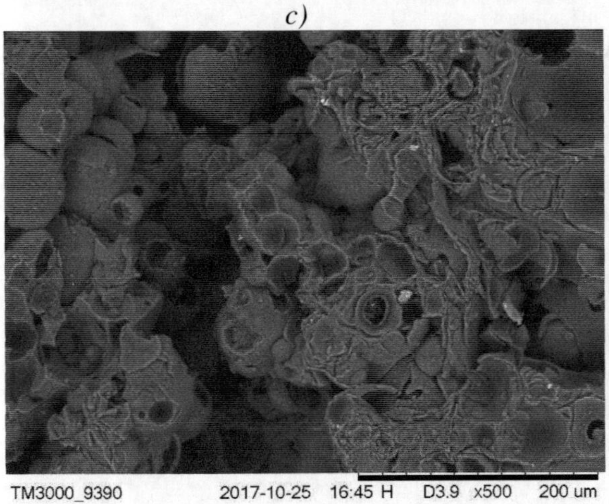
TM3000_9390 2017-10-25 16:45 H D3.9 x500 200 um

e)

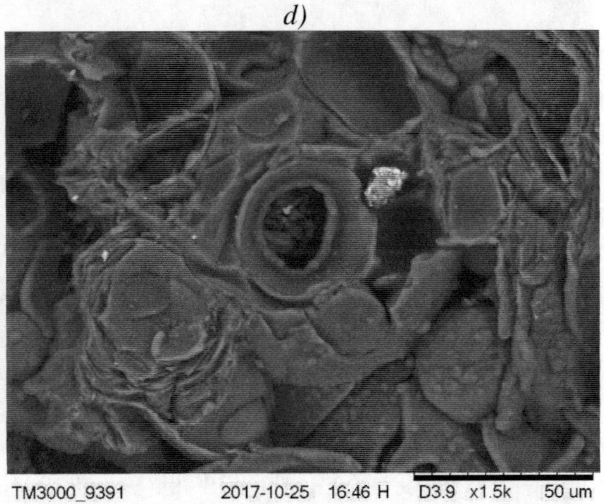
TM3000_9391 2017-10-25 16:46 H D3.9 x1.5k 50 um

f)

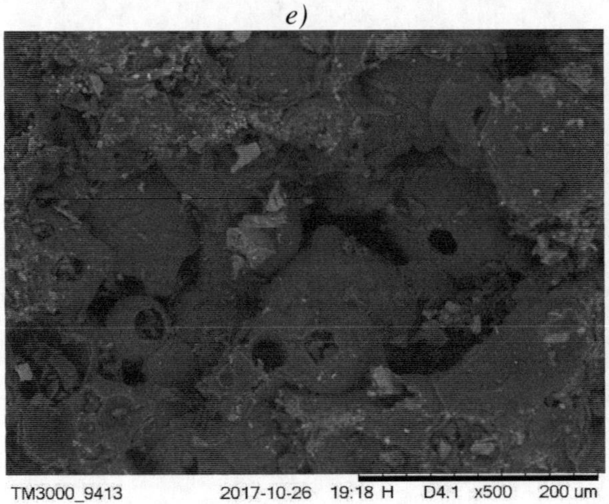
TM3000_9413 2017-10-26 19:18 H D4.1 x500 200 um

g)

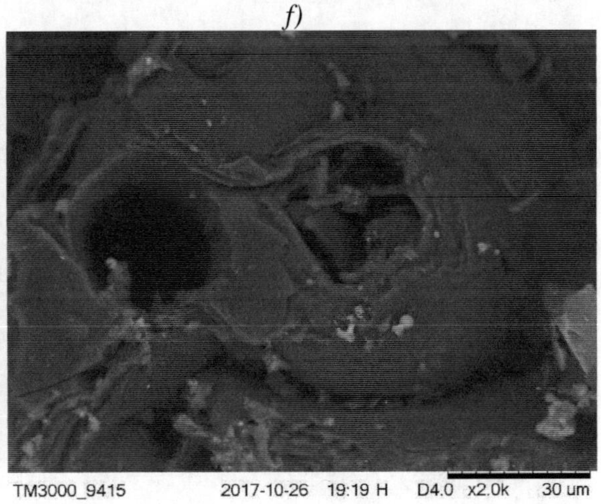
TM3000_9415 2017-10-26 19:19 H D4.0 x2.0k 30 um

h)

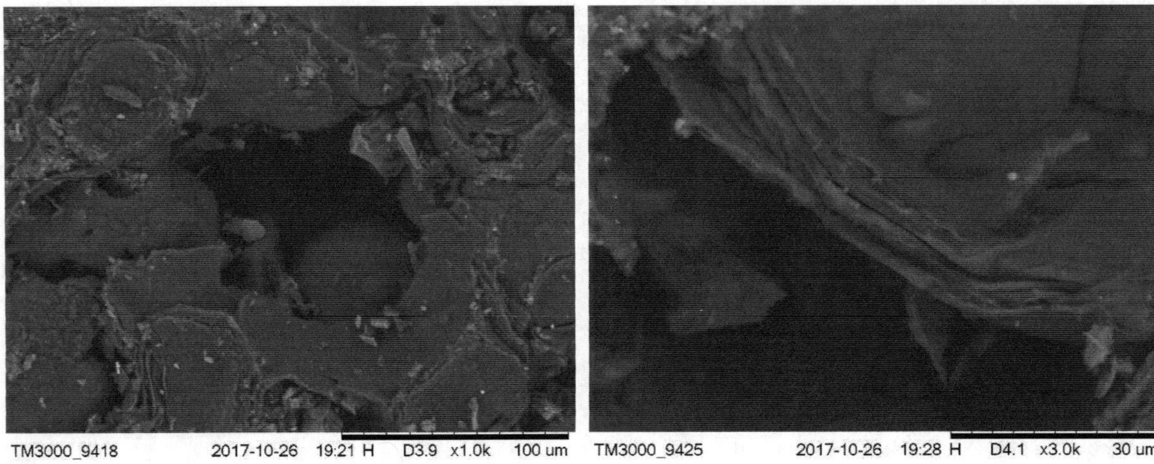

i) *j)*

Figure 5. SEM photographs of foams obtained on the basis of carbon microspheres and coal tar pitches

3. Conclusions

Thus, the proposed technology allows obtaining SCFs with intermediate average values of the thermal conductivity coefficient (10-50 W/m·K) and satisfactory strength characteristics which makes it possible to use them for manufacturing of products including the ones with a complex surface geometry for which excessive heat sink is harmful or does not play a noticeable role (rocket technology and etc.).

4. Acknowledgment

The authors are grateful for the financial support of the Ministry of Education and Science of the Russian Federation within the framework of the Federal Target Program “Research and development in priority areas of development of the scientific and technological complex of Russia for 2014–2020” agreement No. 14.583.21.0057 of July 28, 2016 (Project ID RFMEF158316X0057).

References

[1] Tikhomirov V K 1975 *Foams. Theory and practice of their production and destruction* (Moscow: Khimiya Publ) pp 11-15

[2] Galimov E R, Fedyaev V L, Galimova N Ya, Takhaviev M S and Siraev A R 2017 Mathematical modeling of processes occurring during deposition of sprayed particles of polymeric powder *Journal of Physics: Conf. Series* vol 789 **012007**

[3] Salosina M O 2016 *The thermal conditions of the solar probe taking into account the impact of high-speed dust particles* (Moscow: Proceedings of MAI) p 86

[4] Shutov F A Syntactic polymer foams 1986 *Advances in Polymer Science* pp 63–123

[5] Inagaki M, Qiu J and Guo Q 2015 Review Carbon foam: Preparation and application *Carbon Vol 87* pp128-52

[6] Internet source: Firefly Energy Co. Available at: http://fireflyenergy.com/.(accessed 04.02.2018)

[7] Anghelescu M S 2009 Integrated Engineering Thermal and Mechanical Analysis of Carbon Foam *Ph.D.* 122 p

[8] Chen Y, Chen B, Shi X, Xu H, Shang W, Yuan Y and Xiao L 2008 Preparation and electrochemical properties of pitch-based carbon foam as current collectors for lead acid batteries *Electrochimica Acta* vol 53 pp 2245-49

[9] Kashapov N F, Luchkin G S and Luchkin A G 2010 Vacuum technologies for the application of functional coatings *“Vestnik” of Kazan State Technological University* 2 number pp 340-45

[10] Galimov E R, Shveyov A I, Shveyov I A, Lysanov D M and Galimova N Ya 2016 Development of combined technologies and multifunctional equipment for application of polymer powder coatings *International Journal of Applied Engineering Research* vol 11 number 3 pp 1583-89
[11] Samoylov V M, Danilov E A, Galimov E R, Fedyaev V L, Galimova N Ya and Orlov M A 2017 Production of thermally conductive carbon foams and their application in automobile transport *IOP Conf. Series: Materials Science and Engineering* vol 240 **012062**
[12] Galimov E R, Fedyaev V L, Galimova N Ya, Gimranov I R and Marat M M 2017 Formation of polymeric powder coatings *IOP Conf. Series: Materials Science and Engineering* vol 240 **012014**
[13] Galimov E R, Fedyaev V L, Galimova N Ya, Gimranov I R and Siraev A R 2017 Dynamics of coalescence and spreading of liquid polymeric particles during coating formation *Journal of Physics: Conf. Series* vol 789 **012006**
[14] Shafigullin L N, Shafigullina G R, Sarimov D R and Romanova N V 2017 Analysis of carbon plastics destruction causes based on epoxy binding *Turkish online journal of design art and communication* vol 7 pp 1872-78
[15] Shafigullin L N, Gumerov I F, Ibragimov A R, Shafugullina G R, Nizamova A I and Romanova N V 2018 Aspects of using accelerated weather testing methods for polymeric materials *IOP Conference Series: Materials Science and Engineering* vol 412 issue 1 **012069**

IOP Conf. Series: Materials Science and Engineering **570** (2019) 012024 doi:10.1088/1757-899X/570/1/012024

Improving the efficiency of shaving through the use of wear-resistant coatings

V N Gilman[1,2], I P Balabanov[1,3] and A I Faskhutdinov[1,2,3]

[1] Kazan Federal University, 18 Kremlyovskaya street, Kazan, 420008, Russian Federation

[2] KAMAZ PTC, 2 Avtozavodsky Avenue, Naberezhnye Chelny, 423827, Russian Federation

[3] Kazan National Research Technical University named after A. N. Tupolev –KAI, 10 K. Marks street, Kazan, 420111, Russian Federation

Abstract. The influence of the preparation and use of various types of wear-resistant coatings on the tool shaver of high-speed steel R6M5K5 with a hardness of 60-65HRc for shavinging parts such as gear driven made of steel 20HGNMTA with a hardness of 160-210 HB. The use of hydroabrasive cleaning with ceramic balls made it possible to obtain a clean surface with a rounding of the cutting edge to 7 microns. For the analysis of durability, coatings based on titanium nitride (TiN) and a multi-component coating of titanium and aluminum (AlTiN) were used. The coating on the basis of titanium nitride (TiN) increased the resistance 1.35 times. Multi-component coating of titanium and aluminum (AlTiN) increased resistance 1.61 (AlTiN) times.

As finishing operations for parts Gear driven, manufactured from steel 20ХГНМТА with hardness 156-207 HB, the technological operation of shaving with shaver R6M5K5 high-speed steel with hardness of 62-64HRc is performed.

It is known that the shaver is a gear wheel, the teeth of which are cut by transverse grooves 1.5 mm wide [1-5]. These grooves form a set of blades, which are in constant engagement with the response profile of the blank cut (shave) with her thin chips. Thus the resulting accuracy class reaches 5-6 degrees (Ra 0.63-0.16 μm) [1,3-5,7-9].

The main goal of the tool management process is to increase efficiency and reduce the user costs of a tool for processing by optimizing processes and / or increasing tool durability [2-6]. In the case of shaving operation, the most optimal way to reduce costs is to assume the use of various methods of hardening cutting edges [4, 7, 8–10]. First of all, we are talking about physical vapor deposition in a vacuum (Arc-PVD vacuum arc method) or, in other words, ion-plasma coating [11]. However, taking into account the design features of shavers, the use of this type of hardening has not previously been possible. The problem was the proper preparation of cutting edges and deep cleaning of the grooves between the cutting edges. The presence of burrs on the cutting edge did not allow to form a perfect blade evenly coated [12]. And pollution in relatively deep grooves did not allow to achieve the required purity of the coating process (see Fig.1).

IOP Conf. Series: Materials Science and Engineering **570** (2019) 012024 doi:10.1088/1757-899X/570/1/012024

Fig.1 Shaver cutting edge in the initial state

The use of hydroabrasive cleaning with carriers in the form of ceramic balls (for removing burrs and rounding the edges) and aluminum oxide powder made it possible to solve part of the problems connected with preparing for the coating process. Namely, an evenly cutting edge was obtained, rounded off by an insignificant radius (up to 5-7 microns), while all surfaces were subjected to deep cleaning and activation, which, together with an ultrasonic cleaner, realized the maximum surface cleanliness (see Fig.2).

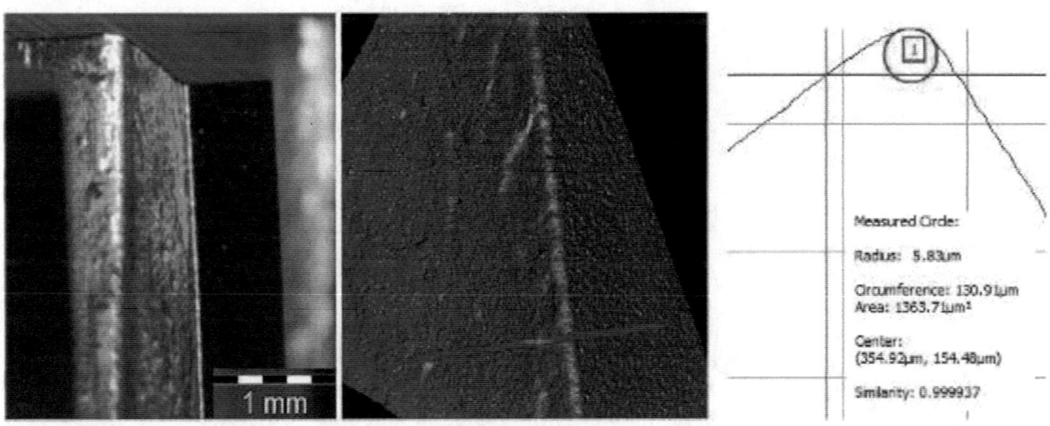

Fig.2 Shaving cutting edge after surface preparation

The next stage of the study was the selection of the optimal type of coating [13-17]. Taking into account the nature of the work of a multi-blade tool, when a small amount of chips is cut in one touch, it can be concluded that the local heating of the cutting wedge is extremely small. That is, as variants of coatings, it is possible to consider both widely known titanium nitride (TiN) with a low threshold value of temperature destruction and oxidation of about 600 ° C [17, 19], as well as all kinds of

IOP Conf. Series: Materials Science and Engineering **570** (2019) 012024 doi:10.1088/1757-899X/570/1/012024

advanced multicomponent coatings based on titanium and aluminum (AlTiN) with temperature indicators from 800-900 ° C [18,20,21,22]. To obtain complete information, experiments were carried out with two types of coatings (see Fig.3).

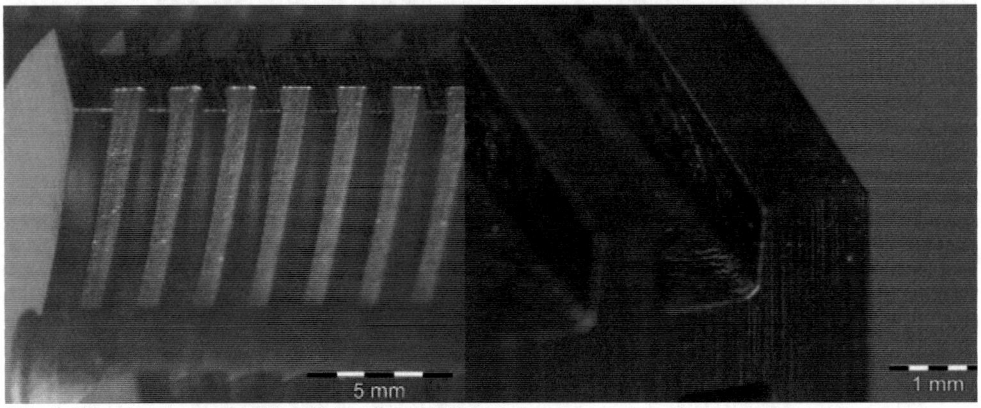

Fig.3 Shaving cutter with TiN and AlTiN coating

The analysis of the application of these shavers in the technological cycle showed an increase in durability 1.35-1.61 times and is shown in the diagram (see Fig.4).

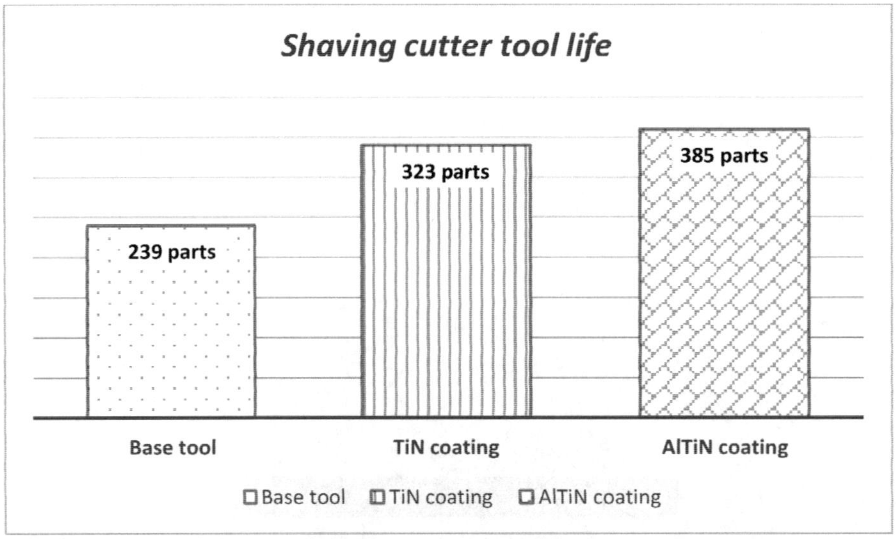

Fig.4 Diagram of change of shaving cutter tool life

In addition to increasing the overall resistance, undoubted increase in the quality of the blank surface on one or two classes was obtained, which in turn, at no additional cost, improved the characteristics of the gear and, ultimately, should increase the durability of the gear. Taking into account that a relatively similar increase in resistance has been obtained for different types of coverage, the choice of a specific type of coverage should be made after calculating the cost of consumables. It is also necessary to provide further work on the selection of the optimal coating, taking into account the greater resistance to abrasion wear.

References

[1] Weiss Denis 2016 Influence of the geometry of the rounding of the cutting edge wear of a tool Machine Park No 12

[2] Faskhutdinov A I, Emelyanov D V and Blinova A S 2015 *Impact of the size of the shear layer on the performance of mechanical processing* Proceedings of scientific articles of the 2nd International youth scientific-practical conference "Progressive technologies and processes" pp 111-15
[3] Balabanov I P and Chermenin A A 2015 *Analysis of simulation systems machine tool systems//results 2015: ideas, achievements* Proceedings of materials of the II Regional student scientific-practical conference with all-Russian participation KNRTU-KAI
[4] V A, Romanov V B, Pivkin P M 2016 *Determination of tool parameters in technological systems by cutting* STIN No 10 pp 17-20
[5] A V Shaparev and I A Savin 2018 *Influence of the State of the Contact Surfaces on the Formation of the Joint of Steel and Brass during Cold Cladding* Solid State Phenomena Vol 284 pp 319-25
[6] Kasjanov S V, Kondrashov A G, Safarov D T 2017 *Regulation of Geometrical Parameters Deviations of Automotive Components Parts through Diagnostic Measurements Organization* Procedia Engineering Vol 206 pp 1508-14
[7] A V Shaparev and I A Savin 2017 *Calculation of Joint Plastic Deformation to Form Metal Compound in Cold Condition* Solid State Phenomena 265 pp 313-18
[8] A G Kondrashov, D T Safarov, A I Faskhutdinov and G K Davletshina 2017 *Single-Turn Worm Mills for Conical Round-Tooth Gears* Russian Engineering Research Vol 37 No 9 pp 812–13
[9] Kas'yanov S V, Kondrashov A G, Safarov D T 2017 *Rapid assessment of wear-resistant tool coatings* Russian Engineering Research Vol 37 Is 11 pp 969–73
[10] Khusainov R M and Khaziev R R 2017 *Modeling of forming technological errors in processing by gear shaping machine* IOP Conference Series: Materials Science and Engineering Vol 240 Is 1 **012045**
[11] Khusainov R M and Sharafutdinov I F 2016 *Methods of assessing the dynamic stability of the cutting process using UNIGRAPHICS NX* IOP Conf. Series: Materials Science and Engineering 134 (2016) **012042**
[12] R M Khusainov and R R Khaziev 2017 *Mathematical Model for Assessing the Accuracy of Processed Gears on Gear Shaping Machines* Procedia Engineering Vol 206 pp 1087-92
[13] Kasyanov S V, Kondrashov A G and Safarov D T 2014 *Research of characteristics of wearproof coating for cutting tools* INTERFINISH-SERIA 2014: International Conference on Surface Engineering for Research and Industrial Applications 62 p
[14] Weiss Dennis Effect of cutting edge rounding geometry on tool wear, Machine Park No. 12, 20163
[15] Berend Denkenaa, Jens Koehlera and Michael Rehe 2012 *Influence of the Honed Cutting Edge on Tool Wear and Surface Integrity in Slot Milling of 42CrMo4 Steel* Procedia CIPR pp 190-195
[16] Aperador W, Amaya C. and Espana, C. 2011 CORROSION RESISTANCE OF TiN/AlTiN (n) MULTILAYERS DEPOSITED ONTO AISI 1045 STEEL. *Dyna-Colombia* **78** pp 183-89
[17] Bemporad E, Sebastiani M, Pecchio C and De Rossi S 2006 *High thickness Ti/TiN multilayer thin coatings for wear resistant applications* Surface & Coatings Technology **201** pp 2155-65
[18] Garcia J, Pitonak R, Weissenbacher R and Kopf A 2010 *Production and characterization of wear resistant Ti(C,N) coatings manufactured by modified chemical vapor deposition process. Surface & Coatings Technology* **205** 2322-27
[19] Hovorun T P, Pylypenko O V, Hovorun M V and Dyadyura K O 2017 *Methods of Obtaining and Properties of Wear-resistant Coatings Based on Ti and N and Ti, Al and N* Journal of Nano- and Electronic Physics **9**
[20] Balabanov I P, Gilman V N, Timofeeva T S and Faskhutdinov A I 2018 *Modeling of the cutting edge rounding influence on the tool life in processing a gear wheel by the Power Skiving method* International Journal of Engineering and Technology (UAE) 7(4) pp 71-73

[21] Kuang J H and Chen W L 1996 Determination of tip parameters for the protuberance preshaving cutters *Mechanism and Machine Theory* **31** pp 839-49
[22] Sergevnin V S, Blinkov I V, Volkhonskii A O, Belov D S, Kuznetsov D V, Gorshenkov M V and Skryleva E A 2016 *Wear behaviour of wear-resistant adaptive nano-multilayered Ti-Al-Mo-N coatings* Applied Surface Science **388** pp 13-23

IOP Conf. Series: Materials Science and Engineering **570** (2019) 012025 doi:10.1088/1757-899X/570/1/012025

Methods of training specialists in the field of sustainable resource management and waste management

I R Gilmanshin[1,2], S I Gilmanshina[1], A I Galeeva[3], E M Gadirova[4] and I A Galeev[1]

[1]Kazan Federal University, Kazan, Russia, 420008

[2] Kazan National Research Technical University named after A N Tupolev, Kazan, Russia

[3]Kazan State Power Engineering University, Kazan, 420066, Russian Federation

[4]Baku State University, AZ 1148, Baku, Azerbaijan

is-er@yandex.ru

Abstract. the method of the university students' engineering thinking development has been studied in the article, the experience of major universities of Germany implementing master programs in the field of sustainable resource management and waste management has been analyzed, the reasonableness of leading educational practices application by the specific topic lectures, the problem situations analysis, the logic of the engineering problems formulation has been substantiated.

High competition in the market of educational services leads to the transition to popular, problem-oriented courses, which are being developed in students along with their professional competencies by professional-engineering thinking.

Professional thinking is native to the whole group noted in the literature on formal thinking associated with a particular profession or groups of professions [1]. Its specificity expresses in its activity and makes it possible to carry out tasks ([2], [3], [4]), problem situations, solved by professionals at the stage of the labor path ([5], [6], [7]). This is the content side of professional thinking.

Formation in students in technical areas, training in a classical university is a long process of gradual provision of a number of interactive intermediate goals. These are contradictory and formulation problems, their justification and proof in explaining technical processes and phenomena (for more details see [1, 5]).

Separately, it should be noted that the experimental and calculated technical tasks are based on interdisciplinary connections, the systematic implementation of which involves training for the creation of complex educational, cognitive and technical problems. In the theoretical and practical classes in the process of these problems, you should also apply various methods of scientific awareness with the gradual complication of the nature of the tasks to perform the exercises. In other words, organized intellectual activity of students (energy management, waste management, sustainable development, waste management) should lead to contradictory problems, practical problems and subjective problems associated with intellectual problems and the desire to get out of this state. We believe that problem situations are certain structural components of the muscular system.

The world's leading educational practices also focus on the formation of professional engineering thinking.

The tendency to tighten competition in the educational services market is also visible in the European market. Increased demands on the quality of university education from applicants and students along with the social demand of society and employers reinforce this process. The main goal of opening master programs at European universities today is the striving of universities for European integration within the Bologna Declaration [8]. The more interesting is the experience of European universities in solving this problem.

In the leading universities of Europe - the universities of Germany, as part of the master's programs, highly qualified specialists are trained to create continuous training of personnel: professional activities, research and development workers, and teaching and research personnel. At the same time, it should be noted that the level of training and the obtained competencies allow graduates to work in comfortable life simultaneously in several areas, ensuring the interconnection of science, education and business.

In the Analytical review of master programs in the field of waste management, sustainable resource management and recycling technologies of leading European universities Ulanova O.V. [8] presents programs in the field of waste management implemented in Universities of Germany:

- . University of Stuttgart (Universität Stuttgart), Department of Waste Management (the first one in German history department of waste management, founded in 1975 by Professor O. Tabasaran) - an international English-language interdisciplinary master's program "Air Quality Control, Solid Waste and Waste Water Process Engineering" (WASTE). In 2003, the first international graduation of 23 undergraduates took place. For 7 years, representatives of various countries of the world have become graduates of the international master's program: Bangladesh, Bolivia, Brazil, Bulgaria, Chile, China, Ecuador, Estonia, India, Russia, Singapore, Thailand, Turkey and many others.
- University of Civil Engineering, Weimar (Bauhaus-Universität Weimar) and the University. Leibniz, Hannover (Leibniz Universität Hannover) since 2001, offers a joint correspondence master's program "Wasserund Umwelt" in German.
- Weimar University has been implementing English-language online-course (M.Sc.) "Environmental Engineering and Management" since 2007, developed jointly with the Asian Institute of Technology in Bankog, Thailand and one of the largest universities in the UK - University of Leeds.
- In 2008, at the Brandenburg Technical University of Cottbus (Brandenburgische Technische Universität Cottbus (BTU Cottbus), Masters course in German was launched: «Energieträgeraus Biomasseund Abfällen».
- Since 2007, unique Masters course in German has been opened at the Technical University Clausthal (TU Clausthal) in German: «Radioactive and Hazardous Waste Management»
- Since October 2010, the «Umwelt ver fahrenstechnik und Recycling» and «Utilities and Waste – Sustainable Processing» masters programs have been implemented at Clausthal University of Technology.
- Since 1999, the Dresden Technical University (TU Dresden) has been studying under the master's program "Abfallwirtschaftund Altlasten" (in German). This unique master program includes 3 modern areas of sustainable waste management (waste management, hazardous waste management and remediation of contaminated soils and industrial areas). At the initial stage, the program was focused primarily on training German masters for solving actual environmental problems in Germany itself, in particular in its Eastern part.

Also in the master's programs leading German universities introduced specialized study of environmentally sound waste management courses (Rhine-Westphalian Technical University Aachen (RWTH), Master «Entsorgungsingenieurwesen», Technical University Hamburg-Harburg

(Technische Universität Hamburg -Harburg), Master programs «Wasser- und Umweltingenieurwesen» and “Energie- undUmwelttechnik”, University of Kassel (Universität Kassel), Master program “Umweltingenieurwesen”, Technical University Clausthal (TU Clausthal), Master program “Geoenvironmental Engineering”).

These masters programs are distinguished by a large amount of laboratory practical work on modern analytical equipment in environmental laboratories, targeted excursions to existing waste processing and disposal enterprises, students undergo practical training in environmental, engineering companies and industrial enterprises with the prospect of subsequent employment.

In the course of studying the process of training under the master’s program “AbfallwirtschaftundAltlasten” at the Technical University of Dresden (TU Dresden), a significant bias towards solving complex technical problems in the field of waste management, application of industrial techniques and installations in the educational process, close interaction with enterprises of the industry was revealed.

The use of leading practices reinforced by conducting problem lectures, in which students learn to communicate with the analysis of problem situations that have arisen in practical activity, the logic of posing engineering problems and their solutions, along with a combination of lecture topic presentation, syllabic tasks will ensure that technical students of technical areas of the classical university will be introduced to creative mental activity on the basis of today’s carried out through the analysis of contradictions and the identification of problems izhenie hypotheses, their justification and proof.

References

[1] I R Gilmanshin 2016 *Ways of engineering thinking formation among technical students in a classical university* VI International Scientific and Practical Conference in the framework of Eurasian cooperation pp 134-37

[2] I R Gilmanshin. 2013 *The role of integrated training centers in the field of energy efficiency in ensuring the promotion of energy service contracts* Information resources of Russia No 3 pp 2-4 Azimov Yu I, Gilmanshin I R, Gilmanshina S I 2016 *IOP: Conf. Ser. Mater. Sci. Eng.* **134** 012003

[3] Gilmanshin I R, Kashapov N F, Gilmanshina S I, Galeeva A I 2016 *IOP: Conf. Ser. Mater. Sci. Eng.* **134** 012007

[4] GilmanshinaS I, Gilmanshin I R 2015 *Formation of students and engineering thinking as a way to create new equipment, technologies, materials* Innovative engineering technology, equipment materials -2015: materials of the international scientific-technical conference pp 264-68

[5] Gilmanshina S I, Gilmanshin I R 2015 *IOP: Conf. Ser. Mater. Sci. Eng.* **86** 12029-32

[6] Gilmanshin I R and et. 2015 *Utilization of biogas polygons of solid household waste through the construction of an energy complex based on the installation of alternative energy* Kazan economic investor No 2 (16) pp 41-45

[7] *Analytical review of master programs in the field of waste management, resource, energy saving and recycling technologies of introducing European universities-2010* http://www.baikal-wasma.de/rus/PDF/MasterProgramm_Waste_Management_2010.pdf

[8] Kashapov N F, Gil'Manshin I R and Konahina I A 2014 *IOP Conf. Ser.: Mater. Sci. Eng.* **69** 012024

[9] Gilmanshin I, Kashapov N, Gilmanshina S and Galeeva A 2016 *IOP Conf. Ser.: Mater. Sci. Eng.* **134** 012007

[10] Kashapov L N, Kashapov N F, Kashapov R N and Denisov D G 2016 *J. Phys.: Conf. Ser.* **669** 012029

[11] Dautov G, Fayrushin I and Kashapov N 2014 *J. Phys.: Conf. Ser.* **567** 012006

IOP Conf. Series: Materials Science and Engineering **570** (2019) 012026 doi:10.1088/1757-899X/570/1/012026

Formation of competencies of the 21st century by means of project activities

S I Gilmanshina[1], I R Gilmanshin[1,2], E V Dyatlova[1]

[1]Kazan Federal University, Kazan, Russia, 420008

[2] Kazan National Research Technical University named after A N Tupolev, Kazan, Russia

is-er@yandex.ru

Abstract. The formation of the most sought-after competencies of the 21st century (creativity, critical thinking, communication and work with information, cooperation) by means of project activities is considered. Developed interdisciplinary integrated courses. The results of the survey of graduate students of the pedagogical bachelor degree are analyzed for their attitudes and opportunities to organize project activities for the formation of competencies of the 21st century.

Today, competences can be viewed as a value, a holistic world view, a person's ability to change in response to modern challenges, while preserving the core of education. The most sought-after competencies in engineering education of the 21st century, which allow developing innovative technologies [1, 2, 3], are creativity, critical thinking, communication and work with information, cooperation.

The ideas of formation of the learning competencies of the 21st century have always aroused interest in the scientific world [4, 5]. It was noted that their formation requires personality-oriented learning. One of the conditions for the implementation of student-centered learning is project activity, taking into account the individual psycho-physiological characteristics of students. In general, the application of the project method in education is widely reflected in the works of Russian and foreign authors [6, 7, 8, etc.]. However, the problem of forming competencies of the 21st century through the system of project activities did not receive a final resolution. Moreover, the project activity is inseparable from the present period of development of education.

Today, in the Russian educational standard of a new generation, project activity is considered as the implementation of a system-activity approach to the formation of universal educational activities. A universal learning activities are a tool to achieve the planned results of general education. Thus, given the significant role of project activities in the implementation of the Russian educational standard of a new generation, the proposed study can be considered as one of the attempts to find ways to form competencies of the 21st century through the organization of a system of project activities based on today's realities.

The purpose of the study: to develop and test a system of project activities for the formation of students' competencies of the 21st century.

IOP Conf. Series: Materials Science and Engineering **570** (2019) 012026 doi:10.1088/1757-899X/570/1/012026

Research methods and experimental base of research. The leading approach to the study of this problem is the system-activity approach, which requires systematic education and the implementation of the acquired knowledge and skills in productive activities. The study involved high school students and graduate students of pedagogical bachelor degree in chemistry.

The results of the study and their discussion. Briefly describe each of these competencies of the 21st century. Creativity is considered as creative thinking, the search for non-standard solutions, the development of new ideas, the manifestation of lateral thinking. Critical thinking is presented as effective thinking, a system of judgments for the formulation of reasonable conclusions, the ability to use various logical techniques (induction, deduction and others), a manifestation of system thinking. Communication and work with information implies the ability to effectively listen and use communication, work with information, evaluate, use and manage information. Cooperation means the ability to be responsible, flexible, compromise, to work effectively in a team, to show respect.

From the variety of modern educational technologies for the development of the ability to be creative, critical thinking, communication and work with information, project technology is most suitable for cooperation. Of course, the project technology in the formation of competencies of the 21st century will dominate others, but this does not mean the complete exclusion of other technologies.

To form students' competencies of the 21st century, their project activities should gradually become more complex, as described in detail in [9]. From observations and research according to the plan provided by the teacher, to complex observations requiring active mental activity; from tasks in which knowledge of one studied topic is applied, to design and research tasks requiring knowledge of various branches of chemistry and related disciplines (biology, physics) [9]. The results of the design work can be presented in the form of a presentation, abstract. At the expense of building in the volume of a school subject a system of project problem tasks, the teacher succeeds in programming the activities of students, according to the logic of scientific research, to gradually form the necessary competencies of the 21st century, the most important of which are critical thinking and creativity. Especially favorable conditions for this extracurricular activities - elective courses or circles.

In order to form competencies of the 21st century, we will consider the previously developed interdisciplinary integrated elective courses, such as “Chemistry and Environmental Protection” and “Basics of Energy Saving”. The course content assumes that students complete project assignments in accordance with the principle “from simple to complex”. For example, in the course “Fundamentals of Energy Saving” [10], the topics of mini-projects are first proposed: “Energy Saving as a Phenomenon of Culture”, “Problems of Nuclear Energy”, “Environmentally Clean Fuel”. Then group projects were carried out, such as “Energy saving in the urban economy”, “Energy saving in agriculture”, “Energy saving technologies in animal husbandry”. In the elective course “Chemistry and environmental protection” [10], in one of the classes, schoolchildren first get acquainted with the concept of “coagulation process in nature”, then carry out an experimental project “The presence and nature of coagulation in lyophilic and lyophobic colloidal systems under the influence of electrolytes”. To do this, students build a search plan, then create conditions for the coagulating action of electrolytes and, in the case of coagulation, determine its result - a change in the number of particles, and, consequently, choose the method of this definition. The course ends with design tasks (“Determination of nitrate content in vegetables”, “Determination of vitamin C content in fruits”, “Secrets of toothpaste” and others), during which students of 21st century form their competencies in students in a relaxed atmosphere.

It should be emphasized that the level of professional thinking and methodical preparation of teachers often limits the possibilities of this work [11, 12, 13]. Therefore, the results of the survey of graduate students of the pedagogical bachelor degree in chemistry were analyzed in terms of their attitudes and opportunities for the organization of project activities of students for the formation of competencies of the 21st century. It was revealed that 12% of respondents poorly represent the essence of the method of project-based learning, and 6% of future teachers do not plan to apply this method at all in teaching chemistry. Among supporters of project training, more than 80% of future teachers prefer individual and pair forms of project activities. According to the results of the survey for the formation of competencies of the 21st century, the implementation of integrated projects in chemistry and ecology (84%), chemistry and physics (41%) is most favorable. According to future chemistry teachers, the undisputed leader among 21st century competencies formed by means of project activities is creativity - 81%, then critical thinking (72%), communication and work with information (72%), cooperation (50%).

Conclusion. The formation of the students' competencies of the 21st century is considered on the basis of the project activities implemented in the course of integrated elective courses (chemistry-ecology, ecology-energy saving). The system of project tasks includes abstract and educational experimental mini-projects, as well as group research project tasks. In general, the system of project activities described in the article seems open to innovation, contributes to the formation of students of not only the competencies of the 21st century, but also the development of interest and creative qualities of the individual.

References

[1] Azimov Yu I, Gilmanshin I R, Gilmanshina S I 2016 *IOP: Conf. Ser. Mater. Sci. Eng.* **134** 012003

[2] Gilmanshin I R, Ferenets A V, Azimov Yu I, Galeeva A I, Gilmanshina S I 2015 *IOP: Conf. Ser. Mater. Sci. Eng.* **86** 12014-16

[3] Gilmanshin I R, Kashapov N F, Gilmanshina S I, Galeeva A I 2016 *IOP: Conf. Ser. Mater. Sci. Eng.* **134** 012007

[4] Zimnya I A *Key competencies - a new paradigm of the result of education* URL: http://quality.petrsu.ru/file/ (Request date: 06/27/2018)

[5] Gilmanshin I R, Gilmanshina S I 2017 *IOP: Conf. Ser. Mater. Sci. Eng.* **240** 012022

[6] Gilmanshina S I, Gilmanshin I R 2015 *IOP: Conf. Ser. Mater. Sci. Eng.* **86** 12029-32

[7] Lazarev V S 2011 *A new understanding of the method of projects in education* Pedagogy No 10 pp 8-9

[8] Matyash N V 2014 *Innovative educational technology. Project training* (M .: Academy) 160 p

[9] Gilmanshina S I, Gayfullina A Z, Burlakova V S 2018 *The Formation of General Research Skills of Students in Conditions of School Profile Camp* Modern problems of science and education No 5

[10] Gayfullina A Z, Gilmanchina S I 2018 *Pedagogical Conditions of Forming Ecological-Chemical Culture of Adolescents by Means of Information Technology* Modern science-intensive technologies No 10 pp 176-80

[11] Gilmanshina S I, Gilmanshin I R, Sagitova R N and Galeeva A I 2016 *The Feature of Scientific Explanation in the Teaching of Chemistry in the Environment of New Information of School Students' Developmental Education* International Journal of Environmental and Science Education Vo 11 Is 4 pp 349-58

[12] Gilmanshina S I, Sagitova R N, Kosmodemyanskaya S S *and et.* 2015 *Professional Thinking Formation Features of Prospective Natural Science Teachers Relying on the Competence-Based Approach* Review of European Studies Vol 7 No 3 pp 341-49

[13] Samigullina G S, Gilmanchina S I, Gaisin I T, Gilmanshin I R, Akchurina I R 2015 Professional and Creative Development of Natural Geographic Course Teachers within the Process of Professional Retraining *International Education Studies* **8(4)**

[14] Kashapov N F and Sharifullin S N 2015 *IOP Conf. Ser.: Mater. Sci. Eng.* **86** 012021

[15] Kashapov R N, Kashapov L N and Kashapov N F 2017 *IOP Conf. Ser.: Mater. Sci. Eng.* **240** 012074

[16] Gil'Manshin I R and Kashapov N F 2014 *IOP Conf. Ser.: Mater. Sci. Eng.* **69** 012010

[17] Kashapov L N, Kashapov N F, Kashapov R N and Semushin V D 2016 *J. Phys.: Conf. Ser.* **669** 012028

IOP Conf. Series: Materials Science and Engineering **570** (2019) 012027 doi:10.1088/1757-899X/570/1/012027

Digital learning technologies in the University education system

S I Gilmanshina[1], I R Gilmanshin[1,2], K G Bendjukevich[1]

[1]Kazan Federal University, Kazan, Russia, 420008
[2] Kazan National Research Technical University named after A N Tupolev, Kazan, Russia
is-er@yandex.ru

Abstract. Digital educational resources developed by the authors for information and methodological support of subject disciplines in the system of university education are considered. The digital learning technologies and the capabilities of the electronic resource "Book" are analyzed.

The idea of using digital technologies in teaching is related to the implementation of the new industrial revolution - Industry 4.0 [1], the development of innovative technologies for recycling and the production of new materials [2, 3] and, as a result, the implementation of the University 4.0 model. Digital learning technologies allow to expand the accessibility of education, enhance the productivity of the learning process, make it possible to build an individual training program taking into account the educational needs of students. The success of the modernization of education also involves the work of teachers in the new information-educational environment [4, 5] with information in digital form.

Digital learning technologies are widely implemented in the Kazan Federal University. Consider as an example the discipline "Didactics of Chemistry", developed at the Department of Chemical Education, the curriculum for the preparation of bachelors of pedagogical education in chemistry. For the information and methodological support of this discipline, the digital educational resource "Didactics of Chemistry" has been developed. Of greatest interest are such topics as "Innovations in the development and education of students by means of chemical education" and "Innovative chemistry training". The content of both topics is presented through the electronic resource "Book".

The resource "Book" allows the developer of an electronic resource to create, like a book with the corresponding chapters and paragraphs, a digital educational sub-resource. At the same time it is possible to include media files, video clips of real-life experiences and projects. In general, it is possible to create an electronic portfolio of students' creative works, which is important for developing their value-semantic competence and engineering thinking based on project-based learning [6, 7]. For the development of critical thinking of students, the technology of smart cards is used, which is presented in the topic "Innovative teaching of chemistry" also through the digital resource "Book". In addition, the resource "Book" made it possible to consider the use of modern digital technologies such as webinars, online schools, conferences, and the Olympiad in training.

Briefly consider each of the marked modern digital learning technologies.

IOP Conf. Series: Materials Science and Engineering **570** (2019) 012027 doi:10.1088/1757-899X/570/1/012027

Webinars are a relatively recent training technology that has become popular over the past three years. Its essence lies in conducting online training sessions on a special online platform. One of the applications of webinars is education and preparation for a unified state exam and competitions in various subjects, including chemistry. On webinar platforms, there is an online board where the teacher can clearly explain the material, it is possible to download presentations, illustrative material, video materials, including video experiments. The developing component of webinars, as noted in [8], is the ability of students to work in a team and be attentive to the study of the material. The advantage of webinars compared to other innovative technologies in the following:

- • unification of students from different cities, settlements of the country, otherwise, expansion of access to the educational process;
- • facilitating the educational process, in other words, enhancing its productivity;
- • relaxed perception of traditional educational information;
- • carrying out visual video experiments.

The technology of online schools used in the preparation for the Olympiads and the unified state exam in all subjects is represented by the following resources.

Online School "Foxford" [9]. Its advantage is that Foxford courses are conducted by teachers from leading universities of the country, members of the jury of All-Russian School Olympiads, experts of the unified state exam and the main state exam. The average teaching experience of teachers for more than 7 years. This online school provides training in all school subjects. Classes are held either in groups or individually in the format of online tutoring on a special platform. The advantage is the availability of an online tutorial for filling gaps in knowledge. The disadvantages include, firstly, the lack of an interactive whiteboard and presentations (during classes, the teacher records in chalk on a traditional blackboard, which, in turn, does not fully comply with the principle of visibility of training), secondly, the lack of a mentor to check homework.

The online school "Lektarium" [10] exists relatively recently - about 2 years. The teachers of this school are university students or young university graduates. Lectarium differs from the previous online school in that information is not always given in a strictly scientific language, but in an accessible, often unconventional form.

Online School "Examer" [11]. This school is completely different from the rest. Examer aims to prepare students for the unified state exam. Preparation takes place in a competitive form. This school has no teachers and mentors who can ask a question. The essence of this school is that the student chooses a subject of interest, creates a personal account. In his private office, theory and practice are available to him for specific exam assignments. To check the correctness of the solution of tasks, the student must go to the "new level", that is, perform new tasks. This school is not very effective for students with a low level of basic knowledge.

The next technology, which is gaining more and more popularity, is the technology of online Olympiads and online conferences, which is very useful for highly motivated students. The advantage of this technology is associated with the possibility of a significant expansion of geography and the number of participants in the educational process, the popularization of science among the younger generation. The tasks of online Olympiads in all subjects correspond to the Russian Federal State Educational Standard and, for example, in chemistry, make it possible to form systematic ideas about chemicals and their transformations; learn chemical language; improve knowledge and develop an interest in the study of chemistry.

The advantages of online competitions include:

• lack of costs for the organization and holding of competitions and conferences;

• a wide opportunity for everyone to participate in the Olympiad movement;

• the olympiad platform is protected from cheating (as soon as the student tries to close the tab with the olympiad, either the attempt is blocked or the penalty points are charged).

A significant drawback of online Olympiads compared to full-time Olympiads is the impossibility of conducting a real experimental tour.

The following are the most popular and successful online Olympiads.

Interregional multi-subject Olympiad of Kazan Federal University for schoolchildren [12]. The qualifying round is held remotely on an online platform using distance learning technologies. Every year, about 2,500 schoolchildren take part in this Olympiad only in chemistry, of which only 45% are following a full-time final stage of the online tour.

The All-Siberian Open School Olympiad [13] was organized in 1962 and is aimed at attracting schoolchildren to the study of natural science disciplines. Schoolchildren from more than 50 regions of the Russian Federation take part in this competition every year. In the 2018-2019 academic year, the Olympiad was conducted in the following subjects: astronomy, chemistry, biology, physics, mathematics, computer science. The All-Siberian Chemistry Olympiad is included in the List of Schoolchildren Olympiads and has the highest level - 1. The All-Siberian Olympiad has three stages: the first is a full-time qualifying, the second is an absentee qualifying, the third is a full-time final. At the second stage, students perform tasks remotely using educational technologies.

The Internet Olympiad "SESC Moscow State University" [14] is held for schoolchildren of 7–10 classes in five subjects: mathematics, physics, chemistry, biology, computer science. The Olympiad is held remotely in three stages: training, qualifying, final. The tasks of the qualifying stage are quite unusual, related to human life and the history of the natural science phenomenon. Winners and winners of grades 9, 10 in one of the subjects are invited to attend the full-time Olympiad "Kolmagorov" with the provision of a hostel. Winners and prize-winners of 9, 10 classes in two subjects are invited to participate in the spring gathering of the SESC of Moscow State University named after Lomonosov out of competition.

All considered digital learning technologies are successfully mastered by students in the process of studying at the university, which allows them to be willing to work with information in digital form in their future professional activity.

Thus, as a result, the use of modern digital educational resources developed by university teachers in support of subject disciplines is used by students to learn online how to apply digital learning technologies in future professional and educational activities.

References

[1] Aleksankov A M 2017 *The Fourth Industrial Revolution and Modernization of Education: International Experience* Strategic priorities No 1 (13) pp 53-69

[2] Azimov Yu I, Gilmanshin I R, Gilmanshina S I 2016 *IOP: Conf. Ser. Mater. Sci. Eng.* **134** 012003

[3] Gilmanshin I R, Ferenets A V, Azimov Yu I, Galeeva A I, Gilmanshina S I 2015 *IOP: Conf. Ser. Mater. Sci. Eng.* **86** 12014-16

[4] Gilmanshina S I, Gayfullina A Z, Gilmanshin I R 2018 *Formation of environmental culture in the conditions of new informational environment (on the example of science disciplines)* Proceedings of INTED2018 pp. 1664-69

[5] Sibgatova K I, Gilmanshina S I, Khalikova F D, Gilmanshin I R, Akchurina I R, Shchaveleva N G, Fassakhova G R 2015 *Peculiarities of pupils and vocational college students' career guidance modeling in the integrated system "school – college – enterprise*" Asian Social Science Vol 11 No 1 pp 386-91

[6] Gilmanshina S I, Gilmanshin I R 2015 *IOP: Conf. Ser. Mater. Sci. Eng.* **86** 12029-32

[7] Gilmanshin I R, Gilmanshina S I 2016 *IOP Conf. Ser.: Mater. Sci. Eng.* **134** 012006

[8] Sternberg R J 2012 *The assessment of creativity: An investment based approach* Creativity Research Journal No 24 (1) pp 3- 12

[9] *We are helping to improve grades and prepare for the exam, OGE, olympiads:* Foxford online school. https://foxford.ru (appeal date 10/26/2018).

[10] Lectarium top preparation for 100 points on the exam in all subjects online: *https://lectarium.ru/courses* (appeal date 10/26/2018).

[11] All for self-preparation for the exam: online school "Examer". *https://examer.ru/* (appeal date 10/26/2018)

[12] Interregional Multidisciplinary Olympiad of the Kazan Federal University for Schoolchildren. *https://admissions.kpfu.ru/mpo#it* (appeal date 10/26/2019).

[13] All-Siberian Open School Olympiad. *http://sesc.nsu.ru/vsesib/* (appeal date 10/26/2018)

[14] Internet Olympiad "SSC Moscow State University". http*://internat.msu.ru/distantsionnoe-obuchenie/internet-olympiad/* (appeal date 26.10.2018)

[15] Kornienko E E, Lapushkina E J, Kuzmin V I, Vaschenko S P, Gulyaev I P, Kartaev E V, Sergachev D S, Kashapov N, Sharifullin S, Fayrushin I 2014 *J. Phys.: Conf. Ser.* **567** 012010

[16] Saifutdinov A I, Fairushin I I and Kashapov N F 2016 *JETP Lett.* **104** 180–185

[17] Luchkin A G, Kashapov N F and Luchkin G S 2013 *J. Phys.: Conf. Ser.* **479** 012019

[18] Denisov D G, Kashapov N F and Kashapov R N 2015 *IOP Conf. Ser.: Mater. Sci. Eng.* **86** 012005

Methods of identifying talent in the field of engineering and natural sciences

S I Gilmanshina[1], K G Bendjukevich[1], I R Gilmanshin[1,2], V A Glushkova[1] and E M Gadirova[3]

[1]Kazan Federal University, Kazan, Russia, 420008

[2] Kazan National Research Technical University named after A N Tupolev, Kazan, Russia

[3]Baku State University, AZ 1148, Baku, Azerbaijan

Is-er@yandex.ru

Abstract. The distinctive characteristics of giftedness in the field of engineering and natural sciences are highlighted. The problems of intellectual giftedness are considered. The application of the projective method of diagnosing divergent thinking to determine endowments in the field of engineering and natural sciences, taking into account its adaptation, is substantiated.

Today, one of the priorities of modern education, in accordance with the requirements of the Federal State Educational Standard, is the identification and development of talented youth. This is due to the need for extraordinary creative personalities capable of ensuring the intensive development of a post-industrial society (for more, see [1], [2], [3], [4]).

Recently, engineering and technological research based on chemical knowledge is highly relevant. In addition, chemistry, thanks to its internal logic, plays a key role in the development of adolescent talent and creative thinking, the ability to explain, analyze, and predict (research [5], [6], [7]). This, in turn, favorably affects the intellectual potential of young people, motivating them in the future to solve important national economic problems.

In scientific works of scientists ([8], [9], [10]), affecting the problems of intellectual giftedness, the ambiguity of the concept of "giftedness" is emphasized. Today, the most relevant are the following definitions of giftedness: a qualitative combination of mental abilities, ensuring the progressive activity of an individual; general expressed abilities of the individual, causing the diversity of opportunities and level of activity; intellectual potential as a holistic characteristic of cognitive abilities and opportunities for the acquisition of knowledge and the formation of competencies; a set of natural instincts and the severity of natural prerequisites; talent and talent, as an internal component of potential, expressed in the productive activity of the individual.

The ambiguity of the concept of “giftedness” primarily indicates the multidimensionality of the problem of studying the phenomenon of giftedness. Giftedness requires an integrated approach in its identification and development, taking into account the psycho-physiological, differential psychological and socio-psychological aspects of the individual. The noted aspects in the aggregate should be taken into account when selecting methods, forms and means of training for gifted young people.

The leading approaches to the study of this problem are the system-activity approach. This approach requires consistency in education and personality formation, proves the futility of the knowledge and skills that are not implemented in productive activities. Within the framework of this study, the system-activity approach is expressed in the development of such pedagogical conditions that would create situations of difficulty and include students in various types of practice-oriented activities in order to identify intellectual talent and their training in engineering and natural sciences.

The identification of intellectually gifted youth is connected, first of all, with the teacher's observation of the student's behavior, his academic performance and interest in the subject being studied and in the disciplines adjacent to him. In addition, it is necessary to take into account that often shyness, isolation and, as a result, problems associated with socialization are characteristic of gifted students. To mitigate the problem of socialization, a gifted person must be surrounded by people who have relatives with his intellect.

According to psychological and pedagogical research ([10], [11], [12]), the methods for identifying intellectual talent are based on two approaches, one of which is based on the method of problem situations, and the other on intelligence testing. Both approaches are aimed at identifying the essential characteristics of the concepts of “giftedness”, “genius”, “talent”, “intellect”, “creativity” and the formation of creative personality qualities.

In general, the following parameters are diagnosed: interest and abilities in demand in a particular activity; prerequisites for general endowments; degree of formation of creative abilities of the individual.

However, according to literary data [10], all methods of identifying giftedness are imperfect, which is associated with the many-sidedness of the concept of “giftedness” and its various types, as well as the insufficiently developed concepts related to the phenomenon of general giftedness.

In this study, an attempt was made to adapt the projective method of diagnosing divergent thinking [12] to identify giftedness in the field of engineering and natural sciences. Since the construction of tasks is universal, this technique easily adapts to the specifics of any natural science or engineering discipline. Considering the fact that by divergent thinking, following J. Guilford, we understand a person’s ability to search for a solution in different directions and, accordingly, to find several solutions to a given problem (this does not consider mental capabilities in general). Otherwise, divergent thinking serves as the basis for creativity (creativity indicators: fluency, flexibility, productivity, development) and provides innovative non-standard ideas and ways to solve them.

Since in engineering and natural science problems it is very valuable to produce several solutions and select the best of them, the diagnosis of divergent abilities can serve as a criterion for identifying intellectually gifted youth.

Based on the above, the author developed a method for identifying intellectual talent, which contains 12 tasks divided into four blocks. The first block of tasks is focused on identifying verbal initiative. The next block is to identify the ability to interpret symbols and signs, as well as the ability to identify and match symbols and pictures with natural science (for example, chemical) objects. The third block is to determine the ease of “manipulating” with digital (symbolic) material. The fourth block integrates tasks, which determine the ease of advancing ideas, abstracting from the content of the objects that are manipulated. Each task is evaluated according to four main parameters: fluency; flexibility; originality; productivity. The results for each block of tasks are summarized, and the total score is determined. Moreover, if some answer is less common than 1% of cases in the sample, then it is recognized as original and an additional 1 point is added to the performance points. Thus, originality acts as a supplement. This technique for identifying intellectual talent can be applied starting from the 9th grade.

The developed methodology was tested during the pedagogical practice at the Municipal Educational Institution "Secondary School No. 9 with in-depth study of the English language" of the Novo-Savinovsky District and the Municipal Educational Institution "Lyceum No. 131" of the Vakhitovsky District of Kazan. The results obtained are well correlated with the data on the success of students' performances at natural science (chemical) competitions at various levels.

References

[1] Azimov Yu I, Gilmanshin I R, Gilmanshina S I 2016 *IOP: Conf. Ser. Mater. Sci. Eng.* **134** 12003

[2] Kashapov N F, Gil'manshin I R, Konahina I A et al 2014 *IOP: Conf. Ser. Mater. Sci. Eng.* **69**

[3] Gilmanshin I R, Kashapov N F, Gilmanshina S I, Galeeva A I 2016 *IOP: Conf. Ser. Mater. Sci. Eng.* **134** 12007

[4] Gilmanshin I R, Gilmanshina S I 2016 *IOP Conf. Ser.: Mater. Sci. Eng.* **134** 012006

[5] Gilmanshina S I, Sagitova R N and et. 2015 *Professional Thinking Formation Features of Prospective Natural Science Teachers Relying on the Competence-Based Approach* Review of European Studies Vol 7 No 3 2015 pp 341-49

[6] Gilmanshina S I, Gilmanshin I R, Sagitova R N and Galeeva A I 2016 *The Feature of Scientific Explanation in the Teaching of Chemistry in the Environment of New Information of School Students' Developmental Education* International Journal of Environmental and Science Education Vo 11 Is 4 pp 349-58

[7] Samigullina G S, Gilmanchina S I, Gaisin I T, Gilmanshin I R, Akchurina I R 2015 Professional and Creative Development of Natural Geographic Course Teachers within the Process of Professional Retraining *International Education Studies* **8(4)**

[8] Khalikova F D, Gilmanshina S I *IOP: Conf. Ser. Mater. Sci. Eng.* **240** 012042

[9] V F Gabdulchakov 2014 *Communicative Core of Interaction and its Influence on Education Results* Procedia-Social and Behavioral Sciences V 146 pp 381-84

[10] Leites N S 1996 *Psychology of gifted children and adolescents* (M .: Academy) 416 p

[11] Sternberg R J 2012 *The assessment of creativity: An investment-based approach* Creativity Research Journal No 24 (1) pp 3-12

[12] Soldatova E L 2005 *Psychology at school* Bulletin of practical psychology of education No 2 (3) pp 119-24

[13] Gavrilova V A, Fazlyyyakhmatov M G and Kashapov N F 2014 *J. Phys.: Conf. Ser.* **567** 012023

[14] Kashapov N F, Nafikov M M, Gazetdinov M X, Nafikova M M and Nigmatzyanov A R 2016 *IOP Conf. Ser.: Mater. Sci. Eng.* **134** 012013

[15] Fazlyyyakhmatov M and Kashapov N 2014 *High Temperature Material Processes* **18** (4) 273–279

IOP Conf. Series: Materials Science and Engineering **570** (2019) 012029 doi:10.1088/1757-899X/570/1/012029

Effects of moisture content on thermal conductivity of thermal insulation materials

A M Gusyachkin[1], L S Sabitov[2], A M Khakimova[1], A R Hayrullin[1]

[1]Kazan State Power Engineering University, 51 Krasnoselskaya street, Kazan, 420066, Russian Federation
[2]Kazan Federal University, 18 Kremlyovskaya street, Kazan, 420008, Russian Federation

kharullin@yandex.ru

Abstract. Insulation materials play an important role in district heating pipeline networks. Thermal losses in pipe insulation systems depend on the thermal conductivity of insulation materials. The thermal conductivity of the insulation materials is influenced by the temperature and moisture content. In this paper the effect of moisture content on the thermal conductivity of rock wool produced by different manufacturers was considered. Experimental studies were performed using the standard guarded hot plate (GHP) method. Higher thermal conductivity is always associated with higher moisture content for all investigated samples.

1. Introduction

Growing population and increasing consumption of energy resources increase the importance of energy saving and rational use of energy [1]. One of the potential sectors that reduce energy consumption is heating networks [2]. The first district heating system appeared in New York in 1877. Since then, they began to spread to many European countries [3]. Heat losses in pipeline networks are dependent on insulating materials. [4,5]. The potential for energy saving is available in the building sector. Energy saving in the building sector also depends on insulation materials [6]. The main characteristics of insulating materials is thermal conductivity. Thermal conductivity depends on temperature and moisture content [7,8]. In paper [8], the effect of moisture content on the thermal conductivity of fibrous insulation materials at different insulation temperatures was investigated. The study was conducted for fiberglass, mineral wool and rock wool at different densities. Insulation temperature values ranged from 14 to 34 °C, moisture content from 0 to 46.6%. Higher thermal conductivity for the investigated densities is always associated with higher moisture content. The maximum increase in thermal conductivity is observed for mineral wool. The increase in thermal conductivity can reach 446% with an increase in the moisture content of 15%. The authors [9] showed that when the mineral wool is moistened, the thermal conductivity can increase up to 24 times. The article [10] showed that when fiberglass is moistened up to 1%, there is an increase in heat losses. The article [11] also determined the thermal conductivity of various insulating materials. Moistened was carried out in the climate chamber. A sample of mineral wool adsorbed a very small amount of water about 2%. Unfortunately, small fluctuations of thermal conductivity values (around 3%) do not fall into the accuracy of the experimental method. In another article [12], it was found that the thermal

IOP Conf. Series: Materials Science and Engineering **570** (2019) 012029 doi:10.1088/1757-899X/570/1/012029

conductivity of fiberglass can increase by 300% with an increase in moisture content by 3%. In the article [13], the effectiveness of insulation of pipes made of fiberglass used in district heating and cooling systems subjected to underground water attack has been investigated. The authors of the article found that the thermal conductivity of insulation can increase from 52 to 185 times depending on the temperature of the surrounding water. The authors of [14] investigated the issue of the operation of piping systems with fibrous insulation under conditions of moisture condensation. It is established that the thermal conductivity of the insulation increases systematically with the penetration of water vapor. The thermal conductivity of wet insulation can increase to 3.51 with a moisture content of 15.1%.

The values of thermal conductivity of insulation materials are usually specified by manufacturers. The values of thermal conductivity are given as a function of temperature. However, when using fibrous insulation materials in heat networks, they may be wetted. Typically, manufacturers do not provide data on the thermal conductivity of insulation materials when they are wetted. Data on the effect of moisture on the thermal conductivity of insulating materials is limited. This is also due to the fact that new manufacturers of insulating materials appear on the market. The purpose of this article is to study the thermal conductivity of modern insulation materials of district heating systems, with different levels of moisture content.

2. Methodology

To achieve the goal of the study, fiber insulation materials had dimensions of 150x150 mm, the thickness of the samples was 40–50 mm (figure 1). For the study, samples from rock wool from three different manufacturers were selected (Table 1.). When measuring the thermal conductivity of insulating materials with moisture content, it is necessary to choose the method of their wetting. For moistening insulation materials, the flooding method, the injection method, the laboratory preconditioning method and the ambient temperature with cold surface are used [15]. This article used the injection method because it is simpler and does not require additional equipment. Moisture was injected uniformly over the entire surface of the samples. Desired amount of moisture content determined by weight.

To measure the thermal conductivity of pipe insulation systems, the standard method is used [16]. Measurements are carried out on a cylindrical heated tube. The temperature on the inner side of the insulation is set by the heated pipe, the outer temperature of the wall insulation is cooled by the laboratory atmosphere. Another standard method is guarded hot plate (GHP) [17]. In this case, measurements are carried out on flat samples. The temperature difference between the faces of the samples is set by the heated plates. One of the plates can also be cooled. This method [17] is simpler because it facilitates sample preparation. It is not necessary to build insulation on a cylindrical tube. In this case, thermal conductivity is measured through flat wall that may give inaccuracy in measurements. However, in article [18] has shown that the thermal conductivities obtained on a flat slab specimen can be used to determine the value of a circular pipe. In this work, the GHP method was used to measure the thermal conductivity of pipe insulation systems.

Figure 1. Samples insulation material.

Table 1. List of test samples ordered by code, densities and thickness.

Material type	Sample code	Density ρ (kgm^{3-1})	Thickness δ (mm)
Rock wool	RW-01	80	40
Rock wool	RW-02	80	40
Rock wool	RW-03	80	40

3. Results and discussion

Measurements of thermal conductivity of insulation materials were carried out for dry samples and at three values of the moisture content of 5%, 9.5%, 13.6%. The average temperature of the samples was 26.7±2.5°C. When installing the insulating material on the pipe insulation systems, it is compacted; therefore, during the tests, the material was pressed to density ρ*. The results of the experiment are presented in table 2 and figure 2–3. The mean deviation of the obtained data is 9.2%. For all samples, an increase in thermal conductivity *k* is observed with increasing moisture content MC. Despite the fact that the samples of the insulation material had the same characteristics (table 1), an increase in the MC content may affect the *k* value in different ways depending on the manufacturer of the insulation material. This feature is more noticeable with MC values > 10%. An increase in moisture content has a greater effect on the RW-1 sample. The thermal conductivity of the RW-1 sample increased by 312.8% with an increase in MC by 13.6%. The effect of MC on the thermal conductivity *k* of samples RW-2 and RW-3 is similar. The differences in the *k* values of the RW-2 and RW-3 samples are in the range of accuracy of the experimental measurement. The thermal conductivity of RW-2 and RW-3 samples increased by 81.3–112.7% with an increase in MC by 13.6%. In article [9], at a temperature of 22°C and a MC of 5–20%, the thermal conductivity of mineral wool was 0.10–0.14 W/m°C. In our studies, the *k* values are in a wider range of 0.047–0.148 W/m°C, while the MC values were in a narrower range of 5–13.6%. In article [8], at a temperature of 24°C and a MC of 5–15%, the thermal conductivity of rock wool was 0.0398–0.0399 W/m°C, which is almost similar to a dry material. This data is not consistent with our data. Perhaps this is due to the fact that the density of the rock wool sample in the article [8] was =46 kg/m^3. However, in the same article [8], at a density of mineral wool of 145 kg/m^3 and for the same values of MC and temperature, the thermal conductivity of mineral wool was 0.0696–0.113 W/m°C, which is closer to our values.

Table 2. Measurement results of *k* of rock wool for different manufacturers with different moisture content levels (MC %) under 26.7±2.5 mean operating temperatures.

Sample code	ρ* (kgm^{3-1})	MC (%)	*k* at 26.7±2.5 (°C)	
			W/m°C	% Change
RW-01	119	0	0.036	
	122	5	0.060	69.0
	120	9.5	0.079	121.5
	120	13.6	0.148	312.8
RW-02	119	0	0.036	
	120	5	0.052	46.0
	118	9.5	0.062	72.7
	121	13.6	0.065	81.3
RW-03	118	0	0.036	
	121	5	0.047	31.5
	120	9.5	0.068	91.0
	120	13.6	0.076	112.7

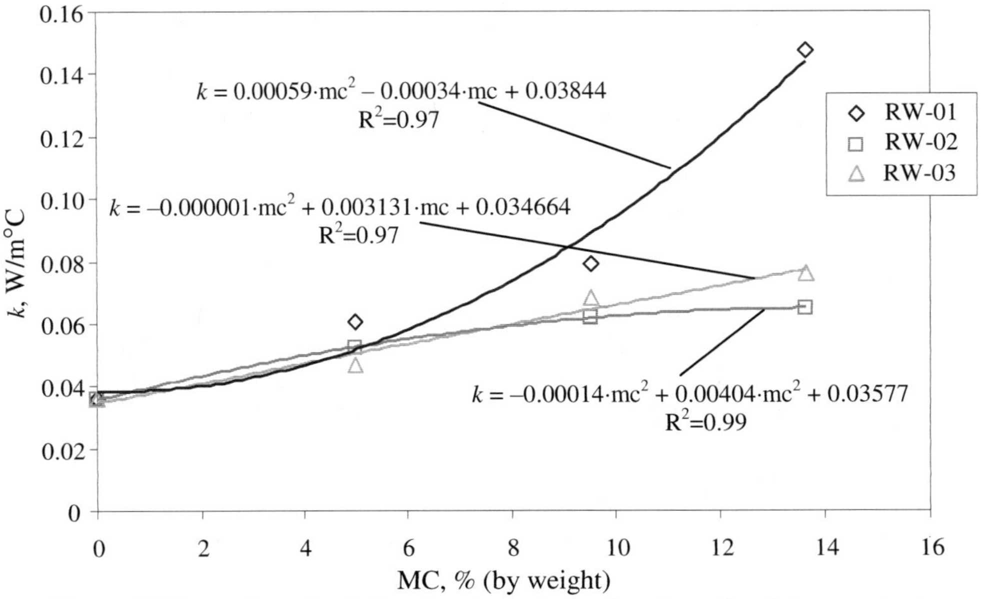

Figure 2. Thermal conductivity of rock wool as a function of moisture content.

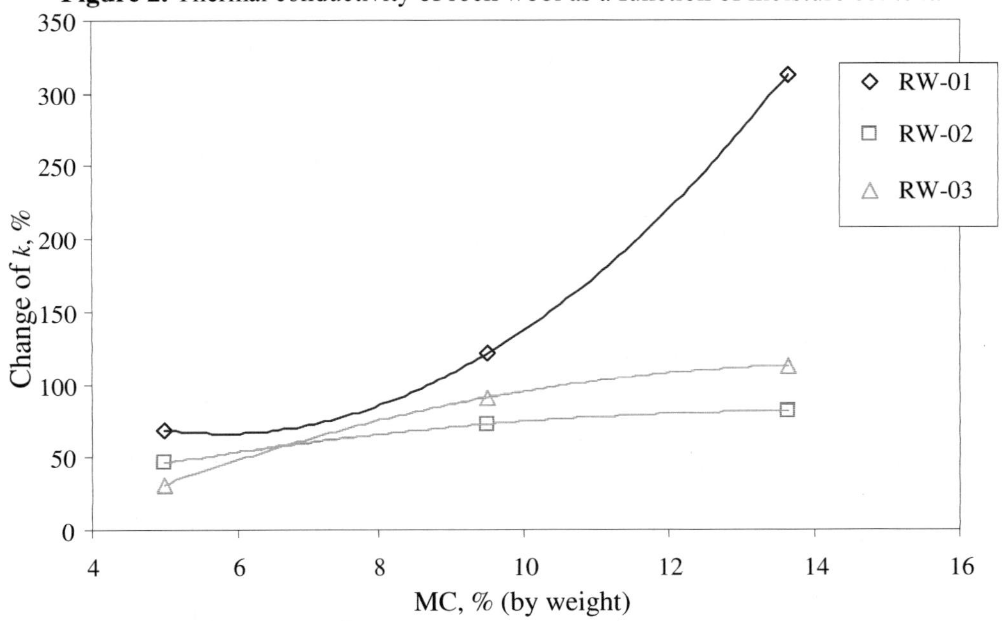

Figure 3. Change of *k* of rock wool, with increase of MC.

4. Conclusion

The thermal conductivity of the insulation material used in district heating networks was investigated. The effect of moisture content on the thermal conductivity of rock wool made by different manufacturers was considered. Higher thermal conductivity is always associated with higher moisture content for all investigated samples. The moisture content of the MC can affect the value of thermal conductivity in different ways depending on the manufacturer of the insulating material. For sample RW-01, thermal conductivity at MC=13.6% increased by 312.8% compared with dry material. For samples RW-02 and RW-03, at MC=13.6%, thermal conductivity increased by 81.3% and 112.7%, respectively.

This study was conducted for one temperature and density of insulating material, that limits the obtained results. As shown in [8], the temperature and density of a material can significantly affect the thermal conductivity. It is also necessary to obtain data on thermal conductivity in a wider range of MC values. Another problem is the possible inaccuracy of determining the thermal conductivity of insulation materials with moisture content using the GHP method [19]. This will be taken into account in our further research.

References

[1] Jamsek M, Dobersek D, Goricanec D and Krope J 2010 WSEAS *Transactions on Fluid Mechanics* **5** 165-74
[2] Dénarié A, Aprile M, and Motta M 2019 *Energy* **166** 267-76
[3] Çomaklı K, Yüksel B, and Çomaklı Ö. 2004 *Applied Thermal Engineering* **24**
[4] Zhang L, Wang Z, Yang X, Jin L, Zhang Q and Hu W 2017 *Energy Procedia* **105** 3369-76
[5] Keçebaş A, Ali Alkan M and Bayhan M 2011 *Applied Thermal Engineering* **31** 3929-37
[6] Schiavoni S, D'Alessandro F, Bianchi F and Asdrubali F 2016 *Renewable and Sustainable Energy Reviews* **62** 988-1011
[7] Berardi U and Naldi M 2017 *Energy and Buildings* **144** 262-75
[8] Abdou A and Budaiwi I 2013 *Construction and Building Materials* **43** 533-44
[9] Jerman M and Černý R 2012 *Energy and Buildings* **53** 39-46
[10] Hedlin C P 1988 *ASHRAE Trans.* **94** 579- 94
[11] D'Alessandro F, Baldinelli G, Bianchi F, Sambuco S and Rufini A 2018 *Construction and Building Materials* **158** 264-74
[12] McFadden T 1988 *Journal of Cold Regions Engineering* **2** 25-34
[13] Chyu M C, Zeng X and Ye L 1997 *ASHRAE* **103** 1136
[14] Zhu W, Cai S and Cremaschi L 2015 *Science and Technology for the Built Environment.* **21** 862–75
[15] Cai S, Cremaschi L and Ghajar A J 2014 *HVAC&R Research* **20** 458-79
[16] EN ISO 8497 1996 Thermal insulation – determination of steady-state thermal transmission properties of thermal insulation for circular pipes
[17] ISO 8302 1991 Thermal insulation –Determination of steady state thermal resistance and related properties – Guarded hot plate apparatus
[18] Koenen A and Marquis D M 2014 *Int. Thermal Conductivity Conf* 74-80
[19] Cai S, Cremaschi L and Ghajar A J 2014 *HVAC&R Research* **20** 458-79

IOP Conf. Series: Materials Science and Engineering **570** (2019) 012030 doi:10.1088/1757-899X/570/1/012030

RANS simulation for the prediction of heat transfer for staggered tube bundle in cross-flow

A I Haibullina[1], K V Chirukhin[1], L S Sabitov[2], A R Hayrullin[1]

[1]Kazan State Power Engineering University, 51 Krasnoselskaya street, Kazan, 420066, Russian Federation
[2]Kazan Federal University, 18 Kremlyovskaya street, Kazan, 420008, Russian Federation

kharullin@yandex.ru

Abstract. In this paper, RANS models for predicting heat transfer for triangular staggered arrangement tube bundle in cross-flow studied. The results of the numerical study were compared with empirical correlation. The tube pitch were p_t/D=1.3. Prandtl numbers were Pr=7.22. The Reynolds number Re based on the maximum average was in the range from 3020 to 6040. The incompressible two dimensional unsteady Reynolds averaged Naviere-Stokes (URANS) equations are employed, with nine different modeling strategies. The best agreement with the empirical correlation was obtained for the SST model. The differences between the numerical simulation and empirical correlation Nusselt numbers are below 12.7%.

1. Introduction

Fluid flow in tube bundles is found in many industrial applications. There are a large number of experimental [1–4] and theoretical studies [5–8] on this topic.

Experimental data on heat transfer and pressure drop were obtained in [1] for cross-flow tube bundles of various configurations for a wide range of Reynolds numbers Re and Prandtl numbers Pr. Correlations are presented for calculating the Nusselt numbers depending on Re and Pr. In article [2], the velocity characteristics with transverse flow of tube bundles using laser Doppler anemometry (LDA) were investigated. In paper [3], heat transfer and pressure drop in the in-line tube bundle with *H* type fin were experimentally investigated. The influence on heat transfer and pressure drop of such parameters as fin height, fin pitch, transverse tube pitch, longitudinal tube pitch and Re numbers are estimated. Experimental studies of heat transfer and pressure drop in cross-flow twisted oval tube bundles were carried out in [4]. Correlations are presented for calculating Nusselt numbers and pressure drops depending on Re.

Reynolds averaged Navier-Stokes (RANS) models are used to simulate turbulence. The authors of [7] analyzed the flow characteristics in a tube bundle with nine different 2D RANS turbulence models. There is no one of the turbulence models that showed superiority over other models. The minimum deviation with experimental data was 46%. The authors of [7] conclude that there is no model that adequately predict flow characteristics in a tube bundle. Another paper [9] also analyzed flow characteristics in a tube bundle with four RANS models. Comparison with experimental data is made. The results show that none of the turbulence models is unable to predict the mean and turbulent quantities reasonably well. However, in the article [10] the standard k-ε and RSM model showed good

IOP Conf. Series: Materials Science and Engineering **570** (2019) 012030 doi:10.1088/1757-899X/570/1/012030

agreement with the experimental data for the velocity profiles in the tube bundle, while the SST *k-w* is not so good. In the same article, a comparison of the heat transfer from the in-line tube bundle with experimental data was made, the deviation was below 20%. In the article [8] a numerical study was conducted using various 3D RANS models. The evaluation of the influence of the transverse and longitudinal pitch of the tube bundles on heat transfer and pressure drop is made for oval tube bundles in cross-flow. Comparison of the obtained results of the RNG *k-ε*,, standard *k-ε*, SST *k-w*, standard *k-w* models with experimental data for oval and round tube bundles was made. The range of numbers Re was in the range from 1000 to 11000, the numbers Pr = 0.7. The best agreement with experimental data for both oval and round tube bundles was obtained with the RNG *k-ε* model.

Over the past two decades, the use of large eddy simulation (LES) to simulate turbulence has increased [5,6,11,12]. In the paper [5] LES, the model gave a good agreement with experimental data when predicting flow characteristics. In another paper [6], both LES and 2D RSM models showed good agreement with experimental data. In the paper [11] authors searched a model of turbulence with lower computational cost comparing SA, *k-ε* and LES models with experimental data. The results show that the two-dimensional SA model gives good agreement with the LES and experimental data. The two-dimensional SA model have lower computational cost in comparison with other turbulence models.

Many researchers continue to explore the possibility of using different turbulence models to predict the flow characteristics for tube bundle in cross-flow. Verification of the results of numerical simulations is mainly carried out with experimental data of the authors [2,12,14]. In these papers [2,12,14], the results of measurements of velocity fields and their distribution in tube bundles are presented. However, data on the ability of 2D RANS models to predict heat transfer in tube bundles in a cross-flow are limited. In this paper, a numerical simulation using RANS models to predict heat transfer in a staggered tube bundle will be investigated.

2. Mathematical model

Computational domain of the mathematical model is shown in figure 1. The tubes are in triangular staggered arrangement (φ = 30°). The tube pitch $p_t/D = 1.3$. Constant temperature $t_w = 18$ °C and non-slip boundary conditions were set on the center tube in the tube bundle. Non-slip and adiabatic wall boundary conditions were set for the another tubes walls. Symmetry boundary conditions were set bottom and top between the tubes. Constant temperature $t_{inlet} = 19$ °C boundary condition were set in the inlet. Outlet pressure were set $P = 101{,}325$ Pa. The incompressible two dimensional unsteady Reynolds averaged Naviere-Stokes (URANS) equations are employed, with nine different modeling strategies. The turbulence models tested in the present study are: Spalart-Allmaras (SA) model, standard *k-ε* model with standard wall function (standard *k-ε* SWF), standard *k-ε* model with enhanced wall treatment (standard *k-ε* EWT), *RNG k-ε* model with standard wall function (*RNG k-ε* SWF), standard *k-ω* model, SST *k-ω* model, transition shear stress transport model (SST), linear pressure-strain reynolds stress model with standard wall function (LPS RSM SWF), linear pressure-strain reynolds stress model with enhanced wall treatment (LPS RSM EWT) [15]. The calculations were performed using Ansys Fluent. For all calculations the SIMPLE algorithm was used. The working fluid was water with constant properties. Prandtl numbers were Pr = 7.22. The Reynolds number based on the maximum average was in the range from3020 to 6040. Expansion factor in the radial direction was 1.2. The number of layers in the near-wall region was 10. The minimum cell size in the near-wall region was $y_{min}/D = 4{,}2 \cdot 10^{-3}$, which is comparable to the minimum cell size $y_{min}/D = 4{,}3 \cdot 10^{-3}$ used in [12].

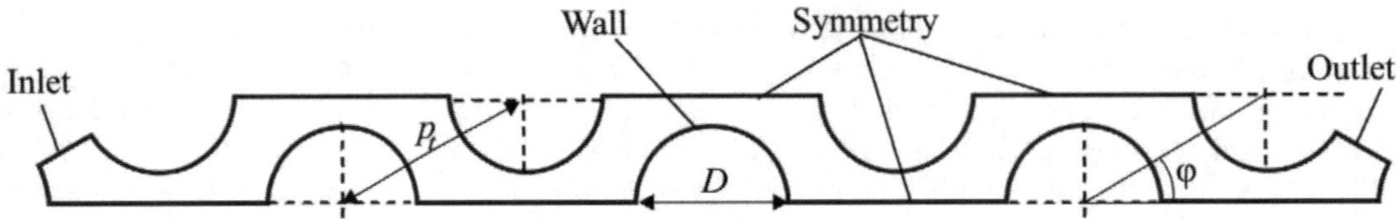

Figure 1. Computational domain.

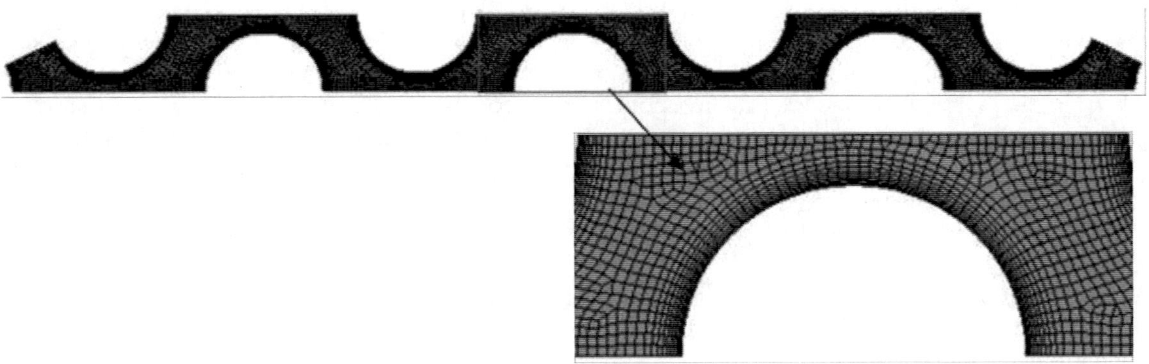

Figure 2. Computational grid.

3. Results

Figure 3–4 shows the Nusselt Nu numbers obtained as a result of a numerical simulation with nine different modeling strategies and calculated using the empirical correlation in a Eq. (1) [1].

$$\mathrm{Nu} = 0.36\mathrm{Re}^{0.6}\mathrm{Pr}^{0.36} \tag{1}$$

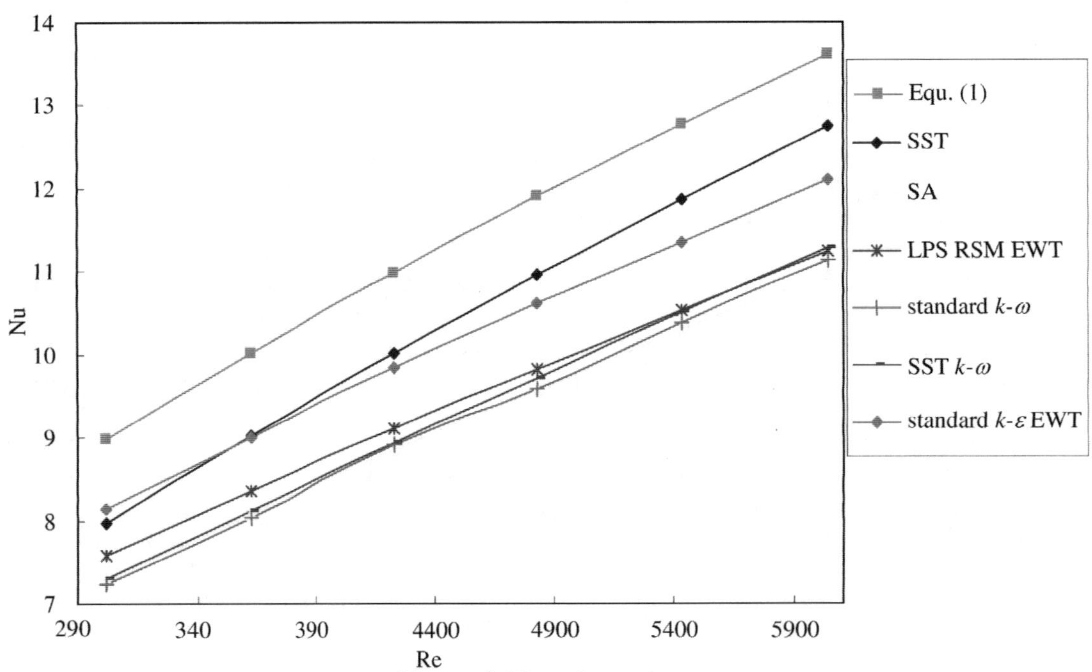

Figure 3. Nusselt number.

The best agreement with Eq. (1) obtained with SST model, the deviation with Eq. (1) is less than 12.7%. Also a good agreement with Eq. (1) obtained with standard k-ε EWT model, deviation is below 12.5%. Standard k-ε EWT model is in better agreement with the empirical correlation with Re less 3700. With increasing Re number, the deviation with Eq. (1) increases. Nusselt numbers obtained using LPS RSM EWT model, standard k-ω model, SST k-ω and standard k-ε EWT model are similar. The differences between results of numerical simulation and empirical correlation were below 18–24%, which is also satisfactory. The value of Nusselt number may be differ due to inaccuracy of empirical correlation and numerical simulation. The authors of the paper [10] investigated heat transfer in a tube bundle using RANS models. The authors of this paper [10] received deviation 25% in comparing Nusselt number with empirical correlation [1] for in-line tube bundle in the range of Reynolds numbers from 10,000 to 50,000. Significant difference with Eq. (1) obtained for LPS RSM SWF model, standard k-ε SWF model, RNG k-ε SWF model (figure 2). The shift of Nusselt number value for numerical simulation and the empirical correlation Eq. (1) was 94–147%.

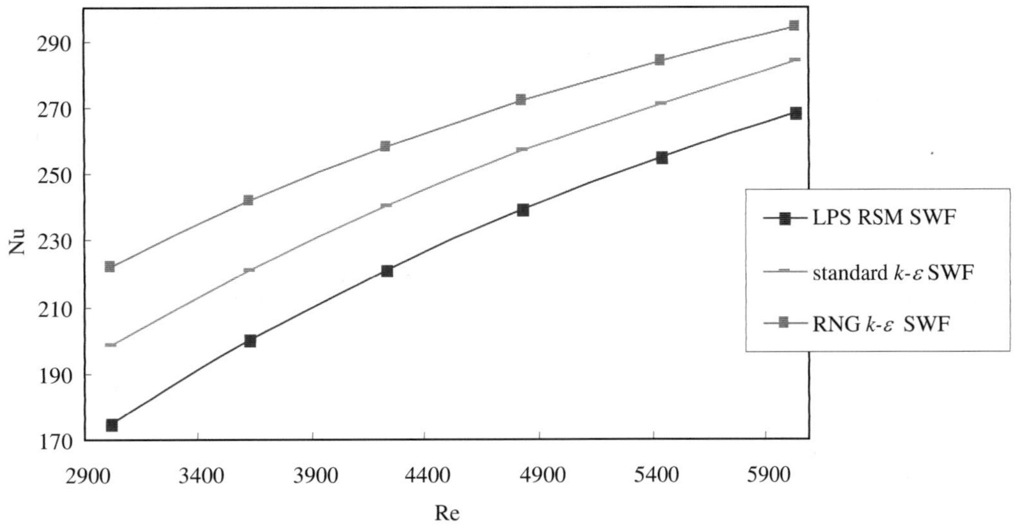

Figure 4. Nusselt number.

4. Conclusion

Two dimensional RANS simulations for predicting heat transfer for triangular staggered arrangement tube bundle in cross-flow with nine different modeling strategies was studied. The best agreement with the empirical correlation was obtained for the SST model and standard k-ε EWT model. The differences between the numerical and empirical correlation Nusselt numbers were below 12.7%. For LPS RSM SWF, standard k-ε SWF and RNG k-ε SWF models agreements with the empirical correlation Eq. (1) were 94–147%. SA model gives satisfactory agreement with empirical correlation (max difference 24%). SA model is more attractive than other turbulence models due to its lower computational cost. However, there is need to proceed further researches of ability of RANS models to predict heat transfer over wide range Reynolds numbers and tube bundle dimensions.

Acknowledgments

The article is executed within the framework of the scientific project 18-79-10136 «Theoretical methods for modeling and developing energy-efficient import-substituting cleaners and deep processing of hydrocarbon raw materials at enterprises of the fuel and energy complex».

References

[1] Zukauskas A, Makarevicius V and Slanciauskas A 1968 *Heat transfer in Banks of tube in crossflow of fluid.* Mintis Vilnius 192 [In Russian]
[2] Balabani S and Yianneskis M 1996 *Proceedings of the IMechE: Part C – J. Mech. Eng. Sci.* **210** 317-31
[3] Li X, Zhu D, Sun J, Mo X and Yin Y 2018 *Exp. Therm. and Fluid Sci.* **96** 146-53
[4] Li X, Zhu D, Sun J, Mo X and Liu S 2019 *Applied Therm. Eng.* **148** 1092-98
[5] Rollet-Miet P, Laurence D and Ferziger J 1999 *Int. J. of Heat and Fluid Flow.* **20** 241-54
[6] Benhamadouche S and Laurence D 2003 *Int. J. of Heat and Fluid Flow.* **24** 470-79
[7] Kulasekharan N, and Prasad B V S S S 2009 *Engineering Applications of Computational Fluid Mechanics* **3** 386-407
[8] Li X, Zhu D, Yin Y, Tu A and Liu S 2019 Parametric study on heat transfer and pressure drop of twisted oval tube bundle with in line layout *Int. Journal of Heat and Mass Trans.* **135** 860-72
[9] Paul S S, Ormiston S J and Tachie M F 2008 Experimental and numerical investigation of turbulent cross-flow in a staggered tube bundle. *Int. J. of Heat and Fluid Flow.* **29** 387-414
[10] Li X, Wu X, and He S 2014. Numerical investigation of the turbulent cross flow and heat transfer in a wall bounded tube bundle *Int. J. of Thermal Sci.* **75** 127-39
[11] Wang Y Q and Jackson P L 2010 *J. of Therm. and Heat Transfer* **24** 534-43
[12] Liang C and Papadakis G 2005 *Engineering Turbulence Modelling and Experiments*. **6** 813-22
[13] Paul S S, Tachie M F and Ormiston S J 2007 Experimental study of turbulent crossflow in a staggered tube bundle using particle image velocimetry *Int. J. of Heat and Fluid Flow* **28** 441-53
[14] Simonin O, Barcouda M 1988 *Proceedings of the Fourth International Symposium on Applications of Laser Anemometry to Fluid Mechanics*
[15] ANSYS FLUENT 2011 ANSYS Inc. Southpointe

IOP Conf. Series: Materials Science and Engineering **570** (2019) 012031 doi:10.1088/1757-899X/570/1/012031

Economy of energy conservation

S R Hajieva[1], E M Gadirova[1], N Sh Sharifova[1], S I Gilmanshina[2], I R Gilmanshin[2,3]

[1] Baku State University, AZ 1148, Baku, Azerbaijan

[2]Kazan Federal University, Kazan, Russia, 420008

[3]Kazan National Research Technical University named after A N Tupolev, Kazan, Russia

Abstract. The economy of energy conservation requires harmonizing the interests of energy producers and consumers. Such an agreement is possible in this case that the resource capabilities of the manufacturer and energy conservation of the consumers are considered simultaneously and on a unified basis. Investments should be directed to where minimal costs of produced or saved energy are ensured. With this approach, the demand for energy resources is provided at the cost-effective manner.

Recently, the topic of energy saving technologies is being considered at the level of international and public policy. Daily questions are discussed about the limitations of natural resources, changes in climate and other problems.

The constant growth of prices and tariffs for energy resources impact on the manufacturing process of each enterprise directly. The solution to this problem lies in one thing-the need to save energy and to undertake activities contributing to this. Energy conservation involves various measures to attract renewable energy, again in the production process. All activities aimed at energy saving are organizational, legal, scientific, economic and technical in nature. Speaking on energy conservation, we also mean the preservation of natural resources that today is more than the actual problem. To date, energy conservation is the main incentive in the economic development of the markets of consumer services and materials. [1, 2]

Alternative energy is a set of promising ways of obtaining, transferring and using energy that are not as widespread as traditional, but they are of interest because of the profitability of their use, with a generally low risk of harm to the environment. There are other types of alternative energy, but the most common types are solar and wind power.

All kinds of solar installations use solar radiation as an alternative source of energy. Sun radiation can be used for the needs of heat supply as well as electricity (using photoelectric elements). The advantages of solar energy include the renewability of this energy source, noiselessness, the absence of harmful emissions into the atmosphere while processing solar radiation in other types of energy. [1,2]

The disadvantages of solar energy are the dependence of the intensity of solar radiation on the daily and seasonal rhythm, as well as the need for large areas for the construction of solar power installations.

IOP Conf. Series: Materials Science and Engineering **570** (2019) 012031 doi:10.1088/1757-899X/570/1/012031

In addition, a serious environmental problem is the use of toxic and toxic substances in the production of photovoltaic elements for solar systems, which creates the problem of their utilization.

One of the most promising sources of energy is the wind. The principle of the wind generator is elementary. The strength of the wind is used to drive the wind wheel. This rotation is in turn transferred to the rotor of the electric generator. First of all, the advantage of the wind generator is that in windy places, the wind can be considered an inexhaustible source of energy. In addition, wind generators, producing energy, do not pollute the atmosphere with harmful emissions. To the disadvantages of devices for the production of wind power can be attributed to the variability of wind power and the small power of a single wind generator. Also, wind generators are known for making a lot of noise, so they should be build away from people's places of residence. [3]

Geothermal energy is considered as the cleanest type of alternative energy. A huge amount of thermal energy is stored in the depths of the Earth. This is due to the fact that the temperature of the Earth's core is extremely high. In some parts of the globe, direct output of high-temperature magma to the surface of the Earth occurs: volcanic regions, hot water sources or steam. Geothermal power supporters offer to use this energy as an alternative source.

Geothermal sources are used differently. Some sources are used for heat supply; the other is to obtain electricity from thermal energy. The advantages of geothermal energy sources include the inexhaustibility and independence from the time of day and time of the year.

The negative side can be attributed to the fact that the thermal waters are highly mineralized, and often also saturated with toxic compounds. The negative side can be attributed to the fact that the thermal waters are highly mineralized, and often also saturated with toxic compounds. This makes it impossible to discharge waste thermal waters into surface water reservoirs. Therefore, the waste water must be pumped back into the underground aquifer. In addition, some scientists-seismologists are opposed to any interference in the deeper layers of the Earth, arguing that this can provoke earthquakes. [4] Other types of alternative energy also exist. But the most common types are already acquainted.

Table 1. Primary production of energy from renewable sources, EU-28, 1990-2015

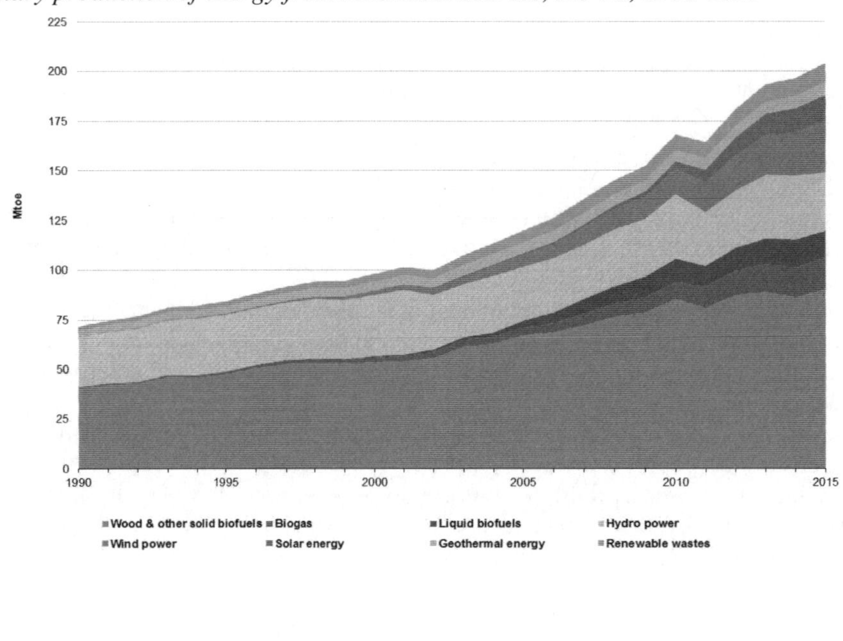

IOP Conf. Series: Materials Science and Engineering **570** (2019) 012031 doi:10.1088/1757-899X/570/1/012031

The use of alternative energy sources is becoming increasingly popular in energy-saving technologies. Solar panels in combination with the use of solar collectors can be used both as an additional and the main source of energy, thus protecting the end user from the necessary dependence in centralized energy networks. Thus, consumption of fossil fuels and energy is reduced. Along with the rapid development of scientific and technological progress in the field of energy conservation technologies, it is often possible to encounter with the notorious human factor stemming from the use of such effective technologies available today. High-tech industries offer us various options for the use and application of energy-efficient technologies, a rich variety of means and methods of energy conservation and options for their use of any energy-consuming industry. As the practice shows, energy efficiency of the projected objects depends not only on the quality of the used equipment, but technology of energy-saving tools play a role in this process. Of course, the design and implementation of such complex systems should be handled by highly qualified specialists who have completed a theoretical and practical training course in the use of these tools and technologies. But as we can see, a number of main tasks are really being solved when designing and implementing a project to increase the energy efficiency of enterprises. The main difficulty appears after putting the object into exploitation. It is not enough to apply energy saving technologies in the production process, it is necessary to use these technologies correctly. As an example, we can cite a common household situation.

Energy-saving technologies should be used in combination with common sense. There is no need to use a working light source when it is not needed, and it is also not worth interrupting its operation with great frequency.

Competent use of energy-saving materials is a practical guarantee of reducing the cost of operation and maintenance of any energy-dependent facilities that require large material costs for maintenance, including thermal power.

In this example, there is only one factor directly affecting energy efficiency, but when it comes to monitoring the energy conservation of industrial enterprises, there are many factors that directly and indirectly affect the energy efficiency of the enterprise as a whole. To improve energy efficiency, a systematic approach is needed, including well-planned and well-defined specific goals for implementation, followed by the development of a basic energy conservation program. The material benefit between the subjects of economic relations deserves no less attention. To date, there are a huge number of ready-made energy-saving solutions, measuring instruments, sensors, and efficient energy-consuming devices. In such circumstances, there is an excellent opportunity to choose energy-effective devices based on their quality and cost.

Table 2. Share of energy from renewable sources in gross final consumption of energy, 2004-2015 %

	2004	2005	2006	2007	2008	2009	2010	2011	2012	2013	2014	2015	2011-2012 average	2013-2014 average	S_{2005}	indicative trajectory				2020 target
																2011-2012	2013-2014	2015-2016	2017-2018	
EU-28	8.5	9.0	9.5	10.4	11.0	12.4	12.9	13.2	14.4	15.2	16.1	16.7	13.8	15.6						20.0
Belgium	1.9	2.3	2.6	3.1	3.6	4.7	5.7	6.3	7.2	7.5	8.0	7.9	6.7	7.8	2.2	4.4	5.4	7.1	9.2	13.0
Bulgaria	9.4	9.4	9.6	9.2	10.5	12.1	14.1	14.3	16.0	19.0	18.0	18.2	15.2	18.5	9.4	10.7	11.4	12.4	13.7	16.0
Czech Republic	6.8	7.1	7.4	8.0	8.6	9.9	10.5	11.0	12.8	13.8	15.1	15.1	11.9	14.5	6.1	7.5	8.2	9.2	10.6	13.0
Denmark	14.9	16.0	16.3	17.8	18.6	20.0	22.1	23.5	25.7	27.4	29.3	30.8	24.6	28.4	17.0	19.6	20.9	22.9	25.5	30.0
Germany	5.8	6.7	7.7	9.1	8.6	9.9	10.5	11.4	12.1	12.4	13.8	14.6	11.8	13.1	5.8	8.2	9.5	11.3	13.7	18.0
Estonia	18.4	17.5	16.1	17.1	18.9	23.0	24.6	25.5	25.8	25.6	26.3	28.6	25.7	26.0	18.0	19.4	20.1	21.2	22.6	25.0
Ireland	2.4	2.9	3.1	3.6	4.1	5.1	5.6	6.6	7.2	7.7	8.7	9.2	6.9	8.2	3.1	5.7	7.0	8.9	11.5	16.0
Greece	6.9	7.0	7.2	8.2	8.0	8.5	9.8	10.9	13.5	15.0	15.3	15.4	12.2	15.2	6.9	9.1	10.2	11.9	14.1	18.0
Spain	8.3	8.4	9.2	9.7	10.8	13.0	13.8	13.2	14.3	15.3	16.1	16.2	13.8	15.7	8.7	11.0	12.1	13.8	16.0	20.0
France	9.4	9.5	9.3	10.1	11.1	12.1	12.5	11.1	13.4	14.1	14.7	15.2	12.3	14.4	10.3	12.8	14.1	16.0	18.6	23.0
Croatia	23.5	23.8	22.7	22.2	22.0	23.6	25.1	25.4	26.8	28.0	27.9	29.0	26.1	28.0	12.6	14.1	14.8	15.9	17.4	20.0
Italy	6.3	7.5	8.3	9.8	11.5	12.8	13.0	12.9	15.4	16.7	17.1	17.5	14.2	16.9	5.2	7.6	8.7	10.5	12.9	17.0
Cyprus	3.1	3.1	3.3	4.0	5.1	5.6	6.0	6.0	6.8	8.1	8.9	9.4	6.4	8.5	2.9	4.9	5.9	7.4	9.5	13.0
Latvia	32.8	32.3	31.1	29.6	29.8	34.3	30.4	33.5	35.7	37.1	38.7	37.6	34.6	37.9	32.6	34.1	34.8	35.9	37.4	40.0
Lithuania	17.2	16.8	16.9	16.5	17.8	19.8	19.6	19.9	21.4	22.7	23.6	25.8	20.7	23.1	15.0	16.6	17.4	18.6	20.2	23.0
Luxembourg	0.9	1.4	1.5	2.7	2.8	2.9	2.9	2.9	3.1	3.5	4.5	5.0	3.0	4.0	0.9	2.9	3.9	5.4	7.5	11.0
Hungary	4.4	4.5	5.1	5.9	6.5	8.0	12.8	14.0	15.5	16.2	14.6	14.5	14.8	15.4	4.3	6.0	6.9	8.2	10.0	13.0
Malta	0.1	0.2	0.2	0.2	0.2	0.2	1.0	1.9	2.8	3.7	4.7	5.0	2.4	4.2	0.0	2.0	3.0	4.5	6.5	10.0
Netherlands	2.1	2.5	2.8	3.3	3.6	4.3	3.9	4.5	4.7	4.8	5.5	5.8	4.6	5.2	2.4	4.7	5.9	7.6	9.9	14.0
Austria	22.6	23.9	25.4	27.2	28.1	29.9	30.4	30.6	31.4	32.3	32.8	33.0	31.0	32.5	23.3	25.4	26.5	28.1	30.3	34.0
Poland	6.9	6.9	6.9	6.9	7.7	8.7	9.3	10.3	10.9	11.4	11.5	11.8	10.6	11.4	7.2	8.8	9.5	10.7	12.3	15.0
Portugal	19.2	19.5	20.8	21.9	23.0	24.4	24.2	24.6	24.6	25.7	27.0	28.0	24.6	26.4	20.5	22.6	23.7	25.2	27.3	31.0
Romania	16.3	17.3	17.1	18.3	20.5	22.7	23.4	21.4	22.8	23.9	24.8	24.8	22.1	24.4	17.8	19.0	19.7	20.6	21.8	24.0
Slovenia	16.1	16.0	15.6	15.6	15.0	20.1	20.4	20.3	20.8	22.4	21.5	22.0	20.5	22.0	16.0	17.8	18.7	20.1	21.9	25.0
Slovak Republic	6.4	6.4	6.6	7.8	7.7	9.4	9.1	10.3	10.4	10.1	11.7	12.9	10.4	10.9	6.7	8.2	8.9	10.0	11.4	14.0
Finland	29.2	28.8	30.0	29.6	31.3	31.3	32.4	32.8	34.4	36.7	38.7	39.3	33.6	37.7	28.5	30.4	31.4	32.8	34.7	38.0
Sweden	38.7	40.6	42.7	44.2	45.3	48.2	47.2	48.7	51.1	52.0	52.5	53.9	49.9	52.2	39.8	41.6	42.6	43.9	45.8	49.0
United Kingdom	1.1	1.3	1.5	1.8	2.7	3.3	3.7	4.2	4.6	5.7	7.1	8.2	4.4	6.4	1.3	4.0	5.4	7.5	10.2	15.0
Iceland	58.9	60.1	60.8	71.5	67.5	69.7	70.4	71.6	72.5	71.7	70.5	70.2	72.0	71.1	55.0	56.8	57.7	59.1	60.9	64.0
Norway	58.1	59.8	60.2	60.1	61.7	64.8	61.2	64.6	65.6	66.7	69.4	69.4	65.1	68.0	58.2	60.1	61.0	62.4	64.2	67.5
Albania	28.1	30.7	32.1	32.8	32.9	31.4	32.0	31.4	35.2	33.2	32.0	34.9	33.3	32.6	:	32.6	33.2	34.3	35.6	38.0
Montenegro	:	35.7	34.8	32.9	32.3	39.4	40.6	40.7	41.6	43.7	44.1	43.1	41.1	43.9	:	27.6	28.3	29.3	30.7	33.0
Former Yugoslav Republic of Macedonia	15.7	16.5	16.5	15.0	15.6	17.2	16.5	16.4	18.1	18.5	19.6	19.9	17.3	19.0	:	23.1	23.7	24.6	25.9	28.0
Turkey	16.2	15.6	14.1	13.3	13.5	14.3	14.2	13.0	13.1	14.0	13.7	13.6	13.0	13.8	:	:	:	:	:	:

(:) not available

S_{2005}: Share of energy from renewable sources in 2005, baseline used for the calculation of the indicative trajectory (in accordance with Directive 2009/28/EC on the promotion of the use of energy from renewable sources

ec.europa.eu/eurostat

As a criterion for assessing efficiency, it is best to apply the minimum cost per unit of energy. Such a criterion corresponds to the interests of both producers and consumers of energy resources. As very often energy conservation is cost-effective, it should be considered a full-fledged alternative to the creation of new energy sources, the construction of new power lines or heating networks. [5]

The effectiveness of energy saving measures can be calculated by considering the different options for energy supply through comparing the cost of production and transportation of energy and at the same time, the cost of conservation the adequate amount of energy when it is consumed. [4, 5] The efficiency of implementing energy-saving measures can be estimated using such an indicator as net present value.

Today, the energy supply is increasing day by day because of alternative sources. The economy of different countries depends on its energy in parallel. Therefore, the growth of energy for different countries is very important.

References

[1] Byers T 1988 *20 design with solar elements*: Textbook (Moscow) 197 p
[2] Aliev R N 2015 *Alternative energy and ecology* (Baku) 368 p
[3] *Project "Renewable energy sources"* Baku 2011
[4] Bokris J O M 1982 *Chemistry of the Environment* (M: Chemistry) 197 p
[5] Petrov K M 1997 *General ecology. Interaction between society and nature* (Moscow: Chemistry), 352 p

IOP Conf. Series: Materials Science and Engineering **570** (2019) 012032 doi:10.1088/1757-899X/570/1/012032

Research of the impact of low-frequency pulse force of lorentz on the process of motion of conducting liquid

V K Ilyin[1], D A Koppen[1], A I Khaibullina[1], O V Ilyin[1], L S Sabitov[2]

[1] Kazan State Power Engineering University ", Kazan, Krasnoselskaya Street, 51, 420066, Russian Federation
[2]Kazan (Volga Region) Federal University, Kazan, Kremlevskaya street, 18, 420008, Russian Federation

l.sabitov@bk.ru

Abstract. The aim of the work was a numerical and experimental study of the influence of the low-frequency pulsed strength of the electromagnetic (EM) field (Lorentz force) on the melt motion GaInSn.

Electromagnetic processes occurring in an induction crucible furnace (ICF) are described by a system of Maxwell equations:

$$\nabla \cdot \boldsymbol{E} = 0,$$
$$\nabla \cdot \boldsymbol{B} = 0,$$
$$\nabla \times \boldsymbol{E} = -\frac{\partial B}{\partial t},$$
$$\nabla \times \boldsymbol{B} = \mu_0 \boldsymbol{J} + \mu_0 \varepsilon_0 \frac{\partial E}{\partial t} \qquad (1)$$

where $\boldsymbol{E}$ [V/m] – electric field strength, $\boldsymbol{B}$ – magnetic induction [Wb / m²], $\boldsymbol{J}$ [A / m²] - current density, t – time, $\mu_0 = 4\pi\ 10^{-7}$ [Gn/m] - magnetic constant, $\varepsilon_0 = 1/(\mu_0 c^2)$ [F/m] – electric constant ($c = 3 \cdot 10^8$ m/sec - constant light speed), ∇ - differential operator (where $\nabla \cdot \boldsymbol{B} \equiv \mathrm{div}\boldsymbol{B}$, $\nabla \times \boldsymbol{B} \equiv rot\boldsymbol{B}$).
Considering the condition

$$\boldsymbol{J} = \sigma_f (\boldsymbol{E} + v \times \boldsymbol{B}), \qquad (2)$$

where σ_f [Cm / m] - magnetic permeability of the medium; v – local conductive velocity
The following two equations for **B** can be derived from the system of Maxwell equations:

$$\nabla \cdot \boldsymbol{B} = 0,$$
$$\nabla \times \nabla \times \boldsymbol{B} = \mu_0 \sigma_f \left(-\frac{\partial B}{\partial t} + \nabla \times v \times \boldsymbol{B} \right).$$

Hereinafter, the index f denotes the parameters of the conducting fluid.
Magnetic induction can also be expressed in terms of the magnetic field vector potentia $\boldsymbol{A}$:

$$\boldsymbol{B} = \nabla \times \boldsymbol{A}. \qquad (3)$$

It follows from the above:

$$\nabla\times\boldsymbol{E}=-\nabla\times\frac{\partial\boldsymbol{A}}{\partial t},$$

as a result of integration we have:

$$\boldsymbol{E}=-\frac{\partial\boldsymbol{A}}{\partial t}-\nabla\cdot\Phi, \tag{4}$$

where Φ – scalar potential.

Considering the equations (1), (2), (3) and (4), an equation for magnetic potentials is obtained:

$$\frac{\partial\boldsymbol{A}}{\partial t}=\frac{1}{\mu_0\sigma_f}\Delta\boldsymbol{A}-\nabla\Phi+v\times\nabla\times\boldsymbol{A}. \tag{5}$$

The last member of the right side of the equation is usually not taken into account, due to the smallness of its effect on the movement of the conducting fluid. For EM fields concentrated in a thin wall layer of the melt, the magnetic induction vector **B** is mainly directed along the axis of the cylindrical volume of the melt. The flow velocity vector v in the near-wall region also has basically only an axial component. Due to the fact that **B** and v are parallel, the result of vector multiplication $v\times\nabla\times\boldsymbol{A}$ is negligible.

Thus, expression (5) can be written as:

$$\frac{\partial\boldsymbol{A}}{\partial t}=\frac{1}{\mu_0\sigma_f}\Delta\boldsymbol{A}-\nabla\Phi. \tag{6}$$

The solution of this equation in two-dimensional cylindrical coordinates is simplified due to the fact that the **A**-vector (the potential of the magnetic field W/m) has only one component — the tangential one. Since $\boldsymbol{A}_z=\boldsymbol{A}_r=0$ и $\nabla\Phi=0$, and

$$\mu_0\sigma_f\frac{\partial A_\varphi}{\partial t}=\frac{1}{r}\frac{\partial}{\partial r}\left(r\frac{\partial A_\varphi}{\partial r}\right)+\frac{\partial^2 A_\varphi}{\partial z^2}-\frac{A_\varphi}{r^2}, \tag{7}$$

where r, φ, z – cylindrical coordinates.

Given the expression (7), the magnetic induction from the equation

$$\boldsymbol{B}=\nabla\times\boldsymbol{A}$$

can be found as:

$$B_z=\frac{1}{r}\frac{\partial(r\partial A_\varphi)}{\partial r}, \tag{8}$$

$$B_r=-\frac{\partial A_\varphi}{\partial z}, \tag{9}$$

Using the expression

$$J=\sigma_f(\boldsymbol{E}+v\times\boldsymbol{B})\text{ и }\boldsymbol{E}=-\frac{\partial\boldsymbol{A}}{\partial t}-\nabla\cdot\Phi,$$

is current density:

$$J_\varphi=-\sigma_f\frac{\partial A_\varphi}{\partial t}, \tag{10}$$

and non-zero components of the electromagnetic force (EMF) in the melt:

$$F_r=J_\varphi B_z, \tag{11}$$

$$F_z=-J_\varphi B_r. \tag{12}$$

Due to the presence of both components of the EMF, there are two types of effects on a conductive fluid - electrodynamic vortex-free and electro-vortex. Non-vortex effect leads to the appearance of a pressure gradient in the conducting fluid, and the electro-vortex to the existence of closed hydrodynamic flows [1].

The result of both effects is the EMF distribution, whose direction is strictly radially in the middle part of the crucible, and closer to the upper and lower edges, where the magnetic field lines are distorted and do not run parallel to the axis, the radial component of the EMF decreases [2] (Fig. 1).

The degree of curvature of the magnetic induction lines depends on the frequency of the supply current. In the case of relatively low frequencies, the EM field penetrates the conductor with almost no distortion and the EMF field vorticity is almost absent. In such systems there are only radial - component EMF (11). At higher frequencies, the field lines follow the contour of the free surface of the melt, concentrating on its outer corners. At the same time, the EMF value in the load corners increases.

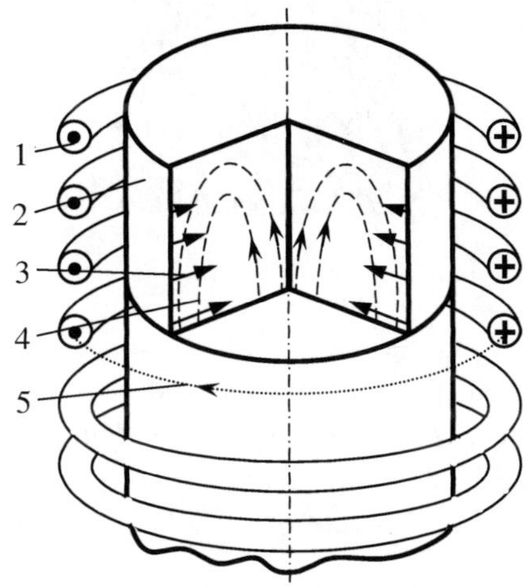

Figure 1. Diagram of the interaction of a magnetic field with a liquid metal in an ICF: 1 - inductor; 2 - melt; 3 - distribution of the radial component of the EMF; 4 - upper vortex zone and the trajectory of the movement of the metal in it; 5 - the direction of the electric current in the inductor

The flow of the described EM processes and the resulting EMF distribution F_r with prevalence in the melt leads to the characteristic azimuthal movement of the liquid: the melt in the middle part of the crucible flows from the periphery to the axis. Further along the axis of the liquid column it is squeezed up to the free surface and to the bottom of the crucible. In the upper and lower parts of the volume, the melt flows to the walls and along them returns to the middle part of the crucible, closing the double-circuit circulation.

As a result of solving equations (7) - (12), the distribution of the Lorentz force density was obtained for the given parameters of alternating current in the inductor. 2

The described interaction phenomena of the primary EM field and the field arising from the induction of eddy currents in the melt can also be described using the equation of induction in dimensionless form:

$$\Pi_t \, \partial\vec{\bar{B}}/\partial\bar{t} = \mathrm{Re}_m \, \mathrm{rot}\left[\vec{\bar{v}} \times \vec{\bar{B}}\right] + \Delta\vec{\bar{B}} \,, \tag{13}$$

where $\overline{B} = B/B_0$, $\bar{v} = v/v_0$, $\bar{t} = t/t_0$ - dimensionless magnetic induction, speed and time;

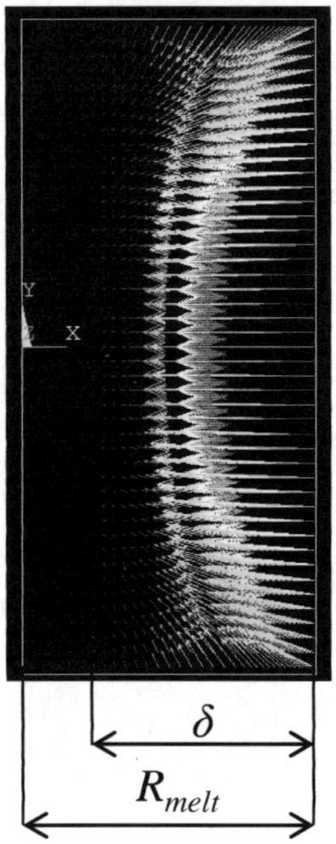

Figure 2 - The distribution of the density of the Lorentz force in the axial section of the melt induced by the action of a single-phase alternating field with a frequency of f_{ac} = 150 Hz and with an amperage in the inductor I_{RMS} = 200 A

$\mathrm{Re}_m = \mu_1\mu_0\sigma_f l_0 v_0$ – magnetic Reynolds criterion;

$\Pi_t = \mu_1\sigma_f l_0^2/t$ – dimensionless criterion that determines the effect of a change in the field in time on its spatial distribution.

For a harmonic change of the field in time, this criterion is converted to a relative (dimensionless) frequency $\hat{\omega}$:

$$\hat{\omega} = \mu_0\sigma_f \omega l_0^2, \tag{14}$$

where $\omega = 2\pi f_{ac}$ – EM field angular frequency

The value $\hat{\omega}$ is closely related to the diffusion of the EM field into the conductor, which, with a sinusoidal change in time of the magnetic field, is determined by the frequency of its oscillations f_{ac} and takes the form of the so-called "skin effect". The essence of this effect is as follows: in the surface layer of a conductor, a standing (in the case of a uniform tangential field) induction wave is formed, the amplitude of which exponentially decreases as it deepens into the surface of the conductor (Fig. 3). The depth of diffusion of an EM field into a conductor is defined as

$$\delta = \sqrt{2/(\mu_1\mu_0\sigma_f\omega)}. \tag{15}$$

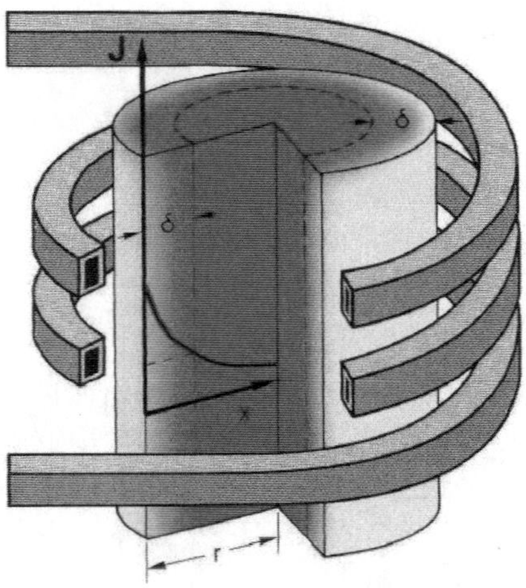

Figure 3. The manifestation of the "skin effect" in a conductor in an alternating magnetic field. *J* is the density of the current induced in the material, *r* and *x* are the radius of the conductor and the distance from the crucible wall

The capabilities of the model for the reproduction of the skin effect in the metal were checked (the error was 4%), which confirms the effectiveness of the developed model of EM processes.

The fluid motion was simulated using a modified *k*-ε turbulence model (taking into account the effect of the presence of a solid wall on the flow — Enhanced Wall Treatment option; the relationship between pressure and velocity was calculated using the SIMPLE algorithm; pressure interpolation calculated by PRESTO).

The results of the analysis of the data obtained are presented in Fig.4.

The functional relationship represented in graph (a) reflects the effectiveness of the application of the EM field to the melt. The reduced form of the curve containing the maximum and, further, the monotonic behavior of the function is characteristic of the interaction of an alternating magnetic field and a conducting material in it.

The obtained dependence of the circulation rate on the AC power supplying the inductor (b) is linear, which also corresponds to theoretical ideas and confirms the adequacy of the model.

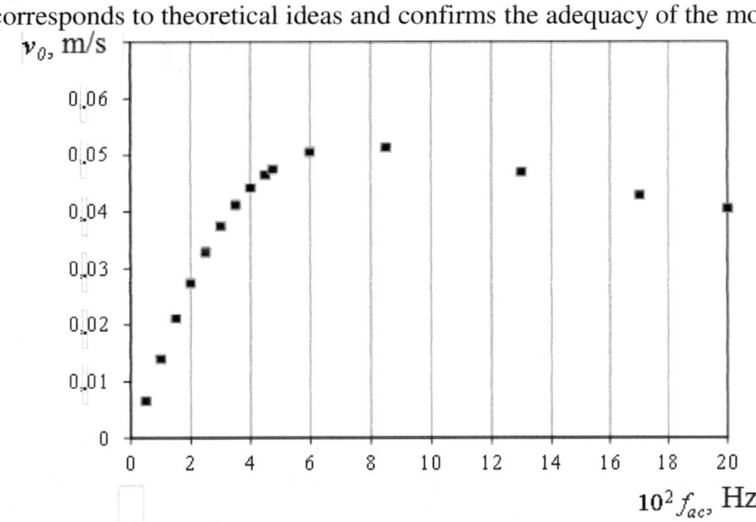

(a)

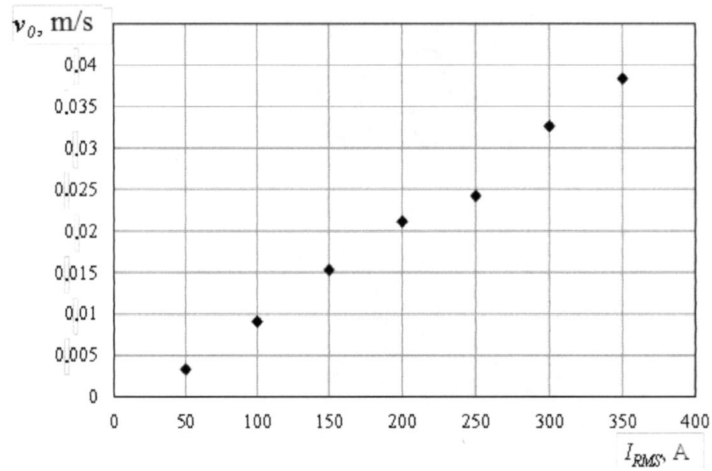

(б)

Figure 4. - Dependence of the melt circulation rate v0 on the frequency (a) and the effective value of the supply current strength (b) obtained using the 2D mathematical model

The analysis showed that the pulse effect with a frequency in the range from 0.5 to 10 Hz has an insignificant effect on the flow - the standard deviation values are close to the value obtained for stationary effects. This may be due to the presence of the frequency of the pulses, in case of exceeding which, the liquid, due to inertness, does not have time to respond with movement to an external effect, and similar modes can be taken as pseudo-stationary [3, 4, 6].

The most significant influence on the motion of the melt was made by pulsations with f_p = [0.05 - 0.2] Hz. In this case, a maximum is observed for f_p = 0.1 Hz. Fig.5

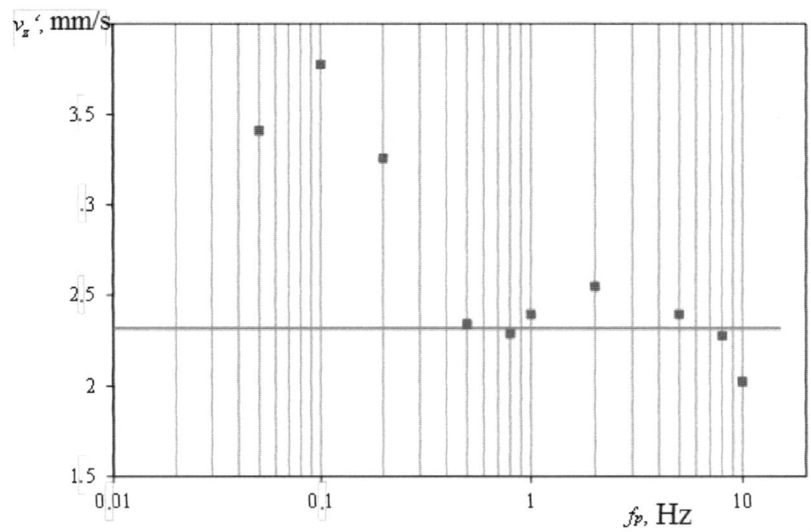

Figure 5. - The effect of the frequency of the pulsed Lorentzian force on the intensity of the pulsation flow velocity. Standard deviations of the axial velocity from its mean value (velocity pulsations - v_z') are compared with a stationary (horizontal line) and pulsating, ψ = 0.25 effects.

Thus, it can be assumed that the presence of a maximum in Fig. 5 is substantiated by the approximation of the frequency of the pulses of the EM field strength to the characteristic frequency of the rotation of the vortices in the flow, which, presumably, increases the efficiency of imposing the pulses. It can also be assumed that the real maximum of the standard deviation for the axial velocity averaged over the profile should be sought near the value $f_p \sim 0.17$ Hz.

Conclusions

The impulsive effect of the Lorentz force on the melt with $f_p = [0.5 - 10]$ does not have a pronounced effect on the flow - the values of the measured axial velocity of the liquid are close to the values obtained for stationary effects.

Under the action of pulses with $f_p = [0.5 - 10]$ Hz, an increase in the magnitude of the pulsation component of the axial flow velocity is observed, the experimental maximum obtained for $f_p = 0.1$ Hz. It is assumed that the maximum effect can be detected when exposed to a stream of pulses with a frequency close to $f_{ch} = 0.17$ Hz.

References

[1] Gelfgat Yu M, Lielausis O A, Scherbinin E V 1976 *Liquid metal under the action of electromagnetic forces* Riga: Zinatne p 246

[2] D Musaeva, V Ilin, E Baake, V Geza 2015 *Numerical simulation of the melt flow in an induction crucible furnace driven by a Lorentz force pulsed at low frequency* Magnetohydrodynamics No 51 (4) pp 771–85

[3] D A Musaeva, V K Ilyin, V Geža, E Baake 2016 *Experimental investigation of low-frequency pulsed Lorentz force influence on the motion of Galinstan melt* St. Petersburg Polytechnical University Journal: Physics and Mathematics vol 2 is 3 pp 193-200

[4] Musaeva D A, Ilyin V K, Gezha V, Baake E 2016 *Experimental study of the influence of the low-frequency impulsive Lorentz force on the melt motion of Galinstan* Scientific and Technical Gazette of the St. Petersburg State Polytechnic University. Physics and Mathematics No 3 (248) pp 39-51

[5] Musaeva D A, Ilyin V K, Sinyavin A A, Gezha V, Bakke E 2013 *Melt movement in an induction furnace crucible under the action of a pulsating magnetic field force* Modern science: research, ideas, results, technologies No 1 (12) pp 147-49.\

[6] Musaeva D A, Baake E, Ilyin V K 2015 *Study of GaInSn Melt Motion under the Action of the Low-Frequency Pulsed Power of an Electromagnetic Field* In the collection: SPbPU Science Week. Materials of a scientific forum with international participation. Institute of Applied Mathematics and Mechanics pp 25-28

[7] S A Farbman, I F Kolobnev 1968 *Induction furnaces for melting metals and alloys* M: Metallurgy 496 p

[8] L L Tyr, M S Shevtsov, L K Chentsova 1965 *Study of the effect of frequency on MHD effects in an induction crucible electric furnace* Electrothermia p 48

[9] L L Tir., A V Svido 1980 *Estimation of melt movement in an induction crucible furnace when designing* Electrical industry. Electrothermia vol 6 No 214 pp 5-7

[10] D A Musaeva, V Gezha, E Baake, V K Ilyin 2014 *Modeling of melt movement in an induction furnace crucible under conditions of low-frequency pulsations of electromagnetic field strength* Materials of the National Congress on Energy pp 139-40

IOP Conf. Series: Materials Science and Engineering **570** (2019) 012033 doi:10.1088/1757-899X/570/1/012033

Simulation as a tool for supporting assembly systems

M Iskandarov, R Khisamutdinov

[1]Kazan (Volga Region) Federal University, 68/19 Peace avenue, Naberezhnye Chelny, 423812, Russian Federation
[2]Kazan (Volga Region) Federal University, 68/19 Peace avenue, Naberezhnye Chelny, 423812, Russian Federation

[1]Iskandarov.MZ@kamaz.ru, [2]HisamutdinovRM@kamaz.ru

Abstract. This article considers the first comprehensive introduction to simulation modeling of assembly systems, and considers the assumption that simulation is a reliable method for testing production and assembly systems. Simulation is an important software tool with which you can perform a large number of experiments with various system parameters. Simulation modeling allows to describe the structure of the system and its processes in a natural form, without resorting to use of complex formulas, equations, and strict mathematical dependencies. Modeling allows to detect and eliminate problems that manifest themselves in the early stages of commissioning the assembly system, which implies significant savings in costs and time [1]. This article provides a part of the theoretical base of the topic and a practical example of using modeling. For practical research software was used Tecnomatix Plant Simulation from the company SIEMENS.

1. Introduction

Considering that modeling is a theoretical and cognitive process based on the repetition of a real production situation, this term is ideally defined as the recognition of objects through other objects that are mostly created artificially. Modeling simulates the structure and behavior of a real system, and is a specific form of the cognitive process. The basic principle is to draw conclusions using modeling within a specific system with objects and their interactions and analyzing the results of the experiment.

Modeling programs, at present, are increasingly used in industrial enterprises [2]. The pressure of competition and customer requirements is steadily increasing, and therefore enterprises need to model production processes. Simulation programs are tools that test the effects of various model decisions, and then estimate production capacity, duration of operations, and other production parameters.

According to the authors R. Debnar and J. Koshtürkak, the main tasks of modeling are:

- Decision support in system design and operation.
- Analysis and optimization
- Forecasting and looking into the future
- Replacing the real system (training, testing, danger, etc.).

Under the simulation model, the authors understand - a mathematical method consisting of a set of logical and mathematical relationships that express the functional or other characteristics of the system [3].

Modeling assembly systems in the workplace helps in planning new and improving existing systems. It is also often used in decision-making processes, such as the size of investments, the cost of planned changes, analysis of production processes, etc. Modeling in assembly systems increases the

IOP Conf. Series: Materials Science and Engineering **570** (2019) 012033 doi:10.1088/1757-899X/570/1/012033

efficiency of the design phase. Currently, simulation accuracy can reach 99%. The accuracy depends on the input parameters and the reliability of the model [4].

Advantages of modeling:

- Testing innovative strategies in a risk-free virtual environment
- Maximum use of production resources
- Reduced investment risk due to fast modeling
- Quick identification of sources of problems in logistic and production areas
- Stock reduction by 20–60%
- Reduced investment costs by 5-20% for the new system
- The fast achievement of positive results and the identification of negative effects [5].

Simulation of assembly areas is currently used in robotized positions or in the area of assembly conveyors [6]. In addition, software is widely used to simulate the movements of personnel, to study the ergonomics of the workplace.

2. Practical results of modeling and analysis in the assembly system

This section discusses an example of designing and analyzing an assembly system consisting of one conveyor, three assembly positions, one accumulator, and one control position (Fig. 1 and 2). To simulate the assembly system, we used software from SIEMENS Tecnomatix Plant Simulation.

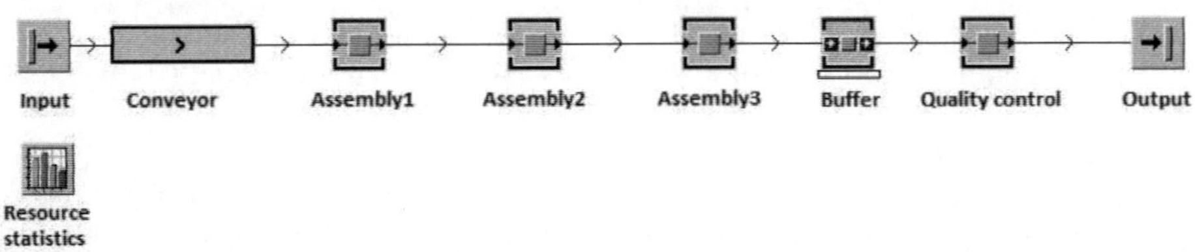

Figure 1. The basic model of the assembly system.

When testing the basic model of the assembly system, it was found that the bulk of the products accumulate after passing through the individual assembly positions at the control point. With the base model, for an eight-hour working day, 159 products were manufactured.

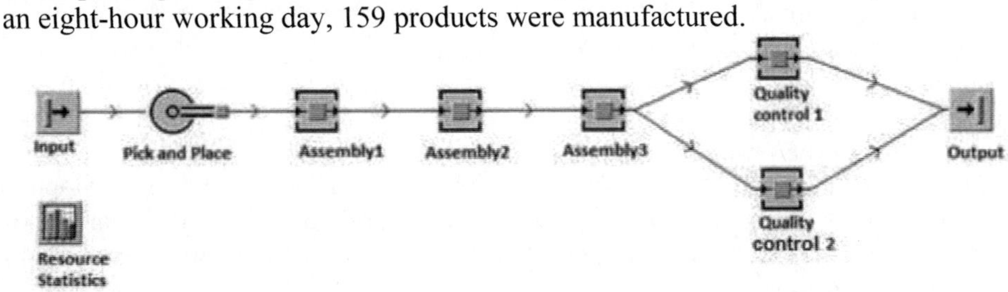

Figure 2. Improved build system model.

After finalizing the model, in the course of analyzing the simulation results, it was found that with the same planned parameters, 409 products were manufactured, which is more than the base model by 250 pieces. It follows from this that productivity increased by 267.78%.

The change in the performance statistics of the assembly systems, before and after tuning, is presented in Figure 3. From the right diagram, you can see that the assembly position at number 3 is not blocked after the model has been refined. The addition of an additional monitoring station reduced the load and allowed an increase in system capacity.

IOP Conf. Series: Materials Science and Engineering **570** (2019) 012033 doi:10.1088/1757-899X/570/1/012033

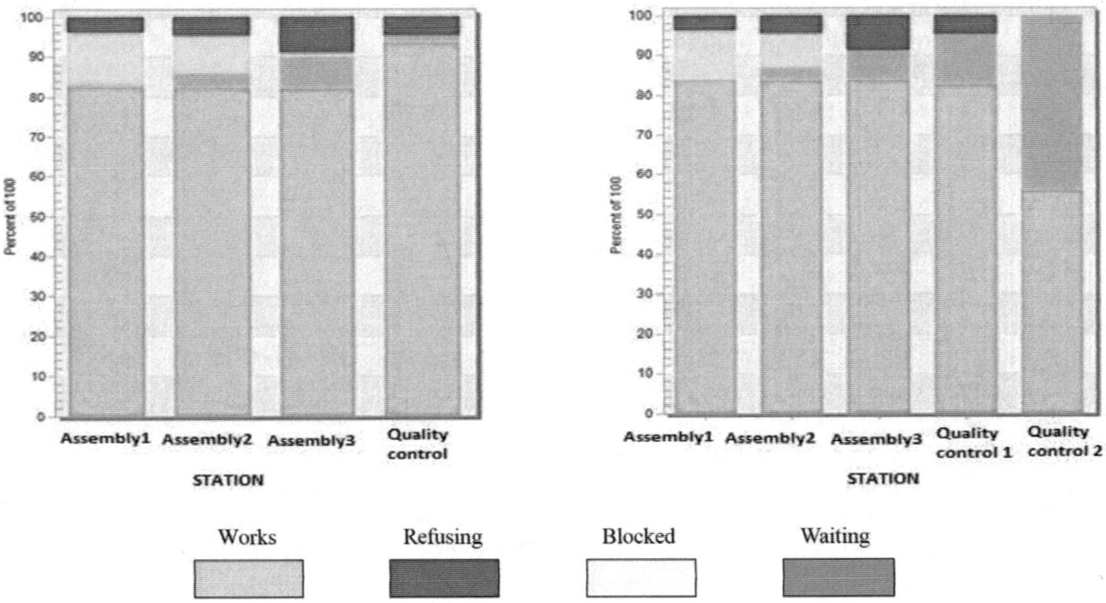

Figure 3. Resource statistics. Base model (left), modified model (right).

3. Results and conclusions.

According to the results of the development of a practical example in the software Tecnomatix Plant Simulation, a number of advantages of assembly modeling were established.

In developing the model, standard objects of the used software were used, which reduces the design time of the assembly system.

Working in a virtual environment and saving intermediate results allowed us to test several versions of the build system and select the most optimal and balanced version. It follows from this that constructive errors can be eliminated before their commissioning.

Tracking and analyzing the output results in the form of diagrams made it possible to increase the capacity of the assembly system by 267.78%, which indicates the possibility of optimizing the production indicators of the assembly system.

Currently, simulation is increasingly used in assembly, especially during the development phase of assembly systems. With the development of IT in the world, modeling is becoming increasingly noticeable, and we can expect that the simulation of the assembly will become one of the main tasks of the software development team.

References

[1] Steffen Bangsow. Manufacturing Simulation with Plant Simulation and SimTalk / Springer-Verlag Berlin Heidelberg 2010. P– 6.

[2] Grechishnikov V.A., Chemborisov N.A., Khisamutdinov R.M., Egorova E.I., Modeling the systems of instrumental support of machine-building production // Stary Oskol: TNT.-2011.-208 p.

[3] TECNOMATIX Plant Simulation, Product overview, SIEMENS (2016)

[4] J. Burieta, Simulácia. In Ipaslovakia (2007)

[5] P. Trebuna, M. Kliment, M. Petrik, Creation of simulation model of expansion of production in manufacturing companies. Modeling of mechanical and Mechatronic Systems MMaMS (2014).

[6] Khisamutdinov RM, Khisamutdinov MR 2015 *Modeling the exchange of data between multi-platform information systems for solving the problems of pre-production* National Association of Scientists No 6-2 (11) pp 72-75

IOP Conf. Series: Materials Science and Engineering **570** (2019) 012034 doi:10.1088/1757-899X/570/1/012034

Technological Support of Plasma Coating with the Plasmatron Power Modulation

A M Kadyrmetov[1], S N Sharifullin[2], G A Suhochev[3] and E V Snyatkov[1]

Voronezh State University of Forestry and Technologies named after G. F. Morozov, Russia[1]
Kazan Federal University, Russia[2]
Voronezh State Technical University, Russia[3]

Saidchist@mail.ru

Abstract. The possible ways of improvement of plasma coating technologies, the results of researches and basic components of engineering technology of this process with the modulation of electric parameters of indirect and remote arc of the plasmatron are presented. Also the integrated processes with the thermal and electromechanical machining are proposed. New high-saving technologies, methods and means of technological support for the required performance indicators of coatings are developed.

1. Introduction

The goal of technological support of plasma spraying and hardening of coatings (PS) on the base of searching and optimization of design and technological solutions (DTS) of the problems of getting the competitive products is achieved, first of all, by the forecasting and evaluating of the technological realizability of advanced PS methods and, secondly, by the technological support of PS design based on these methods.

2. Design methodology of PS with parameters modulation

The methodology includes the actions algorithm on design technologies, means, methods and principles of solution of technological problems, process models.

Structure of the design process includes three stages. The first stage of the technological process design is the statement of the technological task (TT) and the technical requirements (TR) for the coating. In accordance with the operational requirements for the coating and the part parameters, these requirements are divided for: external or small-sized internal surfaces; mating and noncontacting surfaces of parts; surfaces of movable and stationary mates; increased operating loads (alternative loads, impulse loads, working in the conditions of abrasive wear).

During the *Synthesis of Technologies Principles* on the basis of the physical model the necessary physical effects in the modulation of electrical parameters, which form the basis of the chosen version of the technology of plasma spraying and hardening of coatings with parameters modulation are determined [4]. During the *Structural Technologies synthesis* schemes and algorithms of realization of electric parameters modulation variants depending on operational requirements and parameters of a

IOP Conf. Series: Materials Science and Engineering **570** (2019) 012034 doi:10.1088/1757-899X/570/1/012034

detail are created [1]. During the *Parametric synthesis* the dependencies of process criteria from factors are found and the problems of finding the optimal technological modes of spraying and hardening of coatings taking into account operational requirements and parameters of the part are solved.

At each stage of the design, depending on the operational loads and part parameters, the coating technology is selected and the following procedures are performed: choice of the method of PS and its mathematical model (flowchart and calculation scheme); choice of the solution method, including selection of the optimization method; solution; analysis of the results and decision making.

3. Research results

3.1. Forecasting and assessment of technological realizability of prospective methods of PS

Research results of the PS and hardening of coatings improving possibilities show that the expansion of its use is still constrained by such significant disadvantages as the unsatisfactory strength of the coating joint with bottom layer under impulse loads, high residual tension stress in high-thickness coatings, negatively affecting the fatigue resistance, presence of porosity.

The performed analysis of the current state of the PS process researches shows that the refinement of the PS technology by improving its spraying operation is advisable to realize by appending additives to the plasma-forming gas, by simultaneous processing of the sprayed surface and coating with the additional extension arc, by the parameters modulation of the plasmatron, as well as by combining these methods [1, 2]. However, it was found out that in this method theoretical bases of gas-dynamic, thermal and physical-mechanical processes leading to the improvement of coatings quality were practically not studied.

Refinement of the hardening operation of sprayed coatings is advisable to perform by thermo-or electromechanical machining of coatings using spinning roller or ball tool [3]. However, this important practice for plasma sprayed coatings also has not been fully explored.

The improving of the PS technology in the aspect of its development and universalisation is constrained by spreading of this technology for the application of thick coatings (more than 1 mm), as well as for coating on the internal small-sized surfaces of machine parts, for example, on the internal surfaces of the cylinders liners. For these purposes, the process of plasma spraying has not been sufficiently developed.

These unsolved issues in general represent a practical problem, which consists in improving the quality of coatings with minimal thermal impact on the substrate, increasing the resource, reliability, quality and high resistance to the specific operating loads of the machine parts surfaces .

This practical problem predetermines the scientific problem of researches, consisting in creation of theoretical and methodical bases of intensification of gas-dynamic processes and controlling of heat processes of plasma spraying with hardening of coatings on the basis of modulation of electrical parameters and combined influence of electromechanical processing of forming coating for providing with optimal technological conditions to get the required quality of machine parts coating.

The physical model of the dynamization mechanism of the plasma coating processes is based on transformation of the electric energy of impulses at modulation into thermal and gas-dynamic energy, in this manner changing parameters of a plasma jet, particles energy and the efficiency of the process as a whole [4-6]. Analysis of the physical model allows to develop principles of control of processes dynamics by modulation of electrical parameters on the basis of data controlling by the given physical components which consist in the formation of shock or intense acoustic waves in the plasma jet, intensifying energy efficiency to the sprayed material as well as in the formation of surface temperature fields of the base and discrete distributed local areas of coating penetration with optimal parameters.

The structure of complex mathematical model describing the implementation of these physical processes at the dynamization of plasma spraying and hardening of coatings, consists of three interconnected blocks and allows to describe the system "arc-plasma jet-particles-coating" [7].

Theoretic and experimental researches results and possibilities of improvement of plasma spraying and hardening of coatings by dynamization of processes based on the use of modulation of electrical parameters are described in the works [5-12].

3.2. Technological support of PS technology design

Technological support of PS technology design of high-strength wear-resistant metal coatings for machine parts with surfaces of different profiles includes the Providing of the following Components:

- first, the dynamization of PS processes by applying the modulation of electrical parameters;
- second, the systems of controlling the dynamics of these processes.

These components of PS, in turn, include a set of methodological and technical solutions that ensure the implementation of the results of research and, first of all, the description of the scheme PS [2] and its technical support. One of the option of the technical support can be the equipment that includes the installation of PS (Inventor's certificate № 1774828 of Soviet Union, Patents Russion Federation № 2211256 and № 2480533), the installation of thermo-and electro-mechanical hardening and special technical support [10-12]. Special technical support includes devices for coating of crankshafts necks (for example, patents of the RF № 2085301, № 2447951, № 129021), devices for coating of profile surfaces (for example, the "tooth" type surface Patent RF № 2175024), devices for coating of internal small surfaces, cooling systems parts, manipulators, etc.

The working out of the control system over the dynamics of plasma spraying and hardening of coatings includes the developed means of impact on the object and the set of means of collection, or control of information about the object. For an indirect arc the system of control of the dynamics of the spraying process consists in realization of the principles of maximizing pulse power or their intensity at frequencies of 3-5 KHz to increase the energy of the sprayed particles. For a straight arc the system of control of the process dynamics consists in generation of power pulses, which provides, firstly, locally discrete penetration of the coating with the optimal density of distribution of such zones and, secondly, the uniform field of temperatures with minimal spread. The regulation of the modulation parameters can be carried out by the change of capacities, active resistances of the modulator and the frequency of the generator.

Conclusions

On the basis of the research of PS process with the modulation of electrical parameters, the developed technological support and the methodology of their design it can be worked out the optimal options of the technological process, which provide depending on the specified working loads the required level of strength properties of coatings.

References:

[1] Kadyrmetov A M, Stanchev D I and Suhochev G A 2010 Technology of plasma application and hardening of coatings in resource-saving production processes *Hardening technologies and coatings* **7(67)** 29-36

[2] Kadyrmetov A M 2013 Plasma spraying coatings in the mode of modulation of electrical parameters *Repair, restoration, modernization* **10** 23-29

[3] Kadyrmetov A M, Posmetiev V I, Nikonov V O and Posmetiev V V 2013 Evaluation of the quality of plasma coatings, sprayed by a combined method with the running roll, obtained on the basis of computer modelling *The Polythematic network electronic scientific journal of Kuban State Agrarian University* **03(87)** Access mode: http://ej.kubagro.ru/2013/03/pdf/30.pdf

[4] Kadyrmetov A M, Suhochev G A, Maltsev A F and Popov D A 2013 Physical model of mechanisms of dynamization processes of plasma spraying and hardening of coatings by means of modulation of electric parameters *High-tech technologies in mechanical engineering* **10** pp 19-26

[5] Kadyrmetov A M 2013 Intensification of energy exchange in a heterogeneous plasma jet at modulation of electric parameters of the process of plasma spraying *Engineering-Physical journal* **86 (4)** 739-746

[6] Kadyrmetov A M, Smolentsev E V, Maltsev A F and Suhochev G A 2014 Hardening of sprayed plasma coatings by the power impulse modulation of the remote arc of the plasmatron *Herald of the Voronezh State Technical University* **10 (1)** 336-341

[7] Kadyrmetov A M 2013 Theoretical bases and technological supporting of quality of plasma spraying and hardening of coatings by the electrical parameters modulation: *Extended abstract of Doct. Sci. (Eng.)*: 05.02.07, 05.02. 08/A. M. Kadyrmetov; VSUFT. – Voronezh, 32 p

[8] Kadyrmetov A M and Suhochev G A 2009 Features of the process of air-plasma spraying and hardening of coatings *Hardening technologies and coatings* **4 (52)** 17-22

[9] Kadyrmetov A M, Drapalyuk M V, Posmetiev V I and Nikonov V O 2012 Modelling of the process of plasma spraying of coatings on the parts of transport machines in the mode of power modulation of the plasmatron arc *The Polythematic network electronic scientific journal of Kuban State Agrarian University* **10 (84)** Access mode: http://ej.kubagro.ru/2012/10/pdf/19.pdf.

[10] Suhochev G A and Kadyrmetov A M 2008 Experimental studies of control parameters of the process of air-plasma spraying and hardening of coatings//*Hardening technologies and coatings* **11(47)** 53-56

[11] Kadyrmetov A M, Stanchev D I and Suhochev G A 2010 Technology of plasma spraying and hardening of coatings in resource-saving production processes *Hardening Technologies and coatings* **7(67)** 29-36

[12] Kadyrmetov A M, Stanchev D I and Suhochev G A 2010 Equipment for plasma spraying and hardening of coatings with modulation of electrical parameters *Hardening technologies and coatings* **11 (71)** 41-48

IOP Conf. Series: Materials Science and Engineering **570** (2019) 012035 doi:10.1088/1757-899X/570/1/012035

Use of modified molecular sieves in mechanical filters

O R Karataev [1], V F Novikov[2], E S Karataeva[3] and A A Kartashova[4]

[1] Federal State Budgetary Educational Institution of Higher Education (FSBEI HE) « Volga State Academy of Physical Culture, Sports and Tourism», 420010, Republic of Tatarstan, Kazan, Universiade Village, bld. 35
[2] FSBEI HE «Kazan State Energy University», 420061, Kazan, Krasnoselskaya st., bld. 51
[3] FSAEI HE «Kazan (Privolzhsky) Federal University», Engineering Institute, 420021, Kazan, S.Saydashev st., bld. 12
[4] 4420045, Republic of Tatarstan, Kazan, Nikolay Ershov street, 29 A Association "non-profit partnership "Kama innovative territorial production cluster"

ElSKarataeva@kpfu.ru

Abstract. This article discusses the use of modified molecular sieves in mechanical filters of water treatment plants and the separation of the gas mixture, as they have sorbing properties to all heavy metals. The main indicators of the quality of separation is the adsorption isotherm, which demonstrates the structure of the adsorbent, the thermal effect of separation and a number of other physical,chemical and technological characteristics. System analysis of technological parameters of mechanical filters increases the operational period of equipment.

Synthetic zeolites are classified as hydrated aluminosilicates of alkaline elements. They are solid granular materials that have a very porous structure and have a large specific absorbing surface. Their crystal structure is represented by tetrahedra of silicon and aluminum oxides, combined into lace frames with the same cavity dimensions, which are filled with alkaline and alkaline earth metal cations and water molecules. Each pore of the structure of a synthetic zeolite has a specific size from 2 to 15 angstroms [1].

Natural zeolite-containing rocks have adsorption and ion exchange properties. The chemical composition of zeolite-containing rocks is quite complex. It includes ions of Na, K, Mg, Ca, as well as water molecules located in the crystal lattice of natural zeolites.

Zeolites are rock-forming minerals and are fairly widespread in the rocks of the earth's crust. One of the largest deposits of natural zeolites in Russia is the Tatar-Shatrashan deposit, in which about 13.3% of all natural reserves are concentrated. The chemical composition is one of the important indicators of the quality of zeolite-containing raw materials.

The composition of the zeolites of the Tatar-Shatrashan deposit differs from the ores of other deposits. The most important differences are the reduced sodium content (23 times), aluminum (2 times) and increased calcium content (5 times), that is the aluminosilicate component is reduced. This difference in the main components is explained by the presence of different rocks in the section: marly limestone and flasks characterized by specific properties of the mineralogical composition (Table 1) [2].

Table 1

Chemical composition of zeolite-containing rocks of the Tatar-Shatrashan deposit, (% wt.)

Oxides	Productive stratum (in brackets - number of samples)				
	Pack 1 (37)	Pack 2 (83)	Pack 3 (74)	Pack 4 (22)	Average
SiO_2	57,87	58,49	61,45	52,67	56,27
В т.ч.SiO_2	21,55	30,08	27,59	15,94	26,71
TiO_2	0,28	0,25	0,37	0,33	0,30
Al_2O_3	6,26	5,24	5,38	4,94	5,37
Fe_2O_3	2,48	2,21	2,27	2,58	2.30
MnO	<0,01	<0,01	<0,01	<0,01	<0,01
CaO	15,58	15,84	12,49	17,24	14.90
MgO	1,46	1,25	1.15	1,37	1,26
Na_2O	0,14	0,13	0,14	0,20	0,14
K_2O	1,18	1,13	1.36	1,40	1,24
P_2O_5	0,10	0,09	0,12	0,18	0,11
P.p.p. (H_2O+CO_2)	19,73	18,12	15,37	19,15	17,72

As you can see in the table in zeolite-containing rocks the highest content of silicon dioxide and aluminum oxide is observed, which have good adsorption properties and are widely used in a pure form for water purification from related impurities. A high content in zeolite-containing rocks is observed for calcium oxide, which is an inert filler. Zeolite-containing rocks of the Tatar-Shatrashan deposit are characterized by a rather high moisture content, which is located in the mesopores of the material and evaporates when heated to 350-4000C and frees up the pores for sorption processes.

Natural zeolite-containing rocks of the Tatar-Shatrashan deposit were used for wastewater treatment from organic and inorganic impurities (Table 2) [3].

As you can see in the table 2, after filtration of water through zeolite-containing rocks, all indicators show a decrease in the concentration of pollutants, especially it is significant for ammonium ions: from 15.7 to 0.68 mg/l, and for chlorine ions: from 364 to 65.2 mg/l.

Table 2

Results of wastewater treatment with natural zeolites of the Tatar-Shatrashan deposit

Indicator	Water entering the waste treatment facilities, C, mg/l	Water after filter with zeolite, C, mg/l
pH of water	7,66	7,34
Transparency, cm	2,8	1,30
NH_4^+	15,7	0,68
NO_3^-	3,62	3,50
NO_2^-	0,67	0,07
SO_4^{2-}	65,2	35,0
Cl^-	364	65,2
PO_43^-	4,98	4,20
$Fe_{general}$	0,85	0,19
Anionic surfactants*	0,20	0,13

Zeolite-containing rocks of the Tatar-Shatrashan deposit belong to the first group of natural materials according to thermal and acid resistance and have rather good regeneration abilities. They withstand high temperatures and are resistant to aggressive media.

The pore size of zeolites is not constant: it increases with increasing temperature. The adsorption isotherm is a source of information about the structure of the adsorbent, the thermal effect of adsorption, and a number of other physicochemical and technological characteristics. Zeolites have an ordered system of internal pores. Information on the equilibrium absorption is clearly reflected in the adsorption isotherms (Fig.1)

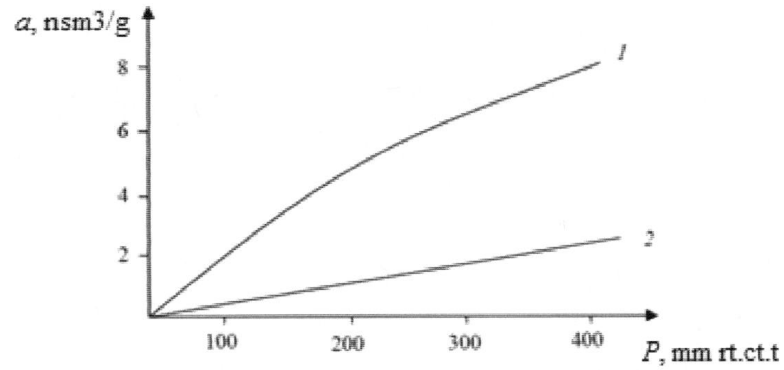

Figure 1. Adsorption isotherms:
1- nitrogen, 2- oxygen

One of the factors in choosing between using zeolites or molecular sieves is the nature of the gas mixture to be separated. Taking into account the stability of zeolites to weakly acid and weakly alkaline media and its mechanical strength and their positive filtration results and sorption results, it can be said that it is useful for water purification with a high content of impurities. They can be used at water treatment plants, as they have sorbing properties for all heavy metals. It is also possible to solve the problem of water filtration at the household level: water purification comes from physicochemical and biological contaminants. Also an important part of water treatment with zeolites is the ability to saturate water with potassium, magnesium and calcium, which has a beneficial effect on the cardiovascular system of the body [4].

References

[1] Karataev O R, Karataeva E S 2018 Mechanical filtration based on selective concentration of particles as an innovative method of water preparation IOP Conference Series: Materials Science and Engineering Vol 412 Is 1 012030.
[2] Shushkov D 2004 Natural zeolites Bulletin of the Institute of Geology of Komi, Scientific Center of the Ural Branch of the Russian Academy of Sciences No 9 pp 23-24.
[3] Brek D 1976 Zeolite molecular sieves M.: Mir 781 p.
[4] Denisov, AV Chernyshev, SA Yakushev 2006 Molecular sieves for gas separation mebran Materials of the COMPETITION OF RUSSIAN INNOVATIONS 16 p.

IOP Conf. Series: Materials Science and Engineering **570** (2019) 012036 doi:10.1088/1757-899X/570/1/012036

Research of efficiency of work of heat-mechanical equipment by the thermography method

O R Karataev [1] and E S Karataeva[2]

[1] Federal State Budget Educational Institution of Higher Education «Kazan National Research Technological University», Kazan city, Karl Marx street, bld.68, Russia
[2] Federal state autonomous educational institution of higher professional education «Kazan (Privolzhsky) Federal University», engineering institute, Kazan city, Saidashev street, bld.12, Russia

ElSKarataeva@kpfu.ru

Abstract. The paper deals with the solution of problems to assess the technical condition of thermal mechanical equipment, on the example of boilers by thermography. As a result of the study of the brickwork of the steam hot-water boiler, thermal imaging images were obtained, from which it can be concluded that the brickwork of the boiler does not perform its direct functions, the insulation is outdated. Thus, constant monitoring allows making current repair of machinery, reducing wear and heating loss to maintain efficiency at the project level.

The solution of the tasks on the assessment of the technical state of the heat-mechanical equipment, to a large extent, is associated with the introduction of effective methods of instrumental control, and in particular, the thermal method of non-destructive testing [1].

A significant proportion of equipment in boilers has a long service life. The condition of thermal insulation of boiler units, at the same time, is of great importance in matters of its efficiency and maintaining coefficient of efficiency at the design level.

In order to determine the temperature state of the heating surfaces, a thermal imaging survey of boilers DKVR 4-13 was conducted.

Modern thermal imaging systems allow you to quickly and accurately identify areas with increased heat loss and determine their boundaries.

Three main tasks in the processing of thermal images of thermal equipment can be distinguished:

1. Identification of sections of structures with increased heat loss. Here, heat transfer resistances R (m2·°C/W) for the base and other characteristic areas are calculated and compared with the required value.
2. Specific heat losses q (W/m2) are determined for all characteristic areas.
3. Determination of economic damage from the identified thermal anomalies.

- determining the plottage of the areas with high heat loss;
- determination of excess heat losses through these areas;
- determination of the amount and cost of overspending energy sources.

High-quality performance of thermal insulation (TI) is due not only to the marginal rate of heat loss, but also the consumption of thermal insulation materials.

The actual heat losses differ from the designed ones, because of the poor quality of the performance of the insulating works, by the complexity of taking into account the actual conditions of

IOP Conf. Series: Materials Science and Engineering **570** (2019) 012036 doi:10.1088/1757-899X/570/1/012036

external heat transfer [2].

Considering that each percentage of excess [3] of heat loss norms is equivalent to overspending of about 300 kg of equivalent fuel per year per 1 MW of installed capacity, the cost of maintaining TI in a proper condition, and therefore systematic control, certainly pay off. The use of a thermal-imaging (non-contact) measuring system allows reducing the laboriousness of work in monitoring the condition of TI, with an increase in the area of measurement unavailable for the contact devices.

TI of the main and auxiliary equipment and pipelines [4] with the heat carrier temperature above 100°C is subject to the tests. When the air temperature in the room is 25°C, the temperature on the surface of TI should not exceed 45°C when the heat carrier temperature is equal to or less than 500°C, when the heat carrier temperature is 501-600°C temperature on the surface of TI should not exceed 48°C. For the objects located in the open air the temperature on the surface of TI should not exceed 55°C with a metal coating layer and 60°C - with other types of the coating layer.

The losses of heat energy by the boiler units in boiler houses make a considerable part from their own needs [5]. Such losses are determined by the formula:

$$Q_{cm} = \sum_{i}^{n} B_i \cdot Q_p^H \cdot \frac{q_5}{100} \cdot 10^{-3}$$

where: *Bi* - consumption of natural fuel by a boiler unit in the estimated month, t;

q_5 - average loss of heat energy by all elements of the boiler units into the environment, in % of the amount of heat of the fuel;

$Q_{r.}^{n}$ – calorific value of the used natural fuel, kcal/kg;

i - number of boiler units.

As a result of the study of setting the steam water-heating boiler unit of type DKVr-4-13 {double-drum, vertically-water tube reconstructed unit} (steam capacity - 4 t/h, steam temperature - up to 194 degrees, steam pressure - 1.3 MPa, efficiency– 88%) by thermography, the thermal imaging pictures were obtained (Fig.1).

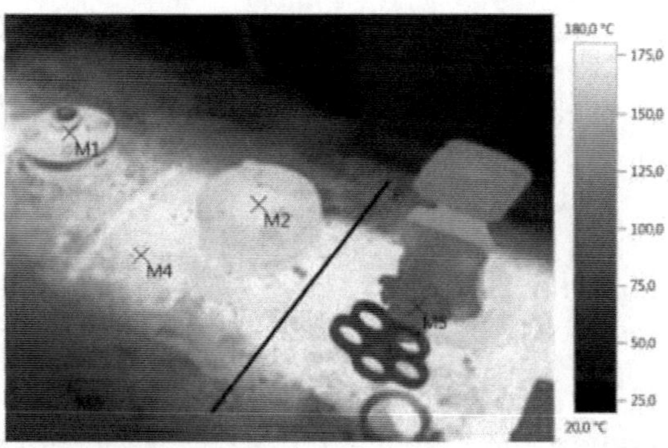

Data: 03.04.2013 **Reflective temperature [°C]: 20,0** **Emissivity: 0,93**

Measured object	**Temperature [°C]**
Measuring point 1	164,3
Measuring point 2	171,1
Measuring point 3	63,1
Measuring point 4	170,1
Measuring point 5	96,2

IOP Conf. Series: Materials Science and Engineering **570** (2019) 012036 doi:10.1088/1757-899X/570/1/012036

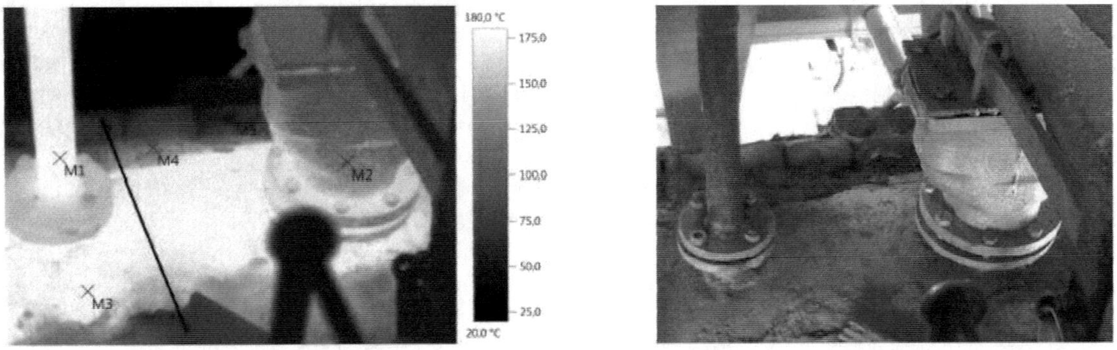

Data: 03.04.2013 **Reflective temperature [°C]: 20,0** **Emissivity: 0,93**

Measured object	Temperature [°C]
Measuring point 1	175,0
Measuring point 2	118,4
Measuring point 3	176,6
Measuring point 4	119,1
Measuring point 5	58,1

Fig.1 Thermogram of TI of the boiler-unit bundle

From the pictures we can conclude that setting of the boiler unit is not fulfilling its direct functions, the thermal insulation is outdated. The temperature on the surface of the boiler in some parts of it exceeds twice or even three times. Constant monitoring of the condition of setting the boiler unit by thermography will allow timely detection of local faults in installation or errors in the project, indicating insufficient thermal protection of individual sections and eliminate defects and, as a result, reduce heat losses.

References

[1] RD {regulatory document} 153-34.0-20.364-00 Methods of infrared diagnostics of heat mechanical equipment.
[2] Konahina I A, Kashapov N F, Gil'Manshin I R, Ganiev R R 2014 Optimization of industrial steam supply and steam-and-condensate farming of machine building enterprise IOP Conference Series: Materials Science and Engineering.
[3] Methodical recommendations on the procedure for conducting thermal control of technical devices and structures used and operated at hazardous production facilities 2007 RD-13-04-2006 LLC STC «Scientific and Technical Center» «Industrial Security», Moscow 24 p
[4] Karataev O R, Ryzhkov D V, Golubeva G N 2015 Evaluation of the technical condition of walls of the building by means of thermal imaging control Bulletin of the Technological University No 3 T 18 pp 182-84
[5] Karataev O R, Karataeva E S 2018 Mechanical filtration, based on elective concentration of particles, as an innovative method of water treatment IOP Conference Series: Materials Science and Engineering Vol 412 Is 1.

IOP Conf. Series: Materials Science and Engineering **570** (2019) 012037 doi:10.1088/1757-899X/570/1/012037

Development of turning-filtering mechanismof centrifuges

E S Karataeva[1] V A Lashkov[2], A A Brodt[3] and M N Akhlyamov[4]

[1]Federal state autonomous educational institution of higher professional education «Kazan (Privolzhsky) Federal University», Engineering institute, Kazan, Saidashev str, 12, Russia
[2,3]Federal State Budget Educational Institution of Higher Education «Kazan National Research Technological University», Kazan city, Karl Marx str, 68, Russia
[4] OOLLC "PLKGROUP", 420141, SalihBatyev str, 19, office 6

e.s_karataeva@mail.ru

Abstract. Analysis of the literature revealed a lack of filtering centrifuges, which include the difficulty of regenerating the filtering surface. The presented constructive solutions for its elimination provide only local, but not complete regeneration of the filtering surface. The authors investigated the regeneration of filter elements when the blades rotate around the axis as a result of the engagement of parts of the rotary device. In order to exclude shock loads in the process of turning, various cam designs are considered, the best of which is a fork cam.

Separation of heterogeneous liquid systems (suspensions) in many enterprises is carried out in the field of centrifugal effect (in centrifuges). Regardless of the principle of operation, centrifuges have a common drawback which lies in the difficulty of regenerating the filtering surface. Incomplete removal of dispersed particles from the filtering surface, their introduction into the pores of the filtering element and compaction by centrifugal forces leads to an increase in the separation resistance.

In this regard, it became necessary to find a technical solution for the effective regeneration of the filtering surface. The literature presents various structures of regenerating units, based on: mechanical removal of the sediment layer, the effect of the hydraulic wedge and hydraulic shock with back pressure pulsations in the additionally formed gap, changing the geometry of the filtering element and breaking the integrity of the compacted layer, etc. Analysis of the literary sources allows for the conclusion that in these technical solutions there is a local, but not complete regeneration of the filtering surface. On the basis of the conducted research, a promising direction was identified to restore the activity of the filtering surface, consisting of individual elements (blades), with the possibility of alternate rotation with acceleration around its own axis (Fig.1.).

IOP Conf. Series: Materials Science and Engineering **570** (2019) 012037 doi:10.1088/1757-899X/570/1/012037

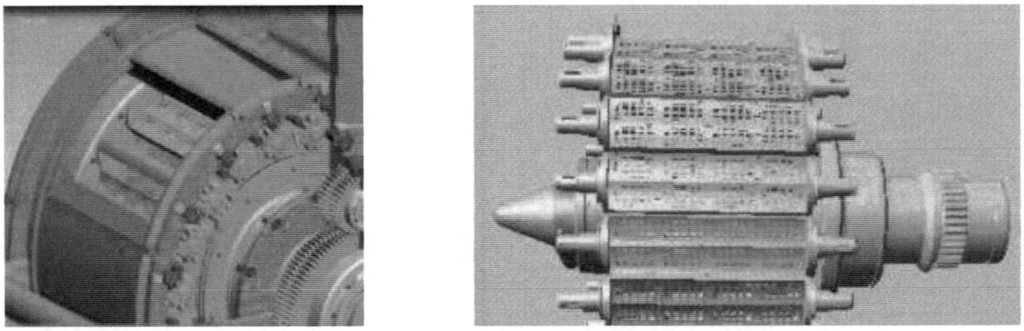

Fig.1. Filtering centrifuge with rotary filtering elements (paddles): a - general view, b - filtering surface

A centrifuge consists of a body and a rotor coaxially placed in it, the cylindrical surface of which is formed by a set of perforated elements. The elements are installed with the possibility of rotation on their axes; for this purpose, toothed gears are fixed on one of the semi-axes of the elements.

A feeder with a loading opening is placed inside the rotor. The rotor is enclosed in a shell with holes for draining the liquid phase. The shell is provided with a toothed sector. The teeth of the sector are engaged with the toothed gears located in this area. The discharge box is provided in the shell.

During operation of the centrifuge, the rotor and the feeder with the shell are driven into rotation. The initial suspension is fed into the feeder, which is thrown onto the surface of perforated blades under the action of centrifugal force, where the separation of the liquid and solid phases takes place. The filtering process continues within one rotation of the rotor relative to the feeder with the shell. At the end of one rotation the toothed gears engage with the toothed sector of the shell.

A rotation with acceleration of the filtering elements around their axes at 1800 takes place, resulting in the sediment is discharged into the discharge window. The unloaded element occupies its working position, on the filtering surface of which the initial suspension is fed. Due to the rotation of the blades at 1800 their filtering surface reorients relative to the direction of the centrifugal forces and their continuous centrifugal regeneration is ensured.

The analysis of the centrifuge operation revealed the shortcomings of the rotary device of the system of the filtering elements regeneration, which are as follows:

- between-centers distance between the rotor and the blade can not be arbitrary and depends on the modulus of the tooth and the number of teeth;
- a module of the toothand the number of teeth should be chosen so that when the last tooth leaves the gearing, the retainer without impact falls on the fixing surface (otherwise, these structural elements collide (Fig.2).

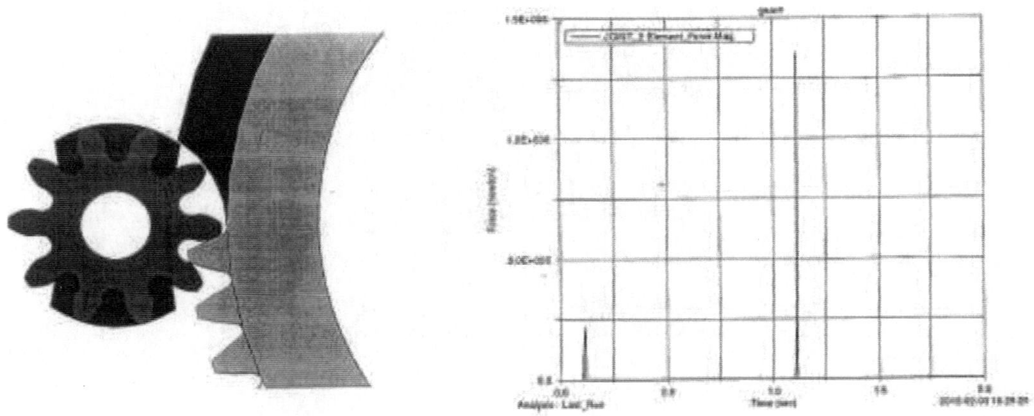

Fig.2. Results of the analysis of the work of a toothed pair in the system ADAMS, fixing the strike of the retainer on the guide

To eliminate the shortcomings, studies were carried out with various parts of the rotary device — cams: a cross-shaped cam with a retainer, a modified cross-shaped cam with a retainer and one pin; a cross-shaped cam with a retainer and two pins; a forked cam.

It was found out when turning the cross-shaped cam with a retainer and one pin:

- there is a rebound from hitting the linear part of the cam to the engagement of the pin and tooth;

- the angle of rotation of the blade is random and depends on the friction on the axis of the blade;

- contact of the fixing surface occurs with a rebound.

To eliminate the rebound when turning with a modified cross-shaped cam with a retainer and one pin, the lead-in part of the cam was changed from straight to radial.

In order to streamline the rotation of the filtering element, an additional pin was inserted into the structure. As a result, the rotation of the elements became more predictable, but the rebound of the fixing device was not eliminated (Fig.3).

Thus, all the considered mechanisms have one common drawback. Different geometry elements that are responsible for the rotation and fixation of the blade lead to the impact interaction of their surfaces.

Still the task in hand was solved by the use of a forked cam. When using this device, it became necessary to determine the geometry line of the axes centers. As a result of mathematical processing, the following relations between the angles were obtained:

$$\beta = \beta_0 - (\beta_0 t),$$

$$\theta = 90 - \varphi_0 + t\varphi_0 + arctg((l/2sin\beta)/(R_c - l/2cos\beta)),$$

$$R = \sqrt{(R_c - l/2\,cos\beta)^2 + (l/2sin\beta)^2},$$

$$\varphi = \varphi_0 - \varphi_0 t,$$

где $l = 10$; $R_c = 78$; $\beta_0 = 90$; $\varphi_0 = 12$.

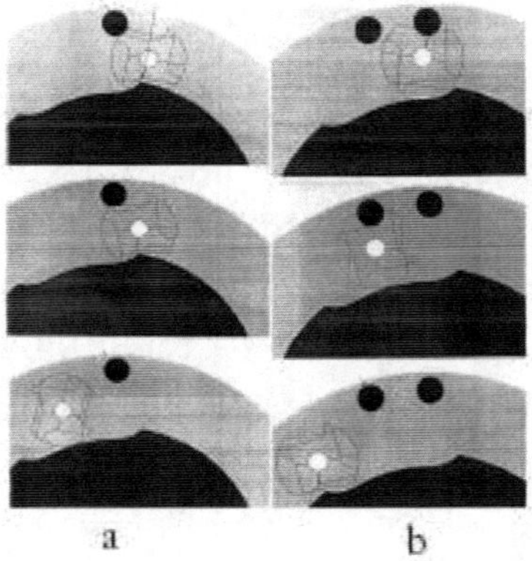

Fig.3. Cross-cam operation: a - with one pin; b - with two pins

The expression for the linear variation of the angle βcan be given by a more complex dependence, providing smooth acceleration at the terminal parts of trajectory

$$\beta = \beta_0 - \beta_0(\sin(t \cdot 180 - 90) + 1)/2.$$

The results of calculations and constructions are presented in Fig.4.

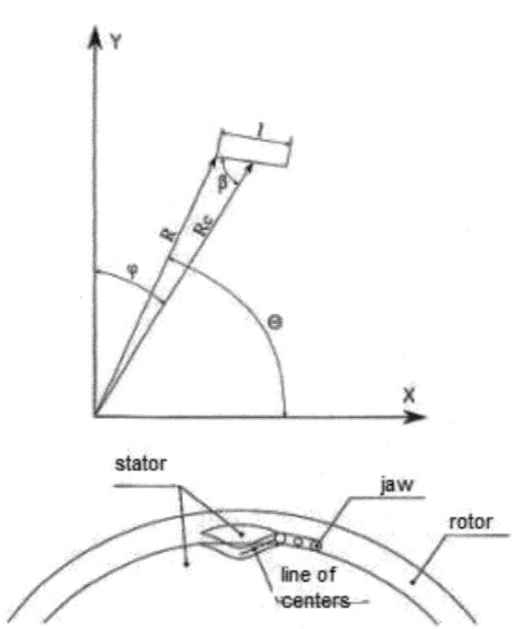

Fig.4. Line drawing of cam centers

The resulting solution is scaled and recalculated to any size (angle of rotation, fork arm, between-centers distance.

References

[1] MukhtarovYaS, SufiyanovRSh, LashkovV A 2012 *The effect of solvent losses on the profitability of the process of dehydration of oil-contaminated soils*Bulletin of Kazan Technological University T15 No 11 pp 197-98
[2] YaS Mukhtarov, RShSufiyanov, NIGdansky, V.A. Lashkov2012 Modeling the process of handling oil-contaminated soils / // Bulletin of Kazan Technological University. - 2012. – T15No11 pp 201-04
[3] YaSMukhtarov, R ShSufiyanov, V A Lashkov 2014*Methodology for calculating the optimal parameters of the functioning of the system for separating the liquid phase from suspensions and high-moisture materials*Bulletin of Kazan Technological UniversityT17 No 3 pp 230-32
[4] Karataev O R, Karataeva ES 2018*Peculiarities of calculating the longitudinal mixing of adsorbent in a fixed layer with account of surface friction*IOP Conference Series: Materials Science and Engineering 2018 Vol 412 Is1 **012031**

IOP Conf. Series: Materials Science and Engineering **570** (2019) 012038 doi:10.1088/1757-899X/570/1/012038

Automation of process of manufacturing products from a mixture of fluids fast polymerized in air using the jet centrifugal method

F G Karikh and L N Shafigullin

Naberezhnye Chelny Institute (branch), Kazan Federal University
Sjujumbike street 10a, Naberezhnye Chelny, 423800 Tatarstan, Russia

misharin_82@mail.ru

Abstract. The developed method provides the extraction of each component from the transported agents into flexible vessels followed by blending in a dosing unit, mixing in a static mixer, spraying the mixture on the surface of ellipsoidal drums, while ensuring the required thickness of the gel coating transported to the line of formation.

1. Introduction

Among the resins polymerized in air, dicyclopentadiene (DCPD) has the widest range of uses. Today, it is polymerized using barrels [1 - 5] by combining two premade mixtures. Typically, the first mixture includes DCPD, and the second one – other components, such as plasticizer, modifier, dye, etc. These mixtures are premade due to a rapid rate of DCPD polymerization and difficulties in providing small amounts of hardener (1-2% wt.). The drawbacks of this method are highly labour intensive preparation of a two-component mixture and a low-speed usage of this mixture.

2. Body text

The following installation has been designed to correct these drawbacks. The units of the installation.

1. Jet centrifugal device. It helps to eliminate the drawback of the method [1] by using the jet centrifugal method to form a stream of droplets with controllable mixture jet velocity and path [6]. The schematic view of the jet centrifugal device is shown in Figure 1 where the fluid (2) is supplied from the flexible vessel (1) through the connection tube (3) to the gland (4) used to seal the connection with a system of the rotation unit (5). The silicone tube (6) provides a leak-tight connection between the rotation unit and the glass atomizer (7). It ensures the fluid outflow through the nozzle (8) which defines the jet (9) diameter.

The path of the resulting droplet stream (10) is determined by an inclination of the nozzle (8) in relation to the axis of rotation (11). The flow rate of fluid depends on two parameters: nozzle (6) diameter (typically 50-200 μm) and distance between the nozzle and the axis of rotation (11) (usually 2-50 mm). The silicon tube (6) has a 30-μm mesh filter to prevent the nozzle (8) from clogging (omitted in Figure 1). It is mounted in the silicon tube (6) at a sharp angle to its axis.

2. Dosing unit. It has been proposed to use a dosing unit for continuous preparation of a mixture of fluids which are not polymerized in air [7]. It controls the flow rate ratio for each component and ensures that no mixture component has a contact with air. Each component is added through the flow passage into the respective dosing chamber on the mounting plate, with the adjustable size of a dosing opening separately for each chamber. No compressor is needed thanks to the flexible vessels that are

IOP Conf. Series: Materials Science and Engineering **570** (2019) 012038 doi:10.1088/1757-899X/570/1/012038

sealed to the inlet fittings of the dosing chambers. The quantity of a prepared mixture in unit time depends on the performance parameters of the jet centrifugal device: its rotational speed and section of the atomizer nozzle.

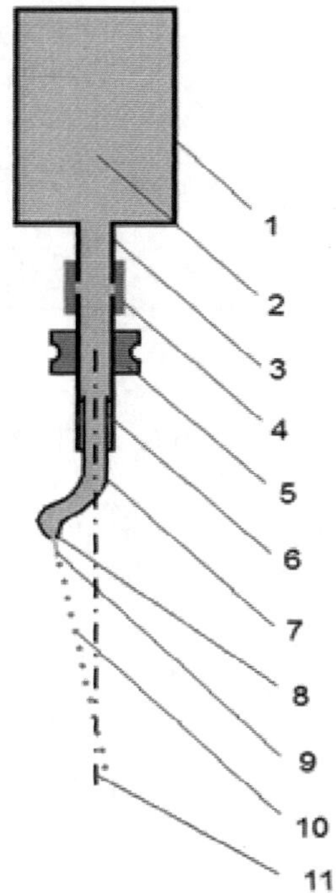

Figure 1. Device for manufacturing products from a mixture of fluids fast polymerized in air using the jet centrifugal method

The components are dosed with a required ratio based on resin consumption per a specified period, sizes of the ring dosing openings are set for each chamber according to pre-compiled tables on the relationship between the component flow rate and the ambient temperature.

3. Static mixer [8]. Figure 2 shows the schematic longitudinal section of the mixer which includes the housing (1) with the tangential inlet (2) for a mixture from the dosing unit. This mixture is added to the mixing cylinder (3) with the cylindrical insert (4) and filler (5), which press tightly against the metal mesh (6) in the annular space (7). It mates with the threaded portion of the cap (8) which presses tightly against the cutoff (9). The cutoff is basically a twisted wedge with the blade (10), and it is used to form a circular vortex in the chamber (11), which is limited by the tapered surface (12) extended to the cylinder (3). As the fluid spirals in the mixing cylinder (3) and contacts the mesh (6) the components are mixed, and the mixture is transferred to the jet centrifugal device through the openings in the profile ring (13) that connects with the fairing (14) with the outlet duct (15) and outlet orifice (16).

A different variant of the static mixer [9] has a conical mixing chamber. It is intersected with a metal cylinder-shaped mesh to provide a higher degree of sealing with linking of the joint in the region of its generator.

IOP Conf. Series: Materials Science and Engineering **570** (2019) 012038 doi:10.1088/1757-899X/570/1/012038

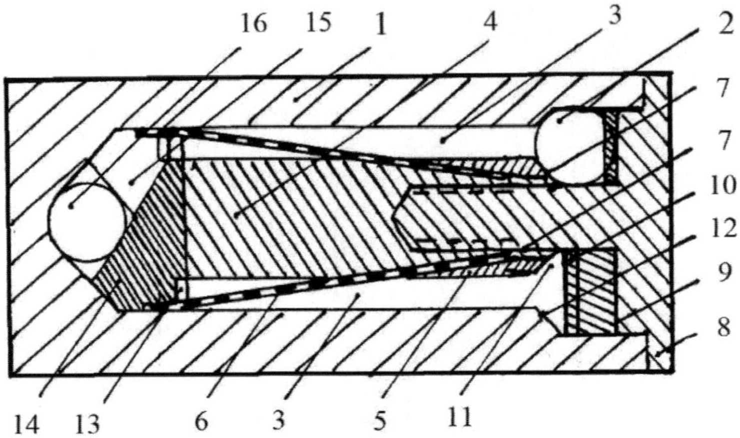

Figure 2. The mixer

4. System of mechanically interconnected rotating drums. In order to extend the application of the method [10] for production of a wider range of products, the method and apparatus [11] have been proposed to apply coatings on the surface of rotating drums with a toroid-shaped surface, which have various diameter ends. The number of drums depends on a product range. For example, Figure 3 shows the three-drum installation where the jet centrifugal device (1) includes two atomizers (2) with opposite jets that are atomized into droplets (4) onto the toroid-shaped surface (5) of the drums (6). These drums are rigidly secured to the rods (7) which are connected via identical bevel gears (8) with the motor (9) to ensure the same rotational speed for all the drums. As the duration of contact between droplets (4) and atmosphere is constant due to the stable distance between the atomizers (2) and wetting area (5), the degree of polymerization of a gel layer formed on the drums (6) is constant. All the drums (6) are positioned in such a way that their ends (10) with a larger diameter are overlapped by the ends (11) with a smaller diameter. The gap between the adjoined pair of drums is minimal, so the surface (12) of each drum remains unwetted. It addresses the following needs: prevention of leakages of polymerized droplets (4) through the gap between the adjoined pair of drums; prevention of environmental pollution; facilitation of throwing the gel polymerized layer off the drums (6); stability of the size and cleanness of the edges (14); facilitation of layer (13) removal from the transporter line (15) during its delivery to the process line with suitable forming equipment.

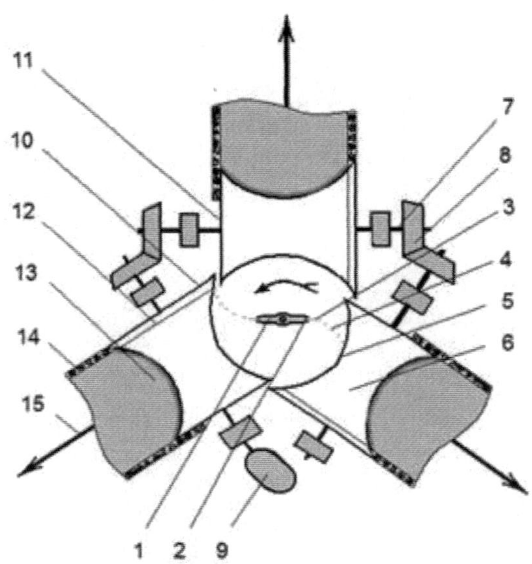

Figure 3. The three-drum installation

IOP Conf. Series: Materials Science and Engineering **570** (2019) 012038 doi:10.1088/1757-899X/570/1/012038

3. Conclusions

1. The installation can operate automatically when its components follow the program which takes into account a change in such variable parameters as ambient temperature, composition of a used fast polymerizing mixture of fluids, rotational speed of drums and atomizer.

2. The number of drums can be increased if it is necessary to expand a range of products.

This work was supported by the research grant of Kazan Federal University.

References

[1] Busygin B M and et al. 2012 *Method for making a two-component system based on dicyclopentadiene* RF Patent 2450028

[2] Romanova N V, Yurasov S Yu, Romanov , Sokolova Yu A, Gumerov I F, Shafugullina G R and Shafigullin L N Producing Polydicyclopentadiene from Lower-Grade Dicyclopentadiene *Research Russian Engineering Research* vol 38 pp 1046-48

[3] Kashapov N F, Nafikov M M, Gazetdinov M X, Nafikova M M and Nigmatzyanov A R 2016 Innovative production technology ethanol from sweet sorghum *IOP Conference Series: Materials Science and Engineering* vol 134 issue 1 8 number 012012

[4] Fedyaev V L, Galimov E R, Galimova N Ya, Gimranov I R and Siraev A R 2017 Dynamics of coalescence and spreading of liquid polymeric particles during coating formation *IOP Conf. Series: Journal of Physics: Conf. Series* vol 789 number 012006

[5] Fedyaev V L, Galimov E R, Galimova N Ya, Takhaviev M S and Siraev A R 2017 Mathematical modeling of processes occurring during deposition of sprayed particles of polymeric powder *IOP Conf. Series: Journal of Physics: Conf. Series* vol 789 number 012007

[6] Karikh F G and Karikh A F 1993 *Method for generating a stream of droplets* RF Patent 1834714

[7] Karikh F G and Karikh A F 2000 *Dosing unit for controlling flow rates of the two fluids* RF Patent 2144693

[8] Karikh F G and Karikh A F 1999 *Static mixer for polymerizing fluids* RF Patent 2135354

[9] Karikh F G 2001 *Static mixer for fast polymerizing fluids* RF Patent 2171135

[10] Karikh F G and Mukhametzyanova G F 2016 *Method and apparatus for making multicomponent mixture of flows of fast polymerizing fluids when exposed to the atmosphere* RF Patent 2644910

[11] Karikh F G, Shafigullin L N and Mukhametzyanova G F 2018 *Method and apparatus for making products from fast polymerizing fluids when exposed to the atmosphere* RF Patent 2644910

IOP Conf. Series: Materials Science and Engineering **570** (2019) 012039 doi:10.1088/1757-899X/570/1/012039

Automation of metal and alloy melting processes using spectral analysis data on the composition of exit gases

F G Karikh[1], L N Shafigullin[1] and G F Muhametzjanova [1]

[1]Naberezhnye Chelny Institute (branch), Kazan Federal University
Sjujumbike street 10a, Naberezhnye Chelny, 423800 Tatarstan, Russia

misharin_82@mail.ru

Abstract. The methodology has been developed for determining the physical chemical parameters of real-time metal and alloy melting to identify the correlation between the intensity of the analytical lines of exit gases and the intensity of the lines which are generated in the workshop laboratory by analyzing the samples cast by the furnace operator at the same time, which enables automation of melting process.

1. Introduction

Today, low temperature plasma is generated mainly by a dual-jet arc plasma generator which serves as a spectrum excitation source for solutions and powder materials, as well as for heat treatment of metals and alloys. From the point of view of spectral analysis, the drawback of a dual-jet arc plasma generator is its inability to control the composition of gas flows that do not reach glowing areas. This drawback is eliminated by using a six-electrode device based on three high-voltage transformers powered from a three-phase network [1 - 7].

2. Body text

The production tests were performed in KAMAZ foundry plant to identify the capability to monitor the melting status in real time by the composition of exit gases. The tests showed that the usage of this device in the electric-arc steelmaking furnace DSP-50 made it possible to correct the alloy composition during its production by re-blending the required components; to reduce the power consumption by 10% and electrode consumption by 3%; to control the temperature of molten metal during melting and furnace downtime; to reduce the losses of metal resulted from burnt-off disperse waste by more than two times; to decrease the “time constant” for the process of chemical composition measurement in a produced alloy up to a few seconds, i.e. by two orders of magnitude; the number of defects was decreased from 22% to 15% in the aluminum alloy casts made in the holding furnace Dozamatic 400 by controlling the modifier content in the melt [8].

However, automation of a melting process is not possible with a high-voltage six-electrode discharge due to a low temperature of this source (4800°C), which cannot excite atoms with an excitation energy of higher than five electron volts, that is typical for most elements. This problem can be addressed by using a six-jet plasma generator [9] as a spectrum excitation source for exit gases. It heats exit gases up to 15,000°C, and it is based on three dual-jet plasma generators, which are powered by a three-phase rectifier or electrocar storage batteries. This electric vehicle also carries a spectrum analyzer between the controlled furnaces equipped with offtakes from the main exhaust duct, and displays to provide a furnace operator with melting status data in real time.

A six-jet arc plasma generator is shown in Figure 1. The plasma forming copper heads (1) are mounted on the nonconductive plates (2), and secured rigidly to the brackets (3) to enable travel along the axes of the heads (1) in the direction perpendicular to the pipe supports (4). Above them there are annularly positioned tube chamber (5) responsible for argon supply to the heads (argon protects electrodes against oxidation), and a chamber (6) responsible for distribution of working gas (air) via the flexible hoses (7). Protective gas is supplied to copper anodes with contacts A1, A2, A3 and tungsten cathodes with contacts K1, K2, K3 via the flexible hoses (8). Above the supports (4), the chamber (9) for addition of cooling water and chamber (10) for water discharge to the passage (19) are positioned axially to the chambers (5) and (6). Cooling water is added to the head compartments from the vertical passage (17). The chamber (10) for water discharge to the passage (19) is near the tube chamber (9). The chambers (9) and (10) are connected with head compartments via flexible hoses (11) and (12). The water flow (19) (heated by the plasma generator) is directed to the radiator to be cooled, and then the cooled water returns to the passage (16). The supports (4) are positioned on the mounting table (20), and between them the tube (14) is rigidly mounted to form an analyzed gas flow or processed powder material (15). The cylinder (13) is mounted axially to the tube (14). This cylinder provides the uniformity of change in angle of convergence of six heads with the help of the articulated joint system (21) which ensures the synchronous change in inclination of the heads in relation to the plane of the table (20). The system (21) includes the plates (22) with the brackets (3) which are mounted loosely on them. The plates are responsible for adjustment of the interelectrode gap between the plasma forming heads in the plasma generator, which are mounted on the nonconductive plates that are secured rigidly to the brackets to enable travel along the axes of the heads in the direction perpendicular to the pipe supports. Above the supports are annularly positioned tube chamber for argon supply to the heads, chamber for distribution of working gas and chamber for water cooling, which are connected to the heads via the flexible hoses. Between the supports (installed on the mounting table), there are rigidly mounted tube which forms an analyzed gas flow or processed powder material, and cylinder which provides the synchronous change in angle of convergence of six heads' axes with the help of the system shown in Figure 1. The system includes the plates with the loosely mounted brackets to ensure synchronous adjustment of the interelectrode gap between each pair of plasma forming heads.

At the first stage, the water cooling and gas supply systems of the plasma generator are started, the water, argon and air flow rates are set during pre-commissioning. At the second stage, the power supply system is activated in the normal operation mode of dual-jet plasma generators: arc discharge is triggered in each head by a momentary high voltage supply to the gap between the electrode and the nearest orifice plate, then the arc column is transferred by an argon stream to the auxiliary counter electrode to form a plasma flux which combines cathode and anode jets into the main plasma flow. At the third stage, the system for separating plasma forming heads is activated to form a plasma dome at an angle of convergence of plasma jets. The angle is defined during the optimization of gas dynamic properties to enable fulfillment of specific tasks when it comes to development of spectrum analysis techniques or development of heat treatment processes for powder materials. At the fourth stage, the system for supply of heat treated powder material or controlled gas flow is started, for example, exit gases for monitoring the status of melting in real time with data recording. At the fifth stage, all the systems are turned off one by one, repeating the plasma generator starting procedure in reverse order.

In order to prevent a significant argon consumption during the operation of six-jet plasma generator, the controller of ratios of two gas flow rates [10] has been developed. It reduces argon consumption by two orders of magnitude (the consumption may remain significant only at the start), and ensures a constant total flow rate of mixed media when their ratios change by a few orders of magnitude, the controller design and operation is simplified.

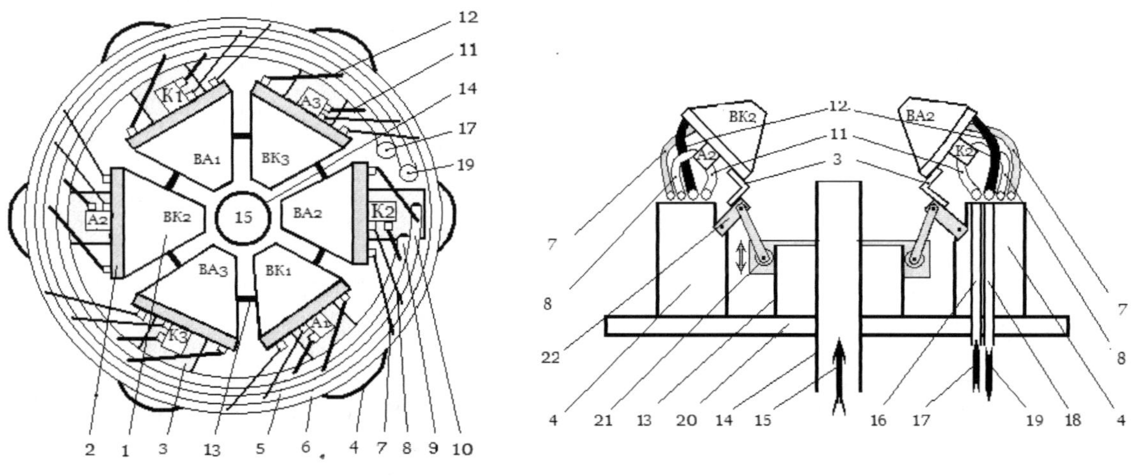

a) b)

Figure 1. Six-jet arc plasma generator: a) top view; b) side view: 1 – copper heads; 2 – plate; 3 – brackets; 4 – pipe supports; 5, 6, 9, 10 – chambers; 7, 8, 11, 12 – hoses; 13, 14 – tube; 15 – powder material; 16, 17, 19 – passage; 20 – mounting table; 21 – articulated joint system; 22 – plate

3. Conclusions

The developed six-jet arc plasma generator can be used to control the quality of a produced alloy in real time, which enables automation of metal and alloy melting processes.

This work was supported by the research grant of Kazan Federal University.

References

[1] Karikh F G 1997 *Device for spectrum excitation* RF Patent 2085871 (IPC 6 G 01 J 3/10).

[2] Fedyaev V L, Galimov E R, Belyaev A V, Dolgov A V and Samoylov V M 2018 Evaluation of interphase boundaries dynamics at the contact of melt with solid material *IOP Conf. Series: Materials Science and Engineering* vol 412 number 012015

[3] Kashapov N F, Dautov I G, Tkachenko L A, Fadeev S A and Shaidullin L R 2018 Calculation of the distribution of temperature in the form based on magnesium oxide for the casting of titanium products *Journal of Physics: Conference Series* vol 1058 issue 1 number 0120629

[4] Kashapov N F Dementyeva J N Kashapov R.N Kashapov L N 2018 Use of polystyrene production wastes in selective laser sintering processes *IOP Conference Series: Materials Science and Engineering* vol 412 Issue 1 number 012093

[5] Galimov E R Shveyov A I Shveyova T V Sharafutdinova E E Belyaev A V Sirotkina L V 2016 Method of predicting the hardness of welded joints *International Journal of Applied Engineering Research* vol 11 Number 3 pp 1603-08

[6] Khisamutdinov RMZvezdin VV Saubanov RuzR Israfilov IH Rakhimov RR Spirin AA 2016 Study of processes of steels surfaces modification with highly concentrated energy flows *IOP Conference Series: Materials Science and Engineering* vol 669 Issue 1 number 012024

[7] Astashchenko V I, Zapadnova N N, Mukhametzianova G F and Shafigullina A N 2017 Key concepts for production of high-quality parts *IOP Conference Series: Materials Science and Engineering* vol 240 issue 1 number 012007

[8] Karikh F G 2004 *The methodology for determining physical chemical melting parameters of metals based on spectral analysis data* Dissertation of the doctor of technical sciences (Chelyabinsk: South Ural State University Publ) p 310

[9] Karikh F G, Karikh A F, Poprotsky V S and Poprotsky R V 2014 *Six-jet arc plasma generator* RF Patent 2529740 (IPC H05H1/26)
[10] Mukhametzyanova G F, Karikh F G and Mukhametzyanov I R 2017 *Controller of ratio of flows from two media* RF Patent 2620631

IOP Conf. Series: Materials Science and Engineering **570** (2019) 012040 doi:10.1088/1757-899X/570/1/012040

Description of a complex technical system of urban electric transport from the standpoint of synergistic methodology

N F Kashapov[1], L S Sabitov[1], A E Auhadeev[2], R S Litvienko[2] and I Z Gatiyatov[2]
[1]Federal State Autonomous Educational Institution of Higher Education "Kazan (Volga Region) Federal University", Kazan, Kremlevskaya street, 18, 420008, Russian Federation
[2] Federal State Budgetary Institution of Higher Education "Kazan State Power Engineering University ", Kazan, Krasnoselskaya Street, 51, 420066, Russian Federation

l.sabitov@bk.ru

Abstract. City electric transport is considered from the standpoint of a synergistic methodology in order to search for modern approaches to solving the actual problem of increasing its efficiency. The properties of the urban electric transport system are identified in their unity and interdependence in terms of synergetics. Methodological interpretations of these properties from the standpoint of an actually existing system of urban electric transport will allow determining its theoretical synergetic model.

Introduction

Improving the efficiency of urban electric transport (UET), primarily energy, is a prerequisite for the modernization and innovative development of the transport system of modern cities, to ensure that the transport needs of the population are met, and also to integrate urban agglomerations into a single transport system. These processes require new methodological approaches to the description of the GET, as a complex technical system with a huge number of internal and external relations with their heterogeneity and structural diversity.

The modern methodological space of science has been enriched with synergetic ideas about the phenomena of reality as complex systems. Such a view of the world, born of the efforts of a number of outstanding scientists of the new time, is reflected in the concept of “Synergetic”, which was developed and formulated as a new scientific paradigm by the German mathematician Hermann Hacken [1]. An attempt to interpret all these methodological bases from the standpoint of a real-life organization of the urban electric transport system will serve to develop innovative solutions to the actual problem of increasing its efficiency.

Materials of theoretical study

The modern city, its production and non-production technologies, where the UET system occupies a central place and is of unifying importance for many, is one of the most problematic objects of synergistic research [2, 3] in the context of the growing urbanization of public life. In the domestic scientific literature [3, 4], attempts are made to describe transport systems in terms of open complex systems and synergetics, and also to determine which theoretical and practical problems can be solved from the standpoint of innovative, positive development of urban transport systems. The most relevant from the standpoint of systemic synergetic methodology seems to be the definition in terms of synergetics of the following properties of the UET system in their unity and interdependence (Figure 1).

1. **The complexity of the organization of the system.** This is one of the most important properties of the system, determining all other qualities and manifestations of its order [5, 6].

The UET system is complicated by the presence and quantity of various types of rolling stock, means of communication, sources and methods of energy supply, production and repair facilities for transport units, etc. It is complicated not only by its technical and technological content, but also by its complex social infrastructure that provides the functions of the transport system, including the human, personnel, professional, and educational factors. The complexity has a number of necessary characteristics.

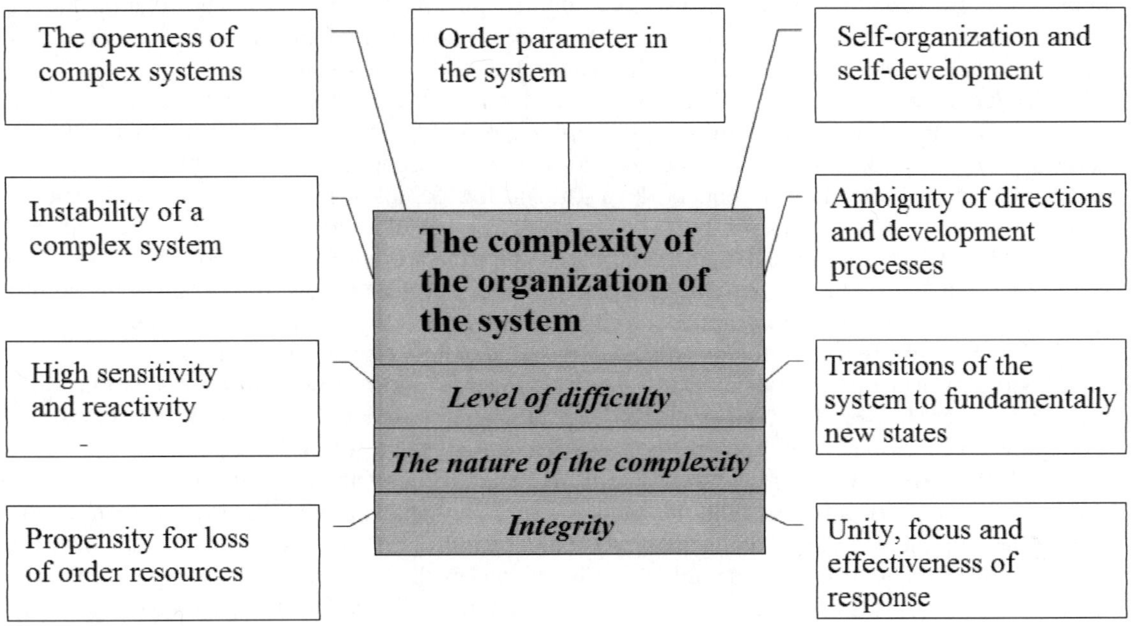

Figure 1 - Synergistic properties of the UET system

a) ***Level of difficulty.*** It consists in a variety of parts included in the system. A systematic approach to urban transport has been significantly developed and presented in the scientific literature [4, 5], there remains the problem of determining the specific level of complexity of this system, which is crucial for its positive innovative development based on the properties of self-organization and self-development. In the UET system of a modern large city there are types of transport that compete with each other, for example, by their availability, comfort, etc.

b) ***The nature of the complexity,*** consisting in the fact that parts of the system and their space-time relationships are not the same, are significantly different, they coexist as "the unity and struggle of opposites". The search for specific ways to optimize such a dialectic unity is the second problem of the complexity of the UET system. The transport system of a city, having a certain level and nature of complexity, may not have in its composition any elements, for example, sufficient energy sources, repair facilities for the production of vehicles and means of communication, institutions preparing personnel, etc. [7-9].

c) ***Integrity*** - a sufficient number of parts included in the space-time relationship. It characterizes the organization as a system capable of displaying the qualities of order. The more complex the system is in its level and nature and, if it is integral in its composition, the more pronounced are the qualities inherent in complexly organized ordered systems. Determining the sufficient integrity of the UET system for its innovative development is the third problem of its complexity.

2. **The openness of complex systems.** This is a property of systems that determines their dependence on external resources and the ability to perceive these resources in the interests of the emergence and development of their internal qualities and their manifestations. The system UET is open to relationships with other technological systems of the city. Moreover, it plays the role of connecting many technologies of the city into a single technological system, ensuring the flow of people - the main energy potential of society in a single stream of raw materials, means and products of their production and non-production activities. However, the whole set of non-transport technologies of the city has an impact on its transport part. These relationships can be conducive to or impede the processes of positive mutual development of city technologies. The relationship of the transport system with other technologies of the city as a source of innovative development is the central problem of the technological development of the city as a whole [10, 11].

3. **Order parameter in the system.** This is the criterion by which order arises and manifests itself — an organization-system is formed with certain properties that correspond to this criterion. In addition, the order parameter or regulation is an external condition and an internal factor in the emergence and subsequent independent development of a complex system. For the UET system, the order parameter is various kinds of positive effects on the life of a city, for example, on "convenience for life" [12]. Order (complex system) and order parameter are born simultaneously. In the chaos of events, the random movement of particles in certain conditions "seizes" the parameter according to which the motions acquire the space-time order and thus the order parameter is realized, fixed in the properties and manifestations of the system. Based on the value of the UET system for people's livelihoods, its innovative development cannot be focused only on random events of choosing the order parameter [13]. Finding the optimal order parameters in the UET system is an urgent task for the technological development of modern cities as a whole at all stages of this process.

4. **Self-organization and self-development.** This formation of the relations of parts of the system suddenly arising, in accordance with the original order parameter, into a space-time organization that corresponds to this parameter, with subsequent complication of the organization, the establishment of new order properties in its new parameters. This is due to the appearance in the system of internal, intrinsic to it, mechanisms and development resources and, of course, due to its openness to perception of external resources and exchange with them [14]. The transport systems of modern cities have a centuries-old history of their origin and development. Now there is no need to talk about the spontaneous generation of the UET system, especially in large cities, in which the development of electric transport is a pressing social problem - in them transport systems have already been "born" and formed in one form or another. The existing organization of the city's transport needs to be improved. Based on the principles of self-development, transport, the transport system, the UET system develops in the direction of improvement, and therefore positive development, thanks to the means inherent in it — internal mechanisms that are different from other technological systems. But this is not only the case - with all the openness towards the surrounding technological environment and considerable dependence on it, the mechanisms and processes of self-development of the UET system are active, largely independent and independent of external influences. However, they can lead to development and to an unfavorable result, both for the system itself and global, socially significant. The problem of analyzing the processes of self-organization and self-development of the UET system consists in knowing the specifics of the internal mechanisms of development and determining the parameters of the order in which development should occur. The result of self-development (positive or negative) depends on a number of other crucial properties of the transport complexly organized system [6, 15].

5. **Instability of a complex system.** In everyday terminology, instability is perceived as a negative quality of any phenomenon. Relatively complex systems with the properties of openness, self-organization and self-development instability is defined by the terms "nonequilibrium", "non-

IOP Conf. Series: Materials Science and Engineering **570** (2019) 012040 doi:10.1088/1757-899X/570/1/012040

equilibrium states", which are also understood as positive properties of systems that provide them with the necessary variability, ability to easily adapt to changing conditions of existence and development. UET systems, the more complex in their composition, nature and integrity, the more they are non-equilibrium. Such a state may have various manifestations, the essence of which requires definition and specification. This may be existing fluctuations and the emergence of new characteristics of the complexity of the transport system. This may be the instability and coexistence of different, possibly contradictory, order parameters; the emergence of new parameters that require selection and consolidation. The non-equilibrium state of the UET arises for internal reasons related to the complexity of the system, and for external reasons related to the processes of the entire technological system of the city, requiring the satisfaction of new order parameters. In this non-equilibrium state, the UET system is faced with the choice of a new path of self-organization and self-development based on existing or emerging additional internal resources. The problem arises of determining and studying the mechanisms and conditions for choosing this path, which depends on the problem associated with the following property of the transport systems [3].

6. **Ambiguity of directions and development processes.** This is the possibility of different lines, directions of development of the system in its given non-equilibrium state. In terms of synergetics, this phenomenon of bifurcation is a split line of development, possibly of more complex splitting. **Bifurcation** arises in connection with the instability and variability not only of the states of the system itself, but to a greater degree of the environment of its existence. It is regarded as a non-linear development at each moment of growing imbalance, its probabilistic and unpredictable nature, at the same time as the ability of the system to choose the direction of development itself, the most adequate to its internal properties and therefore more efficient. The lines of development of the UET system may be different due to changes in the characteristics of its internal complexity associated with changes in the scale of the city, its production and non-production technologies, socio-economic tasks and processes, and other characteristics. In any case, development should proceed along a positive line in its social and humanitarian significance. This is expressed in the problem of finding mechanisms for providing some kind of "stability in instability" - ensuring sustainable positive development of the city and its technologies in general, taking into account the non-equilibrium properties of its technologies as complex open systems [16]. The solution to this problem is largely due to the property of the system, manifested in its high sensitivity and reactivity.

7. **High sensitivity and reactivity.** This is a property of a system that has a complex organization, to react to the smallest influences from the external environment and to the smallest changes in the internal environment, at the same time, it is the ability to give large reactions to these small external or internal changes. In technical terms, this means the following: for small information "at the entrance" to the system, it gives great information - a great result of the response at the output. This property has both a positive side and a negative one, consisting in the fact that a small, but unpredictable, possibly negative, effect can be caused by small impacts on the system [6]. Due to its material "cumbersome" and organizational complexity, the urban transport technical, technological system can be perceived as an inert enough education - sensitive only to strong influences from the general factors of the social life of the city, its general technological basis, and not capable of fast and complete response. on these impacts. However, the methodological approach suggests that a certain level and certain characteristics of the complexity of the UET system provide it with high sensitivity and reactivity to the smallest external influences and changes in internal processes, not to mention the effects of large and strong ones. The methodological approach reveals in this the problem of finding criteria for the complexity of the UET system and the influencing factors that can serve to identify both the positive and negative sides of the state of its hypersensitivity and reactivity. This has not only a theoretical and methodological meaning, but also a practical one, which consists in the need for balanced decision-making on introducing changes to the UET system, often of an administrative-voluntaristic nature. One of the conditions for studying this problem is to refer to the properties of the system, which determine the unity and purposefulness of its response to external influences.

IOP Conf. Series: Materials Science and Engineering **570** (2019) 012040 doi:10.1088/1757-899X/570/1/012040

8. **Unity, focus and effectiveness of response.** This is a special property associated with the presence in the system of coherence mechanisms, in a technical language, the coherence of its parts in their space-time relationships and actions. Consistency - the **coherence** of the parts of the UET system, which are part of its complex organization, is one of the most realistic, practical problems of innovative, positive development of urban transport. Its deepest methodological essence consists not only in the coherence of one part with another, but also the coordinated unity of the space-time actions of the entire set of parts as a whole. The most important thing in this problem is the theoretical search, definition and practical implementation of such coherence. The system is coherent in the relationship of its parts, many times more efficient than non-coherent. Coherence gives unexpected, emergent results [18].

9. **Propensity for loss of order resources.** This is a property of the system associated with its internal complexity and openness towards the external environment. In the language of physics, a complex system is "**dissipative**," that is, prone to dissipation, loss of its resources — "energy, matter, and information" when interacting with other systems. Thus, it loses the level and nature of its organization - the resources of its order. In everyday terminology, dispersion, loss has a negative meaning. Relatively complex systems, a dissipative property means dispersion and loss of resources that go beyond the established order, are not fixed in order, and free up space for acquiring new resources from outside to form a new order with its new parameter. For the innovative development of the GET system this is a favorable condition [3]. It is the replacement of obsolete resources with new ones - in engineering, technology and in the organization of the system. Such a methodological approach requires a theoretical and practical definition of dissipation criteria, its characteristics positive or negative if they lead to the loss of valuable resources due to their disorder, disorganization.

10. **Transitions of the system to fundamentally new states**. These are the processes of transition of the system to states that are radically different by their internal organization, properties, and their external manifestations, which take place under the influence of various factors. In modern scientific methodology, they are referred to as "phase" states of physical and chemical material objects. Depending on the nature of the system and the influences exerted on it, new phase states can be very diverse. On this methodological basis, a practical, real definition of variants of phase states of UET systems with different technological characteristics and development trends is needed. Such states can be certain types of the UET system, characterized by special, significantly different properties. In these states, it is important to determine the real mechanisms of transitions, the corresponding attractors, given the need to acquire positive development directions based on a synergistic view of the development of transport systems in the global unity of all the technologies of life in a modern city.

Conclusion

The above properties of the UET require careful study in theoretical and practical terms from the standpoint of a real organization of the city's transport system, which ultimately will make it possible to obtain a model of an open, complexly organized technical system of UET. This is a serious theoretical and methodological problem, practical approaches to the solution of which require the use of modern information and analytical technologies [18, 19]. The model should be fairly simple, but not as simple as possible, in order not to reduce the synergistic essence and complexity of a truly existing UET system. The authors of this article did not disclose all the properties of the UET as a complexly organized technical system, but a start has been made to rethink the principles of organization, development and management of the UET system in the context of the dialectics of the relationship between its external target management, on the one hand, and the internal properties of self-organization - on the other [3, 6].

References:

[1] Haken G 1980 *Synergetics*, trans. from English by V I Emelyanova, ed. Y L Klimontovich (Moscow: World) p 423

IOP Conf. Series: Materials Science and Engineering **570** (2019) 012040 doi:10.1088/1757-899X/570/1/012040

[2] Chumenkov V P 2012 *The emergence of a new scientific concept and the penetration of its ideas into some areas of natural, technical and humanitarian* science (Moscow: Cosmos) p 428
[3] Aukhadeev A E 2014 *Self-development of the transport system of the modern city: Search for an innovative model of intellectual management* (Moscow: VINITI) p 220
[4] Yembulaev V N 2004 *Theoretical foundations and methods of managing the transportation system of a large city* (Vladivostok: Dalnauka) p 212
[5] Frolkov A I 2007 *System approach in science and technology* (Moscow: Book and business) p 174
[6] Knyazeva E N and Kurdyumov S P 1994 *Laws of evolution and self-organization of complex systems* (Moscow: Science) p 240
[7] Gorev A E 2010 *Fundamentals of the Theory of Transport Systems* (St. Petersburg: SPbGASU) p 214
[8] Khusnutdinova E M, Khamidullina G R, Pavlov P P, Aukhadeyev A E, Litvinenko R S and Khusnutdinov A N 2018 *Training transport specialists based on a contemporary view of self-organization of sophisticated developing systems* IOP Conference Series: Materials Science and Engineering 412(1) 012047
[9] Troitskaya N A, Chubukov A B 2004 *The Unified Transport System* M .: Academy p 240
[10] Aukhadeev A E, Litvinenko R S and Pavlov P P 2017 *Interpretations of the description of the urban transport system in the light of the theoretical and methodological achievements of modern science Transport: science, technology, management* (Moscow: VINITI) No 8 p 29-33
[11] Khristoforov E N and Sakovich N E 2012 *Transport and environment* (Bryansk: BrGSA) p 196
[12] Vukan R Vuchik 2011 *Transport in cities convenient for life*, transr. from english by A S Kalinina (Moscow: Territory of the Future) p 413
[13] Aukhadeev A E 2012 *Methodology of research and solutions to the problems of the technogenic influence of electric transport on the ecological and anthropological situation in the city* (Kazan: GU "NCCJZD") No 1 (11) p 788
[14] Yakimov M R 2008 Transport systems of large cities. Analysis of operating modes on the example of the city of Perm (Perm: PGTU) p 184
[15] Korolkov B P 2011 Thermodynamic bases of self-organization (Irkutsk: IrGUPS) p 120
[16] Aukhadeev A E Litvinenko R S and Rylov Yu A 2017 *Theoretical and methodological guidelines for research and solving problems of managing the development of the urban transport system* Transport Information Bulletin No 8 (266) pp 28-34
[17] Aukhadeev A E 2015 *Contradictions of "emergence" and bioethical problems of managing the development of the transport system of the city* Bulletin of the Scientific Center for Vital Safety No 2 (24) pp 5-9
[18] Gaykovich A I 2001 *Basics of the theory of designing complex technical systems* (St. Petersburg: SIC "MORINTECH") p 432
[19] Sopov V I and Shchurov N I 2005 *Modeling of electric transport systems* (Novosibirsk: NSTU) p 189

IOP Conf. Series: Materials Science and Engineering **570** (2019) 012041 doi:10.1088/1757-899X/570/1/012041

New methodological approach to the development of the theory of electric traction of urban electric transport

N F Kashapov[1], L S Sabitov[1], A E Auhadeev[2], R S Litvienko[2] and P P Pavlov[2]
[1]Federal State Autonomous Educational Institution of Higher Education "Kazan (Volga Region) Federal University", Kazan, Kremlevskaya street, 18, 420008, Russian Federation
[2] Federal State Budgetary Institution of Higher Education "Kazan State Power Engineering University ", Kazan, Krasnoselskaya Street, 51, 420066, Russian Federation

l.sabitov@bk.ru

Abstract. The article presents the scientific hypothesis that the modes of operation of electric traction equipment as part of an automated electric drive of the rolling stock of urban electric transport are formed under the influence of a complexly organized technological process of electric traction and are the topological basis of its structure. The aim of the study was to study the process of converting electrical energy into a targeted mechanical movement of rolling stock in the implementation of all technological conditions of the production and technical system of urban electric transport. The theoretical study was carried out on the basis of a methodological analysis of the theory of electrical equipment, the theory of automatic control and an automated electric drive, the theory of electric traction, the theory of motion construction, a systematic approach and a synergistic natural science paradigm. The main result is a formulated scientific concept, which allows to present the process of electric traction as a system consisting of five levels that are in hierarchical relations and solve each specific problem of building the movement of electric transport. The obtained theoretical system will allow developing the general theory of electric traction of urban electric transport by improving the methodological principles of the formation of operating modes of electric traction equipment when implementing the transportation process technology in the production and technical system of urban electric transport.

Introduction

On the modern urban electric transport (UET) for the implementation of controlled mechanical movement is widely used automated traction drive, improving the energy efficiency of which is an important scientific problem, which has important national economic importance. The implementation of the specified operating modes of the electric traction electrical equipment (TEE) that is included in the electric drive is carried out by means of an indirect control system (driver / driver controller) in an automated one, and when using an automated driving system - in automatic modes. In this case, the modes of operation of the TEE, providing for the implementation of electric traction, are interrelated with the modes of movement of electric rolling stock (ERS) [1 - 3].

The general theory of UET electric equipment, based on the methodological principles of the theory of electric machines, the theory of automatic control and automated electric drive and the

IOP Conf. Series: Materials Science and Engineering **570** (2019) 012041 doi:10.1088/1757-899X/570/1/012041

theory of electric traction, has many methods for calculating rational modes of operation of the TEE, which practice shows, do not always show a sufficient level of adequacy for UET . Obviously, this is due to the high complexity of the processes that determine the formation of the operation mode of the TEE during the implementation of electric traction in the UET system and the lack of a sufficiently elaborated methodology for their interpretation in the sectoral theory [4, 5]. Against this background, many researchers confirm that experienced, well-trained drivers are able to implement rational modes of operation of the feasibility study, providing ERS modes of movement, saving up to 10% of electricity compared to the resulting map and recommended for execution [6 - 8] . All this determines the relevance of the scientific problem of the development of the theory and methods for calculating rational operating modes for a feasibility study of ERS UET based on the development of a methodology for describing the processes that determine the formation of operating modes of a feasibility study as part of an automated electric drive when implementing the UET technological process.

As mentioned above, the issues of the theory of electrical equipment related to the calculation of the technological modes of its operation as an element of the automated electric drive ERS lie in the subject area of the theory of electric traction. This fact determines the possibility of developing the theory and methods for calculating rational operating modes of traction electrical equipment by improving the basic concepts of the formation of electrical traction as a process of converting electrical energy into a targeted mechanical movement of ERS in the implementation of all technological conditions of the production and technical system of UET. For this, the authors formulated a scientific hypothesis based on a system approach and synergetic methodology on the interpretation of electric traction as an open complexly organized process, the topological basis of a multi-level configuration of which is the implementation of TEE operation modes as part of an automated electric drive [9 - 12]. Such a methodological approach will allow us to develop ideas about the formation of the operating mode of traction electrical equipment, the architecture of interactions with other technical, technological and production processes that characterize the implementation of electric traction, which will help to solve the actual scientific problem of improving the efficiency of the automated traction drive.

Methods of theoretical research

According to the scientific hypothesis expressed, the electric traction process, the main task of which is the purposeful mechanical movement of the ERS, implemented by the TEE as part of an automated traction drive, can generally be represented by a set of hierarchically aligned subprocesses. To describe these subprocesses, it is advisable to use the theory of "building motion" proposed by the famous biomechanics of the first half of the twentieth century. ON. Bernstein [13] and has found application in robotics and mechatronics [14]. In accordance with this theory, the realization of controlled motion of objects or systems of various nature, including biological and technical, is carried out by means of its "construction". In this case, the construction of motion occurs in the general case by 5 levels ("A", "B", "C", "D", "E"), which are called "levels of motion construction" [13]. The theory of motion construction determines the mechanisms of motion control and management, which are formed in a special way by levels and have a complex structure. Such a mechanism of "building motion" allows organizing an effective algorithm of "building solutions" in controlling the motion of complex dynamic systems [14].

Based on this theory, the levels of construction of the ERS movement during the implementation of electric thrust can be identified as follows:

Level A ($У_A$) – a special level of building movements, on which movement as such is absent, but at the same time there are processes associated with preparing for movement, i.e. resource support of the ERS process.

Level B ($У_B$) – the level of building motions in the system's own coordinates (specific ERS), i.e. spatial, temporal and power coordination of the mechanical work of the TEE as part of an automated traction electric drive.

IOP Conf. Series: Materials Science and Engineering **570** (2019) 012041 doi:10.1088/1757-899X/570/1/012041

Level C ($\mathbf{У}_C$) – the level of building motions in the surrounding objective space, i.e. spatial, temporal and power coordination of the controlled mechanical movement of ERS in terms of the route of movement (plan and profile of the track, length of the stage, etc.).

Level D ($\mathbf{У}_D$) – the level of construction of the system's movements when interacting with objects of the surrounding space, determined by the ideas about their qualities and properties of mutual relations (based on objective monitoring of parameters through measuring tools), i.e. implementation of a targeted mechanical movement of ERS in terms of the technology of the transportation process (schedule, inter-train interval, travel time, etc.)

Level E ($\mathbf{У}_E$) – the level of construction of the movements of the system when interacting with objects of the surrounding space, based on knowledge of their qualities and properties of mutual relations, i.e. implementation of hypothetical ("idealized") mechanical movements of ERS, satisfying the main forecast parameters of the production process task of the UET, formed by the organizational management system.

Levels of building movement are arranged in accordance with the hierarchy of specific tasks to be solved. In this case, the tasks are solved in parallel, for each level they are continuously formed by the higher level and control the lower level, subject to the conditions of the task of the general process of EPS movement. Thus, the hierarchy of subprocesses, tasks at each level of motion building, entering into interrelationships, form the structure of the process of mechanical movement of ERS.

The process of ERS movement described above can be conditionally represented as the "mechanical" component of the implementation of electric traction. It is obvious that there is also an "electrical" component, which determines the processes of electrical energy conversion, the formation of electrical modes of the TEE, interaction with the environment, the energy system, other ERS, etc. Such processes also have a clear hierarchical structure and, in the general case, can be represented by 5 levels of building an electric thrust system (both on direct and alternating currents) [15].

The analysis of conditionally separated mechanical and electrical components of the electric thrust technological process leads to the conclusion that the main process is also a system of levels related to hierarchical subordination and interconnecting the implementation of various specific tasks that together solve the main problem. The real technological process of electric thrust cannot be unambiguously divided into mechanical and electrical components, but the similarity of their hierarchical structures makes it possible to formulate the main features of the construction of the electric thrust process: *multilevel process; hierarchy of levels; parallelism and interconnection of subprocesses; exchange of energy, material and information resources between levels and the external environment.*

Taking into account the above-listed features, let us formulate a theoretical system (scientific concept) of the levels of construction of the electric thrust process. For the basis we take the provisions of the theory of the construction of motion, since the main task of electric traction is controlled mechanical movement of ERS. This theoretical system of hierarchy of levels of construction of the process of electric thrust is graphically presented in Figure 1.

IOP Conf. Series: Materials Science and Engineering **570** (2019) 012041 doi:10.1088/1757-899X/570/1/012041

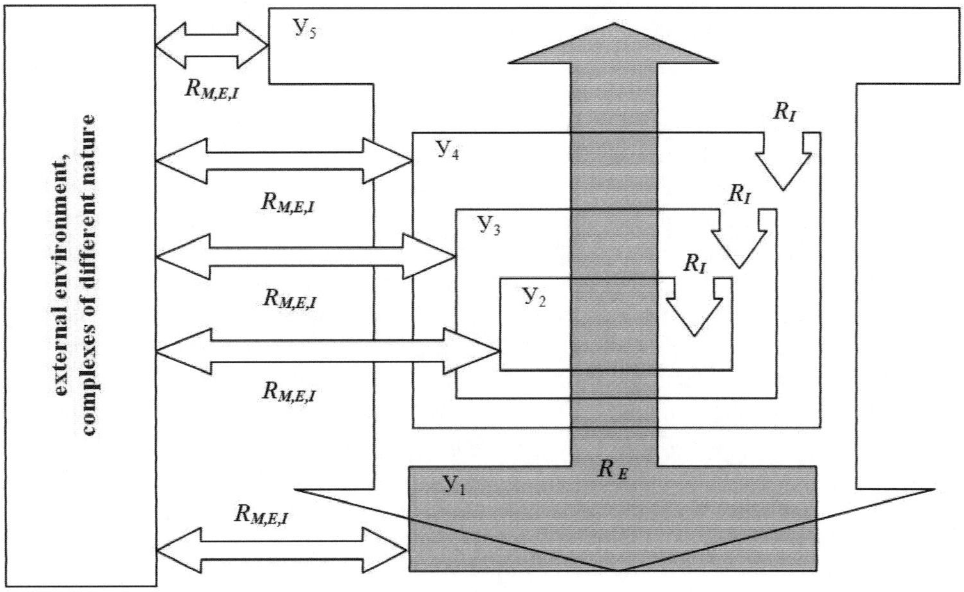

Figure 1. Levels of construction process of electric traction:
У$_1$ –**У**$_5$ – levels of electrical draft process; $\boldsymbol{R_{M,E,I}}$ – exchange of material, energy and information resources respectively

The multilevel process of electric traction is determined by the presence in its structure of several levels (subprocesses), each of which solves a specific specific task, which differs from the others, but in combination with others, aimed at solving the main task of the process. Such a subprocess or a group of subprocesses can form one of the levels of the system. So in the structure of the process of electric thrust it is proposed to distinguish 5 levels:

1. **The level of energy supply (У$_1$).** By analogy with the level of **У**$_A$ building motion is decisive for the entire process of electric thrust. At this level, the formation of traction as such does not occur, but the functional readiness of the process to the activity is ensured by ensuring a stable and reliable energy supply of the transportation process.

Energy supply is determined by the methods, methods and means of receiving, converting and distributing energy in the traction power supply system, its technological level, availability of power reserve, material support, as well as the qualifications of service personnel

2. **The level of implementation of the work of electrical equipment (У$_2$).** By analogy with the level of hydrocarbons **У**$_B$, at this level is considered a subprocess implemented in the system limited by the object's own coordinates (in this case in the ERS system), without taking into account the action of external forces that change the pulse (amount of motion) of the object (but subject to the exchange of energy, matter and substance with the external environment - heating and cooling of the TEE, electromagnetic interactions, changes in humidity, etc.). It is conditionally possible to imagine that the EPS is suspended on cables over rails and, consequently, there are no tangential traction forces or resistance forces to movement external to the ERS. In this case, the electric thrust generated by the TEE, is implemented as a torque applied to the wheel pair or wheel.

The subprocess of this level is characterized by certain algorithms for the operation of the TEE and an indirect control system, which are determined by the passport electrical and electromechanical characteristics of their constituent elements. These algorithms are formed at the stage of designing the EPS and uniquely determine the implemented modes of operation, i.e. under identical external conditions, the EPS implements identical characteristics. This is a feature of this level, which consists

IOP Conf. Series: Materials Science and Engineering **570** (2019) 012041 doi:10.1088/1757-899X/570/1/012041

in the fact that the algorithms for creating electric thrust are unchanged and are determined at the design stage by using a TEC with certain characteristics. Changing the algorithms for the implementation of electric traction requires changing the characteristics of individual elements of the traction drive system.

3. The level of controlled movement of ERS ($\mathbf{У}_3$). At this level, by analogy with $\mathbf{У}_C$ level, a subprocess of interaction of the ERS with the surrounding space and its elements is formed. So when the wheelset interacts with the elements of the path, a tangential traction force $\mathbf{F}_K$, external to the EPS, arises, leading to movement, and the electrical and electromechanical characteristics of the electrical equipment form the ERS traction characteristic. The created $\mathbf{F}_K$ is controlled, which allows to realize the controlled movement of the ERS. ERS control is carried out either by the driver or by the automatic driving system. Different parameters of the surrounding space, expressed in the physical characteristics of the plan and profile of the track, climatic and weather conditions, voltage at the current collector, etc., determine the various implementations of electric traction.

The difference of this level from the next one is that in this case electric thrust, and, therefore, controlled movement is not yet “targeted”, i.e. realizing some goal, in particular - to transport passengers from one point to another, following in the schedule, while the driver's qualifications at this level are not taken into account.

4. **The level of implementation of the technological process UET ($\mathbf{У}_4$).** At this level, by analogy with the level of $\mathbf{У}_D$, a subprocess of interaction of EPS with the surrounding space and its elements is formed taking into account their specific properties and characteristics, for example, the implementation of the required travel speed on the stretch taking into account the movement schedule, ensuring speed limits on individual sections, etc. The movement is no longer just “controlled”, but “targeted”, i.e. aimed at achieving a specific goal - ensuring the transportation of passengers along the route network in accordance with the technology of the transportation process of the UET and the quality requirements of train work. At this level, electric traction depends on the level of professional qualifications of the driver or the ERS driver.

5. **The level of implementation of the production process UET ($\mathbf{У}_5$).** This level of the electric thrust construction system is abstract and corresponds to the $\mathbf{У}_E$ level in the theory of motion construction. Its abstractness is associated with the formation of an "idealized" process of electric traction, which lower levels tend to realize. Conventionally, at this level, a “program is built up”, according to which the electric thrust process is to be implemented, in accordance with the predicted “idealized” process, which continuously controls this process and predetermines it (for example, the predicted level of electricity consumption, passenger traffic, etc.). This subprocess lined up with the strategy of the production process of the transport production UET.

The hierarchy of the electric traction process is determined by the fact that the subprocesses of the lower levels are part of the subprocesses of higher levels (see. Fig. 1). Processes of higher levels “subordinate” to themselves subprocesses of lower ones — they set the parameters and conditions for their implementation of $\boldsymbol{R_I}$. In this case, the higher the level of the system, the more complex the task of implementing electric traction in the UET system, it solves.

So, for example, the subprocess of the controlled movement of ERS ($\mathbf{У}_3$) includes in its composition the subprocess of realization of the work of the electric equipment ($\mathbf{У}_2$) and at the same time is part of the subprocess ($\mathbf{У}_4$) determined by the technology of the transportation process. At the same time, all these subprocesses include a subprocess of receiving, converting and distributing the energy consumed by the EPS (power supply subprocess - $\mathbf{У}_1$), which does not form a movement, but which provides electrical traction for the whole process, and also obey the subprocess of implementing the transport production process strategy ($\mathbf{У}_5$). This subprocess is the highest in this hierarchical system, since determines the requirements and conditions for the implementation of all previous subprocesses.

The parallelism of the subprocesses occurring at each level is due to the mechanism for the implementation of electric traction and the vertical hierarchy of system levels. Levels of a higher order form their subprocesses not at the end of the subprocess of the lower level of the system, but at the

IOP Conf. Series: Materials Science and Engineering **570** (2019) 012041 doi:10.1088/1757-899X/570/1/012041

time of its implementation with continuous interaction, which determines *the interconnection of subprocesses*. At the same time, the implementation of their specific task is built by the higher level on the basis of the task of the lower level, fully defining the conditions and parameters of its implementation (Figure 1).

Thus, the controlled movement of ERS ($\mathbf{Y}_3$) on an elementary section of the path (with certain parameters) is built on the basis of the implementation level of the operation of the EPS electric equipment ($\mathbf{Y}_2$), which generates algorithms and modes of operation of the automated EPS electric traction drive (climatic conditions) and weather conditions, etc.) in which the subprocess $\mathbf{Y}_3$ is built. When executing the EPS technology ($\mathbf{Y}_4$), the movement along the route network while maintaining the driving speed and the inter-train interval in difficult traffic conditions is implemented based on the subprocess of the EPS controlled movement level in the elementary segment, which is formed taking into account the EPS operation conditions. Similarly, the construction of motion for all levels.

The exchange of energy, material and information resources between the levels and the external environment is one of the main conditions for the implementation of the electric traction process in an open complex UET system [16, 17]. Obviously, this condition is strictly implemented only in the general case, and in real conditions, as a rule, only some interactions are obvious. So, for example, energy exchanges (electrical, thermal, mechanical energy, etc.), as well as material resources (including financial and personnel) are obvious [18]. In describing open complex systems, the definition of "information as measures of Order" is used in the synergetic methodology, as opposed to the notion of entropy as a measure of Chaos [19]. By the exchange of information resources, we will understand any interaction leading to the "increased Order" of processes and the system as a whole. It can be various restrictions, conditions and requirements for the parameters of the processes and phenomena (different for each level (subprocess)), expressed also in the regulations, instructions, norms, etc. all that limits ("orders") the numerous multitude ("Chaos") of the options for the implementation of a process or its structure. Conventionally, the structural diagram of the exchange of energy $\boldsymbol{R_E}$, material $\boldsymbol{R_M}$ and information $\boldsymbol{R_I}$ resources between the levels of building electric traction and the external environment is shown in Figure 1.

Conclusion

The theory of electrical equipment requires the development of ideas about the traction process, because It is the modes of operation of the feasibility study that unambiguously determine the modes of movement of the ERS, and electric traction is the point of application for managing the production and technical process of the UET, which is not only a scientific, but also an important practical problem.

In the course of a theoretical study based on the proposed hypothesis, the authors formulated a scientific concept representing a technological process of electric traction, the main task of which is to transform the electrical energy of a source into a targeted mechanical movement of ERS, a set of hierarchically aligned subprocesses, each of which is in the hierarchical subordination of a higher level process and forming the conditions for the implementation of the subprocess of the subordinate, decides on a specific the task of building the movement.

The presented theoretical system will make it possible to develop the general theory of UET electrical equipment by improving the methodological principles of the principles of forming the TEE operation modes when implementing the technology of the transportation process in the UET production and technical system. The identified structural and functional relationships of the implementation process of the feasibility study, which characterize the interaction architecture with other technical, technological and production processes ensuring the implementation of electric traction, will allow determining rational modes and methods of controlling automated traction electric drive, which will solve the problem of increasing energy efficiency in the UET system.

References

IOP Conf. Series: Materials Science and Engineering **570** (2019) 012041 doi:10.1088/1757-899X/570/1/012041

[1] Efremov I S and Kosarev G V 1976 *Theory and calculation of electrical equipment of the rolling stock of urban electric transport* (Moscow: Higher School) p 480

[2] Sleptsov M A, Dolaberidze G P and Prokopovich A V 2006 T*he Basics of Electric Transport* (Moscow: Research Center «Academiya») p 464

[3] Biryukov V V and Porsev E G 2018 *Electric Traction Drive* (Moscow: Research Center "Yurait") p 315

[4] Isaev I P and Golovaty A T 1992 *Rules of traction calculations need to be revised. Lokomotiv* (Moscow: Russian Railways) No 8 pp 6-8

[5] Eiduks Ya 1993 Methods of traction calculations need to be specified Railway of Latvia (Riga: Transport) No 50 pp 8-9

[6] Osipov S I, Gorchakov E V and Isaev I P 1984 *Rational train driving modes and testing of locomotives* (Moscow: Transport) p 280

[7] Rosenfeld V E, Isaev I P and Sidorov N N 1983 *Theory of electric traction* (Moscow: Transport) p 328

[8] Aukhadeev A E 2006 *Modeling of thrust modes of electric power electric transport (Kazan: KGEU) p 176*

[9] Rylov Yu A, Aukhadeev A E, Mukimov A H and Karimov V I 2017 Realization of electric traction in the production and technical system of urban electric transport Modern scientific research and development (Moscow: Science center "Olympus") No. 8 (16) p 487 -491

[10] Hakimov E M 2007 Dialectic of hierarchy and non-hierarchy in philosophy and scientific knowledge (Kazan: FEN) p 288

[11] Rylov Yu A, Aukhadeev A E, Grishina V V, Salikhova A R and Rashitova R A 2017 *Specificity of electric traction of urban transport Herald of Contemporary Studies* (Omsk: Science Center «Orka») No 11-1 (14) pp 250-253

[12] Aukhadeev A E, Litvinenko R S, Rylov Yu A and Hairullin A G 2018 City electric transport from the standpoint of synergetic methodology World of Transport and Technological Machines (Orel: OSU) No. 1 (60) p 67-73

[13] Bernstein N A 1947 *On the construction of movements* (Moscow: Medgiz) p 234

[14] Platonov A K 2010 *On the construction of motions in ballistics and mechatronics* Applied mechanics and motion control (Moscow: Keldysh Institute of Applied Mathematics) pp 127–222

[15] Sopov V I and Shchurov N I 2013 *Electric power supply systems for direct current electric transport* (Novosibirsk: NSTU) p 728

[16] Haken G 1980 *Synergetics, trans.* from English by Emelyanova V I, ed. by Yu L Klimontovich and S M Osovts (Moscow: World) p 423

[17] Aukhadeev A E 2014 *Self-development of the transport system of the modern city: Search for an innovative model of intellectual management* (Moscow: VINITI RAS) p 220

[18] Deev V V, Il'in G A and Afonin G 1987 *Traction trains* (Moscow: Transport) p 264

[19] Prigozgin I and Stengers I 1986 Order from Chaos: A New Dialogue between Man and Nature, trans. from English by Yu A Danilov, ed. by V I Arshinov, Yu L Klimontovich and Yu V Sagnova (Moscow: Progress) p 432

IOP Conf. Series: Materials Science and Engineering **570** (2019) 012042 doi:10.1088/1757-899X/570/1/012042

Consideration of the influence of the reliability of elements of the urban electric transport system on its bandwidth

N F Kashapov[1], L S Sabitov[1], R S Litvinenko[2], A E Auhadeev[2], and P P Pavlov[2]
[1]Federal State Autonomous Educational Institution of Higher Education "Kazan (Volga Region) Federal University", Kazan, Kremlevskayastreet, 18, 420008, Russian Federation
[2] Federal State Budgetary Institution of Higher Education "Kazan State Power Engineering University ", Kazan, Krasnoselskaya Street, 51, 420066, Russian Federation

l.sabitov@bk.ru

Abstract. The article proposes an approach to assessing the capacity of urban electric transport, which is one of the criteria for a stable resource supply of electric traction, taking into account the reliability of the elements of its infrastructure. By introducing the operational readiness factor as an indicator of reliability, the researcher has the opportunity to assess the impact of failures of these components on the efficiency of the electric traction process, as well as to compare the compliance of the transport line infrastructure with the characteristics of the rolling stock used.

Introduction

Capacity is one of the criteria for a stable and reliable resource support of the electric traction process in the urban electric transport system (UETS). For example, the task of guaranteed energy supply of passenger traffic is solved even at the design stage of a traction power supply system, while providing a reserve for increasing the size of traffic, and, therefore, sets the maximum capacity, taking into account the characteristics of the UET traction electrical equipment [1, 2].

The analysis of the existing scientific and methodological apparatus for estimating the throughput of the UET, and its comparison with the methodology for calculating the throughput of railways [3], allows us to make a number of observations:

1. The existing methods for calculating the capacity of route transport lines are rather outdated and do not take into account the current state of the elements of the urban electric transport system (UETS). Publications on this issue refer to the 60-80s of the last century. Studies of the throughput of railways, in contrast to the UET, were updated in the early 2000s, in particular, in 2010, the Instruction on the calculation of the available throughput of railways was approved [3].

2. As a temporary unit of measurement of the throughput capacity of UET transport lines, hours are taken, and the carrying capacity is respectively the transported number of passengers per hour. Since the passenger traffic in the UET S networks varies by the hour of the day, days of the week, months and seasons of the year, taking one hour as a time interval is not correct. It is necessary to evaluate these indicators at least for the daily period in order to increase the accuracy and reliability of the results obtained.

IOP Conf. Series: Materials Science and Engineering **570** (2019) 012042 doi:10.1088/1757-899X/570/1/012042

3. Studies carried out in VNIIZhT in the 60-70s of the last century showed that the level of reliability of technical equipment (locomotives, cars, track devices, signaling systems, etc.) influences the use of railway capacity. The growth of failures led to the fact that they began to be taken into account when calculating the carrying capacity of the corresponding reliability factor$\alpha_н$ [3, 4]

$$N_{nal} = \frac{(1440 - t_{tec\,h})\alpha_n}{I},$$

where$t_{tec\,h}$ – the duration of the free time from the train, provided for the work on the current maintenance of the track, devices, structures, etc.); I – travel interval.

Similar to rail transport, the failures of the elements of the ETS and the infrastructure of the transport line have a significant impact on the efficiency of passenger traffic of [5, 6], so there is an objective need to take into account the effect of reliability on the throughput of the GETS. Consideration of this issue should be carried out in a complex, taking into account the interaction and mutual influence of all components of the passenger transportation process.

Approach to estimating the throughput of the elements of the UETS taking into account reliability.

Approach to estimating the throughput of the elements of the UETS taking into account reliability. Since the reliability of stopping points and intermediate stations of transport lines, as engineering structures, with proper maintenance does not have a significant impact on the efficiency $\alpha_н$ of electric transport, the reliability factor in assessing the throughput of such transport infrastructure elements may not be taken into account. The components of the contact network and the track that are located within the stations (points) must be taken into account when forming the reliability coefficient of the corresponding span.

The expressions for estimating the throughput of the elements of the transport infrastructure of the UETS in the daily time interval, taking into account the influence of failures, with the exception of stopping points and intermediate stations, take the following form

UET ground line carrying capacity

$$P_p = \frac{3600\,V(24 - t_{tec\,h})\alpha_n}{t_p V + \frac{V^2}{2b} + l + l_0};$$

where$V$ - movement speed on the stretch, m / sec;

t_p – driver response time, sec;

b – deceleration during emergency braking, m / sec^2;

l – ngth of electric rolling stock (EPS), m;

l_0 – safety clearance between successive transport units, m

the capacity of the stopping points of the land line UET

$$P_{o.p} = \frac{3600(24 - t_{tec\,h})}{t_1 + t_2 + t_3 + t_4};$$

wheret_1 – времяime spent by EPS for approaching the stopping point, sec;

t_2 – duration of disembarking and landing of passengers, sec;

t_3 – duration of the announcement of the driver about stopping the landing and closing the doors;

t_4 – the duration of the release of the stopping point, s.

thethroughput of regulated land crossings UET

$$P_{reg} = \frac{3600\,(t_з - \Delta t)(24 - t_{tec\,h})\alpha_p}{T_c t_p};$$

where$t_з$ – duration of the green phase of the traffic light, sec;

Δt - the time between turning on the permissive signal of the traffic light and crossing the stop line by the first vehicle;

$t_p = \frac{l_p + l_b}{V_y}$ - the estimated interval of passage of vehicles through the intersection, sec;

V_y - set speed of EPS at intersection;

T_c - traffic light control cycle duration.

bandwidth intersection with a forced control system for a trolley bus

$$P_{pr.reg} = \frac{1200(t_з - \Delta t)(24 - t_{tec\,h})\alpha_n}{T_c t_p};$$

crossing capacity with tram control system

$$P_{pr.reg} = \frac{3600K(24 - t_{tec\,h})\alpha_n}{T_c};$$

tram traffic capacity

$$P_{Yz} = \frac{3600V(24 - t_{tec\,h})\alpha_n}{(l_p + l_k) + t_v V_{c.p} + \frac{V_{c.p}^2}{2b}};$$

wherel_k – distance between serial and shunt contacts;

t_v – driver response time and device response;

$V_{c.p}$ – the speed of passage of the section with a turnout.

Underground driving capacity

$$P_p = \frac{3600(24 - t_{tec\,h})\alpha_n}{t_v + t_{ct} + t_{by} + t_{zy} + t_{cp} + t_p + t_{rez}};$$

whereI_p – interval between metro trains on the stretch;

t_v – time of signal perception by locomotive crew;

t_{ct} – time of service braking, during which the metro train travels the distance from the start of braking to the traffic lights, which changed the signal from prohibiting to allowing;

t_{by} – block time;

t_{zy} – transit time;

t_{cp} – the response time of the alarm, centralization and blocking devices (ACB) from the moment of the departure of the previous EPS from the protective area to the opening of the traffic light;

t_p – the time required for the entire composition to go beyond the site, equal to the length of the metro train;

t_{rez} – reserve (reserve) of the burning time of the traffic light permitting signal, taking into account the accuracy of the metro train.

bandwidth of an intermediate metro station

$$P_{ct} = \frac{3600(24 - t_{tec\,h})}{t_v + t_{ct} + t_{pod\,h} + t_{ost} + t_{yx} + t_{cp} + t_p + t_{rez}}.$$

where$t_{pod\,h}$ – the time required for a metro train to go from the traffic light to its full stop at the platform;

t_{ost} –train stop time on the main tracks of the station;

t_{yx} – time from the moment of starting the metro train after parking to leaving the security area behind the exit traffic lights.

To bring the hourly capacity to the daily interval, a multiplier $(24 - t_{tec\,h})$was introduced, where $t_{tec\,h}$is the duration of the daily time budget allocated for the production of planned work on the current maintenance of the track, devices and facilities. As a rule, these activities are performed during a night break in the work of the UET and range from 4 to 6 hours. In contrast to the railway, to carry out these activities in the UET system, there is no need to look for the possibility of providing technological "windows" in the EPS schedule, which greatly facilitates the work of repair teams. At the same time, carrying out unplanned repairs to eliminate sudden failures of the UET infrastructure can lead to stopping the passenger transportation process at the site, while the railway transport can change the route in the case of freight traffic, or eliminate the backlog due to the subsequent increase in speed.

Consider what physical meaning is embedded in the concept of reliability coefficientα_n. In the Instructions for calculating the available capacity of railways, the reliability coefficientα_n is a coefficient taking into account the reliability of technical equipment (infrastructure and rolling stock), and given fixed values depending on the number of main tracks (single-track or double-track lines) on

IOP Conf. Series: Materials Science and Engineering **570** (2019) 012042 doi:10.1088/1757-899X/570/1/012042

a stretch and type of traction (electric or diesel traction). The author "7" as the indicator of reliability $R(t)$of technical means of railway transport uses the following expression:

$$R(t) = \text{К}_g P(t), \qquad (1)$$

where$\text{К}_\text{г}$ - availability factor (the probability that at any time the technical system is operational and ready to work);

$P(t)$ – probability of failure-free operation of the technical system;

t – duration of a given time interval.

In terms of the theory of reliability of technology, there is no such indicator as the reliability coefficient, but based on formula (1) and GOST [8], it can be concluded that the coefficient of operational readiness is used as an indicator of reliability of technical means of railway transport, since under certain conditions it is a product of availability and probability of trouble-free operation. Under the coefficient of operational readiness of equipment [8], refers to the probability that the product at a given time $t_1$is in working condition and, from this moment, perform the desired function under the given conditions in the interval(t_1, t_2). In connection with the above, the reliability coefficient $\alpha_\text{н}$used in estimating the throughput capacity of the GET transport line will be understood as operational availability$\text{К}_{o.g.}(t_1, t_2)$ factors corresponding to the elements of the GETS transport infrastructure.

In [5, 6, 9, 10] it was proposed to consider the UETS from the standpoint of a systematic approach as a complex technical system consisting of a large number of elements combined into the following subsystems directly involved in the transmission and conversion of electrical equipment into mechanical work of rolling stock environmental conditions:

traction rolling stock subsystem (fleet of traction units (TU);

path subsystem;

traction power supply subsystem (traction substations and contact network).

The proposed systematic approach to the consideration of the UETS makes it possible to solve problems of improving the functioning with regard to the interaction of its components, since, improving only the elements of traction rolling stock without upgrading the supporting transport infrastructure, it will not be possible to achieve an increase in any quantitative indicators of road traffic. It should be noted that some increase in the efficiency of the UETS is possible by increasing the qualification level of full-time specialists [11].

The figure shows a variant of the UETS decomposition using tramcars as traction units for the purpose of subsequent analysis of its reliability and assuming four levels in the structure of the system: the UETS as a whole is located at the initial level, the subsystems of the electric rolling stock, power supply and rail network; the second level is represented by electrical and engineering systems that are part of the subsystems of the 1st level; on the third (lower) level there are elements of the complexes (traction electrical equipment and mechanical equipment).

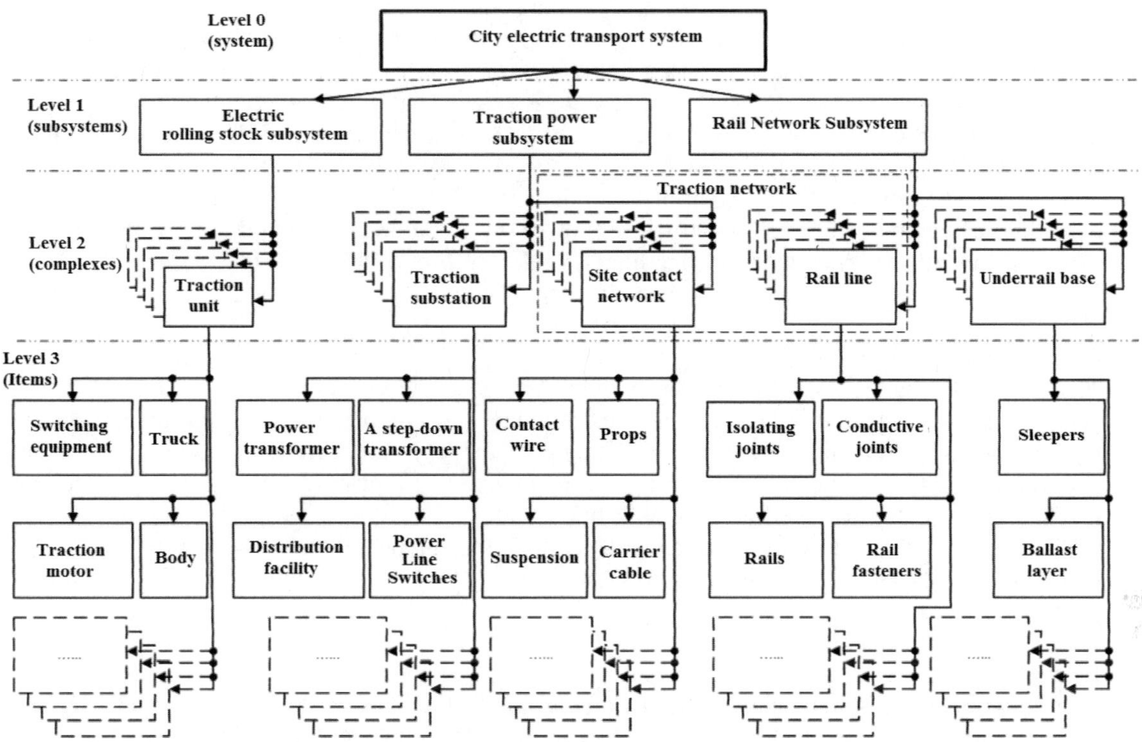

Figure 1. - Decomposition of the urban electric transport system
(option when equipping trams with cars)

Taking into account these provisions, and in order to take into account the influence of the level of reliability of various components of the transport infrastructure on the throughput capacity of the UETS transport line, the formulas for determining the reliability coefficient for various types of UET will take the following form

for trolleybus line

$$\alpha_n^{troll}(t_1,t_2) = \text{K}_{o.g.}^{troll} = \prod_{i=1}^{2} \text{K}_{o.g.i}(t_1,t_2),$$

for tramway line

$$\alpha_n^{tram}(t_1,t_2) = \text{K}_{o.g.}^{tram} = \prod_{i=1}^{4} \text{K}_{o.g.i}(t_1,t_2),$$

for subway

$$\alpha_n^{metro}(t_1,t_2) = \text{K}_{o.g.}^{metro} = \prod_{i=1}^{5} \text{K}_{o.g.i}(t_1,t_2),$$

where

$\text{K}_{o.g.1}(t_1,t_2)$- coefficient of operational readiness of traction substations;

$\text{K}_{o.g.2}(t_1,t_2)$- coefficient of operational readiness of the contact network section;

$\text{K}_{o.g.3}(t_1,t_2)$ - the operational readiness ratio of the rail line section;

$\text{K}_{o.g.4}(t_1,t_2)$ - the operational readiness ratio of the area under the rail base;

$\text{K}_{o.g.5}(t_1,t_2)$- the operational readiness ratio of the used signaling systems.

The reliability indicators used are currently regulated for railway transport and its infrastructure [12-15], but, taking into account the peculiarities of operation, they can be used for UET. Thus, using the operational availability factor as the reliability coefficient of the corresponding components (contact network, traction substations, rail line and rail base) of the electric power exchange system located within the transport line elements (spans, intersections, nodes, etc.) can be corrected the amount of available capacity of passenger routes.

IOP Conf. Series: Materials Science and Engineering **570** (2019) 012042 doi:10.1088/1757-899X/570/1/012042

Similarly, to assess the impact of technical failures on the carrying capacity of the railway section at work [7], the allocation of five farms in the structure that ensure the functioning of the transportation process was proposed. Due to the specific features of the operation of the UET, the subsystems proposed by the authors of this article are somewhat different from the farms of railway transport described in [7], and their number varies depending on the type of EPS. Also, the author of [7] presented separately the locomotive and wagon economy combined into one to consider the reliability of the train's technical equipment. In the case of a UET, it is proposed to consider the impact of failures of the transport line infrastructure (power supply subsystems and rail network) on its carrying capacity, and the reliability (readiness) of the EPS substation from the standpoint of its impact on the carrying capacity of the UET in general.

Conclusions

Thus, by analyzing the impact of reliability of all subsystems in the composition of the UETS on throughput and carrying capacity, you can determine its place in the structure of urban passenger traffic. The proposed approach can be used to identify the least reliable elements, both traction EPS and transport infrastructure, affecting the process of passenger transportation UETS, and if necessary, take measures to improve their efficiency. When creating new passenger routes or upgrading old ones, the described method allows one to practically evaluate the predicted traffic indicators and the compliance of the elements of the infrastructure of the transport line with the capabilities of the EPS planned for use.

Literature

[1] Kisneeva L N, Auhadeev A E, Tukhbatullina D I иEgorova P V 2018 *Energy supply of the electric traction process of the urban electric transport* (Moscow: Modern Science) No 12-1pp 19-21

[2] Rylov Yu A, Aukhadeev A E, Babakuliev R Yu иFazylov S G *Analysis of the motion modes of urban electric transport electric rolling stock* (Moscow: Modern Science)No 1-1 pp 50- 53

[3] Borodin AF 2011 *Instructions for calculating the cash capacity of railways*, (Moscow: Russian Railways) p 305

[4] Levin D Yu and Pavlov VL 2011 *Calculation and use of the throughput capacity of railways* (Moscow: Federal State Educational Institution "Training and Methodological Center for Education in Railway Transport") p 364

[5] Litvinenko RS, Aukhadeev A E and Filina OA 2017 *Investigation of the technical reliability of the urban electric transport system* (Moscow: Transport: science, technology, management) No 8 p 60-71

[6] Pavlov PP, Idiyatullin RG and Litvinenko RS 2017 *On the issue of assessing the reliability of the electric transport system of the city* (Moscow: Transport Information Bulletin) No 5 (263) p 23 - 26

[7] Ivnitsky VA 2008 *Reliability of technical means of railway transport and its connection with the capacity of the directions.* (Moscow: Bulletin of VNIIZhT) p 6- 9

[8] GOSTR 53480 2009 *Reliabilityinengineering. Terms and definitions* (Moscow: Standardinform) p 33

[9] Kisneeva L N, Aukhadeev A E, Tukhbatullina D I иStaroverova U V 2017 *Description of urban electric trance as a complex technical system* (Moscow: Modern Science) No 11 pp 181-183

[10] Rylov Yu A, Aukhadeev A E, Egorova N Yu иUrzhumtsev P S 2017 *Electric traction in urban electric transport system* (Moscow: Modern Science) No 11 pp 195 – 197

[11] Khusnutdinova E M, Khamidullina G R, Pavlov P P, Aukhadeyev A E, Litvinenko R S and Khusnutdinov A N 2018 *Training transport specialists based on a contemporary view of self-organization of sophisticated developing systems* (IOP Conference Series: Materials Science and Engineering) 412(1) **012047**

[12] GOST R 55444 2013 *Railway power supply. The nomenclature of reliability and functional safety indicators* (Moscow: Standardinform) p 14

[13] GOST R 55443 2013 *Railway track. The nomenclature of reliability and functional safety indicators* (Moscow: Standardinform) p 12
[14] GOST R 54461 2011 *Reliability of railway traction rolling stock. Terms and definitions* (Moscow: Standardinform) p 20
[15] GOST 32192 2013 *Reliability in railway engineering. Basic concepts.Terms and Definitions*.(Moscow:Standardinform) p 32

IOP Conf. Series: Materials Science and Engineering **570** (2019) 012043 doi:10.1088/1757-899X/570/1/012043

The approach to the study of the reliability of the electric transport system of the city as a complex technical system

N F Kashapov[1], L S Sabitov[1], R S Litvinenko[2], A E Auhadeev[2], and I Z Gatiyatov[2]
[1]Federal State Autonomous Educational Institution of Higher Education "Kazan (Volga Region) Federal University", Kazan, Kremlevskaya street, 18, 420008, Russian Federation
[2] Federal State Budgetary Institution of Higher Education "Kazan State Power Engineering University ", Kazan, Krasnoselskaya Street, 51, 420066, Russian Federation

l.sabitov@bk.ru

Abstract. The currently existing methods for analyzing and assessing the reliability of a technical system are mainly focused on its individual elements and do not take into account the complexity of their interactions and integration into subsystems of various hierarchical levels. When considering the issues of reliability of electric vehicles, they mainly focus on individual components: the reliability of electric traction rolling stock, the reliability of the power supply system, the reliability of the track infrastructure, etc. The article proposes an approach that allows to consider urban electric transport as a complex technical system consisting of several levels, each of which contains components directly involved in the transmission and conversion of electric power into mechanical work of rolling stock under the influence of the environment. Such an approach will allow, in particular, to assess the impact of the reliability of such individual components on the efficiency of implementation of electric traction in a complex technical system of urban electric transport.

Introduction

In the system of intracity passenger transport, such types of electric transport as the trolleybus, tram and metro are mainly used due to their high productivity and environmental friendliness. The main feature of such transport is the use of electric traction, which provides transmission and conversion of electric power by traction electrical equipment (TEE) into mechanical work of rolling stock through the provision of technological and production tasks of the transport process of passenger transportation and exposure to the external environment [1, 2].

Under the electric transport system of the city (ETSC), in terms of its technical implementation, we will understand the production functioning and developing system, which is a combination of electric vehicles, transport infrastructure, transport enterprises and management systems aimed at the most complete satisfaction of the changing needs of the urban population in time and space on trips inside this city.

ETSC in the complex can be considered as a complex technical system (CTS) [1, 3]. In accordance with [4], by the system we mean a set of elements defined in time and space with known properties and with ordered connections between the elements and properties oriented towards the fulfillment of the main task of the given set. In some cases [5] use the term complex dynamic system - the formation of elements of different nature, which has functions and properties that are missing from each of the elements and is able to function, statically correlating in a certain range with the environment and, due to this, maintain its structure during continuous changes in the interactions of elements according to complex dynamic laws.

ETSC, like any other system, exists among other objects, united by the concept of "external environment". The objects of the external environment are connected with the elements of the system; however, this connection is significantly weaker than the internal connections in the system. Elements are parts or components of a system conditionally accepted as indivisible. The internal structure and properties of an element are not the subject of research and study, only such properties are significant, which manifest themselves in the interaction with other elements and affect the properties of the system as a whole. The division of the system into elements is relative and depends on the task and objectives of the study; otherwise, in other aspects, the element can be considered as a CTS. The combination (integrity) of elements and properties is an essential feature of the system.

A special feature of the CTS is the possibility of dividing it into subsystems - a set of system elements that function independently [5]. In this case, the goals and objectives of the functioning of the subsystem itself, possessing some separateness, are subordinated to the general goals of the system.

Materials of theoretical study

The possibility of the full and effective use of ETSC for the implementation of the passenger transportation process largely depends on the technical condition of its elements and is largely determined by the reliability of their operation. In accordance with [6], reliability is understood as the property of readiness and reliability and repairability properties affecting it, and maintenance support. Previously, the definition of this term [7] was the following: reliability is the property of an object to preserve, within the established limits, the values of all parameters characterizing the ability to perform the required functions in specified modes and conditions of use, maintenance, storage and transportation. Reliability is a complex property, which, depending on the purpose of the object and the conditions of its use, may include reliability, durability, maintainability and persistence, or certain combinations of these properties. The effectiveness of its application largely depends on the reliability of the equipment - the ability to meet the requirements for the service with given quantitative characteristics [6], which depends on a combination of capabilities and readiness for use. Therefore, ETSC should be considered as a single CTS, consisting of a large number of different elements that interact to ensure an efficient passenger transportation process.

Analysis of the domestic literature allows us to conclude that the consideration of the issues of reliability of ETSC is mainly focused on the assessment of the reliability of its constituent elements and does not imply their synthesis into a single CTS. Failures of individual elements do not lead to the loss of efficiency and efficiency of the entire ETSC, but the pace and quality of the performance of the specified functions is reduced, so an objective need arises in assessing the reliability of the entire system as a whole. The following main areas of research on the operational reliability of ETSC elements can be distinguished:

- reliability of electric rolling stock [8 - 11];
- reliability of the traction power supply system [12, 13];
- reliability of the track route [14 - 16].

In addition to the above sources, there is a large amount of literature devoted to the reliability of individual units, assembly units and assemblies of different physical nature and purpose, which are used both as separate elements and as part of technical devices. However, when evaluating the reliability of ETSC, a systematic approach is needed, which involves first calculating the reliability

IOP Conf. Series: Materials Science and Engineering **570** (2019) 012043 doi:10.1088/1757-899X/570/1/012043

indices of simpler subsystems and elements, and then grouping the results obtained in a certain way in order to obtain the characteristics of the whole system.

For the implementation of the process of passenger transportation in the structure ETSC allocate the following technical means [1, 3]:

electric rolling stock (ERS);

track facilities and roads that form the transport network;

plants and depots providing maintenance and repair of ERS;

power supply devices;

communication and blocking devices;

means of organizing and controlling traffic.

Each part of this complex represents a subsystem, and their combination forms a CTS, which is called the urban electric transport system. Figure 1 shows the interaction between these subsystems.

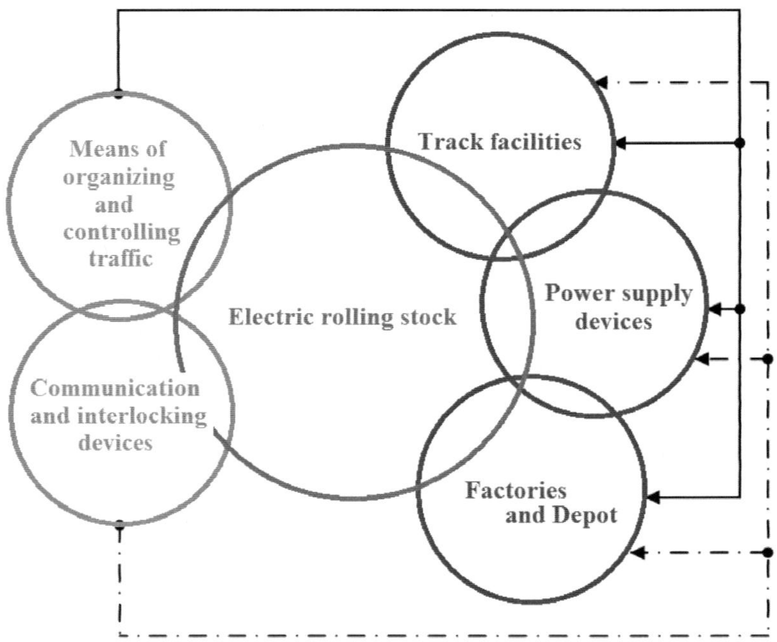

Figure 1 - Diagram of the interaction of subsystems ETS

With reference to the tasks of the ETSC reliability study, as a CTS consisting of a large number of elements, it is proposed to use the following subsystems directly involved in the transmission and transformation by means of a feasibility study of electric power into the mechanical operation of rolling stock under environmental conditions:

traction ERS subsystem;

path subsystem;

traction power supply subsystem.

Further specification and formation of additional subsystems in the composition of the ETSC is not always justified in the analysis of reliability as:

firstly, the required level of reliability of the equipment is laid at the development stage, and then, through proper operation and timely maintenance, it can only be maintained. It is possible to increase the achieved level of reliability only by upgrading the technical object itself. In this regard, it is possible to omit consideration of the issues of reliability of the means of organizing and controlling traffic, since the issues of efficient use of ETS, in this context, belong to the field of theories of electric traction and transport logistics;

secondly, in spite of the fact that ERS is a renewable CTS, the reliability of plants and depots providing maintenance and repair of ERS does not affect the reliability of the system as a whole, but rather depends on the professionalism and training of personnel [17]. Therefore, in the formation of mathematical models, it is possible to accept the assumption that the restoration of the failed elements is carried out within the deadlines set by the regulatory and technical documentation;

thirdly, the mathematical model of any CTS should not be "overloaded" by taking into account all sorts of factors, as this leads to complication of calculations, and often to distortion of the results and the inadequacy of the model to the original. Therefore, when forming the structure and composition of ETSG, it is necessary to be guided by the necessary completeness and reliability of a certain part of the properties (in our case, reliability) of the original, which is essential precisely in this study and in the given formulation of the problem.

Using a systematic approach to assess the impact of reliability on the efficiency and quality of ETSC, it is advisable to apply a structural decomposition method, the essence of which is to divide the system under study into subsystems (components) of a lower hierarchy level, which will replace the solution of one large task with a solution of a series of tasks of less complexity. Figure 2 presents a variant of the ETSC decomposition for the purpose of the subsequent analysis of its reliability, assuming the selection of four levels in the system structure.

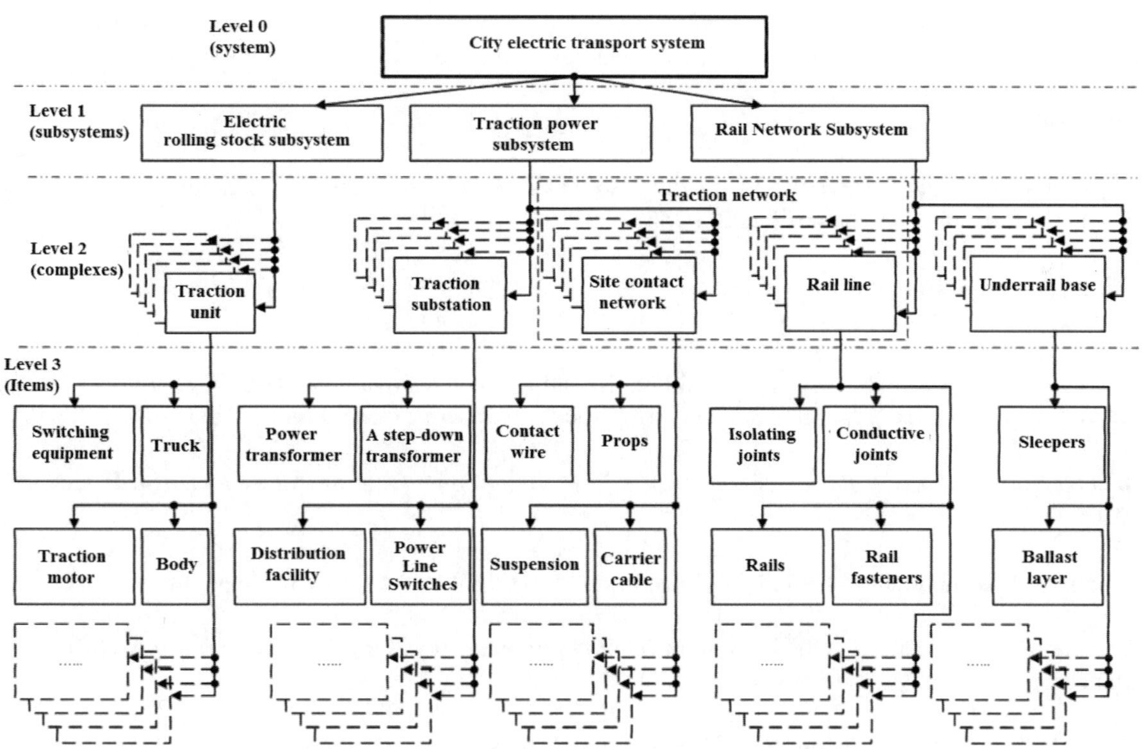

Figure 2 - Decomposition of the electric transport system of the city (option)

ETSC as a whole is located at the initial level; subsystems of electric rolling stock, traction power supply and rail network are located at the first level (in the case of use as a ERS, tram or subway cars); the second level is represented by electrical and engineering systems that are part of the subsystems of the 1st level; on the third (lower) level are the elements of the complexes (TEE and mechanical

IOP Conf. Series: Materials Science and Engineering **570** (2019) 012043 doi:10.1088/1757-899X/570/1/012043

equipment). It is also possible to select components of a lower level in the ETSC elements - assembly units, assemblies, units; however, as mentioned above, the reliability assessment of these technical means (devices) has been studied and studied sufficiently earlier. Therefore, such a degree of detail of the description and the number of levels of the hierarchical structure of ETSC, such as the CTS, can be considered sufficient and convenient for perception, and also corresponds to the modern level of knowledge of a specialist in the field of reliability.

Conclusion

Thus, the task of analyzing ETSC as an object of research into reliability comes down to assessing the reliability of its subsystems, complexes and elements, and the subsequent synthesis of the results obtained. This approach is possible both at the stage of creating a new ETS in a city that uses E for the first time, and in relation to existing megacities in the process of modernizing transport routes by introducing new sections. The quality of the operation of ETSC is estimated by the performance indicator, hence the task of assessing the reliability of the system is reduced to determining its effect on the system's efficiency. For these purposes, it is convenient to use a complex indicator - the coefficient of efficiency preservation. In accordance with [18], the coefficient of efficiency K_{ef} retention is the ratio of the value of the indicator of the effectiveness of a product for a certain period of operation $W(t)$ to the nominal value of this indicator $W_0(t)$ calculated under the condition that no product fails during this period.

Calculated in such a way K_{ef}, it will be possible to estimate the deviation of the calculated parameters of the efficiency of the ETSC functioning from the nominal value, taking into account the reliability of its constituent elements, and, if necessary, take measures aimed at raising it

Literature

[1] Shchurov N I 2003 *Methods and means of saving and increasing energy efficiency in the system of urban electric transport* (Novosibirsk) p 385

[2] Rylov Yu A, Aukhadeev A E, Egorova N Yu и Urzhumtsev P S 2017 *Electric traction in urban electric transport system* (Москва: Modern Science) № 11 c 195 – 197

[3] Kisneeva L N, Aukhadeev A E, Tukhbatullina D I и Staroverova U V 2017 *Description of urban electric trance as a complex technical system* (Москва: Modern Science) № 11 c 181 – 183

[4] Gaykovich A I 2001 *Basics of the theory of designing complex technical systems* (St. Petersburg: "MORINTECH") p 432

[5] Sopov V I and Shchurov N I 2005 *Modeling of electric transport systems* (Novosibirsk) p 189

[6] GOST R 27.002 - 2009. *Reliability in engineering. Terms and definitions* (Moscow: Standardinform) p 27

[7] GOST 27.002-89. *Reliability in technology. Terms and Definitions.* (Moscow: Publishing Standards) p 37

[8] Prokhorov A V 2010 *Reliability of the electrical insulation system of traction motors* (Moscow: Energy Security and Energy Saving) No. 2 p 21-25.

[9] Bachurin N S 2009 *Methods for assessing the reliability indicators of a tram car* (Ekaterinburg: Ural Transport) No. 1 (20) p 36 - 39

IOP Conf. Series: Materials Science and Engineering **570** (2019) 012043 doi:10.1088/1757-899X/570/1/012043

[10] Mikhailov I Yu, Muravlev O P and Fedyanin A L 2016 *Engineering analysis of the operational reliability of trolleybus electrical equipment* (Kemerovo: Bulletin of Kuzbass State Technical University) No. 3 (115) p 85-92.

[11] Hudonogov A M, Olentsevich D A, Lytkina E M and Ivanov V N 2008 *Reliability of asynchronous auxiliary machines for electric locomotives* (Irkutsk: Bulletin of Irkutsk State Technical University) No. 4 (36) p 117-119

[12] P-639 2005. *Recommendations on the main provisions of the study of the reliability of the contact network devices. Organization of Cooperation of Railways* (Warsaw) p 16

[13] Anischenko V A, Kolosova I V 2007 *Reliability of power supply systems* (Minsk: BNTU) p 151

[14] Lysyuk V S and Kamensky V B 2001 *Reliability of the railway track.* (Moscow: Transport) p 286

[15] Lysyuk V S and Semenov V T 2003 *Robust and reliable railway* (Moscow: Academkniga) p 589

[16] Shchepotin G K 2008 *Operational reliability of a railway track* (Ekaterinburg: UrGUPS Publishing House) p 144

[17] Khusnutdinova E M, Khamidullina G R, Pavlov P P, Aukhadeyev A E, Litvinenko R S and Khusnutdinov A N 2018 *Training transport specialists based on a contemporary view of self-organization of sophisticated developing systems* (IOP Conference Series: Materials Science and Engineering) 412(1) 012047.

[18] GOST R 53480 - 2009 *Reliability in engineering. Terms and Definitions.* (Moscow: Standardinform) p 33

IOP Conf. Series: Materials Science and Engineering **570** (2019) 012044 doi:10.1088/1757-899X/570/1/012044

Modern problems of digitalization of agricultural production

N F Kashapov[1] M M Nafikov[1] M KH Gazetdinov[2] SH M Gazetdinov[2] A R Nigmatzyanov[3]
[1]Kazansky (Volga region) federal university. Kazan, Russia
[2]Kazansky state agrarian university. Kazan, Russia
[3]Tatar Institute of Agribusiness Personnel Retraining. Kazan, Russia

arnig76@yandex.ru

Abstract. At the present stage, the most promising form of digitalization of production and management in agriculture is the formation of an integrated digital production complex as an organic merger of technical means of digitalization of production on the basis of precision farming, artificial intelligence management, as well as information and analytical management system. The availability of such a powerful tool as an information and analytical system as part of an integrated digital production complex in combination with the most digitized production processes can simultaneously meet modern requirements for increasing productivity and reducing costs for the production of agricultural products and food, and a sharp increase in the quality of products. It is argued that the functioning of information and analytical systems in some large agricultural formations of the Republic of Tatarstan has provided a qualitative leap in the development of methods and means of digitalization of production and management. It became possible to integrate automated controls up to the level of agro-technological management without a gap between the organizational, economic and agro-technological levels. The main problems to be solved in the development of IAS IDPC are considered. Expert systems, which have been developing rapidly in recent years, seem to be a promising tool for solving such problems.

Introduction. Feature of the present stage of development of rural economics of the Russian Federation is need for modernization on the basis of introduction of all forms of innovations, use of advanced technologies in combination with set of organizational and economic measures. At the same time formation of new technological way which basis the digital technologies which are based on use of electronics, robotics, computer facilities, telecommunications, genetic engineering, etc. [1] are has special relevance. In this regard, in rural economics there are basic changes allowing to implement the concept of creation of digital economy in agricultural industry in recent years.

Research methods. The digital economy should be understood not only as set of means of digitalization of technical training of production, management of agroengineering procedures and organizational and economic management. Digitalization is including, means of increase in flexibility of production, ability of fast reorganization on creation of various coproductions of agro-industrial type, reorganization on broad development of various forms of integration and, first of all, at primary production level, at the level of agricultural enterprises of various forms of ownership and the organization [2]. At digitalization of agroengineering procedures it is necessary to consider that there are fundamental differences in management of engineering procedures for continuous and discrete productions. Methods and means of digitalization of management of continuous engineering procedures are not quite suitable for improvement of discrete productions. During creation of the information and analytical systems (IAS) of management it is required to consider these features as main form of digitalization directly of production for discrete productions [3].

Information technology development and the computer equipment which caused a possibility of development of the built-in control facilities processing equipment and also development of robotics which allowed to automate routine operations on loading and unloading of agricultural machinery and the equipment became a technical prerequisite of creation of IAS.

At this stage of development of economy in the industry of informatization and communication in the Republic of Tatarstan steadily gain gross revenue [4]. So, in 2018 it made 56,7 billion rubles and grew in comparison with 2017 by 1,8 billion rubles, or for 3,3%. The share of a gross income of the industry in a gross regional product of the republic made 3,1%. In general this indicator on the republic tends a steady rise which is described by the following equation:

$y = 19{,}6 + 3{,}4\,t, \quad R^2 = 0{,}941$

3,9 billion rubles are invested in development of fixed assets of the enterprises of the industry of informatization and communication of the Republic of Tatarstan in 2018 that is 25% more than in 2017. Since 2007, the volume of investment into capital investments of the industry of informatization and communications annually grows.

According to the Strategy of social and economic development of the republic till 2030 gross revenue of the industry will grow almost by 8 times to 389 billion rubles that will allow to increase a share of the industry of informatization and communication in a gross regional product of the Republic of Tatarstan to 7%. At the same time the share of information technologies in a gross income of the industry will grow to 75%, the volume of investment into the industry will be 68,6 billion rubles.

It is known that rates of innovative development of agricultural industry in the Republic of Tatarstan remain low, first of all, because of the low level of the technological equipment in many respects determined by technological and technological level of the industry and insufficient qualification of shots [5]. While international and European experience of conducting agricultural works is already directly connected with information technologies, in Russia this direction develops since the beginning of the 2000-th years, but for many reasons rates of its development remain low.

At the present stage of prospect of information technology development in agricultural industry are extraordinary high. In some large integrated formations of the Republic of Tatarstan such as JSC "HK "Ak Bars", CJSC "Agrosila Group", LLC "UK KV Agro" are successfully used new technologies of housekeeping [6] already today. At the moment it is, first of all, programs for calculation and optimization of diets of feeding and mixes of forages for various animals, software products for diagnosis of diseases of animals and crops, information systems of automation of operational accounting, the program for selection of animals, the geographic information systems, accounting information systems considering branch specifics, complex enterprise management systems [7].

Analysis and discussion of results of a research. Introduction and use of the listed software products making IAS basis provided quantum leap in development of methods and means of digitalization of production and management. There was an ability to integrate the automated control facilities up to the level of agrotechnological management without gap between organizational and economic and agrotechnological levels.

At this stage in the integrated formations it is possible to allocate several directions of development of technologies and use of innovations:

– technologies of processing of the soil;
– technologies of cultivation and maintenance of the cattle;
– technologies of drainage and irrigation of the soil;
– technologies of collecting and preservation of products;
– technologies of transportation and product sales.

At the same time the convenient environment for storage and information processing is the cloud computing uniting in themselves license software, hardware, communication channels and also technical support of users [8]. All this provides to specialists of agricultural enterprises a possibility of connection to a system from any workplace in real time, excludes costs of transfer, a razarkhivation, updating of a databank. In LLC "UK KV Agro" in livestock production a common information space on IAS "Seleks" is realized that allows to exchange data through jobs of various services (veterinarians, livestock specialists, appraisers and others) and also to unload via web applications regional reports, breeding document flow, exhibitions, realization of the cattle on племторг.рф, etc.

In the same place the modern software product "Fodder Diets" allowing to make the best decisions in feeding of cattle for increase in profitability of herd is developed. Within this product it is realized: maintaining database of diets and premixes; creation of the database on nutritional value of forages; assessment of the actual diets; calculation of need for sterns; use at the choice of standards of need of animals for nutrients; the analysis and assessment of the received diets.

In CJSC "Agrosila Group"IAS "Veterinary science" which allows to store and use information on all levels of the organization of work of specialists of veterinary service is realized: from a way of introduction of medicine to a consumption of medicines in the enterprise; from diagnosis to one animal to an overall picture of incidence in herd; from planning of preventive processings before preparation of the annual reporting; from survey of a newborn calf to the analysis of the reasons of leaving of animals; from detection of a disease to treatment cost. Use of IAS "Veterinary science" allows to increase efficiency of management decisions in the organization of work of veterinary service and all livestock complex of agricultural enterprise.

Use of information technologies increases productivity and efficiency of administrative work, allowing to solve in a new way many problems [9].

"Exact agriculture" which main objectives are optimization of production for the purpose of receiving the maximum profit and rational use of resources, including natural and also environment protection becomes one of the relevant directions of use of information technologies in agricultural industry. "Exact agriculture" is considered as an integral part of resource-saving and environmentally friendly agricultural industry, it allows to provide control of performance of agrotechnological operations.

We will distinguish from such technologies which became irreplaceable elements on fields in recent years the systems of automatic control of the selkhozyaystvenny equipment, the systems of automatic control of application of fertilizers and quality control of field works, mapping of productivity and some other. Today the equipment equipped with the systems of automatic control carries out operations with high precision, sowing complexes are capable to turn off automatically crops of seeds on overlappings.

Use of modern digital technologies allows to obtain quickly reliable data about the executed amounts of works, about the reaped crop, the spent materials, about a condition of crops of various cultures, and on this basis to make the justified management decisions which directly influence a resulting effect of work of agricultural enterprise on location of a large number of various equipment.

For conducting "exact agriculture" in CJSC "Agrosila Group" use special devices and technologies on the basis of artificial intelligence. For example, for environmentally friendly fight against weeds the "ecorobotix" robot is used, having computer sight for definition of plants and manipulators for weeding, or the swarm of "prospero" robots consisting of small cheap robots who carry out the same task at the same time they communicate among themselves and easily replace some another.

When holding agrochemical actions the system of uniform introduction "Bars-5" which is intended for automatic adjustment of a consumption of liquid is used (fertilizers, means of protection of plants and other). This system is successfully combined with the system of parallel driving "the "Atlas 730" which keeps data of 9 fields, 12 various dosages, etc.

IOP Conf. Series: Materials Science and Engineering **570** (2019) 012044 doi:10.1088/1757-899X/570/1/012044

Now during the agrotechnical works the receivers antennas of global position systems (GPS or GLONAS) installed by any car, the unit, etc. are widely used. They take the bearings of signals from the satellites which are in a zone of reception of information.

The geographic information system (GIS), the software allowing to process and show spatial information, to computerize and make electronic cards is also widely used. The geographic information system allows to process and analyze various spatial data integrated in a digital form.

Except the listed systems when performing agrotechnological operations for actuating of executive parts of machine units onboard and remote sensors are used. By means of remote sensors take temperature, humidity of the soil, define contamination of the soil. Onboard sensors serve for adjustment of norm of seeding of seeds, application of fertilizers, toxic chemicals, water, lime and also for transfer of parameters of the machine unit and determination of its location.

Decision-making, for example, on need of additional application of fertilizers on the concrete site of the field will be based on information obtained by means of global position and geographical information systems, traditional sources and also on the basis of expert estimates of practicians and consultants. The last is especially important for discrete productions of seasonal nature as direct result of it is sharp increase in flexibility and productivity and also a number of the changes of social and economic character which are in details considered in extensive literature concerning creation of IAS.

So, now the most perspective form of digitalization of production and management in agricultural industry formation of the integrated digital production complex (IDPC) as organic merge of technical means of digitalization of production (TMDP) on the basis of the system of exact agriculture, artificial intelligence techniques (II) of management and also means of digitalization of organizational and economic management of IAS IDPC is represented.

In the conditions of IDPC a role organizational and economic the upravlekniya sharply increases, and contents it qualitatively changes in comparison with a traditional form of the organization of production. The special mark on realization of IAS as a part of IDPC is left by current trends in digitalization development - widespread introduction of personal computers, application of a cloud computing of information processing, approach of means of information processing to the end user. At the same time the reliability of information entered into the database increases, fuller and high-quality data processing what the high efficiency of performance of work is result of is provided.

Existence of IAS as a part of IDPC changes also the nature of ensuring production with technical means. The high performance, flexibility and continuity in production of agricultural products can be reached only on the basis of the high-performance robotic equipment which, in turn, is provided with creation of the automated jobs of the expert (ARM-Inzhener, ARM-Agronom, ARM-Zootekhnik, ARM-Ekonomist, etc.) equipped necessary tool to means: the base of resource-saving technologies, programs realizing the automated and robotozirovanny technologies for the direction "Exact Agriculture", "Clever Farm", "Smart Greenhouse", etc.

IDPC as subject to design, has essential specifics in comparison with other types of agrarian formations of various forms of managing created earlier in agricultural industry. These specifics, according to us, relatively do not depend on methods and means of digitalization of production and management and are expressed in the following features:

– complex nature of production of agricultural industry;
– flexibility of productions and management at all levels;
– existence of means of synchronization and coordination of functioning of the IDPC separate elements.

Proceeding from these specifics and also from those general trends in development of agricultural industry mentioned above methods and means of the organization and farm management at the present stage have to undergo certain changes [10].

It is obvious that a set of the functional means focused on creation of IDPC has to be significantly expanded in comparison with earlier developed information management systems the enterprise on the platform "1C: The enterprise" at the expense of tools for such IDPC elements, as a control system of TMDP, II.

Now development of methods and tools essentially for the IDPC new domestic competitive elements only begins. A necessary condition when carrying out these works is system approach to them and the agreement of design decisions on various elements giving the chance of their painless introduction in the future in a uniform system [11].

Its flexibility becomes the major requirement to functional providing IDPC which has to be satisfied during design of structure. At the level of TMDP and II the flexibility is understood as ability to adapt to change of parameters of production and agroengineering procedures and also to production of new types of agricultural products and their engineering procedures (flexibility of working off of management decisions) and can be provided with local means of this level. However the flexibility of functioning of IDPC in general (therefore, and efficiency of economic activity of an object of management) cannot be provided without flexibility at organizational economic level (flexibility of adoption of management decisions). This type of flexibility should be understood as ability of a system in general to adapt to change of a production situation, on the one hand, and to external indignations – with another. This flexibility also has to be provided in IDPC designing process as she significantly mentions methods and implementers of organizational and economic management.

Thus, the specifics of IDPC have significant effect not only on functional decisions, but also on the software of IAS IDPC. Ensuring management of agroengineering procedures and synchronization of events in IDPC assumes, in particular, functioning of IAS IDPC in real time. It means that during design of IDPC careful assessment of critical time intervals has to be made. The developed means have to provide the set time of reaction of a system for indignation that assumes optimization by the program of a system taking into account restrictions for time of performance of these or those operations.

Reliability of functioning of a system of data processing in general and each its element separately becomes extremely important requirement to the software of IAS IDPC. Therefore at design of IAS IDPC effective control over productions on the basis of digital subsystems of management of engineering procedures and receiving the quantitative indices guaranteeing its reliability has to be provided.

One of the most important features of IAS IDPC is approach of means of digitalization of management to the end user – administrative personnel, experts, workers and operators of service. Powerful tools of the interface have to be developed for the organization of effective interaction of the end user with a system (language and program).

During creation of means of the lingware of IAS IDPC there is a number of specific problems. Processing of independent inquiries to a system demands attraction of essentially new methods connected with opportunities of artificial intelligence techniques, first of all, with natural language processing. On the other hand, are required also professional specifics of concrete subject domain as IAS IDPC assumes realization of such tasks as development of the automated systems of collecting, the analysis of the wireless sensors given with application, microelectronic complexes with a digital format of processing and signaling, visualization of data from industrial devices the Internet of things, technologies with use of Big Data and neural networks, etc. focused language tool, reflecting.

Summarizing told above, it is possible to allocate the following main problems which are subject to the decision when developing IAS IDPC:

– design of functional providing IAS IDPC perspective elements;
– ensuring flexibility of production;
– ensuring ability of a system to adaptation at the level of organizational and economic management;
– a possibility of modification and maintenance of a system in the course of its functioning;
– creation of control devices of quality of information support of IAS IDPC and achievement of the set indicators of reliability of a system;

– ensuring functioning of a system in real time;
– design of configurations of the local networks providing functioning of this system;
– design of software of diagnostics of failures and processing of emergencies in networks of the set configuration;
– design of control facilities the database for networks of the set configuration;
– design of languages of communication of the end user with a system taking into account optimization of structure and characteristics of language;
– development of tutorials of the user nonprofessional to work with a system.

It is obvious that only such powerful tool as IAS IDPC in combination with most digitized productions, can meet at the same time modern requirements as on increase in labor productivity and reducing costs for production of agricultural products and food, and on sharp improvement of quality of the made products.

As abroad, and in the Russian Federation works on digitalization of economy are conducted approximately since the beginning of the 2000th years. It is necessary to make a reservation that these works are focused on creation of such object as IDPC. So far in this regard there were two approaches – language and model.

The idea of language approach to creation of IDPC originates from works on digitalization of various processes and is based on use of the language means allowing to describe the created object of management at some level of abstraction. The obvious lack of this method is that the quality of a resultant system (and, first of all, degree of its adequacy to the nature of an object of management) entirely depends on experience and qualification of the developer. Application of this method for creation of the integrated systems (and the more so IDPC in agricultural industry) is connected with the considerable additional difficulties caused by the different nature of components of a system and consequently, and distinction of the language means necessary for their description.

Contrary to language approach the model method of formation of IDPC is based on a possibility of creation of model of a control system on the basis of initial model of an object of management. Making a start from model actually of production, the nature of an object of management can construct the control system adequate by some criterion. The completeness, consistency and adequacy of formal models of an object of management, on the one hand, and control systems, with another, serve as a guarantee of systemacity of a designing process and, finally, to functional and information coherence of components of a resultant system.

The control system model (by whatever method it was constructed – subjectively by the developer or objectively on the basis of model of an object of management) can form a basis for design of all types of providing the automated system. At the same time the model design method allows to raise considerably degree of validity of design decisions during system development.

Process of creation of model of a control system – substantial design of a system – includes such works as the choice of methods of management, distribution of functions of management on levels and elements of a system, determination of structure of the automated functions and also organizational, functional and information interrelations in a system. Here the regulations of functioning of IAS IDPC as the automated system are defined and the choice of implementation methods of artificial intelligence is made.

The complexity of digitalization of a stage of substantial design (and consequently, and creations of model IDPC) is connected with difficulty of formalization of those criteria by which all activity connected with the choice of some options and decision-making is defined. If actually control algorithms rather easily give in to formalization (the best proof to that is a large number of packages of application programs of functional purpose, each of which realizes some formal algorithm), then selection criteria and decision-making at a stage of substantial design of systems are still based mostly on experience, qualification and an intuition of the developer and completely depend on subjectivity of the specific person. Only to the present times the way to objectification of these criteria consists in use of cumulative experience of many experienced experts of agricultural industry.

IOP Conf. Series: Materials Science and Engineering **570** (2019) 012044 doi:10.1088/1757-899X/570/1/012044

The perspective tools allowing to solve the specified problem represent the expert systems which are violently developing in recent years. Huge volumes of information necessary for realization of the automated design, can be organized in the form of knowledge bases that gives the chance to use this information in that indistinct and not completely certain form which is inherent in it. Selection criteria and decision-making by which qualified specialists in design of systems are guided can be described in the form of rules of a logical conclusion.

Conclusions. The solution of essentially new tasks rising today before organizers of IDPC demands application of essentially new means. Apparently, use of methods and artificial intelligence techniques (and, in particular, expert systems) will allow to reach qualitatively new level in digitalization of rural economics.

References

[1] *The future of agriculture* [Electronic resource]. - Access mode: https://www.economist.com/technology-quarterly.

[2] Medennikov V I, Salnikov S G *The main directions of Informatization of agriculture* [Electronic resource] Access mode: http://www.viapi.ru/ publication/full/ detail.php.

[3] Akmarov P B, Knyaseva O P, Tretyakova E C 2018 *Problematic issues of information technology development in Russia* Trudy mezhdunarodnogo simpoziuma «Nadezhnost i kachestvo T 1 pp 264-67

[4] *Rossiya v tsifrakh. 2018: Kratkiy statisticheskiy sbornik* [Russia in figures. 2018: A brief statistical figest] pp 223-238 [Electronic resource]. - Access mode http://www.gks.ru/free_doc/doc_2018/rusfig/rus18.pdf.

[5] Kashapov N F, Nafikov M M, Gazetdinov M Kh, Nafikova M M, Nigmatzyanov A R 2016 *IOP Conference Series: Materials Science and Engineering* **012012**

[6] Kashapov N F, Nafikov M M, Gilmanshin I R, Gazetdinov M Kh, Nafikova M M, Nigmatzyanov A R 2017 *Energy-saving Technologies of Cultivation of Sugar Sorghum* IOP Conference Series: Materials Science and Engineering **012032**

[7] *GeoMation Farm* Access mode: http://www.hitachi-sk.co.jp/products/ geomation/farm/index.html in Japanese.

[8] Kolonejchenko A C 2017 *Information support of management processes in agriculture* Molodoj uchenyj No 15 pp 10-12

[9] Chupina I P 2016 *The competitiveness of products as the object of a targeted strategic development of an economic entity* Eastern European Scientific Journal No 1 pp 59–62

[10] Gazetdinov M Kh, Khabirov R S 2016 *Modernization of the agricultural sector and rural development* Konkurentosposobnost v globalnom mire: ekonomika, nauka, techologii No 3 (15). pp 60-63

[11] Gazetdinov M Kh, Semicheva O S, Nezhmetdinova F T, Gazetdinov Sh M 2018 *Transformation of Work of Rural Population in the Conditions of Modernization of Economy* Towards Productive? Sustainable and Resilient Global Agriculture and Food Systems pp 932-43

IOP Conf. Series: Materials Science and Engineering **570** (2019) 012045 doi:10.1088/1757-899X/570/1/012045

Development of digital and robotic agricultural production in the Russian Federation

N F Kashapov[1], M M Nafikov [1,3], Mansur M Nafikov[2] and A R Nigmatzyanov[3]
[1]Kazan (Volga Region) Federal University. Kazan, Russia
[2]KNITU-KAI, College of Information Technology. Kazan, Russia
[3]Tatar Institute of Agribusiness Personnel Retraining. Kazan, Russia

arnig76@yandex.ru

Abstract. In recent decades, the rural population has increased dramatically. These circumstances necessitate the use of digital, intelligent and robotic. In the agro - industrial complex, these technologies are one of the most important functions of which is intellectual. It is a process to improve the quality of work.

One of the priorities of the development of agriculture in the Russian Federation is the introduction of digital technologies. This trend is particularly relevant in that due to monitoring and certification it is possible to speed up the process of entering into agricultural circulation vacant land. Since 2018, the geographic information system of the agro-industrial complex (GIS AIC RT) has started to work in the Ministry of Agriculture of the Republic of Tatarstan. We developed it and supervise the implementation together with the Ministry of Information and Communication of Tatarstan. With the help of this system, it is possible to effectively use, plan and control all stages of agricultural production. Monitoring of fields, forest belts and forest plantations in Tatarstan began to be carried out using drones. They have found wide application in the farms belonging to Agrosila JSC, Holding Company Ak Bars JSC in precision farming. Thus, GIS AIC RT became an indispensable assistant of local specialists in the village.

Leading automation researchers Carl Osborne and Benedict Fry, exploring the transition of industrial countries to the digital economy, formulated their approach to robotization [1]. In their opinion, "robotization is an automation of a system or task of a level where the need for human labor disappears and human labor is replaced with an automated version." To introduce robotization into production in rural areas, a system of state support is needed for agricultural enterprises to acquire new technology, taking into account the priority of introducing digital intelligent and robotic technologies. A prerequisite is also the preparation of productive livestock and cultivated crops, which, according to various parameters, must correspond to the introduced digital, intellectual and robotic technologies, in general.

Robotics are highly effective when used in animal husbandry for milking cows, mares, goats, etc., for cleaning and removing manure, grooming animals. [2, 3,4,5], in crop production for sizing and inlaying seeds, sowing works, protecting plants from weeds, pests and diseases, for monitoring crops. Nowadays milking robotics can be called one of the most common robotic products in rural areas. A scientist from the Ural State Agrarian University, Ye.A.Skvortsov, argues that the use of robotics is a panacea in terms of increasing productivity, production efficiency, and in terms of solving personnel problems in the industry. Complex work is needed to create and assess the

IOP Conf. Series: Materials Science and Engineering **570** (2019) 012045 doi:10.1088/1757-899X/570/1/012045

prerequisites and the need to introduce robotics, which may include addressing the economic, zootechnical, agronomic, etc. feasibility and feasibility of using innovative techniques based on robotics in agriculture. Such work is conducted abroad. The solution of this complex of questions is impossible without the involvement of industrial science, including agricultural universities, research institutes, technical schools, etc. However, without innovative robotics, qualified personnel it is difficult to imagine an increase in agricultural production and an increase in the competitiveness of domestic agriculture as a whole [6, 7,8].

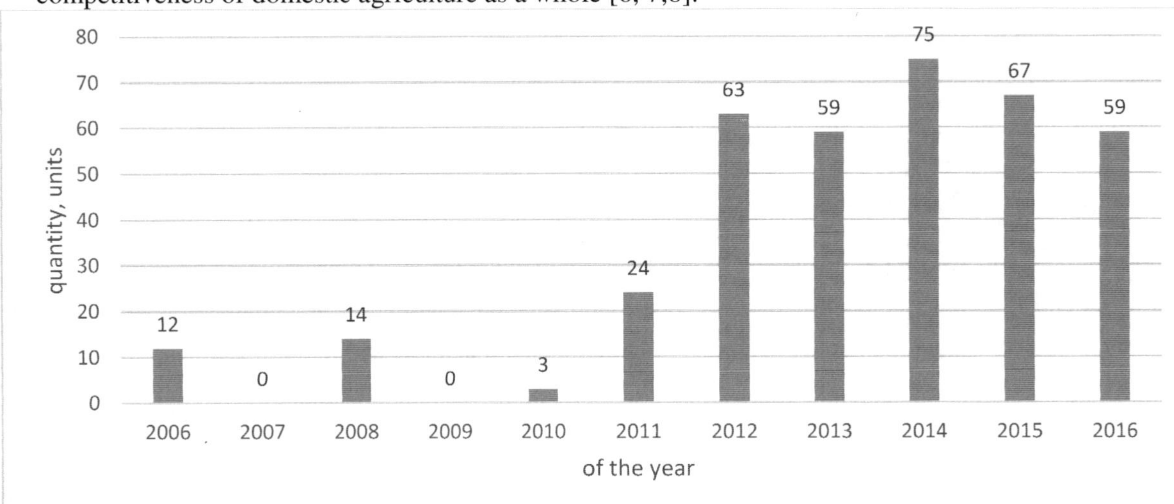

Chart 1. Introduction of robotics in organizations subordinated to the Ministry of Agriculture of the Russian Federation As of the second quarter of 2017, robotics is used in 28 regions, in 103 organizations subordinated to the Ministry of Agriculture of the Russian Federation, mainly dairy production (chart 1).

More active work on the introduction of robotics was carried out in 2015, in subsequent years, the rate of implementation declined slightly. So, in 2016, the number of implemented robotics units compared to 2014 and 2015. amounted to 88.6 and 72.2%, respectively. This is explained by the fact that all of the robotics used in the Russian Federation is imported, and due to the change in the exchange rate and sanctions, it has become almost not available for many AIC organizations.

At the beginning of 2018, the average density of robotization in the world is 74 robots per 10,000 workers in the economy as a whole (Figure 2).

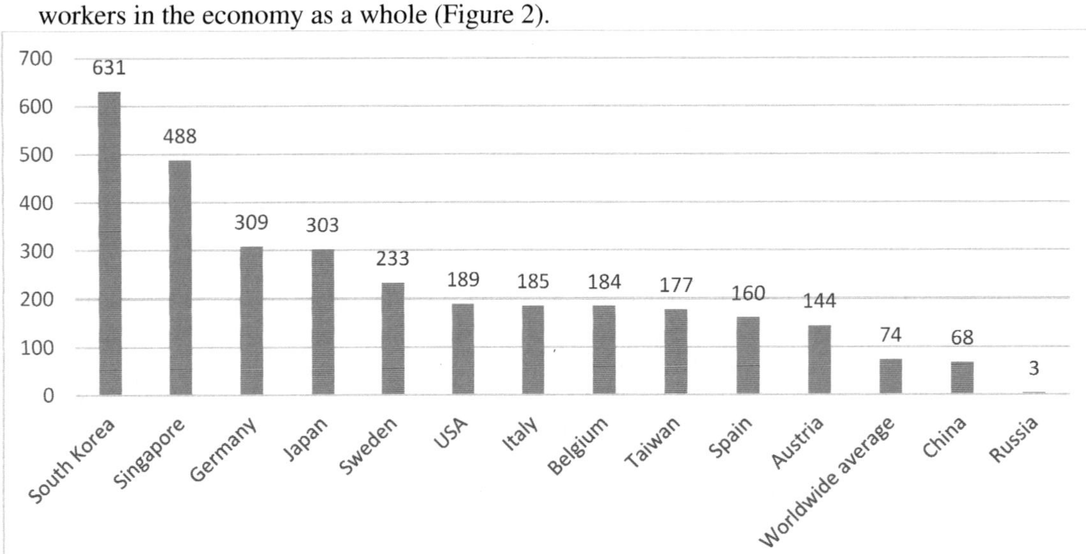

IOP Conf. Series: Materials Science and Engineering **570** (2019) 012045 doi:10.1088/1757-899X/570/1/012045

Chart 2. Use of robots in the countries of the world

According to the International Organization of Robotics (IFR), in terms of the density of robotization, the Russian agro-industrial complex and the country's economy as a whole lag far behind not only the leading countries, but also the world average.

The reasons for the transition of organizations of the agro-industrial complex to robotics are explained by the fact that this process is carried out haphazardly and there are problems of the following order:

A) the creation of domestic-made robotics, taking into account local conditions of production and in accordance with the requests of organizations of the agro-industrial complex. Used in the countryside in terms of sanctions imported robotics;

B) there is a need to issue scientifically based recommendations on the implementation and use of intelligent, digital and robotic technologies in organizations of the agro-industrial complex;

C) the study of the effectiveness of the introduction and use of robotic technologies in rural areas;

D) to develop the RPD and the list of competencies for the training of specialists who are able to master and use robotics;

D) a comprehensive scientifically based assessment of the impact of the use of robotics on the development of rural areas, in particular on the nature and severity of agricultural labor, on the incidence and industrial injuries of workers, on the quality of work, and the level of employment in the rural areas;

E) to form a strategy for the development of production and introduction of agricultural robotics, containing goals and mechanisms for its implementation, taking into account external and internal factors of introducing robotics, observing the principles of its use in the agroindustrial complex, state support for creating agricultural robotics with the involvement of engineering science.

Accordingly, to solve the problems of domestic scientists and developers will require a significant expansion of research programs in the field of robotics with the involvement of industry science, which can be implemented through special programs of the Ministry of Agriculture of the Russian Federation.

References

[1] Frey C B, Osborne M A 2013 *Future of Employment: How Susceptible Are Jobs to Computerization?* Oxford: Oxford Martin School 72 p

[2] Ivanov Yu A 2014 *Directions of scientific research on the creation of innovative technology with intelligent systems for animal husbandry* Vestnik VNIIMZH No 3 pp 4–17

[3] Kormanovskiy L P 2013 *The development of robotization of milking cows* Vestnik VNIIMZH No 2 pp 78–81

[4] Skvortsov E A, Skvortsova E G, Nabokov V I, Krivonogov P S 2017 *Application of milking robotics in the region* Economy of the region No 1 pp 249-60.

[5] Surovtsev V N, Bilkov V A, Nikulina Yu N 2013 *Innovative development of dairy farming in the north-west of the Russian Federation as a basis for increasing the competitiveness of milk production* Economic and social changes. Facts, trends, forecast No 4 (28) pp 143-50

[6] Skvortsov E A 2016 *Personnel aspect of the introduction of robotics in agriculture* Agrarian Bulletin of the Urals No 2 pp 99-106

[7] E A Skvortsov, E G Skvortsova, I G Sandu, G A Iovlev 2018 *The transition of agriculture to digital, intellectual and robotic technologies* Economy of the region T 14 is 3 pp 1014-26

[8] Kashapov N F, Nafikov M M, Gazetdinov M X, Nafikova M M, Nigmatzyanov A R 2016 *Justification of the choice of units for mains-noah soil cultivation of sweet sorghum and their effectiveness* IOP Conference Series: Materials Science and Engineering

IOP Conf. Series: Materials Science and Engineering **570** (2019) 012046 doi:10.1088/1757-899X/570/1/012046

The effect of tillage with various tools on its agrophysical and agrochemical properties

N F Kashapov[1], M M Nafikov[1] , A R Nigmatzyanov[2] and V V Medvedev[2]

[1]Kazan (Volga Region) Federal University. Kazan, Russia
[2]Tatar Institute of Agribusiness Personnel Retraining. Kazan, Russia

arnig76@yandex.ru

Abstract. When analyzing agricultural equipment for tillage and technology, it is observed that despite the introduction of conservation farming technologies (zero, ridge, strip and other methods of tillage), the mechanical method remains prevailing, taking into account the diversity of soil conditions in various soils. - climatic zones of the Russian Federation, the presence of plain, slope and contour farming, the need to combat erosion and preserve its agrophysical and agrochemical properties.

Plowing with dumps is a radical tool in the fight against weeds, pests and diseases and the basis of environmentally safe technologies that reduce the use of chemicals, optimize agrophysical and agrochemical processes in the soil.

Illiterate application of the methods of primary tillage can lead to bogging of crop rotation fields, reduced yields due to waterlogging, insufficient development of the root system of cultivated plants, disturbed gas exchange, reduced microbiota, etc. Deep soil tillage is also necessary on fallow lands. Requirements to the quality of field processing led to the spread of the technology of smooth plowing and a sharp increase in the production of circulating plows, which are designed for smooth plowing without dump ridges and open furrows. Plowing the fields is done in the shuttle mode without splitting the field into pins, which significantly reduces the cost of FCM.

Foreign and domestic firms produce mounted reversible plows (2-6 cases), semi-mounted plows on one support wheel (4-9 cases) and semi-mounted plows on a trolley (7-17 cases) [1].

Gregoire Besson (France) produces a wide range of semi-mounted reversible plows of the original design with a different number of pairs of bodies and the ability to work in the unit with other implements.

Multicase semi-mounted plows consist of two frames (front and rear), pivotally interconnected, which allows you to copy the field relief more smoothly.

In the transport position, the rotary frame of the plow with bodies is fixed in a horizontal position. The hinged front of the plows ensures 110 ° rotation. The toothed rack and pinion mechanism for wrapping the plow is driven by two hydraulic cylinders and ensures work with a constant effort and without jerks during the cycle of transfer of the plow from one working position to another. Plows can be equipped with different types of plow bodies, 11 types of dumps and 6 types of skimmers adapted to all types of soil and vegetation cover.

Kvemeland (Norway) presented mounted, semi-mounted and trailed reversible plows for smooth plowing with a different number of bodies (up to 14 pairs) and a series of plows for driven plowing with a number of bodies from 2 to 12. As a rule, the plows are equipped with devices (rollers, harrows, etc.) for additional crumbling of the soil and leveling the field surface.

On all models of plows manufactured by this company, the hulls are equipped with safety devices (most often spring-backs), which allows them to be used in fields littered with stones. All plows have the ability to change the width of the grip (the width of the grip of one case can vary from 35 to 50 cm).

Depending on the model, the plows of this company can be equipped with a variety of additional equipment: skimmers, saws, knives, rippers, support wheels, plowshares and plow bodies, as well as be equipped with a system that allows you to adjust and adjust the plow in an automatic or semi-automatic mode.

The plants of the Russian Federation produce about 30 models of reversible plows, aggregated with tractors of traction classes 0.9; 1.4; 2; 3; 5 and 6 with the number of buildings from 2 to 7 and a capacity of from 0.2 to 3.7 ha / h [2].

In recent years, in many countries of the world, including the Russian Federation, various methods of soil tillage have been developed and tested in practice. The usual plows are replaced by chisel plows and cultivators. They contribute to the better conservation and accumulation of moisture in the soil, positively affect the physical properties and biological activity of soil microorganisms, prevent the development of water and wind erosion, do not leave developmental furrows and land ridges

After arable land is treated with such aggregates, a stubble remains on the soil surface, which helps to reduce deflating, increase snow accumulation and reduce soil freezing as a result of maximally moistening arable land, reducing the negative impact of water and wind erosions on the soil, and also destroying the “plow sole”. and chisel plows are used to deepen the arable layer in combination with the introduction of higher doses of organic and mineral fertilizers. Sewing treatment, widely used in conditions of insufficient moisture, in steppe areas prone to wind erosion and on sloping lands, loosens the soil (while its most fertile part remains on the surface), cutting weeds and preserving up to 90% of stubble on the surface of arable land. straw [3,4,5].

Subsoiler, if necessary (or on request) can be equipped with compact rollers, mulch disks, tooth harrows, cutters, etc. Along with the main tillage-free tillage, deep-rippers are equipped with mechanical or pneumatic driven bunkers for introducing subsoil and depth mineral tukov[6,7].

In the Russian Federation, more than 30% of cultivated acreage is in arid steppe and forest-steppe areas, and almost all cultivated land in these areas is subject to erosion. The lack of soil moisture is caused not only by a small amount of precipitation, but also by losses to runoff, evaporation, snow blowing, high transpiration of moisture by weeds[8,9,10].

In recent years, disk tools have been recognized in the Russian Federation and some foreign countries, whose working bodies are mounted on individual racks placed on the cross beams of the frame of an agricultural machine or tool[11,12,13,14]. Such working machines are called diskators, discocats, diskators, front disc harrows, and so on. In the Russian Federation, the production of machines of this type is widely spread, they are produced by more than fifty companies and organizations located in various fields.

References

[1] *The state program for the development of agriculture and regulation of agricultural markets for 2013-2020:* Decree of the Government of Russia dated July 14, 2012 No. 717 // Collect legislation of the Russian Federation - 2012. No 32 **4549**

[2] *National report "On the implementation and results in 2014 of the State Program for the Development of Agriculture and Regulation of Agricultural Products for 2013-2020* M.: Federal State Scientific Institution Rosinformagrotech 2015 364 p

[3] A A Borin, A E Loschinina 2015 *Effect of soil cultivation in combination with the use of fertilizers and herbicides on crop productivity of crop rotation* [Vliyanie obrabotki pochvy v komplekse s primeneniem udobreniy i gerbitsidov na urozhaynost kultursevooborota] Zemledelie. – Agriculture No 7 pp 17-20

IOP Conf. Series: Materials Science and Engineering **570** (2019) 012046 doi:10.1088/1757-899X/570/1/012046

[4] A P Shindin and others 2012 *Kukuruza. Sovremennaya tekhnologiya vozdelyvaniya* [Corn. Modern cultivation technology] 152 p

[5] Medvedev V V, Fomin V N, Nafikov M M 2017 *Produktivnost kukuruzy v zavisimosti ot vidov i doz azotnykh udobreniy,srokov ikh vneseniya.* [Productivity of corn depending on the species and doses of nitrogen fertilizers, the timing of its application] Problems of innovative development of the agroindustrial complex:personnel, technology, efficiency. Collection of scientific articles) Vol 11 pp 290-95

[6] M M Nafikov, N M Yakushkin, V N Fomin, I P Talanov 2015 *Vozdelyvanie odnovidovykh i smeshannykh posevov sorgovykh kultur* [Cultivation of single-species and mixed crops of sorghum crops] 248 p

[7] Z P Okazova, D M Mamiyev, A A Tedeyeva 2015 *Ways to increase corn productivity in the conditions of the steppe zone of North Ossetia-Alania* [O putyakh povysheniya urozhaynosti kukuruzy v usloviyakh lesostepnoy zony RSO-Alaniya] Sovremennye problemy nauki i obrazovaniya. - Modern problems of science and education No 5 p 695

[8] Sotchenko V S 2012 *Corn: the main directions in the selection of highly productive hybrids* [Kukuruza: osnovnye napravleniya v selektsii vysokoproduktivnykh gibridov] Niva Tatarstana No 2-3 p 10

[9] Z I Usanova, I V Shalnov, A S Vasilev 2016 *Influence of calculated fertilizer doses and density of standing on corn productivity, removal and economic balance of basic nutrients* [Vliyanie raschetnykh doz udobreniy i gustoty stoyaniya na produktivnost kukuruzy, vynos i khozyaystvennyy balans osnovnykh elementov pitaniya] Zemledelie. Agriculture No 3 pp 23-26

[10] I A Shmalko, V N Bagrintseva 2012 *Urozhaynost kukuruzy pri otvalnoy i poverkhnostnoy obrabotke pochvy* Materialy nauchno-prakticheskoy konferentsii. Selektsiya. Semenovodstvo. Tekhnologiya vozdelyvaniya kukuruzy [Corn productivity in dump and surface tillage. Proceedings of the scientific and practical conference. Selection. Seed production. Corn cultivation technology] pp 220-30

[11] B A Yagodin, Yu P Zhukov, V I Kobzarenko 2004 Agrokhimiya. [Agrochemistry] M.: Mir 584 p

[12] Semina S A 2012 *Efficiency of fertilizer systems in crop cultivation in the forest-steppe of the middle Volga region* [Effektivnost sistem udobreniya pri vozdelyvanii kukuruzy v lesostepi Srednego Povolzhya] Niva Tatarstana No 1 pp 39-42

[13] Ziganshin A A 2001 *Modern technologies and productivity programming* [Sovremennye tekhnologii i programmirovanie urozhaynosti] Kazan: Izd-vo Kazanskogo un-ta 109 p

[14] Dospekhov B A 1985 *Metodika polevogo opyta s osnovami statisticheskoy obrabotki rezultatov issledovaniy* [Field experience methodology with the basics of statistical processing of research results] M.: Agropromizdat 351 p

IOP Conf. Series: Materials Science and Engineering **570** (2019) 012047 doi:10.1088/1757-899X/570/1/012047

The state of the technical level of domestic agricultural machinery

N F Kashapov[1], M M Nafikov[1,2], A R Nigmatzyanov[2] and I G Gajnutdinov[3]

[1]Kazan (Volga Region) Federal University. Kazan, Russia
[2]Tatar Institute of Agribusiness Personnel Retraining. Kazan, Russia
[3]Kazansky state agrarian university. Kazan, Russia

arnig76@yandex.ru

Abstract: Comparative tests of domestic and foreign equipment designed for carrying out soil tillage and sowing, plant protection techniques and the application of organic and mineral fertilizers under the conditions of compliant regulatory documentation, carried out in recent years, have shown that by basic technical and economic and agrotechnical indicators domestic technology is not inferior to foreign counterparts and may well replace them in a crisis relationship.
It has been shown that it has been carried out in recent years under the conditions of the regulatory documentation, It was shown that technical indicators of domestic equipment were not affected by the equipment.

Modern agricultural production is impossible without technology that goes to agricultural producers through the market. In recent years, a significant share of the Russian agricultural machinery market was taken by foreign firms, which led to an increase in the number of foreign machinery in the countryside. Imported equipment is poorly adapted to work in our conditions, often fails and requires the purchase of appropriate consumables, over time there is a need to replace units, purchase spare parts. In addition, prices for spare parts are quite high. In connection with the foregoing, the import substitution of foreign agricultural equipment, as well as its units, spare parts and consumables is an urgent task under sanctions.

In the Russian Federation, large and medium agricultural holdings, small agricultural enterprises, peasant farms (peasant farms), individual entrepreneurs, personal subsidiary farms of individuals (private farms) are engaged in agricultural business (Table 1).

Table 1 - Agricultural organizations of all forms of ownership engaged in agribusiness in the Russian Federation

Farm categories	Number of objects	Area of agricultural land, ga	The proportion of the area sh land in this category of farms, in%
Large and medium agricultural organizations	27779	3867	64,1
Small agricultural enterprises	20404	1201	14,6

Subsidiary farms of non-agricultural organizations	19649	183	1,2
Peasant farms	253331	85	12,9

The largest number (64.1%) of the share of agricultural land is occupied by large and medium agricultural holdings, the other categories of farms are significantly inferior to them.

The availability of agricultural equipment available in farms of all forms of ownership (Table 2) also has a big difference: in personal subsidiary farms and other individual farms of citizens of tractors - 37% of the total, tractor mowers - 61, tractor plows - 36%; in farms and individual entrepreneurs of tractors - 14%, tractor mowers - 12, tractor plows - 22, combine harvesters - 28%; in large and medium-sized agricultural organizations of tractors - 40%, tractor mowers - 22, tractor plows - 34, grain combine harvesters - 59%.

Table 2 - Availability of agricultural equipment in farms of all forms of ownership in the Russian Federation

Farm categories	Tractors		Tractor mowers		Tractor plows		Combine Harvesters	
	number machines	share of all%	number machines	share of all%	number machines	share of all%	number machines	share of all%
Large and medium agricultural organizations	437752	40	49434	22	120341	34	112208	59
Small agricultural enterprises	73421	7	9144	4	24337	7	21821	11
Subsidiary farms of non-agricultural organizations	19649	2	2256	1	6488	2	4002	2
Peasant farms	158827	14	27375	12	76794	22	54037	28
Individual entrepreneurs	411863	37	132065	61	122860	35	-	-

As can be seen from Table 3, the average load of arable land on one tractor, the area of grain crops per one combine harvester, is greatest in small farms, large and medium agroholdings and peasant farms.

Table 3 - The load on agricultural machinery in farms of all forms of ownership of the Russian Federation.

Indicators	Large and medium agricultural organizations	Small agricultural enterprises	Subsidiary farms of non-agricultural organizations	Peasant farms Individual entrepreneurs
The load of arable land on one tractor, ha	153	194	60	105
It is necessary to plant (plant) the corresponding crops for one combine, ha:				
grain harvesting	257	231	118	167
corn harvesting	259	418	198	238
potato harvesting	30	45	62	40
Beet-harvesting (without topper)	129	140	95	96

According to the “Strategy for the development of agricultural engineering for the period up to 2020”, for the effective management of agribusiness in the countryside, the tractor park should be about 610 thousand units, and grain and forage harvesters no less than 147 thousand units. A separate

IOP Conf. Series: Materials Science and Engineering **570** (2019) 012047 doi:10.1088/1757-899X/570/1/012047

subprogram "Agricultural Engineering, Food and Processing Industry" includes the provision of agricultural producers, food processing and processing industry with reliable high-performance agricultural machinery, equipment to reduce overhead costs, as well as capable of working on modern waste-free, energy-saving technologies, which in turn is efficient will affect[1].

Agricultural machines in the field perform a large number of technological operations aimed at changing the state of processing objects. Therefore, the absence of one or another type of equipment violates the performance of the necessary cycle of technological operations for the production of agricultural products. The system of agricultural machinery used in the late 90s. It included about 3 thousand names of agricultural machines and implements, their modifications and devices, which took into account the zonal principle of agricultural production.

At present, the market of agricultural machinery and equipment in the Russian Federation, as well as throughout the world, is mainly oligopolistic, i.e. It has a limited number of large manufacturing enterprises. At the same time, it is changing, since buyers have an unlimited choice, for example, the possibility of acquiring agricultural equipment produced by regional enterprises, as well as analogues from near and far abroad [2].

In order to import substitution in the Russian Federation, mechanisms are being used to support the renewal of the agricultural machinery park through the State Program for the Development of Agriculture and Regulation of Agricultural Products, Raw Materials and Food Markets for 2013-2020, the subprogram “Technical and Technological Modernization, Innovative Development”, the main event “Renovating the Park agricultural machinery "(subsidies to producers of agricultural machinery for reimbursement of the cost of production and real zatsiyu agricultural machinery); financial lease (leasing) through Rosagroleasing OJSC and other leasing companies; credit programs of OJSC “Rosselkhozbank” and other banks; special programs of enterprises that provide discounts, increase the warranty period, etc .; recycling programs; regional programs, including the subsidization of part of the costs of agricultural producers for the purchase of equipment from regional budgets [3].

Considering the global tendencies of the widespread use of direct-seeded seeding complexes, modular modular units and machines for applying liquid organic fertilizers, anhydrous ammonia, and liquid complex fertilizers, it is advisable to start their production at plants in the Russian Federation[4,5,6,7,8].

The analysis has shown that domestic combine harvesters of various classes are at the level of foreign ones or somewhat inferior in terms of specific technical level indicators. However, they have a clear advantage in the implementation of different options for harvesting non-grain parts of the crop, and they are preferable by the price-performance criterion. The payback period of domestic combines is 2-3 times less than foreign ones, since their cost is lower (by 1.5-1.8 times) and the prices for spare parts (by 5-6 times). Foreign combine harvesters are effective in harvesting crops with a yield of more than 30 c / ha, but the areas with such a yield in Russia are less than 20%. Domestic harvesters are suitable for work in all climatic zones. They are more effective in harvesting wet, long-straw and weedy breads, the cost of harvesting grain crops is 20-30% less. Therefore, the potential for import substitution of foreign combines can be assessed as sufficient.

References

[1] *The head of the Ministry of Industry and Trade of Russia on measures to develop domestic engineering* (Electronic resource). URL: http://www.test.lobbying.ru/content/sections/articleid_12023_linkid_122.html

[2] *Agricultural machinery: national agricultural catalog* (Electronic resource). URL http://www.selhoz-katalog.ru/pdf/kat_selhoz_17-2018.pdf

[3] *Russian analogues of foreign agricultural machinery, import substitution of aggregates, spare parts and consumable*s M. FGBNU "Rosinformagrotekh, 2015. – 340 p

[4] N F Kashapov, M M Nafikov, M Kh Gazetdinov, Sh M Gazetdinov and A R Nigmatzyanov 2018 About one approach to the assessment of technical equipment of agricultural enterprises in conditions of economy modernization IOP Conference Series: Materials Science and Engineering,Vol 412 **012038**

[5] N F Kashapov, M M Nafikov, I R Gilmanshin, M Kh Gazetdinov, M M Nafikova and A R Nigmatzyanov 2017 *Energy-saving technologies of cultivation of sugar sorghum* IOP Conference Series: Materials Science and Engineering V 240 **012032**

[6] N F Kashapov, M M Nafikov, M Kh Gazetdinov, M M Nafikova and A R Nigmatzyanov 2017 *Comparative evaluation of different machines for seedbed for sorghum* IOP Conference Series: Materials Science and Engineering V 240 **012033**

[7] N F Kashapov, M M Nafikov, M X Gazetdinov, M M Nafikova, A R Nigmatzyanov 2016 *Justification of the choice of units for mains-noah soil cultivation of sweet sorghum and their effectiveness* IOP Conference Series: Materials Science and Engineering V 134 **012013**

[8] N F Kashapov, M M Nafikov, M KH Gazetdinov, M M Nafikova, A R Nigmatzyanov 2015 *Economic justification of the choice of machines and tools for primary tillage under sweet sorghum proceedings of the International scientific-technical conference* MNTK "IMTOM–2015" Part 1 pp 285-88.

IOP Conf. Series: Materials Science and Engineering **570** (2019) 012048 doi:10.1088/1757-899X/570/1/012048

Influence of roughness on operational properties of details

I I Khafizov[1] **and I G Nurullin**[1]

[1]Kazan Federal University, 18 Kremlyovskaya street, Kazan, 420008, Russian Federation

khafizov@kpfu.ru, opros16@yandex.ru

Abstract. The roughness has significant effect on operational properties of details. For increase in wear resistance and durability of cars and mechanisms, it is necessary to provide appropriate roughness of a surface of details. In this article influence of roughness on some properties of details is considered: on contact deformation, stability against corrosion, wear resistance and on other indicators.

Shows results of long-term operation of production of the machine-building industry of different function that, parameters of a condition of a blanket strongly affects reliability and durability of details of cars and, respectively, products in general. This results from the fact that the majority of operational characteristics of details substantially depend on quality of a surface.

The surface quality of parts is a complex indicator determined by such characteristics as macrogeometry, surface roughness and waviness, the state of the surface layer.

On the finished and machined surface of the part there are always irregularities of different types and sizes, depending on the methods of machining, in contrast to the surface depicted in the drawing [1].

During the cutting process, the metal layer is deformed and the structure changes [2]. The degree of hardening and depth of the hardened layer depend on the processing methods, modes and cutting speed, the geometry of the working edge of the tool and the properties of the starting metal [3].

Deviations from the theoretical form can be of three groups: macrogeometric-deviations from the correct geometric shape; waviness – as a result of vibrations and unevenness of the cutting process; microgeometric, formed as a result of the impact of the cutting edge of the tool on the treated surface.

For increase in wear resistance and durability, it is necessary to provide appropriate roughness of a surface of details. The details with a surface having big roughness are exposed to bigger wear during operation, than the details having a plain surface. It can be explained with the fact that the area of a support of a rough surface is less, than equal as surfaces adjoin among themselves only combs [4].

Also, the details of roughness which are available on a surface and defects promote further concentration of tension which can exceed the strength of metal and be a cause of destruction of a detail [5].

The roughness step taken as a roughness must be very small, relative to the base length of the entire surface. There are three types of object roughness:

1) Initial-occurs as a result of the technological sequence of processing of the product by various abrasives.

2) Operational (ES) - acquired roughness during operation of the part, as a result of wear and working friction.

3) Equilibrium is a type of ES that can be reproduced in stationary friction conditions.

Despite their small size, such irregularities have a significant impact on the performance properties of parts (ESD). The results of the study of the effect of surface roughness on some ESD can be seen in table 1.

Table 1 - Influence of Roughness on Operational Properties of Details

№	Operational properties of details	Influence of roughness on operational properties of details
1	The corrosion firmness	Has a considerable impact. Corrosion of details in atmospheric conditions are formed easier and go quicker on roughly processed surfaces. As a rule, on purely processed surfaces, the corrosion resistance is higher.
2	Contact deformation	Strongly affects contact rigidity of joints of the interfaced details. Reducing roughness and waviness, increase bearing detail surfaces for 80 – 90% is possible and by that to increase contact rigidity.
3	Friction and wear	The value of working wear increases. Wear is affected by the shape of micro-irregularities and their direction. Peaked wear faster than flat-topped.
4	Fatigue strength (up)	Irregularities affect the stress concentration and the formation of fatigue cracks. Insufficient up will lead to rapid breakage, causing failure of the machines.
5	The stability of the stationary landings	When pressing parts is smoothing of asperities leading to a decrease of the actual tensioning. In this regard, a decrease in the strength of the connection of parts is detected with more rough surfaces.
6	Thermal conductivity	There is a decrease in the thermal conductivity of the joints.
7	Tightness of joints (HS)	There is a decrease in contact stiffness and HS, due to a decrease in the actual contact area.

But, also it is necessary to remember that not in all cases the purely processed surface is the most wear-resistant, as the retention of lubricant under different friction conditions depends on the micro-roughness available on the surface. And therefore, depending on specific conditions of friction, it is necessary to pick up optimum roughness of a surface. Parameters of roughness are determined in GOST 2789-73 "Roughness of a surface. Parameters, characteristics and designations" [6].

Analyzing results, it is possible to draw a conclusion that the roughness has significant effect on various properties. Also it is necessary to remember that influence of roughness it is essential also to economy as repair and restoration of worn-out details can be quite often more expensive, than production of the new car. It is possible for this reason, exact determination of value of roughness - one of important problems of metrology.

References

[1] Regov B A and Below P C 2012 Study of the effect of cutting conditions at fine turning on the surface quality *Vestnik RUDN. Series: Engineering research* No 4

[2] Khafizov I I 2013 Ways to reduce the consumption of materials when they are separated by combined methods *Bulletin of Kazan technological University* (Kazan: Kazan State Technical University) No 1 p 208-11

[3] Morandin I O and Volkov R S 2012 Surface quality after cutting *Successes of modern natural science* No 6 pp 76-79

[4] Khafizov I I, Nurullin I G and Sadykov Z B 2018 Application of the method of electroerosion treatment to improve the quality of the treated surface *Materials of IX conference «IMCOM-2018»* vol 432

[5] Khafizov I I, Nurullin I G and Sadykov Z B 2018 Problems of development of electrochemical

production of Russia and possibility of their decision *IOP Conference Series: Materials Science and Engineering* vol 412 **012042**

[6] Khafizov I I and Nurullin I G 2017 Product quality as the main factor of increase of competitiveness (on the example of JSC «Kazan helicopter plant») *IOP Conference Series: Materials Science and Engineering* vol 240 **012039**

[7] Gil'manshin I R and Kashapov N F 2014 Energy service contracts in regional engineering center for small and medium businesses *IOP Conference Series: Materials Science and Engineering*

[8] Nurullin I G and Khafizov I I 2016 The impact of innovation implementation and certification on the company's activities and reputation Modern technologies: topical issues, achievements and innovations: collection of articles of the winners of the IV international. science.- practical Conf. (Penza: ICSU " Science and education») p 44-6

[9] Akhmetov I D, Zakirova A R and Sadykov Z B 2017 New electrode-tool for the combined kerf of electrically conductive materials *IOP Conference Series: Materials Science and Engineering* vol 240 **012003**

[10] Khafizov I I and Galimov A N 2017 IT-strategy and major aspects of quality management on the market of goods and services *IOP Conference Series: Materials Science and Engineering* vol 240 **012038**

[11] Khafizov I I, Saveleva T N and Lyubtsov V S 2017 Estimation of parameters of charge carriers in dielectric materials by CELIV method *IOP Conference Series: Materials Science and Engineering* vol 240 **012040**

[12] Khafizov I I 2017 Ways of decrease in the material consumption in case of their separation by the combined methods *IOP Conference Series: Materials Science and Engineering* vol 240 **012037**

[13] Khafizov I I 2016 Economic efficiency and effectiveness of ways of separating materials electro diamond processing *IOP Conference Series: Materials Science and Engineering* vol 134 **012014**

[14] Kashapov N F, Khafizov I I and Nurullin I G 2018 Influence of introduction of robotics on increase in efficiency of electrochemical production *IOP Conference Series: Materials Science and Engineering* vol 412 **012034**

[15] Fazlyyyakhmatov M G and Kashapov N F 2014 Corona discharge in the process of spraying protective powder coatings on piezoceramic materials *High Temperature Material Processes* vol 18 pp 273-79

[16] Kashapov N F and Sharifullin S N 2015 Hardening of the surface plasma jet high-frequency induction discharge of low pressure *IOP Conference Series: Materials Science and Engineering* vol 86 **012021**

[17] Kashapov N F, Gil'manshin I R and Konahina I A 2014 System analysis of the energy complex of engineering enterprise as a basic tool of effective energy management *IOP Conference Series: Materials Science and Engineering* (Bristol-UK) vol 69

[18] Gubaidullin D A, Kashapov N F and Zaripov R G 2017 Experimental studies for flow velocities of oscillations gas when transiting through resonance in the area of an open end of pipe *IOP Conference Series:Journal of Physics: Conference Series.* vol 789 **012017**

[19] Gilmanshin I R, Kashapov N F, Gilmanshina S I and Galeeva A I 2016 Landfill energy complex based on the renewable energy installations *IOP Conference Series: Materials Science and Engineering* vol 134

IOP Conf. Series: Materials Science and Engineering **570** (2019) 012049 doi:10.1088/1757-899X/570/1/012049

Influence of application of robotic systems on prospects and directions of development of electroerosive processings

I I Khafizov[1], I G Nurullin[1] and N F Kashapov [1]

[1]Kazan Federal University, 18 Kremlyovskaya street, Kazan, 420008, Russian Federation

khafizov@kpfu.ru, opros16@yandex.ru

Abstract. Application of robotization in production is necessary for achievement of the required quality of products. Use of the industrial robot gives an unlimited range of opportunities on processings of preparations. The robotic system can be applied without readjustment to auxiliary operations.

At this stage of development of the engineering industry, the competitiveness of modern machine tools is ensured by a high level of automation and robotization.

The introduction of robotics and automation of work processes in production lead to improved product quality while reducing costs. This approach has a positive effect on the growth of the enterprise and the overall performance of the industry [1].

The use of robotics in production is necessary to achieve the required product quality. Industrial robots have a very high positioning accuracy, and the repeatability provides the ability to achieve the desired level of processing of the product with a decrease in manufacturing defects. With the exception of the human factor in technological processes, the percentage of operating errors is significantly reduced [2].

The technology of EDM has all the characteristics of key technologies. For a limited period of time the production of EDM machines in terms of sales was able to reach the fourth place in the market of Metalworking equipment. Developing in parallel with the production of other types of equipment, it has also become a platform for the introduction of a number of innovations - the processing of nanoobjects, the use of linear actuators and special structural ceramics.

By means of electroerosion processing high accuracy of forming and receiving the set sizes is provided. Also, the warping of parts is eliminated, because of the possibility of processing blanks after heat treatment [3].

Wide possibilities of shaping in EDM allow you to get the solutions that is hard to attain with other processes or possible only with very great cost.

At the enterprises of machine - and instrument-making the most common are the following technologies of electroerosion treatment:

1) Processing complex contur surfaces and holes (EDM copying broaching);

2) Treatment complex contur ruled surfaces (EDM profile cut);

3) Piercing of deep holes of small diameter (electroerosive piercing) [4].

For use in sparsely populated technology EDM machines must be equipped with robotic devices:

- automatic change of the electrode-tool and download the details on the plate-satellites;

- automatic change and refilling of wire electrode tool;
- manufacturing of start holes and removal of cut parts.

On the basis of a database of technological recommendations, leading companies have developed software that allows you to automate the processing technology based on the minimum necessary information (workpiece materials and electrode-tool, geometric characteristics of the processed profiles, the required surface roughness) [5].

And the equipment of electroerosion machines with adaptive protection and control devices helps to improve the reliability and accuracy of processing.

The robotic system can be used without readjustment for auxiliary operations:

- installation-removal of parts;
- marking-marking;
- measurements-control;
- application of adhesive, lubricating or insulating compounds, etc.

Reconfiguration of the robot with the ability to move in space on an additional external axes makes the machine a unique technological equipment.

The most important problems of the development of the method of electrical discharge machining include the following:

- development of methods and tools to improve the accuracy and quality of the treated surface;
- creation of integrated technological systems based on the processes of electrical discharge machining;
- development of scientific bases and appropriate protection technologies in the implementation of robotic systems;
- creation of new and (or) improvement of existing combined methods of treatment with the use of methods of electrical discharge treatment;
- further increase in the level of automation, both in technological preparation and in the production of products [6].

The use of an industrial robot provides an unlimited range of possibilities for electrical discharge machining of metal, wood and polymer.

It should be noted that electrical discharge equipment and technology have formed an important element in modern engineering, whose role in the development of production of high-tech products is constantly growing.

Directions of development of electrical discharge machining, the main objectives of which are the need to improve the accuracy and productivity of the process, at the present stage and in the future can be implemented through the use of new materials and the introduction of robotic systems in production.

Robotizations of production to reduce influence of a human factor and will allow to reach the best quality and quantitative index of products at the expense of the accuracy of work and resistance to influence of external factors.

References

[1] Kashapov N F, Khafizov I I and Nurullin I G 2018 *Influence of introduction of robotics on increase in efficiency of electrochemical production* IOP Conference Series: Materials Science and Engineering vol 412 **012034**

[2] 2018 *Robotics production: scope, pros and cons of the introduction of robots* Personal journal of the head "General Director" Mode of access: https://www.gd.ru/articles/9644-robotizatsiya-proizvodstva Date:27.10.2018

[3] Ogleznev N D 2014 Current state and prospects of development of electroerosion treatment News of Samara scientific center of RAS No 1-2

[4] Saushkin B P 2012 *Electroerosive processing: state and prospects of development*, Part 1 Complex contur electroerosive processing Finval group of companies Mode of access: https://finval.ru/about/articles/elektroerozionnaya-obrabotka-sostoyanie-i-perspektivy-

razvitiya Date:28.10.2018

[5] 2016 *Electroerosive processing LLC "Progress-Metalworking"* Mode of access: https://hitechmetal.ru/publications/eliektroerozionnaia_obrabotka Date:29.10.2018

[6] Saushkin B P 2012 *Electroerosive processing: State and prospects of development Part 2. Insertion of openings Metalworking equipment* "IONEX" Mode of access: http://www.ionexedm.ru/EDM_RITM_77_part02.pdf Date:28.10.2018

[7] Khafizov I I, Nurullin I G and Sadykov Z B 2018 *Application of the method of electroerosion treatment to improve the quality of the treated surface* Materials of IX conference «IMCOM-2018» vol 432

[8] Khafizov I I, Nurullin I G and Sadykov Z B 2018 Problems of development of electrochemical production of Russia and possibility of their decision *IOP Conference Series: Materials Science and Engineering* vol 412 **012042**

[9] Khafizov I I and Nurullin I G 2017 Product quality as the main factor of increase of competitiveness (on the example of JSC «Kazan helicopter plant») *IOP Conference Series: Materials Science and Engineering* vol 240 **012039**

[10] Gil'manshin I R and Kashapov N F 2014 Energy service contracts in regional engineering center for small and medium businesses *IOP Conference Series: Materials Science and Engineering*

[11] Nurullin I G and Khafizov I I 2016 The impact of innovation implementation and certification on the company's activities and reputation Modern technologies: topical issues, achievements and innovations: collection of articles of the winners of the IV international. science.- practical Conf. (Penza: ICSU " Science and education») p 44-6

[12] Akhmetov I D, Zakirova A R and Sadykov Z B 2017 New electrode-tool for the combined kerf of electrically conductive materials *IOP Conference Series: Materials Science and Engineering* vol.240 **012003**

[13] Khafizov I I and Galimov A N 2017 IT-strategy and major aspects of quality management on the market of goods and services *IOP Conference Series: Materials Science and Engineering* vol 240 **012038**

[14] Khafizov I I, Saveleva T N and Lyubtsov V S 2017 Estimation of parameters of charge carriers in dielectric materials by CELIV method *IOP Conference Series: Materials Science and Engineering* vol 240 **012040**

[15] Khafizov I I 2017 Ways of decrease in the material consumption in case of their separation by the combined methods *IOP Conference Series: Materials Science and Engineering* vol 240 **012037**

[16] Khafizov I I 2016 Economic efficiency and effectiveness of ways of separating materials electro diamond processing *IOP Conference Series: Materials Science and Engineering* vol 134 **012014**

[17] Fazlyyyakhmatov M G and Kashapov N F 2014 Corona discharge in the process of spraying protective powder coatings on piezoceramic materials *High Temperature Material Processes* vol 18 Is 4 p 273-279

[18] Kashapov N F, Gil'manshin I R and Konahina I A 2014 System analysis of the energy complex of engineering enterprise as a basic tool of effective energy management *IOP Conference Series: Materials Science and Engineering* (Bristol-UK) vol 69

[19] Kashapov N F and Sharifullin S N 2015 Hardening of the surface plasma jet high-frequency induction discharge of low pressure *IOP Conference Series: Materials Science and Engineering* vol 86 **012021**

[20] Gubaidullin D A, Kashapov N F and Zaripov R G 2017 Experimental studies for flow velocities of oscillations gas when transiting through resonance in the area of an open end of pipe *IOP Conference Series:Journal of Physics: Conference Series* Vol 789 **012017**

IOP Conf. Series: Materials Science and Engineering **570** (2019) 012050 doi:10.1088/1757-899X/570/1/012050

Application of the method of electroerosion treatment to improve the quality of the treated surface

I I Khafizov[1], I G Nurullin[1] and Z B Sadykov [1]

[1]Kazan Federal University, 18 Kremlyovskaya street, Kazan, 420008, Russian Federation

khafizov@kpfu.ru

Abstract. Intensive use of electrical discharge machining is currently determined by the very dynamic development of domestic engineering. The use of electrical discharge machining helps us to achieve the desired results, namely high machining accuracy, low surface roughness and low dependence of performance on the mechanical properties of the processed materials. Thus, the study of processes and tools of electrical discharge machining to improve the quality of machined parts is one of the urgent tasks.

A high degree of sophistication and performance of the manufactured machine tools and the significant results of machining materials in a modern production has allowed to solve a variety of complex tasks.

But the rapid progress of aviation, nuclear and electronic technology, tool production and instrumentation has necessitated the use of materials, machining of which can be made only with great difficulty or practically impossible. These include materials with very high hardness (diamond, magnetic alloy, hardened steel), brittleness (ceramics, quartz, glass, ferrites), viscosity (stainless and heat-resistant steel), as well as materials with magnetic properties. Particular difficulties arise in the shaping of such materials when they need to get grooves, narrow slits, cavities and blind holes of complex shape.

The modern stage of development of materials science is characterized by the emergence of new materials with unique properties. Which, in turn, require new processing technologies to ensure high quality of the treated surfaces. An illustrative example of the state of materials science can be called the aircraft industry, where new materials appear in the first place, and then find application in other industries, including engineering and medicine.

Accordingly, the need for effective machining of complex shapes of materials difficult to cut predetermined the emergence of a number of new methods. These include electrophysical and electrochemical processing methods. The important and necessary methods include electrical discharge machining [1].

The process of electroerosion treatment is the destruction of a metal or other conductive material as a result of local exposure to short-term electrical discharges between two electrodes, one of which is an electrode tool, and the other – the workpiece. Under the influence of high temperatures in the discharge zone there is heating, melting and partial evaporation of the metal.

Modern wire EDM machining is widely used in various industries: Aeronautics, machine tools etc. But even more the spread she received in high-tech fields of engineering, namely the engine.

IOP Conf. Series: Materials Science and Engineering **570** (2019) 012050 doi:10.1088/1757-899X/570/1/012050

Continuous improvement of one of the main indicators of the quality level of engines - specific thrust - is provided, including the use of low-strength, curvilinear-shaped parts and structural elements. With the help cutting process such the surface very difficult, and in some cases and impossible.

On the details of high-tech products there are surfaces located in hard-to-reach places, which can be processed only with the use of EDM operations. In the designs of aerospace engineering parts, titanium and aluminum alloys are most often used in all their diversity. Because of their lightness and strength, they have become indispensable materials of particularly important parts, which are subject to the most stringent requirements, both for the quality of the material and the quality of their processing. By means of electroerosion processing it is possible to carry out easily processing of these alloys with the greatest productivity.

Consider the similarities and differences of electroerosion and electrochemical processing on copying and stitching machines. The similarity is in the kinematic-geometrical scheme of geometry generation:

1) use an electrode tool with a similar design of the working part;

2) the electrode tool is fed in relation to the workpiece;

3) processing is carried out at small gaps between the tool and the workpiece;

4) there is no significant force impact on the workpiece;

5) removal of metal from the workpiece occurs most intensively in places with a smaller interelectrode gap.

Table 1 shows a comparison of the characteristics of electrical discharge machining electrochemical machining.

Table 1-Differences between electroerosion treatment and electrochemical treatment

No	Indicator of difference	Electroerosion treatment	Electrochemical treatment
1	The main difference between the methods is the physical mechanism of metal removal	This is the process of dimensional destruction of the material under the action of an unsteady electric discharge in a liquid dielectric	It is based on high-speed dimensional anodic dissolution of metal under the action of high-density electrolysis current.
2	Different performance of the equipment	More slowly	Anodic dissolution is faster by an order of magnitude
3	Ensuring accuracy	Accuracy is easier to ensure	It is more difficult to ensure accuracy. Careful protection of untreated areas is required.
4	The presence of a defective layer on the treated surface..	There is, but quite insignificant	During electrochemical treatment there is no defective layer.
5	Dissolution of the workpiece at different temperatures	The temperature does not affect the electroerosion treatment	The temperature affects the electrochemical treatment.
6	The presence or absence of wear of the electrode-tool	The wear of the electrode tool in the area of 3-10%	No electrode-tool wear. The electrode-tool is made of cheaper and easily processed material

In our opinion, the main advantages of electrical discharge machining, compared with other types of treatments, include:

- high quality of the received surfaces which do not demand further finishing treatments;
- you can treat the surface with high hardness;
- do not deform thin-walled parts due to lack of mechanical load;
- no noise.

And of course it is worth considering improving the performance and efficiency of electrical discharge machining. For example, the introduction of the electroerosion method in the production allowed "Severnaya Torgovaya Kompaniya" LLC to increase the productivity of processing deep

IOP Conf. Series: Materials Science and Engineering **570** (2019) 012050 doi:10.1088/1757-899X/570/1/012050

holes of small diameter by 3.5 times, cutting Windows by 2 times and processing composite parts, such as forming molds, by 8.3 times, reducing the cost price by 22, 48 and 82%, respectively [2].

The following factors affect the performance of electrical discharge machining:

- modes of electrical discharge machining (process parameters);
- physical and mechanical properties of tool and product materials;
- physical and chemical properties of dielectric fluid.

Thus, to increase productivity, it is necessary to choose the optimal combination of the above factors, allowing to increase the share of useful pulse energy, its power and pulse repetition rate [3].

For example, the performance of electrical discharge machining can be improved by using multi-circuit processing schemes, when one product is processed by several electrode tools isolated from each other (in sections). Each section is connected to the pulse generator autonomously. In this scheme, electrical discharge machining can be carried out several electrical discharges simultaneously. The use of multi-circuit processing scheme helps to increase productivity, but not in proportion to the number of circuits, as these circuits operate from a single supply regulator and in case of violation of the electrical discharge process in one of the circuits, the process stops in all circuits [4].

Electroerosion treatment does not exclude machining, but complements it, taking its specific place, corresponding to its features, namely: the possibility of processing conductive materials with any physical and mechanical properties and display the shape of the tool in the product.

From all this, it can be concluded that electrical discharge machining is important and has the right to be widely used in production as the basis of technologies for forming complex profiles.

References

[1] Khafizov I I, Karatayev R O and Shamsutdinova R Z 2015 Wastewater treatment by electrochemical methods *Bulletin of technological University* No 22 p 21-3

[2] 2018 Performance of electrical discharge machining *«Northern trading company»* Mode of access: http://sev-torg.com/articles-ip-proizvoditelnost-eeo.htm Date:27.09.2018

[3] Sychev E A and Tarapanov A S 2013 Prediction of surface roughness during electrical discharge machining of parts of complex configuration *News Of TulGU. Technical science* No 8

[4] Sarilov V J, Sarilova O A and Budnitz S V 2017 Improving the efficiency of electrical discharge machining of aluminum and titanium alloys on the basis of research processes in the interelectrode interval *Fundamental study* No 6 p 74-9

[5] Khafizov I I 2007 Low-waste finishing separation of scarce materials by the combined method. *Abstract of the thesis for the degree of candidate of technical Sciences* (Voronezh: Voronezh State Technical University)

[6] Khafizov I I 2013 Ways to reduce the consumption of materials when they are separated by combined methods Bulletin of Kazan technological University (Kazan: Kazan State Technical University) No 1 p 208-11

[7] Khafizov I I 2015 Processing methods with imposing of electric field at low- waste division of materials *IOP Conference Series: Materials Science and Engineering* vol.86 Art. No 012013

[8] Khafizov I I and Nurullin I G 2017 Product quality as the main factor of increase of competitiveness (on the example of JSC «Kazan helicopter plant») *IOP Conference Series: Materials Science and Engineering* vol 240 Is 1 **012039**

[9] Gil'manshin I R and Kashapov N F 2014 Energy service contracts in regional engineering center for small and medium businesses *IOP Conference Series: Materials Science and Engineering*

[10] Akhmetov I D, Zakirova A R and Sadykov Z B 2017 New electrode-tool for the combined kerf of electrically conductive materials *IOP Conference Series: Materials Science and Engineering* vol 240 **012003**

[11] Khafizov I I and Galimov A N 2017 IT-strategy and major aspects of quality management on

the market of goods and services *IOP Conference Series: Materials Science and Engineering* vol 240 **012038**

[12] Khafizov I I, Saveleva T N and Lyubtsov V S 2017 Estimation of parameters of charge carriers in dielectric materials by CELIV method *IOP Conference Series: Materials Science and Engineering* vol 240 **012040**

[13] Khafizov I I 2017 Ways of decrease in the material consumption in case of their separation by the combined methods *IOP Conference Series: Materials Science and Engineering* vol 240 **012037**

[14] Khafizov I I 2016 Economic efficiency and effectiveness of ways of separating materials electro diamond processing *IOP Conference Series: Materials Science and Engineering v*ol 134 No 1 **012014**

[15] Kashapov N F, Khafizov I I and Nurullin I G 2018 Influence of introduction of robotics on increase in efficiency of electrochemical production *IOP Conference Series: Materials Science and Engineering* vol 412 **012034**

[16] Khafizov I I, Nurullin I G and Sadykov Z B 2018 Problems of development of electrochemical production of Russia and possibility of their decision *IOP Conference Series: Materials Science and Engineering* vol 412 **012042**

[17] Gubaidullin D A, Kashapov N F and Zaripov R G 2017 Experimental studies for flow velocities of oscillations gas when transiting through resonance in the area of an open end of pipe *IOP Conference Series:Journal of Physics: Conference Series.* Vol 789 **012017**

[18] Gilmanshin I R, Kashapov N F, Gilmanshina S I and Galeeva A I 2016 Landfill energy complex based on the renewable energy installations *IOP Conference Series: Materials Science and Engineering* vol 134

[19] Kashapov N F and Sharifullin S N 2015 Hardening of the surface plasma jet high-frequency induction discharge of low pressure *IOP Conference Series: Materials Science and Engineering v*ol 86 No 1 **012021**

[20] Kashapov N F, Gil'manshin I R and Konahina I A 2014 System analysis of the energy complex of engineering enterprise as a basic tool of effective energy management *IOP Conference Series: Materials Science and Engineering* (Bristol-UK) vol 69

[21] Fazlyyyakhmatov M G and Kashapov N F 2014 Corona discharge in the process of spraying protective powder coatings on piezoceramic materials *High Temperature Material Processes* vol 18 Is 4 p 273-79

IOP Conf. Series: Materials Science and Engineering **570** (2019) 012051 doi:10.1088/1757-899X/570/1/012051

The design of automotive alternator with high efficiency

A A Khafizov, R I Valiev, R S Akhmetshin and Yu I Shakirov

Branch of Kazan Federal University in Naberezhnye Chelny, 68/19 Mira Street, Naberezhnye Chelny, 423810, Russia

almazok75@yandex.ru, rivaliev87@mail.ru, zoia2007@mail.ru, inekashakirov@mail.ru

Abstract. A technical solution aimed at increasing the motor life and efficiency of the automotive alternator to an average of 80% is suggested by replacing the permanent magnets on the stator with electromagnetic ones, reducing the number of gaps and their geometric parameters, as well as eliminating brush contacts.

1. Introduction

The proposed technical solution relates to the field of electrical equipment of the vehicle, in particular the electric generators of the vehicle (alternator). It can be used in the automotive industry.

Famous automotive alternator [1-9], which contains the rotor with pole halves on a movable shaft; a stationary magnetic circuit with the field winding between the rotor and the pole halves; the stator winding of the generator. In this case, the pole halves of the movable rotor are located between the magnetic core with the excitation winding and the magnetic core of the stator with the generator winding.

The disadvantage of it [1] is the low efficiency due to the set of gaps and their geometric dimensions. For example, there is the gap between the rotor and the fixed magnetic core with the excitation system winding and the gaps between the pole halves, as well as the gap between the stator magnetic core with the generator winding and the mentioned pole halves.

2. Problem Statement

The increasing of the service life and the efficiency of the automotive alternator to 80% by the replacement of the permanent magnets on the stator solenoid and reducing the number of clearances and geometry, as well as the exception of the brush contacts.

3. Technical Solution

Automotive alternator includes a movable shaft of the rotor pole halves, a stationary magnetic circuit with the field winding and a stationary magnetic circuit of the stator winding of the generator. The rotating pole halves of the rotor are located in the gap between the mentioned stationary magnetic cores with the excitation winding and the stator with the generator winding. A rotating shaft made of non-magnetic material has support bearings at the ends. The first magnetic core of the rotor with a six-phase winding of the machine pathogen system is situated at one end. In the middle of the shaft there are two three-phase rectifiers with "LC" high-pass filter, the input of which is connected to the multiphase winding, and the output is connected to the excitation winding of the generator. The excitation winding of the generator is located on the second magnetic circuit and is located at the other

IOP Conf. Series: Materials Science and Engineering **570** (2019) 012051 doi:10.1088/1757-899X/570/1/012051

end of the shaft. In this case, the excitation winding with a magnetic core is located inside the cavity of its stationary second stator, represented by a magnetic core with a generator winding.

The winding of the rotor of the machine exciter has a six-phase "star" scheme, symmetrically located in the grooves along the surface of the rotor magnetic circuit. The winding of its fixed stator is connected to a DC source, and the ends of the stator windings of the generator according to the scheme "star" are displayed on the clips on the generator housing. Two three-phase rectifiers are connected to the six ends of the rotor windings, on the terminals of which there is a "LC" high-pass filter. The rectifiers and the "LC" filter are evenly mounted on the shaft of the generator. The general body of the machine exciter system and the generator system is made of non-magnetic material and contains the inputs of the stator excitation winding of the machine exciter, provided for the parameters for connection to the car battery. The output of the stator winding of the generator has a three-phase circuit provided for the parameters for connection to the electrical loads of the car.

4. The principle of operation of the presented generator.

The schematic diagram of the generator is presented in Fig. 1. A battery is connected to the input (1) of the stator winding (2) of the machine exciter system (3). The current flowing through the winding (2) forms a constant electromagnetic field of the stator of the machine exciter system (3). Since the shaft (4) rotates by rotor windings (5), an induction voltage (EMF) is formed in them. The induction voltage is rectified by two three-phase rectifiers then filtered by the "LC" filter (7) into a constant voltage. The rectified voltage is applied to the excitation winding (8). Since the excitation winding is located on a rotating shaft, a rotating constant electromagnetic field is formed, the power lines of which cross the stationary winding (9) of the stator of the generator system (10). As a result, the required three-phase alternating voltage of the generator system is generated in the stator winding. In this case, the generator has two gaps with minimum dimensions, on average 1mm, the excitation system is on the one hand along the shaft, the diodes of the two rectifiers and the "LC" filter are in the middle, and the generator system is on the other side.

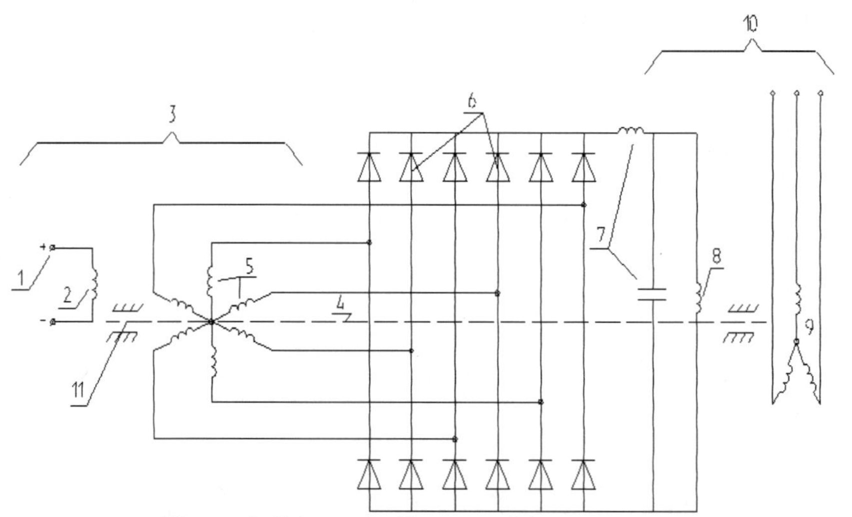

Figure 1. Schematic diagram of the generator.

1 – input, 2 –machine exciter, 3 – machine exciter systems, 4 – shaft, 5 – rotor windings, 6 – diodes, 7 –"LC" filter, 8 – excitation winding, 9 – stationary stator winding of the generator system, 10 - three – phase alternating voltage of the generator system, 11-bearings.

5. Conclusion.

This technical solution ensures the implementation of the goal – to increase the efficiency of the automotive generators on average up to 80% by eliminating permanent magnets, the use of electromagnets, reducing the number of gaps and their geometric dimensions, as well as eliminating brush connections in electrical circuits.

The proposed technical solution is protected by a utility model patent [10] and can be implemented in the automotive industry.

References

[1] Akimov A V, Akimov S V, Leykin L P 2003 *Generatoryi zarubezhnyih avtomobiley* Za rulem p 36
[2] Akimov S V, Bashkirov A V, Runovskiy S K 2002 *Vyichislitelnaya model ventilnogo generatora s razlichnyimi variantami obmotki* Avtotraktornoe elektrooborudovanie.- No **5-6** p 10
[3] Akimov S V, Chizhkov Yu P 2001 *Elektrooborudovanie avtomobiley* Uchebnik dlya vuzov. M. Za rulem 384 p
[4] Balagurov V A 1982 *Proektirovanie spetsialnyih elektricheskih mashin peremennogo toka:Ucheb. Dlya vuzov* M.: Vyisshaya shkola 272 p
[5] Galkin Yu M 1967 *Elektrooborudovanie avtomobiley i traktorov* M.: Mashinostroenie 280 p
[6] Yutt V E 2000 *Elektrooborudovanie avtomobiley* M.: Transport 320 p
[7] Akimov S V, Borovskih Yu I, Chizhkov Yu P 1988 *Elektricheskie i elektronnyie oborudovanie avtomobiley* M.: Mashinostroenie 277 p
[8] Akimov S V, Akimov A V 1995 *Avtomobilnyie generatornyie ustanovki* M.: Transport 118 p
[9] Litvinenko V V 1997 *Elektrooborudovanie avtomobiley VAZ* M.: Za rulem 240 p
[10] Ilin V I, Shakirov Yu I, Akhmetshin R S 2016 *Generator preimuschestvenno dlya avtomobilya*. Patent No 164845

IOP Conf. Series: Materials Science and Engineering **570** (2019) 012052 doi:10.1088/1757-899X/570/1/012052

Electric motor with electronic control unit for changing the polarity of the winding to improve reliability and energy efficiency

A A Khafizov, R S Akhmetshin and Yu I Shakirov[1]

Branch of Kazan Federal University in Naberezhnye Chelny, 68/19 Mira Street, Naberezhnye Chelny, 423810, Russia

almazok75@yandex.ru, zoia2007@mail.ru, inekashakirov@mail.ru

Abstract. A technical solution is proposed that increases the motor life, reliability and energy efficiency of the DC motor by replacing the plate collector with ring ones and introducing an electronic control unit for changing the polarity of the winding.

1. Introduction

The proposed technical solution relates to the field of electrical engineering in particular DC motors.

An electric motor [1-3], consisting of a frame with poles and excitation windings and an anchor with a winding and a plate collector, brush connection and a power source is known. However, such an electric motor has a significant disadvantage, which is the presence of a plate collector, which reduces the service life and reliability, due to the deposition of conductive dust between the plates of the collector, which also limits the power and speed.

2. Problem Statement

The purpose of the technical solution is to increase the reliability and service life of the electric motor by eliminating the plate collector at anchor.

3. Technical Solution

The proposed DC motor consists of a frame with the main poles and with excitation windings on them and an anchor with windings on the shaft. There are two ring collectors with a brush system, through which the anchor winding is connected to a DC source. To change the polarity of the main poles of the excitation windings, an additional unit of the analog-pulse system of electronic switching of the current direction in the excitation windings is introduced. To control the moment of switching the direction and magnitude of the current in the field windings, additional induction or magnetic sensors on the frame between a pair of different-polar main poles are provided. These sensors monitor the position of the armature winding. The number of additional sections of the main poles, additional induction or magnetic sensors, respectively, can be increased [4,5].

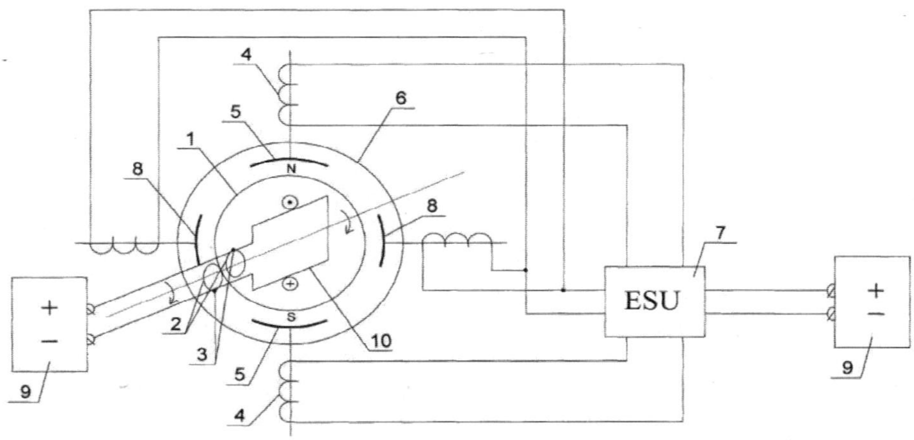

Figure 1. The schematic diagram of the electric motor. 1– anchor, 2 – contact rings of the anchor collector, 3 – brush-pin assembly, 4 – excitation winding, 5 – main pole of the frame, 6 – frame, 7 – analog-pulse unit of electronic switching of the current direction (ESU) in the excitation winding, 8 – additional induction or magnetic sensors, 9 – DC source 10 – armature winding.

4. The Principle of Operation of the Presented Electric Motor

The schematic diagram of the electric motor is presented in Fig.1. The polarity of the poles of the armature (1) is unchanged. It is determined by the direction of the current flowing from the source (9) through the brush-pin assembly (3) from one ring (2) through the winding to the other ring (2) through the brush-pin assembly (3) back to the current source. On the corresponding excitation winding (4) of the main pole (5) of the frame (6), the current passes the direction determined by the block (7). The armature winding (10) will be affected by the electromagnetic force of the desired direction.

At a certain point in time, the direction of the current in the field windings is switched. This switching is performed in addition provided by the electronic switching unit (ESU) in order to switch in a timely manner in the appropriate spatial position of the armature winding relative to the induction or magnetic sensors (8). The sensors are mounted on the frame between the main poles of the excitation windings. EMF is induced on these sensors from passing the corresponding winding with the current of the armature. Induced EMF is a command pulse to change the direction of current in the windings of the main poles of excitation. By switching the direction of the current in the excitation winding of the main poles, the corresponding polarity will be maintained, which determines the direction of the electromagnetic force on the armature winding, due to which the armature rotates in one direction. If it is necessary to change the direction, it is required to change the direction of polarity on the anchor rings, by changing the polarity of the supplied voltage from the current source [6,7].

It is possible to increase the efficiency of the motor by increasing the number of sections of the main poles, sections of the armature windings, respectively, as well as the number of induction or magnetic sensors [8,9].

5. Conclusions

Analog-to-pulse switching of the direction and magnitude of the current is a remote to start and stop the motor. The analog-pulse unit facilitates the start and stop of the rotating rotor, and when starting, the possibility of changing the polarity of the main and additional poles to change the direction of the current in their windings, from which the direction of rotation changes. At the same time, with the achievement of nominal speed, the additional poles act as sensors of the anchor winding position, and from this the ESU unit switches the polarity of the main poles in accordance with the position of the anchor winding under the main pole.

The proposed technical solution is protected by a utility model patent [10] and can be implemented in the industry.

IOP Conf. Series: Materials Science and Engineering **570** (2019) 012052 doi:10.1088/1757-899X/570/1/012052

References

[1] Ivanov-Smolenskiy A V 2004 *Elektricheskie mashinyi*. V 2-h t.: uchebnik dlya vuzov. T.1 / 2 izd., pererab. i dop. Moskva: MEI p 256

[2] Moskalenko V V 2007 *Elektricheskiy privod* M. Izdatelskiy tsentr «Akademiya» p 362

[3] Kuzmenko G P 2011 *Elektrodvigatel postoyannogo toka*. U. A. No 61940 U. MPK Ukraina

[4] Geller B 1973 *Impulsnyie protsessyi v elektricheskih mashinah* M.: Energiya p 440

[5] Moskalenko V V 2007 *Elektricheskiy privod: Uchebnik dlya studentov vyisshih uchebnyih zavedeniy* M.: Izdatelskiy tsentr «Akademiya» p 362

[6] Kireeva E A, Sherstnev S N 2013 *Polnyiy spravochnik po elektrooborudovaniyu i elektrotehnike (s primerami raschetov)* M.: KNORUS p 864

[7] Kolli Ya N, Soboleva L P, Fradkin B M 1972 *Zadachnik po teoreticheskim osnovam elektrotehniki (teoriya polya)*: Uchebnoe posobie dlya elektroenergeticheskih, elektrotehnicheskih i radiotehnicheskih spetsialnostey vtuzov M.: Energiya p 168

[8] Akhmetshin R S, Ilin V I 2016 *Generator preimuschestvenno dlya avtomobilya* Patent No 164845

[9] *Ob'em i normyi ispyitaniy elektrooborudovaniya*: SO 34.45-51.300-97 (RD 34.45-51.300-97). – M.: NTs ENAS, 2001

[10] Akhmetshin R S, Shakirov Yu I, Ilin V I, Akhmetshin M R 2016 *Elektrodvigatel postoyannogo toka s koltsevyim kollektorom* Patent poleznoy modeli RU 163 130 U1

IOP Conf. Series: Materials Science and Engineering **570** (2019) 012053 doi:10.1088/1757-899X/570/1/012053

Operation of brushes with increased service life on traction machines of electric vehicles

A N Khusnutdinov[1], E M Khusnutdinova[2] and B M Butakov[1]

[1]Kazan State Power Engineering University
51 Krasnoselskaya Str., Kazan, 420066

[2]Kazan (Volga Region) Federal University
18 ul. Kremlyovskaya, Kazan, 420008

Abstract. The paper describes how to use a split brush with increased service life. Serial brushes were compared with those proposed by design, reliability and cost characteristics. The authors propose a method of how to improve the design of serial brushes and demonstrate a dynamic performance curve.

Introduction

The operation of electrical traction vehicles of rolling stock includes the periodic replacement of brushes subject to mechanical and electroerosive wear.

Operational investigations show that the service life of a serial brush can significantly vary due to different operational factors. The observations demonstrate that the service life of the brush is found to be much lower compared with the calculated values. This happened because the manufacturer failed to comply with the appropriate workmanship: low-quality materials; deviations from the design (collector geometry excursion due to insulation shrinkage); poor setting of machine black band; poor quality balancing of machine rotating parts, etc. All these factors make the brushes wear faster and reduce their service life.

Long-term observations show that these negative phenomena are pretty much difficult to control, i.e. this approach is ineffective. This requires many efforts to launch more stringent regulations for manufacturing the electrical machines; radical renewal of hardware at maker facilities, etc. It is necessary to search for beneficial solutions aimed at increasing the service life, manufacturability, cost reduction and increase of the operational reliability of the brushes [1, 2, 3].

Description of brushes with increased service life

The recommended brush is fundamentally different in design features and in terms of operation from the existing ones. This brush is composite in design [4], while the existing brushes are solid ones. The chemical composition and proportions of the components of composite brushes and their sizes do not differ from the serial ones. The design values of the electrical machine and the brush holder also remain unchanged. This means that the new type of brushes do not pose any technological risk while in service [5, 6].

The main idea of the project is to increase the service life of electrical machine brushes (additionally from 30% to 44%), depending on their design.

IOP Conf. Series: Materials Science and Engineering **570** (2019) 012053 doi:10.1088/1757-899X/570/1/012053

The brushes experience a continuous wear every time the motor running. Worn brushes cannot be restored and must be replaced along with valuable components:

1) flexible non-insulated braided wire. Material: refined copper.
2) tinned cable lugs. Material: copper.
3) unworn section of brush (the brush wears out 2/3 of its height). Material: graphite, carbon black, copper.
4) materials for connecting the conductive wire with the carbon-graphite section of the brush (caulker, electrically conductive adhesive);

The estimated cost of the not completely used upper section of the brush is 40% of its total cost.

The proposed engineering solution makes it possible to use the upper (not completely used) section of the brush many times by means of a removable liner, which, in turn, can significantly reduce operating costs. The entire process of replacing the replacement liner greatly facilitates the work of the maintenance personnel, and reduces the time of maintenance up to 70% as the personnel does not need to loose and secure the lug bolt of the conductive shunt of the brush.

Fig. 1 shows a serial brush and its main components.

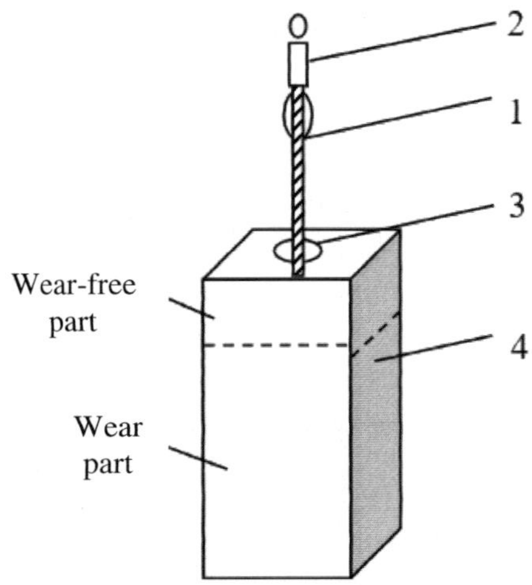

Fig. 1 A serial brush.
1 - tinned lug; 2 - flexible non-insulated braided copper wire;
3 - materials for connecting the conductive wire with the carbon-graphite section of the brush (caulker, electrically conductive adhesive); 4 - carbon-graphite section.

Functionally, the composite brush does not differ from the serial one. The main difference lies in its design. It consists of two sections: the main section that supplies power from the brush holder, and a replacement graphite liner. As the electrical machine is in service, the replacement graphite liner wears out almost completely and during the routine inspection/repair is replaced with a new one [7, 8].

The main section of a new brush includes valuable components described above. As the wear part of the brush when used properly is a replacement graphite liner, its upper section is wear-free not requiring the replacement. Obviously, it is beneficial in terms of saving funds on valuable components, which are waste during the operation of a serial brush. Therefore, the use of a composite brush excludes the possibility of under-utilization of the design service life of the brush.

The wide application of the proposed brush design is determined by the typical brush holder. This unit is designed to provide accurate positioning of the brushes on the collector and reliable contacting. The brush holder shall be so designed that the brush is nearly to be engaged to ensure stable pressing on the brush, regardless of the degree of its wear in height and reliably insulated from the housing. The brushes are held in the brush holder by special sleeves. The technical condition of the brush holder should meet the specifications for their operation.

Fig. 2 depicts a fragment of a serial brush holder. The sleeve of the brush holder 1 receives the upper section of the composite brush 2 and the lower replacement graphite liner 3 sliding along the collector 4 of the electrical machine. The brush is pressed to the collector by a spring 5. This force creates a reliable contact in the contact surface 6 between the main section of the composite brush and the replacement graphite liner. A puller 8 is used to inspect the condition of the brushes and replace the extremely worn bottom replacement section. The tip of the puller 8 goes into the hole 7 of the lower replacement section 3. Then the puller 8 takes the lower replacement section 3 out of the brush holder sleeve 1.

The recommended brush with an increased service life has been successfully tested in actual operating conditions in test trolley buses and trams at Metroelectrotrans, the Municipal Unitary Enterprise of Kazan, as well as in the facilities of Nizhnekamskneftekhim, JSC in Nizhnekamsk and Electrotransport, LLC in Naberezhnye Chelny. These tests confirmed the increased service life and reliability of its operation, both on rolling stock and on synchronous machines. Weekly inspections of motors equipped with composite brushes showed the absence of any tell tales of undesirable processes in motors (sparking, local burnout of collector plates, destruction of brushes, flashing, winding overvoltage, current pulsations leading to dynamic beating of the armature of the traction motor, etc

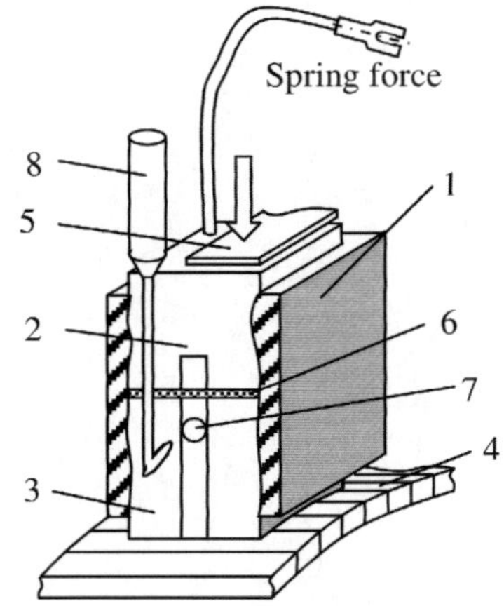

Fig. 2 A schematic of brush holder sleeve with a composite brush:
1 - brush holder sleeve; 2 - upper section of composite brush; 3 - lower replacement liner; 4 - collector; 5 - brush holder spring; 6 - contact surface between sections of composite brush; 7 - hole for replacement graphite liner puller; 8 - puller.

The replacement liners of brushes with increased service life shall be replaced as soon as they are worn out. To improve the switching conditions of new liners, they are recommended to pre-grind to fit the radius of the collector. A grinding matrix serves for this purpose (Fig. 3). Ease of handling and small size shorten the time of the installation of replacement liners that, in turn, minimizes the downtime of the electrical machine.

IOP Conf. Series: Materials Science and Engineering **570** (2019) 012053 doi:10.1088/1757-899X/570/1/012053

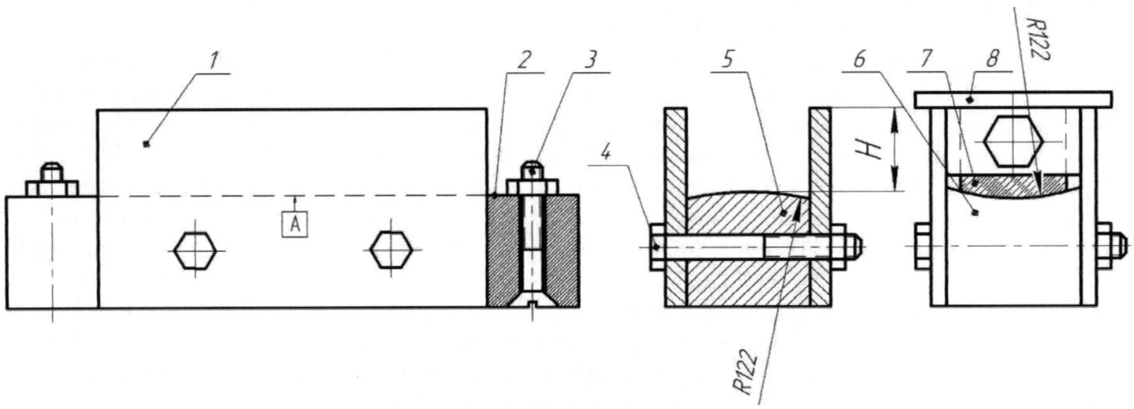

Fig. 3 The grinding matrix
1 - jaw, 2 - abrasive sheet, 3 - screw, 4 - bolt, 5 - matrix (curved), 6 - matrix (concaved), 7 - brush, 8 - brush holder for 2 brushes.

The dynamic performance of the split-type brushes are shown in Fig. 4 [9, 10].

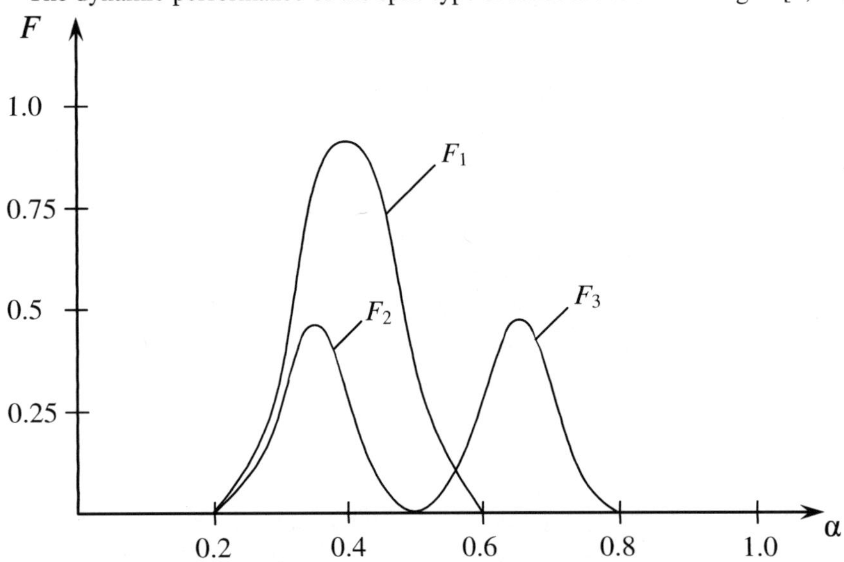

Fig. 4 Dynamic performance of split-type brushes

F_1 – pulse produced by impact of solid brush;
F_2 – pulse produced by impact of upper section of brush;
F_3 – pulse produced by impact of lower liner.

Using split-type brushes makes it possible to:
1. Reduce the impact force of the collector;
2. Spread impact pulses F_2 and F_3 in time;
3. As a result, the brush less detaches from the collector surface;
4. All this improves the switching and reduces the contact resistance (collector – brush);
5. Reducing the impact of mechanical and electrical (switching) factors reduce the wear of the components in rubbing contact, i.e. collector and brushes.

The existing approach of how to operate the brushes is inefficient resulting in:

IOP Conf. Series: Materials Science and Engineering **570** (2019) 012053 doi:10.1088/1757-899X/570/1/012053

- to the low service life of brushes and in some cases it can drop to 44%;
- copper waste;
- significant labor costs in their service.

In this regard, a new approach of operating the brushes of electrical machines is recommended to remove these drawbacks. However, the design of the brush holders of electrical machines and the brand name of brushes remain the same. The new approach of operation of brushes has been confirmed through lab and in-service tests [11]. The way they work on traction engines of urban electric transport (trams and trolley buses), electric and diesel locomotives yield encouraging results.

Summary

Thus, the in-service tests carried out on diesel locomotives, electric locomotives (Kazan branch of the Gorkovskaya Railway), trams, trolley buses (Metroelectrotrans, Kazan) demonstrate the following essential results:

- the service life of brushes used by the new approach as compared with the existing ones is increased by 30% or higher depending on the height of the brush;
- the cost of a new brush is reduced by 28.8%;
- labor costs are reduced to 70%;
- the payback period is zero.

References

[1] Yermuratsky P V, Kosyakin A A and Listvin A S 1995 *Electrical engineering and electronics reference book.* M.: Energoatomizdat, p 353

[2] Electric cars handbook / under general editorship of I. P. Kopylova vol. 1. M.: Energoatomizdat, 1988. p 453

[3] Electric cars handbook / under general editorship of I. P. Kopylova vol. 2. M.: Energoatomizdat, 1988. p 688

[4] Brush assembly of an electrical machine: Patent 64821, the Russian Federation: IPC H 01 R 39/40 / R. G. Idiyatullin; applicant and patent owner R. G. Idiyatullin – No. 2007102195/22; published on 01.19.2007

[5] Litvinenko R S, Idiyatullin R G and Aukhadeev A E 2015 *Simulation of failures of the electrical system and its components at the stage of development*. Electrical and information and systems No 4 vol 11 pp 17-25

[6] Idiyatullin R G, Khusnutdinov A N, Vdovin A M, Popov A V and Kisneeva L N 2012 *Evaluation of the operational reliability of traction generators.* Proceedings of higher educational institutions. Power engineering problems No 11-12 pp 108-114

[7] Filina O A and Idiyatullin R G 2015 *Approximation of the current-voltage characteristic of the brush contact of a composite brush with an increased service life.* Modern trends in the development of science and technology. No. 8-1 pp 134-136

[8] Khusnutdinov A N, Rybakov R B and Allyamova N M 2014 *The use of iterative methods in the design of electrical machines*. Scientific challenge of transport means in Siberia and the Far East. No. 1-2 pp 272-274

[9] Pavlov P P, Litvinenko R S, Mubarakhin M N, Yushin I O and Nigmatullin V M 2008 *A technique for selecting a rational variant of a multifunctional aircraft system.* Economical aspects in aircraft engineering. Vol 51 Is 2 pp 198-204

[10] Filina O A and Tsvetkov A N 2019 *Evaluation of the operational life of direct current motors* IOP Conference Series: Materials Science and Engineering. Vol. 489. Is.1. - Art. No. 012016

[11] Khusnutdinova E M, Khamidullina G R, Pavlov P P, Litvinenko R S and Khusnutdinov A N 2018 *Training transport specialists based on a contemporary view of self-organization of sophisticated developing systems.* IOP Conference Series: Materials Science and Engineering. Vol. 412. Is.1. - Art. No. 012047

IOP Conf. Series: Materials Science and Engineering **570** (2019) 012054 doi:10.1088/1757-899X/570/1/012054

Issues related to energy saving in petrochemical sector and methods for their solution

E M Khusnutdinova, I A Konakhina and G R Khamidullina

Kazan (Volga Region) Federal University
18 Kremlyovskaya Str., Kazan 420008

Abstract. The paper describes the case study of energy saving problems in the petrochemical sector and discusses methods of recovery of low-grade renewable energy resources and approaches to assessing their potential use.

The coordinated action of the power facilities within the industrial undertakings (distinguished below in three main parts) make it possible to efficiently consume fuel and energy resources in the required amounts. These are [1]:

1. Sources of energy resources, i. e. dedicated companies or in-house systems that produce the required types of energy.
2. Systems of transportation and distribution of energy resources between consumers.
3. Consumers of energy resources.

Each of the party in the energy resources producer – consumer system has its own hardware and is characterized by certain indicators of energy and thermodynamic efficiency. This may result in a situation when high efficiency indicators of some of the system participants are leveled by other participants, so that the total efficiency of the heat and power system turns out to be low [5,9,10].

The most problematic in this regard is the stage of consumption of energy resources. Inspection of some petrochemical companies shows that the actual *energy resource* consumption 1.7÷2.6 times exceeds the theoretically required one [1], i.e. targeted use of energy resources is about 43% of the actual necessary costs of production technologies. The main causes for this situation to happen in the domestic industry are:

1) failure to select correct machinery, which in real conditions of operation can not facilitate the power usage effectiveness that correspond to the actaul level of technical development;
2) inadequately arranged coordination between heat and power systems and heat engineering systems;
3) insufficient or inefficient use of secondary medium- and high-grade energy resources;
4) low-grade secondary energy resources, which include heat flows having a temperature [2]:

$t < 90°$ C for liquids;

$t < 150°C$ for gases;

are almost never used in industry and energy systems. However, the temperature level of consumption of heat resources in processes, as a rule, is low (see Table 1), and therefore the output of low-grade secondary energy resources used by the petrochemical undertakings makes up almost 90% of their total secondary energy resources.

Table 1 Heat consumption by a petrochemical undertaking [2]

Consumers	Heat consumption		
	Total consumption of steam and hot water,%	Temperature level of heat absorption	
		Up to 150 °C	Over 150 °C
Process	65	55	10
Heating and ventilation	25	25	-
Hot water supply	10	10	-
Total:	100	90	10

The linking chart of power production and distribution systems typical applied in petrochemical undertakings is shown in Fig. 1.

In the conventional approach, the heat and power systems of industrial consumers are solely considered as a sufficient power source adequate in quality as required by the process regulations [3]. The mode of operation of heat and power systems is subject to the conditions dictated by the consumer. Such an approach often leads to miscalculations while setting up heat engineering systems and heat and power systems, i.e. to the latent or apparent overspending of fuel and power resources, which naturally affects the cost of products.

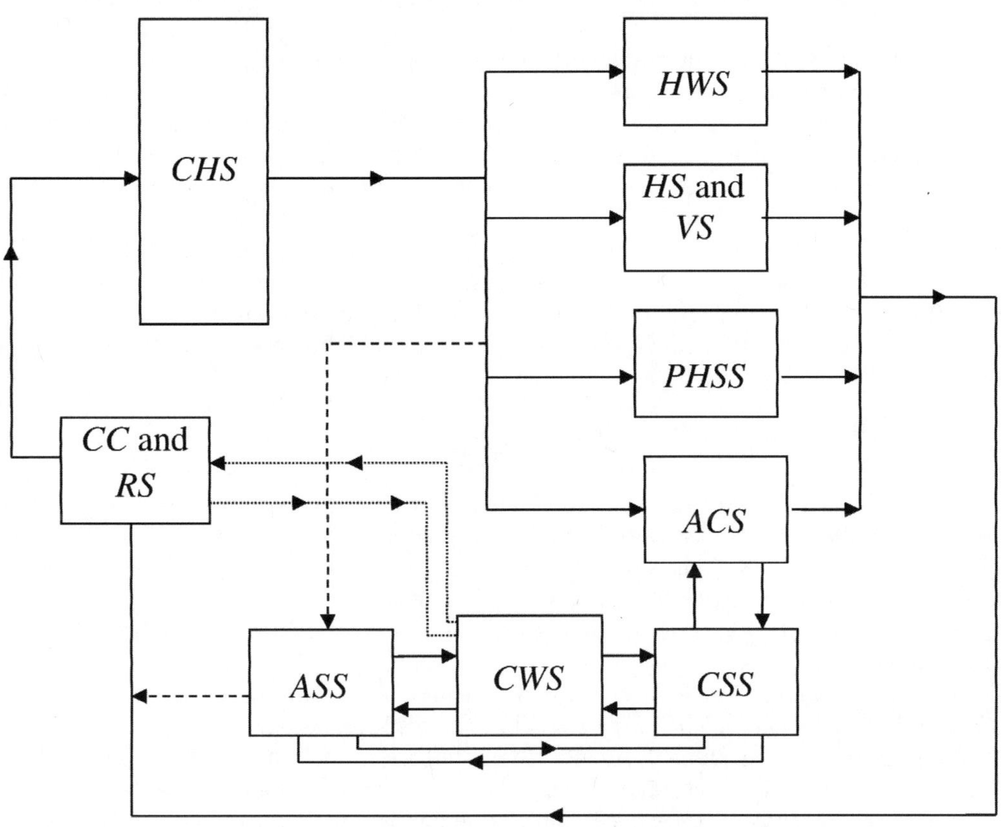

Fig. 1 The linking chart of power production and distribution systems typical applied in petrochemical undertakings

CHS stands for centralized heating system;
HWS stands for hot water supply;
HS and *VS* stands for heating and ventilation systems;
PHSS stands for production heat supply systems;
CC and *RS* stands for condensate collection and return systems;
ACS stands for air conditioning systems;
CSS stands for cold supply systems;
CWS stands for cooling water system;
ASS stands for air supply systems.

In particular, the seasonal fluctuation [1, 3] has a rather strong influence on the overall indicators of the energy efficiency of industrial undertakings. In summer time, the secondary energy resources are usually supplied in excess, and at the same time the people face problems associated with insufficient amount and quality of cooling fluids due to an increase in the temperature of the cooling water. At the stable lowest outside temperatures, on the contrary, the heat energy is consumed to excess because of an increase in heat losses through cladding taking much time to identify and requiring substantial economic costs [8].

The key challenges to be addressed include:

1) ensuring the balance of energy resources of the required characteristics in any period of time for reliable and cost-effective operation of individual units and the entire production association;
2) optimal selection of utilities according to thermophysical and thermodynamic parameters;
3) definition, nomenclature and modes of operation of reserve and accumulating utility sources, as well as alternative consumers of secondary energy resources in the period of their excess supply [4];
4) identification of space for the energy efficiency of a production to grow at the current level of technical development and in the distant future.

Closed heat supply systems are one of the main components of waste-free production systems. The recovery of low-grade heat and its transformation to the required temperature level may return a considerable part of the energy resources, which is usually discharged into the atmosphere back to the process [6-7].

Cold supply systems of an industrial undertaking do not have a significant impact on the supply of secondary energy sources, but can be regarded as consumers of heat when developing recovery measures. In particular, one of the effective solutions for smoothing seasonal fluctuations in the power consumption system in the course of the year is the inclusion of absorption heat transformers in recovery systems.

References

[1] Nazmeev Yu G and Konakhina I A 2002 *Heat power systems and power balances of production facilities*. M: Moscow Power Engineering Institute Press

[2] Grigorov V G, Neyman V K, Churakov S D, Semenyuk L G and Presich G A 1987 *Recovery of low-grade secondary heat resources at chemical plants*. M.: Himiya

[3] Sazanov B V and Sitas V I 1990 *Heat power systems in industrial undertakings*. M.: Energoatomizdat. p 304

[4] Khafizov I I 2013 *Ways to reduce the consumption of materials in their separation by combined methods Bulletin of Kazan Technological University* Vol. 16. No. 1. pp 208-211

[5] Khusnutdinova E M, Konakhina I A and Khamidullina A F *Comparative analysis of the thermal-hydraulic efficiency of intensified heaters of viscous oils and petroleum products.* International scientific and technical conference "Innovative engineering technologies, equipment and materials - 2016" - Kazan, 2016. Part 1. pp 360-364

[6] Khusnutdinova E M, Konakhina I A, Khamidullina G R and Mamonova A O 2018 *Ways to Transfer Oil and Oil Products in Heating Conditions and Methods for Their Enhancement.* IOP Conf. Series: Materials Science and Engineering 412, № 012046

[7] Danilov O L and Garyaev A B 2010 energy Saving in heat power engineering and heat technologies: textbook for universities – M.: publishing house MEI, p 424

[8] Amirova S S, Bulatova V M and Chekunov N I 2010 *Modern ways of energy-saving in the industry of petrochemical complex* / Proceedings of the Samara scientific center of the Russian Academy of Sciences, volume12, №4 (3) pp 540-543

[9] Kapustin M In 2013 *Modernization of the refining and petrochemical industry in Russia* // Himagregaty. – № 4 (December). pp 12–16

[10] *Project: energy-efficient Russia.* [Electronic resource] URL: http:// www.energy2020.ru/energy_russia/about.

IOP Conf. Series: Materials Science and Engineering **570** (2019) 012055 doi:10.1088/1757-899X/570/1/012055

Current issues of resource and energy savings in the domestic oil and gas complex.

I K Kiyamov[1,3], L S Sabitov[1,2], G I Kabirova[1], L Sh Akhtyamova[1] and L Sh Iskhakova[1]

[1]Kazan (Volga region) Federal University, Kremlevskaya str., 18, Kazan, 420008, Russia

[2]Kazan State Power Engineering University, 51 Krasnoselskaya street, Kazan, 420066, Russian Federation

[3]Autonomous non-profit organization of higher education "Academy of Social Education", Isaeva str.,12, Kazan, 420039, Russian Federation

l.sabitov@bk.ru

Abstract. The paper deals with the effective implementation of resource and energy saving policy in the oil and gas complex of the Russian Federation (RF). modern regulatory and legal framework is analyzed. The key features of the new "Energy Strategy until 2035" are identified. The development of the modern oil and gas complex of the Russian Federation is analyzed. Justified the problems of direct and indirect effects of the efficiency of resource and energy saving in the oil and gas industry of the Russian Federation, focusing on the problem of metrological provision of domestic oil and gas industry. Promising technologies in the field of geological exploration, as well as in the field of oil and gas production, aimed at improving the efficiency of the resource and energy saving process in the oil and gas complex are proposed.

The organization of an efficient process of resource and energy savings today is the one of the priorities of the real sector of the domestic economy. This process is considered as one of the main sources of accelerated economic growth. Increasing competitiveness, economic security and sustainable socio-economic development of the country, as well as raising the level and quality of life is not possible without an effective implementation of building energy and resource saving. The key element in this situation is the process of effective modernization of the technological development of the domestic economy through the effective transition to environmentally responsible consumption of the domestic resource base. The results of solving this problem will ultimately determine the place of Russian society among the world's economically developed countries with trans.

Practice shows that, despite the fact that this issue is relevant for the past ten years, up to now a full realization of the potential energy and resource saving in the domestic economy is not involved to the full. Thus, according to the Russian Federation Ministry of Energy (RF), the energy savings potential in the final sectional key industries RF until 2020, an average is about 200 million tons of reference fuel (toe) [1]. The leaders are the electric power industry (42 million toe), housing and utilities (33 million toe) and the oil and gas complex of the Russian Federation (21 million toe) [1]. Paradoxically, the barriers that impede the organization of an efficient energy and resource saving process in the domestic economy have long been known – the lack of motivation, the lack of adequate experience in financing resource and energy saving projects, the insufficient organization and coordination, and the lack of information [2]. Nevertheless, it must be noted that today a fairly extensive regulatory and legal framework has been created in the questions of increasing resources and energy efficiency of the domestic economy. The main documents include: The Federal Law No. 261 of November 23, 2009 "Energy Saving and on Increasing Energy Efficiency, and Amendments to Certain Legislative Acts of the Russian Federation"; the federal and departmental target programs; the government Russian program "Energy efficiency and energy development"; regional and municipal programs of energy saving and energy efficiency increasing and a number of others.

Separately, the "Energy strategy of Russia till 2030" approved by order of the RF Government, dated 13.11.2009 should be selected. It provides the organization of the most efficient natural energy resources utilization, as well as the effective realization of the energy potential in order to ensure sustainable growth of the domestic economy and improving of economic development. This strategy is based on the previous "Energy Strategy of Russia for the period until 2020", and includes such areas of the state long-term policy as the organization of budget efficiency, the organization of energy security and energy efficiency of all sectors of the domestic economy based on principles of environmental policy.

The high volatility of the global energy markets over the past five years has led to the fact that already in 2014 a new draft of the "Energy Strategy until 2035" was approved, which was supplemented and revised in February 2017. Despite the fact that the "Energy Strategy 2035" is successive in relation to the previous one; nevertheless, a number of additions appeared in it [3].

At first, the emergence of new divisions on emerging new challenges for the Russian energy sector, considering the main strategical accents; tax subsections tax, customs and tariff policy, the energy sector; environmental, social policy subsections, human capital in the energy sector and separately added the subsection of the high-level hydrocarbon processing.

At second, the forecast data regard to socio - economic development of the Russian economy, were revised because of the availability of a temporary slowdown in economic activity and, as a result the decline in domestic consumption of energy is projected, and also the average annual decline in the rate of growth in the 2013-2035.

Finally, in the new version of the strategy, the great attention is paid to the process of global energy markets transformation in the post-crisis period. Here are the issues of reducing global primary energy consumption, a shift in global demand towards developing countries, where the presence of the Russian Federation is problematic due to a number of factors, as well as a decrease in demand in traditional for the Russian Federation markets. Noting the important role of non-traditional hydrocarbon sources, also the accelerated development of renewable energy sources [3].

Based on the above, let us disagree with the thesis about the lack of information today aimed at organizing an efficient process of energy and resource saving in the domestic economy. Modern Russian legislation provides for a number of effective methods and mechanisms for stimulating energy and resource saving processes in the real sector of the economy (concessional lending and taxation, financing and co-financing processes for returning and non-refundable energy and resource saving measures, tariff policy, etc.). Certainly, solving the problem of efficient resource and energy saving, as well as increasing energy efficiency is a long-term one, however, for the real sector of the Russian economy, today this is a chance to switch to a model of sustainable advanced development, which in practice will provide:

- increasing the level of welfare of the population on the basis of accelerated economic growth;
- rate of state's energy security on the basis of reducing dependence on export of hydrocarbon resources;
- development of innovative, knowledge-intensive sectors of the Russian economy;
- a comprehensive solution of social and environmental problems in energy-intensive industries;

IOP Conf. Series: Materials Science and Engineering **570** (2019) 012055 doi:10.1088/1757-899X/570/1/012055

- development of a model of socially-oriented businesses in accordance with the concept of sustainable development to it.

The analysis showed that although the foregoing, a number of experts [4,5], the authors of which are solidary work indicate to a number of problems in the Russian economy industry associated with low levels of extraction and processing depth of hydrocarbon resources; reduction in the quantity and quality of exploration ; the significant deterioration of the geological and climatic conditions of new hydrocarbon fields ; a high degree of depreciation of fixed production assets ; the lack of highly developed energy infrastructure in a number of backbone regions engaged in the extraction and processing of hydrocarbons [5] . All these moments predetermine the conduct of further research in the framework of this work, regarding the topical issues of resource and energy conservation in the domestic oil and gas complex.

Modern oil and gas complex of the Russian Federation is a kind of donor of the domestic economy. It plays a significant role both on internal and on external (global) energy markets. For more than twenty years, the development of the domestic oil and gas complex has fully determined the quality and dynamics of the socio-economic development of our country. Today - this is one of the most competitive economy industries in the context of world economic relations. Oil and Gas are the main Russian export goods. In general, the dynamics of the Russian economy development on the state oil and gas complex, because more than 70% of hydrocarbon exports generate more than 40% of federal budget. The efficiency of the Russian oil and gas complex functioning, determines the basis balance payments at domestic economy, the backbone element of supporting the national currency. Its development closely interacts with many sectors of the real sector of the Russian economy, largely determining their development trends in the near future. The oil and gas complex of the Russian Federation is the source of "maximum investment multiplier effects", creating and stimulating, a high level of demand for goods and services of interdependent industries in practice.

For example, in Norway, the multiplier size is 1.65, in the USA is about 2, in Australia is more than 2. The domestic multiplier of the oil and gas industry is practically comparable to the US and is 1.9, which "corresponds to the level of oil-producing industrialized countries" [7]. Today, Russia is one of the leading oil and gas world powers. So by the end of 2015, the Russian economy ranked first place in oil production, ahead of its rival Saudi Arabia by almost 30 million tons [8]. However, by the end of 2016, it went down to the second line of the countries rating in oil production, nevertheless remaining in the TOP-3 countries, the leaders in oil production for the year.

The last ten years, the dynamics of oil prices, is influenced by the economic crisis and is determined by a number of factors, among which the most important are the world growth retardation the development of alternative energy, the technological innovation in the field of oil production, the high volatility of demand and supply in the hydrocarbon markets (Figure 1).

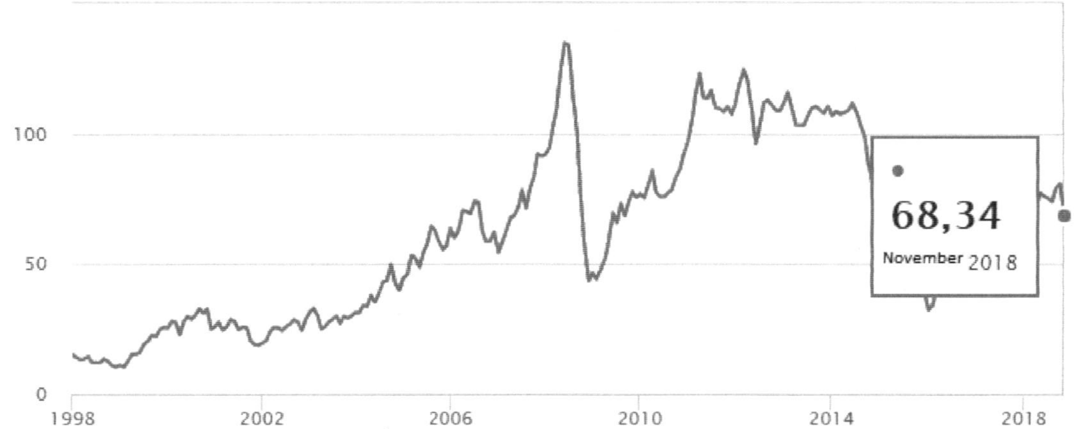

IOP Conf. Series: Materials Science and Engineering **570** (2019) 012055 doi:10.1088/1757-899X/570/1/012055

Fig. 1. Dynamics of prices on futures contract for Brent oil (MOEX, USD for barrel) [10]

During the economic crisis of 2008-2009 the average annual price of oil fell by almost 40 %, exposing the problems and failure of the Russian economy in the system analytics and forecasting the behavior of industrial markets. The fall in oil prices on the world markets triggered the domino effect of the Russian economy. Sharply decreased foreign exchange earnings from the sale of hydrocarbons, the rise in price of money is occurred, the inflow of foreign capital in the form of investment is decreased, the rates of economic growth are reduced and the same with the level of internal revenues. However, in 2011 prices returned to the pre-crisis level, which was a catalyst for the regeneration of placement of funds in recoverable and unconventional sources of hydrocarbons. Thus, according to the results of 2013-2014. the trend of "overproduction" of hydrocarbons was already clearly observed, and in 2015, supply exceeded demand, and led to the fact that in the first half of 2016, the world oil prices reached to a multi-year minimum of $ 28 a barrel mark Brent [10]. However, in the second half of 2016 thanks to the efforts of OPEC countries and the largest oil exporters, the price adjustments occurred, as a result in a barrel of oil at the end of 2016 and at the beginning of 2017 is cost $ 48-57 [10]. At the end of 2017, oil rose in price by 22%, largely agreement by OPEC + contributed to this [10]. In the first half of 2018, world oil prices had a positive trend, reaching a maximum of 82.73 dollars per barrel in September, but the dynamics of October-November of the 2018 year are worrying, since the beginning of October the price of Brent has increased by 28% [10]. Despite the fairly high prices prevailing on the world markets, today, there are a number of problems in the domestic oil and gas complex, the solution of which will predetermine the further development of the Russian economy. Practice shows that these problems are directly or indirectly related to the issues of improving resource efficiency and energy saving in the oil and gas complex of the Russian Federation.

1. The deterioration of the state of the domestic mineral resource base of hydrocarbons, in view of the decrease in the current proven reserves and the rate of their reproduction, which makes it imperative to organize work to convert the country's high oil and gas potential into active hydrocarbon reserves.
2. The growth in the share of difficult -to-recover oil and gas reserves from their total number, which predetermines the introduction of innovative, resource-saving production technologies and the use of modern hydrocarbon solvent-slug injection.
3. Decrease of the average daily production rate of oil wells, due to the high depletion of hydrocarbon fields.
4. The backlog of domestic technologies and the high level of depreciation of the basic production assets of the Russian oil and gas complex.
5. The existing taxation system, as a result of which almost half of the company's revenue goes to paying taxes, which does not reduce the motivation to implement projects in the field of resources and energy saving [11].
6. The strong dependence of the domestic oil and gas complex on the global market for hydrocarbons.
7. Irrational organizational structure of the vertically- integrated companies, implementation of inefficient business processes, the low response rate to external factors, in view of the weak development of IT - technology, violation of uniform standards for storing and transmitting information.
8. Problems of metrological support of the oil and gas complex of the Russian Federation, due to the lack of "modern means of monitoring the parameters of technological processes in the production of oil and petroleum products and their sale abroad" [12]. This problem is due, firstly, to an outdated regulatory framework governing the issues of metrological support for the accounting of oil and petroleum products, as well as sanctions imposed by the European Union and the United States, implying a full-scale ban on the supply of new technology and specialized equipment in the field of extraction, transportation and processing of hydrocarbons.

Of course, each of the above problems is a topic for a separate study, but today we can offer a number of promising areas that contribute in practice to the effective realization of the potential of the resource and energy saving in the Russian oil and gas complex. According to specialists of the National Research University Higher School of Economics [13], within the framework of promising geological exploration technologies , it is necessary to consider "exploration of hydrocarbon deposits in 4D format high-thresolution, and I compulsory visualization of the results "; " exploration of the offshore hydrocarbons deposits in the 3D/4D format based on submarine autonomous self-guiding devices by means of GPS / GLONASS navigation " ; "data acquisition software big data of exploration based on the use of artificial intelligence. "In the field of advanced oil and gas technology, it is necessary "to create underwater robotic systems for hydrocarbon production, which are capable of operating under the conditional of the remote control in the automatic mode"; "To create means of continuous monitoring of the state of hydrocarbon fields reservoirs with the high resolution and computer processing of the received information with the 4D-visualization based on the obtained results" ; "technology development of ice and seismically stable platforms for hydrocarbon production in arctic conditions" ; "development of technologies for drilling wells, secondary opening productive strata " ; "development of hydrocarbon production technologies from unconventional fields under abnormal conditions" and a number of others.

In general, we believe that today the state should pay close attention to the innovation and investment problems of the oil and gas complex, should contribute to creating additional potential opportunities for its participants, create a powerful incentive and help increase the motivation of market participants in addressing the priorities for the strategic development of the domestic oil and gas complex [14]. As priority measures, it is necessary to consider : issues of modernization of the industry through the introduction of innovative technologies in the processes of exploration, production and processing of hydrocarbons , in order to update the basic production assets and improve the efficiency of resource and energy-saving policies of oil and gas companies ; issues of infrastructural nature, aimed at the formation of an integrated transport infrastructure covering the needs of all sectors ; issues of reengineering of main , auxiliary and management 's business processes in the oil and gas sectors corporations ; the question of creating a flexible, adaptive management model of oil and gas companies capable of quickly responding to changes in the external environment in the context of modern economic realities and trends in the development of the world oil and gas market .

Thus, timely decision of the above issues, in practice will contribute to the creation of a powerful economic mechanism, technological and legal mechanism, on the effective implementation of which will depend the further socio-economic development of the Russian Federation, the level and quality of welfare, development of foreign policy and strengthening the country's position in the world economy.

References

[1] Hovak A V The results of the work of the fuel and energy complex of Russia in 2013. Tasks for the medium term (*Electronic Materials)* http://energetika-restec.ru/netcat_files/userfiles/energetika-restec/14-01-2014.pdf

[2] Fatkhutdinov R I, Ayupova A Z, Baimova D I and Galiev G F 2017 Economic Aspects of Energy Saving *Issues of Economics and Law* No *2* p 65

[3] Russia's energy strategy for the period until 2035 (*Electronic Materials* URL : https://minenergo.gov.ru/system/download-pdf/1920/6905521.11.2018.

[4] Ilinsky A A, Shamalov Yu V 2008 Priority strategic initiatives for the development of the oil and gas complex of the Russian Federation *Problems of the modern economy N 4 (28)* (*Electronic Materials)* URL: http://www.m-economy.ru/art.php?nArtId=2220

[5] Larichkin F D, Cherepovitsyn A E, Fadeev A M Problems and opportunities for the innovative development of the oil and gas complex: a strategic vision (*Electronic Materials)* https : // cyberleninka . ru / article / n / problemy - i - vozmozhnosti - innovatsionnogo - razvitiya -

IOP Conf. Series: Materials Science and Engineering **570** (2019) 012055 doi:10.1088/1757-899X/570/1/012055

neftegazovogo - kompleksa - strategicheskoe - videnie
[6] Oil and gas and non-oil revenues (*Electronic Materials)* http : // www . iminfin . com / areas - of - analysis / budget / dohody - fb / neftegazovye - i - neneftegazovye - dokhody ? territory = 45000000
[7] Ovinnikova K N The current state of the oil and gas complex of Russia and its problems (*Electronic Materials)* https : // cyberleninka . ru / article / n / sovremennoe - sostoyanie - neftegazovogo - kompleksa - rossii - i - ego - problemy
[8] Galyautdinov I M Improving the economic efficiency of oil production at a late stage of field development through the introduction of energy-saving measures(*Electronic Materials)* http://old.spmi.ru/system/files/lib/sci/aspirant-doctorant/avtoreferaty/2017/2017-3/avtoreferat_galyautdinov_0.pdf
[9] List of countries for the extraction of oil (*Electronic Materials)* https : // ru . wikipedia . org / wiki /% D 0% A 1% D 0% BF % D 0% B 8% D 1% 81% D 0% BE % D 0% BA _% D 1% 81% D 1% 82% D 1 % 80% D 0% B 0% D 0% BD _% D 0% BF % D 0% BE _% D 0% B 4% D 0% BE % D 0% B 1% D 1% 8 B % D 1% 87% D 0% B 5_% D 0% BD % D 0% B 5% D 1% 84% D 1% 82% D 0% B 8
[10] New collapse in oil prices (*Electronic Materials)* https://www.rbc.ru/economics/21/11/2018/5bf480e99a794758d1158dbf?from=
[11] Timofeev R A, Shlychkov V V 2010 *Optimization of the consumption of fuel and energy resources as a factor in increasing the level of economic reliability of an enterprise* (Kazan: Kazan. state energy Univ.) p 135
[12] Goryunov S M , Muhamethanova L M , Petukhova L and Nikolaeva N G *Problem metrological provision of Russian oil complex* (*Electronic Materials)* https : // cyberleninka . ru / article / n / problemy - metrologicheskogo - obespecheniya - neftyanogo - kompleksa - rossii . pdf
[13] Gokhberg L M, Filippova S P 2014 Forecast of the scientific and technological development of Russia: 2030. *Energy efficiency and energy saving* (Moscow: Ministry of Education and Science of the Russian Federation, National Research University Higher School of Economics) p 27
[14] Kiyamov I K 2012 Features of the formation of innovative potential in the modern economic space of Russia *European Social Science Journal No 5 (21)* p 359-61

IOP Conf. Series: Materials Science and Engineering **570** (2019) 012056 doi:10.1088/1757-899X/570/1/012056

Actual issues of digital transformation of the Russian economy in modern conditions

I K Kiyamov[1,3], L S Sabitov[1,2], G I Kabirova[1], L Sh Akhtyamova[1] and L Sh Iskhakova[1]

[1]Kazan (Volga region) Federal University, Kremlevskaya str., 18, Kazan, 420008, Russia

[2]Kazan State Power Engineering University, 51 Krasnoselskaya street, Kazan, 420066, Russian Federation

[3]Autonomous non-profit organization of higher education "Academy of Social Education", Isaeva str.,12, Kazan, 420039, Russian Federation

l.sabitov@bk.ru

Abstract. The first new kind of inexhaustible resources of modern digital economics - "information" in the article is analyzed; the basics of the concept of "Big Data" are summarized; the promise of concept and "Industry 4.0" is justified; the main conclusions about the prospects of further mutually beneficial cooperation with the People's Republic of China in the development of a modern digital economy are formulated.

«I am well. It's not me who got sick
It's just the world economy.
pregnant with digitalization
and pregnancy is, as you know,
not a disease, but a normal state»
V.V. Putin [1]

The growth rate of economic development in the most countries has slowed sharply in recent decades. In most cases, this is due to run-to-run structural changes, as well as newly emerging imbalances in the global economy. In such circumstances, the issue of finding new sources of further economic development and growth becomes relevant. In the modern world, the digital economy is becoming such a source. Today, an uncompromising struggle for leadership in the field of digital technology is developed, which on practice become one of the key competitive advantages not only individual companies, but also in many countries.

Modern digital economy has created a new kind of resources - information, which is one of the most important factors of successful economic operators at different levels. Information today is a priority product, the resource of which is practically inexhaustible. The advent of the digital economy has led to a number of significant changes in society, namely [2]:

- to the transformation of existing business models;
- to the changing of the standards of the relationship between the seller and the buyer;
- to the changing of corporate culture, focusing on issues of motivation and staff training;

- to the revision of a number of regulations and the level of responsibility of IT departments with the transition to cloud technologies and virtualization technologies;
- to the changes in the company's infrastructure based on new technologies, hardware and software requirements of the environment, key interests of both sellers and buyers.

Day by day increasing flow of information creates the need for a transition to a qualitatively new level of management of economic processes. The information retrieval systems that exist today make it possible for almost completely automate the managerial decision-making processes. All this contribute to a more detailed analysis of the production and economic participants market, not only at the micro, but also at the macro levels of the modern world economy. The emergence of modern digital devices and their active use led to the genesis of the so-called concept of “Big Data”. The amount of data is increasing every day, and this creates new, promising opportunities for all spheres of modern human activity (science, management, healthcare, etc.). The development of the concept of " Big Data is a key element of the modern digital economy. Working with big data allows for review a qualitatively new analysis of socio-economic data, the timely processing of which contributes to the further modeling and forecasting the socio-economic development of market actors.

If the world economy was formed at the expense of oil, machine-building, metallurgical large companies in the framework of the fourth technological order, today the rapid development in the fifth and sixth technological order is provided by companies of the digital sector. They are unique growth points, providing the world economy with digital resources. So, according to the analytical company Brand Finance” in the Global 500 annual report on the most valuable brands in the world published in February 2018, the first ten positions were taken by world companies directly or indirectly connected with the digital technology sector. This fact is undoubtedly the evidence that the modern world economy is moving towards digitalization (Fig. 1) [3].

Fig.1. Rating Top 10 the most valuable (profitable) brands in the world, according to the analytical company Brand Finance” (February, 2018) [3].

The digital economy in modern material production is best characterized by the German concept "Industry 4.0". The previous concept to her the of " Industry 3.0" was mainly focused on the automation of individual production processes through machine tools with numerical control in XX century. Nevertheless, it allowed to increase significantly the productivity growth of the German economy and to become one of the most efficient in Europe [4, p.30]. At the beginning of XXI centuries, in order to increase the competitiveness of industrial enterprises, to intensify the integration of "cyber-physical" (CPS) systems into the factory processes began in Germany, which led to the beginning of a new era in the "Industry 4.0" concept. Today, this concept includes "innovative methods like big data analysis, machine learning, machine vision, industrial Internet of things, virtual reality, augmented reality, three-dimensional modeling, three-dimensional printing, unmanned aerial vehicles and robotics" [5]. These technologies are already widely used in the global industry, and their full-scale implementation in the world economy, in future it may have an effect on performance and labor market comparable to industrial revolutions of the past. According to statistics, companies which are already using CPS today are 25% more profitable compared to competitors (Siemens, Cisco), and analysts at Gartner predict that by 2025 all the major global companies and organizations will become digital [6]. Today, the concept of " Industry 4.0" covers the whole world economy. Except Germany, such countries as the USA, China, and Japan have achieved major successes in the implementation of this concept. In the United States, in 2014, the *Industrial Internet* consortium was formed, combined such large companies as IMB, Intel, AT & T, and several others. In China, the implementation of the "Industry 4.0" concept has opened up the new opportunities for both industrial growth and the development of global e-commerce. A striking example is the Chinese company Alibaba, which develops its business in the b -2-b market (business to business). The "digital platform" of this company in November 2016, just in one day, sold almost 18 billion goods US $ [7]. By 2020, through the flexibility and openness of using digital technology, China plans to reduce the costs of the national economy above the average level, achieved by the key global companies. Japan's leading companies are focused on internal digitalization processes. Developed digital interoperability technologies allow end-to-end processes to be maintained with the key partners in accordance with the horizontal value chain. This ultimately leads to the improvement of the national economy efficiency by increasing productivity, optimizing costs and improving the quality of manufactured products and services.

What are the prospects for the domestic economy in the conditions of modern processes of world digitalization? Today, more than half of industrial Russian companies operate within the framework of the concept "Industry 3.0". The average age of production capacities in the manufacturing industry is about 10-12 years, more than 25% were commissioned until 2000, half of them no longer meet modern requirements for the production of competitive products [4, p. 32]. Certainly there are individual national companies which have achieved success in today digitalization and the key ones are the company IT - industry. The rating of Russian IT companies, according to research by RBC+ analytical company based on the results of 2016, is presented in Table 1 [8]. Studies have shown that today Russia's entry into the global digital ecosystem is constrained by the following factors:

- firstly, this is the technological lag of the Russian economy in relation to the world industrialized countries, such as the USA, Germany, Japan, and China.

Table number 1.

Rating of companies in the Russian IT-market.

Position in the ranking	**Company (company group)**	**Main activities**	**Growth in 2016 compared to 2015,%**
1	National Computer Corporation (incl. Systematika Group	Diversified IT-holding; classical distribution - about 70% *	16.2

	of Companies, OCS, Aquarius Group of Companies)		
2	LANIT (incl. Treolan and CompTek)	Diversified IT-holding; classical distribution - 50%	6.9
3	EPAM Systems	Offshore programming, system integration, consulting	39.9
4	GC Softline (incl. Develonica)	Deliveries of software and computer equipment in retail and wholesale; cloud services	30.7
5	GC "Technoserv" (incl. "Reksoft")	System integration (share of the project supplies - 18.5%)	1,3
6	PJSC "Kaspersky Lab"	Information Security Software Development	14.5
7	ITG (Inline Technologies Group)	System integration (share of project deliveries - 43.5%)	0.6
8	I-Teco (incl. "Servionika")	System integration (share of project supplies - 49%)	3.9
9	Cognitive Technologies	Software Development and Implementation	7.2
10	SAP CIS	Selling and implementing of a software	9.1

In most cases these countries have come into the sixth technological structure, where NBIC technologies are becoming key, in other words NBIK -convergence of such four fundamental knowledge branches as Nano, bio, information and cognitive technologies;

- secondly, there is not enough attention from business and government to the development of digital infrastructure. However, it should be noted that at the end of July 2017, the Prime Minister of the Russian Federation D. A. Medvedev approved the program "Digital Economy of the Russian Federation" according to which, it is planned to spend more than 500 billion rubles for the priorities for the development of the digital economy of the Russian Federation (regulation, information infrastructure, the formation of research competencies and technological reserves, personnel and information security). It will be show how effectively this program will be implemented;

- and finally, one of the most important factors is the lack of a digital culture and a large shortage of specialists with key competences in the field of the “Industry 4.0” concept.

Nevertheless, in spite of all these factors, it can already be said that relating to the economy of the Russian Federation, Industry 4.0 technology in such industries as oil and gas production, electric power industry creates a significant potential for improving efficiency, without a radical transformation of existing business models. For more labor-intensive industries the key opportunities are concentrated in increase the efficiency of the production process by automation, using connected to industrial internet of things sensors and in-depth analytics. The significant effect from the introduction of digital technologies can be obtained in the manufacturing industry due to its high labor intensity and technological backwardness of Russia from the advanced countries. The efficiency gains are possible at all parts of the value chain - from accelerate development and output to market for new products, synchronization of production chains and supplies, completing till the significant increase in the efficiency of planning, production, quality control and level of service maintenance of the final product. Thanks to the modernization of the Russian engineering industry based on the principles of Industry 4.0, you can significantly improve this industry productivity, reducing the backlog of the most industrialized developed countries. According to experts, the annual effect of the introduction of elements of "Industry 4.0" in Russia approximately will amount from 1.3 to 4.1 trillion rubles by 2025. [9].

In general, many experts are agreed that the transition to the digital economy is one of the key GDP growth factors of the Russian Federation. According to the various estimates forthcoming digitalization up to 2025 years will cause from 19 till 34% from the entire increase in the country's GDP [9]. Similar forecasts exist for other countries too. In accordance with the latest research of the McKinsey Global Institute , in the next 5-10 years, the United States and China will become the absolute leaders in the development of digital technologies. In the United States, by 2025, due to the digitization of the economy, the increase in value will be about $ 1.6–2.2 trillion. In China, where the potential for the rapid growth owing to the large investments and increasing of employment decreased each year, the total GDP growth of the country through the introduction of digital technology to 2025 year will be estimated from 7 before 22% [9]. Modern experience of the development of digital technology in China and the United States suggests that these processes in these countries contribute to the improvement competition, increase productivity and manpower qualifications, to price reduction, and easier access to any relevant information. Our country as quickly as possible is necessary to establish a process of mutually beneficial cooperation with these countries. However, due to tense relations with the United States, today, the development of digital technologies, we should pay attention to cooperation with the People's Republic of China. Namely:

- first, it is necessary to pay close attention to the national strategy "Made in China - 2025", strengthen bilateral cooperation in developing the "Industry 4.0" concept, and develop further productive steps to digitize the industry in order to increase the competitiveness of the manufacturing industry of both countries.

- secondly, in order to create a joint digital infrastructure, to ensure the development of joint Russian-Chinese projects in the field of satellite navigation, satellite broadband services, in the development of applications and services supporting the 5G standard, in the field of the Internet of things, Smart city and a number of others.

- thirdly, to create a database in the field of education, through the modern Internet technologies, with the aim of developing a common access to international educational resources. Such a mechanism would create a common platform for the implementation of different types of distance learning and online courses that will surely give real benefits to citizens of both countries.

- and finally, it is necessary to establish cooperation in the development of Smart Energy, covering the development of energy efficient technologies, the development of energy saving industry, the study of the new models of mining, production and distribution of energy using, and the use of new types and sources of energy [10].

In conclusion, we would like to note that the timely and purposeful cooperation with China in the field of digital technologies will contribute to the creation of favorable conditions for the development of Russian Federation competitive economy, the well-being and quality of life of our population, by improving the quality of goods and services, produced Institution in the digital economy based on the use of modern digital technology; the increase awareness and digital literacy of society; as well as the improvement of the quality of public services for the citizens of the Russian Federation.

References

[1] Putin diagnosed the global economy "pregnancy" by digitalization (*Electronic Materials)*: https://www.rbc.ru/economics/25/05/2018/5b07effd9a794708da380dd4?story=5af980859a7947b069a0a9d3 (the date of circulation is 12.04.2018)

[2] Perpelyak A I *The digital economy: new business opportunities* (*Electronic Materials).* URL: https://sibac.info/archive/technic/4(51).pdf (the date of appeal: 06/19/2018)

[3] 2018 *Brand Finance Global 500.* The annual report on the world's most valuable brands (*Electronic Materials)* URL:http://brandfinance.com/images/upload/bf_global2018_500_website_locked_final_spread_03042018.pdf (accessed date: 19/06/2018)

[4] 2017, Potential of competitive release *Expert No 7* (1017) pp 30-35

[5] Business is waiting for digitalization (*Electronic Materials)* URL http: // www.kommersant.ru/doc/2993675 (accessed date 06/03/2017)

[6] The main events of the world IoT. The results of 2017 (*Electronic Materials).* Access

Mode URL: https://iot.ru/promyshlennost/glavnye-sobytiya-iot-itogi-2017-goda

[7] 2016, Who will win: "more data" or materials science. / *Expert No 48* (1009) p 13

[8] The rating of Russian IT companies compiled by RBC + based on 2016 results (*Electronic Materials)* Access Mode: URL http://rt.rbcplus.ru/news/59013cb47a8aa90a969da442

[9] Digital Russia: a new reality (*Electronic Materials).* Access Mode: URL :https://www.mckinsey.com/en/~/media/McKinsey/Locations/Europe%20and%20Middle%20East/Russia/Our%20Insights/Digital%20Russia/Digital-Russia-report.ashx (accessed date 04/12/2018)

[10] Timofeev R A, Shlychkov V V 2010 *Optimization of the consumption of fuel and energy resources as a factor in increasing the level of economic reliability of an enterprise.* (Kazan: Ministry of Education and Science of the Russian Federation, State educational institution of higher professional education 'Kazan State Energy University" pp 35-41

IOP Conf. Series: Materials Science and Engineering **570** (2019) 012057 doi:10.1088/1757-899X/570/1/012057

To a question from About modern state and prospects of development ah Russian oil and gas complex

I K Kiyamov[1,3], L S Sabitov[1,2]*, G I Kabirova[1], L Sh Akhtyamova[1] and L Sh Iskhakova[1]

[1]Kazan (Volga region) Federal University, Kremlevskaya str., 18, Kazan, 420008, Russia
[2]Kazan State Power Engineering University, 51 Krasnoselskaya street, Kazan, 420066, Russian Federation
[3]Autonomous non-profit organization of higher education "Academy of Social Education", Isaeva str.,12, Kazan, 420039, Russian Federation

l.sabitov@bk.ru

Abstract. The work explains the reasons for the high level of energy intensity of the Russian economy. The analysis of the current state of the domestic oil and gas complex is presented. Its key's problems are identified. Further promising directions for the effective development of the Russian oil and gas complex are proposed. As one of the most promising mechanisms to improve the competitiveness of domestic companies' oil and gas complex, organization is invited to consider the effective process and resource and energy efficiency of these companies.

According to the forecast data of the International Energy Agency (IEA), by the end of 2030, the planned energy consumption in the context of the development of the global economy should increase in 40% [1]. According to the study of international consulting company "McKinsey & Company", in the nearest ten - fifteen years, in order to fully provide the world economy with energy, it will be necessary to build about 77% of the energy infrastructure [2]. Energy Ministry data of Russian Federation (RF) show that despite the progress in recent years, progress in reducing the energy intensity of GDP (based on 2007 energy intensity of GDP was reduced to 5.94% in2015 a year) [3], Russian economy is still in a little more effective, because, in the real sector the production of a unit of GDP consumes almost twice as much energy as in the world developed countries (picture 1) [4]. According to experts, an increased level of GDP energy intensity of the Russian economy in comparison with other countries, caused be the complex problems [5].

First, these are established regional sustainable energy systems, regional cluster unions that are interconnected by the supply of energy resources, the structure of production and consumption of energy resources, the result is the problems associated with the process of resource optimization and energy saving.

Secondly, despite the extensive Russian regulatory and legal framework (Federal Law No. 261 "Energy Saving and Improving Energy Efficiency ..."; "Energy Russian strategy for the period till 2035 year"; Russian government program "Energy efficiency and energy development", etc.), control the process, resource, and energy saving, on the regional and municipal levels is still not effective. It is believed that the regulatory functions comprehensively should be implemented at all levels - from the federal to the level of local government.

Thirdly, effective incentive mechanisms for innovative investment and organizational activities of resource and energy conservation are not fully developed. One of the effective method can be energy management, which includes a set of organizational and technical measures needed to improve the energy efficiency of the resources [6]. This mechanism makes it possible to increase the interest of private investors in the financing of resources and energy-saving projects companies in practice.

Finally, one of the key problems today is the absence of fully developed innovation infrastructure in the field of resource and energy conservation. In a number of regions there are diverse innovation centers, IT - parks, technology transfer centers, small innovative enterprises. Nevertheless, we have to admit that in practice, all of them do not fully cooperate effectively with each other, and even more so with other subjects of the resource and energy saving processes on the mutually beneficial use of energy-saving innovations. The failure of these problems reduces the role and importance of the resource and energy saving process in increasing the economic efficiency of the Russian economy. All of this, in practice, does not allow to fully realize the existing potential resources and energy conservation. This figure is quite impressive. So according to the RF Department of Energy the potential savings of final energy in the key industries of RF until 2020, it amounts more than 190 million tons of reference fuel (toe) [1]. Electric power industry (42 mln. Tons of equivalent gas), housing and public utilities (33 mln. Tons of equivalent gas) and the oil and gas complex of the Russian Federation (21 mln. Tons of equivalent gas) are in the lead parts of [1].

Since the object of our research is the oil and gas complex of Russia, we will try to analyze what are the main trends exist today. Currently, the domestic oil and gas complex is a key sector of the Russian economy. This is a kind of backbone element of both the internal (Russian) and external (global) energy market. The last twenty years, this segment of economy fully defines the dynamics and quality of the socio-economic development of the Russian Federation. Oil and gas are high margin export of Russian goods on the world energy market. Oil and gas complex of Russia is a one of the most competitive industries in the context of world economic relations. Its development and operation determines the state of the Russian economy, as almost 70% of exports of hydrocarbon materials generates over 40% of federal budget (Figure 1).

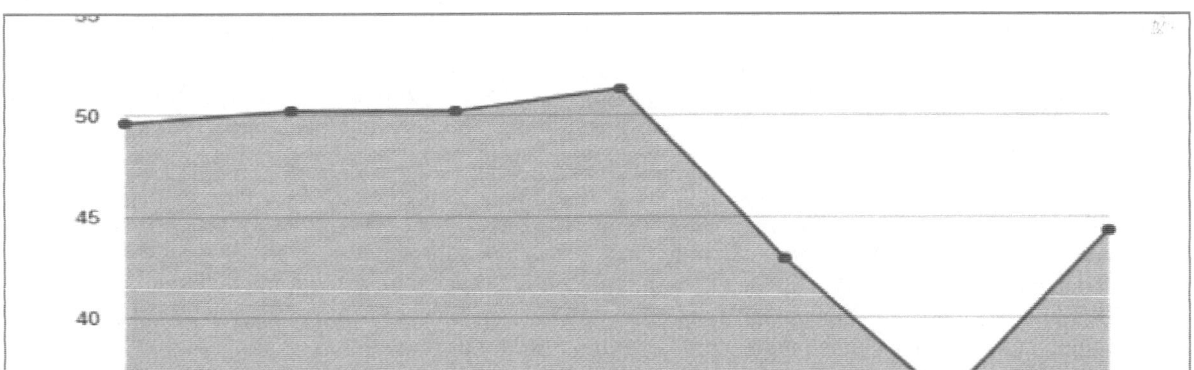

Fig. 1. The share of oil and gas revenues in the Russian budget on years. [7]

The balance of payments of the Russian economy and the rate of the national currency depend on the effectiveness of the domestic oil and gas complex. The development of the domestic oil and gas complex creates a certain demand for the products of related sectors in Russian's economy real sector, determine the prospects for their development in short and medium terms. It is maximally

IOP Conf. Series: Materials Science and Engineering **570** (2019) 012057 doi:10.1088/1757-899X/570/1/012057

connected with industrial companies in the mining and manufacturing industries, mechanical engineering, electric power, and construction. An additional effect is formed through the growth of the tax base, the creation of new jobs, and the increase in real incomes of the population. It provides a powerful impetus for the development of science and high-tech sectors of the domestic economy. This is a peculiar source of investment - multiplier effect.

Nevertheless, the studies have shown a number of significant problems that characterize the current state of the oil and gas complex of the Russian Federation.

1. The problems are associated with the deterioration of the domestic mineral resource base of hydrocarbons produced. This problem is due to a sharp decrease in current proven reserves and their reproduction rates. If there are no questions about the proven reserves of Russian natural gas, the questions of oil supply in Russia give a certain suspicion.

Today, most of the Russian oil-producing regions do not have significant industrial oil reserves, which are necessary to maintain the required volumes of its production for the medium term (10-15 years).

According to the forecast data compiled at the end of 2011, oil reserves will last approximately for 24 years in Russian [9]. Considering that today is the end of 2018 we can talk about a figure not exceeding 20 years.

This fact, from our point of view, makes it necessary to organize work to transfer the high oil and gas country's potential into active hydrocarbon reserves.

2. The problems associated with a significant lag of domestic technologies from foreign analogues, on the background of a high level depreciation of fixed production assets in Russian oil and gas complex [10]. This fact is one of the reason in a growth in the share of hard-to-recover oil and gas reserves from their total number.

3. The problems related to the strong dependence from the conjuncture of the world energy markets. Also, we should not forget about the sanctions imposed by the European Union and the United States in 2014 and still ongoing, which imply a total ban for the supply of innovative technologies, as well as specialized equipment in the field of hydrocarbon production, transportation and processing.

4. Finally, the key issues are associated with the lack of an effective innovation management system in the oil and gas sector, and also the lack of effective mechanisms to stimulate the use of new resource and energy-saving technologies. As a result, our country today is seriously lagging behind in terms of the main indicators of innovation activity from developed oil-producing states. Thus, all of the above-listed problems require the development of further promising areas for the effective growth of the domestic oil and gas complex. We agree with the opinion of many experts, and we believe that the implementation of the following activities can be considered as one of the most popular areas. [11]:

1. Measures aimed at solving problems associated with the increase in oil recovery coefficient. First, it is necessary to increase the effectiveness of work with difficultly recoverable reserves in the fields, with their further inclusion in the industrial turnover. To organize work to identify the latest hydrocarbon fields, including in the implementation of oil and gas potential of the seas on the basis of large-scale geological exploration work. With the help of modern science potential, at each significant field, to organize work to determine the maximum allowable intensification's degree at oil and gas production. To carry out full-scale systemic diversification of the transport infrastructure for the extraction and processing of hydrocarbons. And finally, to make a register of regions that have oil and gas potential to create specialized large-cent oil and gas production under the auspices of the regional authorities.

2. Measures related to the regulatory framework in the oil and gas sector. We agree with the opinion of experts [12] who propose to improve the mechanism of legislative consolidation of the procedure for deductions for replenishing the mineral resources base. As an alternative, it is proposed to consider the issue of creating a specialized fund, the functioning of which could replace all kinds of deductions. It is also proposed to consider the abolition of the flat tax scale in the DPI; to carry out in

accordance with the international classification, an adjustment in the export duty tariffs, namely, to consider a barrel as a unit of measure, and not as a ton of oil; to develop mechanisms of improving the tax system in the field of research geologic exploration and development.

3. Measures related to the implementation of the innovative potential of the domestic oil and gas industry. As we indicated above, it is inappropriate to wait for the foreign investments and innovative technologies from our "foreign colleagues" in view of a tough sanction policy. Studies show [13,14] that domestic oil and gas industry is not fully ready for the implementation of the Energy Strategy 2035. This is primarily due to technological problems, the lack of necessary innovative technologies, including technologies for the development of hydrocarbon fields on the Arctic shelf of the seas (the Yamal Peninsula, Eastern Siberia, the Sakha Republic), which has significant oil and gas potential. First of all, it is necessary to organize large-scale work at the federal and regional levels : to develop effective mechanisms and effective instruments for the implementation of innovative development programs ; to organize transfer of existing innovations in the oil and gas sector; to develop mechanisms for industry's development stimulation through the full realization of the innovative potential oil and gas companies in the field of resource and energy saving; to intensify the work in the direction of increasing the ecological and energy efficiency at oil and gas production processes , their processing and transportation . In conclusion, it should be noted that the analysis of the main problems presented at Russian oil and gas complex allows us to conclude that today, as one of the most available tools to improve the performance and competitiveness of domestic oil and gas companies is the process of resource and energy saving to ensure economic efficiency and the effectiveness of the extraction and oil and gas processing.

References

[1] WEO from IEA: a regular forecast of the development of world energy has been published (*Electronic Materials)* Access mode: http://renen.ru/iea-world-energy-outlook-2017/

[2] *McKinsey Herald* (*Electronic Materials)* The access mode: http://vestnikmckinsey.ru/Services/Doc/2015/f604927a43da4f68ad1e001913766a51/True

[3] Hovak A V The results of the work of the fuel and energy complex of Russia in 2013. Tasks for the medium term (*Electronic Materials)* Access Mode: URL :http://energetika-restec.ru/netcat_files/userfiles/energetika-restec/14-01-2014.pdf

[4] Dynamics of energy intensity of GDP by region of the world. For the oil needle will have to fight (*Electronic Materials)* Access mode: URL :http://expert.ru/expert/2014/21/za-neftyanuyu-iglu-pridetsya-poborotsya/media/ 238775 /

[5] Ovinnikova K N The current state of the oil and gas complex of Russia and its problems (*Electronic Materials)* Access mode: https : // cyberleninka . ru / article/ n / sovremennoe - sostoyanie - neftegazovogo - kompleksa - rossii - i - ego - problemy

[6] Kiyamov I K, Alafuzov I G 2016 The key principles of energy management in energy saving *Bulletin of Economics, Law and Sociology № 1* p 39

[7] The share of oil and gas revenues in the Russian budget began to increase (*Electronic Materials)* Access mode:https://investbrothers.ru/2017/03/28/dolja_neftegazovyh_dohodov_rastet/

[8] Which countries have the most oil and gas? (*Electronic Materials)* Access mode: http://www.aif.ru/money/economy/u_kakih_stran_bolshe_vsego_nefti_i_gaza

[9] Statistics on oil reserves in the world (*Electronic Materials)* Access mode: https://pikabu.ru/story/statistika_po_zapasam_nefti_v_mire_1542369 (the date of circulation is 11/21/2018)

[10] Timofeev R A, Shlychkov V V 2010 *Optimization of the consumption of fuel and energy resources as a factor in increasing the level of economic reliability of an enterprise* (Kazan: Kazan. state energy Univ.) p 135

[11] Mozgova A S *Organizational - economic mechanism for improving the energy efficiency of oil*

and gas enterprises on the basis of an energy audit (*Electronic Materials)* Access mode: http://diss.muctr.ru/media/dissertations/2014/04/%D0%94%D0%B8%D1%81%D1%81%D0%B5%D1 82% D0% B0% D1% 86% D0% B8% D1% 8F_% D0% 9C% D0% BE% D0% B7% D0% B3% D0% BE% D0% B2% D0% B0.pdf
[12] Gizatullin H N, Dyshaev A Yu 1989 *Environmental Performance and Environmental Management Modeling* (Ural Branch of the Academy of Sciences of the USSR, Bash. scientific Center, Sep. econ Research) p 42
[13] Fadeev A M, Cherepovitsyn A E, Larichkin F D 2012 Sustainable development of a new mining region in the implementation of oil and gas projects on the Arctic shelf *Economic and social changes: facts, trends, forecast № 1 (19)* p 33-37
[14] Glushakova V G, Simagin Yu A 2012 *Regional economy. Natural resource and environmental foundations* (Moscow, Knorus) p 320

IOP Conf. Series: Materials Science and Engineering **570** (2019) 012058 doi:10.1088/1757-899X/570/1/012058

The study of the insulating properties of a new material based on straw and isoprene rubber C-nitrous vulcanization

O R Klyuchnikov[1], N A Martianov[2], I K Kiyamov[3], L S Sabitov[3], D A Morozov[4], I R Gilmanshin[3,5]

[1]Kazan National Research Technological University, Kazan, K. Marx Street, 68, 420015, Russian Federation
[2]Kazan Branch of NOA "Rostekhekspertiza", Kazan, Zinin Street, 3, 420097, Russian Federation
[3]Kazan Federal University, 18 Kremlyovskaya street, Kazan, 420008, Russian Federation
[4] Kazan State Power Engineering University, 51 Krasnoselskaya street, Kazan, 420066, Russian Federation
[5] Kazan national research technical university named after A N Tupolev, K. Marx St. 10, Kazan, 420111 Tatarstan, Russia

Abstract. Research has been carried out on the thermal insulation properties of a new composite material based on straw and synthetic rubber of low-temperature vulcanization. The values of heat transfer resistance (R_{cp}) and thermal conductivity coefficient (λ) are determined depending on the humidity of the environment.

The problem of energy conservation has become one of the most pressing for our country. Therefore, the development of new efficient and affordable energy-saving materials is an urgent task. On the fields after harvesting, a significant amount of straw remains, which can become the basis for creating a number of new composite heat-insulating materials. The main mass of straw remaining after harvesting is burned [1]. As is known, a straw is a potential competitive heater, based on a renewable environmentally friendly resource base.

Straw is used as a thermal insulation material in house building, namely, in a clay earthen house on a lightweight foundation [2], for the manufacture of building materials, for thermal insulation of walls, ceilings, partitions, etc. [3]. It is used as organic materials for the manufacture environmentally friendly multi-layer lightweight fire-resistant wall panels [4].

Prospects for the use of straw due to its low specific density $\rho \approx 10{,}2$ kg / m3, fiber structure and good physical and mechanical properties. Due to the low weight of straw blocks, the building does not require a heavy foundation; no lifting mechanisms are required for construction.

The straw has a smooth stem, long (up to 50 cm) internodes, small changes in the stem thickness along the length. With relatively thin walls of the stem, rye straw is the most durable and coarse. After moistening, it acquires plasticity, and when dried, it keeps well the shape that was given to it.

IOP Conf. Series: Materials Science and Engineering **570** (2019) 012058 doi:10.1088/1757-899X/570/1/012058

The combination of elastic-deformation properties of the material and the availability of raw materials, straw - unsaturated rubber can be widely used in the production of new heat-insulating materials in rural areas.

To create experimental samples of the material, we used rye straw, isoprene rubber SKI – 3 and one of the available C-nitrous vulcanization systems generating p-dinitrosobenzene [5].

The heat-insulating characteristics of plates made of straw material - rubber of different thickness 10 mm, 20 mm, 30 mm were studied for the first time according to the method [7, 6] using ITP-MG-4.03 X(Y) «Potok» with 3 density sensors of heat flows and 2 temperature sensors attached to the material through a layer of thermal paste KPT-8 (Figure 1.)

Figure 1. Straw material plate - rubber with heat flux density sensors and temperature sensor

A 1l air thermostat was used. A 40 W incandescent lamp was used as a heat source. The data from the sensors of the density of heat fluxes and temperature on the surface of the material with an error of 10% was recorded in the memory of the device every 20 minutes. within 24 hours (Figure 2).

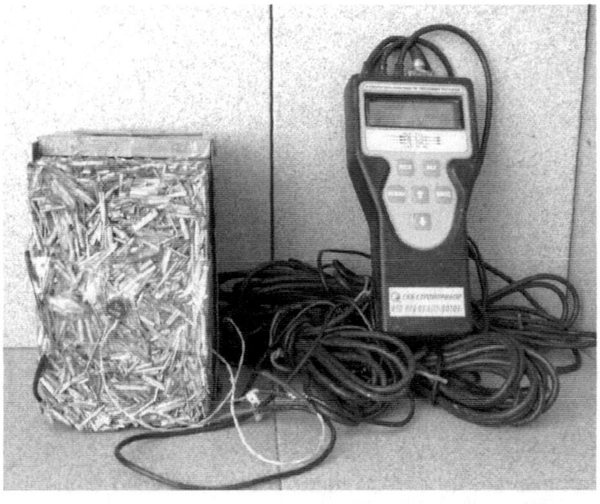

Figure 2. Air thermostat with a device ITP-MG-4.03 X(Y) «Potok».

Heat transfer resistance is determined by the formula: $R_{av} = (T_{in} - T_{out})/ q_{av}$

The arithmetic average values of heat transfer resistance (R_{av}) of plates of different thickness were: $R_{10мм}$= 0,07 м²·°C/W; $R_{20мм}$= 0,29 m²·°C/Вт; $R_{30мм}$= 0,40 м²·°C/W.

Also, the thermal conductivity coefficient of the straw sample was found - SKI-3 rubber according to the GOST method [8] with dimensions of 150x150x9 mm. Using the ITS – 1 device. Value of thermal conductivity λ= 0,046 W/mK.

Figure 3 shows the installation for the study of the coefficient of thermal conductivity λ of a sample based on straw and rubber according to the method of GOST [6].

Figure 3. Study of the thermal conductivity coefficient λ of a composite based on straw and rubber using an ITS device - 1

Also, a sample of 150x150x9 mm was placed in a climate chamber CM-70/75-XXTBX with a set temperature of 22 ° C (Figure 4.). In this chamber, the sample was kept for 1 hour with a humidity of 20%. After that, we measured the thermal conductivity λ, which turned out to be equal to 0.054 W / m K.

Following this, a similar series of experiments were carried out by holding the sample for 6 hours with a change in the relative humidity of 40.60.80%, the results of which are presented in Table 1 (the dependence of the thermal conductivity λ on the sample moisture based on straw and rubber).

IOP Conf. Series: Materials Science and Engineering **570** (2019) 012058 doi:10.1088/1757-899X/570/1/012058

Figure4. Climate chamber CM-70/75-XXTBX

Table 1
The dependence of thermal conductivity λ on the moisture content of the sample based on straw and rubber.

№	T, °C	Humidity, %	λ, W/m K
1	22	20	0,053
2	22	40	0,057
3	22	60	0,059
4	22	80	0,065

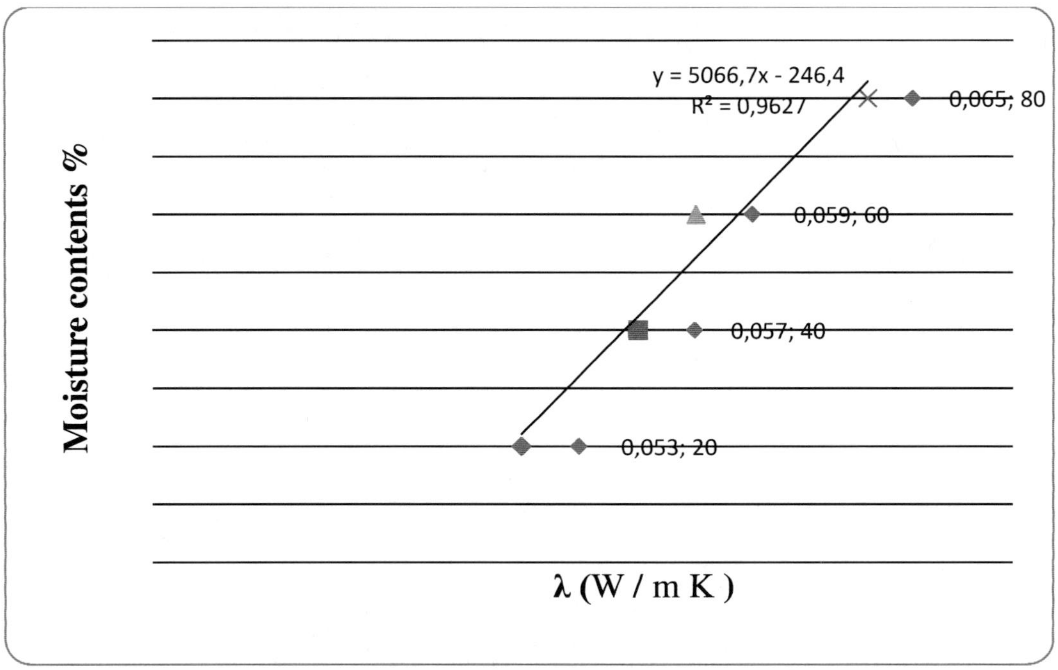

Figure 5. Graph of change of thermal conductivity λ as a function of moisture content in a sample 150x150x9 mm.

As can be seen from the table and graph with an increase in humidity, thermal conductivity coefficient λ does not significantly increase.

A sample based on straw and rubber 150x150x9 mm in size was examined for the influence of the external environment, while the sample was exposed for one year in the open air (Figure 6).

IOP Conf. Series: Materials Science and Engineering **570** (2019) 012058 doi:10.1088/1757-899X/570/1/012058

Figure 6. Sample based on straw and SKI-3 C-nitrous vulcanization rubber after exposure to the external environment for one year.

Under the influence of the external environment, including solar radiation and precipitation, the prototype darkened, and the insulating properties did not significantly change $\lambda = 0.056$ W / m K.

The change in the surface layer of the composition is associated by us with the partial destruction of vulcanized SKI-3, which has in its composition double bonds that are not resistant to ultraviolet, oxygen or air ozone. Promising in this regard may be a material based on straw and rubber, SKEPT-ENB- (DCPD) with a C-nitrous vulcanization system, known for its high heat resistance [9, 10].

Conclusion

For the first time, the values of heat flux density and thermal conductivity of a sample based on straw and rubber $\lambda = 0.04$ (W / m K) were measured experimentally. After accelerated thermal aging, the sample retained its previous form, destruction was not observed, and the thermal conductivity remained at the same level . We see the prospect of research on heat insulating material based on straw and rubber type SKEPT-ENB- (DCPD).

References

[1] Lapin Yun *Comparative assessment of some house-building systems for low-rise buildings* Center "Siberian settlement"

[2] Boldyrev G G, Barvashov V A 2009 *The utility model to the patent 88043* Frame clay residential house on lightweight foundations No 2009116206/22

[3] Khozin V G, Petrov A N, Sannikova V I, Zagoskin S V, Artemenko N F 1998 *Raw mix for the manufacture of insulating material* Patent 2101255

[4] Semenov D K 2015 *A method for automated manufacturing of energy-efficient environmentally friendly multi-layer lightweight fire-resistant wall panels based on organic materials, energy-efficient environmentally friendly multi-layer lightweight fire-resistant wall panel based on organic materials, production line for the automated production of energy-efficient environmentally friendly multi-layer lightweight fire-resistant wall panels based on organic materials* Patent 2549939

[5] Klyuchnikov O R 2018 *C-nitroso-N-oxide vulcanization systems: a monograph* Kazan: Publishing house KNRTU 216 p

[6] *Method of measuring the density of heat flux passing through the enclosing structures 1998* GOST 25380-82 M.: Publishing house of standards 12 p

IOP Conf. Series: Materials Science and Engineering **570** (2019) 012058 doi:10.1088/1757-899X/570/1/012058

[7] Klyuchnikov O R 2014 *The study of the thermal insulation properties of a material based on straw and unsaturated rubber* pp 250-52

[8] *Building materials and products. Method for determining thermal conductivity and thermal resistance in a stationary thermal regime* 2000 GOST 7076-99 M.: Gosstroy of Russia 27 p

[9] I O Klyuchnikov, O R Klyuchnikov, OV Stoyanov 2016 *Study of an elastomeric coating based on EPDM of cold vulcanization* Polymer Science Series V 9 Is 2 pp 157-60

[10] I O Klyuchnikov, O K Klyuchnikov, O V Stoyanov 2015 *Heat-resistant rubber coatings based on EPDM cold vulcanization* Bulletin of Kazan Technological University T 18 No 1 pp 224-25

IOP Conf. Series: Materials Science and Engineering **570** (2019) 012059 doi:10.1088/1757-899X/570/1/012059

Investigation of Properties of Electro-metallized Coatings Deposited by Supersonic Air Stream with Aerosol Fluxing

A V Kolomeichenko [1], S N Sharifullin [2], V N Logachev [1] and A A Izmalkov[1]

Orel State Agrarian University named after N.V. Parakhin, Russia [1]
Kazan Federal University, Russia [2]

Saidchist@mail.ru

Abstract. The article presents the results of studies of the effect of flux components on the adhesive strength, porosity and oil absorption of coatings obtained by supersonic arc metallization (EM) using aerosol fluxing (AF) deposited on 45G2 steel with a speed v = 400-410 m / s. Due to the use of AF in EM, the adhesion strength of electrometallization coatings (EM coatings) increases. The highest value of adhesive strength is 66 MPa. Porosity due to the use of AF decreases by 1.5 - 1.6 times, in comparison with EM coatings without AF. At the same time, the oil absorption of EM coatings using AF decreases 1.6–1.7 times. It is advisable to use EM coatings with AF when restoring worn-out parts for various purpose equipment, as well as for eliminating defects at the production stage.

1. Introduction

One of the common ways to restore worn surfaces of parts is EM. However, wires that provide high physicomechanical properties of applied EM coatings (50HFA, 40X13, PP-PN-100H15, PP-TP-1) have a high cost. In this regard, it is proposed to use an inexpensive welding wire Sv08G2S (with a hardness of the obtained coatings of 300 HV) and to alloy it with EM flux components [1-3].

When conducting research, 45G2 GOST 4543 steel, which is widely used for the manufacture of shaft-type parts, was used as a material for manufacturing samples [4].

In EM it is most expedient to use a flux consisting of the components Na_2CO_3, Na_3AlF_6, $Na_2B_4O_7$ [5 - 7]. Microhardness studies of EM coatings obtained using this flux with various contents of the above components in it were carried out. During the studies, fluxes were established that showed the best results: flux No. 1 (665 HV) $CNa_2\,CO_3$ = 42 g / l, CNa_3AlF_6=6 g/l, $CNa_2B_4O_7$=12 g/l and flux No. 2 (771 HV) CNa_2CO_3=42 g/l, CNa_3AlF_6=6 g/l, $CNa_2B_4O_7$=16 g/l [8]. In this regard, further studies of the physicomechanical properties of EM coatings considered in this work were performed using these two fluxes.

2. Materials and research methods

It is known that one of the most important characteristics determining working properties of various coatings notably determining durability of the parts working areas, is adhesive strength between coating and base (adhesive strength) [6 – 8]. Adhesive strength of the deposited electro-metallized coating (EM-coating) was determined with the pin method according to GOST 15140. The method essence is in destructive force evaluation at pin extension in the direction normally to its end surface with the deposited electro-metallized coating (EM-coating). The sample (fig. 1) consists of a pin and a nut with a gripping device in the shape of an opening.

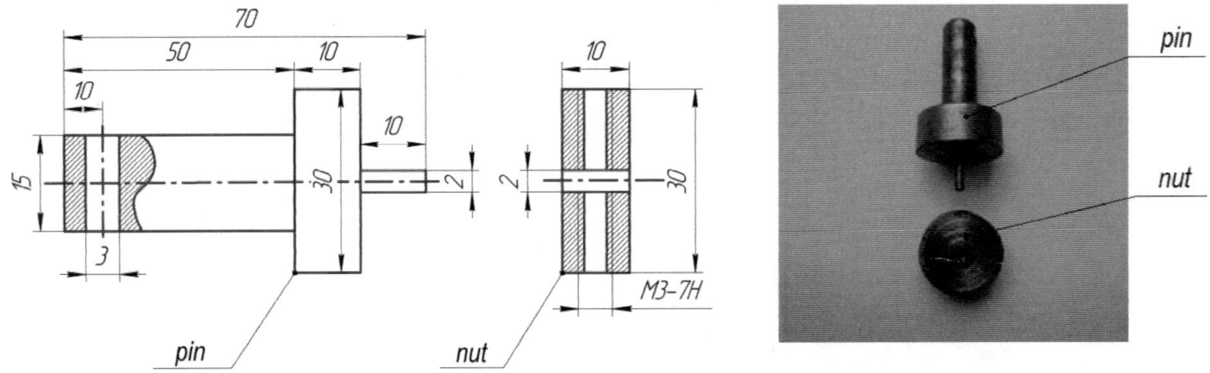

a) b)

Figure 1. Sample for investigation of adhesive strength of electro-metallized coatings (EM-coatings): a) scheme; b) general view

The pin assembled with the nut was set in the sample fixer and preliminary jet-oriental treatment was carried out. The electro-metallized coating (EM-coating) was deposited on the working surface of the sample at its even movement around its radial axis (fig. 2). The investigations on determination of the adhesive strength were carried out on the versatile tensile machine PM-1000.

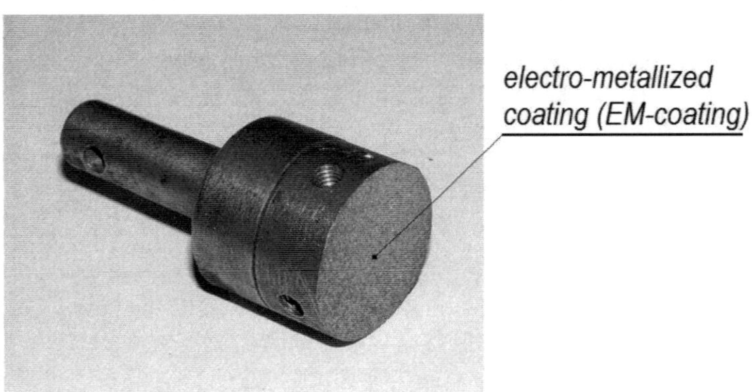

Fugure 2. Sample (in the assembled condition) for determining the value of the adhesion strength of the EM-coating

3. Results and discussion

The comparative analysis of the obtained adhesion strength depending on the applied fluxes and the components contained in them (fig. 3) displayed that the electro-metallized coatings (EM-coatings), deposited with application of aerosol fluxing (AF) (60…66 MPa), it is in 2.1 – 2.2 times higher in comparison with analogous coatings, obtained without aerosol fluxing (AF) (30 MPa).

The increase of adhesive strength of the EM-coatings obtained with aerosol fluxing (AF) application, is explained with deoxidate carbon effect. Under the effect of high temperature of the electric arc thermal dissociation with emission of active elements with C, CO and subsequent carbothermal process take place, i.e. ferrum deoxidation at steel fusion wire in the arc of metal spraying gun occurs. As a result the purer dispersed metal contacts with base, thereby increasing strength of its cohesion with base material [8].

Porousness is one of the main characteristics of the electro-metallized coatings (EM-coatings) [9, 10]. It characterizes density and homogeneity of the deposited metal layer [11]. Porousness of the electro-metallized coatings (EM-coatings) was determined with the method of dye penetrant testing according to GOST 9.302. After electro-metallization and grinding of the external surface R_a = 0.63…0.64 μm the samples (fig. 4) were cooled up to ambient temperature and ungreased with spirit. Then they were washed with distilled water and dried with filter paper.

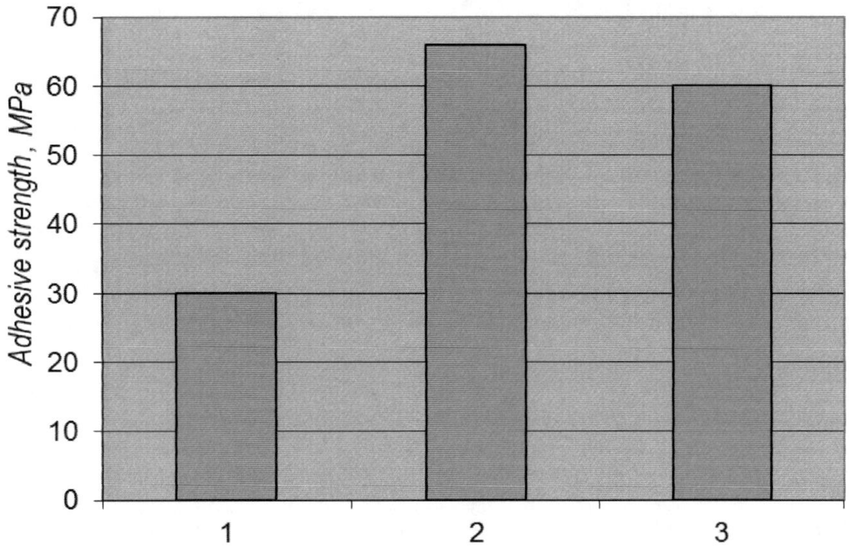

Figure 3. Values of adhesion strength of EM coatings: 1 - Sv-08G2S (without AF); 2 - Sv-08G2S + AF (flux No. 1); 3 - Sv-08G2S + AF (flux number 2)

For pores coloring the samples were put into the solution containing ferricyanide potassium 3 g/l and sodium chloride – 10 g/l, and were held in it for 5 minutes at the temperature of 18…30°C [12, 13]. Washed and dried samples with blue spots areas appeared on the surface were studied under the microscope, 20^x. Porousness of the electro-metallized coatings (EM-coatings) was determined as percentage ratio of the areas of the colored spots to the area of the examined region. The porousness value change at the electro-metallization depending on aerosol fluxing (AF) application and the flux content in it is presented in figure 5.

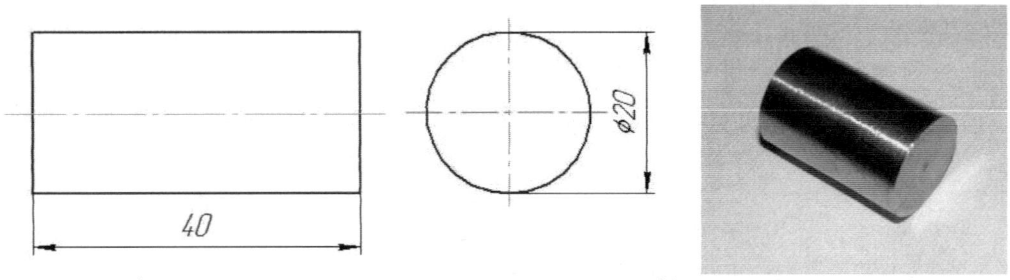

a) b)

Figure 4. Sample for study of porousness and oil absorption of the electro-metallized coatings (EM-coatings): a) scheme; b) general view

The investigations displayed that the application of aerosol fluxing (AF) allows decreasing porousness of the electro-metallized coatings (EM-coatings) in 1.5 – 1.6 times. Tis is explained by the

IOP Conf. Series: Materials Science and Engineering **570** (2019) 012059 doi:10.1088/1757-899X/570/1/012059

fact that because of ferrum deoxidation its oxide content decreases. This provides increase of density and quality of the applied metal layer [15, 16]. Besides, low porousness is connected with high speed of the deposited particles (400 – 410 m/s) and thus, with their higher kinetic energy, which makes it possible to obtain denser and semi-porous electro-metallized coating (EM-coating).

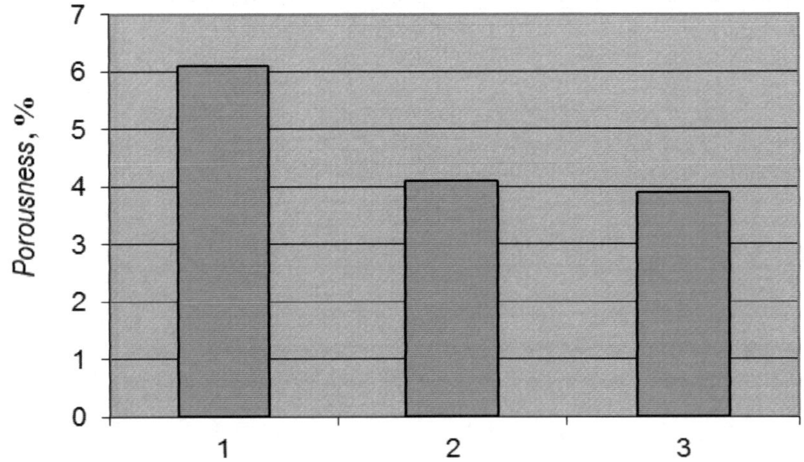

Figure 5. Porosity of EM coatings: 1 - Sv-08G2S (without AF); 2 - Sv-08G2S + AF (flux No. 1); 3 - Sv-08G2S + AF (flux number 2)

Determination of oil absorption of the deposited metal layer was done according to GOST 9.302. The method is based on determining the quality of the oil absorbed with the electro-metallized coating (EM-coating). During the investigation we applied the oil with kinematic viscosity (2.5 – 3.5) 10^{-5} m^2/s at the temperature of 25°C.

To determine oil absorption we used the cylindrical samples with the electro-metallized coating (EM-coating) (fig. 4). They were plunged into the oil heated up to the temperature of 130°C and were kept in it up to the temperature of (20±5)°C. Then they were withdrawn, the oil excess was removed with filter paper and they were weighted. Oil absorption (M) in grams per square meter was calculated using the formula (1):

$$M = \frac{m - m_1}{S}, \qquad (1)$$

where m – mass of the sample after its plunging into oil, g; m_1 – mass of the sample before plunging into oil, g; S – area of the sample surface, m^2.

Oil absorption of the electro-metallized coatings (EM-coatings) is connected with porousness, which is formed during the deposition process [17, 18]. For the parts operating in the conditions of the greasing variation from liquid to boundary, the coatings oil absorption is one of the most important properties that allows improving self-lubrication of the co-working surfaces and decreasing friction coefficient in sliding joint [19]. That way, the wear decreases and the component life in assembly unit increases.

The investigations showed that the oil absorption value of the electro-metallized coating (EM-coating), deposited without aerosol fluxing (AF) application (65 g/m^2), is higher than with aerosol fluxing (AF) application (36 – 39 g/m^2) in 1.7 – 1.8 times (fig. 6).

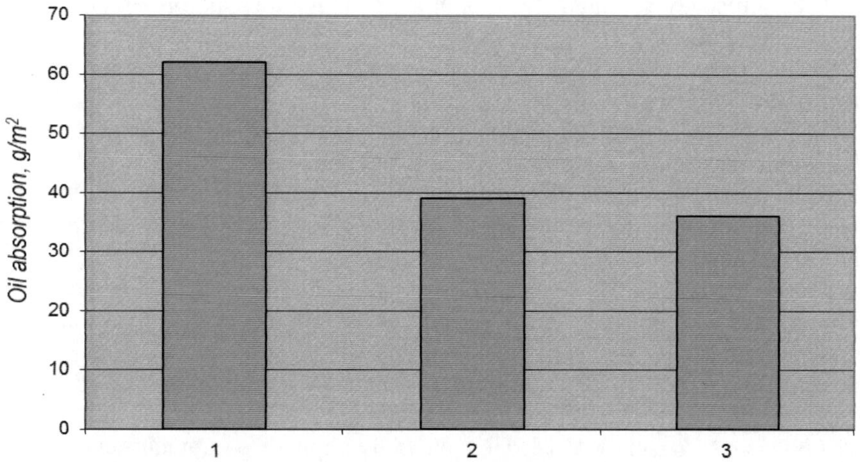

Figure 6. Oil coverage of EM coatings: 1 - Sv-08G2S (without AF); 2 - Sv-08G2S + AF (flux No. 1); 3 - Sv-08G2S + AF (flux number 2)

The decrease of oil absorption of the electro-metallized coatings (EM-coatings), obtained with equal speed v=400 – 410 m/s and with aerosol fluxing (AF) application, is explained with the decrease of their porousness, because of ferrum deoxidation its oxide content decreases and it provides the increase of the density and the quality of the deposited metal layer [20].

Conclusion

Due to the use of AF in EM adhesive strength of EM coatings increases. The highest value of adhesion strength (66 MPa) was obtained on the flux No. 2.

Porousness of the electro-metallized coatings (EM-coatings) at the account of the aerosol fluxing (AF) application decreases in 1.5 – 1.6 times in comparison with the analogous coatings, obtained without the aerosol fluxing (AF) application. Herewith the oil absorption of the coatings, obtained with the electro-metallization method, at the aerosol fluxing (AF) application decreases in 1.6 – 1.7 times. It should be taken into account at prediction of the sliding joints resource, where the parts reconditioned by the electro-metallized coatings (EM-coatings) are used. Divergence in the obtained test results according to porousness and oil absorption is connected with error at calculation of painted area spaces on the electro-metallized coatings (EM-coatings) surface, and with some porousness depth depending on many factors.

To increase physical and mechanical properties of electro-metallized coatings (EM-coatings), it is recommended to apply aerosol fluxing (AF) on the basis of distilled water of the following composition: $CNa_2CO_3 = 42$ g/l, $CNa_3AlF_6 = 6$ g/l, $CNa_2B_4O_7 = 16$ g/l.

References

[1] Kolomeychenko A V, Kravchenko I N, Logachev V N et al. 2015 Improvement of physicomechanical properties of the coverings received by arc metallization *Construction and road machines* **7** 25-29

[2] Matyushkin B A, Denisov V I and Tolkachev A A 2017 Adjustable change in the percentage of elements in the working layer when parts are restored by arc metallization *Works GOSNITI* **127** 166-170

[3] Vorobiev P A, Yusim M Yu, Litovchenko N N et al. 2008 Aerosol fluxing method for arc metallization *Works GOSNITI* **101** 201-204

[4] Tolkachev A A, Denisov V I, Matyushkin B A 2017 The use of arc metallization to restore the cylinder liners of diesel engines of agricultural equipment *Works GOSNITI* **126** 179-182

[5] Litovchenko I N, Denisov V I, , Sparrows P A et al. 2008 Restoration of parts by arc metallization *Technique in agriculture* **2** 28-32
[6] Logachev V N and Litovchenko I N 2014 Electric metallization: ways to improve equipment and technology *Works of GOSNITI* **117** 228-234
[7] Logachev V N 2017 Improving the physicomechanical properties of coatings produced by arc metallization *Technique and equipment for the village*. **3 (237)** 28-31
[8] Logachev V N, Kolomeichenko A V, Kuznetsov Yu A, Solovev R Yu, Denisov V I and Sharifullin S N 2017 Aerosol fluxing in electro arc metallization *IOP Conference Series: Materials Science and Engineering Ser.* **240** 012049
[9] Litovchenko N N, Petryakov B I, Tolkachev A A et al. 2013 Effect of the flow rate of a heterophase flow on the physicomechanical properties of an electrometallization flow *Welding production* **6** 43-47
[10] Gusev V M and Buklakov A G 2010 Influence of air flow on the quality of electric arc spraying *Quality Management in the Oil and Gas Complex* **3** 34-38
[11] Litovchenko N N and Logachev V N 2013 Ways to improve equipment and technology of arc metallization *Tractors and agricultural machines* **11** 52-54
[12] Radyuk A, Gorbatyuk S M and Gerasimova A A 2011 Use of electric-arc metallization to recondition the working surfaces of the narrow walls of thick-walls slab molds *Metallurgist* **55 (5-6)** 419-423
[13] Soloviev R Yu, Vorobiev P A and Litovchenko N N 2012 Metal-carbothermic methods for reducing the oxidation of dispersed metal during arc metallization *Welding production* **3** 43-47
[14] Lyalyakin V P, Murzaev V P, Slinko D B et al. 2017 Physico-mechanical properties of coatings obtained by arc metallization with flux-cored wires *Engineering Technology* **5** 24-28
[15] Murzaev V P, Slinko D B and Yu N Schetinin 2013 Development of resource-saving flux-cored wires for electric arc spraying *Works of GOSNITI* **113** 262-268.
[16] Borisov Yu S, Vigilyanskaya N V, Demyanov I A et al. 2013 Investigation of the dispersion of dissimilar wire materials in the process of electric arc spraying *Automatic welding* **2 (718)** 25-31
[17] Matyushkin B A, Denisov V I and Tolkachev A A 2016 Technological features of arc metallization in the industry of APK *Mechanical Engineering Technology* **10** 32-37
[18] Borisova M Z, Pods N F and Vinokurov G G 2016 Analysis of the structure of wear-resistant coating obtained by arc metallization of cored wire with refractory additives *Science and Education* **2 (82)** 76-80
[19] Borisov Yu S, Vigilyanskaya N V, Demyanov I A et al. 2013 Investigation of the influence of electric arc spraying on the structure and properties of pseudo-alloy coatings *Automatic welding* **12 (727)** 11-17
[20] Lyalyakin V P, Murzaev V, Solovev R Yu and Slinko D B 2017 Physical-mechanical properties of flux-cored wires *Welding International* **9** 729-732

IOP Conf. Series: Materials Science and Engineering **570** (2019) 012060 doi:10.1088/1757-899X/570/1/012060

Flow resistance of ducting with triple-threaded auger inserts in laminar flow regimes

I A Konakhina[1], E M Khusnutdinova [1], A I Kadyirov[2] and E K Vachagina[2]

[1]Kazan (Volga Region) Federal University, 18 Kremlyovskaya Str., Kazan 420008

[2]Institute of Power Engineering and Advanced Technologies, FRC Kazan Scientific Center, Russian Academy of Sciences, 2/31 Lobachevskogo Str., P.O. box 261, Kazan, 420111

Abstract. This paper describes the mathematical model of a laminar Newtonian flow in ducting with triple-threaded auger inserts and delivers the results of the solution to the problem. The values of the flow resistance coefficients have been calculated to demonstrate the purpose of the approach.

Currently, multiple-thread auger inserts have proven as intensifiers of the dead-type convective heat exchange. As is known [1-10], they are preferably to be used for the highly viscous flows, when the discrete roughness cannot satisfy the expectations. In these cases, the entire flow are preferred to swirl in order to create secondary transverse flows being the cause of the intensification of heat exchange. Note that swirling the entire flow increases the flow resistance of the ducting that contain intensifiers, so it is necessary to have power reserves for pumping the fluid. In such circumstances, the need to study how the flow resistance of such ducting depend on the operating conditions of the flow becomes obvious.

Assume that the Newtonian flow is stationary, laminar and isothermal; gravity is negligible; the liquid sticks to the duct walls. Let's take the flow rate through the cross section of the duct as a given one. A general view of the duct is shown in Figure 1.

Calculations and research of hydrodynamic flow fields, which, if the traditional Cartesian coordinate system is in use, are essentially three-dimensional, become simpler when using a special helical coordinate system [11] with variables η^1, η^2, η^3 that are linked with the variables of the Cartesian coordinate system in x, y, z correlations

$$\begin{cases} x = R\eta^1 \cos\left(\eta^2 \pm \kappa\eta^3\right) \\ y = R\eta^1 \sin\left(\eta^2 \pm \kappa\eta^3\right) \\ z = R\eta^3 \end{cases} \qquad (1)$$

where $\kappa = 2\pi R/S$, R is the radius of the duct with a triple-threaded auger insert, S is the screw pitch, $z = R\eta^3$ is the duct axis [11]. In (1), the "+" sign is for the duct, whose spiral ribs are twisted clockwise, and the "-" sign is for the duct with spiral ribs twisted counter-clockwise.

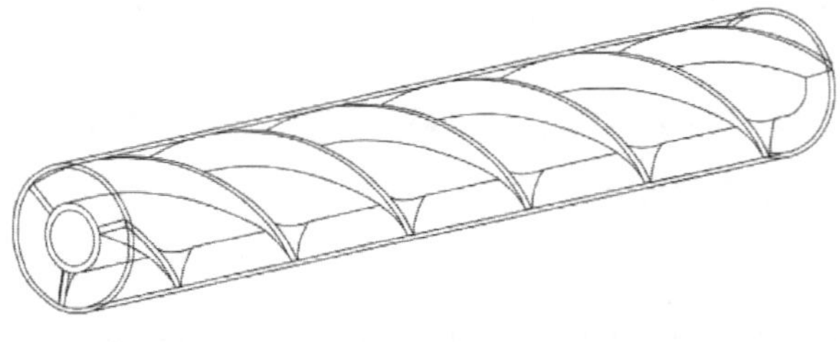

Fig. 1 A general view of the duct with a triple-threaded auger insert

Taking the assumptions into account, the mathematical model yields

$$\mathrm{Re}_R\left(\frac{\partial v^1}{\partial \eta^1}v^1 + \frac{\partial v^1}{\partial \eta^2}v^2 - \eta^1 v^2 v^2 \pm 2\kappa\eta^1 v^2 v^3 - \kappa^2 \eta^1 v^3 v^3\right) =$$

$$= -\frac{\partial p}{\partial \eta^1} + \frac{\partial s_1{}^1}{\partial \eta^1} + \frac{\partial s_1{}^2}{\partial \eta^2} - \frac{1}{\eta^1}s_2{}^2 \pm \frac{\kappa}{\eta^1}s_2{}^3 + \frac{1}{\eta^1}s_1{}^1,$$

$$\mathrm{Re}_R\left(\eta^1\right)^2\left(\frac{\partial v^2}{\partial \eta^1}v^1 + \frac{\partial v^2}{\partial \eta^2}v^2 + \frac{2}{\eta^1}v^1v^2 \mp \frac{2\kappa}{\eta^1}v^1v^3 \mp \kappa\left(\frac{\partial v^3}{\partial \eta^1}v^1 + \frac{\partial v^3}{\partial \eta^2}v^2\right)\right) =$$

$$= -\frac{\partial p}{\partial \eta^2} + \frac{\partial s_2{}^1}{\partial \eta^1} + \frac{\partial s_2{}^2}{\partial \eta^2} + \eta^1 s_1{}^2 \mp \kappa\eta^1 s_1{}^3,$$

$$\mathrm{Re}_R\left(\mp\kappa\left(\eta^1\right)^2\left(\frac{\partial v^2}{\partial \eta^1}v^1 + \frac{\partial v^2}{\partial \eta^2}v^2 + \frac{2}{\eta^1}v^1v^2 \mp \frac{2\kappa}{\eta^1}v^1v^3\right)\right) +$$

$$+\mathrm{Re}_R\left(\left(1+\kappa^2\left(\eta^1\right)^2\right)\left(\frac{\partial v^3}{\partial \eta^1}v^1 + \frac{\partial v^3}{\partial \eta^2}v^2\right)\right) =$$

$$= -\frac{\partial p}{\partial \eta^3} + \frac{\partial s_3{}^1}{\partial \eta^1} + \frac{\partial s_3{}^2}{\partial \eta^2} \pm \frac{\kappa}{\eta^1}s_2{}^1 \mp \kappa\eta^1 s_1{}^2 + \kappa^2\eta^1 s_1{}^3 + \frac{1}{\eta^1}s_3{}^1,$$

$$\frac{\partial v^1}{\partial \eta^1} + \frac{\partial v^2}{\partial \eta^2} + \frac{1}{\eta^1}v^1 = 0,$$

where $\mathrm{Re}_R = \rho_f V_a R/\mu_f$ stands for the modified Reynolds number calculated from the average flow rate $V_a = Q/\Omega$ and the duct radius; Q stands for the volumetric flow rate; Ω stands for the cross-sectional area; $v^i = V^i/V_a\ (i = 1,2,3)$ stands for dimensionless contravariant components of dimensionless velocity $\mathbf{v} = \mathbf{V}/V_a$; $V^i(i=1,2,3)$ stands for the contravariant velocity components $\mathbf{V}$ in the helical coordinate system; $p = PR/(\mu_f V_a)$ stands for the dimensionless pressure; P stands for the pressure; $s_i^{\ j} = (grad\mathbf{v})_i^{\ j}$ stands for the dimensionless components of the velocity gradient tensor.

Single-valuedness conditions can be written as

$$v^i(\delta,\eta^2)=0(i=1,2,3), -\pi/3 \le \eta^2 \le \pi/3,$$

$$v^i(1,\eta^2)=0(i=1,2,3), -\pi/3 \le \eta^2 \le \pi/3,$$

$$v^i(\eta^1,\pm\pi/3)=0(i=1,2,3), \delta \le \eta^1 \le 1,$$

where $\delta = r/R$; r is the smaller radius of the duct with a triple-threaded auger insert.

The components of velocity depend only on two variables η^1, η^2, and the dimensionless pressure can be written as $p = f_0\eta^3 + f_1(\eta^1,\eta^2)$, where $\partial p/\partial\eta^3 = f_0 = const$.

The solution to this problem requires the finite element method and the COMSOL Multiphysics software suite to be used.

For the results of calculations to present easily, we used traditional Reynolds numbers $\mathrm{Re}_h = \rho_f V_a D_h/\mu_f$ with the hydraulic diameter D_h as the defining size. For ducting with triple-threaded auger inserts, the expression D_h takes the form

$$D_h = \frac{4V}{\Sigma},$$

where the volume (V) and the area (Σ) of the wetted surface of a single duct travel with a length S are determined by the formulas

$$V = \frac{\pi(1-\delta^2)R^2S}{3},$$

$$\Sigma = RS\left[\frac{2\pi(1+\delta)}{3} + \left(\sqrt{\kappa^2+1} - \delta\sqrt{\kappa^2\delta^2+1} + \frac{\arcsin\mathrm{h}(\kappa) - \arcsin\mathrm{h}(\kappa\delta)}{\kappa}\right)\right].$$

The coefficient of friction ξ is calculated by the formula

$$\xi = \frac{1}{\mathrm{Re}_h}\frac{2D_h^{\ 2}|f_0|}{R^2}.$$

Fig. 2 shows $\xi\mathrm{Re}_h$ as a function of Re_h for different values of the Reynolds numbers, $\delta = r/R = 0.5$ and two values of $S/(2R) = 4$ (здесь $D_h/R = 0.7295$) and $S/(2R) = 5$ (здесь $D_h/R = 0.7392$).

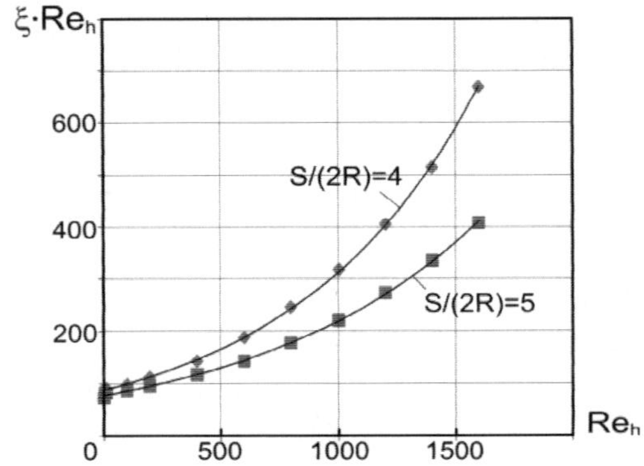

Fig. 2 $\xi\mathrm{Re}_h$ vs. Re_h.

As can be seen from the figure, unlike the flow in a round pipe, where the correlation takes place $\xi\mathrm{Re}=64$, the product $\xi\mathrm{Re}_h$ increases with the increase in Reynolds number and can be described by an exponential dependence of the form: $\xi\mathrm{Re}_h=88{,}088e^{0.0013\mathrm{Re}_h}$ for $S/(2R)=4$, and $\xi\mathrm{Re}_h=77{,}408e^{0.001\mathrm{Re}_h}$ for $S/(2R)=5$.

This work is supported by the Russian Science Foundation under grant 19-11-00220.

References

[1] Bergles A E, Jensen M K, Somerscales E F C and Manglik R M 1991 *Literature Review of Heat Transfer Enhancement Technology for Heat Exchanges in Gas-Fired Applications*, Report GRI 91–0146, Chicago, Gas Research Institute

[2] Kalinin E K, Dreitser G A and Yarkho S A 1990 *Heat-Exchange Enhancement in Channels*, 3rd edition, Moscow: Mashinostroenie

[3] Webb R L 1994 *Principles of Enhanced Heat Transfer*, New York: John Wiley & Sons Inc.

[4] Gortyshov Yu F 2009 *Thermohydrodynamic Efficiency of Prospective Methods of Heat Emission Enhancement in the Channels of Heat-Exchange Devices*. Heat-Exchange Enhancement, Kazan: Tsentr Innovatsionnykh Tekhnologii

[5] Dewan A, Mahanta P, Sumithra K R and Kumar S P 2004 *Review of passive heat transfer augmentation techniques*, Proc. Instn Mech. Engrs. Part A: J. Power and Energy vol 218 pp 509–27

[6] Mitrofanova O V 2003 *Hydrodynamics and heat exchange of swirled flows in channels with swirlers* (An analytical review), Teplofiz. Vysokh. Temperatur, vol 41 no 4 pp 587–633

[7] Nazmeyev Yu G 1996 *Flow dynamics and heat transfer of swirling flows of rheologically complex fluids* (M. Energoatomizdat) p 304

[8] Vachagina E K and Kadyirov A I 2014 *The use of helical coordinate systems*. The Quarterly Journal of Mechanics and Applied Mathematics Vol 67 No 4 pp 553–566

[9] Nazmeyev Yu G and Konakhina I A 1993 *Investigation of the heat transfer intensification in the laminar flow of viscous fluids in helically knurled pipes*. Teploenergetika No 11 pp 59-62

[10] Kadyirov A I, Khalitova G R and Vachagina Ye K 2016 *The hydrodynamic structure of a viscous flow in ducts with intensifiers as auger inserts*. Papers of Akademenergo. No 3 pp 7–16

[11] Khusnutdinova E M, Konakhina I A, Khamidullina G R and Vachagina Ye K 2017 *Intensification of oil and oil product heaters by means of auger inserts*. Interbranch scientific and technical complex IMTOM-2017, Part 1. pp 361-364

IOP Conf. Series: Materials Science and Engineering **570** (2019) 012061 doi:10.1088/1757-899X/570/1/012061

Comparative analysis of friction losses in long pipelines with profiled channel flow area

I A Konakhina[1], E M Khusnutdinova[1], G R Khamidullina[1] and E K Vachagina[2]

[1]Kazan (Volga Region) Federal University
18 Kremlyovskaya Str., Kazan 420008

[2]Institute of Power Engineering and Advanced Technologies, FRC Kazan Scientific Center, Russian Academy of Sciences
2/31 Lobachevskogo Str., P.O. box 261, Kazan, 420111

Abstract. The paper discusses the matter of how friction losses change in long pipelines with profiled channel flow area and describes the method of calculating the reduction of friction losses. The reduction of friction losses in the pipeline components has been analyzed.

The friction calculation is an important component to select the pipe size for pipelining. The friction calculation uses many methods describing all the options for calculating the coefficient of friction, depending on the flow regime, type of fluid and its temperature, as well as on the roughness of the pipe, an equation with variation of its parameters and the introduction of various correction factors [1,9,10].

Heat exchangers are used to stabilize the fluid temperature. This case study used heat exchangers with low rolled fins on the side of high-viscosity oil and screw rolling, which have similar characteristics, and the discrepancy between these indicators was within ±10% [4].

Analysis of how the power consumption changes during the operation of the oil heater is shown in Fig. 1-2 [2]. Fig. 1 shows the analysis of the change in the required electric power N_e, while controlling the temperature conditions of the heat exchanger in the high heat period of operation.

$$\frac{(N_{CК})_i}{(N_{CК})_j}$$

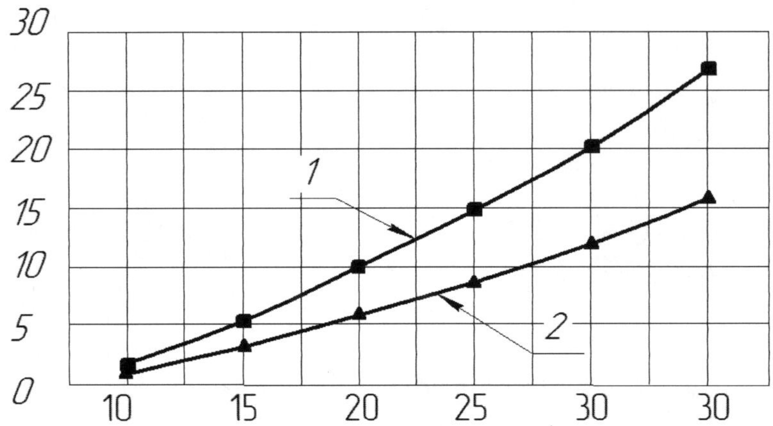

Fig. 1 The dependence of the cost of electrical power for transferring the coolants
1 - screw rolled pipe; 2 - bare pipe

The curve (Fig. 1) has been calculated using the following relationships:

$$N_э = \frac{\Delta P \cdot G}{\rho}; \qquad (1)$$

$$\rho_i = \rho_j; \qquad (2)$$

$$\frac{(N_э)_i}{(N_э)_j} = \frac{\Delta P_i}{\Delta P_j} \cdot \frac{G_i}{G_j}. \qquad (3)$$

Formulas (1) through (3) refer to the following legend: ΔP is the pressure drop in the pipe, G is the water flow, kg/s; ρ is the water density, kg/m^3; i is the index corresponding to the design mode, j is the index corresponding to the base rating

Fig. 2 shows the correlation of the cost of electrical power and the thickness of buildup in bare pipes.

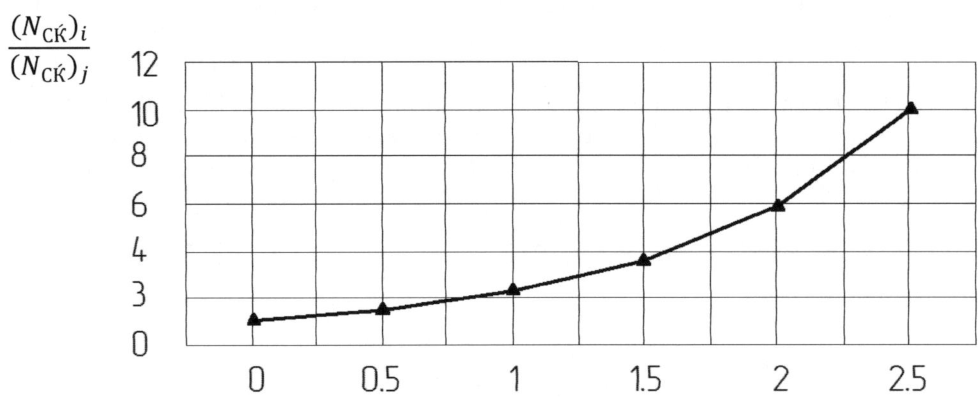

Fig. 2 Cost of electrical power vs. thickness of buildup in bare pipes

The curve (Fig. 2) has been calculated in the approximation of smooth friction using the relations:

$$\frac{(N_э)_i}{(N_э)_j} = \frac{\Delta P_i}{\Delta P_j} = \frac{\lambda_i}{\lambda_j} \cdot \frac{(d_э)_j}{(d_э)_i} \cdot \left(\frac{w_i}{w_j}\right)^2; \qquad (4)$$

$$\Delta P = \lambda \frac{L}{d_э} \frac{\rho w^2}{2}; \quad (5)$$

$$\lambda = 0{,}316/\sqrt[4]{Re}; \quad (6)$$

$$G_i = G_j; \quad (7)$$

$$L_i = L_j. \quad (8)$$

Formulas (4) through (8) use the following legend: λ is the channel resistance coefficient; d_e is the equivalent diameter of the channel, m; *L* is the channel length, m; *w* is the flow rate in the channel, m/s; *Re* is the Reynolds number.

As can be seen from Figures 1 and 2, the relative increase in electric power costs when regulating the temperature conditions of an oil heater in the summer time can reach 18 times the values compared to the winter times, and the narrowing of the pipeline flow area will increase the power costs to 6 times the values [8].

Thus, conventional systems of oil heaters are not capable to achieve a significant energy-saving effect even in the case of using intensified heat exchangers, since all the removed thermal energy is then discharged into the atmosphere [3-7].

It is advisable to solve this problem by cutting the power consumption and transition to fundamentally new systems.

References

[1] Li Long-jion, Xin Ming-dao and Cui Wen-zhi 2004 *An experimental study of the intensification of heat transfer in a screw channel with three-dimensional internal finning*. Therm. Energy and Power. No. 3, pp 270-273

[2] Voropaev A N and Fatkullin P P 2008 *Intensification of heat exchangers of viscous media used at thermal power plants*. Radio electronics, electrical engineering and power engineering. International Scientific and Technical Conference of Students and Graduates. Moscow: Publishing House of Moscow Power Engineering Institute. pp 238 -239

[3] Vachagina Ye K 2004 *Investigation of flow modes of heat transfer in the flow of nonlinear viscous fluids in channels with screw-shaped symmetry*. News of Higher Educational Institutions. Energy Problems. No. 1-2. pp 173-176

[4] Khusnutdinova E M, Konakhina I A and Khamidullina A F 2017 *Comparative Analysis of Thermohydraulic Performance of Enhanced Viscous Oil and Oil Product Heaters*. IOP Conference Series: Materials Science and Engineering. Art. No. 012046

[5] The method of determining the actual loss of thermal energy through the thermal insulation of pipelines of water heating networks of Central heating systems. - M.: NC ENAS, 2007. p 183

[6] Maron B I 2012 *Hydraulics of two-phase flows in pipelines* - M.: LAN, p 256

[7] Nazmeev Yu G and Lavygin In M 2005 Heat exchangers TPP: textbook for universities in the direction of "Power" – M.: Publishing house MEI, p 260

[8] Ibrahimov M K, Novozhilov E V and Subbotin B I 1961 *Heat Transfer and hydraulic resistance in the screw motion of the liquid in the pipe*. Heat power engineering. No. 7 pp 57-60

[9] Gortyshov Yu F 2009 *Thermohydrodynamic Efficiency of Prospective Methods of Heat Emission Enhancement in the Channels of Heat-Exchange Devices*. Heat-Exchange Enhancement, Kazan: Tsentr Innovatsionnykh Tekhnologii

[10] Kadyirov A I, Khalitova G R and Vachagina Ye K 2016 *The hydrodynamic structure of a viscous flow in ducts with intensifiers as auger inserts*. Papers of Akademenergo. No. 3. pp 7–16

IOP Conf. Series: Materials Science and Engineering **570** (2019) 012062 doi:10.1088/1757-899X/570/1/012062

Engineering education in the system of modern innovation

I A Konakhina, G R Khamidullina, E M Khusnutdinova and N F Kashapov

Kazan (Volga Region) Federal University
18 Kremlyovskaya Str., Kazan 420008

Abstract. The paper discusses and reviews key issues of engineering education development as one of the most important component of modern innovative activity in the Russian Federation. The discussion is based on the example of modern educational technologies used in the Institute of Engineering at Kazan Federal University.

As the military tensions continue to grow across the globe and the competition escalates in the global high-tech market, it is very important for our country to ensure the reproduction of intellectual personnel trained not only for highly specialized activities, but also to perform in the field of innovations, which implies creativity and ability to find extraordinary solutions. We cannot afford to lag behind in the development of basic and key technologies providing innovative breakthroughs in the current environment [1-3].

Analysis of social features of engineering activity at the turn of the 20th and 21st centuries reveals the following major trends in this area have emerged over the last 15–20 years [3]:

1. The steady growth in the significance of innovations in the economy, which contributes to the rapid development of technologies and increase their scientific content.
2. The role of small and medium-sized innovative companies in a modern high-tech economy is gradually increasing.
3. The breakdown of centuries-old patterns, when each subsequent generation generally became more educated than the previous one. There has been a reverse trend, and the education system itself has begun to degrade everywhere (a trend that is characteristic not only of our country but noted in many economically developed countries of the world).

The above trends form a number of requirements to the engineering education, which can be regarded as the challenges of modern time to the traditional educational system [1, 3]:

1. The requirements to the basic education of engineers, the quality of their intellectual, volitional and organizational skills become dramatically tougher.
2. The extended responsibility of a modern engineer, who is increasingly being considered all in one as a scientist, specialist, technical expert, and often the leader of small and large groups of projects. All this places high demands on the integrity, versatility and breadth of training an engineer.
3. Training of engineers at the university, at enterprises of various sizes – from small-size companies to major corporations, implemented in the forms of additional education acquires a holistic personal character.

All of the above brings us to the conclusion that the classical concept of engineering education is becoming actual again, but with a pronounced individualization [5-8].

The Institute of Engineering of the Kazan (Volga Region) Federal University introduces the concept of training of engineers qualified for innovation activities. At the same time, innovations are regarded as an aggregate process that combines several stages:

1. Studying the fundamentals of engineering.
2. Forming insights into the nature of scientific work, targeted search and the development of technical ideas.
3. Learning the basis of development of applied technologies and bringing them to industrial use.
4. Studying the methods of work management and control of small groups of diverse professionals.
5. Legal backing of the new product and its commercialization.

The Institute of Engineering introduces a two-level system for training specialists in undergraduate and graduate programs, which differ in the depth and breadth of coverage of the treated stages [11,12].

The undergraduate program is focused on the training of a specialist who is good at background and applied knowledge, as well as the principles of economic management [4].

The graduate program is focused on the development of outside-the-box thinking skills, the formation of a scientific view of the process, and the training of a specialist for the role of performer and leader of small innovative groups [9,10].

Building an innovative infrastructure in the educational institution is an extremely difficult mission and requires in-depth analysis of the scientific and engineering capabilities. For this purpose, the Institute of Engineering has launched a system of R&D laboratories (R&DLab), among them:

- R&DLab for Bioengineering and Life Safety;
- R&DLab for Prototyping (additive process);
- R&DLab for Production and Processing of Construction Materials;
- R&DLab for Plastic Production Process;
- R&DLab Precision Technologies Center;
- R&DLab for Engineering and Management Systems Research

The Institute of Engineering also has a Comprehensive Training Center for Energy Efficiency, which focuses on the development of energy-saving technologies, the study of the scientific background of energy conservation, the development of applied aspects of alternative energy and the advancement of dedicated supplementary education [13-15]. As part of the supplementary education, students have the opportunity to learn the components of robot-operated and fully automated productions, including control systems and setting up of Siemens Sinumerik 840 D sl.

References

[1] Naumkin N I, Senin P V, Maslennikova L V and Maikov E V 2007 *Innovative methods of training in a technical university*. Saransk: Mordovia University Press p 122

[2] Slastenin V A and Podymova L S 1997 *Pedagogics: Innovation Activity*. M.: Individual entrepreneur Magistr Publishing p 456

[3] Yodgorova M O 2018 *Modern engineering education and innovations in training of engineers*. Young Scientist. No 9 pp 147-149

[4] Novikov P M and Zuev V M 2000 *Advanced vocational education: Scientific and practical guide*. M: Rgatis p 266

[5] Kjersdam F, Enemark S 1994 *The Aalborg experiment project innovation in university education*. Aalborg, Denmark: Aalborg University Press

[6] Eyerer P, Hefer B and Krause B 2000 *The reformation of technical education through project orientation education*. Global J. of Engng. Educ., 4, 3, 2812286

[7] Enemark S 2002 *Innovation in surveying education*. Global J. of Engng. Educ., 6, 2, 1532159

[8] Shukshin I E, Vzyatyshev N f and Romankova L I 1996 *Innovative education: ideas, principles, models*.

[9] Sycheva N V 2012 *Essence and mechanism of investment of the innovation process. Siberian financial* No 1 pp 150-55

[10] Panshin I V 2013 *The study of the readiness of Russian regions to create high-tech jobs in the process of innovative modernization Regional economy: theory and practice* No 16 pp 11-20

[11] Gilmanshin I, Kashapov N, Gilmanshina S and Galeeva A 2016 IOP Conf. Ser.: Mater. Sci.

Eng. 134 012007

[12] Khristoliubova V I, Kashapov N F and Shaekhov M F 2016 IOP Conf. Ser.: Mater. Sci. Eng. 134 012017

[13] Kashapov N F, Nafikov M M, Gazetdinov M X, Nafikova M M and Nigmatzyanov A R 2016 IOP Conf. Ser.: Mater. Sci. Eng. 134 012013

[14] Kashapov N F, Gilmanshin I R and Konahina I A 2014 IOP Conf. Ser.: Mater. Sci. Eng. 69 012024

[15] Kashapov L N, Kashapov N F, Kashapov R N 2013 J. Phys.: Conf. Ser. 479 012003

IOP Conf. Series: Materials Science and Engineering **570** (2019) 012063 doi:10.1088/1757-899X/570/1/012063

Features of the formation of interdisciplinary competencies in engineering education

S S Kosmodemyanskaya[1], I R Gilmanshin[1,2], Yu I Azimov[1]

[1]Kazan Federal University, Kazan, Russia, 420008

[2] Kazan National Research Technical University named after A N Tupolev, Kazan, Russia

Is-er@yandex.ru

Abstract. The contradiction between the demands of employers and the insufficiently developed methodology for the formation of over-subject competencies of graduates of technical universities has been revealed. On the example of federal universities, the technology of forming “flexible skills” in mechanical engineering, allowing to be successful regardless of the specifics of a person’s professional activity, is considered.

The socio-economic changes occurring in our country impose new problems and problems on the formation of a new generation of specialists and university graduates to solve. According to the implementation of the Federal Target Program for the Development of Russian Education for the period 2016–2020, the issue of qualified and competitive specialists is currently provided.

Analysis of the existing literature and advanced pedagogical experience on the problem of training qualified specialists - university graduates, reveals a leading contradiction between the demands of a modern customer (represented by society and / or employer) and insufficiently developed methods of forming supra-subject competencies of university graduates.

Modern society, as a customer, needs qualified personnel to develop and implement the latest technologies of private and public business. The most frequent problem is the formation of not only professional, but also supra-professional competencies. This is reflected in the effectiveness of the use of the technology "softskills" - from "soft skills", "flexible skills", allowing to be successful regardless of the specifics of human activity. Psychology often considers these skills as social - the ability to persuade, work in a team, find an approach in dealing with different people, personal development, time management, creativity, etc. . The definition of the existing definitions of this concept characterizes it as a certain complex of non-specialized superprofessional skills governing the successful participation of a specialist in any working process [1].

The etymology of synergy indicates cooperation, assistance and complicity (synergos - Greek, jointly acting). The meaning of this term is currently considered from the standpoint of various fields in dictionaries of foreign words, business dictionaries, economic, sociological, medical, philosophical dictionaries, and dictionaries of a practical psychologist. This term is similar in nature to synergetics, which has an interdisciplinary character. Synergetic studies are associated with the study of self-organization processes of rather complex open systems [2]. Analysis of the definition of a concept indicates that the concept of the combined action of any components is most often used, in which the total effect exceeds the additivity of the actions of these components.

Modernity focuses society on the formation of a new generation of specialists. The report “The Future of Jobs”, presented by the World Economic Forum [3] in Davos (2016), defines the results of research based on 35 competencies that, in various proportions, cover the needs of the largest number of professions worldwide by 2020.

Analysts' forecasts are focused on the following most sought-after competencies included in the top 10: solving complex problems; critical thinking; creativity; people management; emotional intelligence; judgment and speed of decision making; customer focus; coordination skills; interaction; negotiation skills; cognitive flexibility.

We note practical unity in the need to form such a competency plan for students, both for future engineers and for future chemistry teachers. Chemistry and engineering, in our understanding, represent a specific synergy for the formation of a single material picture of the world. For the release of competent specialists in engineering and chemical education, you need the ability to own the above top 10 competencies.

Leading universities of the country for quite a long time took the focus on the training of competent specialists in the field of softskills. In the Kazan (Volga region) Federal University [4], it has been not the first year that the professional development program “Personal Efficiency” has been adapted, which is aimed at the formation and development of softsills as personal, communicative and managerial competencies. This program is based on practical trainings of personal growth and training seminars on the development of key interdisciplinary competencies. The effectiveness of the implementation of this program is determined by the outcome of the students learning the six training modules: self-development and career planning, self-organization skills (time management), intercultural communication skills, conflict management, leadership development and information management.

The Southern Federal University [5] represents the softskills laboratory, which (as well as at KFU) proposes to form an individual program (individual development plan) after analyzing the corresponding testing using a specific course matrix. This matrix contributes to the optimal choice of courses in two ways: through the formation of a comprehensive program for the development of specific competence competencies and / or opportunities to work in the field of those courses that contribute to the development of a wide range of competencies relevant to a particular student.

Based on this context of the development of softsills-competencies, higher demands are placed on graduates of universities. We are interested in the data of the ranking of the best universities in the world - QS World University Rankings [6].

This is an annual (since 2004) large-scale study of the best world-class universities, which is conducted by the British consulting company Quacquarelli Symonds (QS). This project, developed by QS in conjunction with the British Times Higher Education, functioned as the World University Rankings. Since 2010, the single rating has been divided into two: Times Higher Education publishes the World University Rankings in partnership with Thomson Reuters, and Quacquarelli Symonds continues to issue a rating called QS (World University Rankings).

Currently, the founder of QS [7] urged universities to teach students softskills. N. Kvakvarelli noted that as one of the indicators of a university’s performance, a rating company takes into account the gap between the competences that students received in the learning process and those that the employer needs. Research data indicate a lack of softsills-competencies - teamwork, interaction with other people, solving complex problems [7].

Thus, based on the requirements of the employer, the result of the engineering education of a technical university changes also. A graduate engineer (2010-2016) is responsible for work in production, solving tasks close to the type of worker, and / or in the project office, performing the functions of administrator / manager / designer according to the regulations; carries out strict regulations of activities; knowledge of "English with a dictionary"; responsible for certain technological stages of production, etc. For the period 2020-2030, there is a gradual transformation of design engineers into the status of management engineers and business engineers. In this case, the skill of working in a network management team will be demanded [8]; solving routine administrative and managerial tasks through automated control systems and, further, using artificial intelligence; management on the full life cycle of technology; compulsory knowledge of two foreign languages (English and Chinese), the third is desirable, etc.

In Russian universities, the discrepancy between the level of competencies of a graduate and the expectations of employers has been noted for quite some time. Therefore, it is necessary to form students' competencies not only through appropriate leadership training, but also within all professional disciplines, creating a practice context in the classroom, imitating some of the conditions that professionals face in real professional activity [9]. Formation of critical thinking activities is defined as a factor in student readiness for the teaching profession. Many researchers pay attention to the leading role in evaluating student practice.

One can consider engineering thinking ([10], [11], [12]) not as a body of knowledge of specific disciplines, but in the aggregate to see the world as a system, design its elements and structural components and manage them.

Above, we have repeatedly expressed the idea that there is a connection between engineering and chemical-pedagogical education. To summarize the results of our study, data for the last 10 years was used ([9], [13], [14], [15], [16], [17]).

In their teaching activities for the training of engineers ([18]) (Engineering Institute K (P) FU) and future chemistry teachers (Chemical Institute im.B.A. Butlerov K (P) FU) we use the experience of implementation in the educational process project team, which allows to gradually form softskills-competencies. This form of training is quite relevant and most effective for the planned immersion of students in their future professional activities as highly qualified future professionals who successfully solve fairly complex tasks in unusual situations. We train specialists who own softsills-competencies, both for state institutions and for private business organizations. Gradual immersion takes place through the gradual formation of professional, over-subject and interdisciplinary competences with the gradual complication of practical experience in educational laboratories and / or when working with students in basic organizations. The activity of such multi-level project groups is organized at the request of customers and is carried out taking into account theoretical and experimental questions of scientific research of coursework, graduation qualified works of specialists, bachelors and undergraduates. The qualitative and quantitative composition of such variable groups is constantly changing, depending on the goal or experimental task. Logical modeling, forecasting and modeling of various processes and situations in standard and non-standard situations allow students to develop creativity, critical thinking and the ability to work in teams of replaceable staff, form cognitive knowledge and multi-variant solution of a variety of problems. Students acquire the skills of posing and solving problems of scientific research. We note the increased participation of students participating in this form of work in conferences, festivals and seminars at various levels - university, city, regional, Volga, All-Russian and international levels. We also note not only the increased inter-institutional integration within the university, but also inter-university cooperation (and also at the international level).

Thus, the work on this form of training allows future engineers and teachers of chemistry to develop and develop softskills-competencies necessary for more successful involvement of young professionals in their professional activities.

References

[1] Agapov O D 2007 *Socio-phenomenological understanding of interpretation* Scientific notes of the Kazan State University V 149 Book 5 pp 7–23

[2] Khoruzhiy S S 2011 *What is SYNERGEIA? Synergy as a universal paradigm: leading subject areas, discursive communication, heuristic resources* Questions of philosophy No12 pp 19–37

IOP Conf. Series: Materials Science and Engineering **570** (2019) 012063 doi:10.1088/1757-899X/570/1/012063

[3] *World Economic Forum in Davos* [Electronic resource] http://so-l.ru/news/y/2016_01_20_gearmix
[4] *KFU. Competence Development Center Competence Center UNIVERSUM + IMO* [Electronic resource] https://kpfu.ru/zrk/obrazovatelnaya-programma-39personalnaya-245411.html
[5] *Southern Federal University and Center for Career SFU* [Electronic resource] http://softskills.sfedu.ru/
[6] *Analytical portal. ISSN 2310-1792.* [Electronic resource] https://gtmarket.ru/ratings/qs-world-university-rankings/info
[7] *The founder of QS urged universities to train students at softskills* [Electronic resource] https://www.raspissanie.info/osnovatel-qs-prizval-vuzy-obuchat-stud-soft-skills/
[8] *Foresight Competences 2030* [Electronic resource] http://www.myshared.ru/slide/467083/
[9] Khasanova G B 2012 *Requirements of employers to graduates of engineering universities* Bulletin of Kazan Technological University No 20 pp 215-216
[10] Azimov Yu I, Gilmanshin I R, Gilmanshina S I 2016 *IOP: Conf. Ser. Mater. Sci. Eng.* **134** 12003
[11] Gilmanshin I R, Ferenets A V, Azimov Yu I, Galeeva A I, Gilmanshina S I 2015 *IOP: Conf. Ser. Mater. Sci. Eng.* **86** 12014-16
[12] Gilmanshin I R, Kashapov N F, Gilmanshina S I, Galeeva A I 2016 *IOP: Conf. Ser. Mater. Sci. Eng.* **134** 12007
[13] Patlina A S, Popova E D 2017 *Network interaction of universities and schools as a condition for the formation of students softskills* Pedagogical IMAGE No 2 (35) pp 94-102
[14] Kosmodemyanskaya S S, Nizamov I D, Yarullin I F 2017 *Information and educational environment in the system of practices of the future teacher of chemistry* QUID-INVESTIGACION/QUID Sp I 1 pp 373-78
[15] Sakhieva R G, Gilmanshina S I, Gilmanshin I R, Kosmodemyanskaya S S, Akchurina I R, Sagitova RN 2015 *A Portfolio as an Alternative Means of Presenting the University Students Achievements* Asian Social Science Vol 11 No 3 pp 162-67
[16] Gilmanshina S I, Sagitova R N, Kosmodemyanskaya S S and et. 2015 *Professional Thinking Formation Features of Prospective Natural Science Teachers Relying on the Competence-Based Approach* Review of European Studies Vol 7 No 3 pp 341-49
[17] Kosmodemyanskaya S S 2016 *Self-Development of the Future Teacher of Chemistry, through Bilingual Education, in Accordance with New Requirements of Professional Standards* The European Proceedings of Social & Behavioural Sciences Vol XII pp 322-29
[18] Gilmanshin I R, Gilmanshina S I 2017 *IOP: Conf. Ser. Mater. Sci. Eng.* **240** 012022
[19] Abdullin I Sh, Galyautdinov R T and Kashapov N F 2001 *Inzhenerno-Fizicheskii Zhurnal* **74** (5) 104–107
[20] Fayrushin I I, Dautov I G and Kashapov N F *International Journal of Environmental Science and Technology* **14** (12) 2555–2560
[21] Gubaidullin D A, Kashapov N F, Zaripov R G, Tkachenko L A and Shaydullin L R 2017 *J. Phys.: Conf. Ser.* **789** 012017
[22] Khristoliubova V I, Kashapov N F and Shaekhov M F 2016 *IOP Conf. Ser.: Mater. Sci. Eng.* **134** 012017

IOP Conf. Series: Materials Science and Engineering **570** (2019) 012064 doi:10.1088/1757-899X/570/1/012064

Machine elements evaluation diagnostic devices development using compressed air on crankshaft bearings example

A T Kulakov[1], E P Barylnikova[1] and A O Kulakov[2]
[1] Naberezhnye Chelny Institute (branch) of Kazan (Volga Region) Federal University, Russia
10A, Syuyumbike avenue, 423800, Russian Federation, Naberezhnye Chelny, alttrak09@mail.ru, 692401@mail.ru
[2] NP Kamazavtosport, 423800, Russian Federation, Naberezhnye Chelny, ul. Transport passage, 70

Abstract. Devices for diagnosing the degree of wear of crankshaft bearings, suitable for embedding in the on-board system for diagnosing a vehicle during its operation have been developed.

Failures of KAMAZ vehicles often occur due to cranking the connecting rod inserts [4,8,9,10]. The performance of bearings is mainly determined by the lubrication conditions and wear of the crankshaft bearings during operation, which lead to a decrease in pressure in various parts of the lubrication system [5, 6, 7].

The use of diagnostics of the technical condition of the crankshaft bearings involves [3, 7, 11, 12, 13, 14, 15] that during the operation it is possible to find deviations and inconsistencies, wear, changes in the shape and interaction, including changes in the gaps in the connecting rod bearings of the engine crankshaft.

Diagnostics can be carried out using the onboard system installed in the car at the exit from the manufacturer of automotive equipment, and subsequently in road transport enterprises and service centers. Due to this, the resource of the crankshaft and the engine can be used completely by grinding the necks for repair dimensions and installation of inserts of the next repair size, provided maintenance and repair documentation of the manufacturer, preventing the achievement of limit gaps and even more failures in the form of turning liners. To solve this task, a method has been developed for determining the clearance in the connecting rod bearing of an engine crankshaft based on a pneumatic device of differential type used in active control devices [1, 2] (Fig.1). The development is related to the field of engineering for the production of automotive and specialized vehicles, the operation of vehicles and can be used to monitor and diagnose the technical condition of their internal combustion engines

IOP Conf. Series: Materials Science and Engineering **570** (2019) 012064 doi:10.1088/1757-899X/570/1/012064

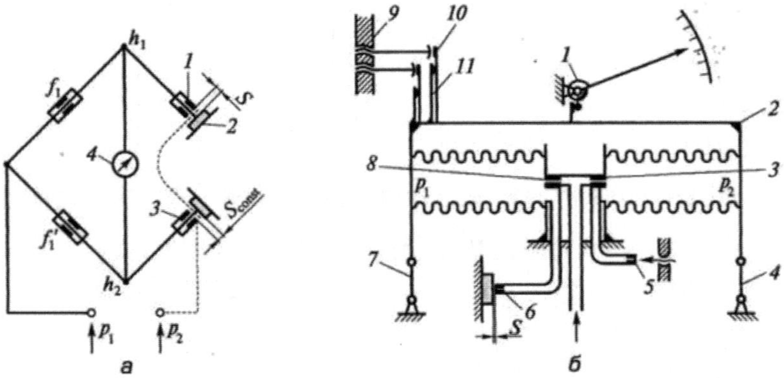

Fig. 1. Differential pneumatic device: a - measuring circuit; b - diagram of the differential bellows instrument [1]:
1 - damper nozzle; 2 - object of measurement; 3 - nozzle; 4 - indicating instrument; 7 - the mechanism; 2 - screed; 3 and 8 - input nozzles; 4, 7 and 11 - flat springs; 5 - counterpressure assembly; 6 - measuring nozzle; 9 - screws; 10 - stops

In the differential circuit (Fig. 1, a) the prepared compressed air is used and two pressures p_1 and p_2 are compared. Bellows, membrane differential pressure gauges reacting to the pressure difference $\Delta p = p_1 - p_2$ in the two branches of the system are used as a pressure meter. Figure 1, b shows the scheme of the differential bellows device. Compressed air from the pneumatic network, passing through the filter and stabilizer unit, under constant pressure flows through the inlet nozzles 3 and 8 in the bellows cavity (metal corrugated springs). From the left bellows air through the annular gap S, formed by the end of the measuring nozzle 6 and the surface of the controlled part, flows into the atmosphere. In this bellows a measuring pressure p_1 is created, the value of which depends on the size of the part being monitored. From the right bellows air flows to the atmosphere through the backpressure node 5, and a constant pressure p_2 is created in the cavity of the bellows. The free ends of the bellows are rigidly connected by a screed 2, suspended on flat springs 4 and 7. The position of the moving system of the device is determined by the difference between the measuring pressure p_1 and some constant back pressure p_2. Figure 2 shows a diagram of the measuring device for determining the clearance in the connecting rod bearing, the section of the first main journal and the first crank pin of crankshaft 1. In the oil channels of the crankshaft, a common air supply channel 2 is made, from which air is supplied to the gap between the crankshaft neck and the inserts 3 through the measuring system of the nozzle-flap at the end of the channel on the surface of the connecting rod neck. The design of the nozzle is made in the form of a screw with an outer diameter of 4 mm, calibrated holes in it with an inner diameter of 1 mm on the spout, recessed by 2 mm, with exhaust air vents inside the crankcase and into the atmosphere. The function of the flap is performed by the working surface of the upper connecting rod liner. Compressed air with a pressure of 0.2 MPa (Fig. 2) from the air preparation unit 5, 6 is supplied through the air line to the transition sleeve 4, the pressure is controlled by a pressure gauge 6, then the air passes through the transition sleeve 4 into the rotating crankshaft and the two nozzles - valve 2, diametrically opposed to the surface of the crank pin, the function of the valve is performed by liners. The volume of air passing through the connecting rod bearing nozzles determines the pressure drop across the diaphragm in the pressure transducer. The pressure drop across the membrane is converted into an electrical signal and the meter indicator readings with a scale calibrated from 0% to 100%.

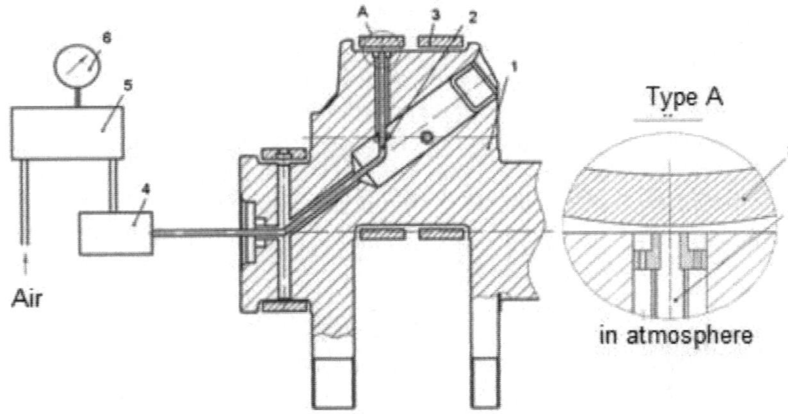

Fig. 2. Diagram of diagnosing the clearance of the connecting rod bearing by measuring air flow: 1 - crankshaft; 2 - air supply pipes to the jets; 3 - connecting rod insert; 4 –transfer coupling; 5 - electronic measuring device; 6 - air preparation unit

Air under pressure of 0.2 MPa is supplied to the connecting rod bearing gap for measurement through throttling devices, the “Sapphire-22D” pressure transducer, a Marposs measuring device (Fig. 3, a) a transition sleeve to the connecting rod bearing gap. These devices are used in the active control when honing cylinder liners.

a) b)

Fig. 3. Device for measuring the gap in the crankshaft bearings during diagnosis: a) measuring unit, air preparation unit, differential pressure gauge; b) receiving choke of the internal combustion engine and transitional coupling for the supply of air

From the measuring unit, the air enters the transitional air supply coupling of the receiving choke (Fig. 3, b), which allows the air to pass from the non-rotating part to the rotating part. Inside the crank pin there are two diametrically opposite holes in the plane of 30 ° from the axis of the crank into which the tubes are inserted. The outlet of the tubes ends with a precise jet with a diameter of 1 mm (Fig. 2). The amount of air entering the connecting rod bearing gap is adjusted by setting an air pressure of 0.2 MPa and adjusting the position of the throttles at the start of the test.

The measuring device measures the pressure drop on the membrane, which increases depending on the increasing flow rate with increases gap.

Calibration of the measuring instrument scale at “100%” of the gap size is carried out by supplying air to the zone of the nominal clearance at the beginning of the engine operation, when the gap between the crank neck and the liner is known. Deformation of the connecting rod bushings with the

IOP Conf. Series: Materials Science and Engineering **570** (2019) 012064 doi:10.1088/1757-899X/570/1/012064

formation of a deflection leads to a decrease, and wear to an increase in the gap with the connecting rod neck, which should be fixed by the device. The control algorithm onboard system must be provided with the same conditions, for which the measurement should be made after the engine stops at a certain position of the crankshaft at the angle of rotation. For the on-board system, the electrical signal must be converted and stored in the on-board computer. Measurements should be made after each engine stop, then superimposed on the mileage data and displayed on the display as a gap value, increase or decrease, intensity, residual life to the limit state. The device with the use of compressed air was used in the diagnosis of the forming of crankshaft connecting rod bearings at the KAMAZ engine factory.

References

[1] PP Volosov 1984 *Aktivnyj kontrol' razmerov* [Active size control] M.: Mashinostroenie 223 p

[2] Dimov YU V 2010 *Metrologiya, standartizaciya i sertifikaciya* [Metrology, standardization and certification] SPb.: Piter 464 p

[3] E P Barylnikova, A A Gafiyatullin, O A Kulakov 2015 *Diagnostika stepeni iznosa podshipnikov dizel'nogo dvigatelya v processe ehkspluatacii* [Diagnostics of wear rate of diesel engine bearings during operation] Problemy funkcionirovaniya sistem transporta pp 68-75

[4] M A Grigor'ev, V A Doleckij 1978 *Obespechenie nadezhnosti dvigatelej* [Engine reliability] M.: Izd-vo standartov 324 p

[5] Denisov A P, Kulakov A T 1986 *Izmenenie uslovij smazki shatunnyh podshipnikov v processe ehkspluatacii avtomobil'nogo dizelya* [Changing the lubrication conditions of connecting rod bearings during the operation of automotive diesel] Dvigatelestroenie No 4 pp 44-46

[6] Denisov A P, Kulakov A T, Gafiyatullin A A, Pankratov D L 2005 *Analiticheskoe issledovanie izmeneniya uslovij smazki shatunnyh podshipnikov v processe ehkspluatacii*[Analytical study of the changing conditions of connecting rod bearing lubrication during operation] Vestnik Saratovskogo gosudarstvennogo tekhnicheskogo universiteta T 3 No 1 (8) pp 69-75

[7] Kulakov A T 1986 *Razrabotka sposoba diagnostirovaniya shatunnyh podshipnikov dvigatelej i prakticheskih rekomendacij dlya snizheniya ih otkazov v processe ehkspluatacii (na primere KamAZ-740)*: diss .... kand. tekhn. nauk : 05.22.10 [Development of a method for diagnosing connecting rod bearings of engines and practical recommendations for reducing their failures during operation (on the example of KAMAZ-740)): diss ... Cand. tech. Sciences: 05.22.10] 173 p

[8] A P Denisov, A T Kulakov 1981 *Analiz prichin ehkspluatacionnyh razrushenij shatunnyh vkladyshej dvigatelej KamAZ-740* [Analysis of the causes of operational destruction of the connecting rod liners KamAZ-740] Dvigatelestroenie No 9 pp 37-40

[9] Kulakov A T 2007 *Povyshenie nadezhnosti avtotraktornyh dizelej putem sovershenstvovaniya processa smazki, ochistki i tekhnologii remonta osnovnyh ehlementov* : disp … d-r tekhn. nauk : 05.20.03[Improving the reliability of automotive diesel engines by improving the process of lubrication, cleaning and technology of repair of the main elements: disp ... Dr. tech. Sciences: 05.20.03] 558 p

[10] Kulakov A T, Denisov A P 2007 *Obespechenie nadezhnosti avtotraktornyh dvigatelej : monografiya* [Ensuring the reliability of automotive engines: monograph] Saratov: Sarat. gop tekhn. un-t 422 p

[11] A T Kulakov, N I Svelichnyj, R T Tazeev, Denisov A P, Sibiryakov P V 2004 Patent na izobretenie No 2221964 *Indikator nerazryvnosti potoka zhidkosti* [Patent for invention no. 2221964. Liquid Flow Continuity Indicator]

[12] Kulakov A T, Gafiyatullin A A, Barylnikova E P, Makushin A A, Kulakov O A *Sposob bezrazbornoj diagnostiki stepeni iznosa podshipnikov dvigatelya vnutrennego sgoraniya* Patent RF No 2560972 [Patent of the Russian Federation No 2560972 "Method for in-house diagnostics of the degree of wear of bearings of an internal combustion engine"]

IOP Conf. Series: Materials Science and Engineering **570** (2019) 012065 doi:10.1088/1757-899X/570/1/012065

Designing of diffusion coatings on bimetallic parts (steel, bronze)

V K Ilyin[1], O V Ilyin[1], L S Sabitov[1,2]

[1]Kazan State Power Engineering University, 51 Krasnoselskaya street, Kazan, 420066, Russian Federation
[2]Kazan Federal University, 18 Kremlyovskaya street, Kazan, 420008, Russian Federation

ilinwk@rambler.ru

Abstract. The article presents the results of theoretical and experimental studies of the formation of diffusion coatings on the bimetallic surface of steel, bronze, analyzed the possibility of obtaining intermetallic compounds found that the mechanism of formation of complex diffusion layers on the bimetallic surface of steel, bronze can be used as a way to improve the tribotechnical and mechanical properties of parts.

The task of predicting the structure and properties of the diffusion layer formed on two dissimilar metals during chemical heat treatment (CHT) is currently relevant and needs to be solved. Therefore, it is interesting to consider the mechanism of diffusion and the formation of a diffusion layer on a bimetallic surface of steel and bronze.

The task of forecasting is complicated by the fact that the saturation is carried out by two or more elements. Experimental data accumulated in this area indicate a high potential for multicomponent coatings. Practically, it is possible to create multi-component alloys with predetermined properties of the surface layer.

There are two possibilities for obtaining such coatings: simultaneous and sequential saturation.

It was established [1] that with simultaneous saturation of the surface with several elements, the depth of diffusion, and the concentration of elements may be insignificant and not have a decisive influence on the change in the microstructure of the diffusion layer. Another option is possible when the content of elements in the surface layer reaches a significant value, as a result of which not only the microstructure of the layer changes, but also the appearance of intermetallic compounds [2].

When forming the composition of the saturating medium, the sizes of the atoms of the saturating elements were taken into account; their diameter should not exceed 15 ... 16% of the corresponding diameter of the saturable metal [3].

In addition, the belonging of the diffusing elements and the metal of the substrate to the same group of the periodic system was taken into account. Mendeleev or to related related groups.

Ferrotitanium and zinc chosen for diffusion saturation meet the above conditions. However, with respect to bimetallic parts, pure titanation or galvanizing has several disadvantages:

- layers obtained on bronze have increased brittleness;
- the formed layers have insufficient thickness;
- deteriorate the physico-mechanical and tribological properties of the part.

IOP Conf. Series: Materials Science and Engineering **570** (2019) 012065 doi:10.1088/1757-899X/570/1/012065

It was established experimentally that due to the fact that two different metals are subjected to saturation in the same reaction space, it is advisable to conduct the saturation process stepwise, taking into account the optimum temperature conditions: the first stage at lower temperatures, the second at higher ones [4].
The diffusion mechanism for stepwise multicomponent galvanizing of a powder mixture of the following composition: zinc Zn, copper Cu, tin Sn, titanium Ti, zinc chloride $ZnCl_2$, aluminum oxide $Al2O_3$ is shown in fig.1.
At the first stage, when t1 = 500 ... 550 °C, atomic zinc appears on the steel surface due to thermal decomposition of the activator and interaction with zinc, which after adsorption by the steel substrate will begin to diffuse deep into the saturated part.

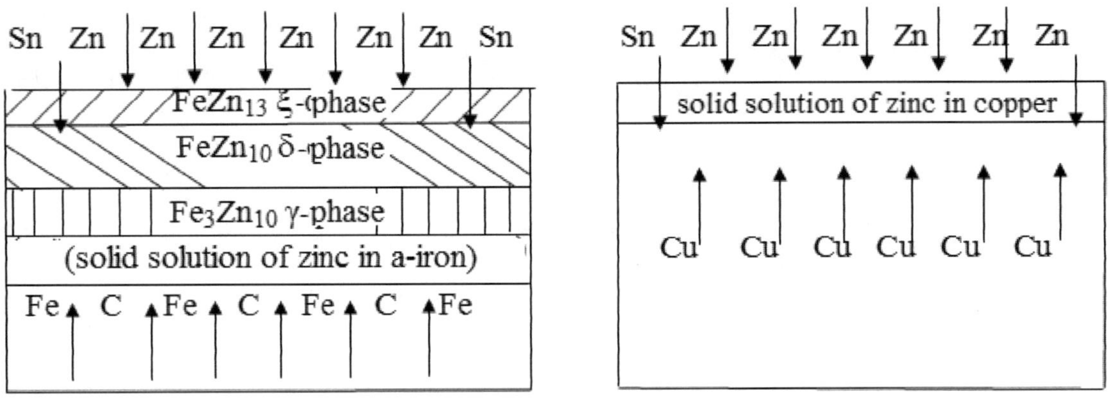

Figure 1. Diagram of the formation of a diffusion layer in the first stage.

Diffusion of zinc occurs predominantly by the vacancy mechanism. The concentration of zinc in the diffusion layer will determine the presence and sequence of phase formation. The process temperature (500 °C) is insufficient to cause a strong counter-current of iron atoms. Therefore, behind the γ- phase, poorer iron δ-phase (FeZn10) and ξ-phase phase (FeZn13) will appear. The sequence of the formation of diffusion layers may be different. It depends on the potential of the saturating medium for zinc, the mass transfer rate and the diffusion coefficient.
The need for the presence of tin in the reaction space due to its ability to increase the depth of penetration of zinc [5].
The mobility of titanium and copper atoms at the first stage (500 ° C) will be low, as a result of which they will not have a significant effect on the diffusion processes occurring in steel.
Considering the processes occurring on the bronze surface, it should be noted that the mobility of zinc in bronze is much lower than in steel. Therefore, the limiting stage of saturation will be the diffusion of zinc in bronze. The diffusion equation in this case is solved under the boundary conditions of the first kind [6]:

$$C(x,\tau)=C_0(\tau)\,[1-erf(x/2\sqrt{D\tau})], \tag{1}$$

where $C(x, \tau)$ – the concentration of the diffusing element at a distance x from the surface of the part at the moment of time τ; $C_0(\tau)$ – concentration of the diffusing element on the surface during the saturation process; $erf(z)$ – Gauss error function.
The diffusion coefficient can be determined by the Arrhenius law:

$$D(\tau)=D_0\exp(-Q/RT), \tag{2}$$

where D_0 – subexponential factor; Q – activation energy; T – absolute temperature; R– gas constant.

As follows from the Arrhenius equation, for the intensification of diffusion processes in copper it is necessary to increase their temperature. Increasing the temperature is also necessary in order to cause a counter-current of copper. According to literary sources [3, 5,], it should be within 650-850 C.
It has been established [5] that the formation of brittle phases in the copper-zinc system occurs because an unlimited amount of a diffusing element (zinc) is supplied to the surface of the solvent metal (copper), which has a sufficiently large diffusion coefficient at the process temperature. Possessing a large diffusion capacity at high temperatures and having an unlimited amount of a diffusing element, copper will have unlimited possibility of saturation. The layer obtained under such saturation conditions cannot be used as an antifriction material due to its low tribological and physicomechanical properties. Therefore, to obtain a certain phase composition of the tribotechnical and physicomechanical properties in the bronze-based diffusion layer, it is necessary to introduce an element into the mixture that would "control" the high-speed modes of diffusion processes and be an alloying element at the same time. In our case, these functions are performed by titanium. Despite the fact that the mobility of titanium at the above temperatures is low, the presence of an activator of $ZnCl_2$ and zinc make possible the following reactions:

$$3Ti+2Zn+2O_2 \xrightarrow{ZnCl_2+Zn} TiCl_2+Zn_2O+Ti_2O_3; \tag{3}$$

$$TiCl_2+2Cu \xrightarrow{ZnCl_2} 2CuCl+Tiат; \tag{4}$$

$$2CuCl+2Fe \xrightarrow{ZnCl_2} 2FeCl_2+2Cu; \tag{5}$$

$$2TiCl_2+2Fe \xrightarrow{ZnCl_2} 2FeCl_2+2Tiат; \tag{6}$$

$$2TiCl_2+2Fe \xrightarrow{ZnCl_2} 2FeCl_2+2Tiат; \tag{7}$$

Reaction (3) goes in saturating mixture. Reaction (4) on a bronze substrate. Reactions (5) and (6) go on a steel substrate.

Titanium oxide formation Ti_2O_3, see reaction (3), leads to slowing down of diffusion processes. However, the presence in the mixture of the activator $ZnCl_2$ makes possible the following reactions:

$$Ti_2O_3+2Zn \xrightarrow{ZnCl_2} Zn_2O_3+2Ti_{ат}; \tag{8}$$

$$Ti_2O_3+Zn \xrightarrow{ZnCl_2} ZnO_3+2Ti_{ат}; \tag{9}$$

Due to reactions (8), (9), atomic titanium additionally appears in the reaction space.

The presence of titanium in the mixture is necessary in order to eliminate the shift of the concentration potential d_c/d_x towards saturating the mixture.

IOP Conf. Series: Materials Science and Engineering **570** (2019) 012065 doi:10.1088/1757-899X/570/1/012065

Based on the above theoretical prerequisites, the mechanism for forming a diffusion layer with a multicomponent thermal diffusion saturation of two dissimilar materials (steel, bronze) in the same reaction space will occur as follows fig. 2.

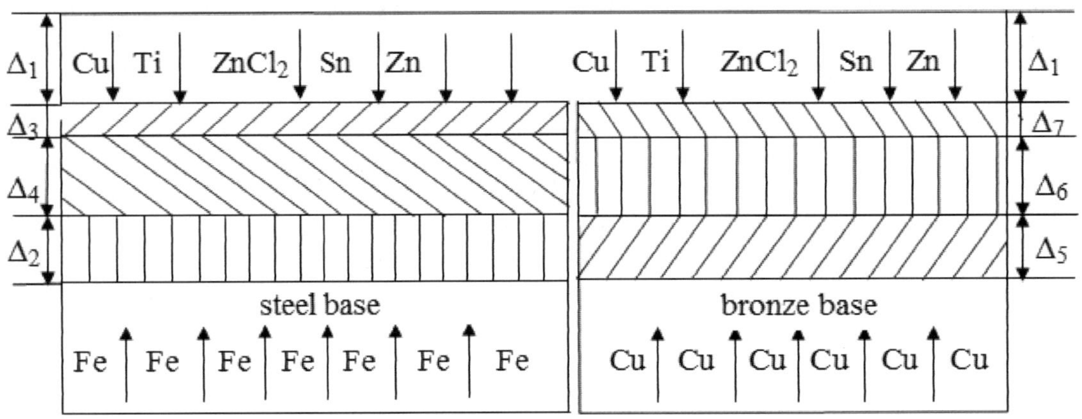

Figure 2. Diagram of the formation of a complex diffusion coating on two different metals (bronze, steel) in the same reaction space.

The location of the main zones (Fig. 2): Δ_1 is the reaction space; Δ_2 - zone of solid solution of zinc in α-iron; Δ_3 - zone enriched with zinc, titanium, and copper atoms; Δ_4 - a zone with a small content of titanium inclusions, enriched with zinc; Δ_5 - zone of solid solution of zinc in copper; Δ_6 is the zone of complex intermetallic compounds of zinc, copper, titanium and tin; Δ_7 is a zone enriched with zinc and titanium atoms.

In the first stage of saturation t=500 °C, in the reaction space, an active process of formation of atomic zinc occurs, which diffuses into the substrate material, forming a solid solution of zinc in α- iron. At the first stage of saturation t = 500 °C, an active process of formation of atomic zinc occurs in the reaction space, which diffuses into the substrate material, forming a solid solution of zinc in α-iron.

When the zinc concentration in iron reaches 20%, the Δ_4 zone will form, the so-called γ-phase (Fe_3Zn_{10}). This is the most solid phase. Its microhardness is within 5047 ... 5390. Further saturation of the steel surface with zinc will cause the formation of compounds consisting of α-phase ($FeZn_7$), δ-phase ($FeZn_{10}$), ξ-phase ($FeZn_{13}$) [7].

The layer of δ-phase is including δ_{1k}, δ_{1n} phases, whose hardness varies widely: 2500-4600 MPa [7]. The Δ_4 zone will be the largest in the diffusion layer. The zinc content in this zone can reach 79% [3]. On the bronze substrate, due to the low mobility of zinc in copper at t = 500 °C, zinc will settle in the form of a thin layer, the thickness of which will depend on the exposure time.

At the second stage, at a temperature of 700–750 ° C, titanium and copper will be more actively involved in the diffusion process. The concentration of atomic zinc in the reaction space due to an increase in temperature will increase, but not significantly, since part of it will be spent on obtaining atomic titanium, see reactions (7), and reducing titanium oxide, see reactions (8), (9). It also follows from these reactions that a decrease in the concentration of atomic zinc below the permissible value will lead to the impossibility of their occurrence. Consequently, a refractory oxide film will form on the surface of the diffusion layer, which prevents the diffusing elements from penetrating deep into the saturable surface. This adverse, at first glance, phenomenon can be used as a mechanism for controlling diffusion processes. The presence of an oxide film allows you to protect the layer from

IOP Conf. Series: Materials Science and Engineering **570** (2019) 012065 doi:10.1088/1757-899X/570/1/012065

excessive embrittlement. However, the amount of titanium alloying in the diffusion layer will also decrease.
In those cases when it is required to obtain a large thickness of the diffusion layer, the diffusion process must be conducted with a sufficient amount of zinc and at the upper temperature limit.
Considering the processes occurring on the steel surface, it is necessary to bear in mind that at the temperatures of the second stage intermetallic compounds of the $(TiFe)Zn_{13}$ type will appear on the surface, which form the surface zone Δ_3.
A layer of zinc, which was deposited on a bronze substrate during the first stage of the process, will diffuse deep into the surface. At first, the Δ_5 zone with the CuZn phase will appear, then, as mutual diffusion occurs, the Δ_6 richer zone with the $Cu5Zn_8$ phase will appear. On the surface of the bronze layer and at a shallow depth there will be inclusions of titanium. These inclusions will be the greater, the more active will be the reaction (8), and (9)

Conclusion

The above mechanism for the formation of complex diffusion layers on a bimetallic surface of steel, bronze during chemical heat treatment (CHT), can be used not only as a way to improve the tribological and mechanical properties of parts, but also as a way to restore bimetallic parts.

References

[1] Filonenko BA 1981 *Complex diffusion coatings* Moscow Mechanical Engineering 136 p
[2] Lakhtin Yu M 1988 *Foundations of Metallography* Moscow Metallurgy 318 p
[3] Dzholabov Yu Sh 1990 *Diffusion metallization in the restoration of parts from copper alloys on the example of the sleeves of the top head of the connecting rod of diesel engines in the conditions of repair enterprises of the agroindustrial complex* Dis. for the Candidate. tech. Science
[4] Shamsutdinova EV, Zueva OS Since 2014 *Nanomaterials and nanotechnologies in power engineering*. Kazan KSPEU 400 p
[5] Ilyin V K 2004 *Repair, restoration, modernization* 5 17-8
[6] Lakhtin Yu M, Arzamasov BN 1985 *Chemical heat treatment of metals* Moscow Metallurgy 256 p
[7] Galin RG 1990 *Development of effective saturating media and diffusion galvanizing technologies using hydrothermal treatment of the starting components* Dis. for the Candidate. tech. Science
[8] Erokhin MN, Kazantsev SP, Chupyatov NN 2017 *Wear resistance of carbide-containing chromium coatings obtained from the gas phase* Vestnik of the Federal State Educational Institution of Higher Professional Education Moscow State Agricultural Engineering University. V.P. Goryachkina 5 81 pp 48-53
[9] Erokhin MN, Chupyatov NN 2016 Wear resistance of precision parts of hydraulic systems restored by carbidochrome coating *Vestnik of the Federal State Educational Establishment of Higher Professional Education Moscow State Agricultural Engineering University* 3 73 pp 41-45
[10] Ilyin V K, Sabitov L S, Haibullina A I and O V Ilyin 2018 IOP *Conf. Ser.: Mater. Sci. Eng.* **412** 012029

IOP Conf. Series: Materials Science and Engineering **570** (2019) 012066 doi:10.1088/1757-899X/570/1/012066

Modifying surface properties of polyamide powders for selective laser sintering

A M Lyakhovich[1], R N Kashapov[1,2], N F Kashapov[1], L N Kashapov[1,3]
[1] Engineering Institute, Kazan Federal University, Kazan, 420008, Russia
[2] Laboratory of Radiation Physics, Zavoisky Physical-Technical Institute, FRC Kazan Scientific Center, Russian Academy of Sciences, Kazan, 420029, Russia
[3] Laboratory of Continuum Mechanics, Institute of Mechanics and Engineering, FRC Kazan Scientific Center, Russian Academy of Sciences, Kazan, 420111, Russia

kashramiln@gmail.com

Annotation. The purpose of this work was determination of the changes in surface properties of Polyamide-12 powder used in Selective Laser Sintering (SLS). The powders of original polyamide Duraform PA^{tm} (3D Systems) were investigated. Also the powders remaining around the product after the sintering process were investigated. The difference in the chemical structure of the Polyamide-12 powder surface layers has been established. The reason is the increased content of oxygen-containing groups in the surface layers. To a greater extent, this difference was revealed for powders left after sintering and intended for reuse. Aliphatic chains are subject to oxidative degradation in the surface layers of the original powder. In the powder remaining after sintering, terminal carbonyl and amine groups are subject to oxidative degradation.

Introduction

The technology of additive manufacturing in the 21st century is rapidly developing. New methods for the additive construction of products based on various physicochemical phenomena are emerging. Relatively old technologies of additive production, discovered in the last century, only strengthen their position in the hierarchy of production processes. One of such methods is the technology of selective laser sintering [1, 2]. In selective laser sintering (SLS) of polymer products, only a part of the powder layer corresponding to the product profile is subjected to laser impact. In this case, there is no need to create supports, since the product is surrounded by "green" powder, which serves as a support. It is the absence of the requirement for the presence of supports and the possibility of using thermoplastic materials that made this technology popular in the field of prototyping functional products. However, when reusing "green" powders, there are difficulties associated with the deterioration of the surface quality of the product, with an increase in roughness, especially on the vertical walls of the product [3], and sometimes with no powder in the structure of the product itself. This leads to deviations in the size of the product from the parameters of the virtual model. In addition, sometimes on the cold metal parts of the working chamber, crystalline structures form in the form of sharp thin needles and condensation of the ultrafine powder fraction, which indicates polymer sublimation processes or the presence of some amount of volatile constituents in the powder [4]. The identified problems raise questions about the impossibility of re-using the “green” powder during the next SLS cycles, which leads to its

IOP Conf. Series: Materials Science and Engineering **570** (2019) 012066 doi:10.1088/1757-899X/570/1/012066

disposal. Understanding the process of changing the properties of the powder in the future would allow the development of methods and equipment for the reuse of powder.

The aim of this work was to investigate the causes leading to changes in the surface properties of polyamide powders used in the selective laser sintering.

Main part

Selective laser sintering of polyamide powders was carried out on SLS Spro 60 HD. This setup is equipped with a continuous CO_2 laser with a power of 70 W and a wavelength of 10.6 microns. Figure 1 shows the internal view of the working chamber. The chamber has three platforms that move in the vertical direction, the right one (1) and the left one (2) are designed for storing the powder and its preheating, the central platform (3) serves as the working area of the building. The roller (4) moves the powder from the platform area (1) or (2) to the working platform and a laser scanning process is performed on the surface of the layer. Before moving the powder into the working area, the powder is heated to a temperature of 135 ° C. In the working area, the temperature of the powder is maintained at 171 ° C. The chamber atmosphere is pre-filled with nitrogen, in the process, the oxygen concentration is 3%.

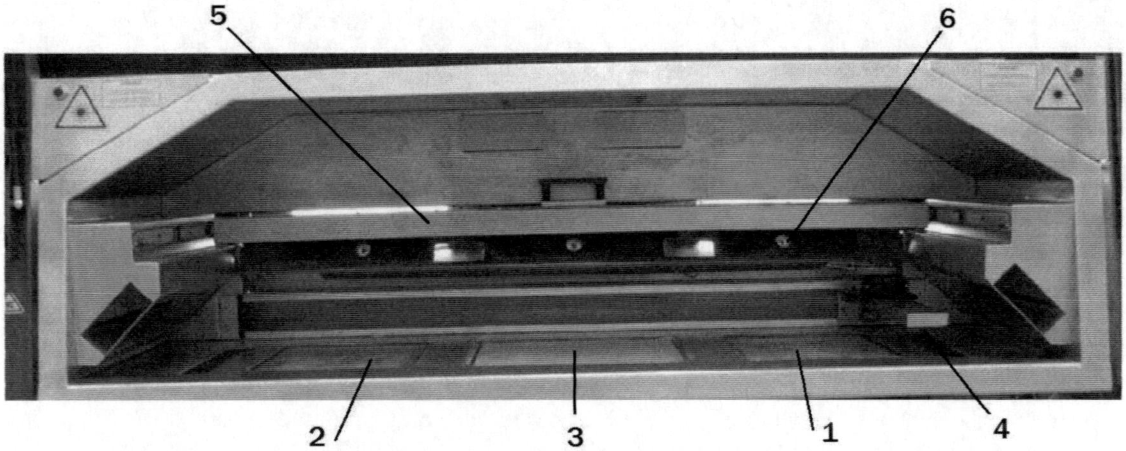

Figure 1. Inside view of a camera for building a SLS Spro 60 HD selective laser sintering machine. 1- right and 2 left powder storage platform, 3-working powder sinter platform, 4- powder transfer roller, 5- slide panel with infrared heaters, 6- IR sensors

Warming up is performed in two ways by using the heaters, located in the pipes moving platforms, and infrared heating elements in the upper extendable panels (5). Temperature control is carried out using infrared sensors (6).

The powders of the original polyamide Duraform PA (3D Systems) (sample 1) and the powder remaining after sintering (sample 2) were investigated. The studies were conducted by X-ray photoelectron, infrared and Raman (Raman scattering) spectroscopy and electron microscopy.

To obtain X-ray photoelectron spectra (XPS), an ES-2401 spectrometer with a magnesium anode (photon energy 1253.6 eV) was used. The power of the X-ray tube was 200 W, the vacuum in the analyzer chamber was $10–6.10^{-6}$ Pa. The spectrometer was calibrated along the line Au4$f_{7/2}$ = 84.0 eV. The value of the binding energy (E_{CB}) C1s- electrons of carbon atoms in the aliphatic hydrocarbon chain is assumed to be 285.0 eV [5]. The accuracy of determining the position of the lines in the C1s- spectrum was 0.2 eV, the decomposition of the spectra was carried out according to the method described in [6].

IOP Conf. Series: Materials Science and Engineering **570** (2019) 012066 doi:10.1088/1757-899X/570/1/012066

IR spectra were obtained on an Excalibur Varian 3100 FT-IR spectrometer in the range of 400-6000 cm^{-1} with a resolution of 1 cm^{-1}. Prepared tablets with KBr for research on transmission. The obtained spectra were normalized to the total intensity.

Raman spectra (RS) were obtained on a Horiba LabRam HR800 spectrometer using a He – Ne laser operating at a wavelength of 632.8 nm, the power consumption did not exceed 20 mW.

The surface morphology of the powders was studied using an EVO 50 Carl Zeiss AG scanning electron microscope.

Results and discussion

Electron microscopy (Fig. 2) showed practically identical size fractions of samples 1 and 2.

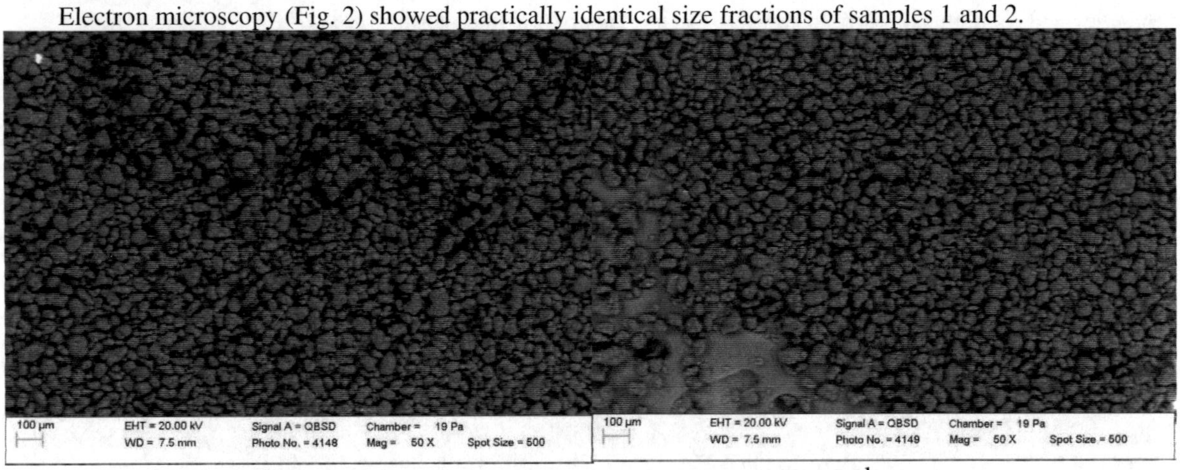

a b

Figure 2 - Microphotographs of polyamide powder: a - sample 1, b - sample 2.

In the IR spectroscopic study of powders, it was found that the functional groups of the powders are similar (Fig. 3) and correspond to polyamide 12 (poly (dodecanamide)) [7]. The authors of [8] for assess the quality of polyamides in the IR-spectroscopic study proposed a method for comparing the intensities of peaks of terminal amine or carbonyl groups and CH2 groups. The amine group in the studied powders is characterized by a frequency of 3304, 0 cm^{-1}, carbonyl - 1639 cm^{-1}, CH_2 - 2920 cm^{-1}. Using the approach proposed in [8], we estimated the ratios of the intensities of the bands of the terminal amine and carbonyl groups to the intensity of the band of the CH_2 group. As can be seen from Table 1, the intensity ratios of the bands of the functional groups for different powders are close and their difference does not exceed 1% for amine and 3% for carbonyl groups, which indicates the proximity of the chemical structure of the powders.

Table 1. The intensity attitude of the bands of the functional groups of polyamide powders according to IR spectroscopy.

Sample	Attitude of intensities	
	$-NH/CH_2$	$=CO/CH_2$
1	0,971	1,10
2	0,967	1,11

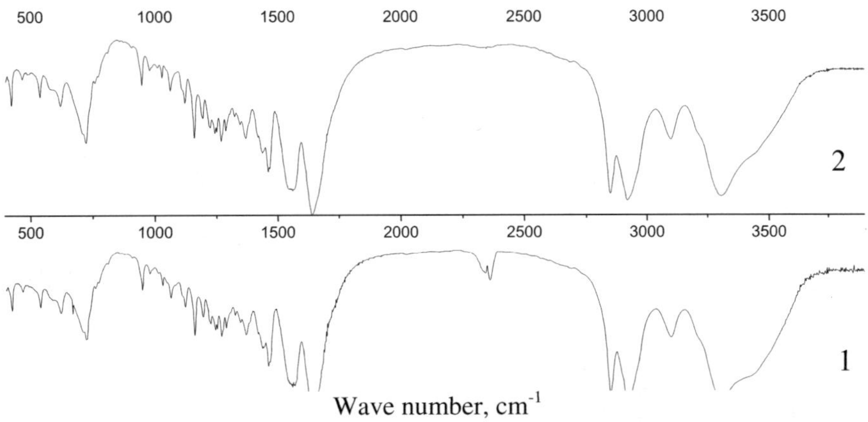

Figure 3 - IR spectra of polyamide powders: spectra 1, 2 correspond to the number of samples.

The study of RS-spectroscopy powders also revealed the similarity of the spectra of all the studied powders (Fig. 4) and their correspondence to the chemical structure of poly (dodecanamide) [7].

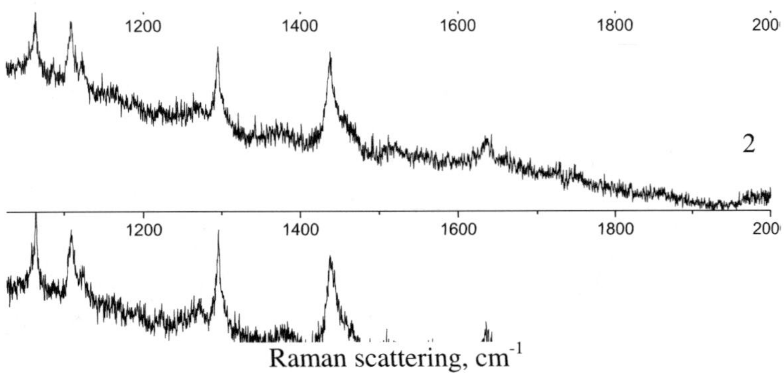

Figure 4 - Raman spectra of polyamide powders: spectra 1, 2 correspond to the sample number.

It should be noted that both IR and RS - spectroscopy in these studies provide almost volumetric characteristics of powders. XPS is a surface research method, the analysis depth of which does not exceed 10 nm [9]. Studies conducted by the XPS method showed a difference in the chemical composition of the surface of the powders (Table 2, Figure 4). As follows from the chemical formula of polyamide 12, the ratio of C: O: N elements (at.%) should be 86: 7: 7. However, an excess of the amount of oxygen in the surface layers of all samples was found, which follows from the O / N ratio (Table 2). The range of content of elements in the surface layers of powders with parallel surveys of different samples (of which there were at least three) of the same sample indicates the presence of powder in the sample with different chemical composition of the surface.

Table 2. The content of elements in the surface layers of polyamide powders (the range is indicated for parallel surveys).

Sample	Elements, at.%			
	C	O	N	O/ N
1	81-90	7-12	3-7	2.3-1.7
2	82-84	10-12	5-7	2.0-1.7

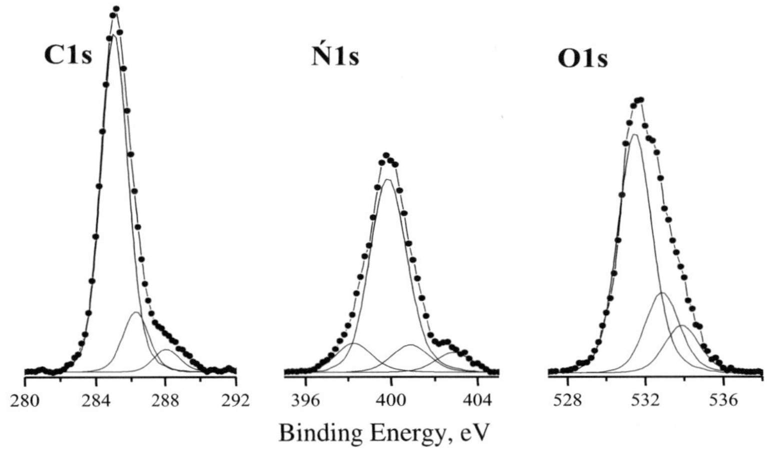

b

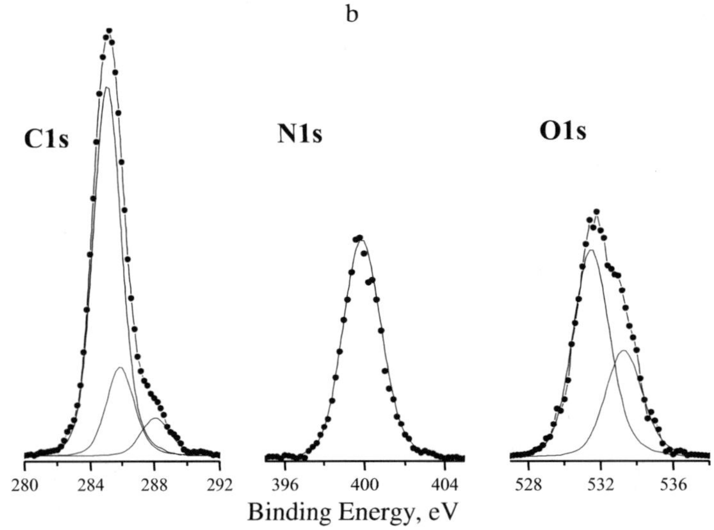

a

Figure 5 - XPS spectra of polyamide powders. Compliance with the sample number: a -1, b - 2.

According to [5], the N1s- and O1s-spectra of polyamide 12 are represented by single lines with binding energy (E_{be}) 399.8 eV and 531.3 eV, corresponding to nitrogen and oxygen atoms in the amide group - (CO) NH. In the C1s spectrum there are four lines with E_{be} 285.0 eV, 285.3 eV, 285.9 eV and 288.0 eV, corresponding to carbon atoms: in the aliphatic chain, in the α-position to the amide group, in connection with the amide nitrogen groups and carbon in the amide group itself, respectively. The relative intensities of (I) lines in the C1s spectrum are 75: 8: 8: 8

Our study on sample 1 found that the C1s- spectrum is represented by three E_{be} lines at 285.0 eV, 285.9 eV and 288.0 eV (Fig. 5 a). The lines with E_{be} 285.0 eV and 285.3 eV are not allowed, as close as possible. The relative intensities of the lines in the C1s spectrum 74: 19: 8. The increase in the second line I and the appearance in the O1s spectrum of the line with E_{be} 533.2 eV indicates the presence of functional groups containing a C – O bond in the surface layers of the polyamide [5]. The appearance of such groups can be explained by the oxidative destruction of the main chain of the polyamide. Even at low (room) temperatures, aliphatic polymer chains are prone to oxidative aging [10].

The study of the powder remaining after sintering (sample 2) revealed that the range of the content of elements during parallel surveys decreased, with an overall increase in the amount of oxygen (Table 2). New lines appeared in the spectrum of B N1s- with E_{be} 398.0 eV, 400.8 eV and 402.8 eV (Fig. 5b). The first line can be attributed to the terminal amino group. The appearance of lines with E_{be} of 400.8 eV and 402.8 eV can be associated with the processes of thermo-oxidative destruction of terminal amino groups during hydrolysis (E_{be} = 531.5 eV in the O1s- spectrum for OH^- ions [11]) and in their interaction with air oxygen. It is possible that the terminal carbonyl groups are also subject to thermal oxidation, as evidenced by the appearance in the O1s- spectrum of a line with E_{be} = 532.4 eV, corresponding to an isolated C=O bond [5]. The ability of aliphatic polyamides to react by terminal carbonyl and amino groups is known [8]. Most likely, layers of powder bordering the contour of the obtained part are subject to thermal oxidation (sintering temperature of polyamide powders is 183 °C). This is indicated by the characteristic redistribution of the line intensities in the N1s- and O1s- spectra in parallel surveys only for sample 2. It is not excluded that the main thermal oxidation occurs after contact of the powders with the atmosphere.

Conclusions

The difference in the chemical structure of the surface layers of polyamide powders used in selective laser sintering, which consists in an increased content of oxygen-containing groups in the surface layers, is established. This difference is revealed for powders left after sintering and intended for reuse. In the surface layers of the original polyamide powder, aliphatic chains are mainly subject to oxidative degradation, while in the powder remaining after sintering, carbonyl and amine end groups are subject to oxidative degradation. When recycling powders, additionally appeared polar functional groups of the surface layers can initiate coagulation of the particles. This leads to the formation of large-sized particles, which degrades the quality of the surface of the finished product.

To avoid deterioration of part surfaces when using "green" (secondary) powder in subsequent cycles, it may be necessary to remove the powder adjacent to the part contour from the process, or create conditions that exclude thermal oxidative destruction of the powder outside the part field. Another alternative is to restore the “green” powder to its original state.

Acknowledges

The work was carried out in the framework of the Base part of the state task of Ministry of Science and Higher Education of the Russian Federation (3.9399.2017/8.9)

References

[1] Lu L, Fuh J, Wong Y S 2001 *Laser-induced materials and processes for rapid prototyping* Kluwer Academic Publishers 267 p

[2] Pohl H, Gosger P 2001 *Three dimensional printing ProMetal - a new technology for the fabrication of tools* Proceedings of Rapid 2001 conference pp 248-52

[3] De Oliveira Setti G , De Silva J V L, De Oliveira M F et al. 2014 *Surface Methods for monitoring the densification of parts in the selective laser sintering process* Rapid Prototyping Journal. Vol 20 Is 2 17106327 pp 157-63

[4] Salazar A, Rico A, Rodríguez J, Segurado Escudero Jv et al. 2014 *Fatigue crack growth of SLS polyamide 12: Effect of reinforcement and temperature* Composites Part B: Engineering Vol 59 pp 285-92

[5] Beamson G, Briggs D 1992 *High resolution XPS of organic polymers* The scienta ESCA 300 database. N.Y.: John Wile & Sons 582 p

[6] 6. Povstugar V I, Shakov A A, Mikhailova S S et al. 1998 Journ. anal. chemistry T 53 No 8 pp 795-99

[7] Kuptsov A Kh, Zhizhin G N 2001 *Fourier spectra of Raman scattering and infrared absorption of polymers* M. Fizmatlit p 169
[8] Karaseva S Ya, Krasnykh E L, Levanova S V, Petrov G G, Sadivsky S Ya 2006 *On the quality of caprolactam and polyamide* Rus. chem. journ. (J. Rus. chem. society of D.I. Mendeleev) vol L No 3 pp 54-55
[9] *Surface analysis by Auger and X-ray photoelectron spectroscopy. Translated from English* 1987 Ed. D. Briggs, M.P. Sich. M .: Mir. 600 p
[10] Kirillova E I, Shulgina E S 1988 *Aging and stabilization of thermoplastics* M. Chemistry 240 p
[11] Nefedov V I, Cherepin V T 1983 *Physical methods for studying the surface of solids* M.: Nauka 33 p

IOP Conf. Series: Materials Science and Engineering **570** (2019) 012067 doi:10.1088/1757-899X/570/1/012067

Features of very heavy crude oil transportation

A O Mamonova[1], O A Plugatyreva[1], E M Khusnutdinova[1] and A N Khusnutdinov[2]

[1]Kazan (Volga Region) Federal University
18 ul. Kremlyovskaya, Kazan, 420008

[2]Kazan State Power Engineering University
51 Krasnoselskaya Str., Kazan, 420066

Abstract. The paper discusses the features of very heavy crude oil transportation in pipelines as exemplified by foreign transportation techniques. Special focus is given on the processes of rheologically complex behavior of waxy crude oils.

Currently, there is a significant increase in the production of heavy crude oil with a viscosity of 30 MPa.s and higher that contains a large amount of paraffin and solidifies at relatively high temperatures. It has a unique chemical composition and is a valuable raw material for the petrochemical industry, which arouses a particular interest [1].

However, very heavy paraffinic oils are characterized by rheologically complex behavior, which creates a number of problems both at the stage of its production, treatment for transportation and during pipeline transportation. The oil becomes thicker so that it is extremely hard to move through the pipes requiring much energy, and sometimes even impossible [2].

Optimization of the oil transfer through pipelines is related to the matter of reducing flow resistance. Improving the rheological properties of heavy and very heavy crude oils and enhancing the oil delivery performance is achieved through physical effects, pretreatment, which allows increasing the flow behavior of the fluid [3].

The most logical and understandable way is to increase the temperature of the oil either immediately before it is injected into the pipeline, or along the entire route of transferring. The paraffins contained in the oil are believed to form crystalline structures, which increase its viscosity with the temperature drop [4]. With the increase in temperature, these structures, in turn, are destroyed and the viscosity of the oil decreases resulting in oil that is more fluid.

This technology has been used for many decades;however, it has its weak points. Firstly, the heat exchange with the environment cools down the oil resulting in a need for periodic heating for further free movement [5].

For heating long pipelines, the SECT (Skin Electric Current Tracing) system developed by Chisso Engineering Co., Ltd., Japan can be used. This system makes it possible to heat up a 30-km pipeline run with a power connection at one end of the system and heat up the entire portion between pumping stations if a power source is available [6].

Secondly, the heating of the oil itself takes a part of the transported fuel, thus increasing the cost of the final product. The lower the ambient temperature, the more fuel will be needed to heat up the oil [6].

An alternative method of transferring heavy oil is to mix it with lighter carbon diluents, which, as a rule, are gas condensate or light (nonviscous) crude oils. As a result, the viscosity of the resulting

IOP Conf. Series: Materials Science and Engineering **570** (2019) 012067 doi:10.1088/1757-899X/570/1/012067

mixture and its pour point are significantly reduced [7]. This is achieved by lowering the concentration of paraffin in the mixture and the content of asphalt-resinous substances in light crude oils, which prevent the formation of a solid crystal lattice of paraffin.

Oil delivery using diluents is widespread in Venezuela, where mixtures of oil of different viscosity are heated up while transferring. In the USA, heavy oils is mixed with light oils during the pipeline transportation. In West Germany, the high pour point oil Libyan oil (30-60%) is mixed with Hassimesaud (30-70%) and Kirkuk (6-7%) in the Strasbourg - Saarbrücken pipeline [8]. In Romania, they transfer waxy crude oils from the Berka field together with diluents (casing-head gasoline).

Special chemicals, called pour point depressants (PPD), are introduced to reduce the pour point, viscosity and yield point of very heavy oils. Typical natural depressants are asphalt-resinous substances contained in crude oil. Adding tar and bitumen rich in these substances can improve the rheological parameters of oil [9].

Treated oil is currently pumped through West European pipelines, such as Ile-de-France, Finnart-Grangemouth and Rotterdam-Rhine.

However, the introduction of solvents leads to a significant increase in the cost of oil delivery, and the addition of depressants is effective only for waxy oils, whose viscosity increases due to a large amount of high pour point paraffin hydrocarbons.

Currently, due to the ever-increasing volumes of very heavy oil transported through pipelines, the issue of reducing flow resistance, energy consumption during transportation, reducing the weight and size of heat exchangers is a challenge. One of the available solutions to this problem is associated with the application of heat transfer intensification, which increases the energy performance of heat exchangers. This offers opportunities for controlling the effective viscosity of oil and significantly reducing the flow resistance in transfer lines [10] by controlling the shear rate in the flow. When choosing the methods of heat transfer intensification, it is necessary to regard not only the efficiency of the surface itself, but also the multipurposeness of heat transfer agents, the ease of manufacturing the surface, the ease of assembling the heat exchanger, and the strength requirements. Of all the methods of heat transfer intensification in pipes, the most regarded is the principle of helical twist, as an efficient and technologically feasible approach.

References

[1] Khusnutdinova E M and Mamonova A O 2017 *Methods of enhancing rheological properties of very heavy oils and petroleum products.* Actual problems of innovative development: collection of papers of the International Conference and Workshop. M.: Impulse. pp 730-732

[2] Sdvizhkov I V 2016 *Characteristics of transferring quick-hardening and high-viscosity oils and oil products through main oil lines* (Graduation thesis / National Research Tomsk Polytechnic University)

[3] Khusnutdinova E M, Konakhina I A, KhamidullinaG R and Mamonova A O 2017 *Ways to transfer oil and oil products in heating conditions and methods for their enhancement.* Proceedings of the 8th International Scientific and Technical Conference "Innovative Machine-building Technologies, Equipment and Materials - 2017". Part 1. pp 364-368

[4] Korshak A A and Shammazov A M 2002 *Fundamentals of Oil and Gas Engineering* (College textbook. Second edition, revised and corrected: — Ufa.: DesignPolygraphService LLC) p 544

[5] Aliyev R A, Belousov V D, Nemudrov A G, Yufin V A and Yakovlev E I 1988 *Pipeline transportation of oil and gas: textbook for higher education.* M.: Nedra, p 368

[6] Starkov N V 1983 *Foreign practices of heat insulation and heating of hot process pipelines.* Transportation and storage of petroleum products and hydrocarbon feed. No. 4 pp 18-21

[7] *Transportation of high-viscosity oils* 2016 (Youth Science Forum: Engineering and Mathematical Sciences: electr. sav. Paper on proceedings of the XXXV International Virtual Student Conference and Workshop) No 6(35)

[8] Kospanov A B, Tyurin A N, Churikova L A 2017 *Features of transportation of high-viscosity oil in the operating conditions of the "hot" pipeline*. Technical science in Russia and abroad: materials of VII International Science Conf. M.: Buki-Vedi. pp 141-144

[9] Gubin V E, Gubin V V 1982 *Pipeline transportation of oil and oil products*. M.: Nedra, p 296

[10] Konakhina I A, Khamidullina G R and Khamidullina A F 2016 *Enhancement of Hydrodynamic Processes in Oil Pipelines Considering Rheologically Complex High-Viscosity Oils.* IOP Conference Series: Materials Science and Engineering. Vol.134, Is.1. - Art. No. 012019

IOP Conf. Series: Materials Science and Engineering **570** (2019) 012068 doi:10.1088/1757-899X/570/1/012068

Modified gypsum-cement-pozzolanic composites reinforced with polypropylene fibers

R Kh Mukhametrakhimov[1], **A R Galautdinov**[1] **and I R Gilmanshin**[2,3]
[1]Federal State Budgetary Educational Institution of Higher Education "Kazan State University of Architecture and Civil Engineering", Kazan, Zelenaya Street, 1, 420043, Russian Federation
[2]Kazan Federal University, Kazan, Russia, 420008
[3] Kazan National Research Technical University named after A N Tupolev, Kazan, Russia

Abstract. The paper shows the effectiveness of the use of a polyfunctional complex modifier for a gypsum-cement-pozzolanic binder (GCPB), which allows reducing the water requirement of the mixture by 35%, increasing the flexural strength of gypsum-cement-pozzolanic composites (GPPC) by 28%, and 25% for compression. The results of experimental studies of the effect of polypropylene fibers on the basic physicomechanical properties of a central processing center are presented. The performed studies have shown high efficiency of polypropylene fibers of the type VSM-II-6 for dispersed reinforcement of the central industrial park. The maximum increases in the values of the flexural strength (by 41.2%) and compression (by 20%) are achieved with their volume content in the amount of 1%. It should be noted that the introduction of the studied fibers with a length of more than 6 mm did not lead to the expected result, which is due to the formation of a heterogeneous structure of the GPPC

Introduction

Products based on gypsum-cement-pozzolanic composites are widely used in construction as a universal building material (dry construction mixtures, gypsum plasterboards, facing tiles for interior and exterior). Along with a lot of positive technical properties, materials and products based on gypsum-cement-pozzolanic binders have considerable brittleness, which is especially true when using thin-walled sheet products, as well as high water absorption, which limits their use in humid environments. To eliminate these drawbacks, various mineral and chemical additives, as well as reinforcing materials, are introduced into the GCPB composition.

The physicomechanical properties of the GPPC depend on the composition of the binder, its activity, water binding relationship, hardening conditions, etc. The main technological factor designed to ensure the possibility of combining the gypsum binder with Portland cement in a single binder composition is the introduction of active mineral additives (tripoli, flask, etc.).

The most typical active mineral additives (AMD) - tripoly, kaolin, blast furnace slag, aluminosilicate, selected in accordance with the procedure described in [1], reduce the concentration of calcium hydroxide in hydrating gypsum-cement mixture only after 4 days of hydration. This effect is caused by hydrolysis of the hydration products of clinker minerals, which are not capable of binding their own calcium hydroxide. As a result, they cannot eliminate the negative interaction with the gypsum substance at the stage of a structured formation. In this regard, the expansion of the system occurs, caused by the formation of ettringite. Thus, the introduction of the gypsum-cement chemical composition leads to the implementation of experimental experimental studies [2,3].

Mineral supplements of the type metakaolin [4,5], ferrosilicon, thermally activated clay [4], microsilica [5], and white soot have a completely different quality. They are characterized by extremely high dispersion, chemical activity and reactivity. This is due to their relatively low required content in the composition of the gypsum-cement mixture [6].

It should be noted that the short setting times for GCPB greatly complicate the processes of forming articles based on it, including gypsum-fiber sheets (GFS). In this regard, it is necessary to consider the features of regulation of the setting time and hardening of the GCPB, analyze chemical modifiers (plasticizing and water-repellent additives) of the gypsum-cement matrix and GFS on its basis to improve their performance and durability.

In the production of materials and products based on GCPs, a wide range of modifying additives is used, such as additives for regulating the kinetics of initial structure formation and hardening, plasticizing additives (PA), water-repellent additives (WRA). At the same time, the nomenclature of modifying additives for cement binders is much wider than for gypsum binders and building materials based on them. It should be noted that the work of modifying additives largely depends on the composition of the binder, the quantity and quality of the compounds formed during the hydration of the binder [7], as well as the alkalinity of the medium.

In [8], the effectiveness of using citric acid to slow down the setting of recycled gypsum plaster (β-hemihydrate calcium) was shown. It was found that citric acid reduces the compressive strength of samples and their hardness. In [9], the effectiveness of the use of superplasticizers based on sulfonated melamine formaldehyde resins for the modification of calcium β-hemihydrate is shown. Introduction in the amount of 0.6% leads to an increase in the strength of the samples by 69%, acceleration of hydration processes, leads to the formation of a denser structure, which increases their water resistance. In [10], it was shown that the use of complex additives containing carbon nanotubes and ultrafine additives leads to an increase in physicomechanical characteristics, a change in the qualitative composition of hydrate tumors, as well as a change in their morphology and size of crystalline hydrates. At the same time, the effectiveness of complex additives should be ensured by the synergistic effect of the interaction of the components of the additive.

In previous studies [11,12], we developed a multifunctional complex modifier, including the Best-GFS regulator for setting and hardening, GP Odolit-K, and DG Ethyl silicate-40, for gypsum-cement-pozzolan mixtures. The effect of the synergistic interaction of the components of the developed complex modifier can significantly improve the water resistance of products based on HRC, their limits of bending and compression strength. In addition, the use of this modifier leads to a lengthening of the dates of the beginning and end of the mixture setting, which will simplify the technological processes of manufacturing and installation of products based on GCPB [13]. In this regard, this multifunctional complex modifier was adopted by us to modify the FIR in these studies.

As is known, dispersed reinforcement of binders based on various compositions allows to increase many of the most important physical and mechanical characteristics: compressive and flexural strength, fracture toughness, crack resistance, etc. [14-16]. The matrix in the reinforced compositions gives the product the necessary shape, creates a monolithic material. Combining numerous fibers in a single whole, the matrix allows the composition to perceive various kinds of external loads (tension, compression, bending, shear, etc.). At the same time, the matrix itself participates in the creation of the bearing capacity of the composition, ensuring the transfer of forces to the fibers [17]. For the calculation of composite materials, in particular concrete, diagram methods [18] are used, which allow a more complete assessment of the stress-strain state of the calculated cross sections taking into account the physical nonlinearity of materials [19].

The use of basalt and glass fibers for the reinforcement of gypsum-cement-pozzolanic compositions is limited due to the available data on their low resistance to chemical corrosion in the medium of the gypsum-cement-pozzolanic stone [20,21]. It should be noted that the solution to this problem is possible by modifying the surface of these fibers [22].

In [23], the effect of polyvinyl acetate and polypropylene fibers on the properties of gypsum composites was studied. It was shown that polyvinyl acetate fibers can significantly accelerate the process of hydration of the binder, increase the flexural strength and impact strength of gypsum composites (compared to polypropylene fibers). It should be noted that the length of the fibers studied in the work was 6-12 mm with their volume fraction in the composition of the gypsum composite up to 1.2%. In this regard, it is of interest to study the effect of fibers of greater length and content on the material properties.

Technical characteristics of various types of fibers, which are used for dispersed reinforcement of products, are given in [16]. Polypropylene fibers according to this classification can be attributed to low-modular, synthetic, since the relative elongation at break is 10-25%, tensile strength 400-770 MPa, average fiber diameter 20-22 µm, length 6, 12, 18, 24, 32 mm and more. Such characteristics of polypropylene fiber provide high impact strength and flexural strength of a composite material.

High technical indicators, chemical resistance and low weight of polypropylene fibers cause a considerable interest in their research as a dispersion-reinforcing material for the GPPC. Of particular interest is the study of the influence of the length and volume content of fibers on the basic physical and mechanical properties of modified gypsum-pozzolanic systems, which can be used in the manufacture of a wide range of building products.

The purpose of the research is to study the effect of polypropylene fibers of various lengths and content in the composition of the gypsum-cement-pozzolan mixture on the physicomechanical properties of modified GPCC.

Materials and Methods

- To carry out experimental studies, the following materials were adopted:
- a) knitting:
- construction gypsum G6BII produced by LLC Arakchinsky gypsum, produced according to GOST 125-79;
- Portland cement Belgorod cement plant PTs500 D0, produced according to GOST 10178-85;
- b) modifying additives:
- AMD - metakaolin produced according to TU 5729-098-12515988-2013, with hydraulic activity of at least 1000 mg / g.
- SE "Odolit-K", produced by LLC "Service-Group" according to TU 5745-01-96326574-08;
- regulator of setting and hardening terms “Best-TB” produced by LLC “Innovative Technologies”;
- Homogeneous mixture of oligoethoxysiloxanes "Ethyl silicate-40" produced by JSC Khimprom, Novocheboksarsk according to TU 2435-427-05763441-2004.
- c) fibers:
- polypropylene fibers of the mark VSM-II, 6, 12, 18, 24 and 32 mm long, produced according to TU5458-001-82255741-2008. Specifications are shown in Table 1.

Table 1 - Technical characteristics of polypropylene fibers of the mark BCM-II

Name	Indicators
Material	High Modulus Thermoplastic Modified Polymer
The average fiber diameter, microns	20-22
Fiber length, mm	6, 12, 18, 24, 32
Tensile strength, MPa	550
Elongation	20%
Modulus of elasticity, MPa	No less 10000

The number of single fibers, ppm / kg	300-600
Fiber surface area, m^2 / kg	150
Melting point, ° C	160

d) drinking tap water that meets the requirements of GOST 23732.

Preparation and testing of samples GPPC. At the first stage, to obtain GCPB, its components were mixed in a dry form with a gypsum: cement: AMD ratio of 76: 20: 4. At the second stage, the rheological properties of GCPB with and without a complex modifier were investigated. The content of polyfunctional complex modifier was 1.5% by weight of the binder. At the third stage, in order to evenly distribute polypropylene fibers in the gypsum-cement-pozzolan matrix, they were fused in the calculated amount of mixing water in the presence of a modifying additive. At the next stage, the components of the gypsum-cement fiber mixture were mixed for one minute, followed by molding the samples.

Tests of samples after their hardening for 28 days in an air-humid environment were carried out on standard beam specimens with dimensions of 4x4x16cm, from a molding mixture of normal thickness according to the method described in GOST 23789-79.

Optical studies of samples of the gypsum-cement-pozzolanic stone were carried out using a Philips XL-30 scanning electron microscope. Optical studies were conducted to study the morphology of hydrated tumors, the surface structure of the studied gypsum-cement compositions and changes in them.

Results and Discussion

The introduction of a polyfunctional complex modifier into the GCPB structure made it possible to reduce its water demand by 35% and to increase the setting time by 19 minutes. At the same time, the flexural strength increases by 28%, and by 25% during compression. The microstructure of the initial and modified gypsum-pozzolanic matrix was studied using scanning electron microscopy with the analysis of the elemental composition of the studied compositions at the indicated characteristic points. Electronic images of the samples are presented in Fig. 1-4.

ig. 1. Electron microscopic image of the original matrix at x500 magnification

Fig. 2. Electron microscopic image of the initial matrix at x800 magnification

IOP Conf. Series: Materials Science and Engineering **570** (2019) 012068 doi:10.1088/1757-899X/570/1/012068

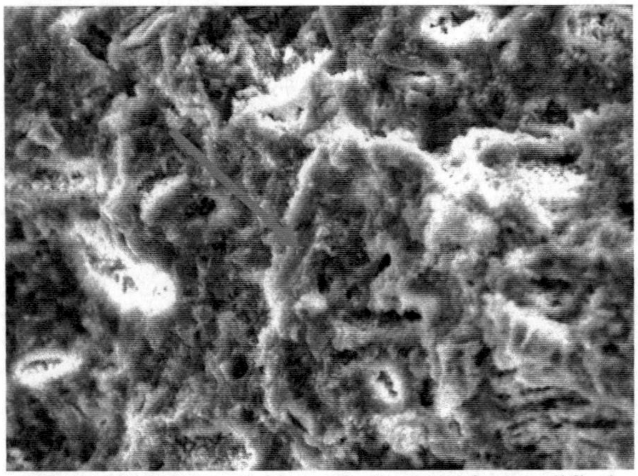

Fig. 3. Electron microscopic image of a modified matrix at x500 magnification

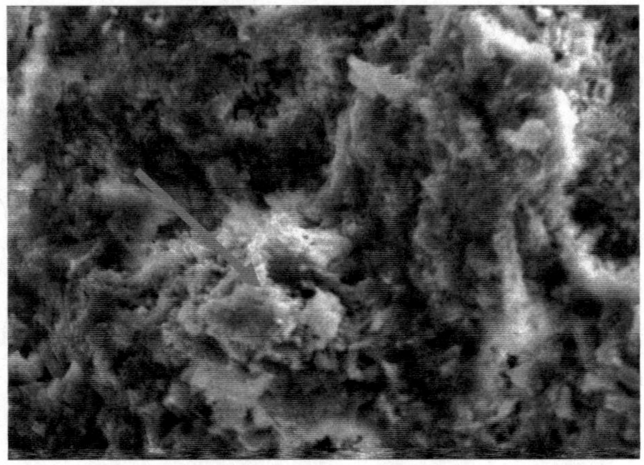

Fig. 4. Electron microscopic image of a modified matrix at x800 magnification

Analysis of the elemental composition of the initial matrix at the indicated points (Fig. 1, 2) allows us to conclude that there are needle-like inclusions of ettringite ($3CaO*Al_2O_3*3CaSO_4*32H_2O$). Electron microscopic images with the elemental composition of the modified GPCP (Fig. 3, 4) indicate the presence in it of the needle forms of calcium monosulfate ($3CaO*Al_2O_3*3CaSO_4*12H_2O$), as well as chains of low-base hydrosilicates ($3CaO$-$2SiO_2$-$3H_2O$), what caused the formation of a more homogeneous crystalline structure and a decrease in the porosity of the modified samples as compared with the initial composition and an increase in their physicomechanical characteristics.

The effect of the volume content of polypropylene fibers of different lengths on the flexural and compressive strength limits of the centrally applied control centers is shown in Fig. 5 and 6.

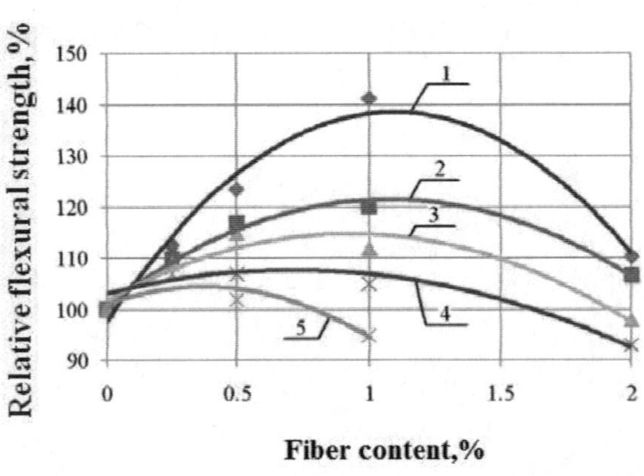

Fig. 5 - The effect of the volume content of polypropylene fibers on the relative flexural strength of the GPPC with a length of: 1 - 6 mm; 2 - 12mm; 3-18 mm; 4 - 24 mm; 5 - 32mm

The following regression equations have been obtained, which characterize the effect of the studied fibers on the flexural strength of the GPPC, the reliability of which is 0.81-0.98:

$R_{flex}(1) = -33.872x^2 + 74.291x + 97.831;$

$R_{flex}(2) = -17.646x^2 + 38.044x + 100.96;$

$R_{flex}(3) = -15.027x + 27.894x + 101.85;$

$R_{flex}(4) = -9.072x + 12.811x + 103.23;$

$R_{flex}(5) = -25.091x + 18.673x + 101.06.$

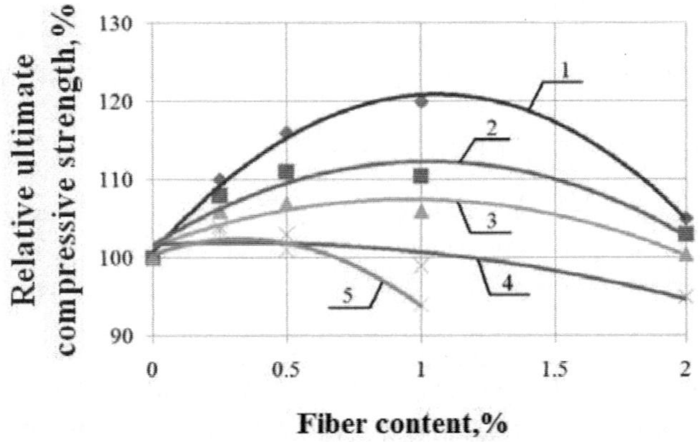

Fig. 6 - The effect of the volume content of polypropylene fibers on the relative compressive strength of the GPPC with a length of: 1 - 6 mm; 2 - 12mm; 3-18 mm; 4 - 24 mm; 5 - 32mm

The following regression equations were obtained, which characterize the effect of the studied fibers on the flexural strength of the GPPC, the reliability of which is 0.82-0.99:

$R_{comp}(1)= -18.045x^2 + 38.117x + 100.78;$

$R_{comp}(2)= -10.174x^2 + 20.833x + 101.68;$

$R_{comp}(3)= = -6.5806x^2 + 12.523x + 101.5;$

$R_{comp}(4)= -2.5211x^2 + 1.4998x + 101.75;$

$R_{comp}(5)= -18.909x^2 + 12.327x + 100.44;$

As can be seen (Fig. 5, 6), the introduction of polypropylene fibers into the composition of the gypsum-cement-pozzolan mixture, depending on the length and their volume content, has a different effect on the flexural strength of the GPCC. The most effective fiber showed a length of 6mm. So, depending on their dosage in the composition of the mixture, the flexural strength increases by 10-41.2%, when compressed by 5-20%. Keeping the fibers with a length of 12 mm to a lesser extent increases the tensile strength of the GPPC when bending (6.7-20 %) and compression (3-11%).

The introduction of fibers of 18, 24 and 32 mm in length with an increase in their content in the composition of the mixture leads to their uneven distribution, difficulty with mixing and the formation of an uneven structure of GPPC. This, in our opinion, causes a decrease in the strength of the samples with an increase in their content in the mixture. The increase of the flexural strength (up to 15%) and compression (up to 6%) is achieved with relatively low dosages of fibers (0.25-0.5%).

Conclusions

1. Analysis of literature data indicates the effectiveness of the use of complex chemical additives for the modification of gypsum-based binders, which, due to the effect of the multifunctional interaction of the components, have greater capabilities in regulating the processes of formation of the structure and properties of finished products. In this regard, a complex modifier was adopted for the modification of GCPB on the basis of "Best-TB" regulator of setting and hardening, "Odolit-K" hyperplasticizer and "Ethylsilicate-40" water-repellent additive
2. Analysis of the elemental composition of the initial gypsum-cement matrix allows to conclude that there are inclusions of ettringite in it. ($3CaO*Al_2O_3*3CaSO_4*32H_2O$). Electron-microscopic images with the elemental composition of the modified gypsum-pozzolanic matrix indicate the presence of acicular forms of calcium monosulfate ($3CaO*Al_2O_3*3CaSO_4*12H_2O$), as well as chains of low-base hydrosilicates ($3CaO$-$2SiO_2$-$3H_2O$), what causes the formation of a more homogeneous crystalline structure and a decrease in the porosity of the modified samples as compared

with the initial composition and an increase in their physicomechanical characteristics.

3. It has been established that one of the effective ways to solve problems aimed at improving the performance properties of products based on gypsum binder is their disperse reinforcement. Analysis of the experience of using various types of reinforcing fibers made it possible to identify polypropylene fibers, as one of the most effective for disperse reinforcement of products based on GPPC.

4. The completed studies have shown high efficiency of polypropylene fibers of the type VSM-II-6 for dispersed reinforcement of GPPC. Thus, the maximum increase in the values of the flexural strength (by 41.2%) and compression (by 20%) is achieved with their volume content in the mixture in an amount of 1%.

5. It should be noted that the introduction of the investigated fibers with a length of more than 6 mm did not lead to the expected result. In our opinion, this is due to the formation of an inhomogeneous structure of GPPC. In addition, the introduction of silicone modifier could reduce the adhesion strength of the fibers with the matrix. Therefore, the influence of the organosilicon modifier and the technology of introducing fibers into the gypsum-cement-pozzolanic mixture is of interest for further research..

Acknowledgements

The work was performed within the framework of the Grant of the President of the Russian Federation for state support of young Russian scientists - PhDs (grant number MK-2509.2018.8)

References

[1] Ferronskaya A V 2004 *Plaster materials and products. Production and use.* Catalog (Moscow: ACB)

[2] Mukhametrakhimov R Kh, Galautdinov A R 2017 The role of active mineral additives of natural origin in the formation of the structure and properties of the gypsum-cement-pozzolanic binder *Bulletin of the Technological University* **20** 60–3

[3] Mukhametrakhimov R Kh, Galautdinov A R 2016 Gypsum-cement-pozzolanic binder based on low-quality raw materials and industrial waste *Bulletin of the Technological University* **19** 56–9

[4] Khaliullin M., Rakhimov R., Faizrakhmanov I 2017 The influence of thermally activated clay additives on the properties of composite gypsum binder *ZKG ZEMENT - KALK - GIPS Int. Bauverlag BV GmbH. Gutersloh* 58–63

[5] Butakova, M.D., Gorbunov S P 2016 Study of the Influence of Complex Additives on Properties of the Gypsum-Cement-Puzzolan Binder and Concretes on its Basis *Procedia Eng.* **150** 1461–7

[6] Alksnis F F 1988 *Hardening and destruction of gypsum-cement composite materials* (Leningrad: stroiizdat)

[7] Amelina, D V, Sycheva L I 2010 The Effect of Chemical Additives on the Properties of Gypsum Binder Advances in Chemistry and Chemical Technology 41–3

[8] Camarini G, Pinto M C C, Moura A G de and Manzo N R 2016 Effect of citric acid on properties of recycled gypsum plaster to building components *Constr. Build. Mater.*

[9] Pundir A, Garg M and Singh R 2015 Evaluation of properties of gypsum plaster-superplasticizer blends of improved performance *J. Build. Eng.* **4**

[10] Tokarev Y, Ginchitsky E, Sychugov S, Krutikov V, Yakovlev G, Buryanov A and Senkov S 2017 Modification of Gypsum Binders by Using Carbon Nanotubes and Mineral Additives *Procedia Eng.* **172** 1161–8

[11] Galautdinov A R, Mukhametrakhimov R X 2016 Improving the water resistance of a gypsum-pozzolanic binder based on low-quality gypsum *Izvestiya KGASU* **38** 333–43

[12] Mukhametrakhimov R Kh, Galautdinov A R, Lukmanova L V 2016 The influence of plasticizing additives on the basic properties of the gypsum-pozzolanic binder based on low-quality and technogenic raw materials *Izvestia KGASU* 382–7

[13] Mukhametrakhimov R Kh, Galautdinov A R, Dikina A N 2016 Manufacturing technology and installation of modified gypsum sheets *Izvestia of Kazan State University of Architecture and Civil Engineering* **2** 194–200

[14] Frantsevich, IN, Karpinos DM 1970 *Composite materials of a fibrous structure* (Kiev: Naukova Dumka)

[15] Izotov V S, Mukhametrakhimov R Kh, Sabitov L S 2010 Experimental studies of the effectiveness of dispersed reinforcement of a stretched zone of concrete bent elements Scientific *Journal of the Voronezh State University of Architecture and Civil Engineering. Construction and architecture* **1** 119–25

[16] Rabinovich FN 2004 *Composites based on dispersion-reinforced concrete. Questions of theory and design, technology, design* (Moscow: DIA)

[17] Zotov, VS, Mukhametrakhimov R X 2014 Modification of the structure of dispersion-reinforced compositions to increase the fracture resistance *Materials of IRTC VIII Academic Readings of the RAACS* pp 220–4

[18] Radaikin O V, Sabitov L S, Kashapov N F, Gilmanshin I R 2018 Current state and prospects of development of diagrammatic methods for the calculation of reinforced concrete structures *IOP Conference Series: Materials Science and Engineering*

[19] Radaikin O V, Sabitov L S, Kashapov N F, Gilmanshin I R 2018 The way to improvement of the method of calculation of rigidity of bending reinforced concrete elements from conventional ferro-concrete *IOP Conference Series: Materials Science and Engineering*

[20] Anushina A S, Urbanov A V, Nemtsev A D, Potapova EN 2016 The effect of fibers on the properties of a gypsum-cement-pozzolanic binder Advances in chemistry and chemical technology 66–8

[21] Batalin B S, Semkova E N, Saraikina K A 2013 Study of the process of interaction of fiberglass with cement stone Master's J. 143–7

[22] Starovoytova I A, Zykova E S, Suleimanov A M, Semenov A N, Mishurova M 2016 Study of the physicomechanical characteristics of nano-modified basalt roving and composite material based on it. *News of KSAU* **3** 217–24

[23] Zhu C, Zhang J, Peng J, Cao W and Liu J 2018 Physical and mechanical properties of gypsum-based composites reinforced with PVA and PP fibers *Constr. Build. Mater.* **163** 695–705

IOP Conf. Series: Materials Science and Engineering **570** (2019) 012069 doi:10.1088/1757-899X/570/1/012069

Development of high-strength cast iron for back-up layer of bimetallic products

G F Mukhametzyanova[1], I R Mukhametzyanov[1] and M S Kolesnikov[2]

[1] Kazan (Volga region) Federal University, Naberezhnye Chelny Institute (branch), Department of Materials, Technologies and Quality, Republic of Tatarstan, Russia, 423812, Naberezhnye Chelny, 10A Syuyumbike Ave.
[2] Kazan (Volga region) Federal University, Naberezhnye Chelny Institute (branch), Department of Mechanical Engineering, Republic of Tatarstan, Russia, 423812, Naberezhnye Chelny, Syuyumbike Ave., 10A

Abstract. The article presents chemical composition of the charge and smelting regimes of the newly designed austenitic-bainitic high-strength cast iron with spheroidal graphite (ABHCSG), the analysis of research results of influence of isothermal hardening modes in pseudo-boiling layer on the mechanical properties and structure of cast iron.

In the modern world it is important to obtain new materials with special properties. A special role among new materials with special properties belongs to layered metal compositions. The use of bimetallic compositions in the production of die tools is also promising in order to save expensive materials and improve their durability [1, 2, 3]. Therefore, the design and study of materials for high resistance bimetallic dies is an urgent task.

Practicability of application of high-strength cast iron with spherical graphite for production of a basic layer is caused by its higher thermal conductivity that allows to improve at operation of stamps a heat sink from the warmed-up working surface of the tool on the weight or the refrigerator established in the basic part of a stamp. This significantly reduces the temperature gradient and the stress level in the contact zone of the die and, as a result, causes an increase in its performance. Working capacity of bimetallic dies increases at reception in a surface layer of the tool of dispersed structure. Namely, due to the lower heating temperature of cast iron than that of alloy steels, when filling the support layer there is an additional internal cooling-down of the top-doped layer. In this case, the grain size is reduced, the separation diffusion of alloying elements is slowed down, which leads to a decrease in dendritic liquation and an increase in the characteristics of heat resistance and thermal and mechanical fatigue of the surface layer of the dies.

There are high-strength cast irons with increased strength, ductility, toughness and heat resistance [4, 5]. However, to obtain them complex heat treatment technologies are used which lead to a high cost of production.

The aim of the work was to develop a method for producing aluminum high-strength cast iron with spherical graphite and austenitic-bainitic structure with a tensile strength $\sigma_B \geq 1200$ MPa by bainitic quenching from the cast state without additional heat treatment [6].

To obtain aluminum high-strength cast iron with spherical graphite and austenitic- bainitic structure in the cast state, the iron is smelted in an induction electric furnace, the liquid alloy when drained into the bucket is modified to obtain graphite inclusions in the castings of a globular shape, when poured into a metal form (chill mould), the liquid alloy in the bucket is further subjected to inoculating

IOP Conf. Series: Materials Science and Engineering **570** (2019) 012069 doi:10.1088/1757-899X/570/1/012069

modification, the castings after crystallization are removed from the mold at a temperature of 900 - 1000°C, moved to a furnace with a temperature of 950-1000°C, kept in the furnace for 10 - 30 minutes, then isothermal quenching is applied to the installation in a pseudo-boiling dispersed carborundum layer in the area of austenitic-bainitic transformation at a temperature of 300°C for 60 minutes, followed by air cooling, while using cast iron containing, wt. %: carbon (C)– 2,5 - 3,2; silicon (Si) – 1,5 - 2,5; aluminum (Al) – 7,2 - 9,0; manganese (Mn) – 0,7 - 0,75; magnesium (Mg) – 0,04 - 0,06; molybdenum (Mo) – 0,7 - 1,2; nickel (Ni) – 0,48 - 0,52; copper (Cu) – 0,49 - 0,52; sulfur (S) – 0,01 - 0,012; phosphorus (P) – 0,06 - 0,08; iron (Fe) – the rest.

Heating of high-strength cast iron is carried out, for example – in an induction furnace MGP – 52, with acidic lining on the charge, consisting of cast iron GOST 805-80, scrap steel and fight graphitized electrodes. Alloying iron Si (silicon), Al (aluminum), Mn (manganese), Mo (molybdenum), Ni (nickel), Cu (copper) is carried out by the corresponding ferroalloys in the furnace for 10-15 minutes before the release of the metal (copper M2 (99.9 wt. % - copper (Cu), other impurities), GOST 859-78, Nickel, H3 (Nickel (Ni), wt. % more than 98.6%, the rest cobalt (Co) - 0.7; carbon (C) - 0.1; sulfur (S) - 0.03; copper (Cu) - 0.6;) GOST 849-70). Inoculating modification is carried out in the bucket modifier ZHKMK-4R at a temperature of 1450-1470°C the following chemical composition, wt. % : silicon (Si) – 49.6; calcium (Ca) – 9.5; magnesium (Mg) 8.6; rare earth metals (REM) – 4.7; iron (Fe) – the rest. As a flux cryolite was used K2 (GOST 10561-73). Gravitizing modification was performed by ferrosilicon FS – 75 (GOST 1415-78). Then in a ladle in a certain way [5], by means of quartz "bell", the modifier ZHKMK–4R is added. Obtaining castings is carried out in a metal form (casting mold), which is removed after crystallization from the mold at a temperature of 900 - 1000°C, then moved to a furnace with a temperature of 950 - 1000°C and kept in the furnace for 10-30 minutes. As a result, castings of cast iron with globular graphite and austenitic structure of the proposed chemical composition are obtained.

After hardening in a chamber furnace of electrical resistance in the temperature ranging from 950 to 1000°C cast iron samples are transferred to the installation of "pseudo-boiling" layer. Isothermal hardening of cast irons was carried out at temperatures of 300°C, 350°C and 380°C for 30, 60 and 120 minutes. Heating of the working medium (carborundum) to these temperatures is carried out by pumping compressed air through the heating elements of electrical resistance (tubular furnace). The temperature in the working space is controlled, for example, by chromel-aluminum thermocouples.

The results of studies about the effect of the duration and temperature of isothermal exposure within 300 - 380°C on the mechanical properties of austenitic-bainitic cast irons are presented in table 1.

Table 1 - Effect of temperature and duration of isothermal hardening on the mechanical properties of austenitic-bainitic cast irons.

Modes of isothermal hardening	Heat treatment parameters				Mechanical properties
	Austenitization		Isothermal hardening		
	T, °C	Time of hardening	T, °C	Time of hardening	σ_B, MPa
Mode № 1	950	30	300	60	1220
Mode № 2	950	30	350	60	1084
Mode № 3	950	30	380	60	962
Mode № 4	950	30	300	90	1210
Mode № 5	950	30	350	90	1076
Mode № 6	950	30	380	90	954
Mode № 7	950	30	300	120	1210
Mode № 8	950	30	350	120	1070
Mode № 9	950	30	380	120	950

IOP Conf. Series: Materials Science and Engineering **570** (2019) 012069 doi:10.1088/1757-899X/570/1/012069

In the study of the influence of isothermal hardening modes of cast iron in a pseudo-boiling layer wedge-shaped samples with a thickness of 55mm (GOST 7293-85) were cast. It was found that the increase in the temperature of isothermal hardening in the range of 300 - 380°C during quenching of iron leads to a decrease in strength by 21%; plasticity by 3.5% and toughness by 20%. At the same time, the hardness in cast iron practically does not change (Table. 1). The increase in isothermal hardening at the cooling in the pseudo-boiling layer over 60 min virtually has no effect on the hardness and other mechanical properties of aluminum austenitic-bainitic cast iron. The maximum strength characteristics (σ_B, $\sigma_{0,2}$) were obtained by isothermal hardening in a pseudo-boiling layer of austenitic-bainitic iron at an isothermal hardening at a temperature of 300°C for 60 minutes according to Mode №1 (Fig. 1).

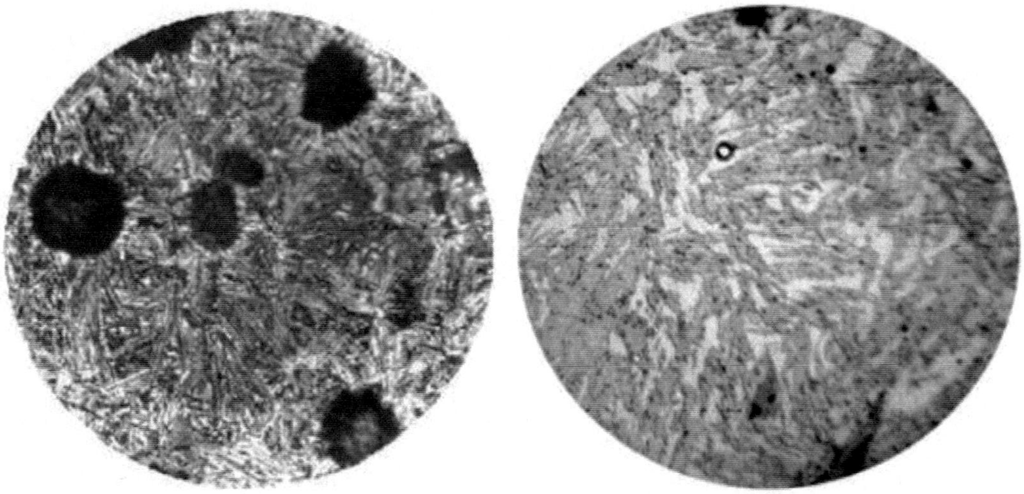

a b

Fig.1 –Microstructure of ABHCSG (mode No. 1): a) x500, b) x1000.

Mechanical tests of cast irons under static loads were carried out on a universal breaking machine. Hardness (HB) was determined according to GOST 24805-81 on the Brinell device; impact strength was determined according to GOST 9454-78 on the pendulum.

The proposed method allows to obtain aluminum austenitic-bainitic cast iron with globular graphite maximum with the following mechanical characteristics: tensile strength σ_B = 1220MPa; yield strength $\sigma_{0,2}$ = 960MPa; plasticity δ = 4,6%; hardness HB 415; impact strength KC =72J/cm^2.

The obtained high strength properties in the proposed austenitic-bainitic cast iron were achieved due to the rational content of graphite-forming elements: aluminum, silicon, nickel and copper, as well as carbide-forming molybdenum in combination with manganese. A positive role is also played by the proposed method of isothermal quenching of iron in a mobile "pseudo-boiling" layer, which provides uniform cooling of iron and as a result – obtaining a homogeneous structure.

References

[1] Golovanenko S A and Meandrov L V 1996 Production of bimetals *M: Metallurgia* 303 p

[2] Kolesnikov M S, Mukhametzyanova G F, Bikulov R A and Astashenko V I 2014 Bimetallic stamps of hot deformation of steel and suspension monoalloyed austenitic-bainitic cast iron. *Foundry operation* Number 3 pp 2 - 4

IOP Conf. Series: Materials Science and Engineering **570** (2019) 012069 doi:10.1088/1757-899X/570/1/012069

[3] Mukhametzyanova G F, Kolesnikov M S and Mukhametzyanov I R 2016 Molding bimetallic dies for liquid forging in the mold from the gas-tight molding sand *Foundry operation* Number 3 pp 26-28

[4] Patent of the Russian Federation 2449043 2012 IPC C22C37/10 and C21D 5/00 Method of heat treatment of spheroidal cast iron / K V Makarenko published 27.04.2012 Bulletin No 12

[5] Patent of the Russian Federation 2250268 2005 IPC C21C 1/10, C22C 37/04 and C21D 5/00 Method of producing castings of half cast iron with austenitic-bainitic structure / K V Makarenko published 20.04.2005 Bulletin No 11

[6] Patent of the Russian Federation 2605016 2016 IPC C21C1/10 and C22C37/04 Method of producing a ductile cast iron / M S Kolesnikov, G F Mukhametzyanova, V I Astashenko and Mukhametzyanov I R published 20.12.2016 Bulletin No 35

IOP Conf. Series: Materials Science and Engineering **570** (2019) 012070 doi:10.1088/1757-899X/570/1/012070

Ranking materials technologies by limiting characteristics of heat-resistant alloys and their longevity in the problems of import substitution

F I Murataev[1], E R Galimov[1] and N Ya Galimova[1]

[1]Kazan national research technical university named after A.N. Tupolev, K. Marx street 10, Kazan, 420111, Tatarstan, Russian Federation

kstu-material@mail.ru

Abstract. The structural strength properties determined by the true stress-deformation diagram and the fatigue characteristics of materials at combined action (tension and bending) under the conditions of temperature change are used to substantiate materials and technologies. The domestic alloy is suggested as an alternative to imported heat-resistant alloy IN-738LC used to manufacture the turbine rotor blades of stationary gas turbine plants (GTP). Expressions to determine the fatigue characteristics according to the criteria of structure stability are presented.

1. Introduction

The operating practice of stationary GTP shows that their parameters temperature, turbine speed and the actual loads in comparison with transport gas turbine engines (GTE) (in particular aviation) are more gentle [1, 2]. In this regard, some companies, for example, "MAN" use cheaper materials in the manufacturing of gas turbines [2-5].

2. Body text

The analysis of turbine rotor blades structure made of CrNi65WMoTiAl alloy and IN-738LC according to the variants of their state after bench-tests and operating time of GTP is performed in order to identify the damage laws and to use the domestic alloys, for example in the repair of imported products. The microcracks with a size of 3…7 μm were detected in the alloy IN-738LC after operating time (Fig. 1, a), cracks [2-4, 6, 9], fatigue striations and breakdown of carbide (Cr, Mo, W)$_6$C into the σ-phase were detected in the CrNi65WMoTiAl alloy (Fig. 1, b). Table 1 shows the comparative data of the element-by-element composition of the blade material under study.

Table 1. The chemical composition of the alloys under study (in volume %)

Alloy grade	W	Mo	Zr	Co	Nb	Ta	Ni	Cr	Ti	Al
IN-738LC [1], USA	2.6	1.7	0.05	8.5	0.9	1.7	61	16	3.4	3.4
CrNi65WMoTiAl, Russian Federation	9.2	4.0	0	0	0	0	65	16	1.4	1.4

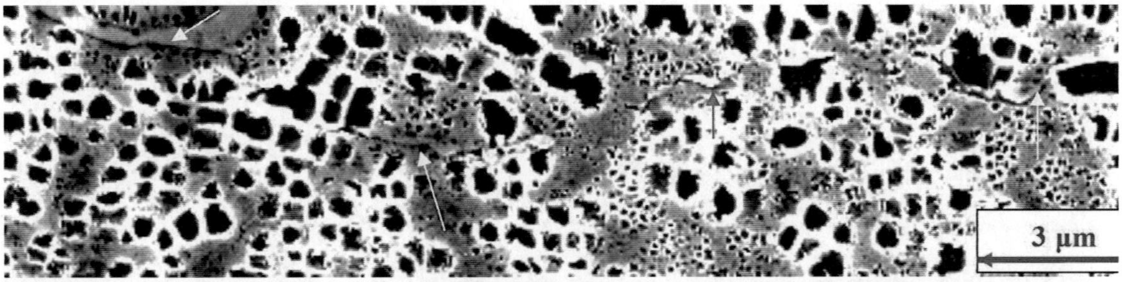

a

b

Figure 1. Metal degradation after operating time in the alloy IN-738LC (a) and in the alloy CrNi65WMoTiAl (b): cracks and breakdown of $(Cr,Mo,W)_6C$ into brittle σ-phase [2, 3, 6] and fatigue striations (right).

Structural damage of the alloy (SDA) expressed by the values of the standard deviations of the reflected ultrasonic signals [2-4, 9] over the sections of the blade airfoil cross-sections, is used to assess the degree of metal degradation. The dependencies (1) – (8) are used for the analysis of IN-738LC alloy damage, the dependence (9) – for the alloy CrNi65WMoTiAl:

$$SDA = 42.896 + 2.4382 \cdot \gamma'_b, \tag{1}$$

- from the size of the grain-boundary γ'-phase;

$$SDA = 64.758 - 1.8067 \cdot \gamma'_{eu}, \tag{2}$$

- from the size of the eutectic γ'-phase (fig. 2);

$$SDA = 41.262 + 0.925 \cdot (Ti,Nb,W) \cdot C; \tag{3}$$

$$SDA = 41.262 + 0.925 \cdot (Ti,Nb,Ta) \cdot C; \tag{4}$$

$$SDA = 47.551 + 1.2171 \cdot \lambda_{II}; \tag{5}$$

$$SDA = 61.68 - 60.50 \cdot (Ta+Co)/Ni; \tag{6}$$

$$SDA = 57.73 - 81.41 \cdot (Mo+Ta)/Ni; \tag{7}$$

$$SDA = 58.49 - 107.3 \cdot (Ta+Nb)/Ni; \tag{8}$$

$$SDA = 67.554 - 0.255 \cdot \Sigma AE \text{ in } M_6C+\gamma', \tag{9}$$

where ΣAE – (+) alloying elements.

The ability of blades metal to resist fatigue at combined action (tension and bending) under conditions of temperature change is determined according to the generalized condition of the limiting amplitudes of the high-frequency component (σ_{a2}) of the form [8, 10, 13]:

$$\frac{\sigma_{a2}}{\sigma_{-1}} = 1 - R_1 \left(\frac{\sigma_{\max_1}}{\sigma_{R_1}} \right)^{n_1} - (1 - R_1) \left(\frac{\sigma_{\max_1}}{\sigma_{R_1}} \right)^{n_2} . \tag{10}$$

Equivalent biaxial stresses are equal to [13]:

$$\sigma_{\mathrm{Re}1} = \sigma_{X1}(1 - \chi_m) + \chi_m (\sigma_{X1}^2 + \sigma_{Y1}^2 - 2\mu\sigma_{X1}\sigma_{Y1})^{0.5} . \tag{11}$$

$$n_2 = B_1 + B_2 \frac{\sigma_{\max_1}}{\sigma_{R_1}} - B_3 \frac{E_k}{E} + B_4 \frac{\mu}{\Psi} + B_5 \frac{m^{(0)}\Psi}{\delta} , \tag{12}$$

where $\sigma_{\max 1} = \sigma_A + \sigma_C$ – sum of actual and thermal stresses; properties are taken at t_{max} (in ^{O}C).

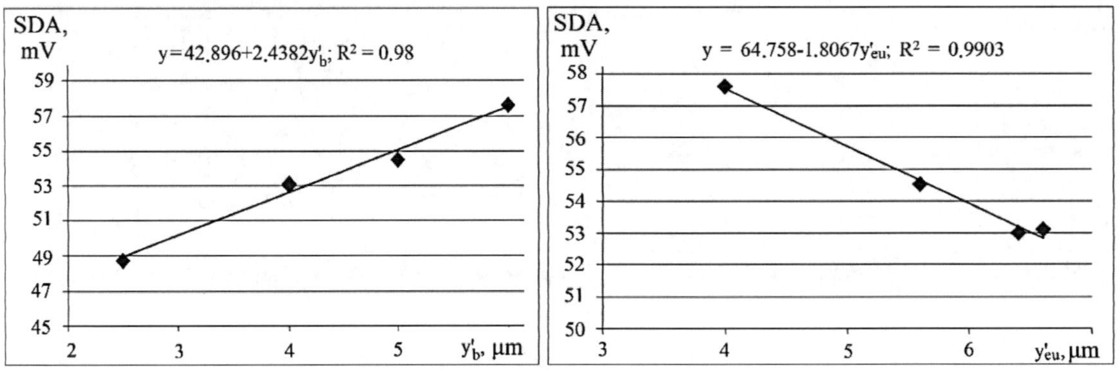

Figure 2. Size effect of the hardening phase located at the dendrites boundary and in the eutectic

Ilyushin's relative module [13]:

$$\mathrm{E_k} = \mathrm{E_k} / \mathrm{E} = \left[1 + \frac{\mathrm{E}}{\mathrm{S_k}} \ln \left(\frac{1}{1-\psi} \right) \right]^{-1.0} . \tag{13}$$

Strain-hardening coefficient [10-13].

$$\mathrm{m}^{(0)} = 0.75 \ln \frac{\mathrm{S_k}}{\sigma_{0.2}} \left\{ \ln \left[\left(\frac{\sigma_{0.2}}{\mathrm{E}} + 0.2 \cdot 10^{-2} \right)^{-1.0} \left(\frac{\mathrm{S_k}}{\mathrm{E}} + \ln \frac{1}{1-\psi} \right) \right] \right\}^{-1.0} . \tag{14}$$

According to the data of simulation modeling (SM) when substantiating the materials choice, the preference should be given to the structural state of alloys with higher values of the characteristics S_K, $m^{(0)}$, ψ. and smaller E_K. At dual-frequency loads the decrease of fatigue life (N/N_1) is controlled by expressions (11, 12, and 16) depending on the inclination angle of the fatigue curve (γ) in the low-cycle fatigue life region (N_1) taking into account the acting stresses (σ_{max1}), values of equivalent stresses (σ_{Re1}), characteristics of strength and plasticity in (12) at the maximum temperature *fn*(t) according to the condition [13]:

$$N_1 / N = 1 - \left(\sigma_{a_2} / \sigma_{-1}^t \right)^{\gamma / n(t)} . \tag{15}$$

The simulation modeling and fatigue tests of CrNi77TiAlB (EI437-BU) and 13Cr11Ni2W2MoV steel (EI961-Sh) were performed under complex thermal and force actions on the material [13] to check the calculated dependences (10-15). When the tensile and bending stresses in the samples (Fig. 3) are combined with the stresses from thermal cycling within the temperature range from 100 to 400°C (tvar 100...400°C) it can be seen (Fig. 4) that the experimental and calculated data obtained by the author [13] are in good agreement.

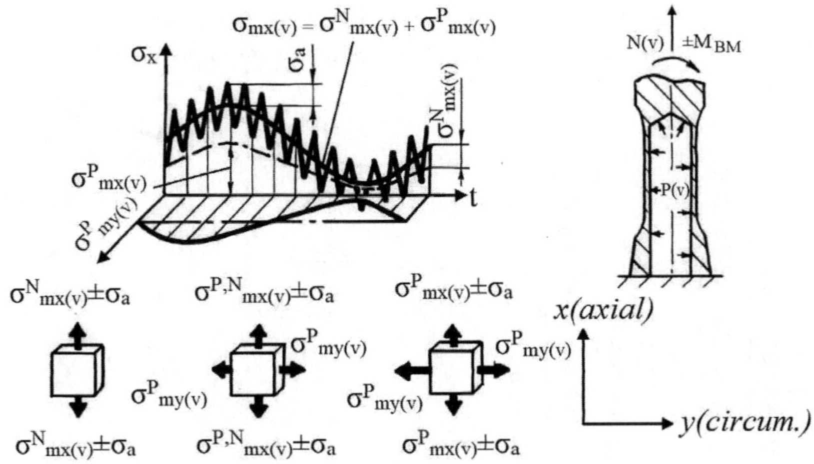

Figure 3. Schemes of combined action of tension (N), pressure (P) and bending (σ_a) at testing of samples

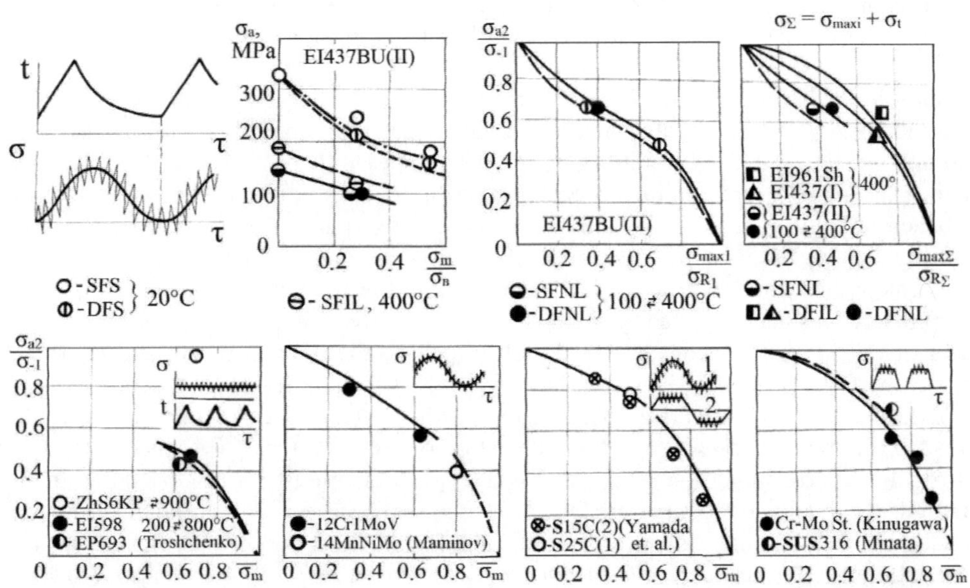

Figure 4. Comparison of experimental and calculated data on fatigue resistance of the author [13] and data of Troshchenko V T [12], Maminov A S, Yamada T, Minata H [16], Kinugawa MSFS – single frequency stress, DFS – dual-frequency stress, SFIL – single-frequency isothermal loading, SFNL – single-frequency non-isothermal loading, DFIL – dual-frequency isothermal loading, DFNL – dual-frequency non-isothermal loading

Other results are obtained by Maminov A.S. (at KAI), at the Institute for Problems of Strength National Academy of Sciences of Ukraine (at t var = 200...900 and 200…800°C) by Troshchenko V T [12] and at the University of Kyoto (Japan at t var = 200...700°C) [16]. The disks of the third stage of the GTE turbine are exposed to severe plastic deformation due to stress concentration in the zone of the hub bolt holes. Because of the constraint of the complex stress-strain state (SSS) [7, 11] and the action of high stresses, significant exhaustion of the plasticity reserve – the relative reduction up to 56% (Fig. 5) – takes place at operating and bench-tests time ~ 10000 hours. Since this process is influenced by both operating [14] and technological heredity, the enterprises have regulations on fixing the current level of properties (CLP) and properties of the initial level (PIL).

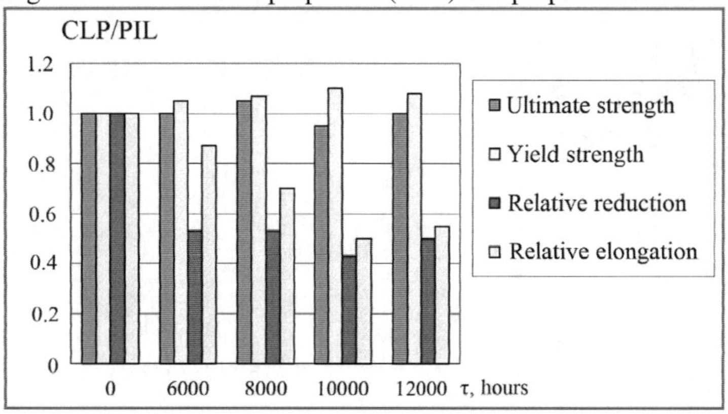

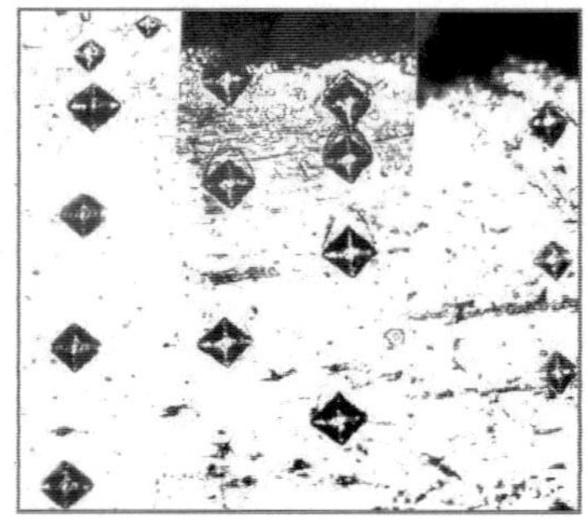

Figure 5. Changes in properties (CLP/PIL) of the CrNi73MoNbTiAl alloy with operating time of disks and measurement of metal microhardness at the hub holes according to the variants: after operation – (a), Restorative heat treatment (RHT) – (b), and operation + RHT – (c)

An increase in the hardness H_{100} of the deformed metal of the hub holes surface by 40…60% (Fig. 5, a) was established [12-15]. The turbine disks under repair are exposed to RHT in the "recovery of metals" mode – as a part of the equilibrium solidus temperature ($T_{OT} \sim 0.35T_{RS}$). As a result, the hardness and plasticity (ψ) are completely restored to the level of PIL and higher including that after repeated bench-tests time (Fig. 5, b and c). Thus, in this part of the research, the criteria for maintainability and assessment of the repair efficiency were established using the available and informative characteristics H_{100} and ψ. This made it possible to repair the turbine disks in engines conducting the RHT instead of replacing them.

The heat treatment modes were optimized with respect to the microstructure and microindentation characteristics performed on the SHIMADZU DUH-211S hardness tester by the Berkovich pyramid. The only indicator (out of 14 analyzed) consistent with the properties of the structural strength of materials is the difference in depth from the indentation of the indenter under constant load, according to the instruction – "creep under constant load – CIT". In our opinion, a more accurate name of this characteristic is the available indentation plasticity (AIP) on the upper area of the diagram: indentation depth – endurance – unloading. Figure 6 illustrates the level of this property by the extent of the areas before indentation of the pyramid according to variants (a-c) of thermal treatment of 13Cr11Ni2W2MoV steel samples according to table 2. From the presented data a good agreement of structure's characteristics and the property ψ with the AIP indicator can be seen.

Table 2. Influence of heat treatment on the steel mechanical properties including the AIP indicator

No. of variant	States of 13Cr11Ni2W2MoV steel		HV_{100*}	σ_B	ψ	AIP
	Double hardening	Tempering	MPa_*10^{-1}	MPa	%	
1 (a)	After operation		~370	1180	41	6.18
2 (b)	(1020 + 1000) °C – oil	560°C	572	1540	55	8.78
3 (c)	(1020 + 1000) °C - oil	640°C	380	1230	72	23.78

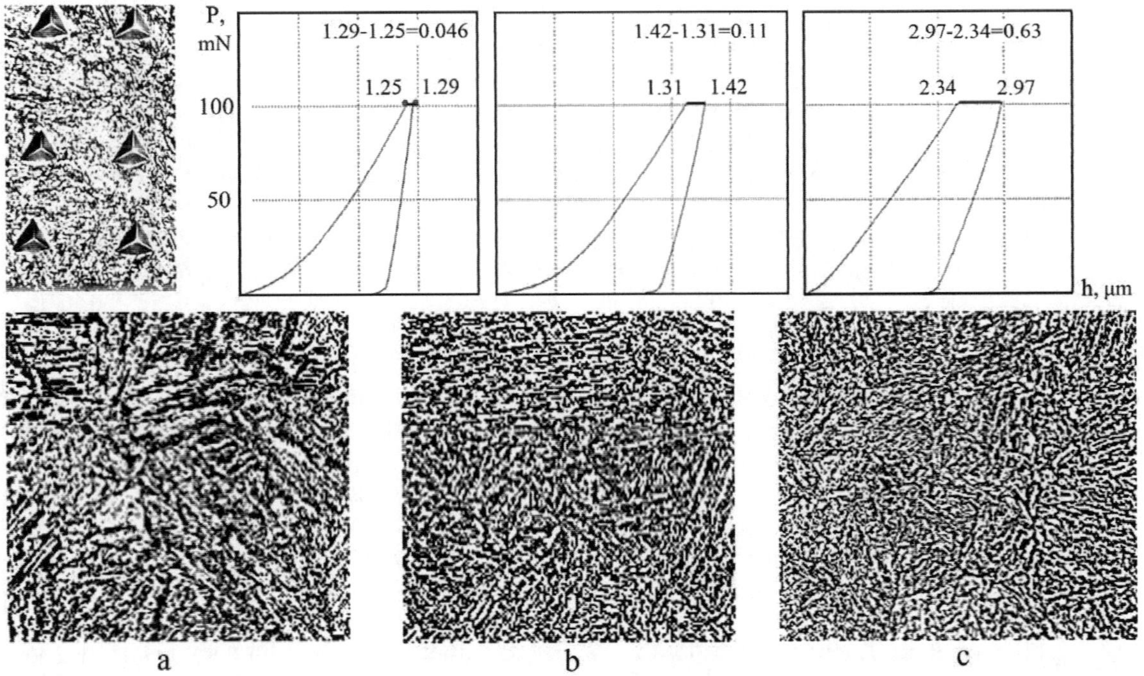

Figure 6. Determination of available indentation plasticity and structural states of 13Cr11Ni2W2MoV steel corresponding to variants of heat treatment (a-c) table 2

Thus, as a result of conducted complex studies the technologies and structural states of materials that provide functionality and long-term operation of complex equipment are revealed. For this purpose, the characteristics of the true deformation curve, the estimated fatigue life, the influence of operating and bench-test times on the parameters of heat-resistant alloy structure (its damageability) are used. In addition, the paper proposes the criterion of "available indentation plasticity" which allows adequately and reliably substantiating the adoption of technological decisions, the choice of the material including the identification of the advantages of the home-grown technologies without destroying the metal.

3. Conclusions

The shape transformation of the globular and cube-shaped particles of the hardening γ'-phase into geometry that is irregular (relief) along the perimeter and their crushing indicates the damage of the alloy structure (up to SDA levels = 0.53…0.57 mV) when microcracks can form in the blade. Increasing the amount of Cr, Ta, Mo, W, Co and decreasing the amount of Ti, Ni in the γ'-phase of the IN-738LC alloy increases the fatigue resistance of the rotor blades with a cast structure. Increasing the sum of alloying elements in $(Cr,Mo,W)_6C$ and γ'-phase of the CrNi65WMoTiAl alloy increases fatigue resistance of the turbine rotor blades. The increase in the distance between the axes of the second-order dendrites (λ_{II}) increases the damageability of the IN-738LC alloy. The possibility of repairing the disks made of the CrNi73MoNbTiAl alloy using the RHT in the recovery mode ($T \sim 0.35T_{RS}$) was confirmed. Effective technologies and structural states of materials were identified namely preference should be given to the states of CrNi65WMoTiAl and CrNi73MoNbTiAl alloys with higher values of S_K, $m^{(0)}$, ψ and smaller – E_K, H_{100} ensuring functionality and long-term operation of complex equipment.

References

[1] Sims C T and Stoloff N S 1995 *Translation Superalloys II: High-temperature materials for aerospace and industrial power* (Moscow: Metallurgiya) 384 p
[2] Murataev F I and Murataev A F 2016 Study of damageability of metal of GTP compressor disks *Vestnik of KSTU* number 4 pp 31-35
[3] Murataev F I, Saitkulov V G, Khusnutdinov Sh N and Trifonov A A 2014 Regularities of composition, structure, and damageability of turbine blades metal after bench-test and operating time *Vestnik of KSTU* number 2 pp 74-78
[4] Murataev F I and Murataev A F 2015 Influence of composition and morphology of intermetallic phase of IN-738LC alloy on damageability of gas turbine blades *Vestnik of KSTU* number 3 pp 43-48
[5] Tarasenko L V 2005 *Heat-resistant cast nickel alloys of equiaxed crystallization* (Moscow, Herald of the Bauman Moscow State Technical University) pp 50
[6] Murataev F I and Klabukov M A 2014 Study of regularities of structure and fatigue damageability of cast blades made of IN738-LC alloy *Vestnik of KSTU* number 3 pp 107-109
[7] Arthur J McEvily 2002 *Metal Failures: Mechanisms, Analysis, Preventions* (Moscow: Tekhnosfera) 416 p
[8] Murataev F I and Zharzhanazi M A 2013 Provision of structural strength of titanium alloys according to criteria of ultimate plasticity and fatigue resistance *Vestnik of KSTU* number 1 pp 50-54
[9] Kantyukov R R, Tameev I M, Saitkulov V G, Khusnutdinov Sh N, Murataev F I et. al. 2014 Application of ultrasonic nanoscopy method to control the blades of gas turbine engines of gas compressor units in production and operation *Proceedings of the XX International Congress «CITOGIC 2014» «New high technologies of gas, oil industry, energetics and communications»* (Kaliningrad: publishing house Ekonomika) pp 178-189
[10] Murataev F I and Khakimov S Sh 2014 Substantiation of microstructure and properties relationships of titanium alloys for prediction problems of fatigue resistance *Vestnik of KSTU* number 3 pp 110-113
[11] Shkanov I N, Braude N Z and Murataev F I 1982 Model of optimization of basic mechanical properties with respect to static and fatigue strength criteria in complex stress state conditions *Soviet Aeronautics* vol 25 Issue 1 pp 79-84
[12] Troshchenko V T 1981 *Deformation and destruction of metals under high-cycle loading* (Kiev: Naukova Dumka) 344 p
[13] Shkanov I N and Murataev F I 1996 Study of fatigue resistance criteria in modeling problems of operating and technological states of engine elements *Izv. VUZ. Aviatsionnaya Tekhnika* number 4 pp 58-61
[14] Murataev F I and Shkanov I N 1999 Substantiation of processing technology of compressor disks according to characteristics of static and cyclic deformation *Izv. VUZ. Aviatsionnaya Tekhnika* number 1 pp 43-47
[15] Murataev F I and Murataev A F 2016 Determination of rupture source of planetary reduction gearbox of HYUNDAI-R170 excavator *Vestnik of KSTU* number 3 pp 87-90
[16] Hideo Minata 1994 Estimation of fatigue life under conditions of varying average stresses and temperature by the example of GTP disk *Departmental special-purpose programme* No 38118 p 26

IOP Conf. Series: Materials Science and Engineering **570** (2019) 012071 doi:10.1088/1757-899X/570/1/012071

Substantiation of domestic material and welding technology for improving properties and competitiveness of pyrolysis furnace coils

F I Murataev[1], E R Galimov[1] and N Ya Galimova[1]

[1]Kazan national research technical university named after A.N. Tupolev, K. Marx street 10, Kazan, 420111, Tatarstan, Russian Federation

kstu-material@mail.ru

Abstract. An alloy of cast pipes produced in the Russian Federation and a welding method for restoring coil sections made of imported pipes damaged by gas corrosion in pyrolysis furnaces are proposed. Substantiation of the alloy choice is carried out according to the research results of composition and structure of the metal of the original pipes and according to the criteria of structural strength using a proven diagram of controlled damageability. Recommendations for optimizing the composition, microalloying, and structure of the proposed alloy of thick-walled cast pipes for the elimination of the deficiencies identified by the research are presented. The alloy advantages by mechanical properties in the temperature range up to 800...890°C are shown on samples of an experimental cast pipe.

1. Introduction

Pyrolysis furnaces coils operate in a wide range of temperatures (400…1125°C) undergoing gas corrosion of the metal with the prevailing reducing medium [1-3]. They can be damaged: in the tube side by C_3+C_4+C_5 hydrocarbons and water vapor at temperatures up to 900°C, in the intertubular space by the NO+NO_2+CO+CH_4+SO_3 flue gases at moderate temperatures in the convection part of the coils and up to 1125°C at the bridge wall (in the radiant section). The coils use imported pipes made of heat-resistant alloy G-X40 NiCrSi 3525. Metal marking in the Russian Federation terminology is 50Cr25Ni35Si2Nb alloy. Damaged areas of the cage (after 10…15 years of operation) are replaced by high-alloy steel cast pipes of domestic production. At the same time, hot cracks occur in the joints of the manual arc welding (MAW) during the formation of a heterogeneous weld [4-6].

2. Body text

The operation of repair sections made of pipes produced in the Russian Federation is only 6…15 months due to their cracking (Fig. 1). This requires the use of new materials functional in tough environments and more efficient welding methods of pipes in the cage during coil repair.

Analysis of pipe fractures indicates that all fractures are brittle, thermal, intercrystalline [1, 2, 4, 6], and from high-temperature gas corrosion. The main part of fractures is across the pipes. Hot cracks from the outer surface resemble in appearance creep cracks [1, 2, 4]. On the inner side of the wall the damages are of the hot cracking nature. Through and non-through cracks of half the wall thickness and more are found on both sides of metallographic sections (Fig. 2).

IOP Conf. Series: Materials Science and Engineering **570** (2019) 012071 doi:10.1088/1757-899X/570/1/012071

Figure 1. Weld seam damaged under repair and pipe after operation

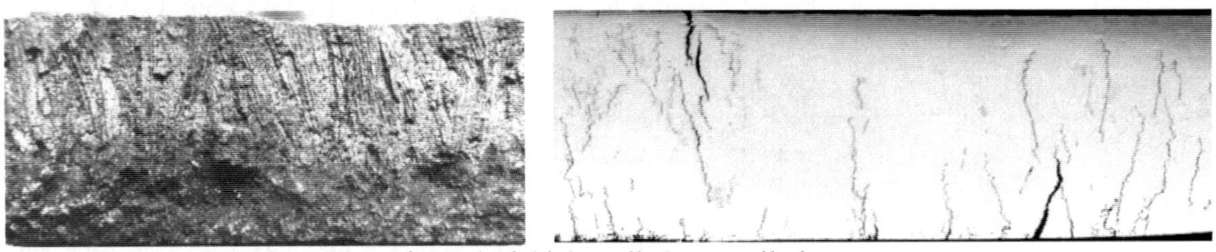

Figure 2. Fracture and cracking of metal of thick-walled cast coil pipes

In order to solve these tasks, integrated studies of the composition of structure and properties of the metal of imported pipes and domestic pipes delivered from Izhevsk and Nizhnekamsk are carried out at higher temperature. The equipment used for the metallography: the METAM-23 research microscope and PS SX150IS (Canon) camera, the X-MET 8000 analyzer, equipment for preparation of metallographic sections: the Remet IPA 30 press, the Remet LSA LS2 polishing pad.

Analysis of the alloys operability according to the composition, technology, and structural state variants was performed based on the characteristics analysis of the stress-deformation diagram [7]: Ilyushin module (1), resistance to tearing-off (2, 3), strain-hardening coefficient (4), condition for normalized metal damageability of metal of cast pipes (5) in the form of expressions:

$$E_K = E_*[1+E/S_K \ln 1/(1-\psi)]^{-1.0}, \quad (1)$$

where: $S_К$ – resistance to tearing-off.

$$S_К = 1233 + 9.27\,\psi \quad (2)$$

or

$$S_К = 75_* m^{(0)} + 955 \quad (3)$$

$$m^{(0)} = 0.75 \ln (S_K/\sigma_{0.2}) \{\ln [l/\sigma_{0.2}/E) (S_K/E + \ln l/1- \psi)]\}^{-1.0}. \quad (4)$$

In the absence of data for Ti alloys:

$$m^{(0)} = 0.75 \ln [(1233 + 9.27\,\psi)/\sigma_{0.2}] \{\ln [l/\sigma_{0.2}/E) (S_K/E + \ln l/1- \psi)]\}^{-1.0}, \quad (5)$$

or

$$m^{(0)} = (S_К - 955)/75, \quad (6)$$

$$\ln 1/1-\psi = \frac{0.5849 S_к}{E_\mu\left(1.071-0.2404\cdot\bar{\sigma}_m-1\right)}. \quad (7)$$

In the absence of data for Ti alloys:

$$\ln 1/(1- \psi) = 0.585\,(75_* m^{(0)}+955)/E_\mu\,(1.071-0.24\bar{\sigma}_m - 1), \quad (8)$$

$$\ln 1/1-\psi = \frac{0.585\sigma_в\left(1+1.35\psi\right)}{E_\mu\left(1.071-0.2404\cdot\bar{\sigma}_m-1\right)}. \quad (9)$$

IOP Conf. Series: Materials Science and Engineering **570** (2019) 012071 doi:10.1088/1757-899X/570/1/012071

In the absence of data and difficulty of using the expressions (3) – (5), you can use the dependency (6) and the graph presented in figure 3. Analysis of experimental and calculated data as well as the operating experience of equipment parts show the effectiveness of using the above characteristics when substantiating materials and technologies [7-9].

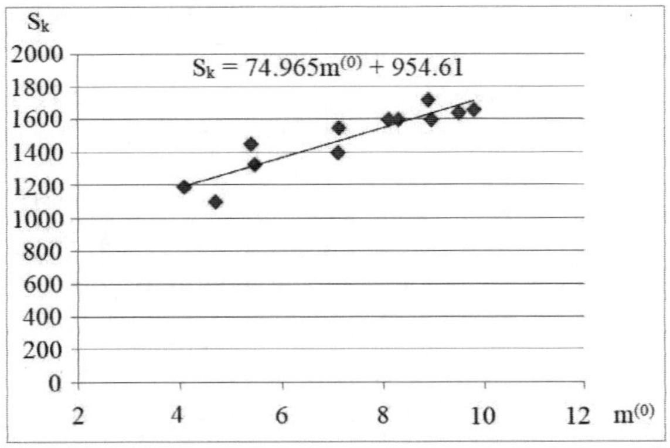

Figure 3. The graph of matching of characteristics resistance to tearing-off (S_k) and strain-hardening coefficient ($m^{(0)}$)

For choosing and substantiating of material with better structural strength characteristics, it is recommended to choose their variants of technology and structure with higher values of ψ, S_K, $m^{(0)}$ and smaller E_K value.

An important but challenging task when substantiating material and technology is finding the optimal ratio of strength and ductility characteristics by searching for a successful structural state. Practically, it is possible to simplify the algorithm of such a solution due to the repeated use of experimentally confirmed dependence (7) in the form of a graph – "a diagram of the limited metal damageability" [7]. Figure 4 illustrates the problem solution of the optimal ratio of ψ and S_K characteristics using the diagram at satisfying the condition when actual properties of the material exceed the values of "the diagram of the limited metal damageability" plotted according to (7) for the conditions of acting stresses ($\bar{\sigma}_m$). That is, it can be seen that when materials represented by the properties are located above the lines, they satisfy the condition (7), and not satisfying materials are located below the lines (solid lines – alloys, dotted ones – steel).

The advantage of imported pipes and pipes produced in Izhevsk, in most cases, is caused by the fundamental difference of the metal compared to the pipes produced in Nizhnekamsk, since they are made of austenitic Fe-Ni alloys 50Cr25Ni35Si2Nb and 45Cr25Ni35NbSi (contain Fe < 37%) and are notable for a large number of the alloying elements Nb and Ni.

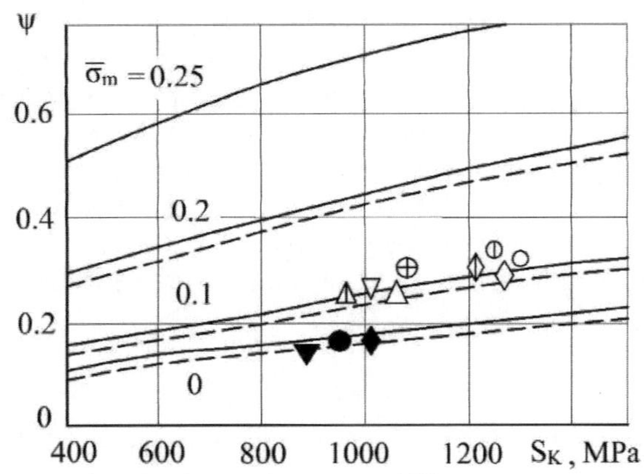

Figure 4. Diagram and curves of limited metal damageability

IOP Conf. Series: Materials Science and Engineering **570** (2019) 012071 doi:10.1088/1757-899X/570/1/012071

Pipes produced in Nizhnekamsk are cast from heat-resistant austenitic steel 45Cr25Ni20Si2. They are less doped and less resistant to high-temperature gas corrosion. The lower content of harmful S+P impurities in imported fusions (0.025%) attracts attention, and this, apparently, also affects the increases in longevity [4, 5]. An order of magnitude smaller copper content in imported and Izhevsk pipe fusions than that in Nizhnekamsk pipes [4], a lower value of λ_2, deserves the closest attention since it can adversely affect the high-temperature gas corrosion resistance [10-13].

Figures 5 and 6 illustrate the nature of metal pipes damage. The metal degrades along the dendrites boundaries from the surface of the pipes (Fig. 5, a) or from the inside (Fig. 5, b). Arms of dendrite with austenite structure (γ), dendrite boundaries, and interdendritic spaces are formed in the crystallization process of high-alloyed steels. Interstitial phases of MC and eutectic ($MC+\gamma^1$) are formed in steels with active carbide-forming components. In the absence of carbide-forming components: Nb, Ti, Ta, and etc because of the high diffusion activity of carbon along the grain boundaries, $Cr_{23}C_6$ is formed and thereby it weakens the degree of γ–doping with chromium (Fig. 5, b).

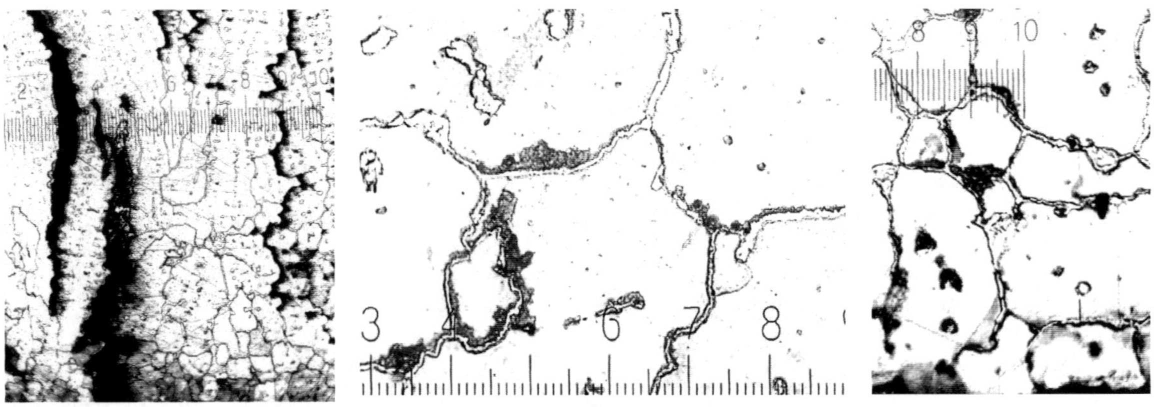

a b c

Figure 5. Types of gas corrosion of cast pipes: dendritic and intercrystalline degradation of the structure of domestic steels and imported alloys

Table 1. Characteristics of the metal composition (in %) and structure after operating time of pipes

Pipe material	C	S	P	S+P	Ni	Cu	Nb	λ_2
50Cr25Ni35Si2Nb	0.45-2.2	0.01-0.015	0.018-0.02	0.025	35.0	0.015	0.80	20
45Cr25Ni35Si2Nb	0.41-2.5	0.08-0.015	0.008-0.026	0.041	31.5	0.012	0.85	30
45Cr25Ni20Si2	0.38-0.43	0.011-0.03	0.02-0.026	0.057	19.0	0.15	0.11	60

If a solid carbide network is formed at the boundaries of metal dendrites, then the boundaries are intensely damaged by creep [3, 6] and intercrystalline corrosion (Fig. 6). Eutectic as well as intermittent chains of carbide $M_{23}C_6$ or MC interfere with the process of degradation at gas corrosion (Fig. 6). Figure 7 shows the damage to metal pipes of different delivery.

The main type of metal degradation of imported pipes (for 10-15 years) and pipes delivered from Izhevsk (for up to 1.5 years) is degradation along the grain boundaries in carbides. Steels of Nizhnekamsk pipes (for 6 month) are intensively damaged by gas corrosion along the dendrites boundaries due to a higher content of Cu (~ 10 times) and impurities (S+P).

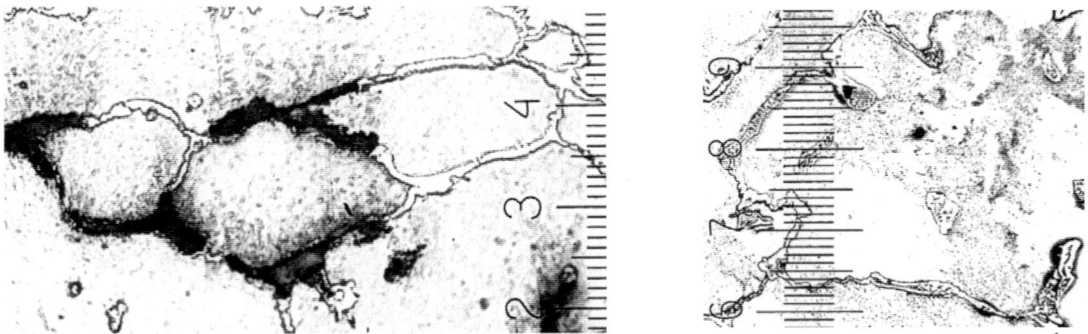

Figure 6. The development of metal corrosion in carbide edging and blocking by eutectic (MC+γ^{I})

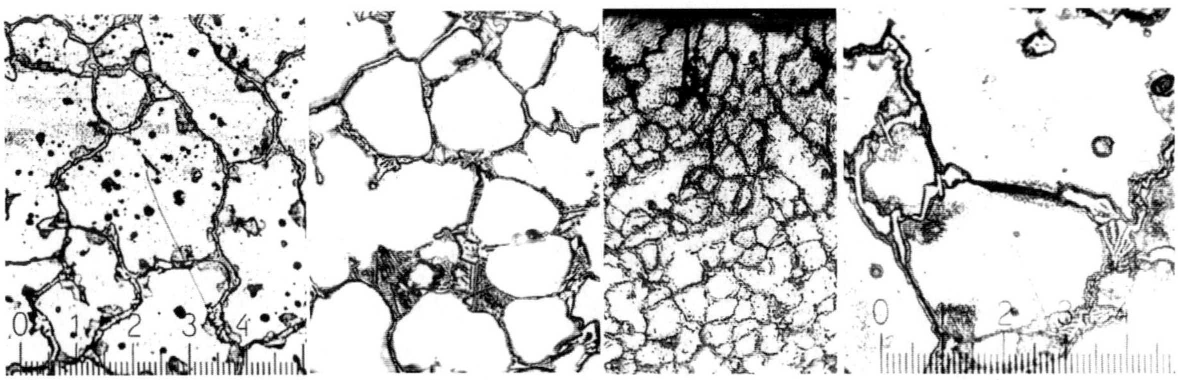

a b c d

Figure 7. Condition of pipes' metal after operation: from Japan (a-c), pipes delivered from Izhevsk (d)

The following materials were proposed to ensure the required service properties: steel 30Cr23Ni7Si and alloys 35Cr24Ni24Nb, 20Cr25Ni25TiAl. The conformity assessment of the test pipes' metal was performed by their properties in the range of operating temperatures. The results are shown in table 2. The advantage of the 35Cr24Ni24Nb alloy in characteristics of high-temperature strength at temperatures of 800°C can be seen.

Table 2. Tensile tests results of metal samples of cast pipes (D85 mm) at temperatures 800-890°C

No.	Steel grade	Tempe-rature, °C	$P_{0,2}$	P_B	$\sigma^t_{0.2}$	σ^t_B	δ^t	ψ^t	$\sigma^t_{B1}/\sigma^t_{Bi}$
			kilogram-force		MPa		%		
1	35Cr24Ni24Nb	800	925	995	311	334	22	30	1
2	30Cr23Ni7Si	890	370	375	127	128	21	54	2.61
3	20Cr25Ni25TiAl	865	-	490	-	165	36	67	2.02
4	45Cr25Ni20Si2	880	625	675	217	234	18	31	1.43
5	50Cr25Ni35Si2Nb	860*	-	680	-	225	22	23	1.48

This advantage is caused by the presence of niobium [12-15] which forms a more resistant to gas corrosion eutectics [(Nb, Cr, Ni) C+γ^{I}] (Fig. 8).

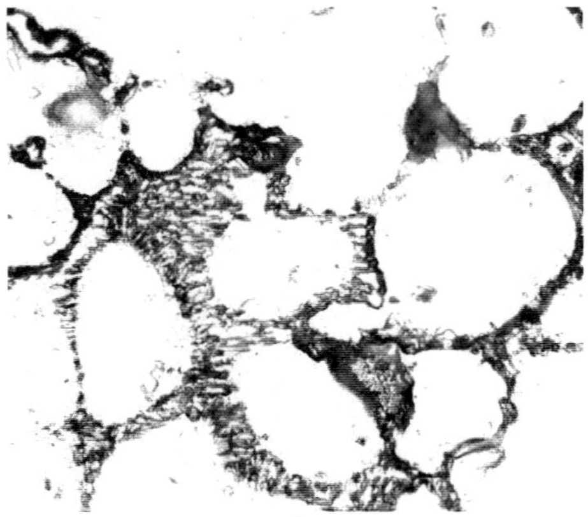

Figure 8. Microstructure of 35Cr24Ni24Nb alloy of pipe's experimental fusion

Eutectic is contaminated with impurities (P+S+Cu); microalloying and refining will allow pipes made of the 35Cr24Ni24Nb alloy to provide better functionality. In addition, it is more resistant to heat and does not burn out at welding, increasing the stability of the welding seam structure in operation [2, 5]. Welding of test joints of pipes delivered from Nizhnekamsk with pipes made of original G-X40 NiCrSi 3525 (50Cr25Ni35Si2Nb) alloy after operation using the method of manual argon-arc welding showed better weldability compared to the standard manual arc welding.

3. Conclusions

1. Substantiation of the alloy grade of the pyrolysis furnace pipes should be agreed with the actual temperature level of the outer and inner wall surface including the furnace specific zone.

2. The most useful alloying elements in relation to operating requirements of pipes are Nb and Ta, which will provide (because of the dispersed arrangement of the MC and the eutectic ($MC+\gamma^1$) at the boundaries of the twisted and cellular configuration of dendrites) long –term heat resistance with low λ_2.

3. When repairing the cages of coils the detailed control and analysis of elements (C, O, N, H) in the pipes after the operation is required.

4. It is necessary to limit the content of impurities (Sn+As+Pb+Sb+Vi+Zn) < 0.015%, S < 0.01%, P < 0.012%, severely limit the Cu impurity and maintain the ratio Nb/C = 2.5…3.8 to increase high-temperature plasticity and improve weldability.

5. It is proposed to replace standard manual arc welding with manual argon-arc welding when repairing damaged pipes sections of cages.

6. The choice of heat-resistant steels and alloys is substantiated by the diagram of the limited metal damageability ($\psi = f\ [S_K]$), lower structural damageability, and dependencies of the properties of structural strength (1), (4), (7) with higher values of ψ, S_K, $m^{(0)}$ and smaller E_K value.

7. An economically alloyed casting heat-resistant alloy 35Cr24Ni24Nb with greater high-temperature strength is proposed.

References

[1] Goritskiy V M 2004 *Metal diagnostics* (Moscow: publishing house Metallurgizdat) 408 p

[2] Murataev F I 2013 Typical and degradation features of welded joints microstructure of structural and special steels *Vestnik of KSTU* number 3 Issue 2 pp 84-89

[3] Pakhomov V S and Shevchenko A A 2009 Chemical resistance of materials and corrosion protection (Moscow: publishing house Khimiya. Kolos-S) 444 p

[4] Murataev F I and Murataev A F 2015 Adaptation of the composition, structure, properties of the metal and welded joints of heat exchangers to the resistance of high-temperature gas corrosion *Vestnik of KSTU* number 3 pp 119-125

IOP Conf. Series: Materials Science and Engineering **570** (2019) 012071 doi:10.1088/1757-899X/570/1/012071

[5] Wegst Z V 1993 *Key to steels. Reference book* (Krasnogorsk: publishing house NovoIM) 581 p
[6] Murataev F I, Klabukov M A and Murataev A F 2016 Degradation of composition and structure of steels and welded joints of vaporization coils in conditions of chemical and electrochemical corrosion *Vestnik of KSTU* number 4 pp 44-49
[7] Murataev F I and Shkanov I N 1999 Substantiation of processing technology of compressor disks according to characteristics of static and cyclic deformation *Izv. VUZ. Aviatsionnaya Tekhnika* number 1 pp 43-47
[8] Shkanov I N, Braude N Z and Murataev F I 1982 Model of optimization of basic mechanical properties with respect to static and fatigue strength criteria in complex stress state conditions *Soviet Aeronautics* vol 25 Issue 1 pp 79-84
[9] Murataev F I and Khakimov S Sh 2014 Substantiation of microstructure and properties relationships of titanium alloys for prediction problems of fatigue resistance *Vestnik of KSTU* number 3 pp 110-113
[10] Sims C T, Stoloff N S and Hagel W C 1995 *Superalloys II. Translation Superalloys II: High-temperature materials for aerospace and industrial power* (Moscow: Metallurgiya) 384 p.
[11] Murataev F I and Murataev A F 2015 Influence of composition and morphology of intermetallic phase of IN-738LC alloy on damageability of gas turbine blades *Vestnik of KSTU* number 3 pp 43-48
[12] Tarasenko L V 2005 *Heat-resistant cast nickel alloys of equiaxed crystallization* (Moscow: Herald of the Bauman Moscow State Technical University) pp 50
[13] Murataev F I and Murataev A F 2016 Structure and properties of structural steel after long-term overheating *Vestnik of KSTU* number 3 pp 82-86
[14] Murataev F I, Saitkulov V G, Khusnutdinov Sh N and Trifonov A A 2014 Regularities of composition, structure, and damageability of turbine blades metal after bench-test and operating time *Vestnik of KSTU* number 2 pp 74-78
[15] Murataev F I and Klabukov M A 2014 Study of influence of structure characteristics on fatigue damageability of cast blades made of IN-738LC alloy *Vestnik of KSTU* number 3 pp 107-109

IOP Conf. Series: Materials Science and Engineering **570** (2019) 012072 doi:10.1088/1757-899X/570/1/012072

Improving the quality of castings made of non-ferrous metal alloys when casting in metal molds

D L Pankratov[1] and R V Gavariev[2]

[1]Kazan Federal University, Naberezhnye Chelny Institute, 423812, Russia, Naberezhnye Chelny, Prospekt Syuyumbike 10A,
[2]Kazan national research technical university named after A.N. Tupolev-KAI,

Gavarievr@mail.ru

Abstract. The article deals with the quality of castings of zinc and aluminum alloys made by means of casting under pressure. The study identifies causes of defects in the outer surface and their influence on the roughness of the castings. The connection between the working surface of the mold and the roughness of the obtained castings is determined. The relationship between operation life of molds and the roughness of the castings is shown. The solution which increases the operational stability of molds as well as the quality of the surface of castings is proposed.

1. Introduction

Pressure casting is one of the most effective methods of obtaining blanks with the highest utilization rate of materials up to 95% [1]. Castings obtained by pressure casting often do not require subsequent machining. Therefore, the casting surface formed in the casting process must meet the requirements specified in the rough draft. The method of pressure casting allows to obtain castings with surfaces corresponding to the purity of not more than Ra=1,6 microns. However, with the wear of the casting equipment, the roughness of the castings increases [2], which will lead to the unfitness of the cast products. Therefore, in order to achieve these values of roughness as well as various other requirements for the surface it is necessary to obtain the desired results.

2. Results

Taking into account the relatively small range of values for regulating the technological parameters of the pressure casting process, due to the opposite of the requirements for obtaining a quality surface, on the one hand, and reducing the porosity inside the casting, on the other, one of the most important parameters is the operation life of the forming surface of the mold [3,4].

The operation life depends on many factors acting simultaneously in the process of obtaining the necessary casting. Therefore, to obtain the most effective technology for the production of high-quality castings by the method of pressure casting it is necessary to find a solution that will reduce the influence of negative factors on the forming surface. One of the possible solutions is coating the forming surface by cathodic-ion bombardment [5,6]. The most suitable option for pressure casting conditions is a coating TiCN-TiN-MoN [7].

IOP Conf. Series: Materials Science and Engineering **570** (2019) 012072 doi:10.1088/1757-899X/570/1/012072

To verify the effectiveness of the protective coating, we carried out a study of the roughness in terms of Ra at three different points on the outer flat surface of the castings of ZAMAK-5 (figure 1A) and aluminum alloy (figure 1B), obtained in the formative inserts of 4Cr5MoSiV material coated with TiCN – TiN – MoN in different periods of operation.

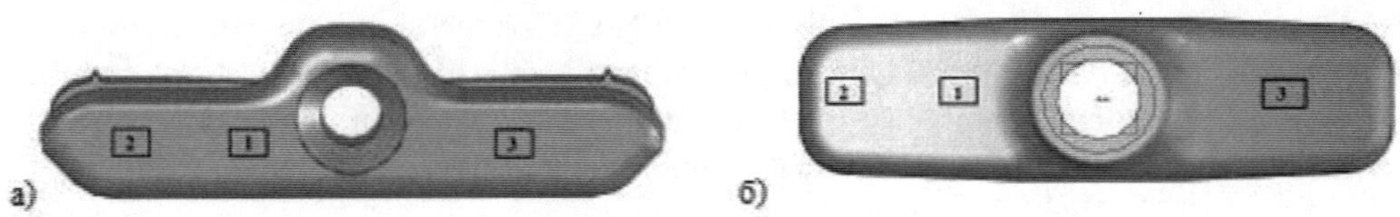

Figure 1. Determining roughness on the resulting casting and the casting of ZAMAK-5; b – casting aluminum alloy

For comparison we investigated similar performance in the casting obtained in the forming inserts of steel 4Cr5MoSiV with nitriding (table. 1).

Table 1. Values of roughness of zinc castings

Processing of molds	No. of the point	Roughness of casting surface Ra, mkm						
		Number of pressing-ins (number of cycles)						
		1	1000	10000	20000	80000	150000	200000
TiCN – TiN - MoN	1	0,547	1,140	1,999	2,109	2,270	3,468	5,867
	2	0,890	1,574	2,224	2,458	2,664	3,736	6,089
	3	0,457	1,325	1,879	1,889	2,005	3,404	5,796
Nitriding	1	1,049	1,726	2,778	3,503	4,798	6,368	-
	2	1,816	2,324	2,861	3,568	4,969	6,758	-
	3	1,593	2,193	2,708	3,373	4,566	6,397	-

According to the data obtained, it was found out that the points lying on the same plane for both aluminum and zinc alloys have different roughness indicators, which can be explained by two factors: the position related to the molding channel, as well as the position related to the structural elements of the forming surface (fillets, holes, low spots, collars and others) [8,9].

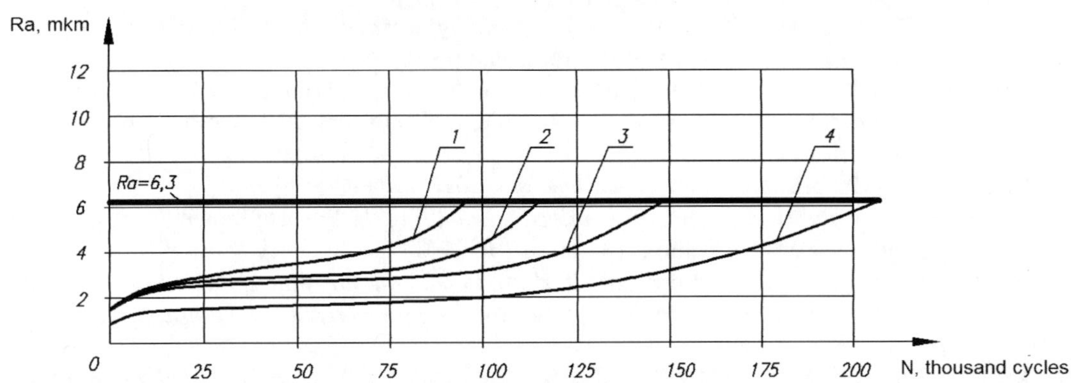

Figure 2. The graphs of surface roughness of castings from a period of operation of the forming panels of steel 4Cr5MoSiV: 1) during casting of the aluminum alloy with nitriding; 2) during casting of the aluminum alloy with coating TiCN-TiN-MoN; 3) in the casting of ZAMAK-5 with nitriding; 4) during casting of ZAMAK-5 with a coating TiCN-TiN-MoN

IOP Conf. Series: Materials Science and Engineering **570** (2019) 012072 doi:10.1088/1757-899X/570/1/012072

It is also revealed that the castings obtained from the forming inserts coated with TiCN-TiN-MoN have less roughness at the points under consideration than the castings from nitriding inserts (Fig. 2).

Figure 2 shows that in the initial period of operation (up to 5000 cycles) all curves have equivalent results, but then there is a sharp divergence [10-12]. The comparison of the castings obtained from forming of the surface coated with TiCN - TiN – MoN (curve No. 2 for the aluminum alloy, No. 4 for ZAMAK-5) and nitriding (curve No. 1 for the aluminum alloy , No. 3 for ZAMAK-5) showed that the change of roughness are similar in both cases. Thus, after 150,000 pressing cycles, zinc castings obtained from nitriding inserts have a roughness Ra=6,3 microns, and castings obtained from inserts coated with TiCN - TiN – MoN - Ra=3,2 microns.

3. Conclusions

Thus, the use of a complex coating TiCN-TiN-MoN, obtained by the cathodic-ion bombardment method, on molds for pressure casting of zinc and aluminum alloys can reduce the growth rate of roughness of end products. At the same time, there is an increase in the operation life of molds with applied wear-resistant coatings in comparison with the traditionally used method of nitriding of the forming surfaces of molds. It is shown that the complex wear-resistant coating TiCN-TiN-MoN, applied to the forming surfaces of molds for pressure casting of non-ferrous metals and alloys based on them can increase the operation life of molds with guaranteed compliance with the quality requirements of castings obtained by pressure casting which in turn reduces the cost of molding in production process.

References

[1] Belopuhov AK 1985 *Tekhnologicheskie rezhimy lit'ya pod davleniem* [Technological modes of pressure casting] : Mechanical engineering 272 p

[2] Goryanov II 1973 *Press-formy dlya lit'ya pod davleniem* [Press-molds for pressure casting]: Reference manual. - L.: Mechanical engineering 256 p

[3] Gavarieva KN, Simonova LA, Pankratov DL andGavariev RV2017 Development of expert systems for modeling of technological process of pressure casting on the basis of artificial intelligence //IOP Conference Series Materials Science and Engineering, Institute of Physics Publishing DOI: 10.1088/1757-899x/240/1/012019

[4] Gavariev RV, Savin IA and Leushin I O 2017 *Improvement of zinc castings surface quality by laminated protective coating* Tsvetnye Metally No 5 pp 84-88

[5] Gavariev R V and Savin I A 2017 *Improvement of surface quality of casting produced by casting under pressure* Solid State Phenomena, pp 988-93

[6] GavarievRV and Savin IA 2018 *Research of the Mechanism of Destruction of Compression* Molds for Casting under Pressure of Color Alloys, Solid State Phenomena Vol 284 pp 326-31

[7] Pankratov DL, Gavariev RV and Gavarieva KN 2016 *Influence of multilayer coatings on the operational stability of molds for injection molding* IOP Conference Series: Materials Science and Engineering Volume 134 Issue 1 **012031**

[8] Gavariev RV, Savin IA and LeushinI O 2016 *Impact of the functional coating on service durability of injection molds for the zinc alloys pressure casting* TsvetnyeMetally No 1 pp 66-70

[9] Deev V B, Prusov ES and Kutsenko AI 2018 *Theoretical and Experimental Evaluation of the Effectiveness of Aluminum Melt Treatment by Physical Methods* La MetallurgiaItaliana V 2 pp 16-24

[10] Gavarieva KN, Simonova LA, Pankratov DL, Shibakov VG and Gavariev RV 2018 *Application of multi-agent system for control of parameters of precision stamping process of bevel gears* IOP Conference Series: Materials Science and Engineering, Volume 412 **012020**

IOP Conf. Series: Materials Science and Engineering **570** (2019) 012072 doi:10.1088/1757-899X/570/1/012072

[11] Safronov NN, Mingaleeva L B and Savin I A 2018 *Optimization of charge material composition in shs process with ferrosilide fabrication from gaseous wastes of metallurgical production* Chernye Metally pp 53-59
[12] Prusov E S, Panfilov AA and KechinV A 2017 *Role of Powder Precursors in Production of Composite Alloys Using Liquid-Phase Methods* Russian Journal of Non-Ferrous Metals Vol 58 (3) pp 308-16

IOP Conf. Series: Materials Science and Engineering **570** (2019) 012073 doi:10.1088/1757-899X/570/1/012073

The concept of quality management systems and monitoring of their effectiveness using "Lean production" concept

O A Plugatyreva and E M Khusnutdinova

Kazan (Volga Region) Federal University
18 ul. Kremlyovskaya, Kazan, 420008

Abstract. An overview of the concepts of the quality management system is given, ways of monitoring their effectiveness are explored using the concept of "Lean Production".

The competitiveness of the Russian economy largely depends on the level of efficiency in the management of industrial enterprises. One of the important directions of increasing the effectiveness of this type of management is a suite of Japanese production management methods that are widely used in leading countries of the world.

In the past decade, Russian industrial enterprises have attempted to introduce quality management methods that include lean production tools. But practice shows that these methods lead to substantial positive results very rarely. Most of the organizations, without delving into the essence and principles of operation of these methods, make typical mistakes that prevent the smooth implementation of new tools for the work:

- low management involvement;
- pursuit of quick results;
- perception of the concept as a set of tools, not a way of thinking;
- embrace the unembraceable;
- the desire to implement "by the book" without adaptation to the conditions of a particular enterprise.

This makes it relevant to study the principles of lean production in the quality management system to identify the causes of the wrong approach of top management to their implementation.

High interest in these methods and concepts is particularly relevant for large industrial enterprises with a legacy of the past in the form of mass production with its large run, used equipment, and excessively large staffs of workers. That is, enterprises that are aware of the need to take urgent measures to maintain the level of competitiveness in the near future.

Quality management systems (QMS) in one form or another exist at each manufacturing company. All QMS are created and regulated according to the standards International Organization for Standardization (ISO). The quality management system is designed to ensure the quality of products or services of the company and to adjust this quality to the customer expectations. At the same time, its main task is not the control of each unit of production, but the ability to make sure that there are no mistakes in the work that could lead to the appearance of defective goods.

The functioning of the quality system is based on a number of general principles that allow you to create products or provide services of a specified quality level. The basic principles of the quality management system are disclosed in GOST R ISO 9000:2015 [1]:

IOP Conf. Series: Materials Science and Engineering **570** (2019) 012073 doi:10.1088/1757-899X/570/1/012073

The first principle is customer orientation. The main focus of quality management is to ensure compliance with customer requirements, as well as the continuous desire to exceed his expectations.

The second principle is leadership. At all levels of the organization, leaders must ensure the unity of the direction and goals of the organization's activities and create an environment in which employees interact with each other to achieve the organization's quality objectives.

The third principle is the workers engagement. Competent employees who are empowered and interact with each other at all levels of the organization increase its ability to create value.

The fourth principle is the process approach. More efficiently and effectively consistent and predictable results are achieved when the organization's activities are understood and managed as interrelated processes that function as a coordinated system.

The fifth principle is improvement. Any successful organization urgently needs continuous improvement in order to be able to preserve and maintain current levels of activity, respond to changes in external and internal conditions, and create new opportunities.

The sixth principle is making decision based on evidence. It is important for the top management of the organization to understand the causal relationships and their possible unintended consequences. Solutions that are based on analyzing and evaluating data and information are more likely to produce the desired results.

The seventh principle - Relationship Management. To achieve sustainable success, organizations manage their relationships with relevant stakeholders to optimize their impact on its performance.

Thus, the quality management system is a proven tool for organizing production activities, and is aimed primarily at increasing the probability of a company achieving its planned results and, ultimately, at improving the quality of a company.

Reliable and accurate data obtained in a timely manner and sufficient for making management decisions play an important role in the management of any enterprise. The introduction of reliable monitoring indicators reflecting the actual stage of each process of the quality management system and the entire management system allows determining and taking timely corrective and preventive measures that will be aimed at achieving goals with minimization of various costs and optimal consumption of necessary resources.

Monitoring is a part of the management of an organization, which is a continuous systematic observation of factors and processes, the values and results of which influence the achievement of goals, which include the analysis and control of the effectiveness of an enterprise's quality management system processes, as well as, continuous updating of data needed to assess the current state of affairs. The main purpose of monitoring should be guaranteed reliable information support, which helps the organization and its managers to make the right management decisions to achieve business goals.

The main problem of the functioning of the quality management system is the lack of competent mechanisms for its monitoring, which makes it difficult to implement the principle of continuous improvement. At the same time, there is a need to minimize and even eliminate any kinds of costs. To eliminate these problems, a special company management scheme is needed, aimed at continuous improvement, which is the Japanese concept of "Kaizen", which is the basis of lean production [2].

Lean production is an integrated approach that includes optimization of processes, provision of management infrastructure and changes in the way of thinking and behavior of the organization's employees. This scheme is aimed at maximum orientation towards the customer [3].

Thanks to the introduction of lean production principles at an enterprise, it is possible to solve basic management tasks: first, minimize costs without reducing the quality level of the final product, second, speed up the production process, avoiding overproduction and oversupply, and debug supply channels.

The main task of the company's management, which seeks to improve product quality by reducing losses, is the creation of a well-developed scheme of actions. It is necessary to develop instructions and rules that ensure the formation of a complete system, increase staff motivation [11,12]. They must have a logical connection with each other. The system distinguishes the following main tools of lean production:

1. Creating a value stream map is an easy to understand graphical scheme of information and material processes that must be implemented in order to provide a customer with a product or service. This map clearly demonstrates the weak points of the stream for the timely detection of current problems in production. After that, an improvement plan is developed.
2. Kanban - notifying employees (through instructions or permits) that it is necessary to start production or withdraw a certain amount of products. This tool is used to plan the cycle of production and sale of goods, from the forecast of demand and setting jobs to employees before the distribution of functions of production capacity.
3. Pull production - within this lean way of organizing production, the quantity of products at each stage depends on the needs of the next stages, and in the long run - on the needs of customers for this product or service.
4. Kaizen - continuous improvement of the value stream, which aims to increase value and reduce costs. In practice, it is expressed in stimulating the initiative of workers [5].
5. SMED (Single Minute Exchange Of Die) - a system for quick reconfiguration of equipment. Machine changeover or tool replacement should be done as soon as possible - within a few minutes or seconds.
6. 5S - a method of creating an ideal workplace and optimizing labor comprising five components:
   - “seiri” or sorting: separating the necessary items from unnecessary ones, and then discarding them;
   - “seiton” or straighten or set in order: distributing to certain places the tools you need to use so that you can quickly and easily find and use them;
   - “seiso” or sweeping: cleaning the workplace, taking care of hygiene and tidiness;
   - “seiketsu” or standardization: a condition that allows you to perform the previous three rules of the methodology;
   - “shitsuke” or sustaining: accustoming employees of an organization to the methodical and precise adherence to technologies, production standards and internal regulations.
7. TPM or Total Productive Maintenance - the principle of effective maintenance of the equipment used in the work, the detection and elimination of its defects before they cause problems. The goal is the most productive and economical use of equipment through preventive maintenance and maintaining it in working condition [4].
8. JIT or Just-In-Time (“strictly on time”) is a technique for the economical use of resources — materials and raw materials. The components required at this stage of production or in a specific operation are delivered to the right moment, but not earlier [13-15].

The essence of the concept of lean manufacturing is formulated simply: to create a competitive company and products (services), you must completely get rid of everything that does not add value to the product, or at least minimize such costs [6-9].

Summarizing the above, we can draw some conclusions. To improve the principles of the company performance, as well as the final product, it is necessary to keep the focus on continuous improvements, focus on the process [10]. You should not spend time and resources on projects that provide only short-term results. Technologies and tools of lean production allow enterprises to achieve their goals and the necessary results in development, as well as continuously monitor the functioning of the QMS. The introduction of lean production techniques is an integral part of modern production. By the standards of world leaders in the introduction of lean production, Russian enterprises are at the very beginning of the journey, and this is an excellent chance to create competitive products that will meet all international quality standards.

IOP Conf. Series: Materials Science and Engineering **570** (2019) 012073 doi:10.1088/1757-899X/570/1/012073

References

[1] GOST R ISO 9000:2015 *Quality management systems. Fundamentals and vocabulary. Intr. 2015-09-28.* M.: IPK Izdatel'stvo standartov

[2] Ohno T 2012 *Toyota production system: moving away from mass production.* M.: IKSI Publ.

[3] Womack J and Jones D 2011 *Lean production. How to get rid of losses and achieve prosperity of your company.* M.: Al'pina Pablisher Publ.

[4] George M L 2007 *Lean production + Six Sigma. Combining the quality of six sigma with the speed of lean production.* M.: Al'pina Pablisher Publ.

[5] Goldsby Thomas 2016 *Lean manufacturing and 6 Sigma in logistics. Guide to optimization of logistics processes.* M.: Grevtsov Publisher, p 416

[6] George Michael 2015 *Lean plus six Sigma in the service sector. How lean manufacturing speed and six Sigma quality help improve business* Michael George. - Moscow: Mann, Ivanov and Ferber. p 868

[7] Russell Jesse Lean Jesse Russell. - M.: Book on Demand, 2012. p 170

[8] Fatkhutdinov R A 2012 *Organization of production*: Textbook. - M.: INFRA-M. p 672

[9] Fox A 2013 *Computational geometry. Application in the design and production of* M. Pratt. - M.: p 282

[10] Womac James 2014 *Lean manufacturing. How to get rid of losses and achieve prosperity of your company* M.: Alpina publisher, p 472

[11] Kashapov L N, Kashapov N F, Kashapov R N 2013 J. Phys.: Conf. Ser. 479 012003

[12] Kashapov N F, Nafikov M M, Gazetdinov M X, Nafikova M M and Nigmatzyanov A R 2016 IOP Conf. Ser.: Mater. Sci. Eng. 134 012013

[13] Dautov G, Kashapov N, Fayrushin I, Dautov I 2013 J. Phys.: Conf. Ser. 479 012014

[14] Kashapov N F and Sharifullin S N 2015 IOP Conf. Ser.: Mater. Sci. Eng. 86 012021

[15] Kashapov L N, Kashapov N F, Kashapov R N and Semushin V D 2016 J. Phys.: Conf. Ser. 669 012028

Investigation and explanation of mathematical tooling for accounting non-economic characteristics during the investment project effectiveness' assessing process

A S Puryaev[1], Zh A Puryaeva[1], A R Kharisova[1], A A Puryaev[2]

[1]Naberezhnye Chelny Institute, Kazan Federal University, 18 Kremlyovskayastreet, Kazan, 420008, Russia Federation
[2]Moscow State University of Geodesy and Cartography, 4,Gorokhovsky pereulok Moscow, 105064, Russian Federation

aidarp@mail.ru

Abstract. In this work we briefly present the study results of the issue of accounting and implementation of specific mathematical tooling for non-economic characteristics in evaluating of projects' effectiveness. There is established classification of events in which using of one or another mathematical apparatus during solving the question of accounting for non-economic characteristics can be observed. It is proposed to study the classification of non-economic characteristics easily formalized and difficult to formalize (parameters) that are used when assessing the effectiveness of projects. There is explained and recommended a certain mathematical tooling for the process of accounting for non-economic characteristics in the assessment on the basis of classification under description.

1. Introduction

As part of the implementation of the research project of the Russian Foundation for Basic Research (RFBR) No. 18-010-00018, we developed a study of foreign and domestic experience on this issue for developing recommendations on the use of mathematical methods and the elaboration of original models for recording non-economic characteristics in the process of evaluating the effectiveness of investment projects. The collection and analysis of material for the period from 1975 to March 2018 based on the Web of Science Core Collection database allowed us to select 168 sources out of them. Among them, 57 sources were selected which directly reflect the subject of the study - the consideration of non-economic characteristics (externalities) in evaluating the effectiveness of investment projects. Originally, to study the domestic experience in this matter, 139 sources from the Russian Science Citation Index (RSCI) database were selected by search queries. After a detailed study of abstracts (annotations), 48 sources remained. The specific results of these studies are presented in articles and submitted to certain publishers for consideration about the possibility of publication. In the frames of this work, 105 analyzed sources were subjected to detailed consideration in the aspect of applying the mathematical method, a tool for recording non-economic externalities in the process of evaluating the effectiveness of investment projects. As a result, 45 cases of application of mathematical tools in evaluating the effectiveness of 41 sources were studied.

2. Aim of research

IOP Conf. Series: Materials Science and Engineering **570** (2019) 012074 doi:10.1088/1757-899X/570/1/012074

Developing of recommendations for the use and elaboration of specific mathematical tooling to account for non-economic characteristics in the process of evaluating the effectiveness of investment projects.

3. Research method

Analysis of the existing mathematical tooling in assessing the feasibility and effectiveness of investment projects (based on the Web of Science Core Collection and RSCI).

4. Resultsanddiscussion

The study of the sources of literature of the foreign Web of Science Core Collection database and the RSCI database on the subject of research allows us to single out the following mathematical methods and their application frequency in evaluating the effectiveness of investment projects (see table).

Table 1. Frequency and percentage of application of certain mathematical tooling to take into account non-economic characteristics when evaluating the effectiveness of investment projects

№№	Mathematicalmethodsandtools	Totalcases	Percentage, %
1	Method of valuation of changes in macroeconomic indicators (*"Cash flow" andcost-benefit analysis* methods (*CBA*)).	22	49
2	Method of multi-criteria optimization based on the use of the Monte Carlo method and data matching	1	2,2
3	Energybalancecalculationmethod	1	2,2
4	Methods for assessing risk and uncertainty:	8	17,9
4.1	*Scenario method (theory of games with "nature")*	*4*	*9*
4.2	*Risk adjustment in the discount rate*	*1*	*2,2*
4.3	*Riskassessmentmatrixmethod*	*3*	*6,7*
5	Methods of fuzzy (vague) sets	4	8.8
5.1	*Efficiency criteria in the scales of fuzzy sets theory in the form of linguistic variable*	*1*	2,2
5.2	*AHP- fuzzy comprehensive evaluation method*	*1*	2,2
5.3	*TOPSIS method in combination with linguistic neutrosophic numbers (LNN)*	*1*	2,2
5.4	*TFIEOWA operator'smethod*	*1*	2,2
6	Methods of the aggregation theory (generalization)	9	19,9
6.1	*Desirability function method by Harrington*	*3*	*6,7*
6.2.	*The method of valuation and the integral indicator (except for the function of desirability)*	2	*4,4*
6.3	*Weightedscoringmethod (ranking)*	2	*4,4*
6.4	*Indicatorevaluationmethod*	2	*4,4*
7	Casesintotal	45	100

The research base contains both foreign sources [1-5] and domestic [6-9], [10]. These links do not list all sources. A complete list of the database under investigation is presented in a separate file of the Internet source [11].

The results of the study suggest the idea of classifying the considered non-economic characteristics (parameters, indicators) into two groups: easily formalizade and difficult to formalize. Easily formalized characteristics are understood as various parameters of a noneconomic nature, the impact of which on the project environment can be assessed in the value scale (monetary units) with a high degree of probability. This will allow to take into account their impact when evaluating the effectiveness of projects directly in the stream of payments. Difficultly formalized characteristics are understood as various parameters of a noneconomic nature, whose impact on the project environment cannot be assessed with a high degree of probability in the value scale (monetary units). To take into account such characteristics, it is necessary to translate the parameters of a different physical nature into another unified normalized rating scale, which will allow evaluating projects based on a generalized criterion and unambiguously choosing the best project. In this case, the category of "effectiveness" and the category of "optimality" become synonymous.

It is fundamentally recommended to use the indicator reflecting the "yield" from the potential as a criterion for determining the effectiveness of a project. I.e. an indicator that would reflect the proportion of the project's compliance with the conditions and restrictions of the decision maker from a potential condition at the time of the assessment. At the same time, the presence of uncertainty in the assessment requires the use of fuzzy sets in the representation of quantitative values of the estimated parameters.

So, economic and easily formalized non-economic parameters are estimated in the value (monetary) scale. In this case, it is possible to take them together in a consolidated stream of payments. Difficultly formalized non-economic parameters (quantitative and qualitative) must be reduced to one single normalized scale, having the limitations set by the decision maker. As far as possible, quantitative parameters should be specified in the form of linguistic variables, each term of which should be represented as a fuzzy set of values. This will allow to take into account at least two parameter estimates with different degrees of membership. It is also proposed to translate economic and easily formalized characteristics into the same normalized scale, having limitations on them (together with difficult to formalize characteristics). The obtained integral final value (as a geometric mean and (or) weighted average) will act as a generalizing quantitatively oriented evaluation criterion, which will have a fraction of the potential value of this criterion. For example, the value of the desirability function is from 0 (the worst) to 1 (optimal, the best) or the radar area fraction, which is a "reflection" of potential probable effectiveness.

An extended article on the subject of the study is under consideration for publication in a journal indexed in the Web of Science Core Collection and Scopus database.

5. Acknowledgements

The reported study was funded by RFBR according to the research project № 18-010-00018.

References

[1] Woods A 2015 Introduction to Monte Carlo Methods.

[2] Mendecka B and Koziol J 2015 Application of the method of data reconciliation for minimizing uncertainty of the weight function in the multicriteria optimization model *Archives of Thermodynamics* **36** 83-92

[3] Romanelli T L and Milan M 2010 Energy performance of a production system of eucalyptus *Revista Brasileira De Engenharia Agricola E Ambiental* **14** 896-903

[4] Platon V, Frone S and Constantinescu A 2014 Financial and economic risks to public projects *1st International Conference 'Economic Scientific Research - Theoretical, Empirical and Practical Approaches', Espera 2013* **8** 204-10

[5] Liang W Z, Zhao G Y and Wu H 2017 Evaluating Investment Risks of Metallic Mines Using an Extended TOPSIS Method with Linguistic Neutrosophic Numbers *Symmetry-Basel* **9**

[6] Kibalov E B and Shibikin D D 2017 Evaluation of the effectiveness of large-scale investment projects in the context of conceptual design *Bulletin of the South Ural State University. Series: Computer technologies, management, radio electronics* **17** 99-108

[7] Epishkina K M 2010 Estimation of public efficiency of rail transport of a megacity *Region: Economics and Sociology* 255-72

IOP Conf. Series: Materials Science and Engineering **570** (2019) 012074 doi:10.1088/1757-899X/570/1/012074

[8] Gert A A, Suprunchik N A, Nemova O G and Kuz'mina K N 2009 Valuation of oil and gas fields and subsoil blocks taking into account uncertainty and risks *Subsoil use of the XXI century* 49-57
[9] Mironyuk S G 2015 Geological dangers of the developed deposits of the eastern shelf of Sakhalin Island: identification and principles of mapping *Scientific and technical collection "News of gas science"* 113-7
[10] Puryaev A S 2017 Alternative evaluation of innovations' effectiveness in mechanical engineering *IOP Conference Series: Materials Science and Engineering***240** 012056
[11] 2018 List of sources for the study of mathematical tools.

IOP Conf. Series: Materials Science and Engineering **570** (2019) 012075 doi:10.1088/1757-899X/570/1/012075

To determination of dependence of parameters of surface cracks in concrete and reinforced concrete volume elements

O V Radaikin[2], L S Sabitov[1], I K KIyamov[1], N F Kashapov[1]
[1]Federal State Autonomous Educational Institution of Higher Education "Kazan (Volga Region) Federal University", Kazan, Kremlevskaya street, 18, 420008, Russian Federation
[2] Kazan State University of Architecture and Civil Engineering

l.sabitov@bk.ru

Abstract. To determine the dependences of the parameters of surface non-through cracks, the problem of tensile bulk concrete elements (in the form of prisms) was considered. One group of elements contained a crack normal to tensile forces, and the other was inclined (the problem was modeled using the finite element method in the Ansys software package). The stress intensity factors of three types were determined, for which a macro (subroutine) was written in the APDL language, the calculated coefficients of the 2nd and 3rd type (K_I, K_{II}) were used for the first time in the considered problems. On the basis of these coefficients, relations between the length, width of the opening, the penetration of cracks are established, the boundaries of the pre-fracture zone are found, and the residual bearing capacity of the elements with a crack is estimated. It is shown that the stress relief area around the crack has a diamond shape. The dependences obtained and the procedure for obtaining the numerical values of K_I abd K_{II} constitute a scientific novelty.

Previously, we obtained dependences between the length, depth, and width of a normal crack opening in a wall panel. The goal of this paper is to find the same relations for the bulk problem of a more general form (using the example of prismatic rods with a single non-through crack).

To achieve the goal, the following tasks are solved:

- modeling of the stress-strain state of bulk concrete elements with a single semi-elliptical crack (with both normal and inclined) by the finite element method;
- stress intensity factors are calculated for cracks with a different ratio of length and depth, the crack opening width is determined;
- the boundaries of the prefracture zone are determined in the frontal region of macrocrack and stress relaxation zone.

Thus, the extension of volumetric concrete elements (in the form of prisms 1x1x1 m), having one surface semi-elliptical crack, is considered. One group of elements contains cracks normal to tensile forces, and the other is inclined (Fig. 1).

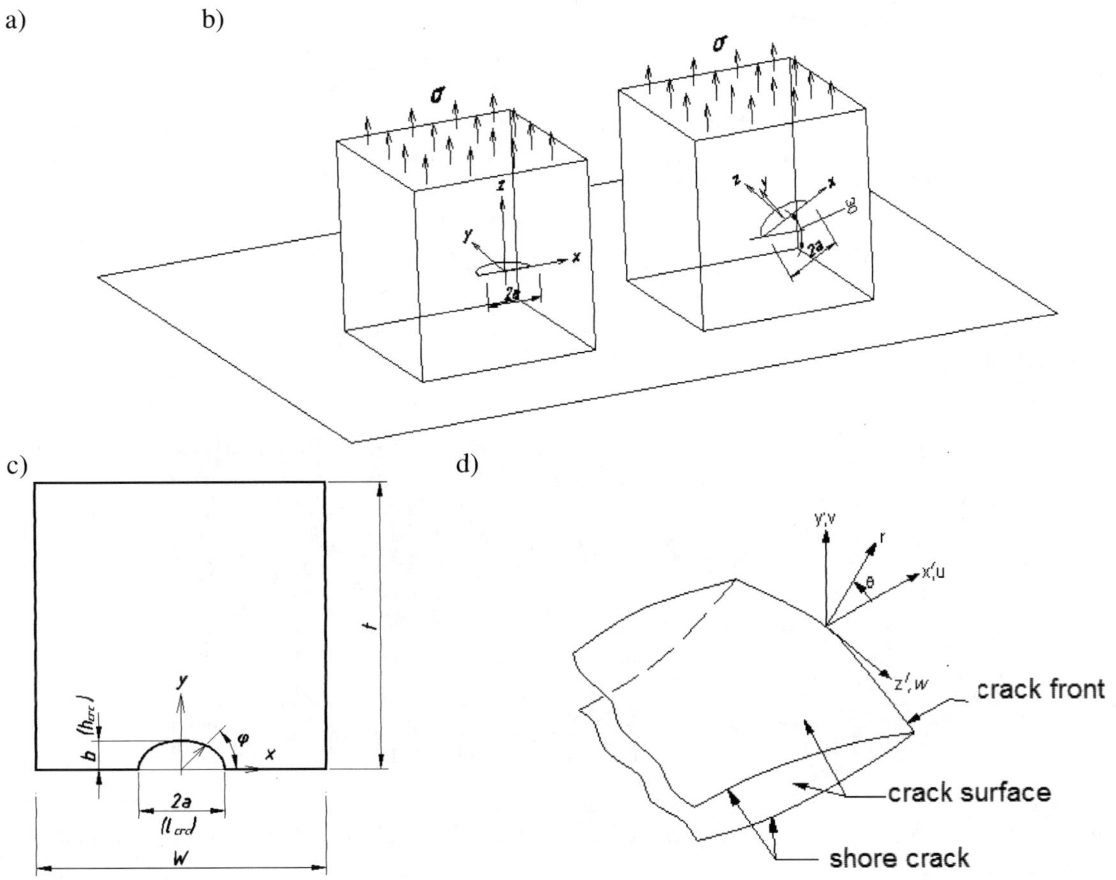

Fig.1. The design model of a bulk concrete element with a lateral crack: a) with a normal break; b) with an inclined crack with an inclination angle of 300; c) a local coordinate system at the center of the semi-elliptical crack; d) local cylindrical coordinate system at an arbitrary point of the crack front.

The research methodology is based on the same assumptions as for the wall panel previously considered in our studies, i.e. the linear mechanics of quasi-brittle fracture for concrete as a uniform elastic material remains valid.

To find the dependencies between the parameters of a non-through crack and determine the stress intensity factors, the finite element method (FEM) is used implemented in the Ansys software package.

Only macrocracks (comparable with the sample cross section) are considered. So their length is $l_{crc} = 2a = 0{,}3\text{м} - const$, burial $h_{crc} = b = 0{,}1...0{,}15\text{м}$ (a,b - ellipse axis). Note that the crack length and its depth (respectively, the major and minor semiaxes of the ellipse) are interconnected by the equation $\frac{x^2}{a^2} + \frac{y^2}{b^2} = 1$ или $\frac{4x^2}{l_{crc}^2} + \frac{y^2}{h_{crc}^2} = 1$, where x,y – coordinates of an arbitrary crack front point.

Moreover, one of the drawbacks of the proposed geometric model of the crack is the limited possible ratio $\frac{2a}{b} = \frac{l_{crc}}{h_{crc}} = 1{,}33...3{,}33$ (in work accepted $\frac{l_{crc}}{h_{crc}} = 1{,}5...3$). Therefore, the FEM of the Ansys complex is applicable only for prismatic or massive elements (columns, dams) and is poorly suitable for calculating bulk beams-walls and slabs. In addition, you must comply with the condition $\frac{2a}{W}<1$, where W - sample width.

The space in the frontal zone of the crack, where, as you know, a feature arises in the fields of displacements and stresses, is broken by special isoparametric 20-node end elements Solid186, which is proposed in the Ansys manual (Fig. 2).

As an example, the sample material is taken from heavy concrete B27.5 with characteristics

$$R_{bt,ser} = 1{,}02\,MPa,\quad E_b = 30200\ MPa,\quad K_{IC} = 0{,}405\ MPa\cdot\sqrt{м}$$

The modulus of elasticity of the material in the simulation does not vary, since the stress concentration zone (pseudoplastic deformations) is small and the nonlinearity of the modulus in it can be neglected, and in the rest of the section $E_b - const$.

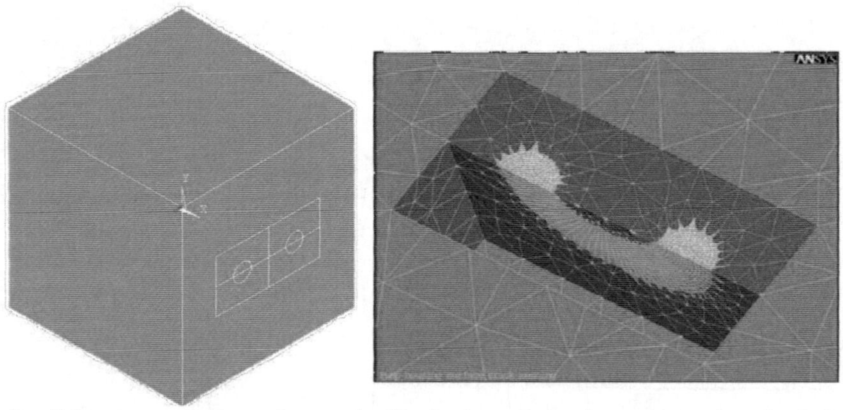

Fig.2. Stage of solid-state modeling of a semi-elliptical crack in the volume (on the left - splitting the concrete element into sections, on the right - a finite-element splitting of the volume around the crack front).

For the above two types of concrete elements (Fig. 1 a, b), the CIN of the normal separation of K_I, transverse K_{II} and longitudinal K_{III} shear (for normal crack only K_I) are numerically determined with a different polar angle of 00≤φ≤1800 (see Fig. 3 , 4a-c) and for different penetration of the crack $0{,}1 \le h_{crc} \le 0{,}15м$. For this, the same approximation algorithm is used as in the plane problem [1], only in a more general form for the formulas [2]:

$$K_I = \sqrt{2\pi}\frac{G}{1+k}\lim_{r\to 0}\frac{|\Delta v|}{\sqrt{r}},\qquad K_I = \sqrt{2\pi}\frac{G}{1+k}\lim_{r\to 0}(A_I + B_I r);$$
$$K_{II} = \sqrt{2\pi}\frac{G}{1+k}\lim_{r\to 0}\frac{|\Delta u|}{\sqrt{r}},\qquad K_{II} = \sqrt{2\pi}\frac{G}{1+k}\lim_{r\to 0}(A_{II} + B_{II} r); \tag{1}$$
$$K_{III} = \sqrt{2\pi}\frac{G}{1+k}\lim_{r\to 0}\frac{|\Delta w|}{\sqrt{r}},\qquad K_{III} = \sqrt{2\pi}\frac{G}{1+k}\lim_{r\to 0}(A_{III} + B_{III} r).$$

$$K_I = \sqrt{2\pi}\frac{G}{1+k}A_I;\quad K_{II} = \sqrt{2\pi}\frac{G}{1+k}A_{II};\quad K_{III} = \sqrt{2\pi}\frac{G}{1+k}A_{III} \tag{2}$$

where Δu, Δv, Δw - the displacement of one crack surface relative to another in the corresponding coordinate axes (Fig. 3d), moreover Δv - esponds to wedging crack shores, Δu - their relative transverse shear, and their antiplane longitudinal shear; A_I; A_{II}; A_{III}, B_I; B_{II}; B_{III} - constant coefficients for linear functions that approximate displacements at the nodes of finite elements on the crack faces; $G = \dfrac{E}{2(1+\mu)}$ - shear modulus; coefficient $k = 3 - 4\mu$ for plane-deformed state, μ – Poisson's ratio.

Note that the normal fracture K_I is determined by the existing analytical dependence [3]:

$$K_I = \frac{\sigma\sqrt{\pi b}}{E_k} F_s\left[\frac{b}{a}, \frac{b}{t}, \frac{a}{W}, \varphi\right] \tag{3}$$

where F_s - approximation function; σ – nominal tensile stresses; E_k – complete elliptic integral of the second kind; t – specimen thickness.

In fig. 3 shows a comparison of the numerical and analytical solutions of K_I in a sample of the first type at a voltage $\sigma = 1MPa$ and ratios $\dfrac{l_{crc}}{h_{crc}} = 2;2{,}4;2{,}73;3$. The discrepancy between the results does not exceed 5%, which is satisfactory.

According to Figure 3 we can draw the following conclusions:

- with a small depth ($b < a$) K_I is maximum, when φ = 900 - the crack with the load growth will develop due to the depth of the central part of the front
- with $b \approx \dfrac{1}{1{,}2}a$ $K_I\big|_{\varphi=0^0} = K_I\big|_{\varphi=90^0}$ - the crack can either be deepened or lengthened by moving its tips along the surface of the concrete;
- with $b \geq a$ K_I maximum when $\varphi = 0^0$ – the crack will lengthen;
- with $b \geq \dfrac{1}{1{,}2}a$ $K_I\big|_{\varphi=90^0, b\geq\frac{1}{1{,}2}a} \approx K_I\big|_{\varphi=90^0, b=a} - const.$

Obviously, the lower two curves of the graph, constructed by the formula (3), have more than one extremum due to the error of the approximation of the function F_s. At the same time, the experimental points on the FEM are described by a smooth curve with one extremum.

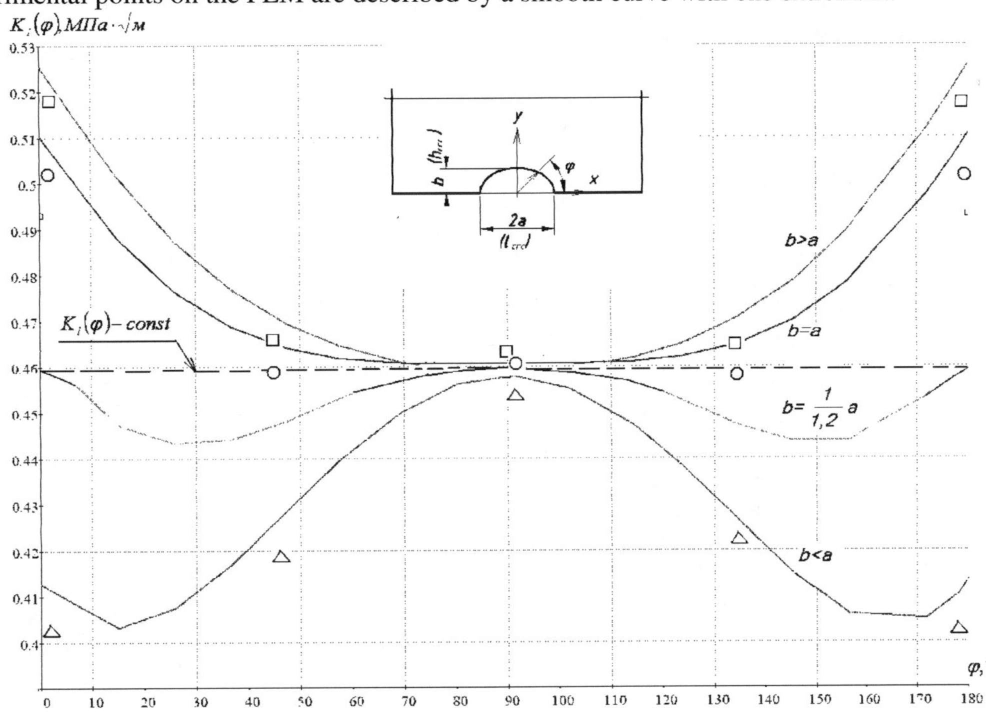

Fig.3. Graph comparing the analytical [3] and the numerical definition of KI (curved lines on the graph correspond to the formula (3));

Numerical solution with $\frac{2a}{b}=\frac{l_{crc}}{h_{crc}}=1{,}5$, ○ – same as with $\frac{2a}{b}=2$, Δ - // - with $\frac{2a}{b}=3$.

Since in the course of its development, a crack can only lengthen and deepen at the same time, $b\approx\frac{1}{1{,}2}a-const$ $(l_{crc}\approx 2{,}4h_{crc})$.Ratio options $b\neq\frac{1}{1{,}2}a$ possible only with an artificially created crack, or when there is some heterogeneity of the stiffness properties of the material along the front of the crack (due to inclusions of coarse aggregate or reinforcement).

Thus with $\sigma=1MPa$ and $b=1/1{,}2a$ $K_I\big|_{\varphi=90^0,b=a}=0{,}46\sigma\sqrt{м}$. Then critical stresses $\sigma_{cr}=\sigma\frac{K_{IC}}{K_I\big|_{\varphi=90^0,b=a}}=1{,}309MPa$ (ONS). The resulting opening width of a normal crack is equal to $a_{crc}=0{,}0423\,\sigma MPa$, mm hen the ultimate disclosure $a^*_{crc}=a_{crc}\sigma_{cr}=0{,}0554$ mm ies in the range of experimental values 0,001…0,06 mm [4].

The above numerical method is used in the work and for inclined cracks. For which, in addition to K_I (there is an analytical formula [3]), K_{II} abd K_{III} are numerically determined (Fig. 4a-b) —data for their calculation are not given in the literature. This stage of modeling and the procedure for finding K_{II} abd K_{III} constitute the scientific novelty of the work and its practical value.

In order to automate the construction of the topological scheme of the model, partition the region around the crack front into finite elements, calculate the CIN, a macro is written in the Ansys language APDL.

To determine the ratio $\frac{2a}{b}$, at which the crack will be lengthened and deepened simultaneously, all three coefficients are taken into account in terms of the intensity of release of the elastic deformation energy (Fig. 4d):

$$G=G_I+G_{II}+G_{III}=\frac{1-\nu^2}{E}\left(K_I^2+K_{II}^2\right)+\frac{1+\nu}{E}K_{III}^2 \quad (4)$$

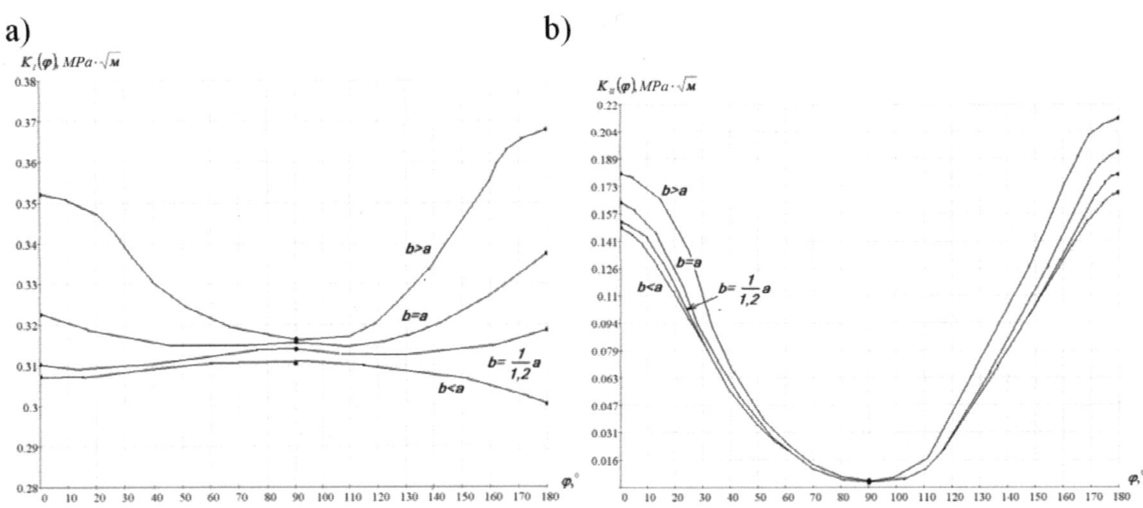

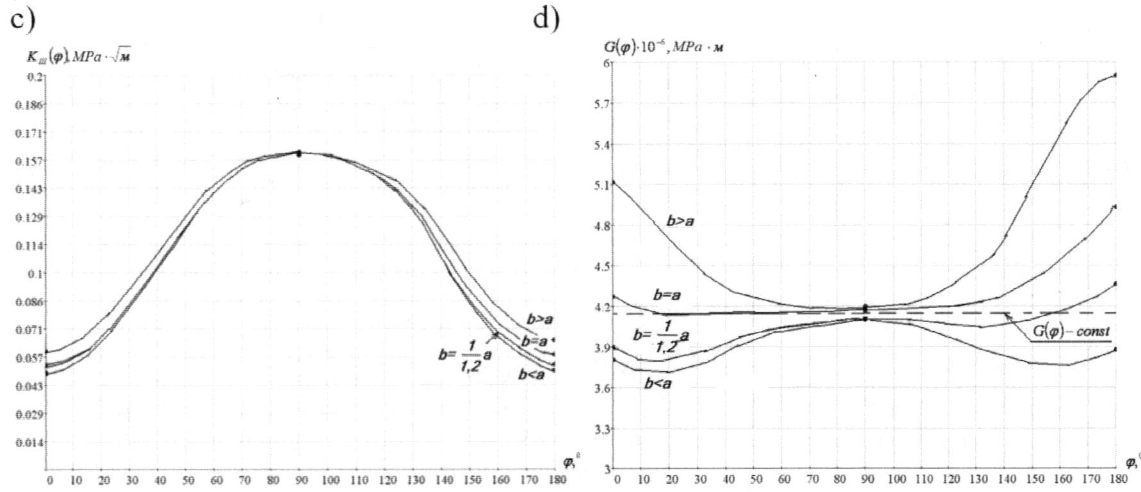

Fig.4. Parameters of fracture mechanics for inclined cracks.

Reasoning in the same way as with respect to a normal fracture, it has been established that the ratio $\frac{2a}{b}$ for an inclined crack is not constant: for the upper half of the crack $(0 \le \varphi \le 90^0)$ $b \approx \frac{1}{1{,}2}a - const$ $(l_{crc} \approx 2{,}4h_{crc})$, and for the lower $(90 < \varphi \le 180^0)$ - $b \approx a - const$ $(l_{crc} \approx 2h_{crc})$. In these arguments, instead of K_I, we consider the generalized characteristic G. For the simultaneous growth of a crack along the entire front, the condition is provided $G(\varphi) \approx G|_{\varphi=90^0, b=a} - const$ (dashed in Fig. 4d), which is possible only if the ratio is variable $\frac{2a}{b}$.

Thus, with $\sigma = 1 M\Pi a$ and $b = \frac{1}{1{,}2}a$ with $0 \le \varphi \le 90^0$, $b = a$ at $90 < \varphi \le 180^0$, the following values of CIN and energy release intensity are obtained (Table 1).

Table 1

Values of CIN and G depending on the polar angle φ

φ	K_I/σ, $\sqrt{м}$	K_{II}/σ, $\sqrt{м}$	K_{III}/σ, $\sqrt{м}$	G/σ, $м$
0^0	0,352	0,180	0,060	$4{,}2 \cdot 10^{-6}$
90^0	0,316	0,004	0,160	
180^0	0,353	0,182	0,061	

In Figures 4a-в -in and the data of Table 1, it can be seen that in the area of the tips of the crack front, a transverse shear mechanism makes a large contribution to the process of its propagation (K_{II} is maximum). On the contrary, in the middle of the crack front, K_{II} is minimal, and K_{III} is maximum, i.e. here the longitudinal shear mechanism (from the crack plane) plays an important role.

The intensity of the release of energy in the moment before the destruction of the wound $G_{cr} = G\frac{K_{IC}}{K_I|_{\varphi=0^0}} = 4{,}83 \cdot 10^{-6} MPa \cdot м$

Critical Stresses (ONS) $\sigma_{cr} = \sigma \frac{K_{IC}}{K_I|_{\varphi=0^0}} = 1{,}15\,MPa$, which is less than for a normal crack of the same length. The width of the inclined crack is equal to $a_{crc} = 0{,}0350\ \sigma MPa$ mm, тогда marginal disclosure $a^*_{crc} = a_{crc}\sigma_{cr} = 0{,}0403мм$, which is also less than that for a normal crack – $a^*_{crc} = 0{,}0554мм$.

hus, the following dependencies are obtained: $\frac{l_{crc}}{a_{crc}} = \frac{300}{0{,}0554} = 5415$ - normal crack $\frac{l_{crc}}{a_{crc}} = \frac{300}{0{,}0403} = 7444$ - inclined crack. These relationships are consistent with the experimental data of the work. [5]: $a_{crc} = (1600..10200)^{-1} l_{crc}$ and on average $a_{crc} = 3190^{-1} l_{crc}$.

From consideration of stress isopoles (Fig. 5), it follows that volumetric tension arises along the crack front. The prefracture zone along the front is similar in shape to a similar region for a planar problem [1], and its size is much smaller than the crack length and the sample size. This is the main condition for the applicability of linear fracture mechanics [6] for the problem under consideration.

а) б)

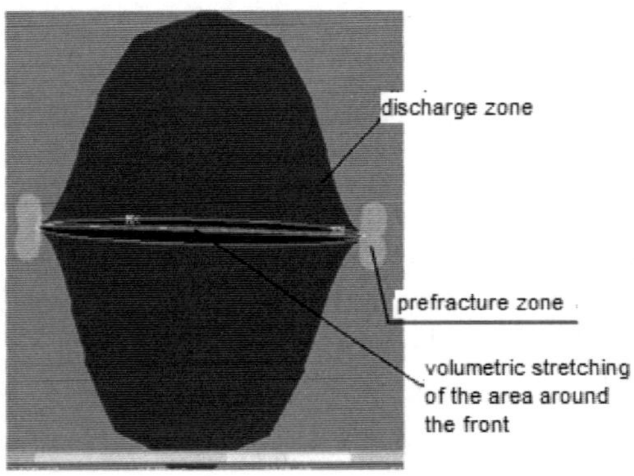

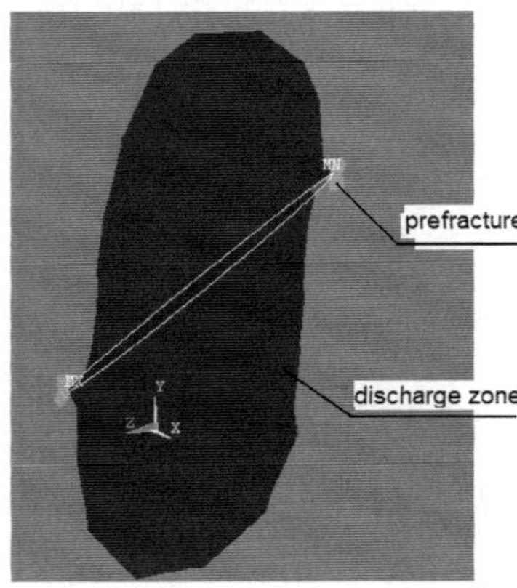

Fig.5. Isopole main tensile stresses (a - cracks of normal separation, b - for an inclined crack with a slope of 300).

Size of the pre-fracture zone at the top of a normal crack for heavy concrete B27,5 равен $d = \frac{K_{IC}^2}{3\pi(1{,}16R_{bt,ser})^2} = 12{,}4мм < l_{crc} = 300мм$, $d << W = 1000мм$. For an inclined crack, this zone in a sample of concrete of the same strength class is approximately 18% smaller than for a normal one and is determined by the formula [6] $d = \frac{G_{cr}E}{3\pi(1{,}16R_{bt,ser})^2} = 10{,}6мм$. Also note that the area of stress relief (Fig. 5) has the shape not of a circle, as is customary in the classical works of Griffith, but of a rhombus with a larger diagonal $2l_{crc}$.

Conclusions

The problem of stretching bulk concrete elements (in the form of prisms 1 × 1 × 1 m) with one surface semi-elliptic crack is considered. One group of elements contained cracks normal to tensile forces, and the other was inclined (the problem was modeled by the finite element method in the Ansys software). The stress intensity factors of three types were determined, for which a macro (subroutine) was written in the APDL language, the calculated coefficients of the 2nd and 3rd types were used for the first time in the considered problems. The relationship between the length, width of the opening, the deepening of cracks: $l_{crc} \approx 2{,}4h_{crc}$, $\frac{l_{crc}}{a_{crc}} = 5415$ ($a_{crc} = 0{,}0554$) - or normal crack; $l_{crc} \approx (2...2{,}4)h_{crc}$, $\frac{l_{crc}}{a_{crc}} = 7444$ ($a_{crc} = 0{,}0403$)-or inclined crack. It is established that the prefracture zone at the top of the inclined crack $d = 10{,}6мм$ 18% less than normal $d = 12{,}4мм$. t is shown that the stress relief area around the crack has a diamond shape.

The relationships obtained, the size of the prefracture zone and the width of crack opening are confirmed by experimental data [4, 5] and have a scientific novelty. Also, the scientific novelty is a procedure for the numerical determination of the coefficients K_{II}, K_{III} for an inclined crack.

Further development of the work should be the integration of the results obtained here and the diagram method for calculating reinforced concrete structures [7, 8].

References

[1] Radaykin O V 2006 *On the assessment of the residual bearing capacity of wall reinforced concrete wall panels by the methods of fracture mechanics* Materials of the 58th republic. scientific Conf., Kazan: KGASU pp 86-92

[2] Sedov L I 1970 *Continuum mechanics* M .: Science 586 p

[3] Murakami Y 1990 *Reference book on stress intensity factors* M: MIR Publishing House 566 p

[4] Pochtovik G Ya, Lipnik V G, Filonidov A M 1977 *Ultrasound concrete defectoscopy in energy construction* M: Energy 121 p

[5] Akhmetzyanov F Kh, Arsentev E Z, Mukhametzyanov I A 2001 *On the regularities of the development of macrocracks in the outer walls of monolithic claydite-concrete in residential buildings during operation* Abstracts of the reports of the 3rd Vseross. conference "New in architecture, design of building structures and reconstruction of NADNR-2001" pp 160-65

[6] Zaitsev Yu V 1991 *Fracture mechanics for builders* M: High School 288 p

[7] Radaikin O V, Sabitov L S, Kashapov N F, Gilmanshin I R 2018 *The way to improvement of the method of calculation of rigidity of bending reinforced concrete elements from conventional ferro-concrete* IOP conference series: materials science and engineering 412 **012058**

[8] Radaikin O V, Sabitov L S, Kashapov N F, Gilmanshin I R 2018 *Current state and prospects of development of diagrammatic methods for the calculation of reinforced concrete structures* IOP conference series: materials science and engineering 412 **012057**

[9] Strelkov Yu M, Radaykin O V, Sabitov L S, Kuznetsov I L 2019 *Comparative analysis of the static work of various types of steel poles of power lines based on computer modeling of the system" support - foundation - foundation soil* Construction mechanics and structural design No 1 (282) pp 71-79

[10] Izotov V S, Mukhametrakhimov R Kh, Sabitov L S 2010 *Experimental studies of the efficiency of disperse reinforcement of a stretched zone of concrete bended elements* Scientific Bulletin of the Voronezh State University of Architecture and Civil Engineering. Construction and architecture No 1 (17) pp 119-25

IOP Conf. Series: Materials Science and Engineering **570** (2019) 012076 doi:10.1088/1757-899X/570/1/012076

To estimate the rigidity of bending reinforced concrete elements based on dependence of length macro trays from load

O V Radaikin[1], L A Sharafutdinov[1], L S Sabitov[2], Yu M Strelkov[2], N F Kashapov[2]
[1] Kazan State University of Architecture and Civil Engineering
[2]Federal State Autonomous Educational Institution of Higher Education "Kazan (Volga Region) Federal University", Kazan, Kremlevskaya street, 18, 420008, Russian Federation

l.sabitov@bk.ru

Abstract. The purpose of this work is to make proposals for calculating the stiffness of normal sections of bending reinforced concrete elements, taking into account the work of the stretched concrete above the macrocrack and stress concentration at its top. In the article, such an account is made by the methods of fracture mechanics using the stress intensity factor K_I. One of the stages of calculation consists in the finite-element modeling of macrocracks in *ANSYS*.

Introduction

In the design of reinforced concrete elements they are allowed to work with macrocracks [1]. At the same time, for structures with a span of more than six meters, compliance with the deflection limitation increases the requirements for the reflection of the actual work of the material.

The issues of calculating the deformations of reinforced concrete bent elements, taking into account the length l of the stretched concrete over the macro-crack M [2], were considered by Ya.M. Nemirovsky. The latter circumstance, and in itself, the complexity of such an approach, as noted in [3], did not allow him at that time to find a wide practical application. In addition, later experiments showed a non-linear relationship between l and M [4].

The proposal of Ya.M.Nemirovsky is developed in our work in the following directions:

1) the use of a theoretically sound approach to fracture mechanics made it possible to obtain an analytical dependence of the crack length on the bending moment, which covers a wider range of parameter values in comparison with the data [2];
2) Our approach additionally takes into account the stress concentration at the tip of the macrocrack, that is, the nonlinear relationship between stresses and strains in the area under consideration.

In the present work, the following prerequisites are used:

a) the macrocracks arising in the stretched zone of concrete of the bent elements - through;
b) short-term static single loading is considered;
c) the beginning of the formation of the first macrocrack condition $M=M_{crc}$ [1];

IOP Conf. Series: Materials Science and Engineering **570** (2019) 012076 doi:10.1088/1757-899X/570/1/012076

d) concrete is a quasi-brittle material [5];
e) the macrocrack in the element is modeled by the Irwin-Ourov mathematical sharp section, and the stress-strain state at its top is determined by the stress intensity factor KI;
f) we consider concrete and reinforcement to be isotropic and elastically deformable materials - the initial connection between stresses and strains is described by Hooke's law. The non-linearity of the deformation of a reinforced concrete element is caused by the spread of a macrocrack in the stretched concrete;
g) deformation of short-term creep and shrinkage are taken into account through the given modulus of deformation $E_{b,red}$, determined according to [1].

Imagine a formula for determining the stiffness of reinforced concrete section in the following form:

$$B_f = E_{b,red} I_{red} \tag{1}$$

where $E_{b,red} = R_{bn} / \varepsilon_{b1,red}$ – reduced modulus of concrete deformation [1], I_{red} – the reduced moment of inertia of a reinforced concrete section relative to its center of gravity, determined taking into account the cross-sectional area of the concrete of a compressed and stretched zone and the cross-sectional area of the tensioned reinforcement with a reduction factor $\alpha_{s2} = E_s / E_{b,red}$:

$$I_{red} = I_b + \alpha_{s2} I_s \tag{2}$$

$$I_s = A_s (y - a)^2 \tag{3}$$

$$I_b = \frac{b(h-l)^3}{12} + b(h-l)\left[x_{ct} - \frac{(h-l)}{2}\right]^2 \tag{4}$$

$$x_{ct} = \frac{\frac{b}{2}(h-l)^2 + \alpha_{s2} A_s (h-a)}{b(h-l) + \alpha_{s2} A_s} \tag{5}$$

where l – crack length; x_{ct} – total concrete height of compressed and stretched zone ($x_{ct} = x_c + x_t$); I_b – moment of inertia of concrete; the remaining designations are taken as in the Manual [1].

Figure 1 shows a diagram of the distribution of stresses and forces in reinforced concrete elements after the appearance of cracks in them.

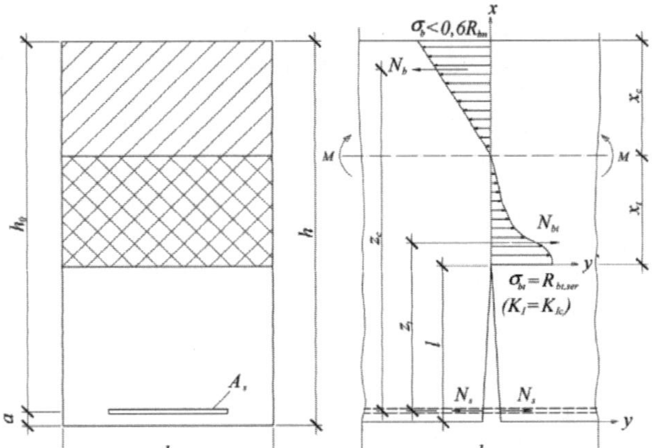

Figure 1. Scheme of the distribution of stresses and forces in reinforced concrete elements after the appearance of cracks in them

We write the equilibrium equations for the considered elements:

$$\sum M_{N_s}=0 \quad => \quad M+N_b z_c-N_{bt} z_t=0 \tag{6}$$

$$\sum Y=0 \quad => \quad N_b-N_{bt}+N_s=0 \tag{7}$$

where the resultant efforts $N_b=\frac{1}{2}\sigma_b x_c b$, $N_{bt}=b\int_{l}^{l+x_t}\sigma_{bt}dx$, $N_s=\sigma_s A_s$ and the corresponding distance between them $z_c=h_0-\frac{1}{3}x_c$, $z_t=l-a+\frac{\int_{l}^{l+x_t}\sigma_{bt}xdx}{\int_{l}^{l+x_t}\sigma_{bt}dx}$. Here x – the current coordinate of the point along the vertical axis with the origin located on the lower stretched face of the concrete.

According to clause "e", the calculated assumptions for stresses in tensile reinforcement should not exceed its yield strength.:

$$\sigma_s=\frac{M-N_{bt}(z_c-z_t)}{A_s z_c}\leq R_s \tag{8}$$

The stress pattern of the concrete of the compressed zone is assumed to be triangular, which is true when $\sigma_b<0{,}6R_{bn}$:

$$\sigma_b=\frac{M}{I_{red,l}}(x-l-x_t),\quad l+x_t\leq x\leq h \tag{9}$$

where $I_{red,l}$ – the reduced moment of inertia of the reinforced concrete section relative to the crack tip, taking into account the cross-sectional area of the concrete of the compressed and stretched zone and the cross-sectional area of the tensioned reinforcement.

The stresses in the concrete of the stretched zone consist of two solutions — singular (first term) and regular (second term):

$$\sigma_{bt}=\frac{K_I}{\sqrt{2\pi(x-l)}}+\frac{M}{I_{red,l}}(x-l),\quad l\leq x\leq l+x_t \tag{10}$$

where K_I – stress intensity factor (σ_{bt} and K_I are determined numerically in the PC *ANSYS*).

The start condition of the macrocrack is written in the form:

$$K_I=K_{Ic},\ \sigma_{bt}\big|_{x=l}=R_{bt,ser} \tag{11}$$

where K_{Ic} - critical stress intensity factor (determined from an experiment on standard samples [6]).

The joint solution of equations (6) - (11) allows determining the stress-deformed state of a bent element taking into account the work of the stretched concrete over the macrocrack and concentration at its top, including the dependence of the length of the macrocrack on the load.

As a result of our theoretical and experimental numerical studies, the following formula was proposed for it:

$$M(l)=M_0\frac{\Phi_I(l)}{\Phi_I(l_{ult})},\ 0\leq l<l_{ult} \tag{12}$$

where l, l_{ult}, – respectively, the current and maximum length of the crack in the section; $\Phi_I(l)$- calibration function, which depends on several factors - it can be determined numerically by the results of modeling the elastic work of an element with a crack under the action of a unit load (M_1=1MH×м) - the values of this function depending on the length of a macrocrack, the class of concrete strength, percent reinforcement and modulus of elasticity of reinforcement were tabulated and tabulated (to reduce the volume of the article we do not give here); M_0 is the maximum experimental value of the bending moment at which the growth of a transverse crack stops, as it rests against the compressed zone of concrete, and the stretched concrete is almost completely shut off from work. The value of this moment M_0 can be obtained by the experimental formula of A.S. Zalesov [7]:

$$M_0 = M_{crc} + \psi R_{bt,m} bh^2$$
$$\psi = 15 \frac{A_s}{bh_0} \frac{E_s}{E_b} \le 0{,}6 \qquad (13)$$

Expressing from formula (12) l, you can build the desired dependence:

$$l = f(M, E_b, \mu, M_0, l_{ult}) \qquad (14)$$

To do this, first calculate the l_{ult} by the formula, which we present here without output:

$$l_{ult} = h - h_0\left(1{,}5 - \sqrt{10\alpha_0 - 2{,}25}\right) \qquad (15)$$

where h_0 – working section height, $\alpha_0 = \frac{M_0}{R_b b h_0^2}$.

Then, using the above tables, I determine the value $\Phi_I(l_{ult})$ and by formula (12) calculate the value $\Phi(l) = \frac{M}{M_0}\Phi(l_{ult})$, according to which the value of the crack length is found from the same tables l.

After that, using the expressions (1) - (5), calculate the stiffness and then determine the deflection of the structure by the already known formulas [1].

To find the calibration function Φ_I, we carried out numerical studies of flexural reinforced concrete elements with a transverse crack. In the course of the experiment, 5 of the most significant factors varied - the plan of the experiment is shown in Figure 2. The simulation was carried out in ANSYS using the methodology [8].

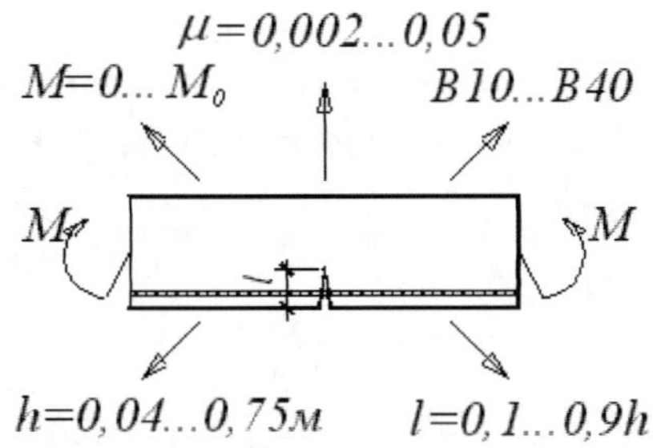

Figure 2. Diagram of numerical studies of reinforced concrete elements

According to the results of the studies carried out in Figure 3, the dependence (14) in dimensionless coordinates is constructed by a solid line for one item. On the same graph, the dots show the experimental data of K A Piradov [4] for a reinforced concrete beam of size $b \times h \times L$ 75×100×840mm from heavy concrete B20. Reinforcement percentage $\mu = 1{,}8\%$. The value of the protective layer of reinforcement a=25mm.

Other beams considered in [4] were also to be compared. As a result, the following discrepancies were found in the results obtained by the author, with the experiments of K A Piradov: no more 8% for μ=0,0089, 10% for μ=0,0180 and 13% for μ=0,0279.

IOP Conf. Series: Materials Science and Engineering **570** (2019) 012076 doi:10.1088/1757-899X/570/1/012076

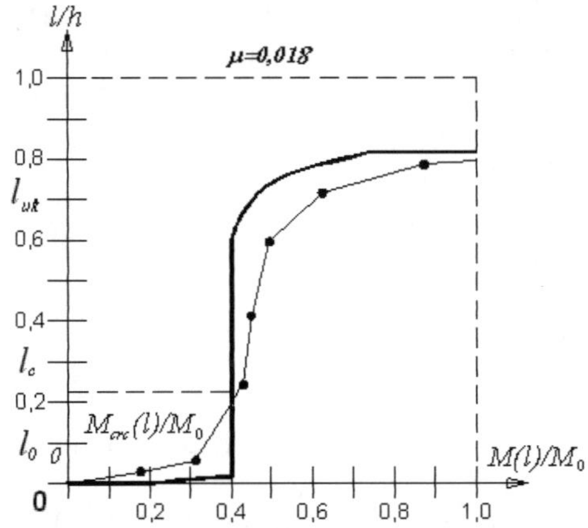

Figure 3. The crack length dependence on the load for a reinforced concrete beam with the percentage of reinforcement μ=0,018: mugs - according to the experiments of K A Piradov [4], the solid line - the results of the author

The volume of the article did not allow to reflect the influence of the initial discontinuities of the material, the prefracture zone at the top of the macrocrack and the coefficient of elasticity of concrete of different strength.

Our subsequent publications will be devoted to this. Further development of the work should be the integration of the results obtained here and the diagram method for calculating reinforced concrete structures [9, 10].

References

[1] *Manual on the design of concrete and reinforced concrete structures made of heavy concrete without prestressing the reinforcement* (to joint venture 52-102-2003) 2005 M: TSNIPromzdany 214 p

[2] M S Borishansky, Ya M Nemirovsky, Yu K Nikolaev, K V Petrova et al1 968 *Study of the stress-strain state of reinforced concrete elements taking into account the work of stretched concrete over cracks and revising, on this basis, the theory of calculating deformations and cracking* Strength and rigidity of reinforced concrete structures: Coll. scientific Art. M: NIIZHB 232 p

[3] A S Zalesov, V V Figarovsky 1976 *A practical method for calculating reinforced concrete structures by deformation* M: stroiizdat 104 p

[4] Piradov K A 1998 *Theoretical and experimental fundamentals of mechanics of concrete and reinforced concrete destruction* 355 p

[5] Akhmetzyanov F Kh 1993 *To features of deformation, damageability, changes in the physico-mechanical characteristics of concrete in structures* News of universities No 8 pp 150-55

[6] *GOST 29167-91 Concretes. Methods for determining the characteristics of crack resistance (fracture toughness) under static* 1992 M: Gosstroy of the USSR p 38

[7] A S Zalesov, E N Kodysh, P L Lemesh, I K Nikitin 1988 *Calculation of reinforced concrete structures for strength, crack resistance and deformations* M: stroiizdat 320 p

[8] E M Morozov, G P Nikishkov 1980 *The finite element method in fracture mechanics* M: Science 256 p

[9] Radaikin O V, Sabitov L S, Kashapov N F, Gilmanshin I R 2018 *The way to improvement of the method of calculation of rigidity of bending reinforced concrete elements from conventional ferro-concrete* IOP conference series: materials science and engineering 412 **012058**

IOP Conf. Series: Materials Science and Engineering **570** (2019) 012076 doi:10.1088/1757-899X/570/1/012076

[10] Radaikin O V, Sabitov L S, Kashapov N F, Gilmanshin I R 2018 *Current state and prospects of development of diagrammatic methods for the calculation of reinforced concrete structures* IOP conference series: materials science and engineering 412 **012057**

IOP Conf. Series: Materials Science and Engineering **570** (2019) 012077 doi:10.1088/1757-899X/570/1/012077

To determination of dependence of parameters of surface cracks in concrete and low-armed wall panels

O V Radaikin[2], L S Sabitov[1], I K Kiyamov[1], N F Kashapov[1]
[1]Federal State Autonomous Educational Institution of Higher Education "Kazan (Volga Region) Federal University", Kazan, Kremlevskaya street, 18, 420008, Russian Federation
[2] Kazan State University of Architecture and Civil Engineering

l.sabitov@bk.ru

Abstract. The dependences between the parameters of non-through normal cracks in the wall "perpendicular plates" panel method were established. The residual carrying capacity of the panel and the size of the pre-fracture zone at the crack tip were determined. The dependences obtained and the proposed method for calculating wall panels constitute a scientific novelty.

Currently in Russia there is an intensification of housing construction, due to the increase in consumer ability of the population in the country in recent years. However, the demand for housing is so great that the cost of its square meter is steadily increasing.

Therefore, there is a problem of reducing operating costs for the repair and maintenance of buildings, as well as the problem of increasing the service life of an already existing fund of houses and a newly erected one.

To solve these problems, it is necessary to develop optimal methods for repairing building elements (fixing cracks in concrete, reinforcing reinforced concrete structures). These techniques require knowledge of the parameters of cracks: their length, width of opening and penetration.

Earlier in the article [1] it was confirmed the presence of a certain dependence between the above values. The authors of [2] emphasize the need to establish relations between the parameters of the crack.

Thus, the purpose of this work is to establish the relationship between the length l_{crc}, width of the disclosure a_{crc} and the depth of the crack h_{crc} in the concrete wall panel.

The panel under consideration is a single-span single-layer supporting wall-wall (aspect ratio $2 \leq L/H < 4$) with constructive reinforcement, which is required for the loads to be perceived only during transportation and installation. Experiments of many authors have established a weak effect of constructive reinforcement on the deformability of reinforced concrete elements

To achieve this goal, the following tasks are solved:

- formed the design scheme of a wall panel with a crack (Fig. 1);
- the stress intensity factor K_I (SIF) is calculated using the known formulas [7];
- according to the method proposed in this paper, the critical length of a crack is determined when it passes from non-through to full;

- an estimate of the residual bearing capacity (RBC) of the panel is given;
- establishes relationships between parameters of a normal normal crack;
- The obtained results are compared with the data of field studies.

The problem of determining the length of a crack during its transition from non-through to through is solved by the method of mutually perpendicular plates (MPP) proposed by the authors, the essence of which is set forth below.

Assumptions research methods:

a) material panel - concrete, outside the pre-fracture zone (which occurs at the crack tip) is elastic isotropic and is characterized by the elastic modulus E_b and Poisson's ratio $\nu = 0.2$, constructive reinforcement is taken into account, if necessary, through the reduced modulus of elasticity;

b) concrete is a quasi-brittle material [3], the characteristic linear size of the zone of pseudo-plasticity (microcracking or pre-fracture) at the top of the crack does not exceed 20% of its length, as shown by studies [4];

c) the size of the prefracture zone at the crack tip is estimated by Irving's plasticity correction, while the stresses of $1{,}16R_{bt,ser}$ and ser are not exceeding the standard concrete tensile strength [4];

d) the macrocrack in the panel is initially specified and its nucleation due to the merger of multiple germinal microcracks is not considered, the concrete with the crack is deformed elastically;

e) динамические эффекты и возможность ветвления макротрещины тоже не учитываются;

f) the length of the crack l_{crc} and the size of the cross section of the panel exceed the characteristic size of the individual elements of the concrete structure l_{ch} ,, it is considered uniform and the scale factor does not affect the determination of the stress intensity factors [5];

g) Only a single short-term vertical loading of the panel (beam-wall) with a uniformly distributed static load q is considered, and the complex-stress state results in a uniform averaging of the main tensile stresses σ_{1m} with a magnitude that is perpendicular to the load q.

The latter assumption is based on the fact that cracks in concrete mainly develop from the action of tensile stresses, since the tensile strength of a material is 8-10 times lower than that of compression and 2-3 times lower than that of shear [6]. In addition , during technical inspections of large-panel houses for their physical deterioration, a multitude of wall panels with vertical rectilinear cracks are revealed, which are obviously caused by stresses σ_{1m}. In the future, this situation will need to be clarified and reviewed s non-uniform stress-strain state arising under the action of vertical force of its own weight, and horizontal wind pressure.

Description of the method of mutually perpendicular plates:

Wall panel dimensions $B \times H \times t$ (Fig.1) with superficial macrocrack length l_{crc} and deepening h_{crc} is represented as two mutually perpendicular plates.

One plate of thickness t is aligned with the front surface of the panel on which the crack lies. This plate simulates the surface layer (flat-stressed state - FSS). Volumes K_I and a_{crc} for this plate with a slit length l_{crc} calculated by Isis formula [7]:

$$K_I^{st}(x) = \sigma_1 \sqrt{\pi x} F_I^{st}(x),\ F_I^{st}(x) = \sqrt{\sec\left(\frac{\pi x}{2B}\right)} \quad \textbf{(1)}$$

$$a_{crc}(x) = \frac{2\sigma_1 x}{E_b t} V_I(x),$$

$$V_I(x) = 1 + 0{,}065\left(\frac{2x}{H}\right) - 0{,}241\left(\frac{2x}{H}\right)^2 + 3{,}76\left(\frac{2x}{H}\right)^3 - 6{,}63\left(\frac{2x}{H}\right)^4 + 4{,}93\left(\frac{2x}{H}\right)^5 \quad \textbf{(2)}$$

With $H >> l_{crc}$ $F_I(l_{crc}) \to 1$ formula (1) degenerates into the Griffith formula for an infinite plane with a cut: $K_I(x) = \sigma_1 \sqrt{\pi x}$. It is valid, for example, for monolithic load-bearing walls erected on the entire height of the building.

nother plate unit thickness simulates the inner layers of the body (flat-deformed state - FDS) and is located perpendicular to the first. In the plate under consideration, K_I is determined already by the Gross formula [7], as for a plate with a lateral notch length h_{crc} :

$$K_I^{gr}(x) = \frac{\sigma_1}{1-\nu^2}\sqrt{\pi x}F_I^{gr}(x),$$

$$F_I^{gr}(x) = 1{,}12 - 0{,}231\left(\frac{x}{t}\right) + 10{,}55\left(\frac{x}{t}\right)^2 - 21{,}72\left(\frac{x}{t}\right)^3 + 30{,}39\left(\frac{x}{t}\right)^4 \qquad \textbf{(3)}$$

Here x – the length of the cut in the plate for the case of flat stress state (FSS) or plate cut length at flat deformed state (FDS) , σ_1 - main tensile stresses.

Note that a given surface crack will become continuous if the plate from the inner layer collapses, that is, when the length of its notch reaches half the thickness of the panel $h_{crc} = 0{,}5t$. This statement can be substantiated as follows: we divide the section of the plate into which the crack deepens into N areas. Let tensile strength be known for each site $R_{bt,i}$ and the law of its distribution is normal. Then, in order for this plate to collapse, it is necessary that the true tensile stresses (including the decreasing cross-sectional area) reach the average concrete strength of the sites: R_{btm} . In this case, exactly 50% of all elementary platforms [8] will be shut down from work - this is equivalent to the cut reaching half the height of the plate section.

It is obvious that in the process of development of a given crack, its length increases simultaneously l_{crc} and deepening h_{crc} . The condition for this isotropic material is equality:

$$K_I^{st}(x) = K_I^{gr}(x) \qquad \textbf{(4)}$$

And from the equation $K_I^{st}(x) = K_I^{gr}(t/2)$ can determine the critical crack length $l_{cr} = 2x$ at the moment when it goes from the non-through to the through (x is the root of the equation).

The value of x can also be determined, as shown in the example below, from the graphs of functions (1) and (3) presented in Figure 2. Where is the desired ratio l_{crc}/h_{crc} or $l_{cr}/0{,}5t$ easily calculated. Knowing this and defining a_{crc} by formula (2), ratios are established l_{crc}/a_{crc} , h_{crc}/a_{crc} .

To estimate the error and preference value of the proposed method, you can use the data [7]. So the error in calculating the SIF by the formula (1) is ± 0.3 with $l_{crc}/H \leq 0$ and ± 1.0 with $l_{crc}/H = 0{,}8$. And according to the formula (3), respectively, ± 0.3 at $h_{crc}/H \leq 0{,}6$. For the opening width of formula (2), the error is ± 1.0 with $l_{crc}/H \leq 0{,}7$.

Below we consider a numerical example of determining the desired dependencies in a concrete wall panel (Fig. 1).

a) б)

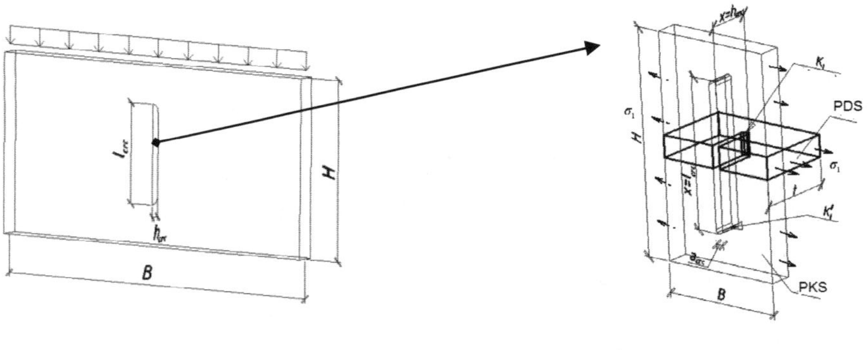

Fig.1. Simulation of a wall panel with a crack by the method of MPP: a - loading scheme; b - scheme of replacing mutually perpendicular plates.

Given: concrete single-layer blank wall panel (constructive reinforcement is not taken into account) with dimensions $B \times H \times t = 6000 \times 3000 \times 300$ mm has a vertical crack of length l_{crc} ($0 < l_{crc} \leq H$) and depth h_{crc} ($0 < h_{crc} \leq \frac{t}{2}$), directed perpendicular to the main tensile uniform stresses $\sigma_1 = \sigma_{1m} = 0{,}1 MПa$ (occur in the transverse direction relative to the vertical compression load). The panel material is heavy concrete class B27.5 with the following indicators of strength and crack resistance [9]: $R_{bt,ser} = 1{,}02\, MPa$, $E_b = 30200\ MPa$, $K_{IC} = 0{,}405\ MPa \cdot \sqrt{м}$ standard tensile strength, elastic modulus and fracture toughness coefficient.

Find: Dependencies l_{crc}/h_{crc}, l_{crc}/a_{crc}, h_{crc}/a_{crc}.

Solution: the design scheme of the panel according to the runway method is shown in Figure 1б.

1. Definition l_{crc}/h_{crc}

Value K_I for two types of plates is determined by the formulas (1), (3) or according to the graphs of fig. 2.

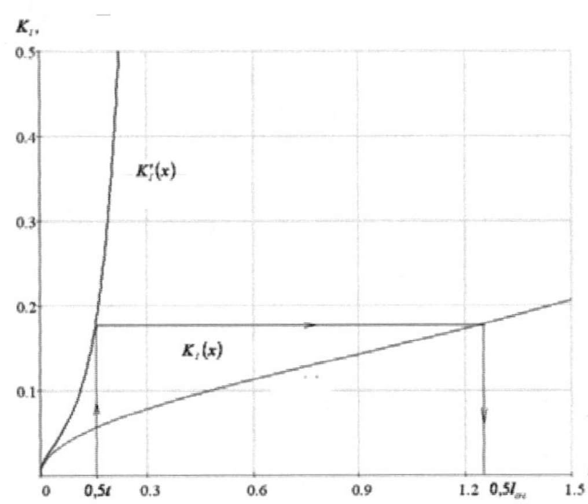

Fig.2. Graphic determination of the dependence of the half-length of a crack on its depth by the stress intensity factors.

According to fig. 2 when buried $h_{crc} = x_1 = 0{,}5t$ determine the critical crack length $l_{crc} = l_{cr} = 2x_2 = 2{,}186м$. Then $\frac{l_{crc}}{h_{crc}} = \frac{2{,}186}{0{,}15} = 14{,}573$. Or taking into account the error of calculations $\pm 1{,}5\%$ will have $\frac{l_{crc}}{h_{crc}} = 14{,}573 \pm 0{,}218$. Note that with smaller depths $h_{crc} < 0{,}5t$ ratio $\frac{l_{crc}}{h_{crc}} \to 1$.

From calculations l_{cr} for panels with different boundary conditions it has been established that the last relation does not depend on voltages σ_1 and ultimately on the material fracture toughness K_{IC}, however, when solving equation (4), it is necessary to observe the condition $K_I^{Гр}\left(\frac{t}{2}\right) \leq K_{IC}$, otherwise, the crack will certainly be through. In our case

$$K_I^{gr}\left(\frac{t}{2}\right) = 0{,}162 \pm 0{,}001\ MPa \cdot \sqrt{м} < K_{IC} = 0{,}405\ MPa \cdot \sqrt{м}.$$

Calculations showed that with $H >> l_{crc}\ l_{cr} \to 2{,}6м$, a $\frac{l_{crc}}{h_{crc}} \to 17{,}336 \pm 0{,}260$ (the length of the critical crack is comparable to a floor height of 3.0 m), and this ratio remains constant regardless of stresses only σ_1, but also on the wall thickness t.

2.Finding l_{crc}/a_{crc} and h_{crc}/a_{crc}

As can be seen from the formulas of Isis (2), the width of the crack opening depends on the stress state and material properties. Therefore, in the limit state, besides the fact that $x = l_{cr} = 2{,}186м$, we take the main voltage equal to the critical

$$\sigma_{cr} = \sigma_1 \frac{K_{IC}}{K_I^{Гр}\left(\frac{t}{2}\right)} = 0{,}1\frac{0{,}405}{0{,}162} = 0{,}25MPa$$ (based on proportion $\frac{\sigma_{cr}}{\sigma_1} \Rightarrow \frac{K_{IC}}{K_I^{gr}\left(\frac{t}{2}\right)}$).

Otherwise voltage $\sigma_{cr} = 0{,}25 \pm 0{,}004\, MPa < R_{bt,ser} = 1{,}02\, MPa$ called the residual strength or residual carrying capacity (RCC). If before it was said that when $l_{crc} > l_{cr}$ the crack will be through, now the condition $\sigma_1 > \sigma_{cr}$ ($\sigma_1 > \sigma_{cr} \cdot (K_I^{gr}\left(\frac{t}{2}\right) = K_I^{st}\left(\frac{l_{cr}}{2}\right) > K_{IC}$) means that the panel in question can be completely destroyed, divided into parts, from the trunk crack. This will happen if the crack accumulates enough elastic energy for its spontaneous growth, otherwise its development will stop (for example: when the crack falls into a compressed force flow, or rests a tip in a hard inclusion - reinforcement or coarse aggregate). In our example

$$\sigma_1 = 0{,}1MPa < \sigma_{cr} = 0{,}25 \pm 0{,}004\ MPa.$$

So, according to (2) we get $a_{crc}(l_{cr}) = \frac{2\sigma_{cr} l_{cr}}{E_b t} V_I(l_{cr}) = 1{,}424 \pm 0{,}014мм$. Thena $\frac{l_{crc}}{a_{crc}} = \frac{2186}{1{,}424} = 1535$, $\frac{h_{crc}}{a_{crc}} = \frac{150}{1{,}424} = 105$. The error is ±2,5: $\frac{l_{crc}}{a_{crc}} = 1535 \pm 38$, $\frac{h_{crc}}{a_{crc}} = 105 \pm 3$. Note that at voltages $\sigma_1 = 0{,}1МПа$ $\frac{l_{crc}}{a_{crc}} = 1535\frac{\sigma_{cr}}{\sigma_1} = 3070 \pm 92$, $\frac{h_{crc}}{a_{crc}} = 105\frac{\sigma_{cr}}{\sigma_1} = 210 \pm 6$. It should also be noted that with increasing stresses σ_1 you can't say that l_{crc} or h_{crc} will decrease because included in the resulting relations a_{crc} there is a non-decreasing function from σ_1.

3. Finally, determine the size of the pre-fracture zone at the crack tip [4]:

$$d = \frac{K_{IC}^2}{3\pi(1{,}16R_{bt,ser})^2} = \frac{0{,}405^2}{3\pi(1{,}16 \cdot 1{,}02)^2} = 12{,}4мм.$$

Consclusionc:

- the penetration of the blind macrocrack in the wall panel is $h_{crc} = (14{,}573...17{,}336)^{-1} l_{crc} \le \frac{t}{2}$ (depending on the height of the panel or the wall section under consideration);
- if the length of the crack $l_{crc} > l_{cr} = (7{,}287...8{,}668)t$, then it will be through,
- if the opening width $a_{crc} > \left(1535\frac{\sigma_{cr}}{\sigma_1}\right)^{-1} l_{crc}$ (depending on the properties of the material), spontaneous growth of a macrocrack is possible, as a result of which a panel or wall section will be destroyed;
- the depth of the cracks in concrete can be determined by the known width of the disclosure of the ratio $h_{crc} = 105\frac{\sigma_{cr}}{\sigma_1} a_{crc}$ (depending on the material properties).
- the size of the pre-fracture zone at the crack tip is within $4мм < d = 12{,}4мм < 75мм$ (lower limit for cement stone, upper for heavy concrete [5]). $d = 12{,}4мм << l_{cr} = 2186мм$, which confirms the above

accepted assumptions about the applicability of linear mechanics of quasi-brittle fracture in the problem under consideration.

The results of the considered example are compared with field observations [1, 10, 11], in which the interval estimate of the desired relations is given. So, in [1] for heavy concrete $\frac{h_{crc}}{a_{crc}} = 100...300$, which corresponds to the above value $\frac{h_{crc}}{a_{crc}} = 105...210$ ($h_{crc} = 105\frac{\sigma_{cr}}{\sigma_1}a_{crc}$).For a three-layer wall panel with an outer layer of heavy concrete [10] $a_{crc} = (1600...10200)^{-1} l_{crc}$ and in average $a_{crc} = 3190^{-1} l_{crc}$, in this paper received - $a_{crc} = (1535...3070)^{-1} l_{crc}$.

Thus, the established ratios of the crack parameters, the size of the pre-failure zone are confirmed by experimental data [1, 5, 10, 11] and, together with the proposed method of mutually perpendicular plates, have scientific novelty.

The further development of the work should be the integration of the results obtained here and the diagram method for calculating reinforced concrete structures [12, 13].

References

[1] Akhmetzyanov F Kh, Arsentev E Z 1999 *On the ratio of the penetration of cracks in concrete to the width of their disclosure on the surface of concrete and reinforced concrete structures* New in Architecture, Design of Building Structures and Reconstruction NACCPR-99 pp 26-28

[2] Orekhov V G, Zercalov M G 1999 *Mechanics of destruction of engineering structures and mountain ranges* M .: ed. DIA 330 p

[3] Akhmetzyanov F Kh 1993 *To the features of deformation, damageability, changes in the physicomechanical characteristics of concrete in structures* Izv. Universities, Building No 8 pp 150-55

[4] Akhmetzyanov F Kh, Radaykin O V 2007 *Towards an assessment of the pre-failure zone of main cracks in damaged reinforced concrete wall panels* Izvestia KGASU pp 46-48

[5] Zaitsev Yu V 1991 *Mechanics of destruction for builders* M .: High School 288 p

[6] V M Bondarenko, P O Bakirov, V G Nazarenko, V I Rimshin 2007 *Reinforced Concrete and Stone Constructions: A Textbook for Building Special Higher Education Institutions* M: Higher School 887 p.

[7] Y Murakami 1990 *A Handbook on Stress Intensity Coefficients* M .: MIR Publishing House 566 p

[8] Radaykin O V 2005 *The effect of the asymmetry of the distribution of the strength of the material on the crack resistance* Materials of the 56th republ. scientific Conf. pp 7-9

[9] Piradov K A 1998 *Theoretical and experimental fundamentals of the mechanics of concrete and reinforced concrete destruction* Tbilisi: ed. "Energy" 355 p

[10] Akhmetzyanov F Kh, Arsentev E Z, Mukhametzyanov I A 2001 *On the regularities of development of macrocracks in the outer walls of monolithic claydite-concrete in residential buildings during operation* New in Architecture, Design of Building Structures and Reconstruction of NASCRR-2001" pp 160-65

[11] Akhmetzyanov F Kh, Arsentev E Z, Radaykin O V, Lizunova N S 2008 *Parameters of cracks in the exterior wall panels of residential buildings with a duration of operation up to 1 year and during the construction of monolithic floors* Abstracts 3 All-Russian Conference "New in Architecture, Design of Building Structures and Reconstruction NACCPR-2007" pp 23-26

[12] Radaikin O V, Sabitov L S, Kashapov N F, Gilmanshin I R 2018 *The way to improvement of the method of calculation of rigidity of bending reinforced concrete elements from conventional ferro-concrete* IOP conference series: materials science and engineering 412 **012058**

IOP Conf. Series: Materials Science and Engineering **570** (2019) 012077 doi:10.1088/1757-899X/570/1/012077

[13] Radaikin O V, Sabitov L S, Kashapov N F, Gilmanshin I R 2018 *Current state and prospects of development of diagrammatic methods for the calculation of reinforced concrete structures* IOP conference series: materials science and engineering 412 **012057**

IOP Conf. Series: Materials Science and Engineering **570** (2019) 012078 doi:10.1088/1757-899X/570/1/012078

Selection of thetoothdepthof the ball end mill based on the analysis of the working conditions of the tool

E A Ryabov, R M Khisamutdinov,O I Yurasova S Y Yurasov

Department of design and technological ensuring of engineering industries, NaberezhnyeChelny Institute of Kazan Federal University, Mira Ave. 423412 13A, NaberezhnyeChelny, Russian Federation.

Evgeniy.ryabov@inbox.ru, docfile@yandex.ru.

Abstract. The working conditions of end radius cutters are analyzed. The length of the chips during processing and the required depth of the tooth are found.

When designing tools with a chip groove the question of its geometric parametersis always acute. So for high-speed steel cutters, the following requirements are imposed on the chip grooves: sufficient chip coverage to avoid jamming; the presence of a sufficient radius of the bottom of the chip groove to eliminate cracks during thermal hardening; the groove must be helical for good chip removal and milling uniformity; the height and shape of the tooth must provide sufficient resistance [1].

For end cylindrical cutters, the shape of the groove is sufficiently studied and conditionally constant in different end sections along the length of its cutting part [2]. For the radius section of the end radius cutters, the choice of tooth depth and shape of the chip groove is still an urgent task. Nowadays, the radius section is obtained by sharpening the end cylindrical cutters using the most simple for universal equipment trajectory of the grinding wheel.

Let us analyze the purpose of tooth depth for various tools that perform finishing work. For the small number of teeth of the end cylindrical milling cutters the recommended depth of the tooth is 0.3 from the district step. The work of broaching which removes a small layer of chipscan be regarded as finishing work too. For this work the tooth depth is selected according to the following conditions $h<0.2*D$ [3]. For the radius section, such values are not applicable due to too large tooth depth values, which will lead to a strong weakening of the design of the cutting part on the spherical section.

Let us consider the geometric conditions of the end radius cutters. End radius cutter is designed for cleaning and finishing of curved surfaces. Preliminary semi-finishing is usually performed equidistant to the final profile of the surface layer by layer or line by line with the formation of scallops. The presence of scallops has little effect on the thickness of the cut layer [4],[5],[6],[7], so we will consider the perfectly equidistant surface as a preliminary one.

The end mill processes surfaces of variable curvature for the largest amount of time, but in order to estimate the approximate length of the chip we shall consider surfaces with constant curvature (Fig. 1). It shows a scheme of layer-by-layer treatment of concave, straight and convex surfaces. In this case, the cut layer is limited form the top by the previous passage of the cutter.

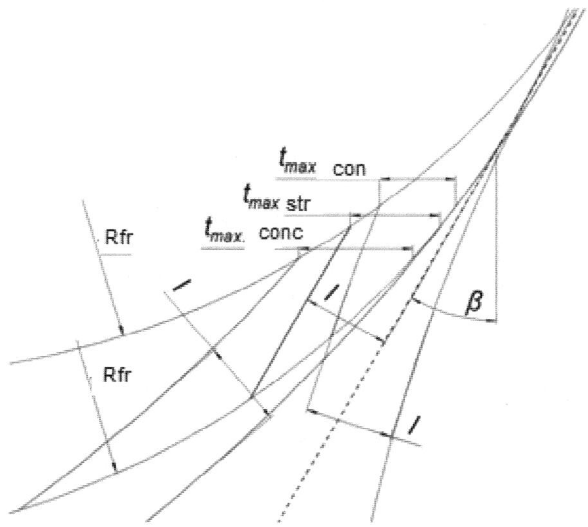

Figure 1. The scheme of treatment of surfaces with constant curvature

It can seen from the figure that at the same value of equidistant l the maximum cutting depth of tmax is achieved when processing a concave surface. Thus, the smaller the radius of curvature of the concave surface, the greater the value of the maximum depth of cut for the considered point of the cutting edge. The minimum radius of curvature of the concave curve is limited by the radius of the tool.

Let us consider theoretically most unfavorable in terms of depth of cut for all points of the cutting edge case (Fig. 2).

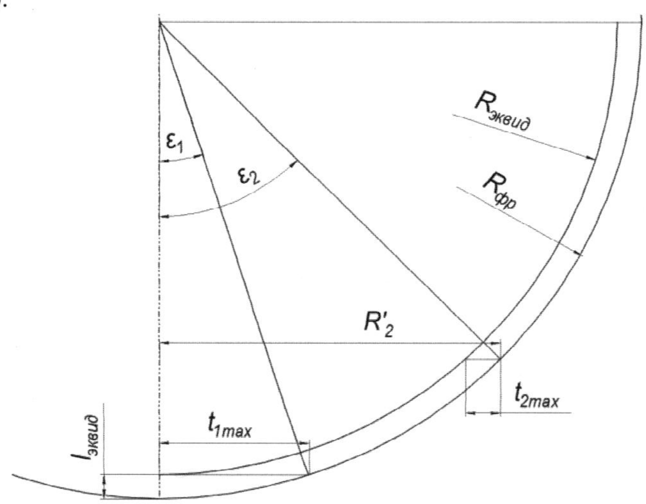

Figure 2.Processing scheme with a maximum depth of cut

In this case, it is a hypothetical surface treatment whose radius of curvature is equal to the radius of the tool, and the treated surface is equidistant by the value of the finishing work allowance. In this case, tmax will be for the case when the cutter operates in a solid material with a full diameter of t1max and a more characteristic case when the depth of cut is variable over timax sections, which will be used as an assessment of the chip placement in the chip groove.

$$t_{1\max} = R'_1 = \sqrt{R_{фр}{}^2 - (R_{фр} - l_{эквид})^2} = \sqrt{2R_{фр} \cdot l_{эквид} - l_{эквид}^2}$$

IOP Conf. Series: Materials Science and Engineering **570** (2019) 012078 doi:10.1088/1757-899X/570/1/012078

$$t_{i\max} = R_i - \sqrt{R_i^2 - 2 \cdot R_{\phi p} \cdot l_{\text{эквид}} + l_{\text{эквид}}^2}$$

In this case the chip length of the AB arc can be found from figure 3 without taking into account the shrinkage and the trochoidal component of the tooth trajectory. At the same time for two adjacent teeth, it is conditionally possible to consider the rectilinear motion of the equidistant plane.

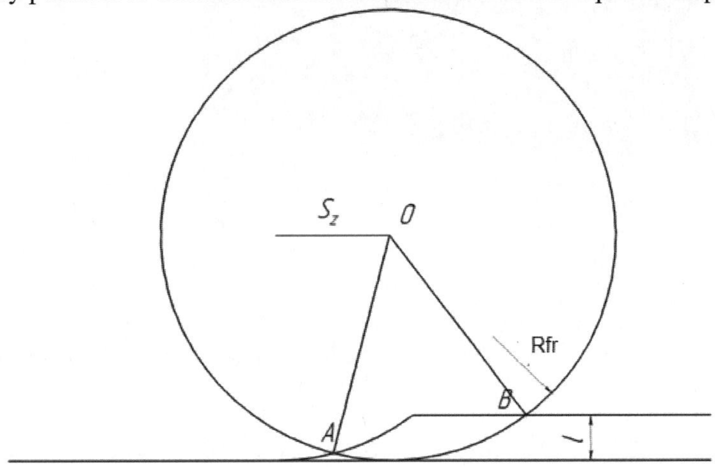

Figure 3.Diagram of the processing with two adjacent teeth

$$AB = (\arcsin(\frac{S_z}{2Ri_{\phi p}}) + \arccos(\frac{Ri_{\phi p} - l}{Ri_{\phi p}})) \cdot Ri_{\phi p}$$

Assuming the condition that the chip descent is close to the normal to the cutting edge [8], the tooth depth will depend on the length of the front surface in the normal section. Taking into account that the depth of the tooth of cylindrical cutters is assigned as 0.45-0.5 of the circumferential pitch[9], it can be concluded that a sufficient depth of the chip groove, and hence the depth of the tooth will be 0.5 of the AB arc.

Thus, the expected height for each diameter along the spherical section in the normal section will be as follows:

$$h_z = (\arcsin(\frac{S_z}{2Ri_{\phi p}}) + \arccos(\frac{Ri_{\phi p} - t_{i\max}}{Ri_{\phi p}})) \cdot \frac{Ri_{\phi p}}{2}$$

However, even with such a depth setting in reality due to the features of the kinematic scheme of sharpening the tool, the depth of the teeth on small diameters will increase (Fig. 4).

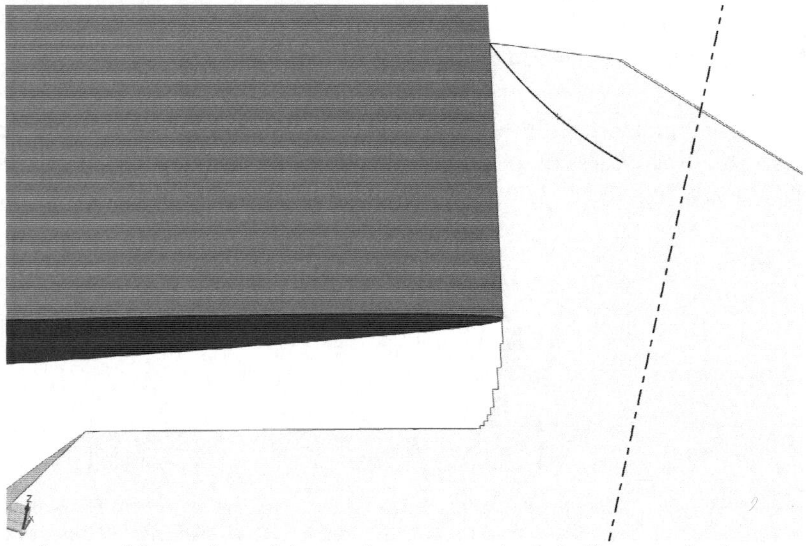

Figure 4.The scheme of formation of the tooth cutter on the i-th diameter in the normal section

References

[1] I I Semenchenko, V M Matyushin, G N Sakharov 1963 *Proektirovanie metallorezhushchih instrumentov* [Design of metal-cutting tools] (Moscow: State scientific and technical publishing house of machine-building literature) 951 p

[2] E A Ryabov, R M Khisamutdinov, S Yu Yurasov 2018 *Direction of chip flow for milling of shaped surfaces by ball end mill* IOP Conference Series: Materials Science and Engineering 412 **012060**

[3] E A Ryabov, S Yu Yurasov 2016 *Creating a rigid model of solid end radius cutter in NX* IOP Conference Series: Materials Science and Engineering 134 **012034**

[4] Safarov D T, Kondrashov A G, Glinina G F, Safarova L R 2017 *Algorithm of calculation of energy consumption on the basis of different model of the production task performed on machines with computer numeric control (CNC)* IOP Conference Series: Materials Science and Engineering. 240 **012060**

[5] E A Ryabov, S Yu Yurasov, O I Yurasova 2016 *Parametric modeling of ball end mills* Russian Engineering Research. Vol 36 No 9 pp 784-85

[6] Liu X, Gao H, Yue C, Li R, Jiang N, Yang L 2018 *Investigation of the milling stability based on modified variable cutting force coefficients* International Journal of Advanced Manufacturing Technology, pp 1-12

[7] Korovin E M, Lunyov A N, Khasanov R Sh 2006 *Optimization of cutting parameters for automatic machines* Russian Aeronautics No 49(4) pp 100-05

[8] Lunev A N, Moiseeva LT 1994 *Calculation of geometry of stamps with small allowances for pieces of the aerodynamic profile* Izvestiya Vysshikh Uchebnykh Zavedenij. Aviatsionnaya Tekhnika No 1 pp 99-102

[9] Turanov A V, Lunev A N, Moiseev V S, Moiseeva L T 2004 *Optimization of trajectory choice for machining GTE blade surfaces* Izvestiya Vysshikh Uchebnykh Zavedenij. Aviatsionnaya Tekhnika No 2 pp 67-72

IOP Conf. Series: Materials Science and Engineering **570** (2019) 012079 doi:10.1088/1757-899X/570/1/012079

Management of elastic displacements of a milling technological system based on the study of the main deformation axes

A R Sabirov, R M Khusainov

Naberezhnye Chelny Institute of Kazan (Volga region) Federal University, Naberezhnye Chelny, Mira 68/19, 423800, the Russian Federation

rmh@inbox.ru

Abstract. The article describes some issues of optimization of technological system parameters and processing parameters to ensure compliance of cutting forces and main deformation axes. As a result there appears a decrease in elastic displacements and an increase in dynamic stability.

1. Introduction
One of the most important factors that affects the efficiency of machine tools is rigidity. The rigidity affects both the machining accuracy [1] and performance, since machining performance is highly dependent on a vibration resistance of the machine and in turn, it is mainly determined by rigidity. In most studies concerning this topic, the focus is on issues of its own and contact rigidity of elements of the technological system. Meanwhile, the system arrangement also influences the rigidity of the entire system; it determines how elements characterized by their own and contact rigidity will perceive loads. This topic is not sufficiently developed in research, but at the same time relevant, because there is a significant reserve for ensuring the accuracy and dynamic stability of machines [2].

2. Optimization of the technological system arrangement by rigidity

2.1. Possibilities of problem solving on optimization of a technological system considering axes of rigidity
Rigidity axes or as they are called in the theory of elasticity axes of principal deformations can be taken as geometrical characteristics, which help to estimate the rigidity of a technological system [3]. The axis of maximum rigidity is a direction in the geometric space of the technological system, where the static elastic deformation is minimal. According to it, the axis of minimum rigidity is a direction in which static elastic deformations are maximal [4].

In terms of computer-integrated "digital" production, finite element modeling (FE modeling) of a technological system can be used to determine axes, which also includes the machine carrier system, cutting and auxiliary tools, the device and the workpiece. The results of finite element modeling are processed by analytical dependencies developed in the theory of elasticity. According to these dependencies with the help of FE modeling the strain tensor of an elementary elastic element (in this case conditional cutting rigidity) can calculate values of principal strains and direction cosines to determine principal axes of deformation [5].

You can solve the following problems, using the results of this calculation:

1. To build the optimal arrangement of the technological system, i.e. to select the relative position of the tool and the workpiece which can be achieved by using the design of the tool or turning axes in five coordinate machines to ensure the maximum rigidity axis approaches resultant cutting forces. This reduces static elastic deformations and increases a dynamic stability. Thus, here is solved the constructive task – the change in the position of axes of rigidity.

2. To change the position of resultant cutting forces. Knowing the arrangement of rigidity axes relative to reference axes of the machine, one can direct the resultant cutting force along the axis of maximum rigidity, then find projections of this vector on the machine coordinate axis. You can implement this method by changing tool parameters and cutting conditions. This method is easier to use in production conditions.

2.2. The solution of the technological problem

Consider the second task on the example of removing the allowance by end mill.

The components of cutting forces along the axes of coordinates depend on technological parameters; in particular, in our case the following analytical expressions are available for components *Fx, Fy* и *Fz* [6]:

$$F_x = P_{cutting} sin\omega;\ F_y = P_{cutting} cos\omega;\ F_z = P_{cutting} sin\omega_0 \quad (1),$$

where $P_{cutting}$ is the resultant of cutting forces, ω_0 is the ascent angle of a helical line of the end mill,

$$\omega = \frac{\pi}{4} + \rho' - \beta \quad (2),$$

$$\text{where } \rho' = \frac{tg(\beta-\gamma)}{tg(\beta-\gamma)+2} \quad (3),$$

$$\text{where } \beta = \frac{\pi}{2} - arctg\frac{1{,}08\sqrt[4]{\left(\frac{\eta E_1 b}{a}\right)}+cos\gamma-sin\gamma}{cos\gamma+sin\gamma} \quad (4),$$

$$\text{where } \eta = \frac{1-\mu_1^2}{E_1} + \frac{1-\mu_2^2}{E_2} \quad (5),$$

E_1, E_2 are elastic moduli of the processed and tool materials; μ_1, μ_2 are Poisson's ratios; γ is the average rake angle of the tool with the rounding of the edge; a, b are the depth and width of the cutting.

Each of projections can be expressed through parameters of a cutting condition or a tool [7, 8].

To solve this problem it is not necessary to determine the exact value of the cutting force. It is enough to take it equal to one. The main thing in this case is to determine relators between projections of the resultant cutting force on coordinate axes, i.e. to specify components of the vector ζ along which the resultant should pass. In our case it turned out that the minimum deformation is ε_2, therefore, components of the vector ζ take the form:

$$\xi 1 = 0;\quad \xi 2 = 0;\quad \xi 3 = 0 \quad (6).$$

Let us determine values *Fx, Fy* и *Fz*:

$$\begin{aligned} F_x &= \xi 1 \cdot \cos(l_1) + \xi 2 \cdot \cos(l_2) + \xi 3 \cdot \cos(l_3) \\ F_y &= \xi 1 \cdot \cos(m_1) + \xi 2 \cdot \cos(m_2) + \xi 3 \cdot \cos(m_3) \\ F_z &= \xi 1 \cdot \cos(n_1) + \xi 1 \cdot \cos(n_2) + \xi 1 \cdot \cos(n_3) \end{aligned} \quad (7).$$

To solve the second problem, it is most convenient to take as an adjustable value the angle of flute helix of the milling cutter ω0, the depth *a* and the width *b* of the cutting.

To determine ω_0 and β, we use expressions for *Fy* and *Fz* from equation (1) and found values of these components from equations (7):

$$cos\left(\frac{\pi}{4}+\frac{tg(\beta-\gamma)}{tg(\beta-\gamma)+2}-\beta\right)=F_x$$
$$sin\omega_0=F_z \quad (8).$$

Solving this equation by numerical methods, for example, using the Levenberg-Marquardt algorithm, we obtain values ω_0 and β. The latter is used to find the ratio between the quantities *a* and *b* from equation (4):

$$\frac{\pi}{2}-arctg\frac{1{,}08\sqrt[4]{\left(\frac{\eta E_1 b}{a}\right)}+cos\gamma-sin\gamma}{cos\gamma+sin\gamma}=\beta \quad (9).$$

This equation is solved by the same method.

3. Conclusions

Thus, we can propose the following method of controlling elastic displacements while cutting through the appointment of technological processing parameters:

1. Finite element modeling of the technological system is the elastic system of the machine, fixture, tool, and parts with the calculation of deformations.

2. Determination of the direction of axes rigidity [5].

3. Setting the direction of the unit vector of the resultant cutting force along the axis of maximum rigidity and the definition of projections of this vector on the axis of the machine (7).

5. Formation of analytical dependencies of cutting force projections through parameters of the cutting condition or the tool (1) – (5).

6. By solving these equations, we obtain necessary technological parameters (in this example, ω_0, *a* and *b*), which provide the necessary direction of cutting force (8), (9).

In this case a maximum approximation of the resultant cutting force to the axis of maximum rigidity is ensured, which causes a decrease in elastic displacements and an increase in the dynamic stability of the technological system.

References

[1] Balabanov I P, Simonova L A and Balabanova O N 2015 Systematization of Accuracy Indices Variance when Modelling the Forming External Cylindrical Turning Process *IOP Conference Series: Materials Science and Engineering* (Bristol: IOP Publishing Ltd) **vol. 86** p 012010

[2] Kasjanov S V, Kondrashov AG and Safarov DT 2017 Regulation of Geometrical Parameters Deviations of Automotive Components Parts through Diagnostic Measurements Organization *Procedia Engineering* (Elsevier) **vol 206** pp 1508-14

[3] Kolesnikov K S, Aleksandrov D A, Astashev V K, etc.. 1994 Dynamics and Strength of Machines. Theory of Mechanisms and Machines. (Moscow: Mechanical Engineering) 534 p

[4] Timoshenko S P, Goodier J N. 1975 *Theory of Elasticity.* (Moscow: Science) 576 p

[5] Sabirov A R, Khusainov R M 2017 Calculation of the Direction of Deformation Axes in a Vertical Milling Machine Zone. Materials of the International Scientific and Technical Conference "Innovative Engineering Technology Equipment and Tools – 2017" (ITM-2017) 2 (Kazan : Foliant) p 228-32

[6] Grechishnikov V A, Petukhov Yu E, Pivkin P M, Romanov V B, Ryabov E A, Yurasov S Yu and Yurasova O I 2017 Trochoidal Slot Milling *Russian Engineering Research.* (New York: Allerton Press) **Vol 37** No **9 pp** 821-3

[7] Grubyj S V, Zaitsev A M 2013The Study of End Mills while Milling Aluminum-Alloy Body Parts. Science and Education No 12 P 31-54

[8] Golovko A N, Kondrashov A G and Yurasov S Yu 2017 Improved Design of a Worm Type Instrument for Final Machining of Evolvent Gear Teeth *Procedia Engineering* (Elsevier) **vol 206** pp 1333-6

IOP Conf. Series: Materials Science and Engineering **570** (2019) 012080 doi:10.1088/1757-899X/570/1/012080

Improving the efficiency of energy output from landfill gas by controlling the morphological composition of MSW

R G Sabirzyanov[1], N A Nurmatova[1], A I Galeeva[3], I R Gilmanshin[1,2] and S I Gilmanshina[1]
[1]Kazan Federal University, Kazan, Russia, 420008
[2] Kazan National Research Technical University named after A N Tupolev, Kazan, Russia
[3]Kazan State Power Engineering University, Kazan, 420066, Russian Federation

Is-er@yandex.ru

Abstract. The change of the elemental composition and increase of the amount of waste, in particular MSW, is urgent environmental and socio-economical problem for Russia today. The main part of MSW is transported to landfills and utilized by inefficient or environmentally unsuitable disposal methods.

The compromise between economic development and environmental conservation is one of the most discussed problems of modern economics and sustainable development of our society.The morphological composition of MSW changes annually due to the growth and changes in the needs of human and industrial waste, and varies depending on the locality.

Mixed MSW are waste that is hazardous to human health, as they are characterized as a multicomponent system, whose components have a complex chemical composition.The presence in the MSW of organic fractions (food, plant, etc.), as well as the content of hazardous type 1 and type 2 materials ("hazardous waste") in them, cause the possibility of the appearance of various diseases due to the ingress of toxic components into the soil and groundwater. [3]

There are 3835 thousand tons of MSW formed annually in Moscow and the Moscow region. The morphological composition of the MSW of the Moscow region and the average morphological composition of the MSW in the world are presented in Fig. 1 and 2. [5]

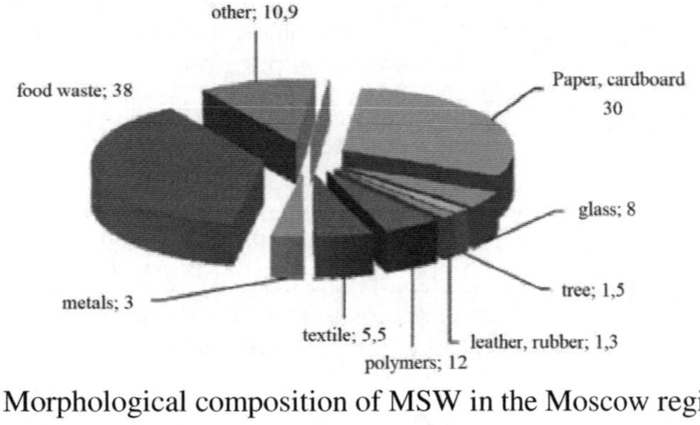

Figure 1. Morphological composition of MSW in the Moscow region,% [2]

IOP Conf. Series: Materials Science and Engineering **570** (2019) 012080 doi:10.1088/1757-899X/570/1/012080

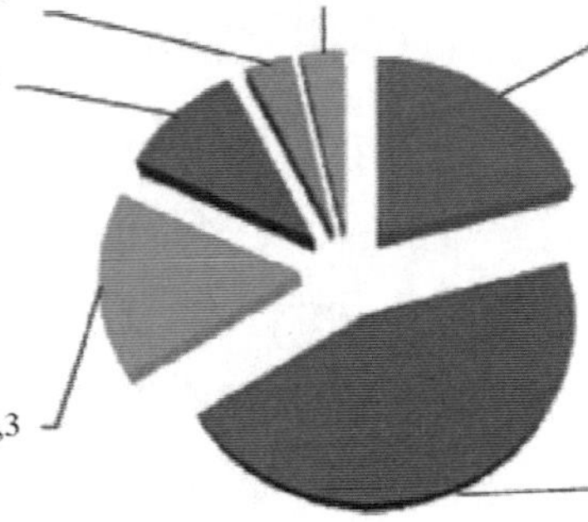

Figure 2. The average morphological composition of MSW in the world,% [2]

Table 1.Factors affecting the component composition of MSW.[2]

Factor	Influence
Temporary	
Stage	The structure of the consumption of goods is subject to constant changes (the emergence of new types of products, packaging, etc.)
Season	In the autumn, as a rule, the content of food waste increases, the specific formation of waste per person in different seasons is also uneven.
Dayoftheweek	The composition of waste on weekends and holidays may differ significantly from that on weekdays.
Territorial	
Climate	The climatic conditions of the area determine the structure of food grown on-site and, accordingly, their consumption.
Transportaccessibility	The development of a transportation access with other territories is reflected on the used goods.
Socio-economic	
Wastesource	Waste of the population is significantly different in composition from waste infrastructure
Populationstructure	Waste from a population with different income levels and employment, have different component composition
Degreeofhouseholdimprovement	Availability of furnace heating, garbage chute, number of storeys affect waste composition
Development of the market of secondary raw materials	The developed market of secondary raw materials helps to reduce the content of useful components in the waste.
Sanitary	
Wastecollectionsystem	The presence or absence of separate collection directly affects the composition of the MSW
Frequencyofexport	Waste composition changes during the week. With prolonged storage of waste, it is possible that they become wet, bloat, and the like.
Prospects for the development of waste management systems	Existence of a waste management strategy and concept implying a change in the waste management system, such as a waste collection system

IOP Conf. Series: Materials Science and Engineering **570** (2019) 012080 doi:10.1088/1757-899X/570/1/012080

MSW includes cardboard, plain paper, leather and rubber, stones, glass and metal, textiles, food waste, wood, polymers and etc. Plastic waste does not decompose for a long time and that's why it is a particular hazard. Rotting communal garbage is the maincomponent for the production of pathogens. Chemicals from the MSW are released into the atmosphere cause allergic, toxic, endocrine diseases. [3, 9, 11]

The number of landfills and unauthorized landfills in Russia is growing every year, occupying more and more areas, creating a bacteriological threat by polluting groundwater, creating a threat to air traffic, attracting large flocks of birds. Also, the formation of landfill gas in large quantities can lead to self-ignition of waste, and harmful compounds (SO2, etc.), which are part of it, poison the environment. [7, 8, 10]

The composition of MSW includes hazardous and toxic substances that threaten human life and health, and since the most part of the waste does not process the sorting, the recycling process must occur through special methods of thermal destruction.

Recycling and disposal of waste is one of the most pressing and requiring special attention problems not only in our country, but throughout the world [8]. This mainly applies to large and rapidly developing densely populated cities, where a huge amount of municipal solid waste (MSW) accumulates. Each year, 140 million cubic meters of MSW accumulate in our country, of which only 3% is recycled, which is unacceptable. The rest of the waste is transported to special landfills for disposal, or to landfills. A distinctive feature of the MSW is that they contain a high percentage of the combustible component contained in such components as rubber, paper, slag, wood, etc. In this regard, people began to think about using waste as an energy source. This idea is the solution to the two most common problems of our time: energy generation and waste disposal. Of course, burning MSW has not only advantages, but also significant negative consequences, difficulties in implementation. The main methods of thermal destruction of waste are incineration, pyrolysis and gasification.

Humidity of waste directly affects the amount of electricity consumed during the process of utilization by thermal methods; therefore, it is necessary to pre-dry the waste to reduce losses.

Table 2. Harmful substances generated during the combustion of MSW. [12]

Substance	mg / m
Flyash, dust	5000
NOx	210
SO2	280
CO	50
HCl	-
Cd,Tl	0.5
Hg	0.06
Pb, Co, Cr, Ni, As, Sn, Bi	60
Dioxins	$2 * 10^{-6}$

Table 3.Comparison of the main thermal methods of disposal of MSW.[4]

	Burning 650 °C	Pyrolysis 450-900 °C	Plasmagasification 2000 °C
Power destruction	70%	90%	99%
The presence of tar and furans	Lot	thereis	Not
Ashandtoxic	30% toxic	10% ash	Not

IOP Conf. Series: Materials Science and Engineering **570** (2019) 012080 doi:10.1088/1757-899X/570/1/012080

resins	resins		
Notsubjectto recycling	Part inorganic waste	Part inorganic waste	-
Sorting waste	Isrequired	Isrequired	Notrequired
Wastevolume	Big	Small	Big
Emissions fluegas	High	Average	Verylow
Sensitivity tohumidity waste	Sensitive	Sensitive	Notsensitive
Quality gasoutlet	Generatorgas (technical)	Ballastedsynthesisgas	High quality received synthesis gas
Output	Heat, electricity	Synthesis gas, liquid fuels electricity, heat	Synthesis gas, liquid fuels electricity, heat

One of the main ways to utilize MSW in the world is to bury it at the close distance area. Under these conditions, waste is subjected to intensive biochemical decomposition with the formation of landfill gas (biogas) [12], which leads to environmental problems. The main components of biogas include not only greenhouse gases (methane and carbon dioxide), but also such toxic compounds as carbon monoxide, nitrogen oxides, hydrogen sulfide, sulfur dioxide. In the process of thermal effects and ignition of waste, carcinogenic compounds benzene and benzapyrene are released. The emission of landfill gases entering the environment has negative effects of both local and global geoecological nature.

Biogas is formed as a result of the anaerobic decomposition of the organic fraction of waste and is 98% composed of methane and carbon dioxide [12]. Of the total amount of methane entering the atmosphere annually, 40-70% is formed as a result of anthropogenic activity, and more than 20% of them are accounted for MSW disposal facilities. It is estimated that about 200 m^3 of biogas is produced from one ton of MSW. At the same time, for the first 15–20 years, when one ton of MSW is decomposed, up to 7.5 m^3 of biogas is released per year. In the future, the intensity of biogas emission is sharply reduced. The dependence of the specific biogas yield for the period of its active stabilized generation is described by the formula:

$$Q = \frac{1.85 G_0 (1 - 10^{kt})}{\left(\frac{59 - W}{13}\right)^4}$$

Organic fractions of MSW are characterized by different periods of decomposition and intensity of methanogenesis. Under these conditions, the qualitative composition of landfill gas acquires a variable, stochastic nature. Existing technologies for utilization of landfill gas are quite sensitive to the stability of the quantitative and qualitative characteristics of landfill gas as fuel. Therefore, experimental methods for determining the specific heat of landfill gas are needed. The work on determining the characteristics of landfill gas as a hydrocarbon fuel of variable composition is carried out jointly by Professor V. Larionov. Experimental determination of the specific heat of combustion of gaseous hydrocarbon fuel is carried out using a diffusion combustion tube of the fuel-air mixture in an air stream.

IOP Conf. Series: Materials Science and Engineering **570** (2019) 012080 doi:10.1088/1757-899X/570/1/012080

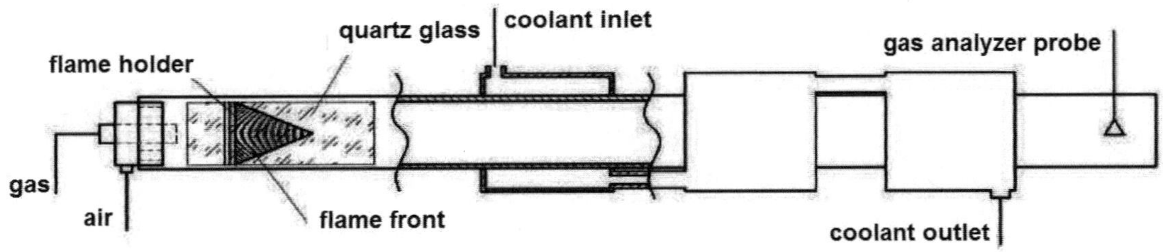

Figure 3. Diffusion combustion tube of the fuel-air mixture in an air stream

The proposed method will allow to determine the dependence of the energy potential of landfill gas on the morphological composition of MSW

Literature

[1] Zilenina VG, Ulanova O V 2017 *The study of the composition of hazardous solid municipal waste (for example hit)* Healthy environment - the basis of regional security. Materials of the first international environmental forum in Ryazan pp 43-49

[2] VI Vorobev, GP Timofeev, VVCharnetsky 2018*Morphological composition of municipal solid waste* Science of the Young is the Future of Russia collection of scientific articles of the 3rd International Scientific Conference of Advanced Developments of Young Scientists: in 6 t. pp 162-68

[3] OD Dnipro, NAKirichenko, D A Solov'ev 2018*Garbage as a socio-philosophical problem* Days of Science Materials of the interuniversity scientific and technical conference of students and cadets on the basis of Kaliningrad State Technical University pp 218-22

[4] R GSabirzyanov, I R Gilmanshin, NF Kashapov, SI Gilmanshina, DR Kraynova2018 *Thermal destruction of production and consumption waste* Proceedings of the IX International Scientific and Technical Conference "Innovative machine-building technologies, equipment and materials - 2018" (ISTC "IMTOM-2018") pp 242-46

[5] Popov V M2012 *Disposal of production and consumption waste* Kursk

[6] IR Gilmanshin, NF Kashapov, SI Gilmanshina 2016*Energy disposal of landfill gas as a way to form a new model of waste management*Problems and Prospects of Innovative Development of the Economy: Materials of the Scientific Forum (XXI International Scientific and Practical Conference) 264 p

[7] I R Gilmanshin, NFKashapov, S I Gilmanshina, AIGaleeva, L S Sabitov 2017*Study of projects for the recultivation of landfills for production and consumption waste in terms of the organization of landfill gas energy utilization* Proceedings of the VIII International Scientific and Technical Conference "Innovative machine-building technologies, equipment and materials - 2017" (MNTK "IMTOM-2017"). Part 1 pp. 220-224

[8] NFKashapov, M MNafikov, IR Gilmanshin, AR Nigmatzyanov, VYaPetrova 2017*Recycling of agricultural production and rural settlements with the production of biogas and organic matter* Innovative machine-building technologies, equipment and materials - 2017 Materials of the VIII International Scientific and Technical Conferencepp 255-59

[9] IR Gilshmanshin, N F Kashapov, AI Galeeva 2016*Energy utilization of landfill gas from landfill gas as a way to form a new model of waste management* Problems and Prospects of Innovative Development of the Economy: Proceedings of the Intern. scientific and practical conf. pp 238-42

[10] Manelis G B 2001*Energy and Ecology, Institute of Problems of Chemical Physics* RAS. Chernogolovka

[11] PetrovV Vand others 2012*Ensuring the functioning of the urban system of environmental monitoring of waste management data production and consumption in the city of Taganrog* Engineering Bulletin of the Don V 23 No 4-2

[12] Bespalov VI, Adamyan RG 2013*Assessment of the conditions of the landfills disposal for the solid waste utilization*Journal "European Applied Sciences V 2 pp 190-94

IOP Conf. Series: Materials Science and Engineering **570** (2019) 012081 doi:10.1088/1757-899X/570/1/012081

Intelligent management of supply chains and outsourcing in mechanical engineering under "Industry 4.0" conditions

R A Sabitov [1], G S Smirnova[1], N U Elizarova[1] and Sh R Sabitov[2]
[1]Kazan National Research Technical University named after A.N. Tupolev – KAI, Kazan, Russia

[2]Kazan Federal University, Kazan, Russia

r.a.sabitov@mail.ru

Abstract. Mechanical engineering industry has recently been seriously affected by achievements and progress of artificial intelligence and by extremely rapid evolution of Industry 4.0. Considering that the market will rapidly change in the coming decades, the only way to preserve and strengthen market positions for enterprises, is the transformation of production processes within the framework of new technological trends and integrated networked cluster ecosystems.

1. Introduction

Technological progress has always influenced business, for example, due to productivity boost because of replacing human labor with machines. Continuous progress in the field of information technologies and related areas should dramatically change both the essence of labor activities and the way of our life in general. The potential power of combination of artificial intelligence technologies and "Industry 4.0" leads to global consequences and challenges that need to be solved fairly quickly and should be overcomed in order to preserve and develop the business. The evident decision is to make the necessary changes in the project management methodologies.

Nowadays an active implementation of intelligent integrated technologies happens in modern mechanical engineering. And it concerns both the production chain and supply chain management [1,2]. Such an interconnection of digital and physical systems has united almost everything: from design and product planning to supply chains and production. The popularity of this approach is due to the short lifecycle of manufactured products, as well as because of high level of customer focus under the conditions of fierce competition.

Digitalization opportunities become important while providing activities of joint production networks which can better adapt to future dynamic markets. With the increasing prevalence of Internet technologies, production systems are transformed into so-called cyber-physical production systems (CPPS). They include a network and autonomous components with the local control of intelligence, which can communicate autonomously with other devices, machines, production modules and products through open networks and semantic descriptions. Thus, traditional rigid hierarchical production will be replaced by a decentralized self-organization, where application of reengineering will allow to create much more flexible production systems.

Besides, for design purposes, dynamic adjustments and reconfigurations of the production system, equipment and software can be performed much more quickly. Also production components can be easily removed or added during product manufacturing. This will allow to better meet the unpredictable demands of the market or to respond flexibly to haltings and failures.

IOP Conf. Series: Materials Science and Engineering **570** (2019) 012081 doi:10.1088/1757-899X/570/1/012081

The goal of this paper is to analyze how intelligent information technologies (IIT), developed within the framework of the "Industry 4.0" projects, and related cyber-physical systems do impact and change principles and methods of supply chain networks managing and outsourcing.

2. Artificial Intelligence and Industry 4.0

Over the past years and decades, computers have constantly transformed both workplaces and society itself. Information technologies have become a vital part of our everyday life and work, changing job descriptions, initiating new business models and processes, and also changing their viability over time. Employees, as well as enterprises, need to be able to quickly adapt to stay competitive and in demand. Changing conditions and markets has always been difficult and problematic for businesses and employees, therefore, to answer the question about the difference of "fourth industrial revolution" from the previous ones it is necessary to understand reasons of wide distribution of the artificial intelligence." Artificial intelligence is not some other type of computer that can do something new. Artificial intelligence tends to imitate and replace human thought processes and create a "learning" virtual computer that can replace human decision-making, i.e. this is an attempt to copy the human thought process by the computer. Artificial intelligence, by means of machine learning, can independently learn and adapt, just like a person [3].

Conception become widely renowned due to the victory of artificial intelligence based program over the "Go" game world champion, although the idea itself and the threat of people replacement by intelligent machines were already discussed in the second half of the last century [4]. However, the "fourth industrial revolution" increases this threat. Due to the wide applying of such technologies as 3D design and printing, Nano-Tech BioTech the labor productivity has grown much. Along with flexibility and convenience, production and supply cost simultaneously goes down [3], what leads to an "economy of abundance" [5].

In the future, company location will no longer be a problem, and two business partners can participate in any venture enterprise, starting production immediately after signing contracts and starting deliveries. At the process of production planning one can reveal the possibility to take into account the potential of outsourcing and supply chain management. Growth of three-dimensional design usage and progress in three-dimensional printing, along with new developments in the field of materials science, undoubtedly, give an unprecedented degree of flexibility to all production innovations.

As an example we can consider the first 3D-printed object in space made for the ISS [6]. Even today, you can already begin production as soon as the product is designed. You no longer need to create new production facilities, since the product can be manufactured on any three-dimensional printer and delivered to the addressee by a definite logistics company. In fact, the need for human participation can be limited to the approval of the design concept together with artificial intelligence that solves all other issues - from fully automated production to the delivery of finished products to client by unmanned cars and drones. Naturally, all these global changes will have a huge impact on the whole business scheme, which will largely be outsourced to artificial intelligence. The workplace of the future will differ significantly from the modern corporate world.

3. The influence of artificial intelligence to project management in mechanical engineering

The transfer of stereotyped repetitive tasks that are performed by deterministic algorithms (for example, accounting) to an artificial intelligence just the first step. Abilities of artificial intelligence systems to learn and self-educate can, in the long run, make it possible to adequately replace an entire army of white-collar workers. Company owners may come up (and many have already come up) to the idea that instead of hiring one or a few hundred people, you can buy one artificial intelligence system that will never get sick, be absent, or complain about working conditions. In this case also there is no need to solve all the complex and often poorly predictable problems arising around the team. System maintenance and upgrading costs for such an artificial intelligence system will be insignificant compared to labor costs and related administrative activities. Cost, as known, was one of factors for

IOP Conf. Series: Materials Science and Engineering **570** (2019) 012081 doi:10.1088/1757-899X/570/1/012081

outsourcing and transferring of manufacturing to countries with low labor costs. [7]. Considering that even the bibliographic work at some university was transferred to artificial intelligence [8], it is easy to imagine that many other similar jobs are under serious automation "threat" by artificial intelligence [9].

If an artificial intelligence system can accurately assess a project, perform tasks, perform outsourcing, for example, like WorkFusion suggests, then, literally no work can be protected from being replaced by artificial intelligence. WorkFusion Platform combines machine learning with crowdsourcing to automate data processing. WorkFusion software is embedded into business processes and it attracts resources of remote employees to perform tasks and uses results to train artificial intelligence, which then takes over the fulfillment of all these tasks.

Deep Learning is a branch of machine learning, based on the model of human brain neurons interaction and in fact is an artificial intelligence, which is not inferior to human accuracy in performing tasks. And it is superior to people by an amount of work it can perform. For project management, this would mean execution of all work types (for example, highly qualified analysis and risk planning and other related tasks) faster and more comprehensively than any of the strongest team of project managers could even dream of. Considering the sad statistics of numerous failures in this area, artificial intelligence can really significantly increase work productivity and reduce its cost.

The methodology of flexible production in combination with artificial intelligence and materials science will allow to organize localized and economical production in a short time and taking into account its specifics. To a large extent, it can be organized without the participation of a human, who, in this case, will only have the auxiliary role of "machine operability" ensuring. How can the sustainability of human labor be improved and counteracted to machine labor replacement? In [13], it is suggested to consider creativity, social intelligence and management of operations as main arguments that can counteract to the replacement of human labor by machine labor. In addition to the new jobs associated with the maintenance and formation of these intelligent supply chains and networked outsourcing, the above arguments ensure that people would retain their jobs, although perhaps in a slightly modified form.

4. Business transformation under conditions of rapid changes

Whenever the economic environment undergoes rapid and dramatic changes, there are winners and loosers. Considering realities of the upcoming challenge, it is necessary to act now to become leaders and drivers of serious market changes. To use capabilities of the artificial intelligence properly, you need to evaluate carefully and critically what kind of business we do represent and where we want to be when the "smart factory" will become an objective reality. Failure or refusal to solve these problems in this case is not a suitable option: competitors will become more active and force other market participants to follow their example or to drop out of the competition.

We can already see this effect even today, but to a lesser degree. Why do clients have to put up with the fact that you can not deliver to them what they want, when they want and how they want. In this case the key to success is to create a value. Only innovative, flexible and customer-oriented competitors of your company will remain on the market. Thus, transforming of your business and yourself is a task that should be started right now in order not to fall behind forever.

Corporate strategy is needed, which, ultimately, will ensure sustainable market success, offering the necessary value to customers. Offers of "value and usefulness" to customers will be the main goal and future distinctive advantage which are necessary for planned results achievement and also for interactive communication with client.

Value and usefulness are much more than just buzz words that identify valuable offers to your clients. Exactly they will be the key factor for future success on the market. Obviously, all products are becoming more and more interchangeable from the technical side. Literally, it does not matter which one gadget, for example, do you buy since all the offers are almost identical.

Decentralized production and outsourcing will be two key functions for successful application of artificial intelligence in Industry 4.0. Today, we already have the necessary tools to solve such

IOP Conf. Series: Materials Science and Engineering **570** (2019) 012081 doi:10.1088/1757-899X/570/1/012081

bottleneck, for example, as a delivery, by implementing artificial intelligence methods into production and logistics. However, in order to truly get around such bottleneck, you need to have localized flexible production with decentralized capacity to work alongside the customer. At the same time, production must be sufficiently "smart" to automatically accept and execute orders using an intelligent system, reducing or completely eliminating the need for human intervention, besides making initial decisions [10-12]. Naturally, at the same time, considered above "lean" and ordered chain of value creation will be very vulnerable to external shocks, what makes viability the main task for the management team, whose functions should significantly differ from today's management.

5. New functions of management in Industry 4.0 environment

The transformation of production industry is called a revolution exactly because not just slight changes are taking place but radical ones: the industry is being restructured from top to bottom. Business models are changing, new companies are born, world-famous brands with a long history are obliterated if they do not have time to join the ranks of digital innovators. Customers have changed their behavior, they want an individual approach, unique products.

The type of required managerial specializations and their relative roles in the management team have been constantly changing since the modern corporations have emerged. And at the beginning there were no positions based, for example, on "equality" or "sustainability", because the need for them evolved only with time. As the business environment evolves, it's necessary to develop, create, change, and replace managerial positions and responsibilities depending on emerging needs.

The actual need for management will be reduced to managerial and business functions, which depend on individual decisions and preferences. This "lean" approach to business management can only be resolved through a combination of artificial intelligence and Industry 4.0, and this is largely dependent on networked outsourcing within the production cluster [2,12]. Creating flexibility for decentralized production will eliminate the need to retain production capacity, which will actually begin to become a burden rather than an asset.

All together this will create a need for new managerial functions, as well as also will change the content and responsibilities list for all departments in company. Naturally, these functions will be largely supported by their own artificial intelligence systems. The human intervention should be carried out only at the very last stage and only if it's necessary and desirable at all [14].

The main idea of Industry 4.0 and CPPS is to create self-organizing and self-adapting dynamic network structures of outsourcing and supply throughout the product life cycle in order to actualize the most flexible individual production with the costs matching to mass-flow production. This trend is already clearly visible in many key sectors of the global economy and political decisions are largely determined by the increasingly tough battle for markets of innovative products and services.

The adoption of Industry 4.0 is actively and effectively hampered by a number of factors, among which, apart from doubts about the digital data safety and requirement in large investments, one can also note insufficient qualification of personnel and customers at all levels, the lack of effective business processes and standards which allow to take advantage of the digital approach. And exactly this approach can provide new opportunities for small and medium businesses. For example, in mechanical engineering clusters, where flexibility and customer focus, combined with standardized networks of own production resources, allows not only to expand the segment of orders from large companies, but also to ensure own accelerated development because of additional investments and refresher training courses. Comprehensive automation, digitalization and intellectualization of main technologies that provide proper functioning of enterprise will ensure guaranteed growth in turnover, production flexibility, productivity and overall company efficiency. The main difficulty lies in the fact that implementation and usage of "Industry 4.0" technologies and cyber-physical systems will take place under the influence of various kinds of uncertain factors, for example, indifferent, targeted and associated with the uncertainty of goals at participants.

Uncertainty analysis of competitor actions can be based on multi-agent systems, various gaming models, methods of fuzzy logic. In these approaches, conflict management principles are

IOP Conf. Series: Materials Science and Engineering **570** (2019) 012081 doi:10.1088/1757-899X/570/1/012081

implemented, as well as taking into account the so-called "soft" factors that are hard to formalize, such as common goals, flexibility, trust, reputation, and so on. The digital revolution in supply chain management and outsourcing can lead to a real growth of business efficiency and to significant improvement of customer services quality.

Often, logistics and outsourcing companies use home-made information systems for the control of production processes and services that leads to significant costs and time losses. It is much more advisable to use a multifunctional renewable platform supported by a serious operator, by creating at the same time a single cooperation space between all participants in the outsourcing and supply chain.

Modeling, optimization and big data analytics allow to form a complete set of technologies for creating an outsourcing network and a digital supply chain, by the identification of process state model for all processes in a real time. At any given time, the digital twin should display outsourcing processes and supply chain status compared to: actual planning data, necessary tools production, component production, transportation, inventory, demand and capacity. The digital twin can be used for making real time decisions, as well as for forecasting and planning of outsourcing. In fact, mechanical engineering company and firms that provide outsourcing and logistics services within the framework of this approach are integrated into a single mechanism for solving tasks of flexible customer-oriented production, while remaining full-fledged market participants at the same time.

For example, if an abnormal situation occurs either at the logistics center or at some cluster enterprise, this deviation can be seen by the risk monitoring tool and transferred to the simulation model to form alternatives. Digital twin simulation can help to reveal impact proliferation for emergency events in system and, besides, provides an effective and almost online adaptation of action plans in accordance with the situation.

6. Conclusion

Decentralized production and distributed outsourcing will be two key functions for the successful application of artificial intelligence in Industry 4.0 for the mechanical engineering. In mechanical engineering clusters, it is preferable to use a multifunctional updateable information management platform and to create a single cooperation space for all participants of the outsourcing network and supply chain. At the same time, simulation, optimization and big data analytics will allow to form a complete set of technologies for creating a flexible network of outsourcing and a digital supply chain, by the identification of process state model for all processes in a real time. And herewith, the digital twin will display and predict the status of outsourcing processes and supply chain compared to: actual planning data, necessary tools production, component production, transportation, inventory, demand and capacity.

References

[1] Bakhtadze N, Smirnova G, Sabitov R and Elpashev D 2017 Identification and simulation models in logistics control systems for production processes and freighting *IFAC-PapersOnLine* Volume 50 Issue 1 pp 14638–43

[2] Gulnara Smirnova, Rustem Sabitov, Boris Morozov, Shamil Sabitov and Natalya Elizarova 2015 To the problem of dynamic modeling and management in an integrated environment of the industrial cluster *IFAC-PapersOnLine* Volume 48 Issue 3 pp 1230-35

[3] Ford M 2015. Rise of the robots: technology and the threat of a jobless future *Basic Books, a member of the Perseus Books Group* New York

[4] Larabell J T 2016 The Rise of the Robots *The New American* 32 pp 10–17

[5] Sadler P 2010. Sustainable growth in a post-scarcity world: consumption, demand, and the poverty penalty *Gower, Farnham, Surrey ; Burlington, VT*

[6] Rainey K 2015. Open for Business: 3-D Printer Creates First Object in Space on Intern *NASA* http://www.nasa.gov/content/open-for-business-3-d-printer-creates-first-object-in-space-on-international-space-station (accessed 11.24.17).

[7] Online resource: https://www2.deloitte.com/content/dam/Deloitte/nl/Documents/operations/

deloitte-nl-s&o-global-outsourcing-survey.pdf (accessed 11.24.17).

[8] Condon W 2013 Large-scale assessment, locally-developed measures, and automated scoring of essays: Fishing for red herrings? *Assessing Writing* 18 100–108

[9] Kashapov N F, Khafizov I I, Nurullin I G and Sadykov Z B Influence of introduction of robotics on increase in efficiency of electrochemical production 2018 *IOP Conference Series: Materials Science and Engineering* 412(1) 012034

[10] Gulnara S. Smirnova, Rustem A. Sabitov , Ekaterina A. Korobkova and Shamil R. Sabitov 2017 Modeling production facility as a dynamic integrated interacting objects system *Procedia Computer Science* 112 965–970

[11] Bakhtadze N N, Smirnova G S, Sabitov Sh R, Sabitov R A, Elizarova N Yu and other 2016 Multi-agent Simulation of SWAP BODIES application in manufacturing supply chain *IFAC-PapersOnLine* 4 № 12 pp 1245–1250.

[12] Smirnova G S, Sabitov R A, Dolgui A, Bakhtadze N N and Sabitov Sh R Design of a Multi-agent System to Manage Relay Intercity Freighting 2016 *IFAC-PapersOnLine* 49 № 12 pp 1656-61

[13] Frey C B, Osborne M A, 2017. The future of employment: How susceptible are jobs to computerisation? *Technological Forecasting and Social Change* pp 114, 254–80

[14] Ruchkina E R, Gilmanshin I R, Kashapov NF, Galeeva A I and Krainova D R 2018 Modeling of the complete life cycle of raw materials in the mathematical apparatus of problem-oriented modification of color networks Petri-DN-networks *IOP Conference Series: Materials Science and Engineering* 412(1) 012059

IOP Conf. Series: Materials Science and Engineering **570** (2019) 012082 doi:10.1088/1757-899X/570/1/012082

Improving the process efficiency of helical gears' toothed rims at the stage of pre-production

D T Safarov [1], A G Kondrashov[2] and I I Khafizov[3]

[1,2] Naberezhnye Chelny Institute of Kazan (Volga region) Federal University, Naberezhnye Chelny, Mira 68/19, 423800, the Russian Federation
[3] Institute of Engineering of Kazan (Volga region) Federal University, Kazan, 420008, the Russian Federation

Abstract. The article reveals modes of increasing the efficiency of preproduction process for manufacturing nodes with helical gears. The increased efficiency is ensured by processing toothed rims of helical gears with end tools on multi-axis CNC machines. The advantage of the proposed processing method is the minimum deviation of actual and theoretical profiles of gear tooth side surfaces.

Introduction

The development and production of new products that include cylindrical spur gears and helical gears (speed reducing gears, transfer boxes, gearboxes) is impossible without working out several options for their design with various test cycles and the selection of the most optimum design. The product design includes some options for its designing and experimental production of its parts. Parts production with simple structural flat, cylindrical, conical elements does not cause significant technological difficulties, but high-quality production of gear rims is impossible without designing and manufacturing a special cutting tool - rams or worm milling cutters [1-3]. Their additional design and manufacture significantly lengthens the process of pre-production and manufacturing pilot runs.

Research objective

Processing gear rims on multi-axis CNC machines without manufacturing special cutting tools can be taken as an alternative solution. Currently, there is a wide range of CAM systems integrated with CAD systems that allow you to generate control program texts for a CNC machine [4-8]. The most famous of them are NX CAM, PowerMill, MasterCam, SprutCam, Adem, etc.

All these software products implement a single sequence of work preceding processing toothed rims, designing the 3D model either in the same integrated CAD software environment or in another, reading 3D model in CAM system. CAM system generates various movement patterns of the selected tool according to universal algorithms incorporated in calculating path-planning module of these programs. The process ends with the formation of a control program code for CNC machine.

As a rule, these universal algorithms implement an equidistant traversal of structural elements' processed surfaces according to previously worked out rules (for example, layer-by-layer removal of the allowance from the periphery to the central point or from the central point to the periphery). The more developed the CAM system is, the greater the number of options for setting the working paths it allows you to perform. Let's consider the most likely ways to handle toothed gear roots. Coordinates of work points can be calculated by all the listed software products in the basic version.

IOP Conf. Series: Materials Science and Engineering **570** (2019) 012082 doi:10.1088/1757-899X/570/1/012082

The first is the transfer of rough milling which should eliminate the main allowance from the root, leaving it on the sides. The allowance is sampled by a spherical end or a keyway cutter (Figure 1a). Then comes the fine milling. It can be made with a modular end mill. This variant of processing assumes the shortest possible processing time of the wheel root, but the modular milling machine must be specially manufactured for parameters of the gear, its manufacturing cost may be comparable to the cost of manufacturing the worm gear-cutting milling machine. Another optional variant is the traversal by a spherical end mill (Figure 1 b). Unlike a modular disk or end mill, a spherical end mill is an inexpensive and affordable tool. The smaller the depth of cut, the better the surface of the side gears, but the longer the process is. In any case, to achieve an acceptable quality of the lateral surface of gears the processing time of toothed rims will be significant. In some cases it can reach 4-5 hours or more.

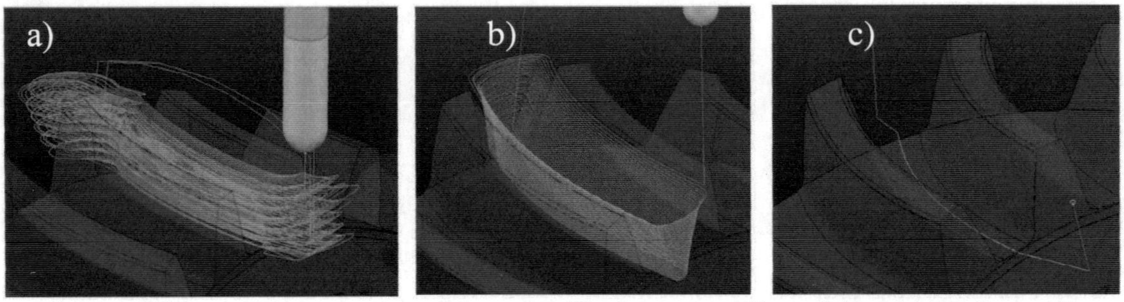

Figure 1. The movement patterns of the end mill: a) rough milling of the root, b) fine milling, c) processing the root of the gear wheel.

The intermediate option among milling options discussed above (modular milling and spherical end milling) is, for example, Sandvik Coromant's offer in the form of "patented InvoMilling process" [9] (figure 2).

This process by the duration of its processing can compete with the milling of gear rims of modular milling cutters. It should be noted that although this solution is ready-made and universal, the cost of the mill body and quick-change milling grades exceeds the cost of manufacturing not only a worm, but also a modular cutter. Its high price is compensated by the versatility of processing gear wheels of a different module with the same tool. In addition to the need to purchase special tools and plates, an enterprise implementing this method should purchase a specialized software product from this company. In addition, the implementation of this processing method requires a 5-axis CNC milling machine, the cost of which is maximum among CNC machining center line. Thus, such a solution can be acceptable for enterprises that process gears in small batches only on condition of their work with significant profitability.

Figure 2. Processing cylindrical gear teeth with special cutters of Sandwik CoroMill 161 and CoroMill 162 [9].

IOP Conf. Series: Materials Science and Engineering **570** (2019) 012082 doi:10.1088/1757-899X/570/1/012082

The work with the above CAD, CAM systems is implemented by various production services of the enterprise. The mathematical model is created by the designer on gears in the CAD system, then he sends it to the control design department of CNC machines in which another expert decides how to process the rim and adjusts one or another variant of equidistant processing of the mathematical model in CAM system. Then, depending on the model of the machining center, he generates the text of the control program and sends it to the workshop for processing. When changing design parameters of wheel rims during the development of the product design or arising deviations in the process of manufacturing gears, the transfer of information from one department to another can lead to errors in design, increasing the duration of getting information.

Results and problem solution

To reduce the number of errors and speed up the process of transferring information from one service to another is possible by transferring the program development functions for CNC machine to the designer on gears. This requires a software product that allows you to generate a control program from a single side, which indicates the necessary structural data of the gear rim of the wheel and data processing wheel on a CNC machine.

For many enterprises the most acceptable option is processing gear rims on the 4-axis machining center. Only the end milling cutter can be the cutting tool for a stack with 4 controlled coordinates for machining helical gears. To reduce the complexity of processing, the main allowance from the root is removed with a mill of the maximum possible diameter. Since the gear profile is formed by an involute called the engagement line rolling along the main circumference, the side surface can be machined with an end mill, the axis of rotation of which is perpendicular to the engagement line and the forming edge is on the side surface of the tooth (Figure 3 b, c).

There has been developed the software product, the input data of which are gear wheel, tool, transition parameters — cutting-in and an overtravel of a tool, safe tool lifting as well as machining process parameters — cutting speed and tool feed amount. The software product calculates coordinates of points of tool's working strokes; the sequence of individual steps of the algorithm is performed taking into account the methodology described [10]. The output of the product is the control programming text with CNC. The program is designed for experts on gears. Approbation of the technique was performed on a 4-axis CNC milling machine when processing a cylindrical helical gear from an aluminum alloy (Figure 3). The results of measuring the machined gear showed acceptable deviations of the actual profile of gear's side surface from the theoretical one.

a) b) c)

Figure 3. Processing the gear rim of a horizontal wheel: a) a semi-fine milling, b) milling the left side of the root, c) milling the right side of the root.

Currently, the software product and processing technology by the end tool of gear wheels are being implemented at the experimental site for manufacturing gear shafts and wheels of "KamAZ".

References

[1] Kondrashov A G, Safarov D T, Faskhutdinov A I and Davletshina G K 2017 Single-turn worm mills for conical round-tooth gears *Russian Engineering Research* Vol 37 No 9 pp 812-13

[2] Balabanov I P, Gilman V N, Timofeeva T S, Faskhutdinov A I 2018 Modeling of the cutting edge rounding influence on the tool life in processing a gear wheel by the Power Skiving method *International Journal of Engineering and Technology(UAE)* Vol 7 No 4 pp 71-73
[3] Golovko A N, Kondrashov A G and Yurasov SYu Improved Design of a Worm Type Instrument for Final Machining of Evolvent Gear Teeth *Procedia Engineering* Vol 206 pp 1333-36
[4] Kravchenko K 2012 The development in CAM system the Power MILL control program for processing a spiral bevel gear on a five-axis machine with CNC *CAD and scheme* No 8 pp 80-83
[5] Safarov D T, Kondrashov A G, Safarova L R and Glinina G F 2017 Energy Planning in Production Shops with Numerically Controlled Machine Tools *Russian Engineering Research* Vol 37 No 9 pp 827–34
[6] Grechishnikov V A, Petukhov Y E, Pivkin P M., Romanov V B, Ryabov E A, Yurasov S Y and Yurasova O I 2017 Trochoidal slot milling *Russian Engineering Research* Vol 37 No 9 pp 821-23
[7] Khusainov R M, Yurasov S Y and Kazargel'dinov R R 2017 Preparations for production in unigraphics NX software *Russian Engineering Research* Vol 37 No 4 pp 363-66
[8] Golovko A and Petrov S 2018 Definition of the initial information for selecting the cutting tool for automated production preparation *MATEC Web of Conferences* Vol 224 **01103**
[9] InvoMilling 2014 *The revolution in processing toothed and helical gears* The catalogue of products of Sandvik Coromant 8 p
[10] Kasjanov S V, Safarov D T 2018 Mathematical modeling of the technological accuracy index deviation structure of the automobile parts *IOP Conference Series: Materials Science and Engineering* 412(1) **012039**

IOP Conf. Series: Materials Science and Engineering **570** (2019) 012083 doi:10.1088/1757-899X/570/1/012083

Bevel gears monitoring methods developmentin the total contact patch terms

D T Safarov, A G Kondrashov

Naberezhnye Chelny Institute of Kazan (Volga Region) Federal University, Naberezhnye Chelny, Mira Ave. 68/10, 423800, Russian Federation

Abstract. The article describes the main provisions of the method, which allows to perform the control of the gear rims of bevel gears according to the shape and relative position of the total contact patch. In comparison with the standard, the developed method allows to evaluate the conformity of the teeth of the ring gear with its considerable displacements relative to the calculated point of the gear transmission.

1. Introduction.

Conical spur gears in the construction of heavy vehicles are used in the differential mechanism, allowing to redistribute the torque from the wheels, when cornering and in the conditions of wheel slip when the vehicle is moving on dirt roads.

To ensure the durability of the differential operation, it is important to ensure the required level of contact patch over all teeth of bevel gears included. Irregularity of the area and location of contact spots on the teeth of the rims of the conical wheels can lead to uneven loading of individual teeth and their intense local wear. In some most unfavorable operating conditions of the car, their breakdown is possible.

The difficulty of ensuring a given spot size of the contact bevel gear in the differential mechanism lies in the fact that it is unregulated. Many factors influence the contact patch size, beginning with the design stage, for example, cutting tool profile provision [1,2], and at the stage of cutting tools wear [3], processed layer state of the blank [4] and other factors. As a result of the above factors, the teeth of the ring gear receive different geometrical deviations, leading to deviations of the location and area of the gear contact patch. Received deviations of the location and area of contact spots can't be compensated or redistributed by means of adjusting the position of gears in the process of assembling the differential. Consequently, the operations of bevel gears are imposed special requirements of ensuring the total contact patch on (picture1).

Under workshop conditions, the main method for assessing the size of the contact patch is a visual method of comparing the resulting contact patch with the allowed forms. Measurements and recording of its parameters are not provided. A visual comparison of the acceptable forms and contact spots location by experienced and qualified personnel ensures the availability of product, but it does not allow to register the quantitative values of this indicator, to monitor the trends of its changes, and therefore effectively manage the process of manufacturing wheels.

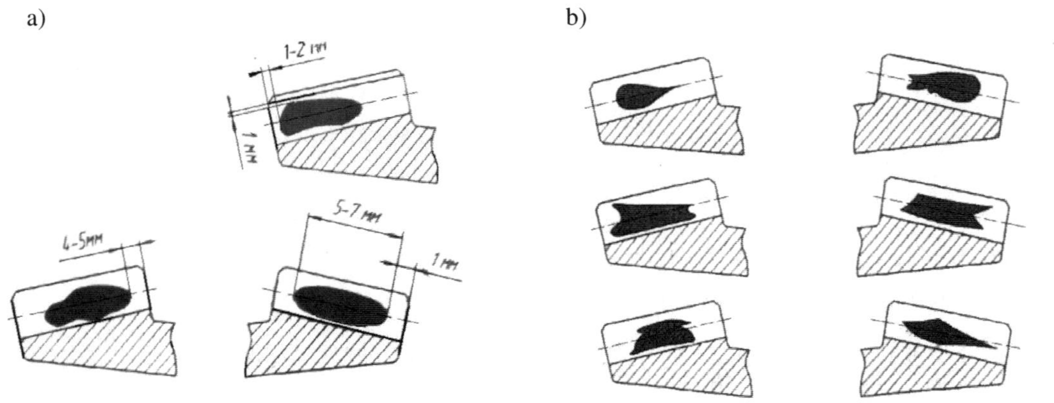

Figure 1. Permissible forms of contact spots a) permissible location of the contact patch, b) types of permitted forms of contact spots.

2. Theoretical part

The contact spot requirements for bevel gears are established by a number of regulatory documents [5, 6]. Regulatory requirements are formulated in tolerance form (Figure 2 a). Standard requirements [1] are applied to the total contact patch dimensions. Its size is determined as a percentage of the length of the tooth "the ratio of the distance between the extreme points of traces of fit to the length of the tooth" ($\frac{a}{b} \times 100$), the height of the tooth "the ratio of the average height of the traces of fit to the average height of the tooth of the corresponding active lateral surface" ($\frac{h}{h_a} \times 100$) (Figure 2 b).

Determining the size of the total contact patch implies its initial location in the coordinates of finding the calculated point (at the intersection of the average height of the tooth and forming the separating cone), i.e. in the conditions of already adjusted process of a crownwork of a cogwheel crown.

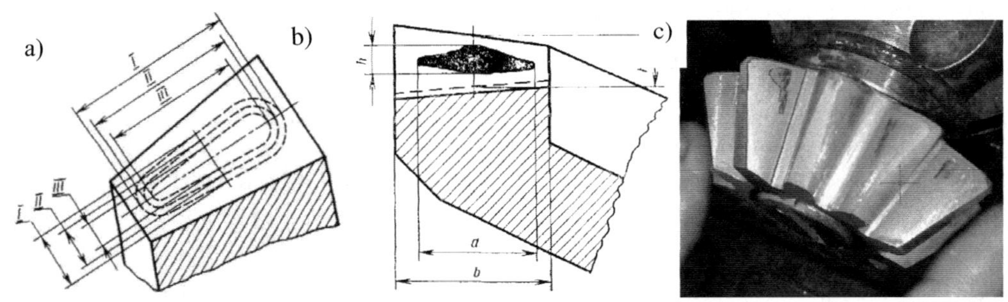

Figure 2. a) Normalization of the size of the contact patch in a bevel gear transmission. I - the largest limit size of the contact patch; II - nominal size of the contact patch, III - the smallest limit size of the contact patch [1], b) the standard scheme for estimating the size of the total contact patch [2], c) an example of a photo of the contact patch.

In practice, even in a well-established process of teeth processing, the total contact patch can significantly shift relative to the calculated gearing point (Figure 2c), and in some cases, may go beyond the boundaries of the teeth. Under these conditions, direct adherence to the method of calculating and rationing the dimensions of the total contact patch by the standard method is the risk of declaring unfit products suitable. Thus, there is a need to develop a more universal method for estimating the total contact patch.

The developed method is given additional modern requirements concerning its incorporation into the quality system of the enterprise [7], providing a collection of quantitative data on controlled characteristics, the identification of diagnostic components of the process of tooth processing [8], suitable for evaluating the result of processing as a result of mathematical modeling of the process [9]. As a result, to ensure the above requirements, in the developed method, the form of the total contact patch and its relative positioning are separately normalized. To establish the requirements of the contact patch shape, similar to the standard method, t the minimum and maximum limits of its boundary are set (Figure 3 b). In contrast with it, the nominal or ideal position of the total contact patch is given as an ellipse. For the considered bevel gear, according to the test results, its optimal position was found, shifted from the calculated transfer point towards the flat end (Figure 3 a). The minimum and maximum dimensions of the ellipse are set depending on the length of the tooth along the pitch circle and the average height of the tooth.

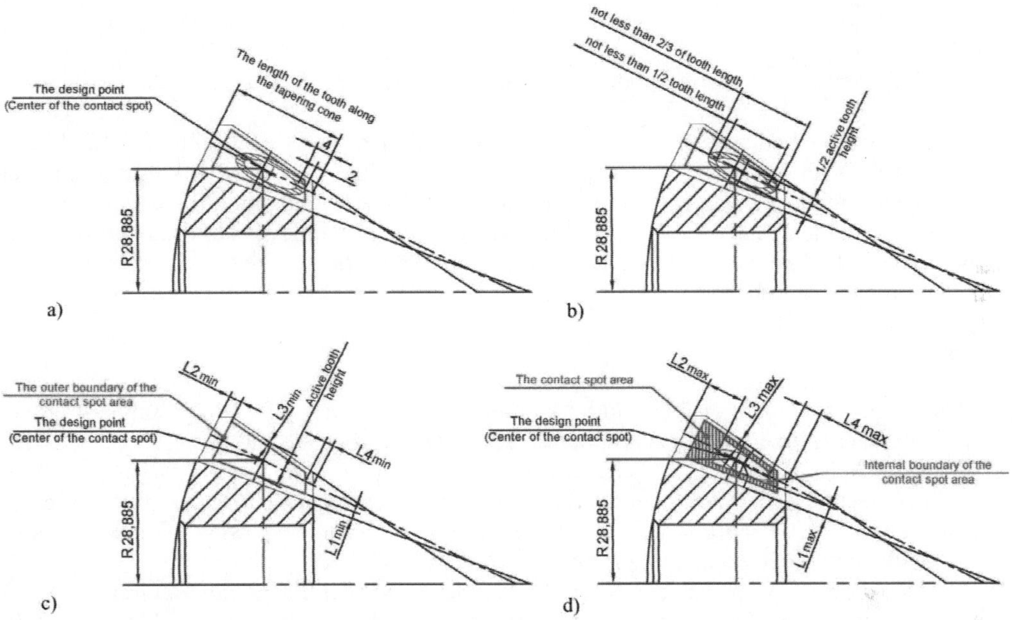

Figure 3. Normalization of the shape and relative position of the total contact patch a) nominal location, b) the allowable dimensions of the total contact patch, c) the maximum allowable boundaries of the location, d) the minimum allowable boundaries. Its relative location is normalized by establishing the minimum and maximum allowable areas for finding contact spots. The minimum allowable area is an area that limits the minimum allowable shape of the contact patch (Figure 3 g). The boundaries of the maximum allowable zone are set to prevent the total contact patch from escaping to the tooth boundaries under the maximum operating load (Figure 3c). The lines limiting the minimum and maximum areas of the total contact patch are equidistant to the corresponding opposite tooth profile. The position of the lines is normalized in the form of linear dimensions from the corresponding boundary of the zone of finding the total contact patch up to the borders of the gear tooth. They take values from 1 to 3 mm.

Thus, the assessment of compliance of the total contact patch is simultaneously performed on a group of indicators: % L,% H - normalizing the shape of the total contact patch along the tooth and its height, as well as indicators L1, L2, L3, L4 - setting restrictions on the relative location of the contact spot. For each of these indicators, the upper and lower deviations are calculated.

Finding the actual values of these indicators is carried out according to the graphical processing of photographic images of contact patches formed as a result of erasing ferric acid pigment on the lateral surfaces of the ring gear teeth (ls) as a result of rolling the processed gear and the reference gear (Figure 2c).

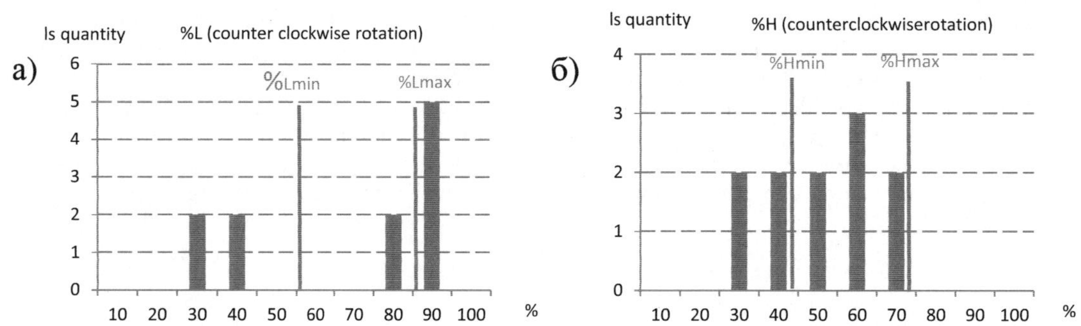

Figure 4 Histograms of the distribution of indicators a)% L and b)% N of the bevel wheel in the process of setting up a gear rolling machine.

Correctly performed adjustment should ensure that all six indicators (% L,% H, L1, L2, L3, L4) hit all the side surfaces of the ring gear in the tolerance field when the bevel gear is run-in on a gear-cutting machine. The technique has shown its high efficiency in terms of maintaining the statistical office [10].

Thus, the developed method of controlling bevel gears is more versatile and allows not only to perform the control of the ring gear for any deviations in the position of the total contact patch, but also to control the setup of the machine.

References

[1] Kondrashov A G, Safarov D T and Faskhutdinov A I and Zairov B F 2017 Single-turn worm mills for conical round-tooth gears *Russian Engineering Research* **37(9)** pp 812-13

[2] Golovko A N, Kondrashov A G and Yurasov S Y 2017 Improved Design of a Worm Type Instrument for Final Machining of Evolvent Gear Teeth *Procedia Engineering* 206 pp 1333-36

[3] Kas'yanov S V, Kondrashov A G and Safarov D T Rapid assessment of wear-resistant tool coatings *Russian Engineering Research* 37(11) pp 969-973

[4] Astashchenko V. I., Zapadnova N N, Mukhametzyanova G F and Shafigullina A N 2017 Key concepts for production of high-quality parts *IOP Conference Series: Materials Science and Engineering* 240(1) **012007**

[5] GOST 1643-81 Osnovnye normy vzaimozamenyaemosti. Peredachi zubchatye cilindricheskie [Basics of feeding and interchangeability. Transmission toothed cylindrical] M.: Izd-vo standartov 45 p

[6] GOST 9368-81 *Peredachi zubchatye konicheskie melkomodul'nye* [Transmission gear conic fine-molded]. M.: Izd-vo standartov 26 p

[7] Safarov D T, Fedorov K A and Ilyasova A I 2016 Algorithms development of making special techniques in APQP manufacturing process of automotive components *IOP Conference Series: Materials Science and Engineering* 134(1) **012036**

[8] Safarov D T, Kondrashov A G and Kasyanov S V 2019 Informative value of measurements for quality management of auto parts *Lecture Notes in Mechanical Engineering* pp 1657-66

[9] Kasjanov S V and Safarov D T 2018 Mathematical modeling of the technological accuracy index deviation structure of the automobile parts *IOP Conference Series: Materials Science and Engineering* 412(1) **012039**

[10] Safarov D T and Kondrashov A G 2018 Technique for monitoring the indicators of the total contact spot of spur gears of the truck's differential gear *IOP Conference Series: Materials Science and Engineering* 457(1) **012014**

IOP Conf. Series: Materials Science and Engineering **570** (2019) 012084 doi:10.1088/1757-899X/570/1/012084

Four-point technique development for measuring spherical surfaces of machine parts

D T Safarov[1], A G Kondrashov[2] and I I Khafizov[3]

[1,2] Naberezhnye Chelny Institute of Kazan (Volga Region) Federal University, Naberezhnye Chelny, Mira Ave. 68/10, 423800, Russian Federation

[3] Engineering Institute of Kazan (Volga Region) Federal University, Kazan, 420008, Russian Federation

Abstract. The article presents the theoretical foundations of the four-point measurements of spherical surfaces of parts method, allowing to increase the efficiency of spherical structural elements of ball tips of various sizes control. Compared to the measurement method on the coordinate measuring machine, the developed technique can be used at workplaces of forming spherical elements. The technique ensures the efficiency of measurements and increases the efficiency of carrying out the cutting tool settings.

1. Introduction.

Traction is considered to be the most important knot of the car from the viewpoint of its safety, when being driven. Traction are a mandatory element of the steering system, it can also be included in the vehicle suspension. The behavior of the vehicle during maneuvering and cornering depends on its technical condition. It ensures the transmission of forces and the prescribed laws of displacement between the various connecting levers. Spherical mates are the main elements that allow this class of parts to perform the function of assignment. Structurally, they are made in the form of a spherical support and a spherical finger. One of the most important key characteristics of a spherical finger is the diameter of the sphere. The correctness and durability of the operation of a spherical coupling depends on the value of this indicator determined as a result of simulating operational loads by the size of the gap in a spherical mate [1]. As a rule, the parameters of the spherical surface of these parts are measured on coordinate measuring machines. Measurements on the coordinate machines are distinguished by high accuracy, but low efficiency and high labor intensity.

As a rule, at workplaces, instead of CMM measurements, measurement of a spherical surface radius by the two-point method (micrometer) is used, these values differ from the values of the diameter of a spherical surface measured by the coordinate method. Since there is an error associated with the position of the measuring tips of the device during the two-point measurements. The difference can be up to 0.02 mm. from the actual size of the spherical surface. Generally accepted tolerances on spherical surfaces allow 0.05 mm, this value can reach up to 1/3 of the tolerance field. Using the two-point method instead of measuring the diametral size of a spherical surface can make the process of timely amending a CNC lathe program dependent on the qualifications of a controller or adjuster, and can potentially lead to the appearance of inappropriate products for the diametrical size of a spherical surface, an excessive gap in the pairing and, as a consequence, increased wear on the whole pairing.

IOP Conf. Series: Materials Science and Engineering **570** (2019) 012084 doi:10.1088/1757-899X/570/1/012084

At present, the radius of the spherical surface of the ball pin and the position of its center in an enterprise is measured only in the laboratory on the IOTA 2204 coordinate measuring machine (Figure 1 b).

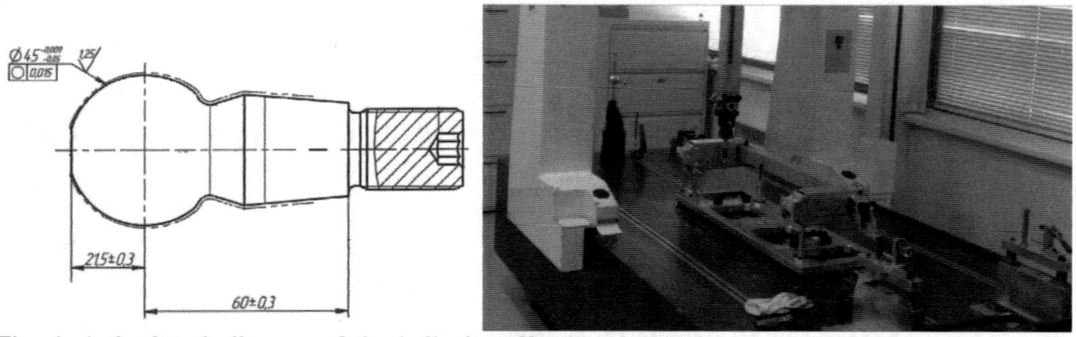

Fig. 1 a) the key indicators of the ball pin, affecting the performance of the function of official purpose, b) the IOTA 2204 measuring machine in the process of measurement. This is a rather complicated and time-consuming procedure. According to internal regulatory documents of an enterprise, 1-2 out of 100 manufactured parts are exposed to measurement. As a consequence, the rapid assessment of values for the management of these indicators in the process is impossible [2, 3].

2. Theoretical part

The task was set to improve the performance of measurements of these indicators, with the organization of measurements near workplaces of forming a spherical surface. This problem can be solved by applying the four-point method of measuring a spherical surface coordinates. To find the diametrical size of a sphere, it is necessary to know the position coordinates of any four points lying on the sphere. To obtain such information it is necessary to apply to the following measurement scheme: the finger is installed on a special support before contact with four recording devices (indicators), registering the size value along the Z coordinate with the previously known position coordinates of the measuring tips along the X and Y axis (X_1, Y_1, X_2, Y_2, X_3, Y_3, X_4, Y_4) - figure 2. The values of the coordinates along the

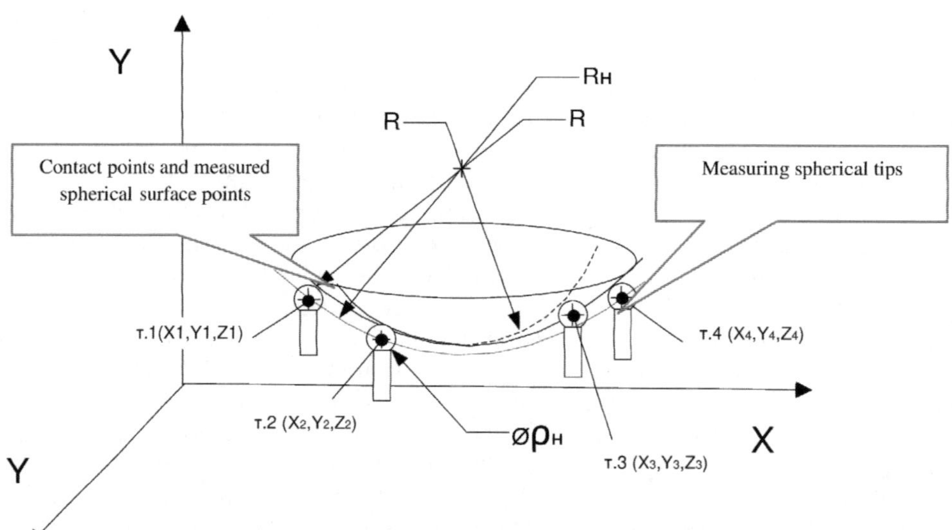

axis Z - (Z_1, Z_2, Z_3, Z_4) obtained in the process of measurement must be inserted into a system of four equations with four unknown parameters of a spherical surface (1).

Fig. 2 Diagram of four-point measurements of the radius and coordinates of its center.

IOP Conf. Series: Materials Science and Engineering **570** (2019) 012084 doi:10.1088/1757-899X/570/1/012084

In this case, the system of equations will have one solution, in the form of the coordinates of the position X_0, Y_0 , Z_0 , R_H, the center of the spherical element of the tip of the measuring indicator and the radius of the spherical surface tangent to the measurement points.

$$\begin{cases} (X_1 - X_0)^2 + (Y_1 - Y_0)^2 + (Z_1 - Z_0)^2 = R_H^2 \\ (X_2 - X_0)^2 + (Y_2 - Y_0)^2 + (Z_2 - Z_0)^2 = R_H^2 \\ (X_3 - X_0)^2 + (Y_3 - Y_0)^2 + (Z_3 - Z_0)^2 = R_H^2 \\ (X_4 - X_0)^2 + (Y_4 - Y_0)^2 + (Z_4 - Z_0)^2 = R_H^2 \end{cases} \quad (1)$$

The found values of the coordinates of the sphere center of the measuring tips X_0, Y_0, Z_0 are the desired coordinates of the center of the measured spherical surface of the ball pin X_, Y_, Z _.

To find the value of the radius of the spherical surface R, it is necessary to subtract from the found radius $R_{\_H}$ the value of the known diameter of the spherical tip $\rho_{\_H}$ (2).

$R = R_{\_H} - \rho_{\_H}$ (2)

To use the methodology, it is necessary to automate calculations for finding solutions to the system of equations. The simplest method of finding a solution to a system of equations is to use the built-in search for solutions add-in in the standard Excel package. To test the methodology, the above formulas of the equations were entered into the program cells. The actual values of the coordinates are obtained by measuring the spherical surface of a known diametrical size. The measurements were carried out when installing the part in the universal globes vice with the single-coordinate altimeter of the MahrDigimahr 817 L firm, similar to the method of finding the diametrical size in [4-8]. The measured values of the indicators are highlighted in red in table 1.

3. Practical results

The main condition for successful solution of the equation is that the coordinates of all four points should not lie in the same plane (this condition must be taken into account when making the body of a special measuring device). After that, a solution was found to the system of equations.

Table1.Table of search results for a system of equations by a superstructure "Search for solutions" package Excel

c conventional notation	Diameter of measuring tip, mm. .			
Ppoint number	1	2	3	4
x xi	-10,000	9,000	10,000	-9,000
y yi	10,000	9,000	-10,000	-9,000
z zi	10,001	9,000	10,003	9,002
	The results of solving the equation - the coordinates of the radius of the measured sphere, mm			
x x0	0,000000	0,000000	0,000000	0,000000
y y0	0,000000	0,000000	0,000000	0,000000
z z0	28,500000	28,500000	28,500000	28,500000
RH	23,286262	23,286262	23,286262	23,286262
	The square of the radius of the sphere center points			
RH ^2	542,25	542,25	542,25	542,25
	Sphere radius, mm			
R	20,286262			

The obtained values of the radius of the sphere and the coordinates of the position of its centers coincided with their theoretical values.

Thus, it has been established that the developed four-point measurement technique is suitable for finding characteristics of the accuracy of a spherical surface provided that a special measuring device is developed, for example, using inductive linear displacement sensors, and measurement data processing instrumentation.

References

[1] Safarov D T and Kondrashov A G 2018 *Stand for monitoring the operational parameters of conjugations "ball support - Body of the tie-rod end" of automotive components* IOP Conference Series: Materials Science and Engineering Vol 289 **012018**

[2] VasilievV A, Odinokov S A and Serov M M 2016 *Modern methods for the quality management of high-rate melt solidification Russian Metallurgy* Metally Issue 13 pp 1321-24

[3] Safarov D T, Fedorova K A and Ilyasova A I 2016 *Algorithms development of making special techniques in APQP manufacturing process of automotive components* Vol 134 Issue 1 8 **012036**

[4] Kondrashov A G, Safarov D T and Kas'yanov S V 2013 *Diagnosticheskie izmereniya geometricheskih parametrov prostranstvenno-slozhnyh detalej avtokomponentov odnokoordinatnym vysotomerom* Kontrol'. Diagnostika No 8 pp 60-64

[5] Kondrashov A G, Safarov D T and Kas'yanov S V 2013 *Izmerenie otnositel'nogo polozheniya i radiusa cilindricheskih ehlementov detali priborom dlya izmereniya vysot* Promyshlennye ASU i kontrollery No 3 pp 18-24

[6] Safarov D T, Kasyanov S V, Kondrashov A G 2019 *Informative value of measurements for quality management of auto parts* Lecture Notes in Mechanical Engineering pp 1657-66

[7] Safarov D T, Kondrashov A G and Khafizov I I 2018 *The method of diagnosing machine systems by measuring the accuracy of manufactured parts* IOP Conference Series: Materials Science and Engineering Vol 412 No 1 **012065**

[8] Kasjanov S V, Kondrashov A G and Safarov D T 2017 *Regulation of Geometrical Parameters Deviations of Automotive Components Parts through Diagnostic Measurements Organization* Procedia Engineering V 206 pp 1508-14

IOP Conf. Series: Materials Science and Engineering **570** (2019) 012085 doi:10.1088/1757-899X/570/1/012085

Promising methods for enhancing the performance of boiler plants of centralized, decentralized and individual type of heat supply

T R Safin, I A Konakhina and G R Khamidullina

Kazan (Volga Region) Federal University
18 Kremlyovskaya Str., Kazan 420008

Abstract. The paper covers the methods of how to enhance boiler plants for centralized, decentralized and individual type of heat supply. The performance analysis uses technical and economic efficiency indicators, fuel and other types of energy resources saving.

Boiler plants in our country are used as sources of thermal energy to provide heating, ventilation and hot water supply. Most often, such a source supplies heat to various groups of consumers distributed over long distances and connected by means of extensive heating networks. In the last 10-15 years, individual heat supply systems designed to provide heat to a single consumer have become more popular, as well as decentralized systems that supply heat to several consumers united geographically on a fairly small area (for example, skid-mounted boiler plants).

The challenge of how to enhance the performance of boiler plants in terms of energy efficiency is actual and significant for the economy of our country due to the fact that they share 60% of total energy saving sector, while their fuel consumption is 80% [1-3]. Its solution requires actions that will contribute to energy conservation and can give a significant effect on fuel economy in the whole country.

Centralized boiler plants.

The main items of energy consumption at centralized boiler plants are:

- heat supply to consumers as steam of industrial characteristics and hot network water providing the heating, ventilation and hot water supply to the consumers;
- discharge of heat to the atmosphere along with fuel combustion products (exhaust gases);
- loss of heat energy due to external cooling of heat exchange hardware and incomplete combustion of fuel in the boiler;
- heat release to the atmosphere with gases removed during the process of deaeration.

Actions that will reduce dead outlay at boiler plants are primarily aimed at reducing heat release to the atmosphere with gases discharged to the atmosphere (exhaust and deaeration gases). Below are two actions:

1. Heat recovery of the discharged gases in condensation heat exchangers [4].
2. The use of heat pump plants [5-8]. This is one of the most successfully promoted actions for the recovery of low-grade secondary energy resources in the energy sector, industry facilities and housing and utility infrastructure. The heat pump plants can be used in boiler plants to increase the

potential of secondary energy resources to a higher temperature level at which their recovery becomes most efficient [14,15].

Both of these actions are competing with each other, because they are able to displace the heat load of the same units, so a special study is required to select one of them or optimize the combined solutions. The method of temperature matching is the most descriptive approach to determining the allowable loads of heat recovery units (Fig. 1) [1, 5, 10].

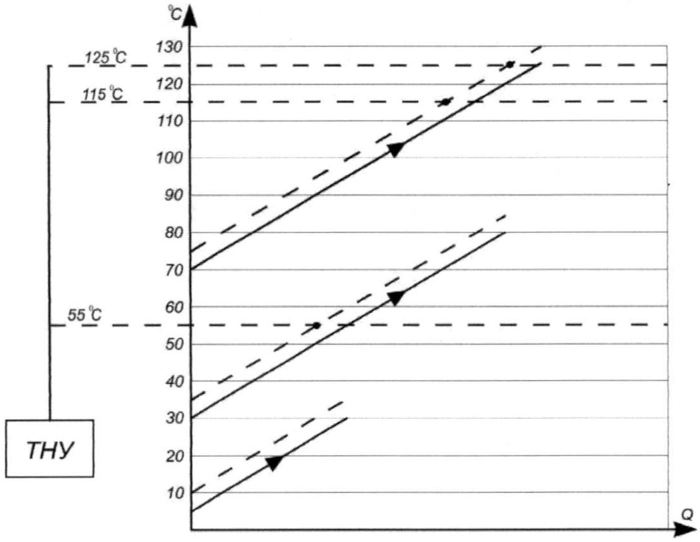

Fig. 1 Determination of allowable loads of heat recovery units using the method of temperature matching

The recovery of secondary energy resources is possible for the following processes [5]:
1) heating of raw water for the makeup system of the boiler plant (Fig. 2, *a*);
2) heating of chemically purified water upstream of the deaerator (Fig. 2, *b*);
3) heating of network water (Fig. 2, *c*).

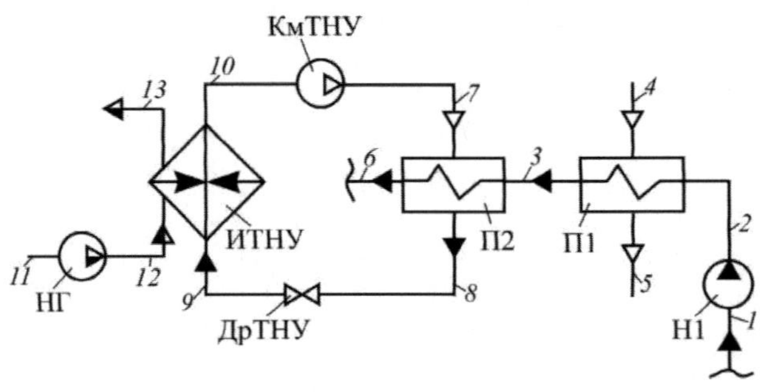

a)

IOP Conf. Series: Materials Science and Engineering **570** (2019) 012085 doi:10.1088/1757-899X/570/1/012085

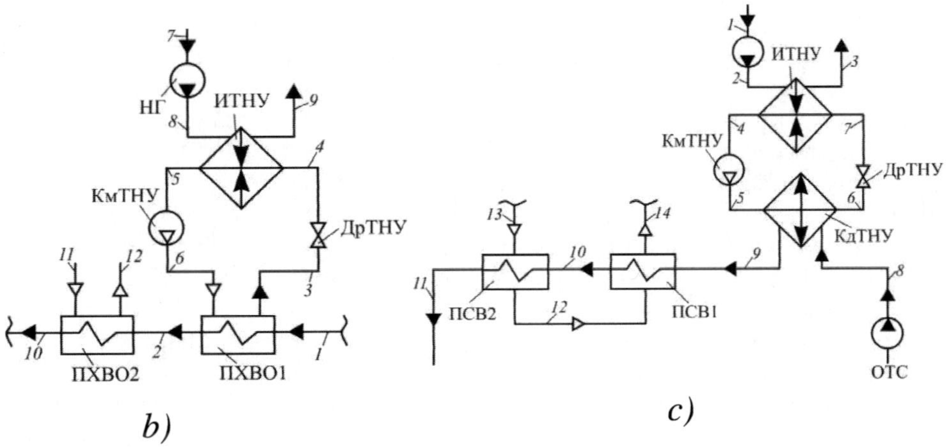

Fig. 2 Possible options of heat recovery of secondary energy resources included in the revised layout of the boiler plant
a) raw water heating; *b)* chemically purified water heating;
c) network water heating

The acronyms on Fig. 2 have the following meaning: KmTNU, ITNU, KdTNU DrTNU stand for respectively, compressor, evaporator, condenser, throttle of heat pump plant; P1, P2 stand for make-up heat exchangers; PHVO1, PHVO2 stand for preheaters of chemically purified water; PSV1, PSV2 stand for raw water heaters, N1, NG stand for pumps; OTS stands for return line of heat network.

In addition to the above actions, the recovery of secondary energy resources for heating the deaerator could be an option.

Boiler plants of decentralized and individual type of heat supply.

For decentralized boiler plants and individual heat supply, the choice of actions is narrower. In this case, heat pump plants have no competitors. Moreover, using of solar collecting panels and batteries in the installations makes it possible to synthesize hybrid plants [8].

Analysis of decision effectiveness

The most important technical and economic indicators of the performance of boiler plants are the cost and profitability of the heat energy produced (with regard to environmental requirements) and, as a result, minimization of the payback periods of power plants.

The main criteria for meeting these requirements are [7, 9-12]:

1) achieving the maximum possible fuel utilization efficiency in a power plant (the ratio of net energy to fuel energy) [13];

2) the maximum possible reduction in capital costs and timelines of construction of power plants.

Based on the calculation of these indicators, the fuel utilization efficiency can be increased for boiler plants by 5-7% with a payback period of 1-2 years.

References

[1] Konakhina I A and Nazmeev Yu G 2002 *Heat power systems and power balances of production facilities*. M: Moscow Power Engineering Institute Press

[2] Tsvetkov V V 1980 *Organization of steam supply of the industrial firms* (M.: Energy) p. 206

[3] Levin M S 1971 *Use of the fulfilled and secondary steam and condensate* (M.: Energy) p.144

[4] Konakhina I A, Safin T R and Gilmanshin I R 2013 *Performance enhancement of fuel-using installations due to external waste heat recovery units* Proceedings of the International Scientific-Technical Conference on Innovative Engineering Technologies, Equipment and Materials. Part 2. – Kazan: Foliant, LLC. pp 209-213

[5] Voropaev A N 2007 *Performance enhancement of cooling system for technical oils used in compressor stations and cogeneration plants: abstract*. dis. ... Ph.D. in Engineering Science. National Research University "Moscow Power Engineering Institute", Moscow.

[6] *A new generation of heat pumps for heat supply purposes and the efficiency of their use in a market economy*. Proceedings of the meeting of the Heat and Centralized Heat Supply subsection of Research and Engineering Board of OAO Unified Energy System of Russia, Moscow, 2004.

[7] Frolov V P, Shcherbakov S N, Frolov M V and Shelginsky A Ya 2004 *Performance analysis of heat pumps used in centralized hot-water supply systems*. Energsberezheniye. No. 2.

[8] Konakhina I A, Khamidullina G R and Khusnutdinova E M 2015 *Hybrid approach in centralized heat and power supply systems*. Proceedings of the International Scientific-Technical Conference on Innovative Engineering Technologies, Equipment and Materials. Part 1. – Kazan: Foliant, LLC. pp 334-337

[9] Israfilov Z K and Kashapov N F 1991 *Journal of Engineering Physics* 60 pp. 364–368

[10] Konakhina I A 2011 *Thermodynamic analysis of collecting and condensate reusing system* (IGEU Bulletin) pp. 11-14

[11] Tabunshchikov Yu A 2005 *Guide on the estimation of economic efficiency investment into the energy saving measures* (M.: AVOK-PRESS) p. 120

[12] Dautov G, Dautov I, Fayrushin I and Kashapov N 2013 *J. Phys.: Conf. Series* 479 012014

[13] Kashapov N F and Sharifullin S N 2015 IOP Conf. Ser.: Mater. Sci. Eng. 86 012021

[14] Fazlyyyakhmatov M and Kashapov N 2014 High Temperature Material Processes 18 (4) pp 273–279

[15] Gubaidullin D A, Kashapov N F, Zaripov R G, Tkachenko L A and Shaydullin L R 2017 J. Phys.: Conf. Ser. 789 012017

IOP Conf. Series: Materials Science and Engineering **570** (2019) 012086 doi:10.1088/1757-899X/570/1/012086

Project activity as a stage of continuous engineering and technical education

R N Sagitova[1], I R Gilmanshin[1,2], A R Akchurina[1]

[1]Kazan Federal University, Kazan, Russia, 420008

[2] Kazan National Research Technical University named after A N Tupolev, Kazan, Russia

Is-er@yandex.ru

Abstract. The distinctive characteristics of the project activity as a stage of continuous engineering and technical education are highlighted. An example of an educational and research project on the corrosion resistance of aluminum is considered, and the competences are defined. The application of project training for the formation of subject competencies and early vocational guidance of students in the field of engineering education has been substantiated.

The modernization of national higher education and the educational policy pursued by the European Union stimulate innovative processes in engineering and technical education. One of them is the project activity of students as a stage of continuous engineering and technical education.

Although the ideas of developing students 'interest in engineering and technical sciences through the development of creative personality traits are not new and have always aroused the interest of scientists. Problems in physics, electrical engineering, chemistry for the development of students' creative abilities have been developed. However, the formation of the experience of engineering and technical creativity requires a new look at design training as a stage of continuous engineering and technical education.

Project activities are inseparable from the present period of development of education ([1], [2]) and science ([3], [4], [5]). Today, in the Russian educational standard of a new generation, project activity is considered as the implementation of a system-activity approach to the formation of universal educational activities. And universal educational activities are a tool to achieve the planned results of school education and the goals of early vocational guidance, in particular on engineering and technical activities.

Thus, taking into account the aforementioned and significant role of project activities in the implementation of the Russian educational standard of a new generation, the proposed study can be considered as one of the attempts to find ways to develop students' interest in engineering and technical sciences by organizing a system of design activities on corrosion processes and energy saving measures based on the realities of today.

IOP Conf. Series: Materials Science and Engineering **570** (2019) 012086 doi:10.1088/1757-899X/570/1/012086

Project activities outside school hours is possible through the implementation of educational research and research projects.

To implement a research project requires, as a rule, the partnership of the school and the university, which is not always possible.

For the implementation of an educational and research project, most often there is enough modern school equipment - a set of micro-laboratories with a set of reagents and a set of digital laboratories.

Modern digital laboratories are equipped with a whole set of sensors for instrumental analysis: pH, electrical conductivity, temperature, gas volume sensors with temperature control, spectrophotometric sensors, etc. All this allows to conduct educational and research projects at a sufficiently high level.

Chemical engineering in the field of corrosion processes and energy saving is necessary for a successful engineering activity.

In the school chemistry course, much attention is paid to metals, their structural features, corrosion resistance and protection from chemical and electrochemical corrosion. Aluminum is an interesting metal from this point of view.

Aluminum in its position in the electrochemical series of the voltage of metals is an active metal ($E^0_{Al^{3+}/Al} = -1{,}662\ B$),but due to the dense air and waterproof oxide film, it is sufficiently resistant to corrosion. In addition, aluminum can react with acids and alkalis.

In front of a group of students of the Lyceum. The Lobachevsky Kazan Federal University was tasked with the implementation of an educational and research project entitled "Investigation of the stability of aluminum with respect to solutions of strong acids and alkalis".

The implementation of the project included the following action plan:

1) formulation of the problem and analysis of literature data;
2) planning and conducting an experiment;
3) analysis and interpretation of experimental data;
4) preparation of the report and presentation;
5) project protection.

Below are the results obtained by schoolchildren in the course of teaching and research activities.

1). When using a solution of sodium hydroxide and a solution of hydrochloric acid with a concentration of 0.1 mol / l, there are no visible signs of a reaction for 15 minutes.

2). When using sodium hydroxide solution and hydrochloric acid solutions with a concentration of 3 mol / l, signs of a chemical reaction appear: gas bubbles are released.

3). Using the gas volume sensor, the dependence of the volume of the released gas (hydrogen) on time was obtained. The results are presented in the form of a graphical dependence (see Fig. 1).

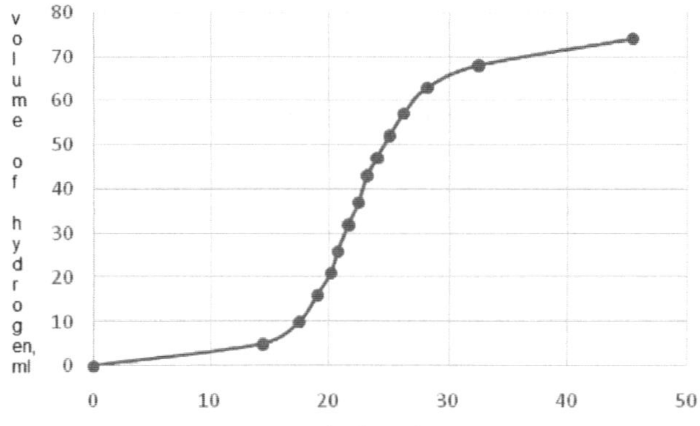

Fig. 1. Dissolution of aluminum in hydrochloric environment

IOP Conf. Series: Materials Science and Engineering **570** (2019) 012086 doi:10.1088/1757-899X/570/1/012086

On the graph of the dissolution of the plate of aluminum in hydrochloric acid can be divided into three areas:

1) Shallow area at the beginning of the graph: a protective film protects aluminum from rapid destruction, gas is released slowly;
2) Sharp rise, almost linear dependence: the surface of aluminum is cleared from the oxide film, the area of the cleaned surface increases, the rate of hydrogen evolution also increases, the gas is emitted rapidly;
3) Aluminum surface decreases, the reaction rate decreases, the rate of hydrogen evolution stabilizes and becomes almost constant (the dimensions of the plate during the experiment have not changed much).

Conclusion: the presence of a flat area at the beginning of the curve confirms the presence of a protective film of aluminum.

Similar experiments were carried out for the reaction of aluminum with alkali.

Next, we will compare the stages of research activities and formulated universal educational actions for students (see Table 1).

Table 1

Stages of research	Formed versatile learning activities		
	regulatory	cognitive	communicative
Problem statement and literary data analysis	Setting goals, the transformation of practical tasks in the cognitive, prediction, goal-setting.	Search and analysis of information, logical operations (comparison, classification, systematization, synthesis, analysis); ability to formulate a research task	The ability to formulate a task, build speech statements, argue your position
Experiment Planning and Implementation	Ability to correlate their actions with the planned results; evaluation and control of their actions volitional self-regulation.	The ability to build simple and complex algorithms based on logical actions, choose a rational solution	Ability to build speech statements, learning interaction, cooperation
Preparation of the report and presentation.	Control, evaluation and correction; possible new goal setting	Comparison, classification, synthesis, construction of conclusions (analysis and synthesis)	The ability to formulate a speech statement, to argue their position

In general, with the implementation of this educational research project, the subject competences are formed:

- the definition of the concepts of "chemical reaction rate", "corrosion", "chemical corrosion", "electrochemical corrosion";
- chemical properties of aluminum;
- the application of scientific research methods, the observation of natural science phenomena, the conduct of experimental research.

These competencies are the basis of any engineering activity in the field of mechanical engineering and contribute to the early professional orientation of students.

IOP Conf. Series: Materials Science and Engineering **570** (2019) 012086 doi:10.1088/1757-899X/570/1/012086

References

[1] Gilmanshin I R, Gilmanshina S I 2016 *IOP Conf. Ser.: Mater. Sci. Eng.* **134** 012006
[2] Gilmanshina S I, Gilmanshin I R 2015 *IOP: Conf. Ser. Mater. Sci. Eng.* **86** 12029-32
[3] Azimov Yu I, Gilmanshin I R, Gilmanshina S I 2016 *IOP: Conf. Ser. Mater. Sci. Eng.* **134** 012003
[4] Gilmanshin I R, Kashapov N F, Gilmanshina S I, Galeeva A I 2016 *IOP: Conf. Ser. Mater. Sci. Eng.* **134** 012007
[5] Gilmanshin I R, Ferenets A V, Azimov Yu I, Galeeva A I, Gilmanshina S I 2015 *IOP: Conf. Ser. Mater. Sci. Eng.* **86** 12014-16
[6] Israfilov Z Kh, Kashapov N F 1991 **60** 364–368
[7] Galyautdinov R T, Kasparov N F and Luchkin G S 2002 *Inzhenerno-Fizicheskii Zhurnal* **75** 170–173
[8] Kashapov L N, Kashapov N F and Kashapov R N 2014 *J. Phys.: Conf. Ser.* **567** 012025
[9] Kornienko E E, Lapushkina E J, Kuzmin V I, Vaschenko S P, Gulyaev I P, Kartaev E V, Sergachev D S, Kashapov N, Sharifullin S, Fayrushin I 2014 *J. Phys.: Conf. Ser.* **567** 012010

IOP Conf. Series: Materials Science and Engineering **570** (2019) 012087 doi:10.1088/1757-899X/570/1/012087

Changes in the CoCr alloys surface relief during plasma electrolytic treatment

L R Saleeva[4], R N Kashapov[1,2], L N Kashapov[1,3], N F Kashapov[1]

[1] Engineering Institute, Kazan Federal University, Kazan, 420008, Russia
[2]Laboratory of Radiation Physics, Zavoisky Physical-Technical Institute, FRC Kazan Scientific Center, Russian Academy of Sciences, Kazan, 420029, Russia
[3]Laboratory of Continuum Mechanics, Institute of Mechanics and Engineering, FRC Kazan Scientific Center, Russian Academy of Sciences, Kazan, 420111, Russia
[4]Kazan State Medical University, Kazan, 420012, Russia

kashramiln@gmail.com

Abstract. Development of dentistry is strongly correlated with the new treatment method development and construction of metal, polymer and ceramic materials. Metal-ceramic crown is a common dental prosthesis. The aim of work was determination the causes of the ceramic coating spalling from the cobalt chrome metal frame. It is proposed to replace the standard sandblasting method to plasma electrolytic treatment. The burning of single microdischarges occurs during plasma-electrolytic process. The formation of the microwells and increasing surface roughness observed in cathode mode. This is due to local melting of the surface. The treated surface does not contain impurity elements that can cause the formation of microcracks.

Introduction

The work is devoted to the study the plasma-electrolyte surface treatment processes of cobalt-chromium alloys used in dentistry. The frame of the metal-ceramic crown is cast from these alloys. The metal frame is lined with dental ceramics, and thus we will get the construction of metal-ceramics. Before applying the ceramic coating, the surface of the frame must be treated with a sandblasting method to increase its roughness. The increase in the roughness parameters contributes to the increase in the adhesion of ceramics to the metal. The metal-ceramic crown is widely used because of its versatility. It is durable and can be used to restore posterior teeth. It is aesthetic and can be used by the smile line. Ceramics has a big palette of shades, which allows you to choose an individual color corresponding to the shade of the patient's teeth. The simplicity of the method of application of ceramics, which does not require special technical devices other than a conventional muffle furnace, and the above-mentioned advantages have caused a wide spread of the metal-ceramic crown. However, as statistic shows, there are cases of chipping ceramic coating from the surface of the metal frame. It was previously established [1] that one of the most likely reasons is the presence of nano-and micro-sized particles of electrocorundum, which is used in sandblasting. High-pressure steam generators are used to remove these particles, but this is not enough to remove electrocorundum particles. These particles are a source of internal stresses and microcracks, as they lie on the border of metal and ceramics. Therefore, the simplest solution is the selection of a new simple non-mechanical method of microrelief formation. Non-mechanical treatment methods include electrophysical and electrochemical processes, laser and ultrasonic treatment. Another important requirement is the geometry of the processed frames, repeating the shape of the human tooth. Considering all the above methods, we can conclude the following:

- laser processing requires expensive equipment;
- electrophysical and electrochemical methods produce anodic dissolution over the entire surface, and the electroerosion process requires a counter-electrode, the control of which is very difficult;
- microplasma processing takes place in a vacuum condition and requires expensive equipment.

The solution to the problem can be found in the use of low-temperature atmospheric pressure plasma initiated from a conventional low-voltage power supply up to 300 V. One possible source of such plasma is gas discharges with liquid electrodes. This method of processing is carried out when the voltage is applied to the electrodes in the electrolyte solution, as a result, a vapor-gas discharge burns around one of the electrodes [2]. The peculiarity of this discharge is that it burns between the solid and the liquid. The solid electrode can be both anode and cathode. The combustion discharge on the cathode is cleaning and polishing workpiece, as under certain conditions it is possible to apply coatings [3]. The anode process on metals that have a tendency to form oxide protective film with high resistance during anodic polarization, anodic oxidation occurs in conjunction with the burning of micro-arcs, this process called micro-arc oxidation [4]. On metals that do not have this property, there is an erosion of the metal and its dissolution. The main difference of this method from the anode process of plasma-electrolyte treatment is that the treatment is carried out only by the interaction of plasma with the surface of the product, and anodic dissolution does not occur under the influence of electrochemical reactions.

The work is devoted to the study of physical and chemical processes of plasma-electrolyte treatment of cobalt chromium alloys, consists of the combination of two processes: the erosion of the metal and its anodic dissolution [5]. Erosion destruction occurs under the influence of a large number of independent micro-electric discharges, which are initiated at the boundary of the liquid and the metal electrode at different times and continue to develop independently of each other. Anodic dissolution of the metal is due to electrochemical reactions. It is necessary to determine the conditions under which the formation of the surface micro-relief happens.

Main part

The implementation of the plasma-electrolyte process can occur in two modes: anode or cathode polarization of the processed frame.

To study the process of plasma-electrolyte treatment, the appearance of micro-discharges and their effect on the surface of the processed metals was made an experimental setup, the functional scheme of which is shown in figure 1. It consists of a power supply system 1, the electrolytic bath 2, the fixing system of the electrode pattern 3, the oscilloscope 4, shunt, 5, voltmeter 6, ammeter 7, the thermocouple 8 and the specimen mounting 9.

The main studies were carried out on samples of cast cobalt-chromium alloy firm Starbond CoS with the following composition and characteristics: Co - 59%, Cr - 25%, W - 9%, Mo - 3.5%, Si - 1%, other components (C, Fe, Mn, N) - maximum 1.5; the limit of the tensile strength (Rp 0.2) Of 650 MPa, a tensile strength of 910 MPa, ultimate elongation 8, the modulus of elasticity of 200 GPa, the Vickers hardness of 280 HV 10, a density of 8.8 g/cm3, solidus-liquidus interval 1305 - 1400°C, the temperature of the casting 1500 - 1550°C, the thermal expansion coefficient of 14.0 µm/m °C. The frameworks of metal-ceramic crowns in dentistry are made from this type of alloy. The electrolytic bath is filled with electrolytes of the required concentration and composition. As a working fluid were used the aqueous solutions of NaCl, Na2CO3, the concentration of which was 1 - 5% by weight.

The electrical power system is a DC (direct current) power supply with a continuously adjustable voltage. The power supply consists of a diode bridge and a laboratory autotransformer adjustment type 1M with a voltage range from 1 to 240V, 50 Hz. Also was used pulsating current source obtained after a two-half-period rectification. The change of the current and voltage form at the

IOP Conf. Series: Materials Science and Engineering **570** (2019) 012087 doi:10.1088/1757-899X/570/1/012087

moment of discharge ignition was determined using an oscilloscope FLUKE 105 SCOPEMETER SERIES II, time scan varied from 5 NS to 60 seconds. Relative measurement errors did not exceed 0.025%. To prevent changes in the size of the anode area, a dielectric protective cover was put on, which leaves open only a strictly defined part of the electrode. The area of the electrode on the surface of which was initiated the process of burning plasma was 0.8 cm^2, the area of the counter-electrode 34$см^2$.

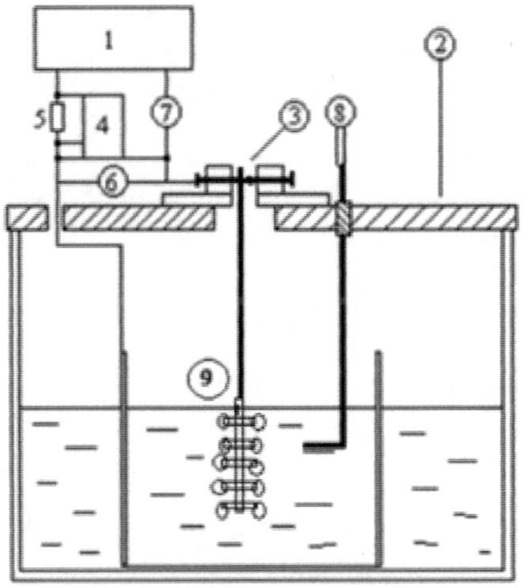

Figure 1 - Experimental setup

The main working parameters of the plasma-electrolyte process are: voltage on the electrodes, discharge current and current density on electrodes.

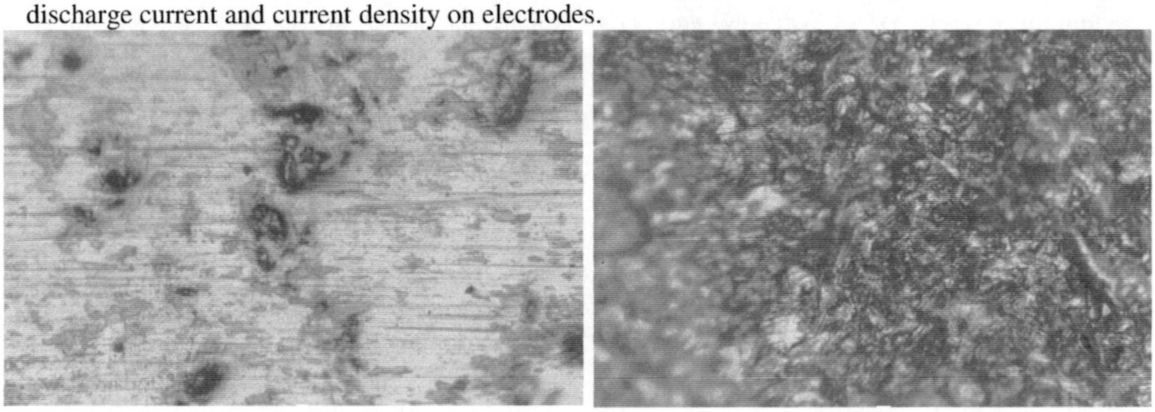
a b

Figure 2 - Image of the surface with magnification × 20: a - after machining with a mill, b - after sandblasting.

Voltage measurement at the anode was carried out using a digital universal measuring device APPA 105, the relative measurement error is 0.1%. With a digital multimeter, APPA 305 was carried out the temperature measurement of the electrolytic cathode with the help of the chromel-copel thermocouple. The electrolyte temperature was measured by a mercury thermometer.

Current-voltage characteristics were measured as follows. The voltage on the electrodes was increased stepwise from 0 to 240-250 V with a step of 5-10 V and exposure for 5 seconds with continuous recording of the voltage between the electrodes, the current and the electrolyte temperature.

The temperature of electrolyte reached 970 °C for two minutes and didn't change.

After plasma-electrolyte processing, the samples were examined using an optical microscope Leica EZ4D, which allows to obtain images of opaque objects, as well as to produce microphotography of these surfaces. The survey was conducted at three points of the sample.

IOP Conf. Series: Materials Science and Engineering **570** (2019) 012087 doi:10.1088/1757-899X/570/1/012087

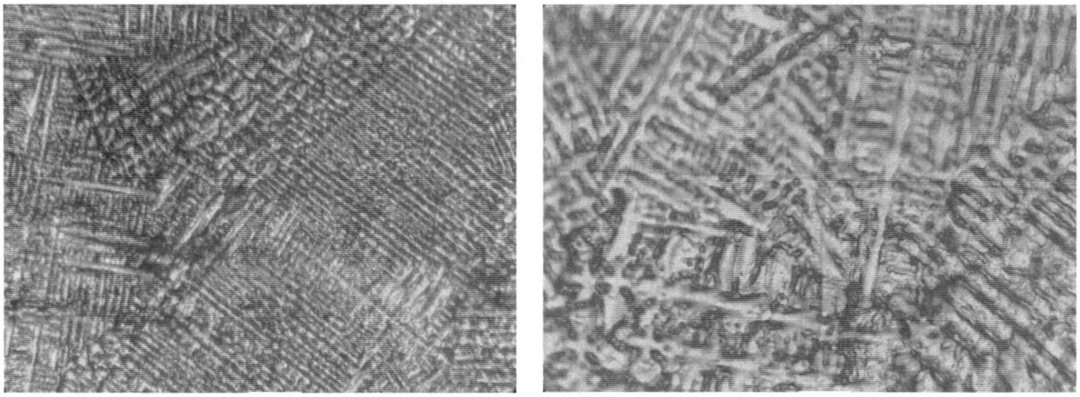

a (×20) b (×40)

Figure 3 - The image of the cobalt-chromium alloy surface after plasma-electrolyte treatment at anode mode: a – ×20 magnification, b – ×40 magnification.

Consider the anode mode of plasma-electrolyte treatment, the current-voltage characteristic of which is shown in figure 4 a. The use of two different polarizations showed that in the anode mode when the electrolyte temperature is above 70ºC, anode dissolution of the electrode prevails, and power of discharge is not enough to form local irregularities and grooves. Figure 3 shows the images obtained on an optical microscope, where we can observe the dendritic structure of the alloy. Similar images are obtained by chemical etching of samples for metallographic analysis. At small processing times, the use of "cold" electrolyte temperature 22ºC and the value of the applied voltage U = 230 V observed the formation of local depressions and holes, under the action of single water hammer formed by micro-discharges. However, intensive heating of electrolyte under the action of gas discharges leads to an increase of temperature, which in turn leads to the disappearance of water hammer. Thus, the anode mode of plasma-electrolyte treatment is inefficient for the formation of the micro-relief surface of cobalt-chromium alloys. This mode of plasma polishing will be effective for reducing the roughness of clasp dentures.

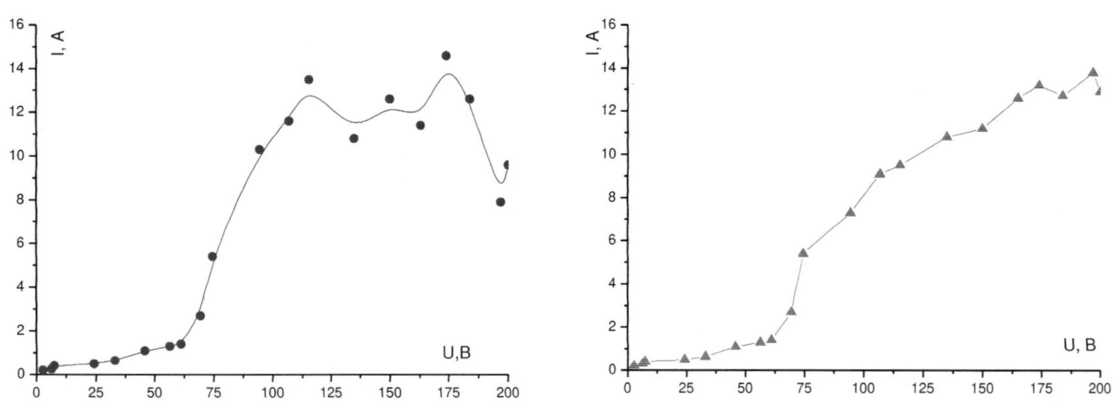

Figure 4 - CVC of the plasma-electrolyte process: a – anode mode, b – cathode mode.

Consider the cathode mode of plasma-electrolyte treatment, the current-voltage characteristic of which is shown in figure 4 b. In the cathode mode is observed more intense of the light effect of single micro-discharges. The thermal energy released by micro-discharges is sufficient to melt the active electrode. Therefore, an important task was the need to select the values of voltage and discharge current for local melting of the surface at which the formation of micro-holes and increase the surface roughness will occur. The lack of a process of dissolution of the electrode during the cathodic polarization eliminates polishing effect observed in the anodic regime. Figure 5 shows surface images obtained at U= 194 V, I = 12.6 A, processing time 40 seconds.

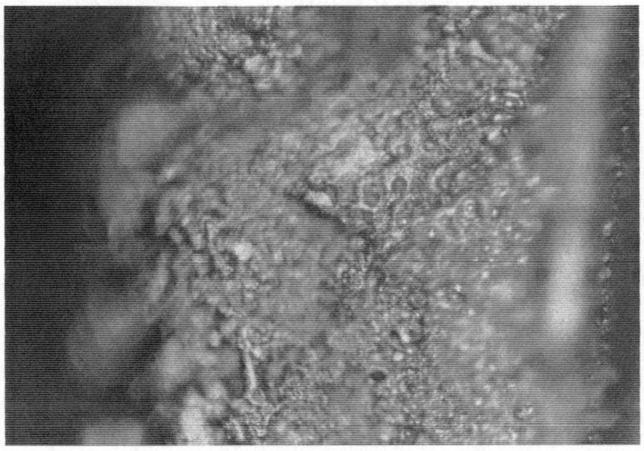

a (×20)

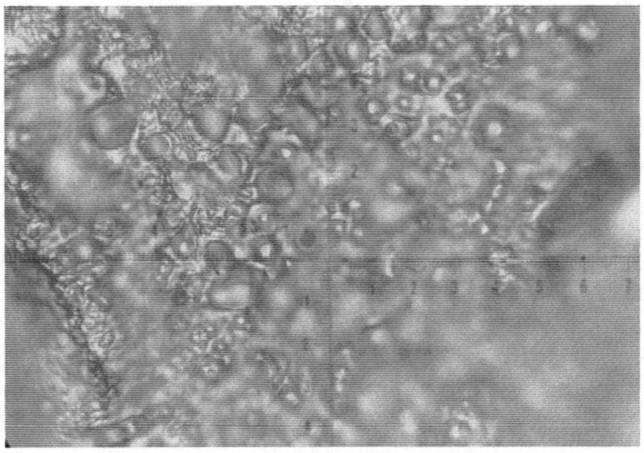

b (×40)

Figure 5 - Image of the cobalt-chromium surface after plasma-electrolyte treatment at the cathode mode: a – ×20 magnification, b – ×40 magnification.

Under these conditions, local melting of the surface and the formation of micro-holes and grooves ranging in size from 40 to 120 microns were observed. This microrelief is suitable for subsequent application of a ceramic coating. The analysis of the volt-ampere characteristics of the two modes shows that the anode mode is characterized by the presence of a maximum in the region of 130 V, fixed fluctuations in the amplitude of the current in the range from 110 to 180 V should be also noted. In the case of the cathode regime, there is a smooth increase in volt-ampere characteristic (VAC) with the presence of an inflection zone in the region of 60 – 75 V. This area is characterized by the beginning of the combustion of micro-discharges.

Conclusion

On the basis of the conducted researches, it is established that the formation of microholes is observed at the cathode mode, as a result of local melting from single micro-discharges. A comparison of anode and cathode modes showed the advantages of using the latter. This is due to the lack of anodic dissolution of the surface and a higher rate of electrical erosion. However, a more intense heat without control of the surface temperature may lead to the electrode melting. In the future, it is necessary to investigate the thermal effect on the cobalt-chromium alloys microstructure under these conditions. In general, we can conclude that this new method effectively solves the problem of replacing the sandblasting method for processing metal frames of dental prostheses. The obtained data can be used to create specialized dental equipment.

Acknowledges

The work was carried out in the framework of the Base part of the state task of Ministry of Science and Higher Education of the Russian Federation (3.9399.2017/8.9)

References

[1] Kashapov L N, Kashapov N F, Kashapov R N 2014 *Research of cobalt chromium alloy surface morphology after sandblasting* IOP Conference Series: Materials Science and Engineering Vol 69 Is 1 **012017**

[2] Kashapov L N, Kashapov N F, Kashapov R N, Denisov D G 2016 *Plasma electrolytic treatment of products after selective laser melting* Journal of Physics: Conference Series Vol 669 Is 1 **012029**

[3] Kashapov L N, Kashapov N F, Kashapov R N 2013 *Research of the impact acidity of electrolytic cathode on the course of the plasma-electrolytic process* Journal of Physics: Conference Series Vol 479 Is 1 **012011**

[4] Ebrahimi S, Bordbar-Khiabani A, Yarmand B 2019 *Enhanced optoelectronic performance of plasma electrolytic oxidized monocrystalline silicon using rGO incorporation* Materials Letters Vol 239 pp 151-54
[5] Barooghi B, Sheikhi M, Amiri A 2018 *Effect of nano-hydroxyapatite and dutycycle on the structure and corrosion performance of plasma electrolyte oxidation coatings in simulated body fluid on Ti–6Al–4 V* Proceedings of the Institution of Mechanical Engineers, Part C: Journal of Mechanical Engineering Science Vol 232 Is 23 1 pp 4229-36

IOP Conf. Series: Materials Science and Engineering **570** (2019) 012088 doi:10.1088/1757-899X/570/1/012088

Development of a polymeric fuel tank with high barrier properties

L N Shafigullin, A M Sotnikov, N V Romanova, E S Shabaeva and D R Sarimov

Naberezhnye Chelny Institute (branch), Kazan Federal University
Sjujumbike street 10a, Naberezhnye Chelny, 423800 Tatarstan, Russia

misharin_82@mail.ru

Abstract. The choice and justification of the polymeric material is made for the fuel tank. It is established a possibility of linear polyethylene application with technology of fluoration as material for production of the fuel tanks with high barrier properties, that is with low permeability in relation to hydrocarbons. The main technological operations for the production of plastic fuel tanks carried out by rotational molding are described.

1. Introduction

The plastic has become the main material used by manufacturers of fuel tanks. In Europe, 90% of all fuel tanks are made of plastic. In the US, more than 75% of automotive fuel tanks are made of laminated plastic with a barrier layer.

The plastic fuel tanks are made from the polyethylene of average and high firmness - strong and light material which allows producers of cars to reduce significantly the lump of the vehicle and to achieve more economic consumption of fuel.

The use of plastic for the production of fuel tanks also makes it possible to make the car safer in the event of an accident. Unlike metal fuel tanks, many plastic tanks can bend and flatten, and not burst and crack with a gasoline spill. It allows preventing outflow of gasoline which can cause a fire and explosion.

The use of plastic for the production of fuel tanks allows vehicle designers to optimize the space used for placing of the fuel tank, thanks the possibility to give to tank the most specific form. The plastic can be formed around objects of any shape. In addition, the plastic fuel tanks are resistant to corrosion, whereas in the case of manufacturing tanks from steel for ensuring the same level of corrosion resistance it is necessary to use a noncorrosive steel that leads to increase in expenses.

2. Body text

The plastic as a material has several advantages over metals, such as low weight and ease of processing. Thus, it becomes the preferred material for the components of the fuel system of the car. Production of multi-layer plastic fuel tanks has gained considerable momentum over the past ten years [1-7].

The main disadvantage of polyethylene as a tank material is high permeability in relation to hydrocarbons, due to the low degree of polarization of polymer macromolecules and weak bonds between the chains of macromolecules. In the absence of through defects in the tank wall, its tightness is entirely determined by the permeability P of the material in relation to gases diffusing through the wall, which, as it is known, represents of the solubility C of a gas by the diffusion coefficient D:

$$P = C \times D.$$

IOP Conf. Series: Materials Science and Engineering **570** (2019) 012088 doi:10.1088/1757-899X/570/1/012088

The permeability of the polymer depends on the nature of the diffusing gas (chemical composition of the fuel), its solubility in the polymer and the surrounding conditions. The diffusion is accelerated by increasing the difference in concentrations of the diffusing substance on both sides of the tank walls and by decreasing their thickness. The solubility of gas more in amorphous zones of polymer also decreases with increase in degree of crystallinity. The presence of alcohols – methanol and ethanol, introduced to increase the octane number, in the composition of the fuel mixture leads to a significant increase in the diffusion of fuel from the tank cavity to the outside. The presence of alcohols – methanol and ethanol, introduced to increase the octane number, in the composition of the fuel mixture leads to a significant increase in the diffusion of fuel from the tank cavity to the outside. The increasing requirements for environmental protection impose increasingly stringent requirements for the emission of hydrocarbons from the vehicle as a whole, and from the fuel system, the main element of which is the fuel tank.

As the monolayer tank from polyethylene cannot provide necessary balance of mechanical durability, low gas permeability in relation to hydrocarbons and processability at low cost, to reduce the diffusion of hydrocarbon it was developed a number of methods for improving the barrier properties of fuel tanks walls (shareware - barrier technologies). These methods are a fluoridation and a sulfonation.

The fluoridation consists in treating the inner surface of the tank with a gas mixture containing nitrogen (99%) and fluorine (1%), as a resulting in the fluorine atoms replace the hydrogen atoms in the surface layer of the polymer. Depending on the reaction conditions, = *CHF,* = *CF2* or *–CF3* groups are formed, and the molecular structure of the surface layer of HDPE becomes more densely packed compared to the initial polymer, therefore, the solubility of hydrocarbons in the surface layer decreases, it is lead to a corresponding decrease in permeability. The main parameters of the process are the partial vapor pressure of fluorine, the contact time of the working gas with the walls of the tank and the temperature. The amount of fluoride in HDPE should be such as to provide a reliable barrier without degrading the mechanical characteristics of the tank. An important factor is also the structure of the surface: if it has a so-called wrinkled appearance, the probability of cracking is reduced. The fluorinated polymer is not “absolutely” impenetrable, and hydrocarbon vapors diffuse through it in small amounts – 1-2%.

The sulphonation means by itself that sulfur trioxide SO3 is introduced into the cavity of the tank, after its manufacture, and as a result of the reactions taking place, a barrier layer of sulfo groups is formed in the surface layer of polyethylene. The neutralization of the reaction by products is carried out with ammonium hydroxide, the excess of which is removed after washing with water. Here, the mechanism for reducing permeability is the same as in the case of fluoridation. The lack of sulfonation is the dispersion of permeability values when processing different tanks.

Modern production of plastic fuel tanks is carried out by rotational molding – a production process which is used for processing the polyethylene in plastic products [8-10]. This process makes it possible to design plastic fuel tanks of complex shape, which will be formed around the free space of the vehicle. The rotational molding allows the use of a variety of workpiece shape, which consists of several parts, so that the product is easily extracted from the tooling.

Rotational molding consists of four main operations (Figure 1):

1. The filling the blank form with powder resin at 50 ° C;

2. The closing the form, heating up to 300 ° C, holding. In the process of heating the form with the workpiece rotates around its own axis and also around the central axis of the equipment, using centrifugal force. During the heating process, the plastic melts and abuts against the walls of the mold;

3. A cooling the mold by using fans. Air is blown through the walls of the mold, and the workpiece is cooled to a temperature of 70 ° C.

4. The hardening of polyethylene and removal of the product from the mold.

The rotational molding is a universal method of processing polymeric materials, which allows producing high-quality competitive products for various applications.

For rotational molding, special grades of polyethylene are used: PE 6432 R (32604) [11], PE 6433 R (32604) [11], PE 6438 R [11], Borealis Borecene Compact RM7406, ROTOPOL UR 644.

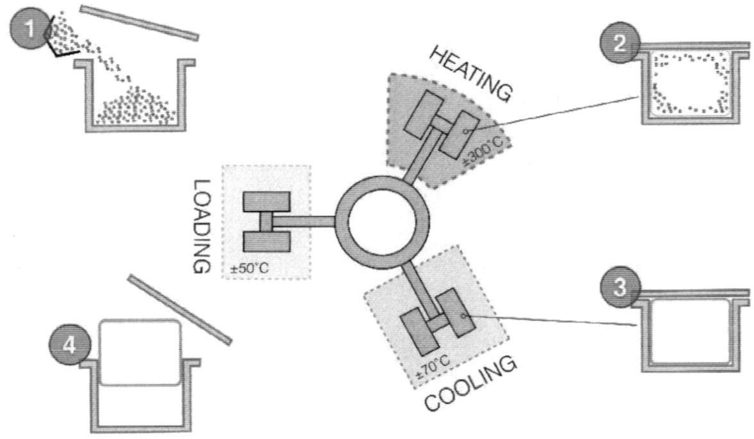

Figure 1. Rotational operation

Analysis of the quality characteristics of polyethylene raw materials for the manufacture of fuel tanks, suitable for rotational molding, revealed the absence of competitive domestic material in price and quality. To improve the properties of the raw materials, it is necessary to use various additives, that is, to carry out an additional process of compounding, and then grinding for the required fraction of rotational molding.

3. Conclusions

Based on the analysis of open sources, it has been established that plastic fuel tanks are made of medium and high density polyethylene. It is established that the main disadvantage of polyethylene as a material for a tank is high permeability with respect to hydrocarbons, due to the low degree the polarization of polymer macromolecules and weak bonds between the chains of macromolecules. This disadvantage is eliminated by methods of improving the barrier properties of the walls of the fuel tanks.

This work was supported by the research grant of Kazan Federal University.

References

[1] Shafigullin L N, Muhametzjanova G F, Muhamatdinova A A, Sotnikov A M and Shafugullina G R 2018 Usage of polymer fuel tanks in the automotive industry *IOP Conference Series: Materials Science and Engineering* vol 412 Issue 1 number **012071**

[2] Official site Group of companies Kuraray [Electronic resource]. - Access mode: http://www.eval.eu/ru/- (accessed 21.09.2018).

[3] Kashapov N F, Nafikov M M, Gazetdinov M X, Nafikova M M and Nigmatzyanov A R 2016 Innovative production technology ethanol from sweet sorghum *IOP Conference Series: Materials Science and Engineering* vol 134 issue 1, 8 number **012012**

[4] Fedyaev V L, Galimov E R, Galimova N Ya, Gimranov I R and Siraev A R 2017 Dynamics of coalescence and spreading of liquid polymeric particles during coating formation *IOP Conf. Series: Journal of Physics: Conf. Series* vol 789 number **012006**

[5] Fedyaev V L, Galimov E R, Galimova N Ya, Takhaviev M S and Siraev A R 2017 Mathematical modeling of processes occurring during deposition of sprayed particles of polymeric powder *IOP Conf. Series: Journal of Physics: Conf. Series* vol 789 number **012007**

IOP Conf. Series: Materials Science and Engineering **570** (2019) 012088 doi:10.1088/1757-899X/570/1/012088

[6] Internet resource: Fuel tank: construction, types, operation. Available at: https://autodromo.ru/articles/toplivnyy-bak-ustroystvo-vidy-princip-funkcionirovaniya. (accessed 12.10.2017)
[7] Internet resource: Construction of present-day vehicle fuel tanks. Available at: https://auto.today/bok/4303-iz-chego-sostoit-sovremennyy-toplivnyy-bak.html. (accessed date: 11.10.2017).
[8] Internet resource: Rotational molding of polymer fuel tanks. - URL: http://www.rototech.it/index.php?evt[page-displayItem] & id=11 - (accessed 23.09.2018).
[9] Internet resource: Rotational molding: new ideas]. Available at: http://polymery.ru/letter.php?n_id=3046&cat_id=2. (accessed date: 25.10.2017)
[10] Internet resource: Rotational molding. Available at: http://www.rototech.it/index.php?evt[page-displayItem]&id=11. (accessed date: 13.10.2017)
[11] Official website of PJSC "Nizhnekamskneftekhim" [Electronic resource]. - Access mode: https: //www.nknh.ru/products / plastics/polyethylene / (accessed 27.10.2018).

IOP Conf. Series: Materials Science and Engineering **570** (2019) 012089 doi:10.1088/1757-899X/570/1/012089

Promising processing technologies of glass-filled polyurethane materials

L N Shafigullin[1], N V Romanova[1], A T Gabdrakhmanov [1], E S Shabaeva[1] and D R Sarimov [1]

[1]Naberezhnye Chelny Institute (branch), Kazan Federal University
Sjujumbike street 10a, Naberezhnye Chelny, 423800 Tatarstan, Russia

misharin_82@mail.ru

Abstract. The review of modern methods of glass-filled polyurethane materials processing is presented. A possibility to use of glass-filled PU in the manufacture of interior parts with different surface textures (instrument panels, door trim, armrests) has been revealed.

1. Introduction

A hundreds of brands of elastic and rigid gas-filled materials are produced on the basis of synthetic resins, rubbers and high-molecular compounds, which are used in all industries and construction. At present, a large number of studies on the development of new polyurethane compositions (PU) were investigated due to a decrease in their cost, simplification of technologies, reduction of toxicity in production, and the possibility of operating in a wide temperature range. It is possible to vary the properties of the obtained PU in a wide range due to the large number of initial components [1-7].

2. Body text

Currently, the following technologies are used for the manufacture of glass-filled polyurethanes [8] (Table 1): Long Fiber Injection Molding (LFI); Fiber Composite Spraying (FCS); Reinforced Reaction Injection Molding (RRIM); Structural Reaktion Injection Molding (SRIM); Resin Transfer Molding (RTM).

Table 1. Description of production methods for glass-filled PU

Method	Basic operations	Advantages
LFI	cutting of fibers, mixing components, stitching fibers and pouring them into the into the socket of the disclosed forms; finishing the product	minimum waste; possibility of combining with cellular cardboard; variable content of fibers and their length; technologies for obtaining the outer surface (staining in the form, film and textile materials)
FCS	cutting of fibers, mixing components, spraying polyurethane, introducing of fibers from the outside into the spray jet and applying separate layers; curing the coating in air; final processing	only the lower half of the form is necessary, the form carrier is not needed; the form can be made of artificial resin (low temperature and mechanical stress); suitable for small batches and large molded products; very high flexural modulus, low mass; structural construction with variable layer

		thickness and its types (monolithic, foamed, filled, unfilled); variable content of fibers and their length; technologies for obtaining the outer surface (staining in the form, film, textiles)
RRIM	polyol with pre-mixed fiber is mixed with isocyanate in the mixing head, poured into the form and there reacts; finishing of products	short cycle times due to high activity components; amenable to spraying surface; high temperature resistance; excellent impact strength
SRIM	before forming and cutting the glassmat.; manual planting; pouring into an open form, followed by pressing; finishing of products	purposeful orientation of fibers for high freedom at adjustment of durability parameters; simple changeover-flexible and low-cost technology; high mechanical properties
SCS	spraying by the reaction mixture of the upper and lower sides of the pre-assembled "sandwich" (fibers, remote glassmats); laying in the form; finishing of products	very light products with high flexural rigidity are produced when using cellular cardboard or similar materia, low consumption of polyurethane due to uniform stitching.
RTM	cutting glassmat; laying in the form by hand; fill in a closed form, where necessary, with vacuum support; finishing of products	targeted orientation of fibers (insoluble binder); high mechanical performance; higher fiber content (up to 70%)

Due to the development of tools for mechanization and automation of labor, cheaper hardware and software tools, LFI and FCS technologies produced by glass-fiber PU [2–8] are widely used in mechanical engineering.

The physical-mechanical and operational properties of glass-filled PU products manufactured according to LFI and FCS technologies were studied [2-7]. Tested two types of samples of glass-fiber PU, manufactured by LFI technology. The first sample consisted of: the front layer - polyurethane film; reinforcing layer (frame) - glass-filled rigid polyurethane; intermediate (connecting) layer - semi-rigid PU. The second sample is glass-filled rigid polyurethane. A sample of glass-fiber PU, made by FCS technology, was a rigid polyurethane.

The results of studies of physico-mechanical and operational properties of glass-filled PU, are presented in tables 2 - 3.

Table 2. Physico-mechanical and performance properties of products from glass-fiber PU obtained by LFI technology

Name of the indicator	Test result		
	Material 1	Material 2	Material 3
Frost resistance	No change in appearance	No change in appearance	No change in appearance
Impact resistance at low temperature: impact with a ball weighing 883 g from a height of 0,724 m immediately after the frost test	Sample №1 No destruction of products: cracks, through holes, chips. Sample # 2 is cracked	No destruction of products: cracks, through holes, chips	No destruction of products: cracks, through holes, chips
Impact resistance at low temperature: impact by	No destruction of products: cracks,	-	-

IOP Conf. Series: Materials Science and Engineering **570** (2019) 012089 doi:10.1088/1757-899X/570/1/012089

a 225 g ball from a height of 0.400 m immediately after the frost test	through holes, chips		
Heat resistance (change in the geometric dimensions of the part,%)	0,09 (no peeling and swelling of the face coating)	0,06 (no peeling and swelling of the face coating)	0,04 (no peeling and swelling of the face coating)
Apparent density, kg / м^2	1036,71	1116,52	535,80
The level of odor, points, not more	2,0 (allowed to use)	-	-
Breaking stress, MPa -in tension	32	47	11,5
Breaking stress, MPa - in bending	45	90	19
Oil-benzine resistance	The lack of stratification and dissolved samples	The lack of stratification and dissolved samples	The lack of stratification and dissolved samples
Material resistance to temperature: +100 °C for 12 h; +45 °C for 16 h; -40 °C for 8 h	-	Lack of detachment of the facial layer, changes in appearance, deformation, delamination	Lack of detachment of the facial layer, changes in appearance, deformation, delamination
Heat conductivity coefficient, W/ (m*K)	-	-	0,148

Table 3. Physico-mechanical and performance properties of products from glass-fiber PU obtained by FCS technology

Indicator	Value	Method
Breaking stress at bending, MPa	49,4	GOST 18564-73 [8]
Shore D hardness	60-69	GOST 24621-91 [9]

3. Conclusions

The test results revealed high values of the physicomechanical and operational properties of glass-filled PU. The results obtained allow us to recommend the use of glass-filled PU as an alternative to traditional fiberglass materials in the production of interior parts with different surface textures (dashboards, door panels, armrests) [10].

This work was supported by the research grant of Kazan Federal University.

References

[1] Gottlieb E M, Galimov E R, Zenitova L A and others 2016 Polyurethanes: synthesis, properties and application in mechanical engineering (Kazan: Kazan publishing house UN-TA) p 149

[2] Shafigullin L N, Yurasov S Yu, Shafigullina G R, Shafigullina A N and Zharin D E. 2017 Sound-Absorbing Polyurethane Foam for the Auto Industry *Russian Engineering Research* vol 37 Issue 4 pp 38-40

[3] Shafigullin L N, Gumerov I F, Gumerov A F, Lakhno A V, Yurasov S Yu and Shafigullina G.R. 2017 Glass-Filled Polyurethane Materials Produced by Fiber Composite Spraying *Russian Engineering Research* vol 37 Issue 11 pp 925-28

[4] Shafigullin L N, Astashchenko V I, Romanova N V, Ibragimov A R, Shafigullina G R and Shafigullina A N 2017 Investigation of physical-mechanical and performance properties of glass fiber reinforced polyurethane materials used in the machine building industry *IOP Conference Series: Materials Science and Engineering* vol 240 Issue 1 012063

[5] Kashapov N F, Nafikov M M, Gazetdinov M X, Nafikova M M and Nigmatzyanov A R 2016 Innovative production technology ethanol from sweet sorghum *IOP Conference Series: Materials Science and Engineering* vol 134 issue 1 8 number 012012

[6] Fedyaev V L, Galimov E R, Galimova N Ya, Gimranov I R and Siraev A R 2017 Dynamics of coalescence and spreading of liquid polymeric particles during coating formation *IOP Conf. Series: Journal of Physics: Conf. Series* vol 789 number 012006

[7] Fedyaev V L, Galimov E R, Galimova N Ya, Takhaviev M S and Siraev A R 2017 Mathematical modeling of processes occurring during deposition of sprayed particles of polymeric powder *IOP Conf. Series: Journal of Physics: Conf. Series* vol 789 number 012007

[8] GOST 18564-73 1973 Rigid cellular plastics. Static bend test (Moscow, Izdatelstvo Standartov Publ.).

[9] GOST 24621-91 1993 Plastics and ebonite. Determination of indentation hardness by means of a durometer (Shore hardness) (Moscow, Izdatelstvo Standartov Publ.).

[10] Online resource: http://poly-tex.ru/fiberglass/glass-roving/ (accessed: 03.07.2018)

IOP Conf. Series: Materials Science and Engineering **570** (2019) 012090 doi:10.1088/1757-899X/570/1/012090

The technological stresses of carbon-filled plastics arising during the pressing of prepregs

L N Shafigullin, N V Romanova, D I Israfilov and G D Shakirova

Naberezhnye Chelny Institute (branch), Kazan Federal University
Sjujumbike street 10a, Naberezhnye Chelny, 423800 Tatarstan, Russia

misharin_82@mail.ru

Abstract. The technological stresses of carbon-filled plastics based on epoxy resins arising during the pressing of prepregs were investigated.

1. Introduction

Carbon-filled plastics - composites based on high-strength carbon fibers - are the most promising type of composite materials. They are different by high specific characteristics, strength and rigidity, low temperature coefficient of linear expansion, resistance to aggressive media. Carbon-filled plastics are widely used in aviation, rocket and space technology, in the automotive industry, in the manufacture of sports equipment and in other areas [1-3].

One of the problems in the production of polymer composite materials is the presence of residual stresses and strains. The reason for their occurrence is the incompatibility of deformation at different points of the product, due to the uneven non-stationary temperature field and the characteristics of the thermomechanical behavior of the material. In order to reduce the residual stresses, they seek to lengthen the cooling stage, equalizing the temperature by the volume of the product and at the same time achieving significant stress relaxation [4].

Residual stresses are not characteristic for polymer composites, but technological stresses at the moment of the end of the manufacturing process are characteristic. These stresses are not residual in the literal sense of the word; it will be change over time, the evolution of residual stresses will occur due to relaxation processes in the polymer material. The result of this change depending on the ratio of the characteristic times of relaxation of the material at temperatures of storage and operation, as well as the designated period of use can be very significant.

Thus, there is the problem of describing the thermomechanical behavior of polymer composites, which is important due to the need to predict the phenomena and patterns of technological processes flow the production of products, accompanied by thermal relaxation transitions [5].

2. Body text

In the process of the pressing of prepregs on the basis of epoxy resins, technological stresses arise when overheating or pressing duration, especially when using an aluminum mold. In this regard, thermal studies were carried out on carbon-filled plastics obtained by pressing into steel (sample 1) and aluminum (sample 2) molding tools. Carbon-filled plastics samples were obtained using a T-700 type of carbon fiber impregnation with epoxy binder (based on bisphenol-A) and hardener - dicyandiamine in methyl ethyl ketones subsequent by removal of the solvent by the drying method [6, 7].

IOP Conf. Series: Materials Science and Engineering **570** (2019) 012090 doi:10.1088/1757-899X/570/1/012090

The method of pressing is based on the ability of the material to transform by heating up to a viscous state and consists in formation of a product from a melt in a closed volume by creating irreversible deformations in the material. The fixation of the specified dimensions and product configuration occurs due to the flow in the material of chemical curing reaction.

The formation of the product by pressing occurs at a temperature of 150 °C for 30 minutes of pressing. These technological characteristics determine the course of various stages of the pressing cycle, as well as the quality indicators of finished products.

It is known that in order to achieve the most homogeneous by weight of the degree of curing in the finished product, it is necessary to quickly achieve and effectively maintain a set temperature of the material of throughout all the volume of the product at the curing stage. Under this condition the finished product does not produce a thermal voltage, which can cause its destruction after the end of the pressing cycle [4].

As a result of the pressing of prepregs, sample 2 was deformed after extraction and destroyed by application of the minimal load.

It is known that as a result of the wrong selection of heat treatment modes, there is often a continuity violation of the material, resulting in the destruction of the structure. But even in cases where the destruction does not occur, the presence of residual stresses and strains can reduce the performance of finished products. One of the main reasons for the formation of a residual stress-strain state is the "freezing" of deformations during in the hardening (polymerization, crystallization, glass transition) of the material, caused by a sharp increase in relaxation times as the temperature falls below T_g.

It is known that the DSC method registers not only the effect of deformation on the thermophysical properties of a glassy polymer, but also special types of molecular motion due to its physical aging. These changes in the mechanical properties during the physical aging of a glassy polymer are also accompanied by a significant change in its thermophysical properties [7].

DSCs were recorded using a Netzsch DSC 204 F1 Phoenix differential scanning calorimeter of the heat flux. DSC analysis was performed in a dynamic heating / cooling mode at a rate of 10 ° C / min in a stream of argon at a rate of 50 cm^3 / min.

In fig. Figure 1 shows the DSC curves for carbon-filled plastic samples obtained in different molds.

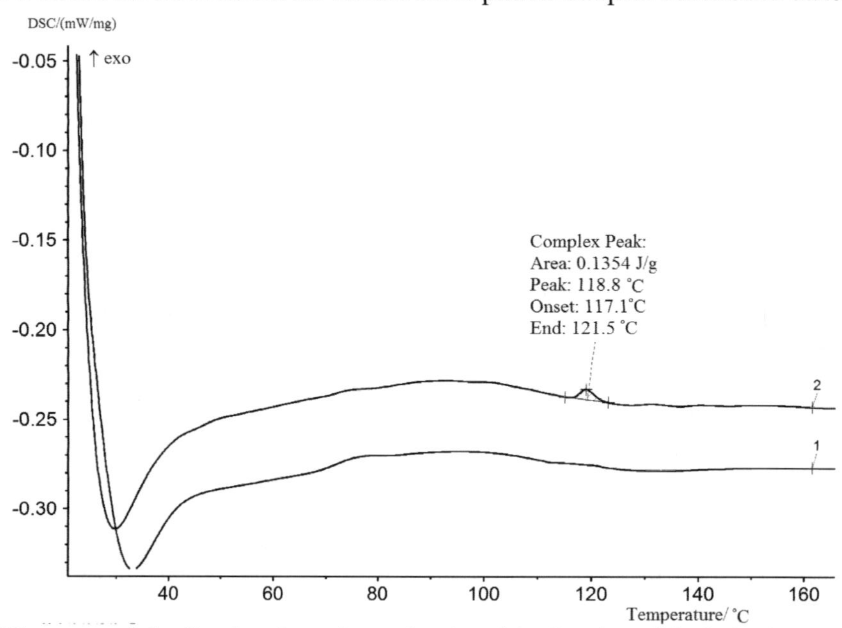

Figure 1. DSC-curves of the first heating of samples 1 and 2 of carbon-filled plastic

The glass transition temperature of carbon-filled plastics is characterized by an area of 100-120 °C. The glass transition temperature of the studied samples is 113 °C (Table 1).

Table 1. The results of thermal studies

Parameter	Sample 1	Sample 2
$T_{1,5}$, °C	300	85
T_5, °C	367	240
T_{max}, °C	429	425
Δm_{Tmax}, %	20	25
T_g, °C	113	113

The deformed polymer (Fig. 1, sample 2) contains an exothermic DSC peak at 118.8 ° C and is characterized by a exo-thermal effect of 0.1354 J/g, located in the region of the glass transition temperature (T_g), while the non-deformed polymer has no peaks in the region glass transition temperature. It can be assumed that the presence of an exothermic peak testifies to the relaxation of the stress of carbon-filled plastics.

TGA was performed using a Netzsch TG 209 F1 Iris thermogravimetric analyzer. Heating up 550 °C was carried out at a rate of 10 ° C / min in conditions of constant blowing by argon. One of the advantages of thermogravimetric method is that it allows carrying out an analysis of thermal decomposition of epoxy binder and evaluating the content of inorganic filler in polymer composite material.

While carrying out the thermal analysis, the following parameters variations have been defined:

-mass of a researched specimen as a function of heating temperature,

-rate of a specimen mass loss as a function of heating temperature [9, 10].

The decomposition temperatures corresponding to 1.5% ($T_{1,5}$), 5% (T_5) and maximum (T_{max}) of the mass loss of the samples, the mass loss at the moment of reaching the maximum rate of decomposition (Δm_{Tmax}) (table 1) were determined from the thermograms of the studied samples.

Analyzing these results of TG treatment (table 1), it can be noted that the low-temperature first stage for the studied samples is characterized by an area of 30-300 ° C, while sample 2 loses about 6% of its mass, and sample 1 only 1.5% of mass.

The second stage of the process is characterized by intense decay in the temperature range of 310–550 ° C, a significant mass loss of all samples (up to 40%), which indicates the active process of thermal decomposition and decomposition. Moreover, T_5 and T_{max} of decomposition of sample 2 is lower than these temperatures characteristic of sample 1, it is explained by the acceleration of these reactions by decomposition products formed in the first stage. So, T_5 of decomposition of sample 2 is lower by 127 ° C, and T_{max} of decomposition is 4 ° C lower than the decomposition temperature of sample 1.

Thus, thermogravimetric studies have shown that the destruction of the deformed sample of carbon-filled plastics obtained in an aluminum mold is faster than standard carbon-filled plastics, it can be explained by the formation of decomposition products in the pressing process, which accelerate the thermal degradation.

3. Conclusions

1. It is shown that in the process of the pressing of prepregs on the basis of epoxy resins, technological stresses arise when overheating or pressing duration, especially when using an aluminum mold.

2. The DSC method revealed the presence of an exothermic peak in the region of the glass transition temperature, testifies to the relaxation of the stress of carbon-filled plastics.

This work was supported by the research grant of Kazan Federal University.

References

[1] Vasiliev V V, Protasov V D, Bolotin V V and other 1990 *Composite materials: Handbook* (Moscow: mechanical engineering) p 512

[2] Bazhenov S L 2014 *Mechanics and technology of composite materials: scientific publication* (Dolgoprudny: publishing House "Intellect") p 328

[3] Galimov E R, Samoylov V M, Danilov E A, Fedyaev V L, Galimova N Ya and Orlov M A 2017 Production of thermally conductive carbon foams and their application in automobile transport // *IOP Conf. Series: Materials Science and Engineering* vol 240 issue 1 p 1-4

[4] Basov N, Kazankov Yu and Ljubartovich V A 1986 *A calculation and design of equipment for production and processing of polymeric materials: textbook for universities* (Moscow: Chemistry) p 488

[5] Matveenko V P, Smetannikov O Yu, Trufanov N Ah and Shardakov I N 2009 *Thermomechanics of polymeric materials in conditions of relaxation transition* (Moscow: FIZMATLIT) p 176

[6] Volynskii A L and Bakeev N F 2014 *The role of surface phenomena in structural-mechanical behavior of solid polymers* (Moscow: FIZMATLIT) p 536

[7] Romanova N V, Shafigullin L N, Shafigullina G R and Sarimov D R 2017 Analysis of carbon plastics destruction causes based on epoxy binding *TURKISH ONLINE JOURNAL OF DESIGN ART AND COMMUNICATION* vol 7 p 1872-78

[8] Sokolova Y, Romanova N, Shafigullin L and Sokolova A 2018 Carbon fiber reinforced plastic technology *Journal of Physics: Conference Series* vol 365 Issue 3 number 032020

[9] Shafigullin L N, Romanova N V, Romanov V B, Yurasov S Yu, Sarimov D R and Sokolova Yu A 2018 Thermal Properties of Carbon Plastics *Research Russian Engineering Research* vol 38 pp 1042-45

[10] Romanova N V, Shafigullin L N, Kokina T M, Ibrahimov A R and Shafigullina GR Thermal Studies of Polymeric Foams Used in Engineering *Helix* vol 8(1) pp 2244 – 48

IOP Conf. Series: Materials Science and Engineering **570** (2019) 012091 doi:10.1088/1757-899X/570/1/012091

Quantum chemical calculations of the vibrational spectrum of aliphatic nitro compounds. Cluster analysis of torsion oscillations

R M Shaikhullina[1], **G M Hrapkovsky**[2] **and M M Shaikhullina**[3]

[1]NaberezhnyeChelny Institute of Kazan Federal University,68/19, Pr.Myra, NaberezhnyeChelny, 423810, Russian Federation
[2]Kazan State Technological University, 68, K. Marx St., Kazan, 420015, Russian Federation
[3]Institute of Physics, Kazan Federal University, 16, Kremlevskaya St., Kazan, 420111, Russian Federation

raviya1@yandex.com

Abstract. The data of theoretical analysis of vibrational spectra in the long-wave range of different isomers of methyl nitrite, methyl, ethyl, n-propyl nitrate, as well as the electron density distribution obtained within the B3LYP Density Functional Theory method with the 6 - 31G(d) basis set are presented. The cluster analysis of charge distribution on carbon, oxygen, nitrogen atoms and torsion oscillation frequencies in the range of 10-400 cm-1 is carried out.

1. Introduction

Quantum chemical calculations are widely used to study the structure, molecular spectra, and reactivity of organic compounds The use of these methods are of considerable interest for the calculation of vibrational spectra of nitrogen-containing compounds which are widely used in various industries, as a component of high-energy substances. Earlier, the vibrational spectra and the structure of a number of aliphatic nitro compounds were studied using quantum chemical calculation within the B3LYP *Density Functional Theory* method with the *6-31G(d)* basis set. As a result, energetically most favorable conformers, as well as the spectral characteristics of the structure and conformational states of molecules were discovered [1-6]. As is known, a long-wave IR spectroscopy is an ideal method for studying the intramolecular and intermolecular displacements, under conditions when the danger of chemical decomposition is practically absent [7].

In our work, the statistical processing and cluster analysis of torsion oscillation frequencies in the long-wave IR spectral range (10-400 cm - 1) are carried out of the following nitro compounds: *cis* - (C) and *trans* (T)-isomer of methyl nitrite (MNI), *trans-* isomer of methyl nitrate (MNA), *trans* - and gauche (G) -isomer of ethyl nitrate (ETN), as well as a series of isomers (TG, TT, GG, GG`) of n-propyl nitrate (PNA). According to the earlier studies [3,4,6] the distinguished conformers are the most energy-efficient.

IOP Conf. Series: Materials Science and Engineering **570** (2019) 012091 doi:10.1088/1757-899X/570/1/012091

Special attention is paid to the analysis of torsion oscillations – τ CO and τ NO with frequencies in the range of 10-400 см$^{-1}$

For the calculations we applied a widely used method of *Density Functional Theory* B3LYP with the *6-31G(d)*-basis, which allows to achieve high accuracy of calculations of both thermodynamic functions and spectral characteristics.

2. Calculations

All calculations within the *Density Functional Theory* (DFT) were performed using the *Gaussian 03*-software package [8]. For DFT, Becker's three parameter exchange functional was used in combination with the Lee-Yang-Parr correlation functional (B3LYP) [9,10]. The standard 6-31G (d) basis was applied. The agreement obtained stationary points and minimum potential energy is checked by calculating the Hessian matrices and analysis of the absence of imaginary frequencies. To correct systematic errors due to the limited size of the basis set of functions, force constants were multiplied by the correction scaling factors. Conversion of force constants calculated in Cartesian coordinates to internal coordinates and its scaling were performed using the program [11].

Statistical processing of torsion oscillation frequencies and electron density distribution on atoms was carried out using the standard program of Stat Graphics Plus 5.0. The following cluster analysis options were used: method-Nearest Neighbor, Squared Euclidean.

3. Results and discussion. The influence of the electronic density distribution on the C, O and N-atoms on the τ CO and τ NO torsion oscillation frequencies

The torsion oscillations onset is associated with the presence of a potential barrier that inhibits free rotation around the chemical bond CO and NO.

The values of the barriers and the frequency of torsion oscillations are sensitive to the intramolecular effect determined by the chemical structure and the conformational state of the molecules. In particular, the electronic nature of the relationship between the groups of CH_3 and NO (MNI), CH_3 and NO_2 (MNA), CH_2 and NO_2 (ETN, PNA), as well as the distribution of q charges on C, O, N atoms of these groups, will determine the height of the barrier of rotation, hence the torsion frequency.

Cluster analysis of the calculated τ CO and τ NO frequencies and q charge distribution on the C, O and N atoms is shown in Figure 1. The first cluster is formed from the spectral and electronic data of the cis- and trans- isomer of MNI, the second cluster – from the trans-isomer of MNA, while the third one is composed from spectral and electronic data of ETN isomers, as well as the considered series of PNA isomers.

It is noteworthy that the low frequencies of τ CO in the range of 70-120 cm^{-1} are typical for compounds with a high electronegativity of the O-atom and a low electronegativity of the C-atom, i.e. compounds forming cluster 3 (ETN and PNA).

At the same time, higher τNO frequencies in the range of 220-380 cm^{-1} are typical for the methyl nitrite isomers with the lowest positive charge of the N atom, i.e. compounds that make up cluster 1.

The obtained correlations between the charge distribution on the C, O, N atoms and the τ CO and τ NO oscillation frequencies of the simplest considered representatives of the series of aliphatic nitro compounds will help further while interpretation of more complex nitrogen-containing macromolecules.

IOP Conf. Series: Materials Science and Engineering **570** (2019) 012091 doi:10.1088/1757-899X/570/1/012091

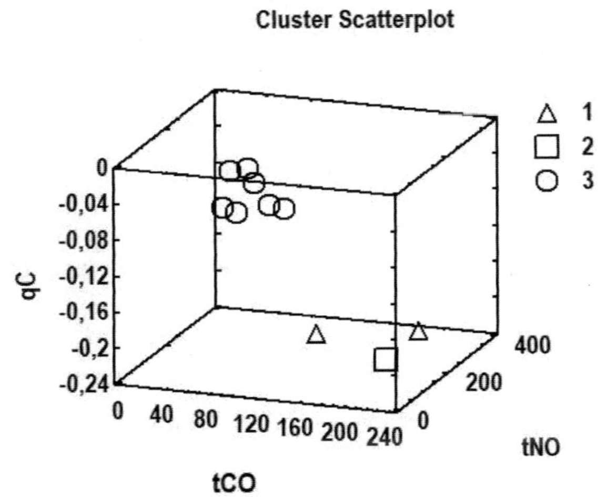

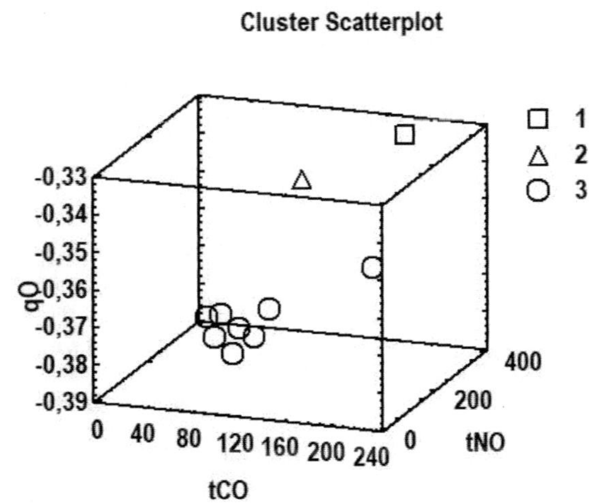

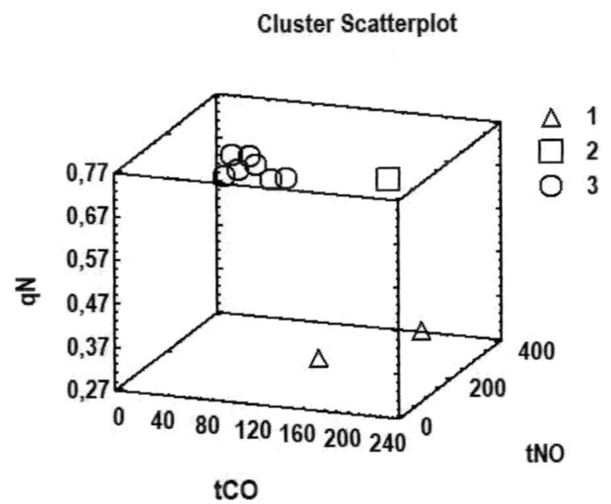

Figure 1.
Cluster analysis of torsion oscillations of τCO, τNO and distribution of electric q charge on the C, O, N atoms in a series of rotating isomers of aliphatic nitro compounds.
1 cluster - methyl nitrite,
2 cluster - methyl nitrate,
3 cluster - ethyl, propyl nitrate
(q - in electron units, frequency CO and NO -in cm^{-1})

References

[1] Shlyapochnikov V and Khrapkovsky G M, Shamov A G 2002 Structure and vibrational spectra of mononitroalkanes *Izv. An.Ser.Chem.***6** 863-70

[2] Joao Bosco PDa Silva, Nivan B. da Costa and Mozart N, Ramos RuiFausto 1996 Vibrational spectra and structure of the cis and trans conformers of methyl nitrite: an ab initio MO study*Journal of Molecular Structure* vol 375 pp 153-80

[3] Shaihullina R M, Khrapkovsky G M and Zvereva E E, Mazilov E A2010 Quantum chemical study of molecular structure and vibrational spectra of methylnitrite*Butlerov communications* **19(3)** pp 10-20

[4] Shaihullina R M, Khrapkovsky G M, Zvereva E E 2015 Quantum-chemical study of the molecular structure and vibrational spectra of methyl nitrate and ethyl nitrate *Butlerov communications* **42(5)** pp 152-61

[5] Gong X D, Xiao H M 2001 Studies on the molecular structures, vibrational spectra and thermodynamic properties of organic nitrates using density functional theory and ab initio methods *Journal of Molecular Structure (Theochem)* **572** pp 213-21

IOP Conf. Series: Materials Science and Engineering **570** (2019) 012091 doi:10.1088/1757-899X/570/1/012091

[6] Shaikhullina R M, Hrapkovsky G M, Shaikhullina M M 2018 N-propyl nitrate vibrational spectrum analysis using DFT B3LYP quantum-chemical method *Journal of Physics: Conf. Series* **1015** 032125 pp 1-7
[7] Finch A, Gates P, Radcliffe K 1973 Application of long-wave spectroscopy in chemistry (Moscow: Mir) p 284
[8] Frisch MJ, Trucks G W at all. 2004 *GAUSSIAN 03, RevisionB.05; Gaussian, Inc.: WallingfordCT*
[9] Beck A D 1988 *Phys. Rev* A vol 38(6) p 3098
[10] Lee C, Yang W, Parr R G 1988 Development of the Colle-Salvetti correlation-energy formula into a functional of the electron density *Phys. Rev.*B **41(2)** 785
[11] Sipachev V A 2001Local centrifugal distortions caused by internal motions of molecules *J. Mol. Structur.* **567** pp 67-72

Assessment of the impact of technological parameters of hot stamping on the cost of the product

V G Shibakov, D L Pankratov, R S Nizamov and L M Nizamova

Kazan Federal University, Naberezhnye Chelny Institute (branch), 423812, Russia, Naberezhnye Chelny, Syuyumbike Ave., 10A
E-mail: E-mail:vladshib50@gmail.com, pankratovdl@gmail.com, nizamovrs@yandex.ru, nizamovalm@yandex.ru

Abstract. The main parameters of the technological process that have the greatest impact on the cost of forgings obtained by hot stamping are established

The main direction of providing competitive advantages of stamping production is to improve the quality and implementation of a flexible pricing policy by controlling the cost of production, which is directly related to the cost of the production process, i.e. the technological process of manufacturing forgings. This dictates a new version of the IATF16949 standard and the development strategy of the world's leading automotive companies.

The aim of the study is to establish the process parameters that have the greatest impact on the cost and quality of forgings. In accordance with [1, 2], in this kind of analysis, it is reasonable to use a systematic approach when bulk stamping is considered as a man-machine system. The main elements of this system (table 1) are: M - the material to be processed; B - dimensional blank (semi-finished product) obtained from the material being processed; E - equipment on which the forging is formed; T - the sequence and modes of technological effects on the blank before, during and after its shaping; T - a tool necessary for the implementation of the TP; E – environment; P - production personnel, designing and implementing the process in production conditions

Table1 - The matrix of significance of the influence of system parameters {M, B, E, T, T, E, P} on the cost and quality of forgings

System elements MBETTEP	Significance levels	Parameters affecting quality and costs in the production of forgings
Material		
The chemical composition of the material	+▲5	- alloying element; - selectivity for each element.
Physical and mechanical properties of the material	+▲3	-macrostructure; -ultrasound; - non-metallic inclusions; - magnetic properties; - landing in a hot and cold state; - carbon-free layer; -strength;

		-austenitic grain; - shock viscosity; - hardenability
Blank		
Surface quality group	+▲5	- turning; - surface finish group.
Sizes of metal	+▲4	- measured and multiple length; - profile; - non-standard profile sizes.
Accuracy	+▲4	- accuracy group; - accuracy grade; - curvature; - roughness.
Edge processing	+▲2	- deburring at the ends; - chamfering at the ends.
Marking	+▲2	- butt end marking; - painting the butt end.
Additional processing	+▲2	- heat treatment; - etching.
Packaging	+▲3	- preservation; - packaging materials ; - slinging materials.
The volume of the original blank	+▲5	material costs
Way of cutting	+▲5	- metal consumption rate; - the need for additional processing (heating); - way of punching
Equipment		
Type of equipment	+○4	- energy carriers; - energy consumption; - rental area; - amortization;
Productiveness	+○3	machine time;
Maintenance	+●2	factor of repair complexity
Load	+●3	load factor
Technology		
Heating		
Type of heating	+▲4	- type of energy carrier
Heating temperature	+▲Δ ● 3	- power consumption
Shaping		
Number of technological transitions	+▲2	- the time of formation; - power consumption.
Closed or open die forging	+▲5	- the consumption of materials
Lubrication (type)	+▲Δ 3	-the cost of lubrication; - lubricant consumption;
Stamping overlays and allowances	+▲4	- consumption of basic material
Tool		
Persistence	+▲4	- the distribution of the cost of equipment on the

		number of manufactured products;
Preheating	+▲ Δ 4	- type of energy carrier;
Staff		
Labor intensity	+▲4	[3]
Qualification	+▲●3	- category
Environment		
Working conditions	+▲ ● 2	- provision of norms
Total ▲	78	
Total Δ	10	
Total ○	7	
Total ●	13	

Experts, based on the possibility of controlling the parameters, divided them into: ▲ - controlled preventively (at the design stage of technology); Δ - controlled by the setting; ● - controlled adaptively; ○ - not controlled. For ranking the indicators of the target importance of the process factors are used: 1-not important; 2-less important; 3-not important enough; 4-important; 5-very important. Engineers and economists of forge and press production, who have experience in designing technological processes of die forging and calculating the cost of more than 10 years, were chosen as experts.

Each element of the system is characterized by a variety of parameters that, to varying degrees, affect the cost and quality of forgings. So, prices for metal products are defined as the sum of the base price and markups for the surface quality and properties of rolled products [4, 5]. Additional requirements for the source material are determined based on the requirements for the characteristics of the finished product [6].

The values of the single indicators of the set {M, B, E, T, T, E, P} are usually governed by GOST, industry standards, guidance materials, recommendations of professional reference books, as well as other information sources, and are usually specified not by numbers, but by ranges . For each individual enterprise, the set is limited by the actual capabilities, however, it should be supplemented as it develops in each of the areas (introduction of new materials, highly efficient technological processes, automation, etc.). According to the results of the evaluation of the influence of the technological parameters of the system on the cost, their share in its value was established.

System parameters:

- technology formed at the design stage, is ~ 72%
- regulable by setting in process, is ~ 9%
- regulable adaptively, is ~ 12%
- unregulable ~ 7%

Conclusion

The variant of preliminary assessment of quality and cost of forgings of the developed technological processes of hot stamping is offered. The influence of the main technological parameters of hot forging on the cost and quality of forgings is estimated.

References

[1] Kashapova L R, Pankratov D L, Shibakov V G, Vinogradov A L *Providing the quality of shaped machine elements at the design phase of the sheet stamping technological process*

[2] V G Shibakov, D L Pankratov, R Khairullin 2017 *Systems Analysis of Process Control Capabilities for Forging Accuracy* Solid State Phenomena Vol 265 pp 1110-15

[3] P 50-72-88 *Recommendations. Unified system of technological documentation* The order of registration of documents used in the standardization of technological processes (operations)

[4] http://www.metalcourier-online.com

[5] http://www.metalindex.ru

[6] V G Shibakov, D L Pankratov, R S Nizamov and R A Hairullin 2018 *Surface quality assurance of ring gear during precision stamping of gearwheels* IOP Conf. Series: Materials Science and Engineering 412 **012073**

IOP Conf. Series: Materials Science and Engineering **570** (2019) 012093 doi:10.1088/1757-899X/570/1/012093

Justification of requirements to quality and accuracy of blankfor precision stamping

V G Shibakov, D L Pankratov, R S Nizamov, L M Nizamova and R A Khairullin

Kazan Federal University, NaberezhnyeChelny Institute (branch), 423812, Russia, NaberezhnyeChelny, Syuyumbike Ave., 10A

E-mail: E-mail:vladshib50@gmail.com, pankratovdl@gmail.com, nizamovrs@yandex.ru, nizamovalm@yandex.ru,guanava@mail.ru

Abstract. The dispersion of blank masses depending on the accuracy class of the initial metal rolling is estimated. The influence of the method of the initial blank segments on the stamping process is described.

The accuracy of mass and geometric parameters of the used blanks is of great importance for precision stamping [1]. In this regard, the study was conducted which aim was to reveal the influence on the result of stamping of the dispersion of blankmasses with different surface quality and it ends.The deviation of the volume of the stamped blank and the quality of the surface of the blankends, affecting its centering in the die, mainly depends on the type of rolled products used and the method of its cutting into dimensional blanks. According to the accuracy of hot-rolled steel round is distinguished according to GOST 2590[2]. To improve the quality and accuracy of the surface of the rods according to GOST 14955-77[3], turning, grinding, lug calibration is carried out, and therefore the accuracy of h8-11 is achieved.

In some cases, for precision stamping blankis used (figure 1, b) of rolled conventional precision, which is ground part of the length l_i≈1⁄2 l, and it is used for making of the untreated part of the forging cutting. Figure 1 and table 1 show the types of blankand information on the mass deviation ΔGbill of the blank, depending on the accuracy of the rolled products used.Mass varies from 0.021 to 0.176 kg.The effect of variation in the mass (volume) of the ΔGbilblankon the variation of the deformation force is shown in Figure 2.Simulation is carried out in QForm V8 for metal standard-accuracy B1.For blank with a diameter of Ø60, the following diameter tolerances are accepted: +0.5 and -1.1mm.Based on the tolerances, the volume of the accepted blank is calculated for the minimum V_{min} and the maximum V_{max} possible tolerances.

From the graphs of deformation forces (Figure 2), obtained by simulating the process of stamping from blanks of various volumes at 1200 ° C, it is clear that the stamping forces in the final transition for the nominal V_{nom} and the maximum V_{max} possible blank volume are relatively close and they are 9.1MN.The reduction of the force to 8.2 MH for the minimum V_{min} of the possible volume of the blank is associated with a decrease in the width of the backwater on the side of the ramming bridge.

The method of cutting rolled products on dimensional blanks also affects the mass accuracy.The most efficient in labor costs and common in mass production is cutting on press shears. For the case of cutting on press shears, the maximum permissible deviations of the shape and size of the blanks are presented in Figure 3.The mass difference between the minimum and maximum possible deviations of the shape of the blankis ~ 0.1 kg. In this case, we can observe a significant distortion of the geometry of the end parts of the blank.

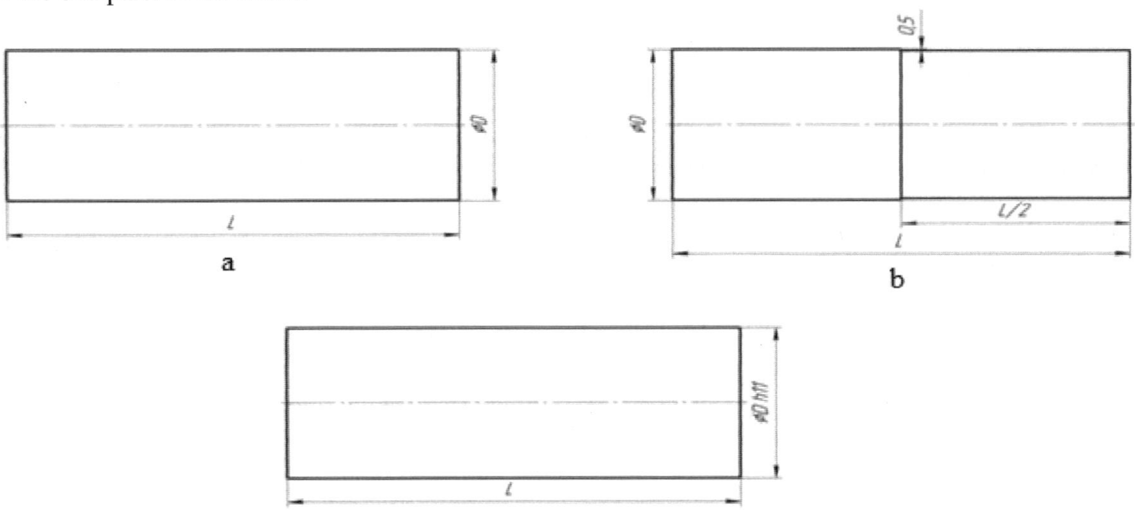

Figure 1 - Types of used blanks: a - according to GOST 2590, b - partially turned from a blank, GOST 2590; c - in accordance with GOST 14955-77 according to h11 quality

Table 1 - Deviation of the mass of the blankdepending on the accuracy of the blank

Blank	Deviation ΔGbil weight of the blank from the nominal value			
Ø60×150mm	Accuracygroup / Blank type	fig.1, a	fig.1, b	fig.1, c
	A	0,110	0,109	0,021
	Б	0,154	0,152	
	В	0,176	0,175	

In the case of cutting the blankon a disc cutting machine, the accuracy of the metal feed is higher, in addition, the crushing and mowing of the cutting ends of the blank are minimized. Deviations of the shape and size of the blankfor this case are shown in figure 4. The difference in weight between the minimum and maximum possible deviations of the form is ~0.02 kg.

The influence of the state of the end surface of theblank on the stability and axisymmetric semi-finished with the lees (the first transfer stamping) was evaluated in the simulation in QFormenvironment (figure 5). As follows from the simulation results, during the draft of the chopped blank due to a significant distortion of the geometry of the ends during the draft, an uneven flow of the metal of the deposited part of the blank occurs, leading to distortion (eccentricity) it relative to the die axis[4].

IOP Conf. Series: Materials Science and Engineering **570** (2019) 012093 doi:10.1088/1757-899X/570/1/012093

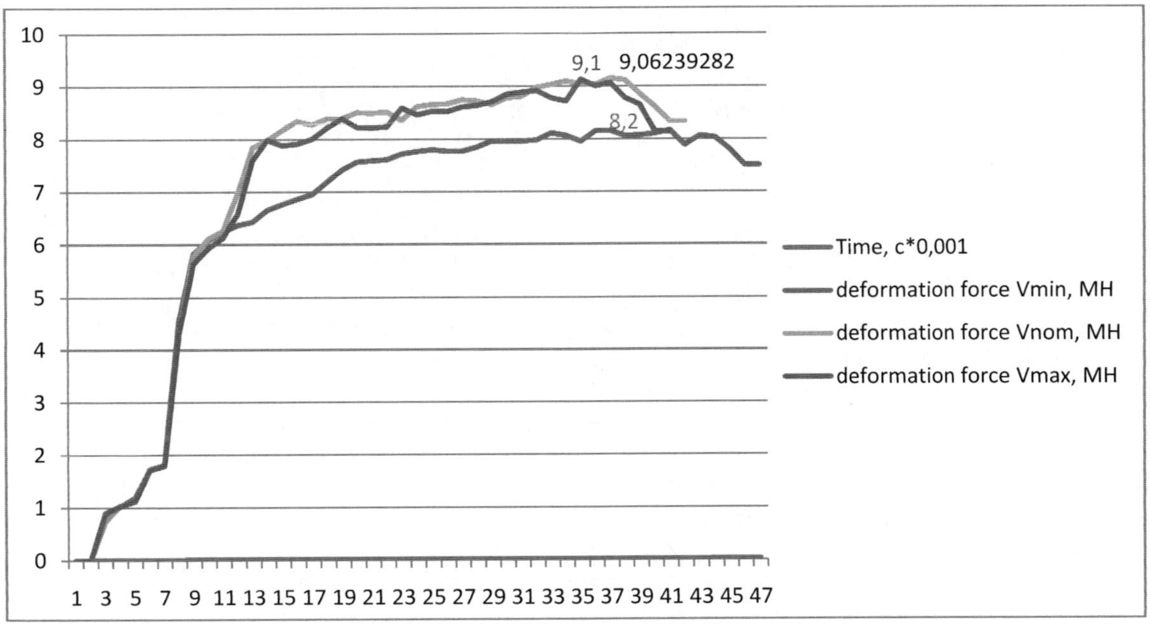

Figure 2- Graphs of stamping force changes for different volumes (V) of the blank

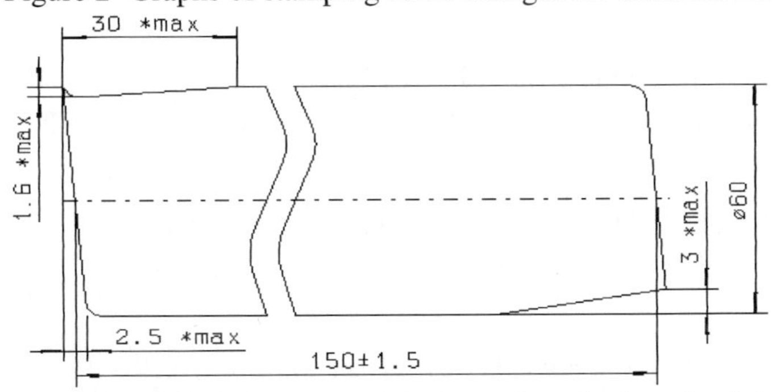

1. Dimensions with an asterisk (*) apply to both ends of the blank
2. Dimensions are given in cold condition

a) b)

Figure 3- The blankobtained by the operation of the segment (cutting) on the press shears: a) schematic image, b) the actual state of the end in the cutting zone.

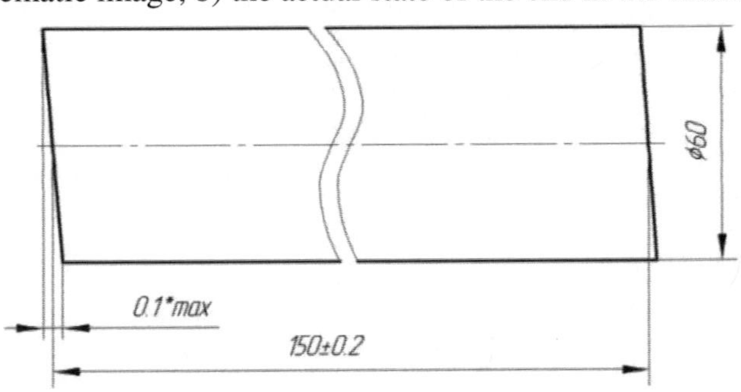

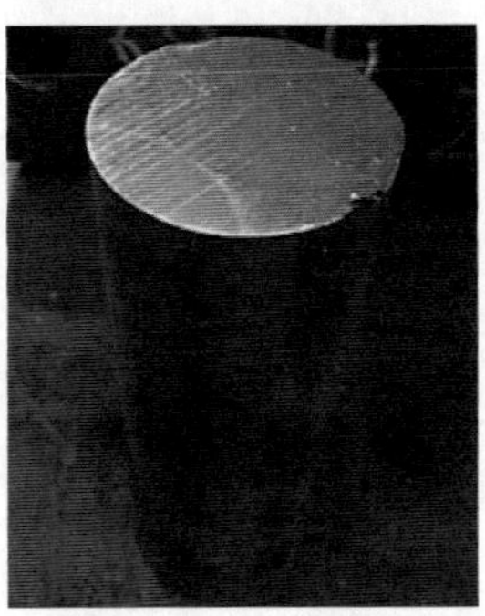

a) b)

Figure 4- The blank obtained by the operation of the segment (sawing) on the disk cutting machine.

IOP Conf. Series: Materials Science and Engineering **570** (2019) 012093 doi:10.1088/1757-899X/570/1/012093

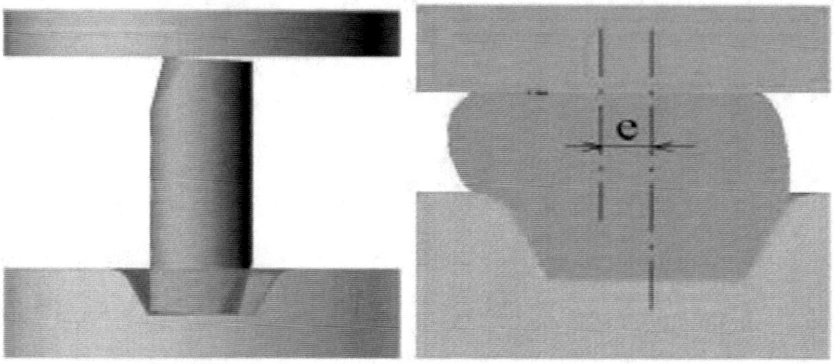

Figure 5 - Distribution of metals in the sediment of chopped blanks: e–eccentricity of the deposited part of the blank relative to the die axis.

Smaller deviation on the bevel of the blank end for the case of cutting on saws provides better stability of the blank during the operation of precipitation, which minimizes the risk of uneven distribution of the metal (e≈0) in the formation of transitions and resulting differences in the height of forging[5]. In turn, this directly affects the accuracy of the product obtained by precision stamping[6].

Conclusions

1. The shape of the ends of the blankshould provide a unique laying of the deposited semi-finished product in the stream of the next stamping transition, which is achieved by cutting on the saws.

2. The minimum deviation of the blank weight is achieved by using a sharpened rolled and cutting dimensional blanks on saws.

References

[1] VG Shibakov, DL Pankratov, RS Nizamov and RA Hairullin 2018*Surface quality assurance of ring gear during precision stamping of gearwheels* IOP Conf. Series: Materials Science and Engineering 412 (2018) **012073**

[2] GOST 2590-2006 Hot-rolledsteelround

[3] GOST14955-77 Quality round steel with special surface finish. Technical conditions.

[4] D LPankratov, RS Nizamov and I Zh Kharisov *Improvements in the process of boss bar upset forging into a horizontal forging machine with the aim of joint knuckle forging quality improvement*

[5] Khairullin RA, Nizamov R S, Pankratov DL, Shibakov VG 2017*The choice of a rational system of transitions in precision stamping of the gear crown* Proceedings of the VIII International scientific and technical conference "Innovative engineering technologies, equipment and materials 388 p

[6] VGShibakov, DL Pankratov, RA Khairullin 2017 Systems Analysis of Process Control Capabilities for Forging Accuracy. Solid State Phenomena Vol 265 pp 1110-15.

IOP Conf. Series: Materials Science and Engineering **570** (2019) 012094 doi:10.1088/1757-899X/570/1/012094

Revisiting the calculations techniques of the optimal planning methods for the local authority activities (on the Republic of Tatarstan example)

A M Shikhalev [1], I A Akhmetova [2], D P Vorontsov [1] and A F Khananova [1]

[1] Kazan Federal University, 18 Kremlyovskaya street, Kazan, 420008, Russian Federation, Engineering institute
[2] Kazan Federal University, 18 Kremlyovskaya street, Kazan, 420008, Russian Federation, Institute of management, economy and finance

128_dmitri@mail.ru

Abstract. The indicators of Savrushinsky local municipal structures development in the Republic of Tatarstan (RT) are defined on the basis of creation, receiving and the analysis of the production function (PF). For the PF creation were collected the input statistical data for 2013 - 2017 on revenues of the studied local municipal district, fixed assets volumes and the staff number. After the logarithmation and receiving the simple multiple regression equation with the potentiation method was received the PF in a multiplicative look that allowed to estimate the degree of intensive and extensive development of the studied municipal structures and also to define its ratio. For the mathematically correct results obtaining beforehand was made the component analysis of future equation, and after its receiving was estimated the residuals autocorrelation existence. As a result was specified that the Savrushinsky local municipal office in the studied period was mostly developed due to the production efficiency increase, than from the production scale growth in which increase consists the positive experience of this municipal structures.

The research purpose is the ratio of innovative character development and scale development (due to the mechanical production scale expansion) for the studied municipal office assessment. However before the PF receiving as variant of the proposed goal achievement it is necessary to make the component analysis for the purpose of prevention in using the general view (1) PF equation according to [1] the existence of the explanatory variables possible collinearity which was realised according to the requirements [2].

$$X = f(K, L), \quad (1)$$

where X is the enterprise revenues (roubles), K is the fixed assets volume, roub., and L is the management staff number.

The component analysis purpose is the compliance degree ascertaining to the complete correlation matrixes requirements among the operating variables [2] for the case if the required PF type look like (1): 1) the X dependent variable must have the strong correlation with the explanatory variables (the coefficient of correlation has to be more than 0.70; 2) the independent variables K and L shouldn't correlate with each other more than on 0.70 [2]. As at the time of the research there were opened only the last five years data the normality test due to the clear reasons for the studied statistical sets elements (see tab. 1) wasn't possible to realise, so we calculated the nonparametric coefficients of pair rank correlation [3] and received the following results: 1) the relations between the dependent variable and the capital and staff - $\rho_{X-K} = 0{,}90$; $\rho_{X-K} = -\,0{,}87$ – were rather high on in modulus exceeding the boundary value 0.70; 2) on the relation between the independent variables between themselves $\rho_{K-L} = -\,0{,}87$ that exceeds in modulus the value 0.70 (but on a few) and nevertheless is proposed to be considered further.

Next, for the PF receiving are required the data given in the tab. 1:

Table 1 - Input data for the Savrushinsky local municipal office.

Counter i	Years	Income, roub., X	Fixed assets (capital) K, roub.	Employees, people, L
		j = 1	j = 2	j = 3 = n
1	2013	2754,8	207,9	7
2	2014	2998,5	228,1	7
3	2015	3531,8	250,4	6
4	2016	3779,3	268,9	6
5 = m	2017	3713,7	322,5	6

After the preliminary logarithmation of the data provided in tab. 1 we have received:

$$\ln X = \ln A + \alpha_1 \ln K + \alpha_2 \ln L. \quad (2)$$

Then, after potentiation from (2) we received a required multiplicative variant:

$$X = A \cdot K^{\alpha 1} \cdot L^{\alpha 2} = 9202{,}3005 \cdot K^{0{,}2272} \cdot L^{-\,1{,}2267}, \quad (3)$$

where A is the neutral technical progress coefficient of; α_1 and α_2 are the elasticity coefficients on for the fixed assets and work (the equation (3) is received as a result of the A.M. Shikhalev author's computer program using).

As the regression equation type (3) connects the the average values it is easy to be convinced that its left-hand and right members are the almost identical: 3355.62 ≡ 3326.51 (a difference is less than 1%). Moreover the reliability of the received PF (3) equation according to Fischer is 90.6% whereas the equation(coefficients) significance on Student's coefficient has an insufficient size, i.e. 57.1% and 88.4%. At the same time the multiple correlation coefficient R is 0.97and the determination coefficient $R^2 = 0.95$ (i.e. about 95% of all really operating factors are included); the general relative error of approximation is only 1.40%. However the Durbin-Watson's criterion is 3.02 instead of expected variant which proposed to be at about 2.0.

The calculations in the author's computer program for the autocorrelation (a/c) received the following results: $r_\alpha^{calc} = -\,0.695$; $d_\alpha^{calc} = 3.022$, which in comparison with the tabular data in [4. - Appendices 7, pp. 133 and 5, p. 132] allow to make the conclusions that: $r_\alpha^{calc} = -\,0.695 > r_\alpha^{table} = -\,0.753$, therefore, autocorrelation here is presented. On the other hand for such conclusion specification it is necessary to use for example the Durbin-Watson's criterion at the same 5% p-level [5].

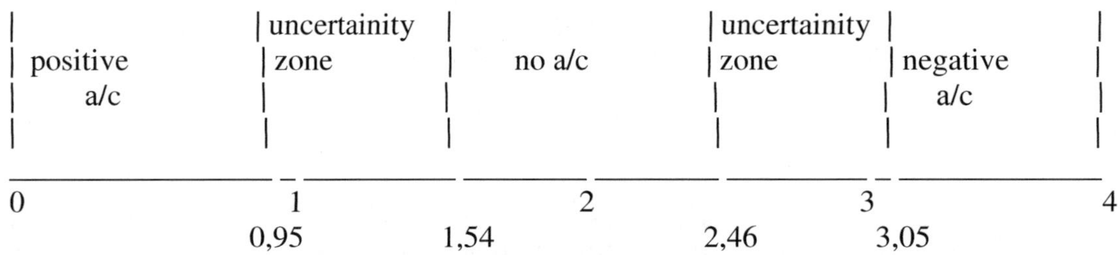

Fig. 1. The graphic interpretation of the autocorrelation residuals existence assessment

However, though the calculated value received $d_\alpha^{calc} = 3.022 < 3.050$, they differ from each other within also no more than 1% which is possible to neglect according to fig. 1.

For the estimation of the goal degree achievement it is necessary to calculate firstly the next data:

1) The income in the studied organisation increased in the $X^* = (X_{2017}/X_{2013})$ = (3713.7 roub. / 2754.8 roub.) = 1.3481 times;

IOP Conf. Series: Materials Science and Engineering **570** (2019) 012094 doi:10.1088/1757-899X/570/1/012094

2) The fixed business assets increased in $K^* = (K_{2017}/K_{2013})$ = (322.5 roub. / 207.9 roub.) = 1.5512 times;

3) The workers number (production staff) in the $L^* = (L_{2017}/L_{2013})$ = (7.0 peop. / 6.0 peop.) = 1.1667 times, and we could express the PF (3) as the Cobb-Douglas function:

$$X = a \cdot K^{\alpha} \cdot L^{(1-\alpha)}, \quad (4)$$

where $\alpha = \alpha_1 / (\alpha_1 + \alpha_2) = 0.2272 / [(0.2272 (-1.2267)] = -0.2273$; then $(1 - \alpha) = [(1 - (-0.2273)] = 1.2273$, and the PF expression in the Cobb-Douglas form function (4) will take a form:

$$X = a \cdot K^{-0,227341} \cdot L^{1,227341}. \quad (5)$$

For the research purpose achievement which is the verification of a working hypothesis about the internal reserves dynamics (negative or positive) of the studied enterprise we need to determine the particular resources effectiveness for the capital productivity (E_K) and for the labour productivity (E_L) with the following formulas (6) and (7):

$$E_K = X^* / K^* = 1,3481 / 1,5512 = 0,8691; \quad (6)$$

$$E_L = X^* / L^* = 1,3481 / 1,1667 = 1,1555, \quad (7)$$

whereas it was already noted, X^* is the incomes growth (as the organic part of the total production output) for 2013 - 2017; K^* is the increase of productive assets volume for the same period and L^* is the staff number rise for the people which were working for the same period.

Further on the basis of the such type (5) function the generalized efficiency factor E could be calculated as the geometric average of particular indicators as the average dynamic is estimated (taking into account (6) and (7)) in a look

$$E = E_K^{\alpha} \cdot E_L^{(1-\alpha)} = 0,8691^{-0,227341} \cdot 1,1555^{1,227341} \approx 1,23 \text{ (times)} \quad (8)$$

Then the average production scale in dynamics also could be determined as the geometric average proceeding from the resources growth rate on formula (9):

$$M = K^{*\alpha} \cdot L^{*(1-\alpha)} = 1,5512^{-0,227341} \cdot 1,1667^{1,227341} \approx 1,09 \text{ (times)}. \quad (9)$$

Therefore, the general incomes growth of the studied production (organization) from 2013 to 2017 in 1.09 times was concerned mostly with the production scales increase whereas the rise by 1.23 times is received due to the production efficiency increase. Then the ratio of scale growth and the effectiveness increase could be expressed (with the expressions (19) and (20)) as their relation like: $S = E / M = 1.23/1.09 = 1.128 \approx 1.13$ times. In other words, during the 2013 - 2017 generally the local municipal office growth rate concerned with the production efficiency share was higher than a share related to its scale rise that could be estimated as the generally positive variant for the studied production (organization).

References

[1] Kolemayev B A 2002 *Mathematical economy*: *The textbook for higher education institutions* (Moscow: UNITY-DANA) p 299

[2] Yeliseyeva I I, Kurysheva S V, Kosteeva T V, Babaeva I V and Mikhailov B A 2004 *Econometrics : The textbook* (Moscow: Finance and statistics) p 240

[3] Shikhaev A M, Khomenko V V and Alyautdinova G R 2016 Evaluation of optimal location of logistics centres in the Republic of Tatarstan on the basis of detailed statistical analysis of expert estimations *The Rev. of Econ., the Law and Sociol.* vol 4 pp 113 – 22

[4] Gromyko G L 1999 *General theory of statistics: Practical work* (Moscow: INFRA-M) p 119

[5] Valeev N N, Aksyanova A V and Gadelshina G A 2010 An analysis of time series and prediction: manual (Kazan: Kazan National Research Technological University Press) p 150

The multicriteria estimation of the Russian Federation region districts promising for the logistic objects placement (on the example of the Republic of Tatarstan)

A M Shikhalev [1], I A Akhmetova [2], D P Vorontsov [1], G R Hamidullina[1], N K Kashapov[1] and O N Rozhko [3]

[1] Kazan Federal University, 18 Kremlyovskaya street, Kazan, 420008, Russian Federation, Engineering institute
[2] Kazan Federal University, 18 Kremlyovskaya street, Kazan, 420008, Russian Federation, Institute of management, economy and finance
[3] Kazan National Research Technical University named after A. N. Tupolev - KAI, 10 Karl Marx street, Kazan 420111, Russian Federation, Institute for Aviation, Land Transportation and Power Engineering

128_dmitri@mail.ru

Abstract. We propose to analyse the multicriteria problem decision making process on the logistic objects placement example and are planning to solve the task in the following steps. We will try to find the alternative optimum to the analysed criteria and ranging the other alternatives beyond the optimum variant which will have the biggest conditional scores volume. The next stage is based on the decision-making person actions which are concerned with the maximizing the rationality in the quasioptimum variants choice. After that basing on the conditions with the big alternative number set the optimum variant could be added by the decisionmaker with the quasioptimum variants (decisions) in three stages. First of all it is necessary to establish the threshold value size in the interactive mode for the next calculations. On the second item of the process the manager mark the significance level for the subsequent set decomposition and then clustering it with the display of the districts multitude (like the images) on a spectral colors manifold (i.e. prototypes). And at last at the third step we propose the adjustment of quasioptimum alternatives number basing on the first choice being guided by the colors of its classes.

It is required to provide the optimum distribution of the Republic of Tatarstan (RT) districts which offer the greatest promise for the logistic objects (LO) creation there that as formulated like a management problem is designated in a tuple (1) with the symbol “t”. Author's approach is based on the preliminary solution of multicriteria decision making task (MC DMT) as the one of the most reliable mathematical tools which is demanded for a wide range of management tasks [1], with the subsequent additional mathematical procedures application [2]:

$$\langle t, X, R, A, F \rangle, \quad (1)$$

where “t” is the studied management problem; $X = \{x_i\}$, $i = 1, n = 43$ (districts of the RT) is the possible management decisions (alternatives) manifold; $R = \{r_j\}$, $j = 1, m = 24$ is the multitude of the requirements (criteria) taken into consideration [3]; “A” is the list of scales criteria (in the input data the criteria are given in interval and binary linguistic scales); “F” is the display of the alternatives manifold to the criteria list in the two-dimensional array $C = \{c_{ij}\}$ form which elements in the [3] are presented in interval and linguistic scales. The MC DMT solution is proposed for an optimum alternative x^{opt} finding which could be meet to the all stated criteria at the same time.

As the additional mathematical procedure for the decision-maker actions is used the fuzzy sets (FS) analysis for the manager immediate ensuring in finding the additional to an optimal solution alternatives manifold (like the quasioptimum subsets) x^{kwopt}.

Together both subsets will create an attractive decision's subset and the research objective will be in the following formal form:

$$x^{promise} = x^{opt} \cap x^{kwopt}; \ |x^{promise}| = |x^{opt}| \ \ |x^{kwopt}| \text{ or } k^{promise} = k^{opt} \ \ k^{kwopt}. \tag{2}$$

For the MC DMT (1) model was chosen the model given in [4]. The criteria according to [4, 5] beforehand were structured in the form of the two-level (close to prof. K. Ishikav's variant) objectives tree (OT) and weighed: $\Omega = \{\omega_j\}$, $j = 1, m = 24$. Then basing on [2, 4] are found a local priorities vector U = $\{u_{ij}\}$ and a global priorities vector $V = \{v_i\}$ with the additive convolution (3):

$$v_i = \Sigma\, u_{ij} \cdot \omega_j \text{ (in the conditional scores)} \tag{3}$$

$v^{opt} = \max \{v_i\}$; $i^{opt} = k = 26$ – is the number from list X:

$x^{opt} = x_{26}$ = "Tukayevsky district" = 70.40 scores = x_1.

The x^{opt} multitude look like the single-point: $k^{opt} = 1$. Other areas are ordered from the x_1. For the x^{kwopt} assessment we will use the FS analysis. Fuzzy sets (FS) are the couples like the $\tilde{A} = \{ < \mu_A(x) / x > \}$ type where "x" is the fuzzy set carrier, $\mu_A(x)$ is its accessory degree, and $\mu_A(x) / x = [0; 1]$. $\tilde{A}$ fuzzy sets carrier: SA = = $\{x \mid x \in X \;\&\; \mu_A(x) > 0\}$. FS depth is $h(\tilde{A}) = \max \mu_A(x) = a_i^{max} = 70.40$, and $\mu_A(x) = a_i / a_i^{max} > 0$ Therefore $S_A = X$ with the elements in accessory degrees $[x_1 = 1{,}000;\ x_{43} = 0{,}151]$. The optimal solution is found in the form of x^{opt} = "Tukayevsky district" = 70.40 scores = x_1.

At the 1st stage the decision-maker only states the β i.e. the boundary value of a set level

$$X^{\beta} = \{x \mid x \in X, \mu (x)/x \geq \beta\} \tag{4}$$

For example, with the $\beta = 0.80$ $k^{promise} = 3$ (see tab. 1, column 4) only a few districts are selected; if $\beta = 0.50$, then with $k^{promise} = 16$ are selected a lot of districts. Let β be defined like $\beta = 0.60$, then $k^{promise} = 10$.

At the 2nd stage the decision-maker sets the FS decomposition level α on T. Saati's scale [3]:

$$A = \bigcup_{\alpha} \alpha \cdot A_{\alpha}. \tag{5}$$

With the taking into account (4) and (5) we made the clustering by the author's method [6], see tab. 1, column 6. An algorithm of clusters formation: at $i = 1$ $d_1 = 1$ forms cluster. For $i > 1$:

$$d_{i+1} = \begin{cases} b_{i+1} = b_i \text{ at } \alpha_{1+1} < \alpha^{crit}; \\ b_{i+1} = b_i + 1 \text{ at } \alpha_{1+1} \geq \alpha^{crit}. \end{cases} \tag{6}$$

At the second stage it is necessary to form the clusters (see tab. 1), for example with the defined $\alpha^{crit} = 5$ ("essential distinctions" on a double quadtree scale). On the 3rd stage and for the method expert opportunities realizing the decision maker must to organize a coloring as the result of spectral colors display (7) to the created classes (see tab. 1, column 8). Let's use the new linguistic variable (LV) Zwet = $\{z_1, z_2, z_3, z_4, z_5, z_6, z_7\}$ for example for the decreasing histogram columns coloring.

$$\tau_{color}: A \rightarrow B \tag{7}$$

Basing to the surjection definition τ_{color} display has the injective character relatively to the B prototypes (elements of a range (7)), and as the images A acted the clusters of districts and the each cluster with $\beta = 0.60$ could be corresponds with its own color.

TABLE 1 - THE RESULTS OF THE LEVEL SET DETERMINING, CLUSTERING AND COLOURING

№	**District's name**	**Scores**	**Grade of membership**	**Weights, %**	**Class №**	**Elem. of cluster №**	**Color Z**
1	*2*	*3*	*4*	*5*	*6*	*7*	*8*
1	Tukayevsky	70,40	1,000	5,08	1	1	red
2	Almetyevsky	65,19	0,926	4,71	2	1	orange
3	Nizhnekamsky	61,66	0,876	4,45	2	2	orange
4	Vysokogorsky	54,70	0,777	3,95	3	1	yellow
5	Arsky	51,29	0,729	3,70	4	1	green
6	Laishevsky	46,50	0,661	3,36	5	1	azure
7	Mendeleevsky	46,11	0,660	3,33	5	2	azure
8	Zelenodolsky	46,01	0,654	3,32	5	3	azure
9	Bugulminsky	45,84	0,651	3,31	5	4	azure
10	Mamadyshsky	43,20	0,614	3,12	6	1	blue

IOP Conf. Series: Materials Science and Engineering **570** (2019) 012095 doi:10.1088/1757-899X/570/1/012095

11	Bavlinsky	38,77	0,551	2,80	7	1	-

In the column 8 table 1 in the decision-maker actions at the 3rd step it is possible to note as an example at least two variants. Variant 1: the "blue column" (Mamadyshsky district) could be estimated by the manager as the unpromising. Then $k^{promise} = k^{opt} + k^{kwopt} = 1 + 8 = 9$ (the Republic of Tatarstan districts). Also is possible the 2nd variant: all "blue columns" which are belonging to the No. 5 class could be marked by the decision-maker as the unpromising. Then $k^{promise} = k^{opt} + k^{kwopt} = 1 + 4 = 5$ (the Republic of Tatarstan districts). Then as the promising cases (or as the "efficient", see [7]) could be accepted the $k^{kwopt} = 8$ districts or the 4 districts. There also could be other variants. Now it is becoming apparent the singleton set x^{opt} role: the optimum set could be not only single-pointed if at a clustering (6) to the 1st class will also belong another (others) nearby located districts with the also assigned to them according to display (7) "red color". So, the proposed problem (2) is solved.

Therefore, basing on the column 8, table 1 the decision-maker here offers the two total variants i.e. it is possible to estimate as the attractive either the first 8 districts, or only the 4 from the 10 earlier selected. This means that the MC DMT solution with the fuzzy analysis using and the decision-maker actions in noted three steps (the first two are formalized) provides the quite objectified way for the proposed objectives pursuing (2).

References

[1] Carver S J 1991 Integrating multicriteria evaluation with geographical information systems. *Intern. Journ. of Geographical Information Systems* vol 5 (3) pp 321–39

[2] Salakhutdinov R Z and Ismagilov I I 2005 *The operations modeling and decision making in the economy on the fuzzy sets theory basis* (Kazan: Kheter) p 86

[3] Rozhko O N, Khomenko V V, Shikhalev A M and Yakimov I M 2016 *The transport and logistics infrastructure of the Republic of Tatarstan development* ed O N Rozhko (Kazan: Publ. House "Fan" of the Republic of Tatarstan Academy of Science) p 78

[4] *Rating of the relative solvency of the Russian Federation territorial subjects* 2001 (Moscow: Rating center of the joint-stock company "AK&M")

[5] Saati T and Kearns K 1991 *Analytical planning. Systems organisation* (Moscow: Radio and communication) p 184

[6] Shikhalev A M, Panasyuk M V and Burilov A R 2014 Application of the Forfold Tables Method for Analysis of Dynamics of Social Systems *Mediter. Journ. of Soc. Sc.* (Rome: MCSER Publishing) vol 5 No 18 pp 335-39

[7] Khafizov I I 2012 Economic viability and effectiveness of materials division methods with the electrodiamond processing *The Rev.of the Kaz. Technolog. Univ.* vol 15 No 17 pp 292-95

IOP Conf. Series: Materials Science and Engineering **570** (2019) 012096 doi:10.1088/1757-899X/570/1/012096

Experimental study of the work of a temporary structure of a dismountable bar system

G N Shmelev[1], L I Khaidarov[1], L S Sabitov[2,3], L R Gimranov[1], N F Kashapov[2]

[1]Kazan State University of Architecture and Engineering, 1 Zelenaya street, Kazan, 420043, Russian Federation

[2]Kazan (Volga region) Federal University, 18 Kremlevskaya street, Kazan, 420008, Russian Federation

[3]Kazan State Power Engineering University, 51 Krasnoselskaya street, Kazan, 420066, Russian Federation

l.sabitov@bk.ru

Abstract. Mobile spatial bar designs are widely used in the construction of temporary buildings and structures, such as scenes, stands, awnings, pavilions, etc. Such structures are statically indefinable and have gaps in the nodes of the connection. To study the actual work of such systems, an experimental study of the construction of the stage was carried out and a comparative analysis of the results of testing and calculation was performed.

Introduction

When conducting private and public events, buildings from collapsible bar systems are in demand. The advantage of these systems is the ability to assemble structures of various shapes, quick installation and disassembly due to special node connections and ease of elements. It is necessary to note the features of such structures compared to capital ones:

- lack of foundation and installation on an unprepared base;
- significant compliance of node compounds;
- short period of operation.

Collapsible buildings were used during the FIFA 2018 World Cup, including erection of the stage in Kazan (figure 1). In connection with the installation of the structure on an unprepared base, during the period of operation, the technical condition was monitored with control of efforts and movements in the most stressed areas. After the event, a series of building tests were performed by loading the poles below with jacks to study its actual work.

IOP Conf. Series: Materials Science and Engineering **570** (2019) 012096 doi:10.1088/1757-899X/570/1/012096

Figure 1. General view of the construction of the stage

1. The construction of the stage

The construction of the stage on which the test was performed was located on the bank of the Kazanka river and served as a stage for performances of creative teams and installation of the screen and sound equipment for broadcasting matches.

The main volume of the structure is assembled from the modular bar system Layher Allround. For the installation of the screen and the canopy device, additional collapsible metal trusses were used. The whole construction along the contour was sheathed with mesh-banner material. The plan and section of the construction are shown in figures 2, 3.

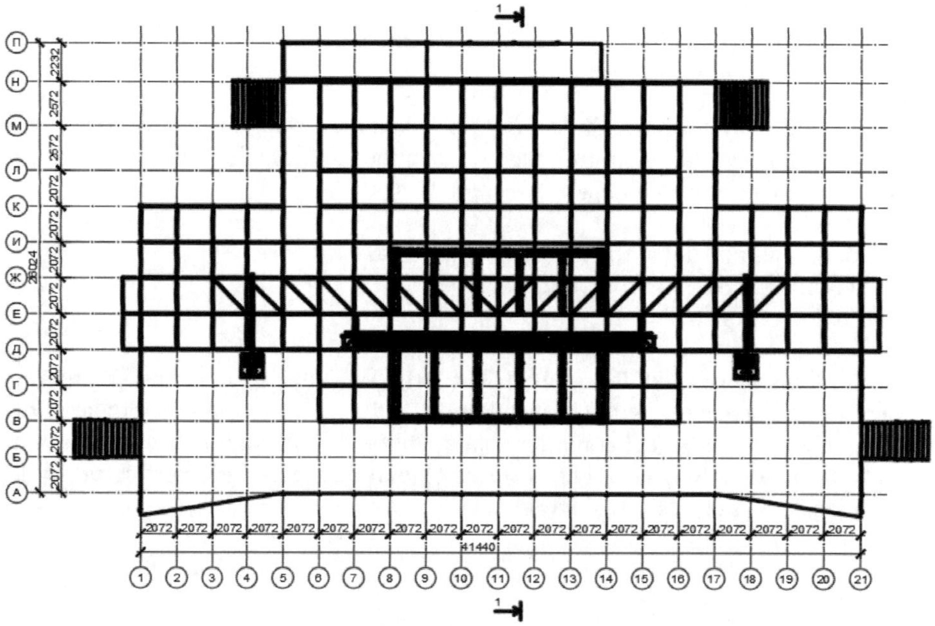

Figure 2. Stage construction plan

IOP Conf. Series: Materials Science and Engineering **570** (2019) 012096 doi:10.1088/1757-899X/570/1/012096

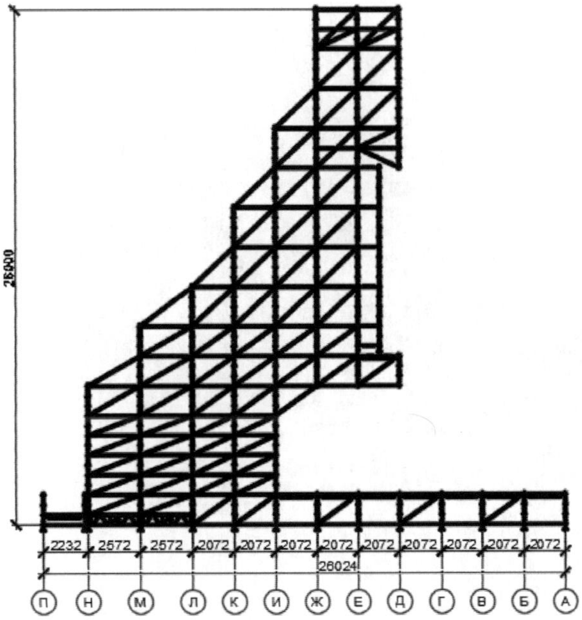

Figure 3. Cross-section of the construction of the stage

2. System Layher Allround

Layher Allround is a modular system of scaffolding assembled from core elements into a three-dimensional structure with a wedge-node connection (figure 3). The main elements of the system are standard, cross-beam, brace and screw support jack. The system is assembled in mutually perpendicular directions, but it is possible to connect not at right angles by using additional building clamps.

Standard, cross-beam and brace are made from pipes with a diameter of 48.3 mm and a wall thickness of 3.2 mm of steel grade C235. Standard s are made 0.5… 3.0 m long with a gradient of 0.5 m, to which, in increments of 0.5, round flanges are welded with eight holes for fastening cross-beams and braces. Assembly of the standards in height is performed by the sleeve connection. Cross-beam and brace have special heads with a fixed wedge at the ends. Head connection with bar of cross-beam is welded, of brace - riveted, allowing rotation of the element in a vertical plane.

The manufacturer applies to the output of the documents on the calculation of the elements and the carrying capacity of the elements and node connections.

3. Numerical simulation of the support lift

To determine the forces and movements when lifting the standard, a calculation was made using the finite element method in the Lira-SAPR PC. The volumetric model is presented in figure 4.
The calculation took into account the following loadings:
- own weight of the bearing elements;
- the weight of the screen and sound equipment;
- the weight of the ballast installed to prevent tilting of the structure;
- load on the support from the jack.

IOP Conf. Series: Materials Science and Engineering **570** (2019) 012096 doi:10.1088/1757-899X/570/1/012096

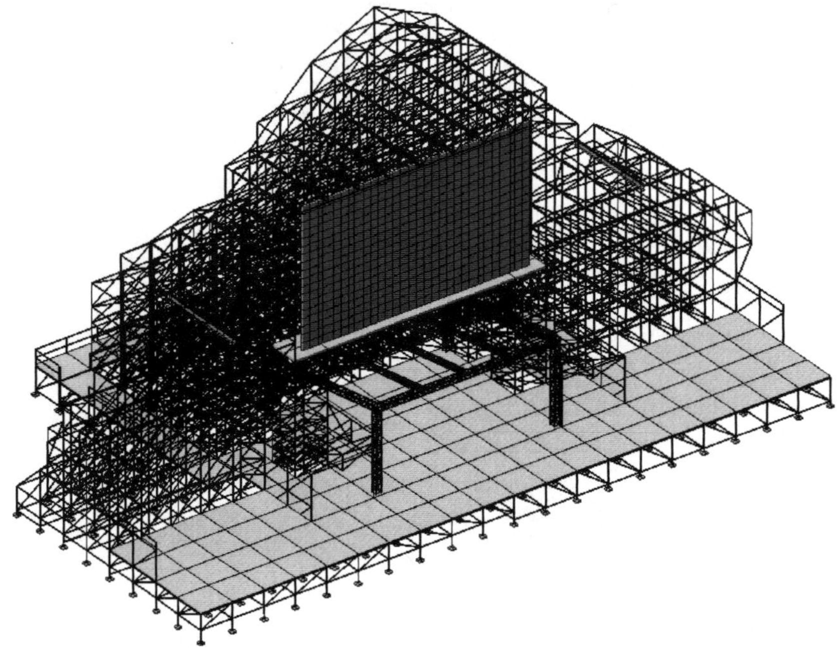

Figure 4. Spatial view of the calculated model of the construction of the scene

Under the distribution boards under the supports, the base soil is modeled in the form of silty sand with a strain module of 18 MPa.

Vertical cross-beams are modeled as bars with absolute hinges at the ends and with eccentricities in accordance with the node design. The cross-beam element model is made with elastic hinges at the ends in the vertical and horizontal planes. In fact, when rotated in the vertical plane, the node cross-beam- standard works in a non-linear fashion. In the calculation, a linear work schedule was adopted due to the fact that only minor moments occur in the cross-beams with the considered loads, and the initial part of the work schedule curve differs little from a straight line.

Stiffness elements were assigned in accordance with the material and the geometric dimensions of the cross section, with the exception of vertical cross-beams, for which the manufacturer has established linear stiffness, taking into account the gaps, turning in the joints and bar bending due to the transfer of force with eccentricity.

According to the results of the calculation, changes in the forces in the elements, movement in the loaded and neighboring standards were calculated.

4. Test

The purpose of the test was to determine the stress-strain state of the elements of the structure during loading and unloading of one and two adjacent supports. The experiment examined the raising and lowering of standard in the K / 13 axes.

4.1. Test method. The rise of the standard was carried out with a manual hydraulic pump with two jacks through the bottom row of cross-beams (figure 3). At the same time, the maximum load on the standard was 3.6 tons. After the lifting by turning the screw, the lifting height of the standard of the standard decreased, the hydraulic jacks were unloaded and the standard lowered to the hanging position.

IOP Conf. Series: Materials Science and Engineering **570** (2019) 012096 doi:10.1088/1757-899X/570/1/012096

Figure 5. Scheme of the standard's rise by jacks

To measure the voltages in the elements, strain gauges and National Instruments equipment were used: the NI SCXI-1000 chassis with the NI SCXI-1521 and NI SCXI-1317 modules. The stresses in the elements were saved in a spreadsheet for further processing and determination of the effort. In each element under consideration, two strain gauges, symmetrically arranged in one section, were installed. The layout of the strain gauges is presented in figure 6.

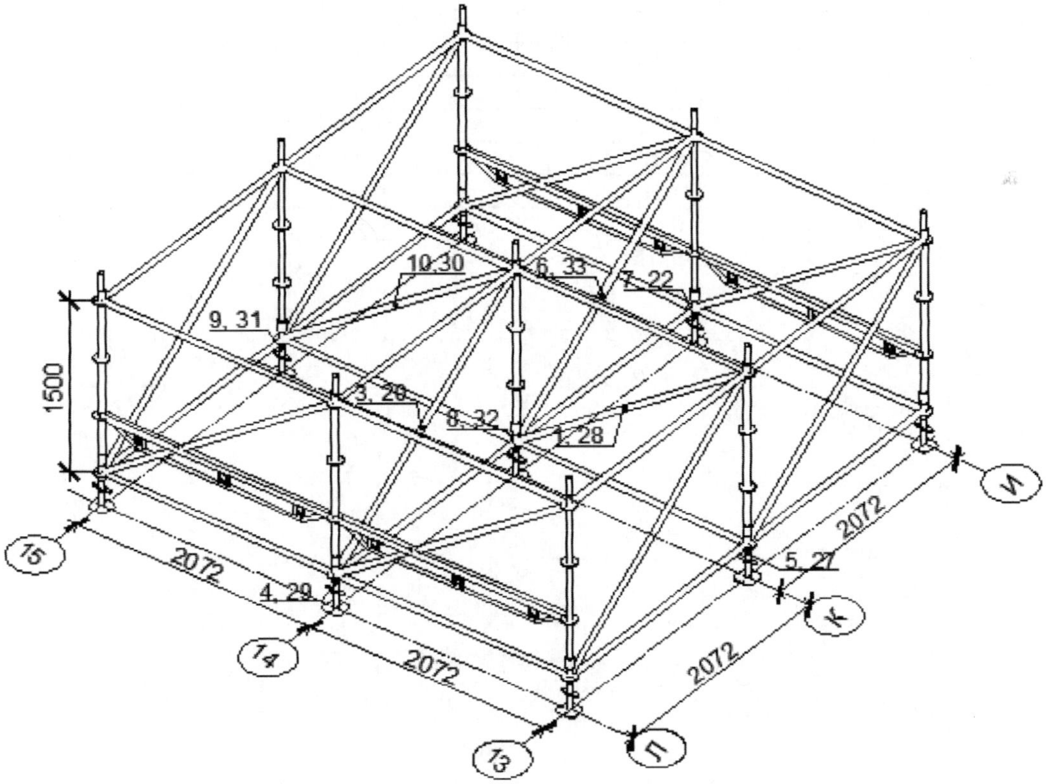

Figure 6. Layout of strain gauges

IOP Conf. Series: Materials Science and Engineering **570** (2019) 012096 doi:10.1088/1757-899X/570/1/012096

Vertical displacements or movements were measured in standards in the "K / 13", "K / 14" and "K /15" axes at the level of the lower tier of cross-beams. The measurement was performed by mechanical deflection of 6-PAO. Instrument readings were recorded in the journal at each stage of loading.

4.2. Test results. Displacements during lifting are presented in table. 1, change of efforts in elements in a graphic form - on figures 7-8.

Table 1. Vertical displacements of supports during testing

Load, kg		Vertical displacements of standards in axes, mm		
		K/15	K/14	K/13
Loading	0	0.00	0.20	0.02
	500	0.01	0.60	0.02
	1000	0.02	1.42	0.02
	1500	0.04	2.22	0.03
	2000	0.08	3.28	0.05
	2400	0.10	4.08	0.10
	2800	0.13	4.80	0.13
	3200	0.18	5.48	0.18
	3600	0.21	6.08	0.20
Unloading	3200	0.21	5.50	0.20
	2800	0.20	4.95	0.20
	2400	0.18	4.35	0.20
	2000	0.13	3.78	0.13
	1500	0.11	3.05	0.12
	1000	0.09	2.09	0.12
	500	0.06	0.42	0.08
	400	0.02	-0.45	0.08
	300	0.01	-1.00	0.00
	200	0.01	-1.20	0.00
	100	-0.01	-1.35	-0.02
	0	-0.01	-1.4	-0.03

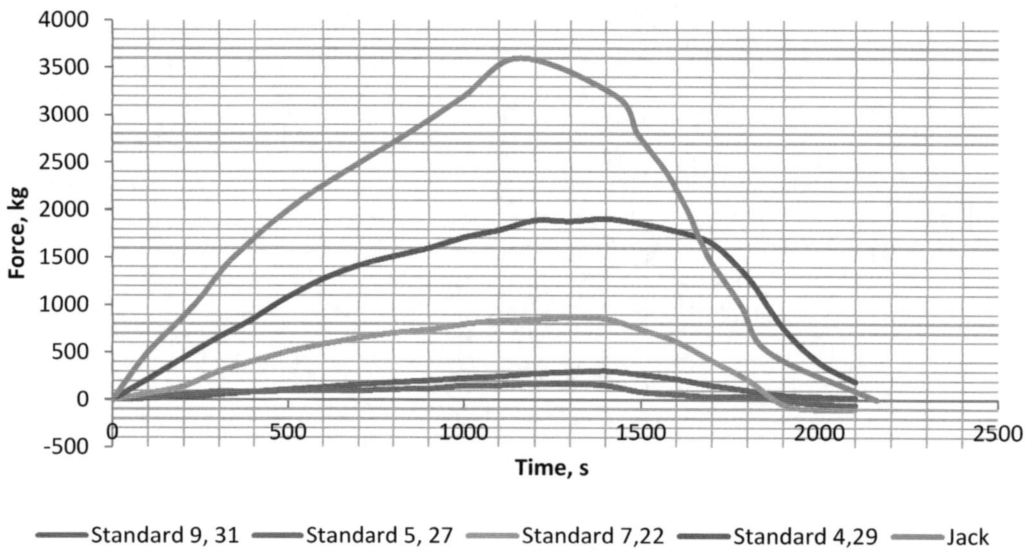

Figure 7. Change of support reactions under standards

IOP Conf. Series: Materials Science and Engineering **570** (2019) 012096 doi:10.1088/1757-899X/570/1/012096

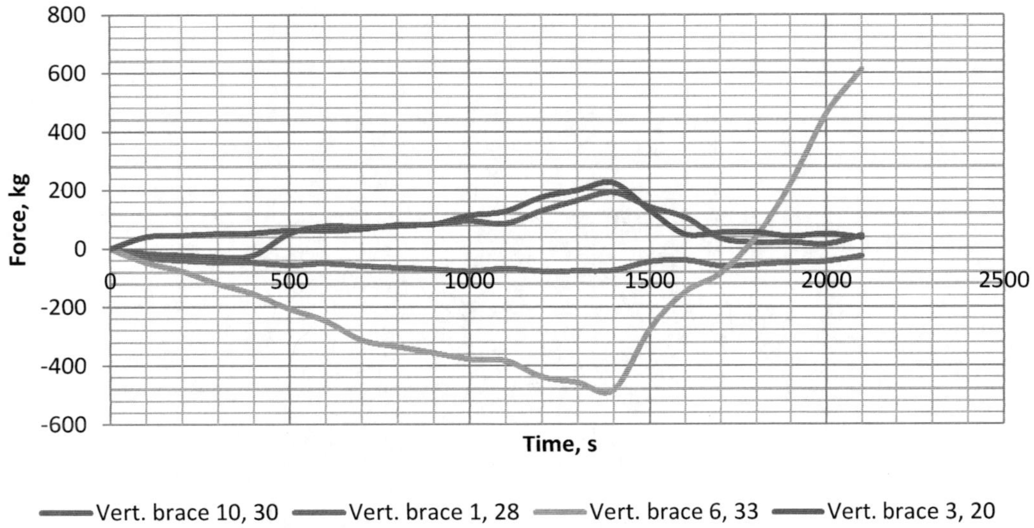

Figure 8. Effort changes in vertical braces

According to the results of the experiment, it was determined that when loading the standard in the K / 14 axes with a load of 3,600 kg, unloading of the neighboring standards occurs in the axes I / 14, K / 15, K / 13 and L / 14 by 3400 kg, i.e. 94% of the load is distributed to adjacent supports. When lifting the considered standard in the launch element below the loading unit after the separation from the screw jack, a tensile force of 186 kg appeared, which is equal to the support reaction under the stand before the test. In fig. 9 you can see the uneven operation of the vertical braces. One of the main reasons for this is the gaps in the nodes of the connection elements. When lowering the standard to the hanging position, the vertical displacement was 1.4 mm from the initial position.

5. Comparison of calculation end experiment results

For a visual comparison of the results of the calculation and experiment, the change in the forces and displacement during the raising and lowering of the K / 14 vertical brace is presented in figures 9-10.

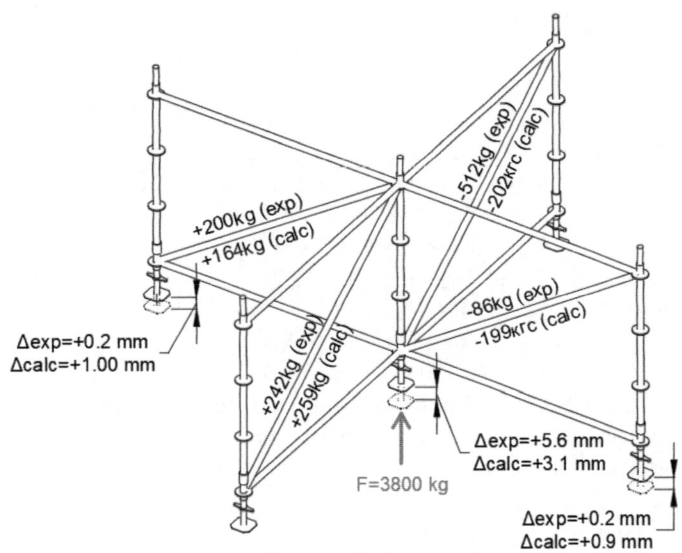

Figure 9. Changes in effort and movement when raising standard according to the results of experiment and calculation

IOP Conf. Series: Materials Science and Engineering **570** (2019) 012096 doi:10.1088/1757-899X/570/1/012096

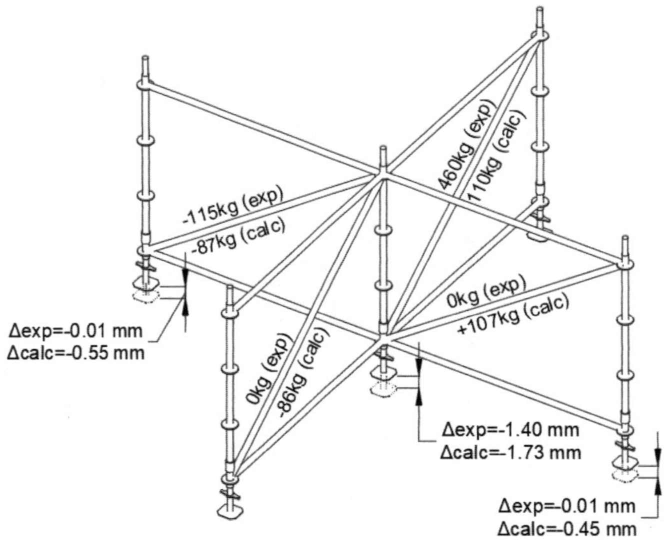

Figure 10. Changes in effort and movement when lowering standard according to the results of experiment and calculation

Conclusion. According to the results of the experiment and calculation, you can see a significant difference in the distribution of efforts in the elements of the structure. One of the important factors is the presence of gaps in the joints. In this regard, it is necessary to further study the actual operation of structures from collapsible bar systems, taking into account the gaps in the nodes of the connection for their correct modeling and determination of their stress-strain state.

References

[1] Liu C, He L, Wu Z and Yuan J 2018 Experimental and numerical study on lateral stability of temporary structures *Archives of civil and mechanical engineering 18* (Elsevier) p 1478-90

[2] Cimellaro G P and Domaneschi M 2017 Stability analysis of different types of steel scaffolds *Engineering structures* 152 (Elsevier) p 535-48

[3] Peng J -L, Ho C -M, Chan S -L and Chen W -F 2017 Stability study on structural systems assembled by system scaffolds *Journal of Constructional Steel Research* 137 (Elsevier) p 135–51

[4] Yuan X, Anumba C J and Parfitt M K 2016 Cyber-physical systems for temporary structure monitoring *Automation in Construction* 66 (Elsevier) p 1–14

[5] Zhang H and Rasmussen K J R 2013 Monitoring System-based design for steel scaffold structures used advanced analysis *Journal of Constructional Steel Research* 89 p 1-8

[6] Chandrangsu T and Rasmussen K J R J 2011 Structural modelling of support scaffold systems *Journal of Constructional Steel Research* 67 (Elsevier) p 866–75

[7] Prabhakaran U, Beale R G and Godley M H R 2011 Analysis of scaffolds with connections containing looseness *Computers and Structures* 89 (Elsevier) p 1944–55

[8] Peng J L, Yen T, Kuo C C and Chan C L 2009 Analytical and experimental bearing capacities of system scaffolds *Journal of Zhejiang University Science A* 10(1) p 82-92

[9] Crick D and Grondin G Y 2008 Monitoring and Analysis of a Temporary Grandstand *Structural Engineering Report No.275* (Edmonton: Department of Civil& Environmental Engineering of University of Alberta) p 48-98

[10] Godley M H R and Beale R G 2001 Analysis of large proprietary access scaffold structures *Structures and Buildings* 146 p 31-39

[11] Chan C L, Zhou C H, Chen W F, Peng J L and Pan A D 1995 Stability analysis of semirigid steel scaffolding E*ngineering structures* vol 17 No 8 (Elsevier) p 568-74

IOP Conf. Series: Materials Science and Engineering **570** (2019) 012097 doi:10.1088/1757-899X/570/1/012097

Numerical study and calculation of the structures of the stands with canopies on the wind impact, taking into account the terrain

G N Shmelev[1], L S Sabitov[2,3], I A Galimullin[1], L I Khaidarov[1] and A S Antonov[1]

[1]Kazan State University of Architecture and Engineering, 1 Zelenaya street, Kazan, 420043, Russian Federation
[2]Kazan (Volga region) Federal University, 18 Kremlevskaya street, Kazan, 420008, Russian Federation
[3]Kazan State Power Engineering University, 51 Krasnoselskaya street, Kazan, 420066, Russian Federation

gn.shmelev@mail.ru

Abstract. Operation of stands with canopies is possible under different landscape conditions, including conditions in areas with a slope or in areas with various obstacles that change the profile of the wind flow and, accordingly, affect the distribution of wind pressure acting on the surface of the canopy. Determination of the actual wind loads on the stands with a canopy is an important task due to the lack of appropriate schemes in the SP 20.13330.2016 "Loads and impacts". The relevance of this problem is associated with the complexity of tests in specialized wind tunnels, as well as with the rapid development of computational fluid dynamics. In this regard, a numerical study of the aerodynamics of these structures was made in the software package ANSYS Fluent for the analysis of flows of liquids and gases. The article presents a comparison of the results with the schemes of wind pressure for awnings from SP 20.13330.2016 "Loads and effects" and Eurocode "Wind effects". The obtained results were used for verification of similar calculation technologies and specialized software systems, addition of regulatory documents and for "adjustment" of newly commissioned wind tunnels (in order to optimize and improve the efficiency of tests).

1. Introduction

Currently, the stands with a canopy are often used as temporary entertainment facilities. Wind is one of the main loads for them, so it is the actual wind loads on the stands with a canopy that are the subject of the study. According to the norms of SP "Loads and impacts" it is expected to use the results of large-scale model tests in special wind tunnels that are able to simulate the boundary layer of the atmosphere. However, such tests are extremely difficult, and for the largest models they need to be carried out abroad. In addition, the method of experimental modeling of wind flows and the impact on the facilities has its own errors and limitations [1], [2].

Recently, there has been a rapid development of computational fluid dynamics (CFD), improving the technological processes of calculations of wind impacts on buildings and structures, constantly growing computing power of personal computers [3], [4]. Leading foreign scientific research and design companies often combine testing and "numerical" experiments [5], [6], [7], [8]. In the future, the role of mathematical modeling, as shown by experience in neighboring industries (such as aerospace) and tasks (construction mechanics), will only increase [3], [7], [9].

From the programs of computational fluid dynamics to perform the necessary calculations, the ANSYS Fluent complex was used. In many ways, the choice is connected both with the possibility of relatively accurate modeling of wind effects, and with the visual process of numerical modeling.

2. Numerical experiment with a calculation model without an obstacle

2.1. Formation of the calculation model

As a prototype object, a temporary structure was taken, installed specifically for the so called "high diving" competitions of the XVI World Aquatics Championships in 2015. The structure of the stands with canopies was located on the banks of the Kazanka river to accommodate spectators. Variable parameters in the calculation model of the stand with canopy are only three – the angle of inclination of the canopy (α) – the base value – 10 degrees, the slope angle (β) – the base value of 15 degrees, the length of the horizontal projection of the slope (h) – the base value – 1h, that is equal to the height of the stand with the canopy (figure 1). C_1 and C_2 – aerodynamic coefficients of wind pressure on the canopy. In the calculations for the basic schemes without obstacles the following values were used: the values of the angle of the canopy (α) of 10, 20 and 30 degrees, the angle of the slope (β) of 0, 15, 30, 45 degrees, length of the horizontal projection of the slope (h) – h, 10h and 20h. Since the length values are taken not in absolute values, but relative to the height of the stand with a canopy, the results of the calculations will be applicable for stands with canopies of different sizes.

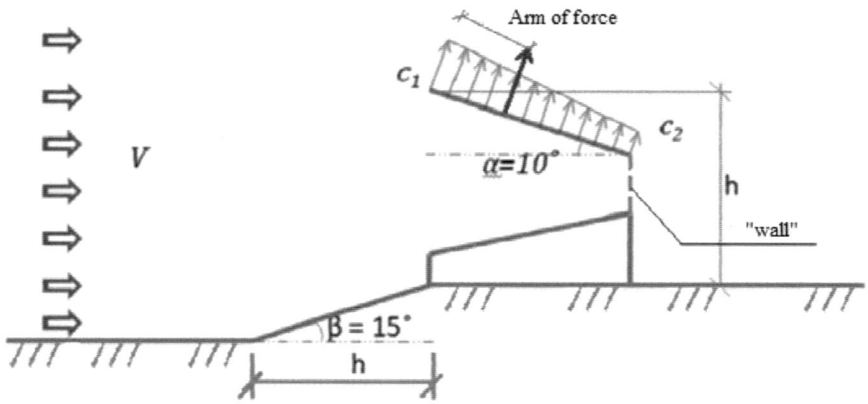

Figure 1. Calculation model of the stand with a canopy.

2.1.1. Basic schemes. The calculations were carried out according to four main schemes, which differed in the direction of the wind (runs in front or behind the stand), as well as the presence or absence of a "wall" between the main part of the stand and the canopy. Scheme 1 – the wind runs in front of the stand, the wall is missing. Scheme 2 – wind runs in front of the stand, the wall is present. Scheme 3 – wind runs behind the stand, the wall is missing. Scheme 4 – wind runs behind the stand, the wall is present.

2.1.2. The distribution of wind pressure. As a result of the calculation in the ANSYS Fluent software package, it is possible to obtain a calculation model with a realistic curvilinear distribution of wind pressure on the canopy. For ease of use in further calculations, the distribution of wind pressure is transformed to a rectangular trapezoid from the condition of equality of the module and the arm of the reduced force. S_{gen} or C_f – the average value of the distributed load factor in this case will be equal for both the calculated and transformed models.

2.2. Selection of comparative values

Review of literature showed that in SP "Loads and impacts" there are diagrams of buildings, open on one side and ones with fully open canopies. At the same time Eurocode "Wind effects" produces examination of stand-alone canopies with barriers, where φ is the barrier degree. To compare the results of numerical modeling with the normative literature, the average pieces of data, most approximate to our models, from each of the sources were taken.

2.3. Results of calculation

The results illustrating figures were constructed as follows. The first line corresponds to the change of the canopy angle parameter (α), the second – to the change of the slope angle parameter (β), and the third – to the change of the length parameter of the horizontal projection of the slope (h). The number in the upper left corner of each segment of the figure is equal to the corresponding value of the parameter to be changed. The numbers on the edges of the diagrams correspond to the boundary values of the aerodynamic coefficients for the corresponding schemes. According to the results, as the angle α increases, the values of C_1 and C_2 begin to increase, while the angle β increases, the values of C_1 and C_2 decrease (figure 2). For the design scheme 4, due to the location of the wind flow behind the podium and the presence of the wall, the C_f values are extremely small, compared to the other three schemes (figure 3).

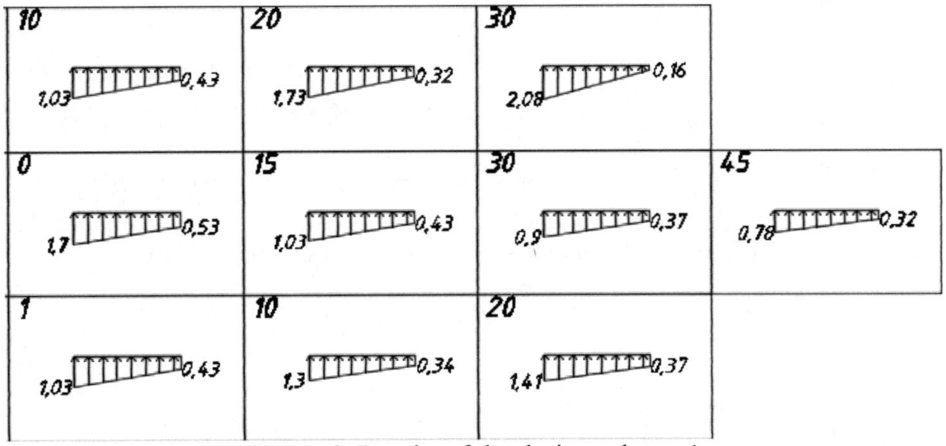

Figure 2. Results of the design scheme 1.

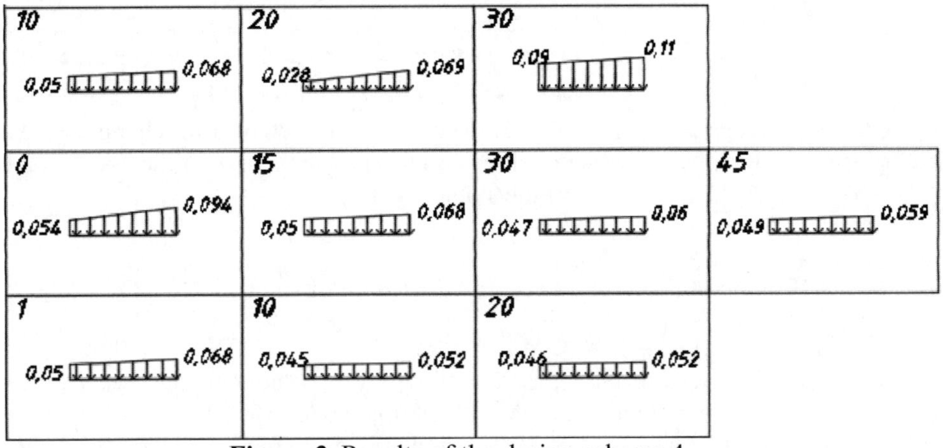

Figure 3. Results of the design scheme 4.

2.4. Comparison of results with normative documents

For each of the considered cases, the average Cf values were calculated, graphs were plotted to allow a visual comparison with the normative documents. For the change of the angle α values of SP "Loads and impacts" almost coincide with the calculated ones. At zero slope angle β, the calculated value is in absolute value greater than the value in SP "Loads and impacts" (figure 4). This is the only such case in all the calculations. For scheme 2, the calculated values are always less than SP "Loads and impacts" and Eurocode "Wind effects". In the case of scheme 3, in some cases, the calculated values of aerodynamic coefficients are greater than in Eurocode "Wind effects" (figure 5). For the calculation scheme 4, as for the calculation scheme 2, the calculated values are always less than the values from the normative documents.

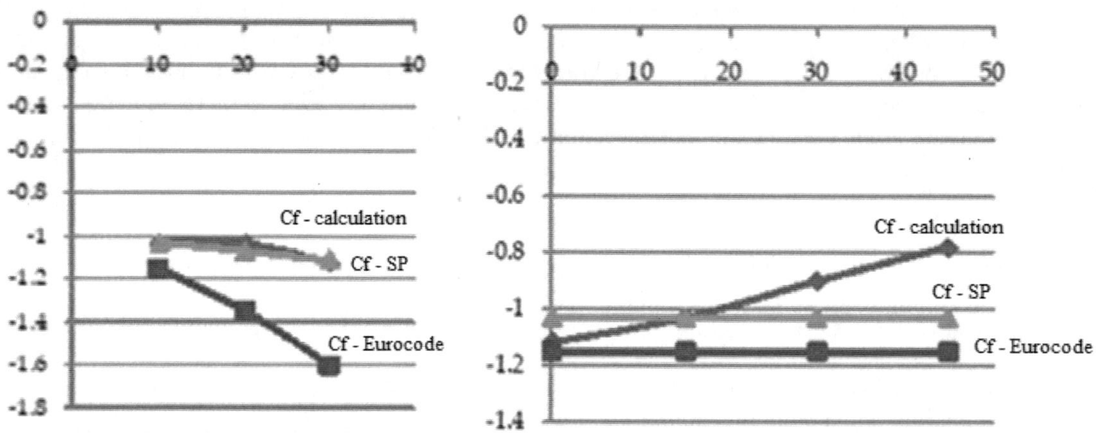

Figure 4. Graphs comparing the results of the calculation scheme 1 for changing the angle of inclination of the canopy (left) and the angle of slope (right).

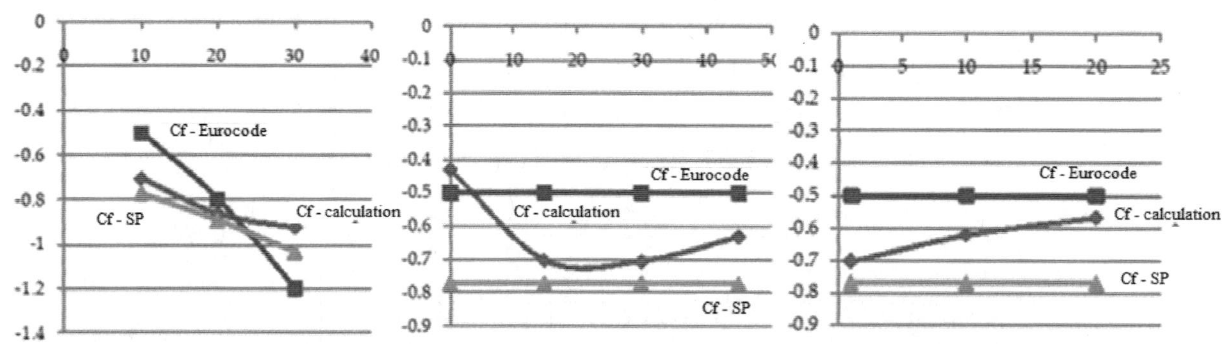

Figure 5. Graphs comparing the results of the calculation scheme 3 for changing the angle of inclination of the canopy (left), the angle of slope (center) and the length of the horizontal projection of the slope (right).

3. Numerical experiment with a calculation model with an obstacle

3.1. Formation of the calculation model

A rectangle with variable values of the sides was chosen as a computational model of a conditional obstacle in the wind path. The second part of the calculation model was again a rostrum with a canopy, similar to the temporary structure at the competitions in "high diving" XVI world Championships in water sports in 2015 (figure 6). Variable parameters in the calculation model of the podium with the

IOP Conf. Series: Materials Science and Engineering **570** (2019) 012097 doi:10.1088/1757-899X/570/1/012097

presence of obstacles are also three – the height of the obstacle (H), the width of the obstacle (b), the distance between the obstacle and the stand (d). The basic value of the parameters – 1h, that is equal to the height of the stand with a canopy. In the calculations for the scheme with an obstacle the values were the following ones: the obstacle height (H) – 0.5 h, h, 2h, the obstacle width (b) – h, 2h, 5h, the distance between the obstacle and the stand (d) – h, 2h, 5h, 10h. Since the values of the dimensions of the obstacle and the distance between the obstacle and the stand are taken not in absolute terms, but relative to the height of the stand with a canopy, the results of the calculations will be applicable for stands with canopies of different sizes.

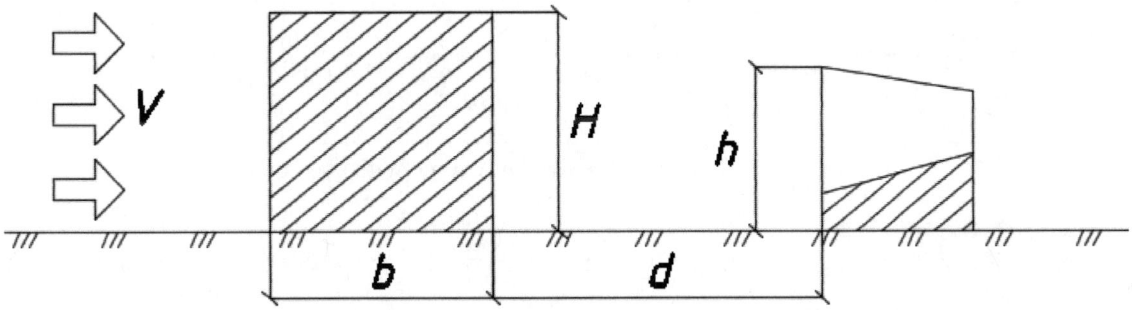

Figure 6. Calculation model of the stand with an obstacle.

3.1.1. The distribution of wind pressure. As for the previous numerical experiment, for ease of use in further calculations, the distribution of wind pressure is transformed to a rectangular trapezoid from the condition of equality of the module and the arm of the reduced force. C_{gen} or C_f – the average value of the distributed load factor in this case will be equal for the calculated and transforned models.

3.2. Results of calculation

The figure illustrating the results was constructed as follows. The first line corresponds to the change in the height of the obstacle (H), the second to the change in the width of the obstacle (b), and the third to the change in the distance between the obstacle and the podium (d). The number in the upper left corner of each segment of the figure is equal to the corresponding value of the parameter to be changed. The numbers on the edges of the plots correspond to the boundary values of the aerodynamic coefficients for the corresponding schemes. It can be seen that with increasing of the angle α the values of C_1 and C_2 start to increase too, and with the increase of the angle β the values of C_1 and C_2 decrease. For the design scheme 4, due to the location of the wind flow behind the stand and the presence of the wall, the C_f values are extremely small, compared to the other three schemes. The analysis of the results showed that the calculation scheme with the presence of an obstacle is characterized by a sharp decrease in C_f – in comparison with earlier calculations without obstacles, and with the normative documents SP "Loads and impacts" and Eurocode "Wind effects" (figure 7).

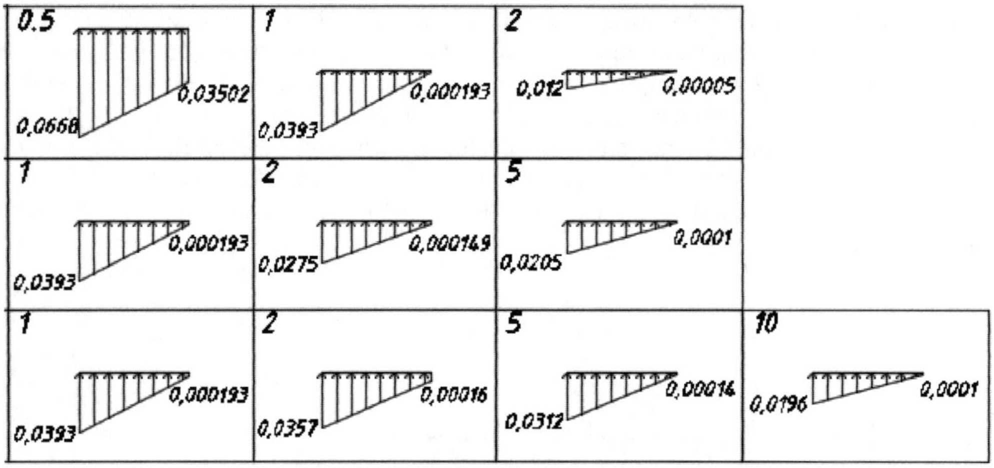

Figure 7. Results of the design scheme with an obstacle.

4. Conclusion

A numerical study of the distribution of wind pressure on the stand with a canopy was done depending on: the angle of inclination of the canopy, the angle of inclination and the length of the horizontal projection of the slope. The results were compared with the values specified in the normative documents SP "Loads and impacts" and Eurocode "Wind effects". Changes in the terrain and the angle of the canopy affect the distribution of wind pressure, but the results are less than the values specified in the SP "Loads and impacts". It is established that with wind influence behind a stand with a canopy the loading is more when the tribune is open whereas with wind influence in front of a stand with a canopy the loading in some cases can be more when the stand is closed with wall. This is important to consider when installing and dismantling a temporary structure. A numerical study of the influence of obstacles on the distribution of wind pressure on the stand with a canopy was done, depending namely on its height, width and distance between the stand with a canopy and an obstacle. It is established that the presence of an obstacle leads to a sharp decrease in aerodynamic coefficients in comparison with both normative documents and calculations without obstacles.

References

[1] Hoffman J and Johnson C 2006 *A new approach to computational turbulence modeling.* (Computer Methods in Applied Mechanics and Engineering, Vol. 195, No: 23-24) p 2865-2870
[2] Hu H, Yang Z and Sarkar P 2011 *Dynamic wind loads and wake characteristics of a wind turbine model in an atmospheric boundary layer wind* (Ames: Springer-Verlag 2011) p 1277-9
[3] Ahmad S, Muzzammil M and Zaheer I 2011 *Numerical prediction of wind loads on low buildings* (Aligarh: International Journal of Engineering, Science and Technology Vol. 3) p 59-60
[4] Gordeev V N, Lantukh-Lyashenco A I, Pashinskiy V A, Perelmuter A B and Pichugin S F 2007 *Loads and impacts on buildings and structures* (Moscow: IASW) p 195-222
[5] Overend M and Zammit K 2006 *Wind loading on cladding and glazed façades* (Nottingham: International Symposium on the Application of Architectural Glass, ISAAG 2006) p 1-3
[6] Mendis P, Ngo T, Haritos N, Hira A, Samali B and Cheung J 2007 *Wind Loading on Tall Buildings* (Australia: EJSE Special Issue: Loading on Structures (2007)) p 41-2
[7] Ghosh A K 2004 *International workshop on wind engineering and science.* (New Delhi: Elite Publishing House Pvt. Ltd.) p 107-112
[8] Yuan X, Anumba C J and Parfitt M K 2016 *Cyber-physical systems for temporary structure monitoring Automation in Construction 66* (Elsevier) p 2–6
[9] Liu C, He L, Wu Z and Yuan J 2018 *Experimental and numerical study on lateral stability of*

temporary structures Archives of civil and mechanical engineering 18 (Elsevier) p 1475-89
[10] Van Uffelen G M 2009 *Wind-induced building interference: increase of wind loads on existing buildings after erection of new high-rises* (Florence: EACWE 5) p 1-3

IOP Conf. Series: Materials Science and Engineering **570** (2019) 012098 doi:10.1088/1757-899X/570/1/012098

Characteristic features of the cooling capacity of aqueous polymer solutions

T V Shveyova [1], G F Muhametzyanova [2], V I Astashchenko [2], N N Zapadnova [3] and E A Zapadnova [2]

[1] Branch of Kazan Federal University in Naberezhnye Chelny, Service of Transport Systems Department, Tatarstan Republic, Russia, 423812 Naberezhnye Chelny, Syuyumbike Avenue, 10A

[2] Branch of Kazan Federal University in Naberezhnye Chelny, Department of materials, technologies and quality, Tatarstan Republic, Russia, 423812 Naberezhnye Chelny, Syuyumbike Avenue, 10A

[3] Technical College n.a. V.D. Potashov, Tatarstan Republic, Russia, 423827, Naberezhnye Chelny, Motornaya Str., 13A

Abstract. For the study of quenching medium a special sample is developed that provides high reproducibility of thermal processes occurring on the surface of the cooled product. The results of the cooling method of aqueous solutions with different polymer content are given. The application of thermogravimetric analysis for the study of physical and chemical processes occurring during heating of a synthetic quenching medium is considered. The temperature intervals of endo-exothermic reactions as a result of oxidation of the polymer component of the solution are shown. The reasons for the formation of two extrema on the cooling capacity curve of aqueous polymer solutions are indicated.

Thermal and surface deformation treatments are the key factors in the system of formation of the structure, properties and reliability of metal products in operation [1, 2]. The most difficult tasks in the field of heat treatment include issues related to the processes of quenching cooling of metal products. A key role in solving this problem is given to the quenching medium. Thanks to its competent and well-equipped application, it is possible not only to control the cooling process, and, therefore, the structure formation of steel during the heat treatment of metal products, but also to solve issues of environmental nature, the cost of production and technological safety. The analysis of literature and patent research indicate that the management of the hardening process is a very difficult task, but the effect of its competent solution is really high. It is known [3-6] that the quenching media must have a strictly defined cooling capacity in the specified temperature ranges of the quenched product, maintain their properties for a long period of operation, be environmentally safe, etc.

Quenching media based on water-soluble polymers, widely used since the middle of the XXth century are of particular interest, as they are devoid of drawbacks that exist in traditional quenching media such as water and oil. Depending on the content of the polymers in water, it is possible to change the cooling ability of the environment from the cooling rate in water to cooling rate in oil [4 - 11].

On the basis of the analytical review of information sources [3-12] it is established that for all synthetic media a certain pattern is clearly traced, expressed in the presence of two maxima on the dependence of their cooling capacity (Fig. 1). However, there is not enough available information about the causes of high and low cooling rates inherent in them, and they do not give a clear answer to this question. Typically, these dependences are associated with the temperature intervals of the boiling point of the quenching fluid that can be changed by hardening in media under pressure [12] or by introducing various additives [8 - 11]. Due to these additives, the temperature range of the film and bubble boiling is changed and, accordingly, an indispensable cooling rate is achieved in a given temperature range.

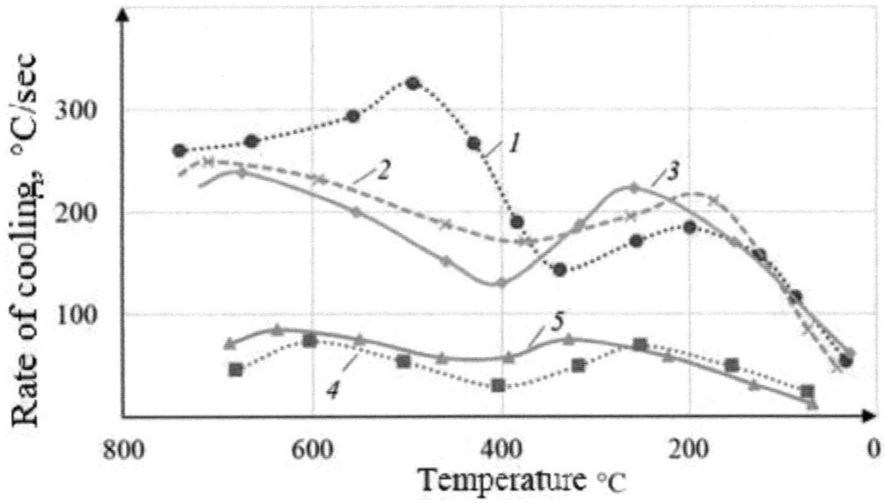

Fig. 1 – Cooling capacity of aqueous solutions "Osmanil-E2 "(1)," PK – 2 "(2)," Tosol – K "(3,4) and" Kamgidrol-ZAK " (5) at different polymer content (wt. %): 1 – 6%; 2 - 0,65%; 3 – 32,5%; 4 – 50%; 5 - 18%.

It should be noted that such additives themselves in high temperatures during quench cooling of the metal are able to undergo a physical-to-chemical reactions: decomposition, oxidation, destruction, melting, recovery etc. When dipping the hot product into the quenching medium at the interface "metal-liquid", these endo-exo-effects create a thermal effect on the surface. The temperature gradient between the surface of the product and the liquid is determined by the amount of heat that is released or absorbed as a result of such reactions. Depending on the type and concentration of the additive in the liquid medium, the level of the thermal effect also changes. Finding out the reasons for the low and high cooling rates on the dependence $V_{охл} = f(T\ °C)$ and establishing the ability to control these values is an important task for specialists engaged in heat treatment of metal products.

Purpose of Work.

Study of the cooling capacity of aqueous polymer solutions and justification of the extreme values of the cooling rate at this dependence.

Materials, Research Methods and Results.

To achieve this goal, the paper has investigated aqueous solutions of liquid "TOSOL-K" [9, 10], which are most often used for bulk and surface hardening of parts. "TOSOL-K" is a concentrated liquid, which contains ~50% of the polymer with the presence of anti-corrosion, anti-foaming and stabilizing additives. Polyglycol – nonylphenol ethoxylates is used as the polymer, and the sodium nitrite and triethanolamine are used as corrosion additives. To ensure the manufacturability of the environment during operation and storage in its composition there are silicone defoamer, castor oil and potassium benzoic acid in small quantities. The study was subjected to aqueous solutions of liquid "TO-

IOP Conf. Series: Materials Science and Engineering **570** (2019) 012098 doi:10.1088/1757-899X/570/1/012098

SOL-K", in which the polymer content was 5%, 15% and 32.5%, as well as 50% (liquid concentrate). The polymer content in the solution was determined by density. The cooling capacity of quenching solutions [12] was estimated by temperature regulation of the inner surface of a specially designed copper sample with a wall thickness of 1.7 mm (Fig. 2A). Thermal calculations have found that the temperature difference between the outer surface of the cooled sample and the controlled area at a specified distance from the surface does not exceed 11°C.

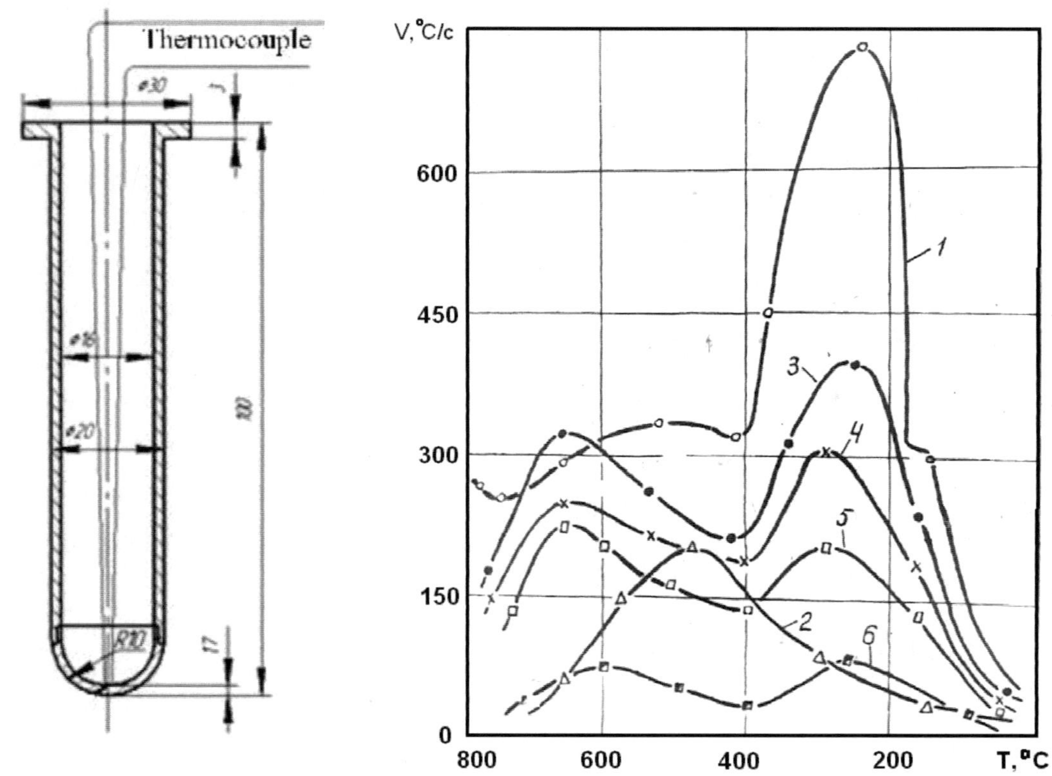

Fig.2 – The cooling rate of the copper sample (a) in water (1), MZM-16 oil (2), 5%-s (3), 15% (4), of 32.5% (5) and 50% (6) aqueous solution of "TOSOL-K" (b).

Using the proposed design of the sample, a high reproducibility of the processes occurring on the surface was achieved and, accordingly, it is possible to constant the accuracy of determining the cooling capacity of liquid media. The results of the study of aqueous polymer solutions of different concentrations has shown that with an increase in the content of nonylphenol ethoxylates in water, the cooling capacity decreases (Fig. 2 b). Thus, an increase in its content in the solution from 5 to 50% reduces the cooling capacity of the quenching medium in the perlite transformation temperature range from 300-320°C/s to 70-75°C/s, and in the martensitic transformation temperature range – from 400°C/s to 80°C/s. it is Noteworthy that there are two maxima on the curves of $V_{охл} = f(T\ °C)$: one in the range of 600-680°C, the second – at 280-300°C. The answer to the reason of formation of two peaks on the curve is performed on the thermogravitometric analyzer TG209F1/ris (f. Netzsch, Germany). Using this method, information on the behavior of the quenching liquid "TOSOL-K", containing 50% of the polymer, when heated from 20 to 800°C was got (Fig. 3).

As a result of the study, the dependence of the mass change on the heating temperature (HT), the change in the rate of mass loss on the temperature (DT), and the curves of temperature change (T) and its differential version (DCT) are obtained. Special attention should be paid to the dependence of DCT, which characterizes the endo-exothermic reactions occurring when the quenching medium is

IOP Conf. Series: Materials Science and Engineering **570** (2019) 012098 doi:10.1088/1757-899X/570/1/012098

heated. It is established that in aqueous solution based on the studied polymer, there are two effects: endothermic – in the temperature range 110 - 170°C and an exothermic asymmetric – in the range of parator 280 - 510°C. It was found that the endothermic peak is associated with boiling water, and its end at 170°C corresponds to the lower boiling crisis – the minimum temperature of stable bubble boiling. The specified temperature determined by the method of dripping [13] corresponds to 157°C [14].

The exothermic peak with a maximum at 400 °C is associated with intensive oxidation of the polymer component under the influence of temperature and sodium nitrite, the decomposition of which is known to occur at 380°C. Comparison of the cooling curves and the differential curve of the temperature change (DCT) of the aqueous solution with the content of nonylphenol ethoxylates shows that the decrease in the cooling capacity in the temperature range of 300-600°C is associated with the exothermic effect arising during the oxidation of the polymer component of the quenching medium (Fig. 3). On the DCT curve, the maximum of this effect corresponds to 400°C, which is explained by a decrease to a minimum value of the cooling rate at a temperature of 400°C (Fig. 2, Fig. 4).

Thus, as a result of thermogravimetrical analysis of polymer quenching media, it can be stated that due to the exothermic reaction, accompanied by heat generation, two bright peaks were formed on the curve of cooling rate – one of which is at the temperature of the beginning of this reaction and the second one – at a temperature of completion of the reaction.

Conclusion

1. An effective method for determining the cooling capacity of quenching media has been developed and tested. It is established that the cooling rate of the solutions decreases with increasing content of nonylphenol ethoxylates in the water.

2. The possibilities of thermogravimetrical method for the study of quenching media based on water-soluble polymers are shown. The technique allows to register physical and chemical processes occurring during heating of quenching liquids and establish endo-exothermic effects and temperature intervals of these reactions.

3. On the cooling rate curve of aqueous solutions "TOSOL-K" two extreme values were revealed – at 600°C and 300°C, which are formed as a result of polymer oxidation at high temperature quenching medium. This reaction is observed in the range of 500-300°C and is accompanied by an exothermic effect with a maximum heat release at 400°C. The beginning and the completion of the exo-reaction correspond to the maxima of the cooling rate of the quenching liquid under study

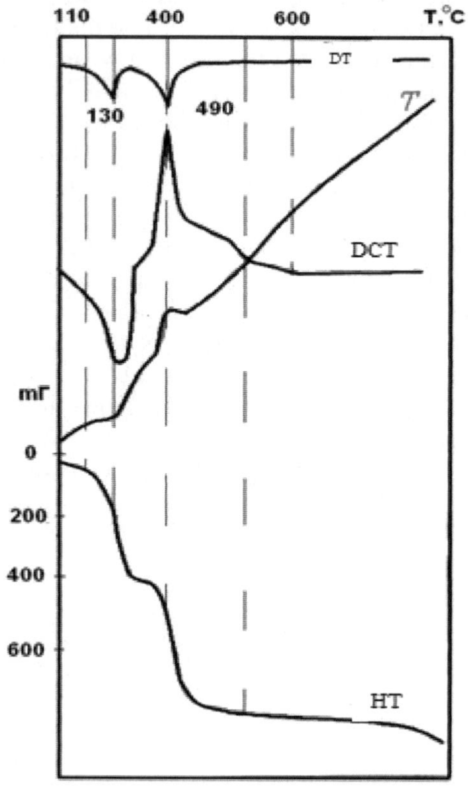

Fig. 3. - Thermographical scheme of liquid concentrate "TOSOL-K": T – temperature change in time; HT – dependence of mass change of the body temperature; DT – dependence of the rate of mass loss of temperature; DCT – differential curve of the temperature change

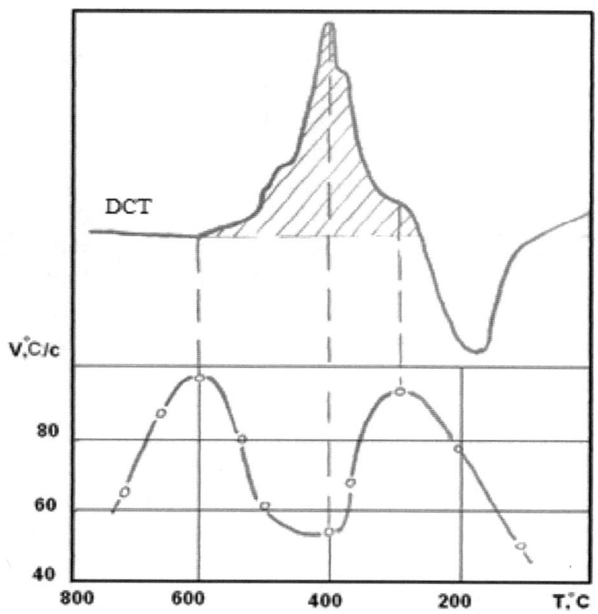

Fig. 4. - The influence of the exothermic effect during the heating of the quenching medium on the cooling ability of the concentrate "TOSOL-K".

Reference

[1] Astashchenko V I, Zapadnova N N and Mukhametzianova G F 2017 Key concepts for production of high-quality parts *IOP Conference Series: Materials Science and Engineering* vol 240 is 1 **012007**

[2] Shveyov A I, Astashchenko V I, Shveyova T V and Shveyov I A 2013 High level forming of properties in steel products *World Applied Sciences Journal* 24(9) pp1148-50

[3] Lyutyi V 1990 The quenching agent Ref. ed. / ed. by Maslennikova S B / Polish transl. *Chelyabinsk: Metallurgy, Chelyabinsk branch* p 192

[4] Petrash L V 1960 Quenching media L: *Mashgiz* p 106

[5] Prihodko V S 1977 Cooling media for hardening *M: Mechanical Engineering* p 32

[6] Astaschenko V I, Rudnitskiy N M and Ustilovskiy S Ya 1984 Application of polymer quenching media in the automotive industry M: *NIINavtoprom* p 55

[7] Hina M L, Vasilkov V F and Kobzov I S Quenching media based on water-soluble polymers 1978 *Metallurgy and heat treatment of metals* Number 8 pp 70-73

[8] Tolstousov A B and Bannyih O A 1981 New quenching medium based on water-soluble polymer *Metallurgy and heat treatment of metals* Number 2 pp 5-8

IOP Conf. Series: Materials Science and Engineering **570** (2019) 012098 doi:10.1088/1757-899X/570/1/012098

[9] Astashchenko V I, Ionkina N P, Yansen G I, Sorokin I. E and Gedberg M G 1982 Use of the "TOSOL-K" quenching fluid for the bulk quenching of steels *Metal Science and Heat Treatment* V 24 № 5-6 pp 377-380

[10] Patent 1016377 RF 1983 MKI S21D-1/60 The quenching medium / V V Nikolaev, V I Astaschenko, G I Yantsen et al. publ. 07.05.1983 Bul №17

[11] Shevchenko S Yu 2014 Comparative assessment of technological capabilities of polymer quenching media [Electronic resource] *Science and education: Electronic scientific- tech. Journal of Bauman* No. 3 pp. 339 - 356 website. – URL: http://technomag.bmstu.ru/doc/700099.html (date of access 25.01.2019)

[12] Kobasko N I 1980 Steel quenching in liquid media under pressure *Kiev: «Naukova dumka»* p 203

[13] Pletneva N A i Ryabinder P A 1946 Regularities of evaporation of liquid droplets in spheroidal state *Physical chemistry* Volume 20 issue 7 p 961

[14] Astaschenko V I, Yantsen G I i Ionkina N P 1981 The influence of the concentration of the boiling of the quenching liquid "TOSOL-K" cracking in steels *Automotive industry* Number 5 pp 29-30

IOP Conf. Series: Materials Science and Engineering **570** (2019) 012099 doi:10.1088/1757-899X/570/1/012099

The criterion of non-stationary cyclic fatigue of bodies, taking into account the presence of an incubation period of destruction

Latif H. Talibly [1,2]

[1] Azerbaijan National Aviation Academy, Baku
[2] Institute of Mathematics and Mechanics of the National Academy of Sciences of Azerbaijan, Baku

ltalybly@yahoo.com

Abstract. A functional of nonstationary cyclic fatigue of bodies was constructed under the assumption that the destruction process occurs in time after the deformation process.
Damage conditions are derived, which determines the number of non-stationary loading cycles, in which damage and cyclic fatigue occurs in the body, which makes it possible to find the non-stationary number of loading cycles before failure.

Introduction

Fatigue or cyclical durability is understood here as a violation of the continuity of the body as a result of the occurrence and accumulation of damage in it during a cyclic change of plastic deformation. The criterion of fatigue failure presented in [1] takes place in the case when the processes of cyclic deformation and cyclic destruction begin simultaneously. In some cases, experimental studies using infrared spectroscopy and acoustic emission methods have shown that damage accumulation begins after a certain incubation time or number of loading cycles after the start of the deformation process. The incubation period significantly depends on the stress amplitude and temperature and is commensurate with the pre-failure period [2,3]. This fact requires the derivation of new ratios of long-term strength and cyclic fatigue. Taking into account the above noted, in [4] theories of deformation and long-term strength of viscoelastic bodies with arbitrary loads in time were developed. These theories also apply to viscoelastoplastic bodies under monotonic loads.

Materials of theoretical research

We will use the concept of accumulation of cyclic damage. Let from the solution of the boundary value problem of plasticity theory at each point $x=(x_1,x_2,x_3)$ of body at any n-cycle in the general case of cyclic loading are known $\sigma_{ij}(x,n)$ $(i,j=1,2,3)$ – the greatest differences between the maximum and minimum values of the components of the stress tensor. In addition, let the temperature field be set $T(x,n)$ in n-cycle, which is considered monotonous with respect to n. In further entries of the argument x we omit. Intensity $\sigma_t(n)$ of components $\sigma_{ij}(n)$ will be: $\sigma_t(n)=\left(\frac{3}{2}s_{ij}(n)s_{ij}(n)\right)^{1/2}$, where $s_{ij}(n)=\sigma_{ij}(n)-\sigma(n)\delta_{ij}$, $\sigma(n)=\sigma_{ij}(n)\delta_{ij}/3$, δ_{ij} – Kronecker symbols. We have: $0\le n\le N_*$, where N_* is the desired number of nonstationary loading cycles with an arbitrary change in intensity σ_i from n, for $\sigma_t=\sigma_t(n)\ne const$. Let now the degree of damage in the case of non-stationary cyclic loading is characterized by $\Pi(n)$. Taking into account the normalization we take: $0\le\Pi(n)\le 1$. While maintaining the initial natural state of the body $\Pi(n)=0$. In addition, we assume that $\Pi(n)=0$ with incubation numbers of loading cycles $0\le n\le N'$, где $N'=N'(x_\alpha)$– The desired number of non-stationary loading cycles before damage points (x_α) of body with arbitrary dependence $\sigma_t=\sigma_t(n)\neq const$ and monotonous temperature $T=T(n)$. At the numbers of loading cycles $N'\le k\le N_*$ parameter Π uniquely determined by parameters $\sigma_t(n)$, $T(n)$, moreover, it is a functional, continuous with respect to $\sigma_t(n)$, $T(n)$. In case $0\le\Pi<1$, what takes place in the interval $0\le n<N_*$, body condition is solid. Destruction points (x_α) body occurs at loading numbers $N_*(x_\alpha)$ for $\Pi(N_*)=1$.

Consider the functionality of the form

$$\Pi(n)=H(n-N')\left[\varphi_1(\sigma_t(n),T(n))+\int_0^n M(n-k)\varphi_2(\sigma_t(k),T(k))dk\right], \tag{1}$$

where $H(n)$– the Heaviside unit function, which is used here to eliminate the negative part of the functional (1). Number N' determined from the condition $\Pi(N')=0$:

$$\varphi_1(\sigma_t(N'),T(N'))+\int_0^{N'} M(N'-k)\varphi_2(\sigma_t(k),T(k))dk=0. \tag{2}$$

On the condition

$$\lim_{\sigma_t\to 0,T=T^0}\varphi_2(\sigma_t,T)=0, \tag{3}$$

where T^0– the initial body temperature, at which there are no initial stresses and deformations, the functional (2) characterizes the damage, the properties of which are given above. Condition (3) is dictated by the property Π, where $\Pi=0$ while maintaining the initial natural state of the body. The ratio (2) determines the number of non-stationary loading cycles N' which begins the process of accumulating damage in the presence of σ_t, T. We call condition (2) a condition of cyclic damage. The number of non-stationary loading cycles N_* s from (1) provided $\Pi(N_*)=1$:

$$\varphi_1(\sigma_t(N_*),T(N_*))+\int_0^{N*} M(N_*-k)\varphi_k(\sigma_t(k),T(k))dk=1. \tag{4}$$

The relation (4) is a condition of cyclic fatigue or cyclic durability at given σ_t and T.

We now define the functions φ_1,φ_2 and M. They are assumed to be characteristic functions of the material. Therefore, when determining them, you can use the results of experiments that are represented by the ratios [3]: $N_1=N_1(\sigma_0,T_0)$, $N_0=N_0(\sigma_0,T_0)$, где $T_0=const$, $\sigma_0=const$. Parameter σ_0 is a double amplitude of stresses in the case of cyclic loading according to the law of symmetric cycles of stresses, N_1 and N_0– experimentally determined characteristic functions of the material: N_1– the number of stationary loading cycles preceding the damage process, N_0– the number of stationary cycles to failure at various constants σ_0 and T_0. From relations (2) and (4) with constant σ_0 and T_0 (in this case $N'\to N_1$, $N_*\to N_0$) define functions φ_1 and φ_2:

IOP Conf. Series: Materials Science and Engineering **570** (2019) 012099 doi:10.1088/1757-899X/570/1/012099

$$\varphi_1(\sigma_0,T_0) = -\frac{\int_0^{N_1} M(\xi)d\xi}{\int_{N_1}^{N_0} M(\xi)d\xi}, \qquad \varphi_2(\sigma_0,T_0) = \frac{1}{\int_{N_1}^{N_0} M(\xi)d\xi}. \tag{5}$$

As we see, φ_1 and φ_2 are connected with functions N_0, N_1 and M. The properties of functions N_i $(i = 0,1)$ are: $N_i(\sigma,T) > 0$, $\partial N_i / \partial\sigma < 0$, $\partial N_i / \partial T < 0$. n this case, the choice of function $M(n)$ ictated by the fulfillment of condition (3). We apply the average theorem to the integral in the second formula (5). We will get

$$\varphi_2(\sigma_0,T_0) = \{N_0(\sigma_0,T)[1 - A(\sigma_0.T_0)]M(\theta N_0(\sigma_0,T_0))\}^{-1},$$

where $0 \le \theta \le 1$, $A(\sigma_0,T_0) = N_1(\sigma_0,T_0)/ N_0(\sigma_0,T_0)$, $0 \le A(\sigma_0,T_0) < 1$.

rom the obtained ratio, taking into account it follows that condition (3) will be satisfied if $M(n)$ will have n^m, where $m > -1$. Based on this, we take: $M(n) = M_0 n^m$, where $M_0 = const > 0$, $m = const > -1$. Given this formula for $M(n)$ from relations (5) we define φ_1 and φ_2. After replacement σ_0 on $\sigma_t(n)$, T_0 on $T(n)$ in the resulting expressions for φ_1 and φ_2, convert functional (1) to mind:

$$\Pi(n) = H(n - N')\left[-\frac{N_1^{1+m}(\sigma_t(n), T(n))}{N_0^{1+m}(\sigma_t(n), T(n)) - N_1^{1+m}(\sigma_t(n), T(n))} + \right.$$

$$\left. + (1+m)\int_0^n \frac{(n-k)^m dk}{N_0^{1+m}(\sigma_t(k), T(k)) - N_1^{1+m}(\sigma_t(k), T(k))}\right]. \tag{6}$$

Conditions (2) and (4) are:

$$\frac{N_1^{1+m}(\sigma_t(N'), T(N'))}{N_0^{1+m}(\sigma_t(N'), T(N')) - N_1^{1+m}(\sigma_t(N'), T(N'))} = (1+m)\int_0^{N'} \frac{(N'-k)^m dk}{N_1^{1+m}(\sigma_t(k), T(k)) - N_1^{1+m}(\sigma_t(k), T(k))}, \tag{7}$$

$$\frac{N_0^{1+m}(\sigma_t(N_*), T(N_*))}{N_0^{1+m}(\sigma_t(N_*), T(N_*)) - N_1^{1+m}(\sigma_t(N_*), T(N_*))} = (1+m)\int_0^{N_*} \frac{(N_*-k)^m dk}{N_1^{1+m}(\sigma_t(k), T(k)) - N_1^{1+m}(\sigma_t(k), T(k))}, \tag{8}$$

Define unknown constant m. With $\sigma_t = \sigma_0 = const$, $T = T_0 = const$ from (6) follows $(N_* \to N_0,\ N' \to N_1,\ \Pi \to \Pi_0)$:

$$\Pi_0(\sigma_0,T_0,n) = H(n - N_1(\sigma_0,T_0))\frac{n^{1+m} - N_1^{1+m}(\sigma_0,T_0)}{N_0^{1+m}(\sigma_0,T_0) - N_1^{1+m}(\sigma_0,T_0)}. \tag{9}$$

Relation (9) is the equation of experimental damage accumulation curves depending on the number of stationary loading cycles n for $\sigma_0 = const$, $T_0 = const$. Let such curves be known from experience. In this case, (9) is used to determine the constant m. Parameter m an be determined using the results of other experiments. For example, let the cyclic loading of an experimental sample occur according to the following program: $T = T_0 = const$, $\sigma_t(n) = n\sigma_0$, where $\sigma_0 = const$, specified doubled voltage amplitude at the first symmetric cycle. At the time of destruction we have: $\sigma_t(N_*) = N_*\sigma_0 = \sigma_b$, where σ_b is a double voltage amplitude in the cycle N_*, which can be found experimentally. Wherein $N_* = \sigma_b / \sigma_0$. We take into account marked in the ratio (8):

$$\frac{N_0^{1+m}(\sigma_b, T_0)}{N_0^{1+m}(\sigma_b, T_0) - N_1^{1+m}(\sigma_b, T_0)} = (1+m)\int_0^{\sigma_b/\sigma_0} \frac{\left(\frac{\sigma_b}{\sigma_0} - k\right)^m dk}{N_0^{1+m}(k\sigma_0, T_0) - N_1^{1+m}(k\sigma_0, T_0)}.$$

The resulting ratio also allows you to define constant m with known functions N_0 и N_1.

Relation (7) is a condition of cyclic damage and determines the desired number of non-stationary cycles N' which begins the process of damage accumulation, relation (8) is a condition of cyclic durability and determines the desired number of non-stationary cycles N_* to failure with arbitrary dependence $\sigma_t \sim n$ and monotonous $T = T(n)$. In the case $N_1(\sigma_t, T)/N_0(\sigma_t, T) \equiv A = const$, and also in the case of $N_1 \equiv 0$ condition (8) gives the same result, which coincides with the fatigue condition presented in [1].

$$(1+m)\int_0^{N*} \frac{(N_* - k)^m}{N_0^{1+m}(\sigma_t(k), T(k))} = 1.$$

Under the condition $N_1 / N_0 = A = const$ cyclic damage condition (7) goes to the ratio

$$(1+m)\int_0^{N'} \frac{(N' - k)^m dk}{N_0^{1+m}(\sigma_t(k), T(k))} = A^{1+m},$$

kinetic equation (6) - to the relation

$$\Pi(n) = \frac{H(n - N')}{1 - A^{1+m}}\left\{- A^{1+m} + (1+m)\int_0^{n} \frac{(n-k)^m}{N_0^{1+m}(\sigma_t(k), T(k))}\right\}$$

Now using the results of fatigue experiments, which are presented in [3] for steel grade EI 437 at a temperature of 1073K, we define the functions N_0, N_1 and constant m. One of the variants of the approximation of functions N_0 and N_1 maybe the following:

$$N_i = N_{is} \exp\left[\beta_i\left(1 - \frac{\sigma_t}{\sigma_s}\right) + b_i\left(1 - \frac{T}{T_s}\right)\right], \ (i = 0,1). \tag{10}$$

At the same time we consider $\sigma_t \neq 0$. Here $\beta_0, b_0, \beta_1, b_1$ are constants; $\sigma_s = const$ - stresses, $T_s = const$ - temperature, which are dimensionless parameters, and are selected from the range of variation σ_t and T as their greatest values, N_{0s} and N_{1s} - the number of loading cycles to failure and to the onset of damage in an experimental sample with $\sigma_t = \sigma_s$, $T = T_s$. For quantities σ_s and T_s will take: $\sigma_s = 600$ MPa, $T_s = 1073$K. It turned out: $N_{0s} \approx 1{,}4 \cdot 10^9$ cycle, $N_{1s} \approx 4{,}1 \cdot 10^8$ cycle, $\beta_0 \approx 1{,}8$, $\beta_1 \approx 1{,}6$, $m \approx 0{,}64$. Note that according to the formulas for N_0 and N_1, when taking into account the found constants, in this case can be taken: $N_1 / N_0 \approx 0{,}35$. The reverse recalculation of the cyclic strength curves showed a stable deviation of the calculated data from the experimental ones, which was about 6%.

If the duration of k-cycle is $t_k - t_{k-1}$ $(t_0 = 0)$, then time t_* and number of cycles N_* before destructionas well as time t' and number of cycles N' before the formation of injuries in the body are related according to the ratios:

$$t_* = \sum_{k=1}^{N*} (t_k - t_{k-1}),\ t' = \sum_{k=1}^{N'} (t_k - t_{k-1}).$$

Finally, we note that the fatigue ratios presented take into account a physically important phenomenon — the influence of the history of cyclic loading on the cyclic strength of bodies.

Conclusions

The criteria of non-stationary cyclic damage and durability are presented, which, in the general case of variable loads, allow to determine the number of non-stationary loading cycles, respectively, before damage and before the destruction of bodies.

References

[1] Moskvitin V V 1981 *Cyclic loading of structural elements* M.:Science 344 p
[2] Goldman A Ya 1988 *Prediction of deformation and strength properties of polymeric and composite materials* L:Chemistry 168 p
[3] Ivanova V S, Ragozin Yu I, Vorobyev N A 1967 *Patterns of destruction of metals at static and cyclic loads* In kN. Thermoplasticity of materials and structural elements is 4 pp 277-85
[4] Talibly L Kh 1990 *On the problem of deformation and destruction of viscoelastic bodies in the presence of a temperature field* Izv. USSR Academy of Sciences Solid State Mechanics No 2 pp 127-39

IOP Conf. Series: Materials Science and Engineering **570** (2019) 012100 doi:10.1088/1757-899X/570/1/012100

Predicting the time of corrosion damage to a plate with a deep double-sided external undercut under stretching

L H Talibly [1,2], F B Imranov[3], A M Jafarova[4]

[1] Azerbaijan National Aviation Academy, Baku, Mardakan Prospect, 30, AZ1045, Azerbaijan
[2] Institute of Mathematics and Mechanics of the National Science Academy of Azerbaijan (NSAA), Baku, B.Vagabzade street, 9, AZ1141, Azerbaijan
[3] Institute of Physics of NSAA, Baku, G. Javid Avenue, 131, AZ1141, Azerbaijan
[4] Institute of Systems Management of NSAA, Baku, B.Vagabzade Street, 9, AZ1141, Azerbaijan

ltalybly@yahoo.com

Abstract. The statement is formulated and the solution of the problem of determining the time of corrosion destruction of an elastic homogeneous plate with a deep bilateral external recess is given. A plate of corrosive material is subjected to bilateral stretching in a corrosive environment in a direction perpendicular to the axis of the recesses. The well-known solution of the problem of the stress state of this plate, obtained by H. Neuber, is used. The following cases are considered: a) the effect of stress, which occurs in the plate when it is stretched, on the corrosion process is negligible; b) the effect of plate voltage on the corrosion process is significant. In both cases, the dependences of the change in the area of the narrowest cross-section of the plate on time are found. The time of complete corrosion wear and the time of corrosion cracking of the plate before its complete wear are also determined.

Introduction

It is known that the combined effect of mechanical stress and aggressive environment on structural elements leads them to corrosion-mechanical destruction. The process of corrosive destruction of bodies has a significant effect on the stress that occurs in them under the influence of external factors (force, temperature, etc.) [1-6]. Corrosion cracks propagate perpendicular to the direction of action of the tensile stress during elastic and non-elastic deformations [7]. Significant works are devoted to the theoretical definition of the corrosion strength of structures, among which works of a general theoretical nature can be noted [2, 4, 5, 8–11], having applied values and works devoted to the determination of the corrosion strength of a specifically considered structure [12–16].

The study was carried out with the support of SOCAR in the framework of the project 12LR-MMA

This article attempts to theoretically predict the time of corrosion damage to a plate with a deep symmetrical two-sided undercut with its elastic stretching in an aggressive medium in the direction perpendicular to the axis of the undercut. Plate material is considered homogeneous and isotropic. Designs in the form of such plates are used in many industries - in engineering, construction, transport, shipbuilding, etc. Under the corrosion destruction of the plate, we understand its corrosion wear and corrosion cracking in an aggressive environment under stress. In this case, the corrosion wear of the plate will be called the process of reducing its thickness due to the separation

IOP Conf. Series: Materials Science and Engineering **570** (2019) 012100 doi:10.1088/1757-899X/570/1/012100

of corrosion products from the action of aggressive environment and tensile force. The corrosion cracking of the plate is the discontinuity of the plate under the conditions of an aggressive medium and the specified force. The process of corrosion wear of the plate can be continued until its thickness completely disappears, and its corrosion cracking can take place at any time. The purpose of this article is to determine the time of complete wear and the time of cracking of the considered plate from the properties of an aggressive medium, the effective tensile force, the mechanical and geometrical parameters of the plate.

Materials of scientific research

A plate of thickness h with a deep symmetrical two-sided undercut is subjected to bilateral stretching in a corrosive environment with a force P (Fig. 1).

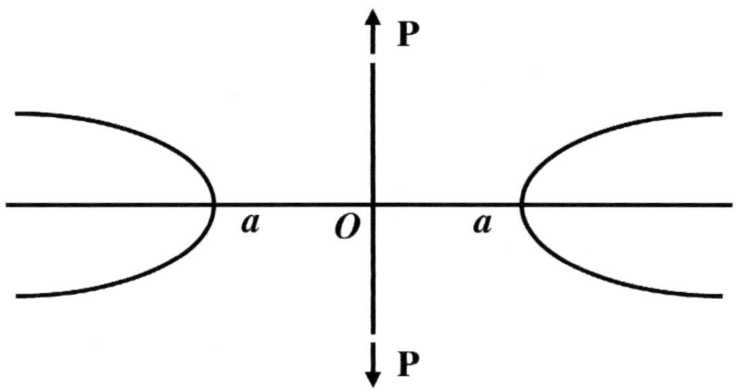

Fig.1

In this case, it deals with a construction with a stress concentrator. In [17] it is shown that with an increase in the depth of the undercut, the concentration coefficient approaches a limit value that does not depend on the depth of the undercut; only the curvature at the bottom of the undercut has a significant effect on the concentration coefficient. The shape of the undercut in the rest of the plate has little effect on the concentration ratio. Therefore, in order to facilitate computational procedures, the undercut is taken in the form of a hyperbola.

In [17], the problem of determining the stress components that arise in the plate under consideration in the absence of the influence of a corrosive environment was solved. In the process of corrosion, the plate sizes are subject to change. Let be ρ- undercut curvature radius, a- half the width of the narrowest part of the plate. In the process of corrosion values ρ, a, h depend from time t: $a = a(t)$, $\rho = \rho(t)$ and $h = h(t)$. As a characteristic size, we take the value $s(t) = 2a(t)h(t)$. . t characterizes the change of the narrowest section of the plate in the corrosion process. Wherein $ds/dt < 0$. We assume that in the process of corrosion the ratio a/ρ remains constant, which is the case, for example, in the case of ρ, a in the process of corrosion changes one common parameter $\lambda(t)$:$a = a(t) = a_o\lambda(t)$; $\rho = \rho(t) = \rho_o\lambda(t)$. Here $a_o u \rho_o$ are constant. Function $\lambda(t)$ satisfies the condidions $\lambda(o) = 1$. $d\lambda(t)/dt < 0$.Согласно [17], the maximum stress in the plate under consideration occurs at the bottom of the undercut and has the expression:

$$\sigma_{\max} = \frac{P}{s(t)} F\left(\frac{a}{\rho}\right), \tag{1}$$

IOP Conf. Series: Materials Science and Engineering **570** (2019) 012100 doi:10.1088/1757-899X/570/1/012100

$$F\left(\frac{a}{\rho}\right)=\frac{2\left(1+\frac{a}{\rho}\right)\sqrt{\frac{a}{\rho}}}{\left(1+\frac{a}{\rho}\right)arctg\sqrt{\frac{a}{\rho}}+\sqrt{\frac{a}{\rho}}}. \quad (2)$$

The quantity a/ρ is called the undercut curvature [17]. At the initial moment of the corrosion process from the formula (1) follows:

$$\sigma_{\max}(0)=\frac{P}{s_0}F\left(\frac{a}{\rho}\right), \quad (3)$$

where s_o- the initial area of the narrowest cross-section of the plate.

Consider the following independent cases of interest.

1. The case of insignificant influence on the corrosion process of a plate of mechanical stress. In this case, the influence of a mechanical stress on the rate of the corrosion process can be neglected and the equation for $s(t)$ can be taken as [2]:

$$\frac{ds(t)}{dt}=-k \quad (4)$$

Here, the value k of the experimentally determined characteristic of the system "material-aggressive medium" is the corrosion rate. The solution of equation (4) with the initial conditions $s(0)=s_o$ is:

$$s(t)=s_o-kt. \quad (5)$$

With full plate wear $s(t)=0$. In this case, from equation (5) follows

$$\frac{s_0}{k}=t_0. \quad (6)$$

Here t_0 - time of complete wear of the plate in the absence of the influence of mechanical stress on the corrosion process. This time is the largest of all times of existence of the plate.

Let t_0^b - time cracking plate in the case when you can neglect the effect of stress on its corrosion. It will be natural if we accept by the condition of plate cracking the condition of equality of the maximum stress and the strength of the plate material:

$$\sigma_{\max}\Big|_{t=t_0^b}=\sigma_b. \quad (7)$$

It is obvious $t_0^b<t_0$ and volume $s(t_0^b)=s_o-kt_0^b$ is non-zero. Using (1) in (7), we obtain the formula for time t_0^b:

$$t_0^b=\frac{s_0}{k}-\frac{P}{k\sigma_b}F\left(\frac{a}{\rho}\right). \quad (8)$$

Using the condition of the existence of cracking time ($t_0^b>0$), from (8) define the force limit P:

$$P_{\lim}<\frac{s_0\sigma_b}{F\left(\frac{a}{\rho}\right)}.$$

Calculate the ratio t_0^b/t_0, where t_0^b and t_0 determined by formulas (8) and (6) respectively:

IOP Conf. Series: Materials Science and Engineering **570** (2019) 012100 doi:10.1088/1757-899X/570/1/012100

$$\frac{t_0^b}{t_0} = 1 - \frac{P}{s_0 \sigma_b}. \qquad (9)$$

We take into account formulas (3) in (9). Will have a relationship t_0^b / t_0, expressed through relationships $\frac{\sigma_{\max}(0)}{\sigma_b}$: $\frac{t_0^b}{t_0} = 1 - \frac{\sigma_{\max}(0)}{\sigma_b}$.

2. The case of a significant effect of mechanical stress on the corrosion process of the plate. In this case, according to [2], for the rate of the corrosion process we will use the equation:

$$\frac{ds(t)}{dt} = -k - n\sigma_{\max}(t). \qquad (10)$$

Here, the values of k and n are considered to be characteristics of the "metal-aggressive environment" system and, for each system, are determined by experiments on corrosion destruction in accordance with the method presented by [2].

Note that the dependence of the rate of the corrosion process on the mechanical stress, generally speaking, should be nonlinear. However, for a certain system "metal - aggressive environment" this dependence is close to linear [2]. At the same time, equation (10) can be used as a first approximation to the solution of the problem of the corrosion resistance of materials.

We take into account formula (1) in equation (10) and determine the differential dt through the differential ds. After some transformations we get:

$$dt = -\frac{s(t)}{ks(t) + nPF\left(\frac{a}{\rho}\right)} ds. \qquad (11)$$

Integrating equations (11), we arrive at a relationship between time and characteristic size — the area of the narrowest section of the plate in the process of corrosion:

$$t = -\frac{1}{k}(s - s_0) + \frac{n}{k^2} PF\left(\frac{a}{\rho}\right) \ln \frac{s + \frac{n}{k} PF\left(\frac{a}{\rho}\right)}{s_0 + \frac{n}{k} PF\left(\frac{a}{\rho}\right)}. \qquad (12)$$

Let $t_{0\sigma}$ there is a time of complete wear of the plate ($s = 0$) in the case when the effect of mechanical stress on the corrosion process is significant. Taking $s = 0$ in (12), we determine the time $t_{0\sigma}$:

$$t_{0\sigma} = \frac{s_0}{k} + \frac{n}{k^2} PF\left(\frac{a}{\rho}\right) \ln \frac{\frac{n}{k} PF\left(\frac{a}{\rho}\right)}{s_0 + \frac{n}{k} PF\left(\frac{a}{\rho}\right)}. \qquad (13)$$

Taking into account (3), formula (13) can be rewritten in the form:

$$t_{0\sigma} = \frac{s_0}{k} - \frac{ns_0}{k^2} \sigma_{\max}(0) \ln\left(1 + \frac{1}{\frac{n}{k}\sigma_{\max}(0)}\right).$$

For comparison, the time to complete wear of the plate in question in cases of dependence ($t_{0\sigma}$)and independence (t_0)corrosion process from mechanical stress, determine their attitude. For this purpose, we will use formulas (13) and (6). In this case, we obtain the ratio:

IOP Conf. Series: Materials Science and Engineering **570** (2019) 012100 doi:10.1088/1757-899X/570/1/012100

$$\frac{t_{0\sigma}}{t_0} = 1 + \frac{n}{ks_0} PF\left(\frac{a}{\rho}\right) \ln \frac{\frac{n}{k} PF\left(\frac{a}{\rho}\right)}{s_0 + \frac{n}{k} PF\left(\frac{a}{\rho}\right)}. \qquad (14)$$

Formula (14) using (3) turns to

$$\frac{t_{0\sigma}}{t_0} = 1 - \frac{n\sigma_{\max}(0)}{k} \ln\left[1 + \frac{1}{\frac{n}{k}\sigma_{\max}(0)}\right] \qquad (15)$$

From the formula (15) follows the obvious inequality: $t_{0\sigma} < t_0$.

Now $t_{0\sigma}^b$ there is a time cracking plate in an aggressive environment under the action of tensile force P. Time $t_{0\sigma}^b$ determine the condition:

$$\sigma_{\max}\big|_{t=t_{0\sigma}^b} = \sigma_b. \qquad (16)$$

Replace in the formula (1) t through $t_{0\sigma}^b$, The resulting expression will be taken into account in (16) have:

$$\frac{P}{s\left(t_{0\sigma}^b\right)} F\left(\frac{a}{\rho}\right) = \sigma_b. \qquad (17)$$

Formula (17) when using (3) goes to: $\sigma_{\max}(0)\dfrac{s_0}{s\left(t_{0\sigma}^b\right)} = \sigma_b$. From here we get:

$$s\left(t_{0\sigma}^b\right) = s_0 \frac{\sigma_{\max}(0)}{\sigma_b}. \qquad (18)$$

Now replacing in (12) $s(t)$ on $s(t_{0\sigma}^b)$by the formula (18) determine the time $t_{0\sigma}^b$:

$$t_{o\sigma}^b = \frac{s_0}{k}\left(1 - \frac{\sigma_{\max}(0)}{\sigma_b}\right) + \frac{ns_0}{k^2}\sigma_{\max}(0) \ln\left[\frac{\sigma_{\max}(0)}{\sigma_b} \cdot \frac{1 + \frac{n}{k}\sigma_b}{1 + \frac{n}{k}\sigma_{\max}(0)}\right]. \qquad (19)$$

Formula (19) determines the corrosion cracking time of the plate under consideration, depending on the characteristics of the "metal-aggressive environment" system, the plate material strength, the initial area of its narrow cross section and the maximum stress occurring at the initial moment of the corrosion process.

Time relation $t_{0\sigma}^b$ and t_0 by using (19) and (6) in:

$$\frac{t_{0\sigma}^b}{t_0} = 1 - \frac{\sigma_{\max}(0)}{\sigma_b} + \frac{n}{k}\sigma_{\max}(0) \ln\left[\frac{\sigma_{\max}(0)}{\sigma_b} \cdot \frac{1 + \frac{n}{k}\sigma_b}{1 + \frac{n}{k}\sigma_{\max}(0)}\right].$$

The last relation implies an explicit inequality $t_{0\sigma}^b < t_0$.

Note that the time to cracking of the plate under consideration can also be determined by the phenomenological formula derived in [9] using the concept of accumulated corrosion damage.

Conclusions

IOP Conf. Series: Materials Science and Engineering **570** (2019) 012100 doi:10.1088/1757-899X/570/1/012100

The problem of determining the time of corrosion damage of a plate with a deep two-sided external undercut when it is stretched in an aggressive medium in a direction perpendicular to the axis of the undercut is set and solved. In cases where a) the effect of the stress that occurs in the plate when it is stretched, the corrosion process is negligible and b) the effect of the plate voltage on the corrosion process is significant, the time of complete corrosion wear and the time of corrosion cracking of the plate are determined.

References

[1] Evans Yu R 1962 *Corrosion and oxidation of metals. Theoretical foundations and their practical application* Moscow: Mashgiz 256 p

[2] Gutman E M 1981 *Mechanochemistry of metals and corrosion protection* Moscow: Metallurgy, 281 p

[3] Ulig G G, Revie R U *1989 Corrosion and fight with it. Introduction to corrosion science and technology* Leningrad: Chemistry 456 p

[4] Ovchinnikov I G, Naumova G A 2000 *Strength calculations of complex core and pipeline structures, taking into account corrosion damage* Saratov: SSTU 207 p

[5] Lokoshchenko A M 2016 *Creep and durable metals in corrosive environments* Moscow: Fizmatlit 504 p

[6] Semenova I V, Florianovich G M, Khoroshilov A V 2002 *Corrosion and corrosion protection* Moscow: Fizmatlit 336 p

[7] Zhukov A M 1954 *Strength and plasticity of brass in ammonia medium* Izv. Academy of Sciences of the USSR No 4 pp 67–72

[8] Ovchinnikov I G 1988 *On the methodology for constructing models of structures interacting with aggressive media* Durability of materials and structural elements in aggressive and high-temperature environments. Interuniversity scientific collection pp 7-21

[9] Talybly L Kh 2003 *On determining the time to corrosion fracture of metals* Transactions of National Academy of Sciences of Azerbaijan, ser. of physical-technical and mathematical sciences. issue mathematical and mechanics pp 239–46

[10] Talibly L Kh 2011 *About one case of derivation of the simplest formula of corrosive destruction under constant load* Technology of machines, mechanisms, from nano-to macro-level. Materials of the 13th International Conference St. Petersburg V 2 pp 423-26

[11] Talybly L Kh, Narimanov S V 2012 *On one formula for corrosion damage under voltage* Heavy engineering No 8 pp 39-41

[12] Dolinsky V M 1967 *Calculation of loaded pipes prone to corrosion* Chemical and petroleum engineering No 2 pp 9-10

[13] Lokoshenko A M, Agakhi K A, Fomin L V 2013 *Bending creep in bending in aggressive media* Problems of mechanical engineering and machine reliability No 4 pp 70-75

[14] Pronina Yu G 2009 *Uniform mechanochemical corrosion of a hollow sphere made of Prandtl material under the action of constant pressure* Vestnik SPSU Issue 1 pp 113-22

[15] Sedova O S, Pronina Yu G, Kabrits S A 2018 *Corrosion of spherical elements under the action of pressure and uneven heating* Deformation and destruction of materials No 1 pp 2-6

[16] Talibly L Kh, Gunbataliev E Z 2013 *About one theoretical solution of the problems of corrosion destruction of a ring sector fixed at the end when bending by shearing force* Mechanics - mechanical engineering No 2 pp 23-25

[17] Neuber G 1947 *Stress Concentration* Moscow - Leningrad: Gostekhizdat 204 p

IOP Conf. Series: Materials Science and Engineering **570** (2019) 012101 doi:10.1088/1757-899X/570/1/012101

Determination of rational parameters of the assembly process of one-piece products which are based on modeling

A M Valiev, D L Pankratov

Kazan Federal University, Naberezhnye Chelny Institute, 68/19 Mira avenue, Naberezhnye Chelny, Tatarstan, Russian Federation, 423812

AjMValiev@kpfu.ru

Abstract. The article is devoted to the development of methods for determining the rational parameters of the assembly of one-piece products. Based on the simulation of the open-source process of cylindrical samples, mathematical models were obtained that allow accurate calculation of the assembly parameters and ensure high quality of one-piece products.

The main advantage of the use of one-piece products is the possibility of combining the properties of various steel grades in one solid product. Methods for obtaining compounds of parts by means of plastic deformation are based on the production of a one-piece product at a temperature below the melting point of the metals being welded. Determining the rational parameters of the assembly process of one-piece products is one of the difficult tasks of their manufacture by means of plastic deformation.

Figure 1 shows the scheme of the proposed method of assembly of axisymmetric one-piece metal products with a barrel shape of the connected surfaces. As the connected elements of one-piece product, the shaped blank 2 and the rod introduced into it 1 act [1, 2].

At the first stage, the rod 1 in the matrix 4 (Fig. 1, a). is put into the preheated forging temperature shaped blank 2 to a depth of l_{in}At the second stage, the compound is kept for a time τ for heating the part of the rod that was put into blank in order to reduce the yield strength of its material. After that, at the third stage, the embedded part of the rod is put off by the value of Δl, due to which the annular gap 3 is filled with the material of the rod (Fig. 1, b). And at the fourth stage, the final assembly is performed by a ring tool 5 on ΔH_d (Fig. 1, c). Figure 1 (d) shows a 3D model of the assembled elements.

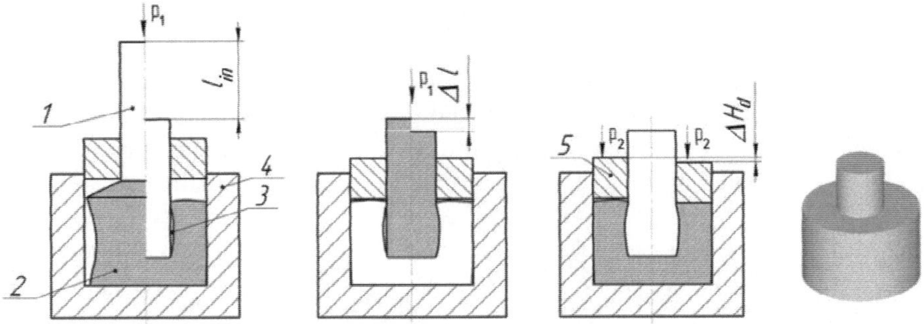

a) b) c) d)

IOP Conf. Series: Materials Science and Engineering **570** (2019) 012101 doi:10.1088/1757-899X/570/1/012101

Figure 1-Diagram of the method of assembly of one-piece products
Note: In the figure, the deformable object is marked in gray at the appropriate stage

Achievement of the specified indicators of quality of assembly of one-piece products, such as the accuracy of the geometric dimensions of the product and the strength of one-piece products depends on the exact calculation of the geometric parameters of the shaped blank [3, 4].

Figure 2 shows the main geometric parameters of the process of open firmware of cylindrical blank with height $H_0$and diameter D_0 (contour 1) by cylindrical punch with diameter d with a taper angle α of the end face. Due to the deformation the blank is reduced to a height of H_K. The flashing process is always accompanied by the formation from the side of introduction of the punch of utiny at the end of the blank, which can be quantitatively described by the angle φ.

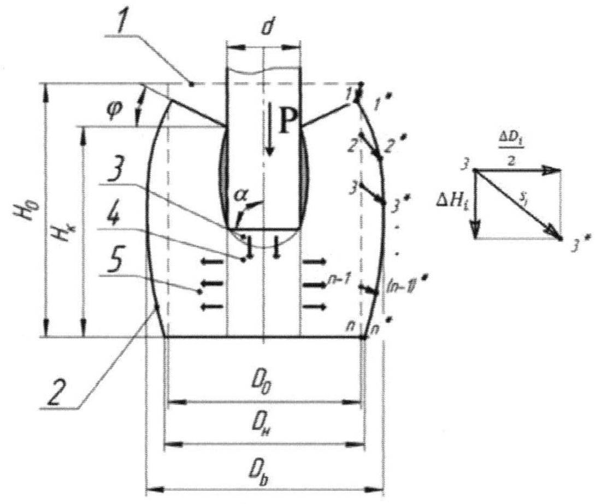

Figure 2 - Parameters of form modification of the cylindrical blank with open firmware

In the process of deformation, the lateral surface of the blank takes the form of a barrel, which has a diameter D_b (contour 2). To accurately determine the displacement of the metal of the blank with open firmware to determine the final shaping of the blank, the values of D_b and H_K are not enough. In this regard, on the side surface of the blank, we denote points i equidistant from each other and enter the vector $\vec{s}_i\left(\frac{\Delta D_i}{2}, \Delta H_i\right)$ displacements of i – points from their initial position 1, 2, ... n into the final 1*, 2*,... n*. The components ΔD_i and ΔH_i are the values of the i – point displacement in the radial and axial directions, respectively. The value D_i is the diameter of the blank at the i - th point, and the value H_i is the distance of the i - th point from the bottom base of the blank. A characteristic feature of shaping the blank with open firmware is the possibility of forming a closed annular gap θ_3.

As a result of the research carried out using simulation modeling, mathematical models have been developed that relate the parameters of the final shaping of the blank ΔD_i, ΔH_i, ΔD_b, ΔH_K, φ, θ_3 with the following variable parameters of geometric dimensions before deformation: d/D_0 is the relative diameter of the rod; H_0/D_0 is the relative height of the blank and α is the angle of the base of the rod.

Figure 3 shows the developed algorithm for calculating rational parameters of the assembly process for one-piece products using the above mathematical models

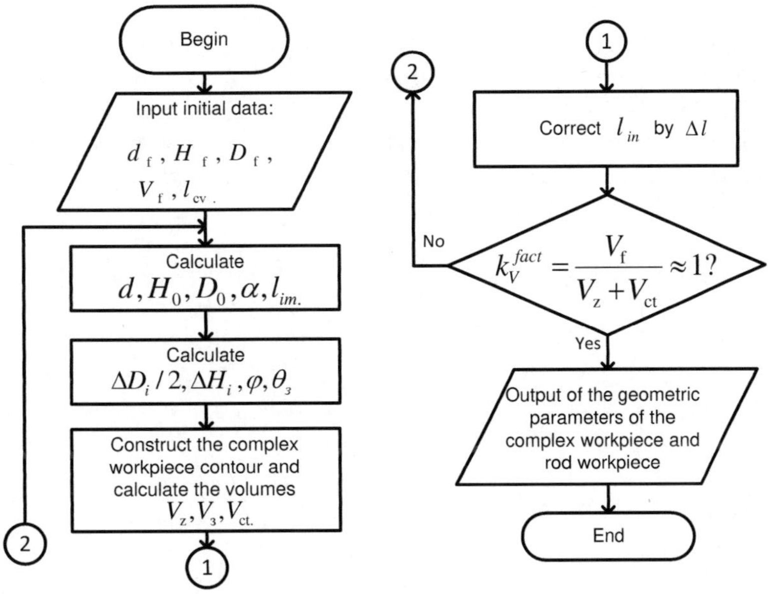

Figure 3 - Algorithm for calculating the geometrical parameters of the shaped blank and rod

After entering the initial data, the diameter D_0 and height H_0 of the cylindrical blank, the values α and l_{in}In the calculation, it is assumed conditions that the desired contour of the cylindrical blank after the firmware fits the contour of the finished one-peace product. This condition is expressed by the following system of relations:

$$\begin{cases} d = d_f \\ H_к = k_V \cdot H_f \\ D_b = D_f \end{cases} \qquad (1)$$

Where H_f и D_f – height and diameter of the body part of the one-peace product after assembly; d_f – diameter of the rod part of the one-peace product after assembly; H_f и f – dimensions of the cylindrical blank after the introduction of the rod; k_V – volume correction factor (on the first cycle of the algorithm calculation k_V=1). Due to the fact that there are many such contours, the contour is selected with the maximum possible relative value of the volume of the annular gap$\theta_3 = \theta_{3\,max}$.

Next, the values of displacement of the lateral surface points in the radial $\frac{\Delta D_i}{2}$ and axial ΔH_i directions are calculated, as well as the value of the angle of the upper end surface φ. The obtained values are used in the construction of the contour of the shaped blank. The length of the embedded part of the rod l_{in}. is adjusted by the value $\Delta l = \frac{\theta_3}{100\%} \cdot l_{in}$.

Then the actual value of the correction factor k_V^{fact} is calculated

$$k_V^{fact} = \frac{V_f}{V_z + V_{st}}, \qquad (2)$$

where V_f, V_f и V_{st} – the volumes of the finished product, the shaped blank and the rod, respectively.

The calculation with the output of the design results stops at values $k_V^{fact} \approx 1$ with a deviation of ±0.01, corresponding to the range of values for accurate hot forging. For large deviations, the calculation starts anew until condition (2) is satisfied, but already with the value$k_V = k_V^{fact}$ in (1).

The developed mathematical models of modification blank forms allow us to determine the rational process parameters and ensure the high quality of the one-piece products. The work makes a significant contribution to the study of the production of one-piece products with plastic deformation with the required geometric accuracy and joint strength.

References

[1] A V Shaparev, I A Savin 2017 *Calculation of Joint Plastic Deformation to Form Metal Compound in Cold Condition* Solid State Phenomena Vol 265 pp 313-18
[2] Valiev A M, Shibakov V G and Pankratov D L 2016 *Russ. Engin. Res.* 36 p 146
[3] Valiev A M, Shibakov V G and Pankratov D L 2016 *Russ. Engin. Res.* 36 p 142
[4] Simonova L A, Valiev A M, Pankratov D L, Valiev R F 2014 *Automated system for design and technological preparation of the production of composite metal products with firmware* Basic Research No 9 (8) pp 1697-02

IOP Conf. Series: Materials Science and Engineering **570** (2019) 012102 doi:10.1088/1757-899X/570/1/012102

The prospects of modern quality management approaches integration (on the example of the Republic of Tatarstan)

D P Vorontsov [1], A S Pirogova [1], A M Shikhalev [1], G R Khamidullina[1], N F Kashapov[1], A F Khananova[1] and O N Rozhko [2]

[1] Kazan Federal University, 18 Kremlyovskaya street, Kazan, 420008, Russian Federation, Engineering institute
[2] Kazan National Research Technical University named after A. N. Tupolev - KAI, 10 Karl Marx street, Kazan 420111, Russian Federation, Institute for Aviation, Land Transportation and Power Engineering

128_dmitri@mail.ru

Abstract. In the research basing on the quality management modern approaches and the practice of the Republic of Tatarstan leading industrial enterprises are estimated the possibilities of energy management systems improving (with intensive introduction of modern industrial equipment) through the integration of the quality management system. In our study is made an analysis of present energy management systems functioning indicators and demonstrated the further need for these two systems integration in the modern management.

1. Introduction

Recently it's possible to see the widespread introduction of robotic systems at the modern industrial enterprises for the purpose of productivity and quality increase in and cost cutting. The wide scales of implementation lead to the growth of power losses that is fraught for the ecology [1, 2] and development of the enterprise, resulting in need of the qualitative organization of work and technical equipment maintenance. The enterprises aim at balance of customers' satisfaction and systematic operation when asking to quality management. The existing approaches to quality management can be divided into three units: statistic methods, verbal directivity tools and the complex methods integrating the two previous units. In the research is analyzed the complex methods as the methodology of the integrated quality improvement like the most effective variant in the most real industry cases.

2. The problem definition

The Quality Management System (QMS) in the current business often could be defined like the firm stability and success guarantor and here it is necessary to determine the QMS [3] model. The QMS model is the set of the principles, methods, requirements to different aspects and processes of activities, the criteria determining the level of perfection of these processes and methods of their assessment which in total define all processes of organization activity directed to achievement of the required results on quality.

The modern quality management methods are the FMEA, QFD, "6 sigma", kaizen, "Lean production", "5C", CRM, TRM, benchmarking, "20 keys" [4]; the suitable choice depending on the firm purposes and the management strategy could propose to produce the high-quality services production. Quality policy is one of the key organizing methods for the company goals achievement. The variant of some noted above complex methods realisation and the current major quality policy tools could be presented in such management instrument like the energy management. An experience of the industrial organizations and firms from the such countries like the Netherlands, the USA, etc. which earlier implemented the energy management systems demonstrate the average annual growth of energy efficiency at least in 2-3% per year, the compatible results are in the agriculture sector [5], etc. Actually in the case of energy management methodology using is realized the systems approach to energy control in the organizations, but for many Russian companies the energy management implementation is mostly a trendy factor or it is concerned with the legislation requirements and in less degree based on the readiness to increase energy efficiency [6].

The GOST R ISO (national) standard 50001 sets the requirements for the energy management system development and improving [7], but the real application of the standard could be ineffective without its quality management system implementation. Obviously, the most effective way here could be concerned with the integrated QMS using. The basis for the integration with the ISO 9001:2015 standard requirements could be the such additions in the quality model elements like the "organizational knowledge" and "knowledge management" as the cumulative knowledge and abilities of all firm staff which could be practically used like the complex concerned with the knowledge and abilities of certain experts and knowledge in the organization information funds, i.e. the certain level of knowledge allowing to create the most qualitative product in shorter periods. The application of risks estimation methods according to the 9001:2015 standard requirements here also could be analysed as the way of the preventive actions efficiency increasing [8], etc. Moreover, the novelties above in the standard version exclude the necessity for the special “system approach” [9] term marking and multiply its using in the text.

The energy management (EM) realising as the integral system made for the specific firm needs and conditions usually allows to gain the multiplicative effect and to increase energy efficiency [10]. According to the Strategy of development for the Republic of Tatarstan (one of the leading Russian Federation regions) before the 2030 the one of the major purposes is “the “smart economy” improving”; the implementation and commercialization of new materials, products and technologies [11], [12], etc. Basing on the previous analysis we could note that the local large firms often spend the resources (and power) with the small profit [13], [14]; then managers employ the experts to develop saving measures like the energy production with trigeneration technologies, or use some alternative sources, etc. [15]. The energy management implementation is a labor-consuming process but the known practice demonstrates certain results in the total productivity growth. Looking on the statements noted above it is necessary to analyse the possibilities of such approach using.

3. Problem solving

Let’s look at the leading firms and enterprises results and available characteristics of its quality systems and policies analyzing the Republic of Tatarstan practice. PJSC «Nizhnekamskneftekhim» is the one of the Europe and the Commonwealth of Independent States petro chemistry production leaders [16]. As it is noted in its open documents the energy saving is the one of its most advantageous activity directions. The important management instruments here are the operations on certification for compliance with the international standards requirements and the enhancement of integrated quality control system. Three energy management programs spared 350 million kWh; the total energy capacity of production is reduced more than on 43%.

JSC «Kazanorgsintez» is the one of the largest Russian Federation chemical firms [17]. At the heart of quality work are the principles of the maximum directivity on the supplier and the special attention to the ecology issues. In 2015 it was signed the program of reliability augmentation lines with the electro-and heat supplies and the WTF equipment for 2016-2020; also are implemented the projects basing on the increasing efficiency of secondary production using methods, etc.

The «KAMAZ» Group of Companies is the largest country autocorporation in the Russian Federation. The company notes that the quality management system is an instrument of competitive production and service achievement satisfying the customers waiting and providing the basis for the stable financial position. It was established the creation by the all company's subdivisions the medium-term program "Actions for the cost cutting on energy carriers of rather adopted annual budget", reducing the expenses to 5% annually [18], etc.

Analyzing the possible mathematical approaches for the task solving here we could propose that choosing among the possible calculation methods [19, 20] or the well-distributed in the quality management different variants of diagrams and the block diagram [21] etc. we could conclude that in this case in will be better to use firstly the such available decisions like the simple expert rating estimation basing on the noted firms open data sources.

See below the results of quality management policy status analysis for all these cases (Table 1).

TABLE 1 - AN ANALYSIS OF THE MODERN QUALITY MANAGEMENT APPROACHES USING

№	**An analysis of the modern quality management approaches using**			
	Methodology	**«Nizhnekamskneftekhim»**	**«KAMAZ »**	**«Kazanorgsintez»**
1	FMEA	5	1	5
2	QFD	4	1	4
3	«6 sigma»	3	3	5
4	«Lean production»	1	1	1
5	Kaizen	2	1	4
6	«5 C»	1	2	4
7	CRM	4	1	4
8	TRM	1	1	4
9	Benchmarking	1	1	1
10	"20 keys"	3	3	4
	Total	25	15	36

The quality management instruments using on these firms basing on its open documents, etc. is estimated in 5 scores scale: "1" - the system is implemented, "2" - implementation result is neutral, "3" - implementation will not bring big changes, "4" - it is recommended, "5" - implementation is necessary. It's possible to make the next conclusions here: JSC «KAMAZ» actively uses the modern approaches of quality management, the policy of activities and quality is effective; JSC «Nizhnekamskneftekhim» successfully uses the modern quality tools; JSC «Kazanorgsintez» needs the using of suitable SMK. The next step is the analysis of the ISO 50001 standard (national version) using in these organizations in the table below, basing on the possible companies open data sources.

TABLE 2 - GOST R ISO 50001 "ENERGY MANAGEMENT SYSTEMS" STANDARD USING

Status	**GOST R ISO 50001 "Energy management systems" standard using**		
	«Nizhnekamskneftekhim»	**«KAMAZ»**	**«Kazanorgsintez»**
Used	+		

Status	GOST R ISO 50001 "Energy management systems" standard using		
	«Nizhnekamskneftekhim»	«KAMAZ»	«Kazanorgsintez»
Not used			+
Planning		+	
Remarks	Implemented and used standard GOST R ISO 50001:2011	GOST R ISO 50001 implementation was planned for 2015	The GOST R ISO 50001 implementation is recommended

From these tables it is visible that the JSC «Kazanorgsintez» rather needs the ISO 50001 standards implementation; «KAMAZ» PTC not fully adheres to the ISO 50001 standard and only some of its subsidiaries implemented this standard; PJSC «Nizhnekamskneftekhim» successfully uses the standard. Thus, here it is almost clear that the JSC «Kazanorgsintez» needs an implementation of the modern integrated quality management approaches and the acceptance of the ISO 50001 standard. The «Nizhnekamskneftekhim» company successfully uses the most of all modern methods of quality policy, energy systems and resource-saving. «KAMAZ» PTC Group of Companies is the company which adhering the successful quality policy respectively the using of the "Systems of Energetic Management" standards is quite expedient.

4. The conclusions and the further researches directions

The using of the most famous energy-saving innovative technologies at the moment is obviously technically and economically profitable. The prospects of the energy management integrated system implementation and the modern quality management approaches are predetermined by the existence and effective interaction of the harmony subsystems. Such integrated system assumes the existence of a competitive quality policy with the aspects of technical elements maintenance due to the environment threats.

Looking at the attractive markets and industry sectors with the speedy growth in the Russian Federation we could note that based on the signed and realised wide government policy, etc. the growth of industrial electrotechnics production and using in the case of current business challenges raises an issue about the new requirements for the high-quality control and production management. As we have seen above the integrated quality management systems and energy management implementation in the Russian conditions could give the completest spanning of production control and due to this also an increase in the productivity growth which is the one of the main task of modern management [22].

5. Acknowledgments

Authors expressed their gratitude to the Department of Quality Management members, and to the members of the Department of Technical Physics and Energy, especially to the Gilmanshin I. R. (vice-director for the scientific activities), Engineering institute, Kazan Federal University; to the colleagues in the Department for Automobile Engines and Service, Kazan National Research Technical University n.a. A.N. Tupolev for their substantial remarks in the discussions of this research materials.

References

[1] Kashapov N F, Gilmanshin I R, Gilmanshina S I and Galeeva A I 2016 Landfill energy complex based on the renewable energy installations *IOP Conference Series: Mat. Sc. and Engin.* (Bristol: IOP Publishing) **vol 134** DOI: 10.1088/1757-899X/134/1/012007 p 1

[2] Kashapov N F, Gilmanshin I R, Gilmanshina S I and Galeeva A I 2017 Instrumental research

method of qualitative composition of landfill gas in the surface layer of landfills *IOP Conference Series: Mat. Sc. and Engin.* (Bristol: IOP Publishing) **vol 240** DOI: 10.1088/1757-899X/240/1/012021 p 1

[3] Milova V M 2014 *Total Quality Management* (St. Petersburg: Publishing House GUAP) p 213

[4] Gorbunova O I and Guseva I K 2016 *The instruments and methods of quality management* (Irkutsk: BSU Publishing House) pp 47-73

[5] Kashapov N F, Nafikov M M, Gilmanshin I R, Gazetdinov M Kh, Nafikova M M and Nigmatzyanov A R 2017 Energy-saving technologies of cultivation of sugar sorghum *IOP Conference Series: Mat. Sc. and Engin.* (Bristol: IOP Publishing) **vol 240** DOI: 10.1088/1757-899X/240/1/012032, p 1

[6] Tretyakova M V 2015 The energy management system as a tool for improving energy efficiency of the organisation energy complex Russia, *Vestnik Universiteta* **vol 13** (Moscow: The State University of Management Publishing) pp 60-65

[7] GOST R ISO 50001:2012 "Energy management systems" URL: eitus.bstu.ru/shared/attachments/124979/ (available at 11.02.2019)

[8] Chaika I I 2014 The ISO 9000:2015 family standards *The continuous improvement of the organizations' activities seminar materials* (Moscow: The Scientific Training Center of quality management systems and certification, National University of Science and Technology MISIS Publishing) pp 1-5

[9] Kashapov N F, Konahina I A, Gil'manshin I R and Fahreev N N 2014Mini-Central heating and Power Plant (CHP): the choice of the optimal structure and modes of operation *IOP Conference Series: Mat. Sc. and Engin.* (Bristol: IOP Publishing) **vol 69** DOI: 10.1088/1757-899X/69/1/012008 p 5

[10] Russian companies experience in energy management implementation, State information system URL: https://gisee.ru/articles/energy_management/59822/ (available at 19.02.2019)

[11] The Republic of Tatarstan Strategy of social-economic development before 2030, Ministry of Economy of the Republic of Tatarstan URL: http://mert.tatarstan.ru/eng/strategiya-sotsialno-ekonomicheskogo-razvitiya.htm (available at 12.02.2019)

[12] Vorontsov D P, Shikhalev A M, Mingafina R Ia, Ialalina V A and Rozhko O N 2018 Interregional logistic centers in the modern regional development: an estimation of needs and prospects (on the example of the Sviyazhsk interregional multimodal logistics center) *Sec. Intern. Conf. of Social Sc. Proc. NORDSCI*, Book 2 **vol 1** DOI 10.32008/NORDSCI2018/B2/V1/29 pp 277-85

[13] Shikhalev A M, Vorontsov D P and Akhmetova I A 2014 The creation of the estimation of internal and external resources ratio in the management of the extensive and intensive firm development *Proc. of Int. Multidis. Sc. Conf. on Soc. Sc. and Ar. SGEM 2014* Book 2 **vol 4** (Sofia: STEF92 Technology Ltd) DOI: 10.5593/sgemsocial2014B24 p 706

[14] Shikhalev A M, Vorontsov D P Akhmetova I A and Khamidullina G R 2015 "Advanced firms in the regional development: real situation and the possibilities of growth (on the example of the JSC "TATNEFT" oil company)," Proceedings of International Multidisciplinary Scientific Conference on Social Sciences and Arts SGEM 2015. Bulgaria, book 2**, vol 3**, pp 1023-1030

[15] Shugoreva E The main consumed resources and energy saving technologies URL: http://altenergiya.ru/energosberezhenie/energosberegayushhie-texnologii.html (available at 12.02.2019).

[16] The energy saving PJSC "Nizhnekamskneftekhim" URL: https://www.nknh.ru/about/energy/energy_saving/ (available at 04.02.2019)

[17] PJSC "Kazanorgsintez" URL: https://www.kazanorgsintez.ru/index.php (available at 21.01.2019)

[18] PTC "KAMAZ" Group of Companies URL: https://www.kamaz.ru (available at 23.01.2019)

[19] Kashapov N F, Konahina I A, Gil'manshin I R and Fahreev N N 2014Mini-Central heating and

Power Plant (CHP): the choice of the optimal structure and modes of operation *IOP Conference Series: Mat. Sc. and Engin.* (Bristol: IOP Publishing) **vol 69** DOI: 10.1088/1757-899X/69/1/012008 p 5

[20] Kashapov N F, Fazlyyyakhmatov M G, Khayritonov Kh A, Lazarev D K and Lazarev V K 2015 Practical experience of using ultrasound flowmeters at the measurement associated petroleum gas *IOP Conference Series: Mat. Sc. and Engin.* (Bristol: IOP Publishing) **vol 86**, DOI: 0.1088/1757-899X/86/1/012020 p 2

[21] Kashapov N F, Sagdiev R K, Denisov E S, Evdokimov Yu K and Fazlyyyakhmatov M G 2014Phased array based ultrasound scanning system development *IOP Conference Series: Mat. Sc. and Engin.* (Bristol: IOP Publishing) **vol 69**, DOI: 10.1088/1757-899X/69/1/012012 p 4

[22] Kashapov N F, Khafizov I I, Nurullin I G and Z B Sadykov 2018 Influence of introduction of robotics on increase in efficiency of electrochemical production *IOP Conference Series: Mat. Sc. and Engin.* (Bristol: IOP Publishing) **vol 412**, Is 1 Art № 0120345 DOI: 10.1088/1757-899X/412/1/012034 p 1

IOP Conf. Series: Materials Science and Engineering **570** (2019) 012103 doi:10.1088/1757-899X/570/1/012103

Highly effective composite materials based on waste of technogenic sulfur

A A Yusupova[1], L R Barayeva[1], R T Akhmetova[1], A I Khatsrinov[1]and A A Bobryshev[2]

[1]Kazan National Research Technological University, 420015, Kazan, st. K. Marksa, 68
[2]Naberezhnye Chelny Institute (branch), Kazan Federal University
Sjujumbike street 10a, Naberezhnye Chelny, 423800 Tatarstan, Russia

alsu16rus@yandex.ru

Abstract. The materials developed on the basis of technogenic sulfur of silicate and chloride zinc of akivator can be used as road materials (tiles, paving stones) for arranging the territory or inner rooms of enterprises for the machine-building complex in areas with aggressive media. X-ray, rheological and quantum chemical studies have shown that when sulfur interacts with zinc chloride, a system of strong thermodynamically stable compounds is formed, causing the formation of an effective material.

1. Introduction

Technological processes of the machine-building complex widely use a different spectrum of aggressive media. Meanwhile, even such effects as air saturated with moisture can relatively quickly - in 10-15 years - to lead the whole building in an emergency condition. Many structural elements of buildings - foundations for acid pumps, some sections of floors, etc. - need of repair or restoration annually, and sometimes more often. The application in the conditions of an aggressive environment is possible not for every material.

At the same time, a large number of sulfur - containing composite materials are known which are distinguished by a number of positive properties: high chemical resistance in corrosive environments with operating parameters, high mechanical strength, low cost and availability of materials [1-3].

2. Body text

The possibility of obtaining sulfur concrete on the basis of sulfuric acid was investigated in this work. The development of technologies for utilization of sulfuric technogenic waste and their processing into sulfuric composite materials is a highly topical task, because it allows obtaining efficient composite materials and solving environmental problems of modern production, which currently has high environmental safety requirements.

Zinc chloride was used as an activator of sulfur and silicate component of the composition. Its electronic structure, in particular, the presence of vacant p- and d-orbitals, will allow activating the initial components [4].

To obtain a sulfur concrete on the basis of sulfur and sulfur waste, the following components were used as the starting components: sulfur (S) - waste from the Nizhnekamsk refinery with a basic substance content of 99.98% by mass. (GOST 127-93); zinc chloride $ZnCl_2$ (GOST 4529-78); sulfur waste of a metallurgical enterprise (sulfur cake) (chemical composition, wt.%: elemental sulfur (S) - 50-58%, aluminum oxide (Al_2O_3) 1.5-3.0%, ferrous sulfate ($FeSO_4$, $Fe_2(SO_4)_3$) - 4.5-15%, calcium oxide (CaO) - 0.3-0.5% silicon oxide (SiO_2) - 12-25% magnesium oxide (MgO) - 0.1-0.2%, water (H_2O) -0.36%).

For testing samples of compositions of metallurgical production were made on the basis of sulfur and sulfur waste. The obtained samples were sent for physico-mechanical tests for sulfur concrete.

The compositions were also investigated by the methods of physicochemical analysis. X-ray diffraction studies were carried out on a DRON-3 diffractometer (Cu Ka-radiation).

As the results of X-ray analysis (Fig. 1) showed, the sulfur cake consists of a crystalline phase, represented by rhombic sulfur, thauzite $Ca_3Si(OH)6CO_3SO_4(H2O)_{42}$, aqueous ferrous sulfates - rhombic glaze and remarite, quartz SiO_2 and calcium sulfate $CaSO_4$.

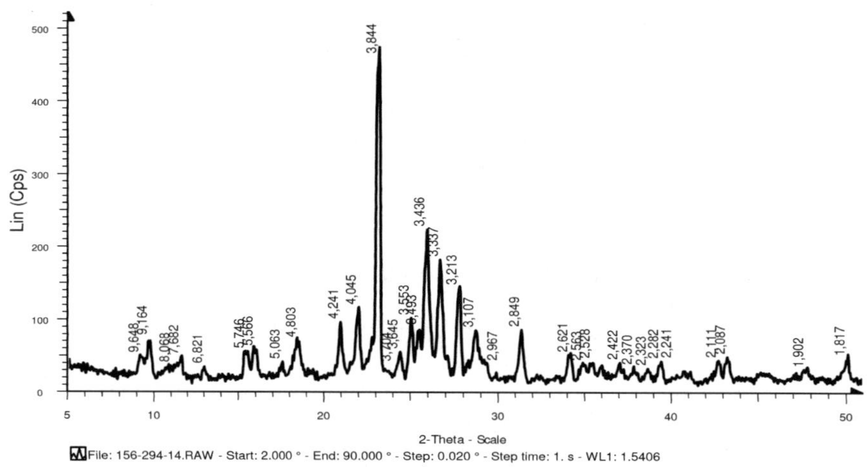

Figure 1. Sulfur cake diffractogram

According to emission spectral chemical analysis, the composition of the cake is represented mainly by the following compounds, wt.%: Sulfur - 50; C, 0.76; Ca 0.5; Al, 3.0; Mg - 0.2; Mo - 28.0; Fe ≥15; Si - 12.0; Co - 1.6; Ti is 0.1; Na - 0.2; K - 0.2.

Since this waste contains a crystalline rhombic sulfur and, at the same time, mineral components - calcite and sulphates - in an amount close to the technological one (mass-to-sulfur: filler ratio is 1:1) are assumed that it is possible to obtain composite materials from it for construction purposes. To compensate for the lack of the sulfur component, pure sulfur in the amount of 10 and 20% of the mass was added to the initial composition. At the same time, the particles of the mineral component were completely enveloped by the sulfur melt, the alignment of the shape of the samples was observed, which acquired a more dense structure.

It is known that the viscosity of a sulfur melt can significantly decrease with the introduction of some electrophilic additives, such as chlorides of aluminum, iron, titanium and zinc. Getting into the sulfuric melt, they interact with sulfur, promoting the shift of the electron density in the sulfur ring, lower the sulfur-sulfur bond energy and intensify the opening of the sulfuric molecule with the formation of a large number of short radicals. The viscosity decreases sharply from 0.025 to 0.003 Pa s [5, 6].

For a detailed analysis of the products of the interaction of sulfur with a zinc chloride activator, X-ray phase analysis was used. On the radiograph, the appearance of zinc sulfide crystalline zinc sulfide (ZnS) cubic syngony and zinc oxysulfate ($Zn_3O\ (SO_4)_2$) monoclinic syngony, which is a product of zinc sulfide oxidation, is observed. Also recorded are the reflections corresponding to the cyclocere S_{12} of the orthorhombic modification, formed, apparently, as a result of the "stitching" of a huge number of short-chain radicals. This indicates some polymerization of short-chain radicals with the formation of polymeric sulfur, which will positively affect the performance properties of the sulfide material [7-10].

In addition, x-ray phase analysis was made by calculating the quantitative ratio of phases of zinc sulfide 8%, 11% of the product of the oxidation of oxysulfate zinc, 11% of cycloseris S_{12}, the unreacted zinc chloride was 2.1%. Considering the amount of injected $ZnCl_2$ (25%), almost all of its amount reacted with sulfur to form zinc sulphide ZnS. The resulting sulfide has a cubic syngony and refers to sphalerite found in nature. Thus, when sulfur interacts with zinc chloride, a system of strong thermodynamically stable compounds is formed.

Quantum-chemical studies using the Priroda program confirmed the activating effect of zinc chloride. The activation energy of opening octasers using zinc chloride is reduced by 30 kJ/mol relative to the opening of cyclooctasers without an additive (Fig. 2). This fact proves the activating effect of the additive on the sulfur component, leading to the opening of cyclic sulfur molecules at lower temperatures, which ultimately leads to a reduction in process costs.

The S_8 molecule is very symmetrical and flexible. In the process of thermal disclosure of this molecule, the formation of various spatial rearrangements is possible. With the addition of $ZnCl_2$, the process of disclosing the sulfur molecule is much easier in two stages.

Zinc chloride significantly reduces the viscosity of the sulfur melt in a wide temperature range, it has a positive effect on the wetting and impregnating properties of the latter.

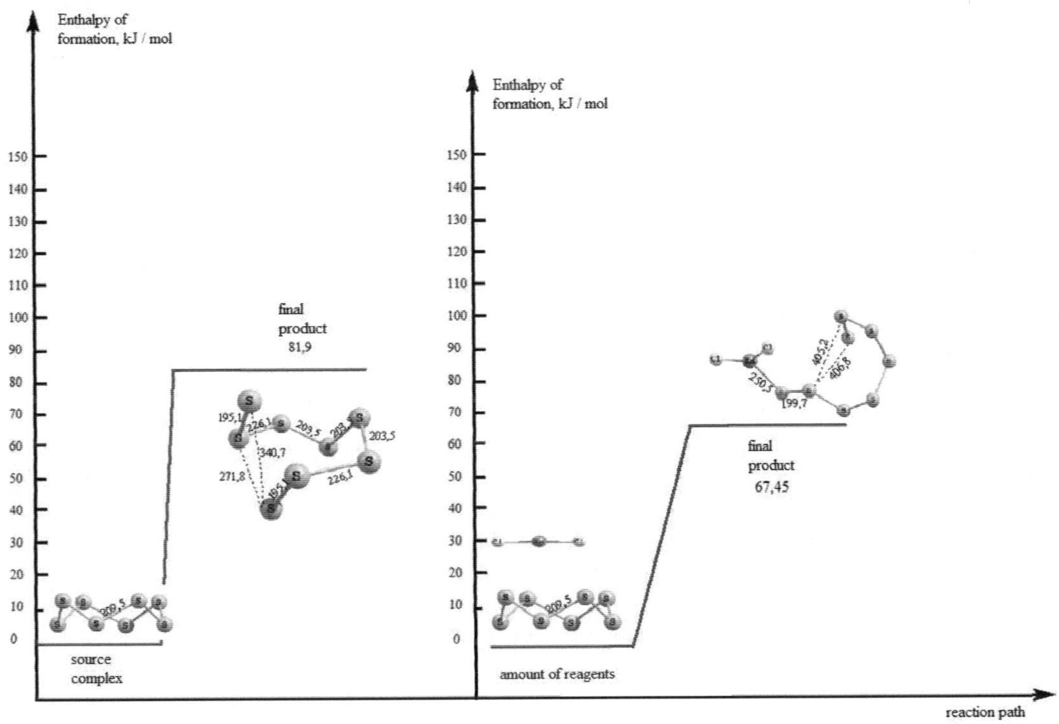

Figure 2. Disclosure cyclooctanone by thermal activation (on the left) and in the presence of the activator of zinc chloride (on the right)

To improve the water-resistant properties, prevent the dissolution of sulphides existing in concrete and the subsequent destruction of the material, the obtained samples were dipped into the sulfur melt for a few seconds, removed from the melt and allowed to cool. The protective layer of hydrophobic sulfur formed on the surface reliably protected the obtained products from dissolution. The obtained samples were tested for mechanical and operational properties - compressive strength (30-35 MPa),

IOP Conf. Series: Materials Science and Engineering **570** (2019) 012103 doi:10.1088/1757-899X/570/1/012103

water absorption (not more than 1%), density (2.15-2.80 g/cm^3), resistance to aggressive media (0.85-0.97).

On the basis of the data obtained, a schematic flowchart of the technology of sulfuric composite materials from sulfur cake was proposed.

3. Conclusions

The developed materials can be used as road materials (tiles, pavers) for arranging the territory or interior of enterprises of the machine-building complex in areas with aggressive media.

References

[1] Korolev E V, Proshin A P, Erofeev V T, Khrulev V M and Burned V V 2003 *Sulfur-based building materials* (Penza: PGUAS) p 372

[2] Akhmetova R T, Yusupova A A , Baraeva L R, Sabakhova G I, Pervushin V A, Khatsrinov A I, Medvedeva G A and Akhmetova A Yu 2015 *Astringent* RF Patent 2555166

[3] Groshin A P, Korolev E V and Kalinkin E G 2005 Structure and properties of modified sulfur binder *Stroit Mater* vol 7 pp 6-9

[4] Roldugin V I 2008 *Physical chemistry of the surface* (Moscow: Dolgoprudny: Publishing house "Intellect") p 568

[5] Fedyaev V L, Galimov E R, Belyaev A V, Dolgov A V and Samoylov V M 2018 Evaluation of interphase boundaries dynamics at the contact of melt with solid material. *IOP Conf. Series: Materials Science and Engineering* vol 412 number 012015

[6] Galimov E R, Bobryshev A N, Lakhno A V, Voronov P V, Galimova N Ya and Sharafutdinov R F 2015 New approaches for evaluating rheological models in composites. *IOP Conf. Series: Materials Science and Engineering* vol 86 number 012001

[7] Jusupova A A, Akhmetova R T, Baraeva L R, Medvedeva G A and Mezhevich Zh V 2018 Development of low-waste technology for the processing of sulphur into polysulphides and materials based on them *IOP Conf. Series: Materials Science and Engineering* vol 450 number 62007

[8] Yusupova A A, Akhmetova R T and Shafigullin L N 2017 Sulfur Composite Materials Based on Sulfide Containing Industrial Waste *Materials Engineering and Technologies for Production and Processing III Solid State Phenomena* vol 265 pp 33-8

[9] Yusupova A A, Khatsrinov A I and Akhmetova R T 2018 Activating effect of aluminum chloride in the preparation of sulfur concrete from sulfur and silica *Inorganic Materials* vol 54 No 8 pp 809–14

[10] Yusupova A A, Erofeev V T, and Bobryshev A A 2018 Activation of sulfur and opal-cristobalite-tridymite phase in sulfur concrete technology *IOP Conf. Series: Materials Science and Engineering* vol 463 number 042033

IOP Conf. Series: Materials Science and Engineering **570** (2019) 012104 doi:10.1088/1757-899X/570/1/012104

About Modeling Kinematic Parameters of Deposition from Composite Electrochemical Coatings for Restoration of Details for Agricultural Machines

S Yu Zhachkin[1], S N Sharifullin[2], N A Penkov[3] and A A Zavrazhnov[4]

Voronezh State technical University, Russia [1]
Kazan Federal University, Russia [2]
Military Training and Research Center of the Air Force «Air Force Academy named after Professor N.E. Zhukovsky and Yu. A. Gagarin», Russia [3]
Michurinsk State agrarian University, Russia [4]

myth_np_nikit@mail.ru

Abstract. Theoretical aspects of process of drawing of composite galvanic coverings on the basis of a chromic matrix on current-carrying surfaces are considered by method of galvanic contact plating (GCP). The analytical dependences connecting parameters of physic mechanical characteristics of a covering with technological modes of sedimentation of coverings and design features of the equipment, used in process are resulted.

1. Introduction

Standard galvanic reduction of parts restored by chromium plating, successfully used for such a long time, despite its ease of use in recent years does not meet the requirements of modern technology and technologies for the quality of coatings used in the restoration of agricultural machinery parts [1, 2]. In this regard, in recent years have been widely developed and implemented in the production of modernized methods of restoration of chrome plating parts.

Taking into account the above, and based on the existing theoretical and practical groundwork on the problem of restoring parts by chromium plating with simultaneous activation of the cathode, a previously developed method for restoring parts was proposed, which consists in combining the process of galvanic coating and its simultaneous machining during deposition with the possibility of regulating the rate of recovery of a particular part surface area, depending on its wear [1]. The method is called galvanic contact deposition (GKO).

2. Materials and research methods

There was a research to determine the operating parameters of the processing method applicable for experimental sleeves. They were conducted on experimental shells, made of steel 30 HGSA TU 14-1-950-74.

Axial residual stresses in chromium coatings was determined by the method of backgroundmaterial with continuous grazing of intense layers from the surface which was carried out on the machine UKOONT [3, 5]. The unit is designed to measure the relative strain and consists of a strain gauge amplifier 8ANCH-7M and recording potentiometer KSP-4. Primary transducers were strain gages 2FKPA-10-200 V.

Residual stresses in the studied layer were represented as the following sum [1]:

$$\sigma_h = \sigma_{hv} + \sigma_{hs}$$

where σ_{hv} – linear component determined by the deformation of the sample which cutts from the part; σ_{hs} – nonlinear component, calculated according to the curve of deformation during the etching of the sample in the study depth.

3. Results and discussion

Plastic deformation of the deposited coating crystal is always accompanied by its elastic deformation [4, 9]. Consequently, the dimensions of the body at the final moment of its loading differ from its dimensions determined when the load is removed.

To represent the logarithmic degree of deformation it is necessary to calculate the integral of an infinitely small increment of the considered size of the body or its element related to its value at each given moment of deformation

$$\delta_x = \int_{x_i}^{x_d} \frac{dx}{x} = lnx \int_{x_i}^{x_d} = ln\frac{x_d}{x_i} \quad (1)$$

On the other hand, the expression of the degree of deformation as the ratio of the size increment to the initial size is of great interest [8].

Due to the fact that the tool forces under its influence on the deposited coating is extremely small compared to the mechanical properties of the coating, it can be assumed that in this case there is a small deformation [3] (the degree of deformation is less than 0.01). For small deformations:

$$\delta = \varepsilon \quad (2)$$

and, accordingly

$$\varepsilon_x + \varepsilon_y + \varepsilon_z = 0 \quad (3)$$

Multiplying all the terms of the equation for the degree of deformation by the volume V of the deformable body, we obtain

$$V\delta_x + V\delta_y + V\delta_z = 0 \quad (4)$$

and for small deformations:

$$V\varepsilon_x + V\varepsilon_y + V\varepsilon_z = 0 \quad (5)$$

Thus, the sum of the multiplication of volume by logarithmic degrees of deformation (crumpled volume) in mutually perpendicular directions is zero.

The elementary displaced volume is defined by the expression:

$$dV_{cz} = F_z dz \quad (6)$$

where F_z – the area of normal cross sections of the deposited coating at each specific moment of the deformation process. Then

$$V_{cz} = \int_{z_i}^{z_d} F_z dz \quad (7)$$

In the case of plastic deformation of elementary volume $F_z = V/z$ [1]. Then

$$V_{cz} = V \int_{z_i}^{z_d} \frac{dz}{z} \quad (8)$$

where Z_i – is the initial height of the deposited layer;

Z_d -the height of the deposited layer after deformation.

After integrating, we can obtain:

$$V_{cz} = Vln\frac{z_d}{z_i} = V\delta_z \quad (9)$$

Such algorithm of layer-by-layer deformation to determine the degree and rate of deformation was't chosen by chance. It was presented in [1, 2] and experimentally tested in [3].

From the expression above (7), taking into account, that $Y_iZ_i = F_{ih}$ and $Y_dZ_d = F_{dh}$ follows:

$$\frac{x_d}{x_i} = \frac{Y_i Z_i}{Y_d Z_d} = \frac{F_{ih}}{F_{dh}} \quad (10)$$

where F_{ih} and F_{dh} are the areas, respectively, normal to the x-axis of the body sections before and after deformation.

Taking into account the above, it is possible to express the degree of deformation and displacement volumes not only through the linear dimensions, but also through the cross-sectional areas normal to the coordinate axis, in the direction of which the degree of deformation and displacement volume is considered:

$$\delta_x = ln\frac{X_d}{X_i} = ln\frac{F_{ih}}{F_{dh}} = -ln\frac{F_{dh}}{F_{ih}} \quad (11)$$

$$\varepsilon_x = \frac{X_d - X_i}{X_i} = \frac{\Delta x}{X_i} = \frac{F_{ih} - F_{dh}}{F_{dh}} = -\frac{\Delta F_x}{F_{dh}} \quad (11a)$$

Similarly, we can get expressions to calculate the degrees of deformation in the directions of the coordinate axes Oy and Oz.

In general, we can write:

$$\delta = -ln\frac{F_d}{F_i};\ \varepsilon = -\frac{\Delta F}{F_d} \quad (12)$$

Conclusion

Theoretically developed and experimentally investigated the dependence of tool pressure on the stages of recovery, implemented by GKO technology, which allows to obtain high-quality thick coatings (in particular chrome) with a given physical, mechanical and, as a consequence, operational properties.

References

[1] Zhachkin S Yu et al. 2002 Cold galvanocontact restoration of parts *VSTU* 138 p
[2] Astahov M V, Sorokina I I et al. 2012 *Prospects for the use of polymer composite materials with the addition of nanocrystalline aluminum oxide powders in the design and repair of agricultural machinery* Technology of metals **12** pp 18-21
[3] Zhachkin S Yu, Penkov N A et al. 2012 *Restoration of hydraulic valves of agricultural machinery by composite galvanic coating based on chrome matrix* Works GOSNITI **109** part 2 pp 89-93
[4] Cherkez M B et al. 1958 *Chrome plating and iron plating* Mashgiz 88 p
[5] Kudryavcev N T et al. 1979 *Electrolytic metal coatings* Chemical 352 p
[6] Silin S S, Hrulkov V A, Lobanov A V and Rikunov N S et al. 1984 *Deep grinding of difficult processing to machine materials* Engineering 64 p
[7] Shluger M A et al. 1985 *Electroplating in mechanical engineering* Engineering 240 p
[8] Shermergor T D et al. 1977 *Theory of elasticity of micro-inhomogeneities environments* Science 400 p
[9] Landau L D, Livshitz E M et al. 1987 *Theory of elasticity* Science 248 p.
[10] Adler U P, Markova E V, Granovskii U V et al. 1976 *Experiment planning in serch of optimal conditions* Science 279 p

IOP Conf. Series: Materials Science and Engineering **570** (2019) 012105 doi:10.1088/1757-899X/570/1/012105

Particular properties of operation of stationary lead-acid batteries at power distribution facilities

A D Ziganshin[1], L S Safiullin[2], A V Bogatkin[3], P P Pavlov[1], I Z Gatiyatov[1] and A E Auhadeev[1]
[1]Kazan State Power Engineering University, Kazan, Krasnoselskaya Street, 51, 420066, Russian Federation
[2] Kazan (Volga Region) Federal University, Kazan, Kremlevskaya street, 18, 420008, Russian Federation
[3]LLC "Akku-Fertrib", Kazan, Ibragimova Avenue, 58, 420066, Russian Federation

Abstract. The article provides an analysis of the features of operation of stationary lead-acid rechargeable batteries at power plants as a source of power for operational DC circuits. The aim of the study was to compare widely used lead-acid battery systems with free electrolyte with dryfit and AGM technologies. The result of the study is to determine the significant advantages of sealed batteries in the implementation of gas evolution and recombination processes. On the basis of the conducted research and the revealed modern trend in the implementation of hardware for automation and telemechanics based on low-maintenance and maintenance-free equipment, conclusions have been made about the feasibility of using sealed lead-acid batteries as sources of operational direct current at power distribution facilities.

Introduction

At present, the highest requirements are placed on the quality and reliability of equipment used at power distribution facilities. To power the operational circuits of the protection, automation and alarm relay, electromagnets for switching off and on switching devices, operational direct current is used predominantly. The current source is mainly the batteries of the lead-acid system manufactured according to the classical technology with liquid electrolyte, which is predetermined by the breadth of the model range, relative simplicity of production and unpretentiousness to operating conditions [1, 2].

The peculiarities of operation of lead-acid batteries with liquid electrolyte should include the need to adjust its level and density and a significant amount of gas evolution during charging, which makes it necessary to install the battery in a separate capital room equipped with forced supply and exhaust ventilation [1-8].

Study of the features of operation of stationary lead-acid batteries

Lead acid battery (LAB) as any other chemical source of current is a system of unstable equilibrium. All processes occurring in it are associated with its slow "destruction", expressed in irreversible loss of capacity.

In order for the processes of "destruction" to proceed with a characteristic speed for this type of batteries and their mode of operation, i.e. in order for the battery life to be predictable, it is necessary

IOP Conf. Series: Materials Science and Engineering **570** (2019) 012105 doi:10.1088/1757-899X/570/1/012105

to ensure the correct operating conditions as well as charge / discharge parameters, which must comply with the modes permitted by the manufacturer [9, 10].

A promising solution to this problem is the use of sealed LAB with their installation in a capital building with natural ventilation. At the same time, the use of modern technologies allows placing LAB in a separate room, as well as in a room with electrical and switching equipment.

Sealed LAB have made a breakthrough in the production technology of lead batteries for industrial use. Thus, in 1957, a battery with a bound electrolyte was developed, the so-called "non-spillable" battery, and the technology itself was named *dryfit* and was patented. The idea of creating a non-spinning battery is based on the principle of thickening the electrolyte to a jelly-like state and sealing the internal volume with an overpressure valve, which is installed in the filling hole to provide one-way communication with the external environment to release a small amount of gas. Later in the 70s of the last century, another electrolyte binding technology was invented by impregnating separators with good capillary properties. This technology is called *AGM* (from Absorbent Glass Mat), and glass fiber sheets are used as separators [10, 11].

The common property of all sealed LAB is that there is no need to dilute the electrolyte during their operation. Overpressure valves are installed during production and do not disassemble during the entire service life of the batteries. Another important quality of the batteries of the sealed design is low gas emission, which significantly softens the requirements for the rate of air exchange at the location of the battery.

However, when building industrial power installations with a high power battery, it is impossible not to take into account the fact of the occurrence of a side reaction of electrolysis of water when charging stationary lead-acid rechargeable batteries.

It is known that when a recharge current flows through a battery, electrolysis of water produces gaseous oxygen (at the positive electrode) and gaseous hydrogen (at the negative electrode). If no special measures are taken, then both gases are released into the surrounding space of the battery in an amount determined by the Faraday electrolysis law, which in terms of 1 Ampere-hour of electricity is 0.037 g or 0.418 l of H_2 (hydrogen) and 0.299 g or 0.209 l of O_2 (oxygen) as a result of decomposition of 0.336 g of H_2O (water). Hydrogen is a potential threat: it is an explosive combustible gas whose mixture with air becomes dangerous by fire, starting with a 4 percent concentration by volume [10, 11].

To remove hydrogen, it is necessary to ensure adequate air exchange of the location of the battery due to natural or forced ventilation. The air exchange rate is determined by the standard GOST R MEK 62485-2-2011 [5].

In a sealed LAB in a bound electrolyte, a mechanism of internal recombination of gases with the formation of water is implemented. The conditions for the reaction are the presence of recombination channels: microcracks in the gel or electrolyte-free pores in the glass fiber separator, through which gaseous oxygen is delivered to the negative electrode, where it enters an electrochemical reaction with hydrogen, accompanied by heat evolution H2 + ½O2 → H2O + heat.

The recombination reaction coefficient in systems with a bound electrolyte is 98–99%, which corresponds to an approximately 100-fold reduction in gas evolution compared with classical structures in which the electrolyte is not bound [10, 12].

In accordance with GOST R MEK 62485-2-2011, the minimum air exchange rate of the location of the battery is calculated by the formula:

$$Q = 0.05 \cdot \mathrm{n} \cdot I_{gas} \cdot C_{\mathrm{rt}} \cdot 10^{-3},\ [\mathrm{m^3/h}]$$

where I_{gas} – gassing current for sustained or accelerated charge [mA / Ah],

C_{rt} – capacity of 10-hour discharge of lead-acid elements to a voltage of 1.8 V at a temperature of 20 ° C,

n – number of cells in the battery.

The values of the gas emission current I_{gas} for battery cells depending on the charge mode and the state of the electrolyte are presented in Table 1.

IOP Conf. Series: Materials Science and Engineering **570** (2019) 012105 doi:10.1088/1757-899X/570/1/012105

Table 1.
Values of gas evolution $I_{\text{газ}}$ for LAB with liquid electrolyte (antimony content <3%) and sealed LAB (with overpressure valve) depending on the charge mode

Charge mode	**Gassing current I_{gas}, мA/A·h**	
	LAB with liquid electrolyte	Sealed LAB
Charge support	5	1
Accelerated charge	20	8

The minimum area of the input and output vents for natural ventilation is calculated by the formula:

$$A \geq 28 \cdot Q, [\text{sm}^2].$$

In the immediate vicinity of the battery, it is necessary to observe a safe air gap, where there should be no spark-forming or hot devices. The required size of the exclusion zone, measured by the straight distance from the battery valves, depends on the type and capacity and is calculated by the formula:

$$d = 28.8\sqrt[3]{I_{gas}} \cdot \sqrt[3]{C_{\text{rt}}}, [\text{mm}].$$

For a monoblock battery, the gas output is determined by the number of elements connected in series in a single housing, connected by a single ventilation system or through a tube. When calculating the safety distance, the value C_{rt} is determined by multiplying the capacity of the monoblock by the number of elements in it.

Based on the above formulas, it can be established that for LAB with internal recombination of gas with a capacity of up to 1000 A·h the size of a safe air gap in any charge mode does not exceed 1 meter. Thus, the creation of an exclusion zone of 1 meter in natural ventilation will guarantee the safe operation of sealed LAB, which allows them to be installed in a common room with electrical and switching equipment [9, 10].

Conclusion

In the industry of automation and telemechanics at the objects of electricity distribution, to a greater or lesser extent, batteries are produced in all three of the above-listed technologies. In this case, the vast majority of batteries used as a source of operating direct current, have a classic design with free electrolyte. The current trend in the development of technical equipment for automation and remote control involves a gradual transition to low-maintenance and maintenance-free equipment [10-12, etc.]. To achieve this goal, it is advisable to use current sources, based on sealed lead-acid batteries, which will reduce operating costs while maintaining the required level of reliability.

References

[1] STO 70238424.29.220.20.001-2009 Battery pack power plants. Organization of operation and maintenance. Norms and requirements (Moscow: NP "Invel") p 80

[2] SO 153-34.20.501-2003 Rules for the technical operation of power plants and networks of the Russian Federation (Moscow: SPO ORGRES) p 161

[3] GOST R IEC 60896-22-2015 Stationary lead-acid batteries. Part 22. Types with control valve. Requirements (Moscow: Standardinform) p 27

[4] GOST 26881-86 Stationary lead batteries. General technical conditions (Moscow: Publishing house of standards) p 167

[5] GOST R MEK 62485-2-2011 Batteries and rechargeable batteries. Safety requirements. Part 2. Stationary batteries (Moscow: Standardinform) p 29

[6] GOST R MEK 8961-95 Lead-acid stationary batteries. General requirements and test methods. Part 1. Open types (Moscow: Publishing house of standards) p 17

[7] GOST R MEK 60896-11: 2002 lead-acid rechargeable batteries. Part 11. Ventilated types. General requirements and test methods (Moscow: Standardinform) p 33

[8] GOST R MEK 60896-21: 2004 lead-acid rechargeable batteries. Part 21. Types of batteries with valve regulation. Test methods (Moscow: Standardinform) p 31

IOP Conf. Series: Materials Science and Engineering **570** (2019) 012105 doi:10.1088/1757-899X/570/1/012105

[9] Makarenko S V 2014 Separate issues of calculations and operation of stationary lead-acid batteries *Electrotechnical market* (Moscow: Elek) No 3 pp 70-71

[10] Skrotskaya O S 2014 The problem of placement of control batteries at the stations. Gas emission and ventilation *Bulletin of reports VII MNPK of the "Transzhat-2014"* (Rostov-on-Don: RGUPS) pp 119-23

[11] Mantsev A V 2015 Review of new battery technologies. *Warehouse complex* (Moscow: Technology) No 4 pp 16-17

[12] Igoshin I N 2018 Sonnenschein PowerCycle batteries - for difficult operating conditions *Communication herald* (Moscow: IRIAS) No 11 pp 3-4

IOP Conf. Series: Materials Science and Engineering **570** (2019) 012106 doi:10.1088/1757-899X/570/1/012106

Minimum density of irrigation of the rotor of a film centrifugal heat and mass-exchange apparatus

N H Zinnatullin[1], G N Zinnatullina[1], A I Haibullina[2], L S Sabitov[3], N.F. Kashapov[3]

[1]Kazan National Research Technological University, 68 Karla Marks, street, Kazan, 420015, Russian Federation
[2]Kazan State Power Engineering University, 51 Krasnoselskaya street, Kazan, 420066, Russian Federation
[3]Kazan Federal University, 18 Kremlyovskaya street, Kazan, 420008, Russian Federation

haybullina.87@mail.ru

Abstract. For anomalous-viscous centrifugal film based on the energy and moment approaches, equations for the critical fracture radius and minimum irrigation density are obtained. Experimental results are more consistent with data from the moment approach.

1.Inroduction

Centrifugal devices, in which the film flow of a liquid is realized, are widely used in chemical technology for carrying out hydromechanical, heat and mass transfer processes [1, 2]. The basis of these devices is the rotor, which is a shaft with nozzles in the form of a flat disk, cone, sphere or other curvilinear bowl. These devices process materials that characterize different rheological properties: viscous liquids, abnormally viscous liquids, viscous-plastic media, etc. [3, 4].

Effective operation of centrifugal film apparatus is possible only if a stable film of liquid is formed. Possible ruptures of the liquid film and excessive irrigation reduce the efficiency of the technological processes, since this either reduces the working surface, or because of the increase in the film thickness, its resistance increases. Therefore, the study of the behavior of a centrifugal liquid film remains relevant from the point of view of determining the minimum irrigation density.

The phenomenon of rupture of a liquid film must also be investigated from the standpoint of dispersion, since it is accompanied by the formation of surface jets and their disintegration into droplets on the edge of the nozzle [5].

Known work on the study of the rupture of a liquid film with an isothermal flow in a gravitational field [6-8]. When studying the rupture of a liquid film, two approaches are used: energetically and forcefully. With the force approach, an equilibrium of the acting forces is formed at the critical point of rupture. In the energy approach, it is assumed that the liquid-gas-solid system chooses such a situation when the energy minimum of the system is possible. If a bursting flow is energetically more favorable for such a system, a discontinuity of flow will be observed, and vice versa.

The force method did not find application: the calculated critical fracture radius is much smaller than the experimental one; energy analysis of the breakdown of a liquid film for a gravitational field gives a satisfactory agreement between the calculated and experimental data. To calculate minimum irrigation density Γ_{min} Hobler offers a formula:

$$\Gamma_{min} = \frac{\mu}{4}\left[\frac{2}{3}\frac{\sigma_3(1-\cos\Theta_0)237{,}4}{\rho g}\left(\frac{\rho^2 g}{\mu^2}\right)^{2/3}\right]^{3/5} \tag{1}$$

where ρ, μ – density and dynamic viscosity of a fluid, respectively [kg/m^3], [Pa·sec]; σ_3 – surface tension, respectively, at the interface of the solid-liquid, solid-gas, liquid-gas [N/m]; Θ_0 - - wetting angle [rad.].

Khobler's ideas were developed in Mikilavinan's works for a horizontal film of liquid moving under the action of a tangential shear stress on a free surface.

Known work on the study of the rupture of a liquid film - the appearance of a "dry spot" - with non-isothermal flow of the medium [9, 10].

It should be noted that in all these works only rheologically simple media were studied, i.e. only Newtonian fluids. In industrial vehicles, more complex rheological media are mainly processed.

The rupture of a liquid film is obviously a consequence of hydrodynamic instability. However, the theory of hydrodynamic instability does not contain information about the conditions for the occurrence of “dry spots”, streams. It does not take into account a very significant parameter - wetting angle Θ_0.

Consider the steady-state isothermal flow of anomalous-viscous fluid flowing in the form of a thin laminar film along a rotating flat nozzle (Fig. 1).

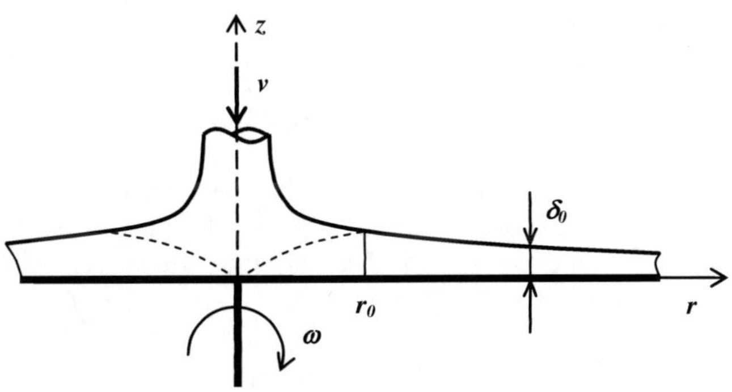

Fig.1. Fluid flow pattern

Orthogonal Coordinate System υ, φ, z rotates together with the disk, the axis of fluid flow is symmetrical.

The rheological equation of state is represented as a power relation (“power law”), which for a one-dimensional flow has the form:

$$\tau_{zr} = k\left(\frac{dW_r}{dz}\right)^n \tag{2}$$

where τ_{zr} — shear stress; k and n – rheological fluid constants; W_r - adial velocity For anomalous-viscous pseudoplastic fluid $n < 1$.

IOP Conf. Series: Materials Science and Engineering **570** (2019) 012106 doi:10.1088/1757-899X/570/1/012106

In the axial region (shock region), a three-dimensional flow of a liquid occurs, starting with r_0, a two-dimensional flow occurs due to centrifugal forces; therefore, the computational region begins with $r > r_0$.

Works [11, 12] are devoted to determining the length of the initial section r_0.

Consider the total energy of a liquid film element having a surface area $S_0 = 2\pi r \Delta r$. If we allow the rupture of this element, we will have (Fig. 2):

$$S_0 = S_1 + S_2 \tag{3}$$

where S_1 and S_2 - respectively wetted and dry areas.

The full energy E_0 of element S_0 could be written

$$\mathrm{E}_0 = \mathrm{E}_\kappa + \mathrm{E}_\sigma \tag{4}$$

where E_κ and E_σ - respectively, the kinematic and surface energy of the liquid film.

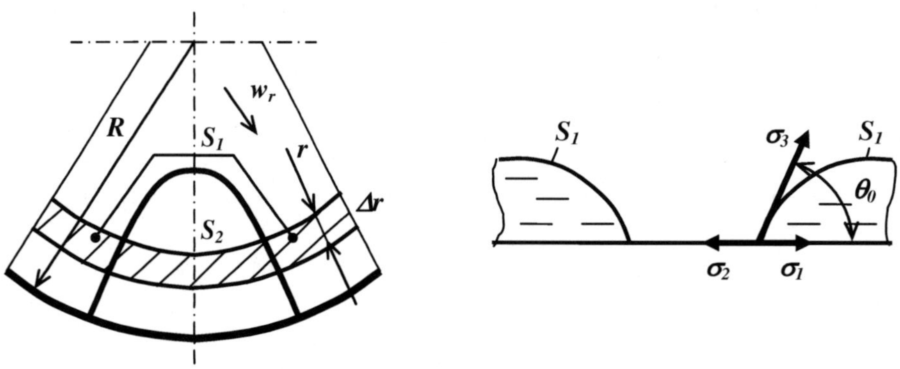

Fig.2. Liquid film rupture pattern

As shown by preliminary calculations for cases where a rupture of a centrifugal liquid film is possible, the flow can be considered one-dimensional [13, 14]. Then, assuming the similarity of the velocity profile in the film continuous and discontinuous flow, the kinematic energy can be represented as:

$$\mathrm{E}_\kappa = \int_0^{\delta_0} \frac{W_r^2}{2} \mathrm{dm} = \int_0^{\delta_0} S_1 \rho \frac{W_r^2}{2} \mathrm{d}\delta \tag{5}$$

Taking into account the velocity profile obtained in [14], we obtain:

$$\mathrm{E}_\kappa = \frac{\rho S_1 n^2}{(2n+1)(3n+2)} \left(\frac{\rho w^2 r}{K} \right)^{2/n} \cdot \delta_0^{\frac{3n+2}{n}} \tag{6}$$

To simplify the mathematical calculations we introduce dimensionless complexes:

$$\bar{\delta}_0 = \frac{\delta_0}{R};\ \bar{r} = \frac{r}{R};\ \frac{S_1}{S_0} = X;\ \frac{S_2}{S_0} = 1 - X;\ J = \frac{\Gamma w R}{\sigma^3};\ \Phi = \frac{w^n K R}{\sigma^3};\ z = \frac{J}{K};\ W_{l_y} = \frac{\rho w^2 R^3}{\sigma^3}$$

IOP Conf. Series: Materials Science and Engineering **570** (2019) 012106 doi:10.1088/1757-899X/570/1/012106

It is easy to show that in the presence of a film break, the irrigation density will be expressed as:

$$\Gamma = \rho \cdot r \frac{\Delta r}{S_1} \text{ or } \Gamma = \Gamma_F / X, \quad \Gamma_F = \frac{\rho r}{2\pi R}$$

Hereinafter, the index F will mark the parameters defined without taking into account the rupture of the liquid film. Liquid film thickness δ_0 taking into account the introduced dimensionless complexes will have the form:

$$\bar{\delta}_0 = \left(\frac{2n+1}{n}\right)^{\frac{n}{2n+1}} \left(\frac{Z_F}{X}\right)^{\frac{n}{2n+1}} \left(\frac{\Phi}{W_{l_y} \bar{r}}\right)^{\frac{n+1}{2n+1}} \tag{7}$$

We present equation (6) to a dimensionless form:

$$\bar{\mathrm{E}} = \frac{\mathrm{E_K}}{\sigma_3 \cdot S_0} = \left(\frac{n}{3n+2}\right)\left(\frac{2n+1}{n}\right)^{\frac{n+1}{2n+1}} \cdot W_{l_y}^{-\frac{n}{2n+1}} \cdot \left(\frac{\Phi}{\bar{r}}\right)^{\frac{3n+1}{2n+1}} \cdot Z_F^{\frac{3n+2}{2n+1}} \cdot X^{-\frac{n+1}{2n+1}} \tag{8}$$

The surface energy in the framework of the considered approximation (no adsorption) at the phase interface can be represented as [15]

$$\mathrm{E}_\sigma = S_1\sigma_3 + S_1\sigma_1 + S_2\sigma_2 \tag{9}$$

According to Young's law [16], which includes only the thermodynamic properties of the interface, one can write:

$$\cos\Theta_0 = (\sigma_2 - \sigma_1)/\sigma_3$$

Then, taking dimensioning into account, we get:

$$\bar{\mathrm{E}}_\sigma = \frac{\mathrm{E}_\sigma}{\sigma_3 S_0} = X(1-\cos\Theta_0) + \frac{\sigma_2}{\sigma_3} \tag{10}$$

Total energy of the element S_0 in dimensionless parameters it looks like:

$$\bar{\mathrm{E}}_0 = \left(\frac{n}{3n+2}\right)\left(\frac{2n+1}{n}\right)^{\frac{n+1}{2n+1}} \cdot W_{l_y}^{-\frac{n}{2n+1}} \cdot \left(\frac{\Phi}{\bar{r}}\right)^{\frac{3n+1}{2n+1}} \cdot Z_F^{\frac{3n+2}{2n+1}} \cdot X^{-\frac{n+1}{2n+1}} + X(1-\cos\Theta_0) + \frac{\sigma_2}{\sigma_3} \tag{11}$$

Since the condition $\frac{d\overline{E_0}}{dX} = 0$ for $\frac{d^2\overline{E}_0}{dX^2} \rangle 0$ corresponds to the minimum energy of the system, then from (11) we have:

$$-A_1 W_{l_y}^{-\frac{n}{2n+1}} \cdot \left(\frac{\Phi}{\bar{r}}\right)^{\frac{3n+1}{2n+1}} \cdot Z_F^{\frac{3n+2}{2n+1}} \cdot X^{-\frac{3n+2}{2n+1}} + (1-\cos\Theta_0) = 0 \tag{12}$$

где $A_1 = \left(\frac{n+1}{2n+1}\right)\left(\frac{n}{3n+2}\right)\left(\frac{2n+1}{n}\right)^{\frac{n+1}{2n+1}}$. From equation (12) we obtain the value of X:

IOP Conf. Series: Materials Science and Engineering **570** (2019) 012106 doi:10.1088/1757-899X/570/1/012106

$$X = \left[A_1 \frac{W_{l_y}^{-\frac{n}{2n+1}}}{(1-\cos\Theta_0)} \left(\frac{\Phi}{\bar{r}}\right)^{\frac{3n+1}{2n+1}} \right]^{\frac{2n+1}{3n+2}} \cdot Z_F \qquad (13)$$

If X varies from zero to unity, then the flow over the disk is characterized by ruptures of the liquid film, i.e. dry areas S_2. Given that the case X = 1 corresponds to the beginning of the gap, from equation (13) we obtain the critical radius:

$$\bar{r}_K = \left[\left(\frac{1}{2\pi A_2}\right)^{3n+2} \cdot \frac{J^{3n+2}}{\Phi W_{l_y}{}^{n}} \right]^{\frac{1}{3n+1}} \qquad (14)$$

where $A_2 = \left[\left(\frac{2n+1}{n}\right)^{\frac{n}{2n+1}} \left(\frac{3n+2}{n+1}\right)(1-\cos\Theta_0) \right]^{\frac{2n+1}{3n+2}}$

From the physics of the phenomenon, it follows that the case X = 1 with $\bar{r}=1$ corresponds to the minimum irrigation density of the rotor apparatus. Then from equation (14) we get

$$\Gamma_{\min} = \frac{2\pi}{wR} A_2 \sigma_3 \left(\Phi W_{l_y}{}^{n}\right)^{\frac{1}{3n+2}} \qquad (15)$$

Consider another approach for determining the rupture of a centrifugal liquid film. As shown by numerous experiments, usually the initial point of rupture of the film is placed along the nozzle radius to the periphery. At some point, an equilibrium is established and, at the same time, a roller of liquid forms on the border of the dry and wetted areas. This equilibrium state is characterized by equality of not forces, but moments of forces, made up for a roller [Fig. 3].

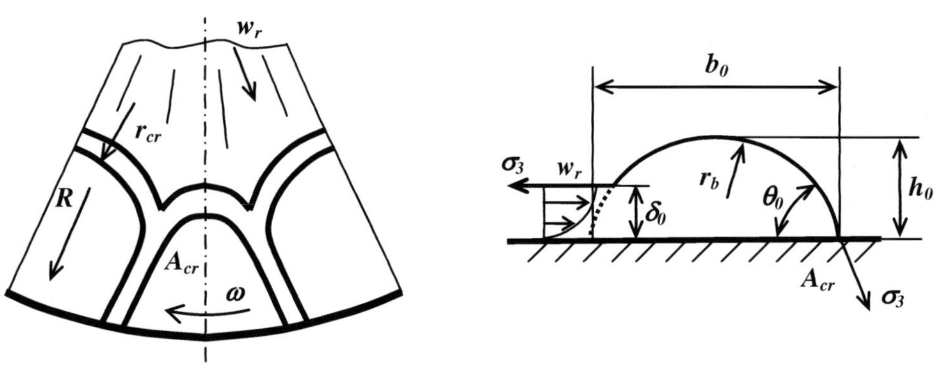

Fig.3. Liquid film rupture pattern

The equation of moments for the equilibrium state of the roller, taking into account the surface, centrifugal forces and the force of the velocity head, we obtain:

$$M_{\sigma} + M_{\omega} + M_{w} = 0 \tag{16}$$

$$M_{\sigma} = \sigma_3 \delta_0$$

$$M_{\omega} = -\int_0^{h_0} \rho b_0 \omega^2 r h dh$$

$$M_{w} = -\int_0^{\delta_0} \rho \frac{W_r^2}{2} \delta d\delta$$

We believe that the roller has a cylindrical surface with a radius r_b, with r_b << R and

$$h_0 = r_b (1 - \cos\Theta_0)$$

For determining M_w the velocity profile is used for the case of one-dimensional fluid flow from [14]. As a result, the equation of moments for the roller can be represented as:

$$\sigma_3\delta_0 - A_3 \delta_0^2 \overline{W}_r^2 \rho \left(\frac{wr}{4}\right)^2 - A_4 r_b^3 \rho w^2 r = 0 \tag{17}$$

where $A_3 = \dfrac{(21n^3 + 16n^2 + 5n + 2)(2n+1)}{4(3n+1)(3n+2)(n+1)^2}$, $A_4 = (\sin\Theta_0 - \Theta_0 \cos\Theta_0 - \frac{1}{3}\sin^3\Theta_0)$

To determine the roller radius r_b using the Laplace equation [17], according to which the relationship between the average overpressure under the roller and the curvature of the roller can be represented as:

$$p_{вс} = \sigma_3 / r_b \tag{18}$$

Determine the excess pressure $p_{вс}$ as the sum of pressures from velocity head and centrifugal force:

$$p_{вс} = p_{wc} + p_{цб} = \frac{1}{\delta_0}\int_0^{\delta_0} \frac{\rho w_r^2}{2} d\delta + \frac{1}{h_0}\int_0^{h_0} \frac{1}{b_0}\int_0^{b_0} \rho b w^2 r db dh \tag{19}$$

Integration of equation (19) with regard to the velocity profile and the roller configuration allows determining the radius of the roller r_b from (18) in the form:

$$r_b = A_5 \left(\frac{\sigma_3}{\rho w^2 r}\right)^{1/2} \tag{20}$$

where $A_5 = \left[\dfrac{2(1-\cos\Theta_0)}{(\Theta_0 - \sin\Theta_0 \cos\Theta_0)}\right]^{1/2}$

The joint solution of equations (17), (20) and (7) allows to determine the radius of rupture of a liquid film:

$$\bar{r}_K = \left[A_6 \left(\frac{\rho w r}{\sigma_3} \right)^n \left(\frac{w^n K R}{\sigma_3} \right) \frac{1}{W_{l_y}^{\ 0,5}} \right]^2 \tag{21}$$

In the dimensionless parameters we get:

$$\bar{r}_K = A_6^2 (2n)^{2n} J^{2n} \Phi^2 \frac{1}{W_{l_y}}$$

where $A_6 = \left(\frac{2n+1}{2\pi n} \right)^n \frac{1}{A_4^{2n+1}} \left(\frac{(\Theta_0 - \sin\Theta_0 \cos\Theta_0)}{2(1-\cos\Theta_0)} \right)^{\frac{3}{2(2n+1)}}$

Given that the case $\bar{r}_K = 1$ corresponds to the minimum irrigation density from equation (21) we get:

$$\Gamma_{\min} = A_7 \sigma_3^{\frac{2n+1}{n}} \rho^{\frac{1}{2n}} / \left(w^2 R \right)^{\frac{2n-1}{2n}} K^{\frac{1}{n}} \tag{22}$$

where $A_7 = \left(\frac{2}{2n+1} \right) \cdot A_4^{\frac{2n+1}{n}} \left(\frac{\Theta_0 - \sin\Theta_0 \cos\Theta_0}{2(1-\cos\Theta_0)} \right)^{\frac{3(2n+1)}{n}}$

Let us compare the equations obtained for the discontinuity radius based on the energy and moment methods. The ratio of the right parts of equations (14) and (21) gives:

$$\frac{\bar{r}_K^{\,M}}{\bar{r}_K^{\,Э}} = A_8 \left[J^{3n-2} \Phi^3 \frac{1}{W_{l_y}} \right]^{\frac{2n+1}{3n+1}} \tag{23}$$

where $A_8 = A_7^2 (2\pi)^{2n} (2nA_2)^{\frac{3n+2}{3n+1}}$

Calculations show that $\frac{\bar{r}_K^{\,M}}{\bar{r}_K^{\,Э}}$ may be either more or less than one. In experimental conditions, in the range of change of technological parameters in the operating conditions of these devices, the condition $\frac{\bar{r}_K^{\,M}}{\bar{r}_K^{\,Э}} < 1$ hard to reach. Therefore, to analyze the operation of centrifugal film apparatus, we propose the moment method. In this case, the energy method should be looked at as an analysis of the causes of the gap.

We carried out experimental studies on a flat disk for a 90% aqueous solution of glycerol (Fig.4) and for a 2.5% aqueous solution of CMC (Fig.5).

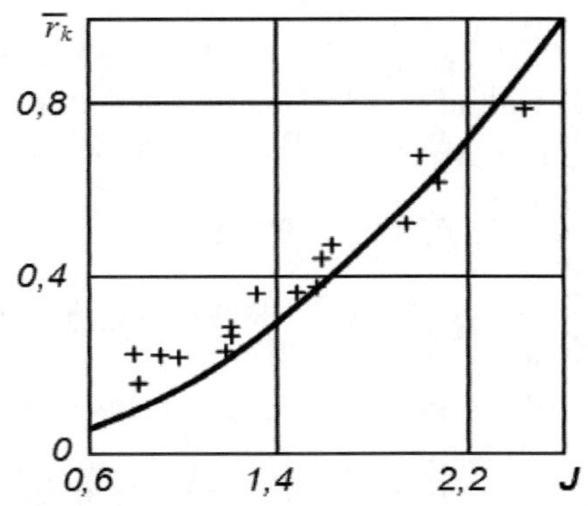

Fig.4. The dependence of the dimensionless critical radius of the dimensionless complex J: with n=1, К=1,8·10^{-1} Нс^n m^2, ρ=1,23·10^3 kg/m^3, R=12,5·10^{-2} m, σ=48,7·10^{-3} N/m, Θ_0=1,223 rad,

$\alpha = \frac{\pi}{2}$ rad, q=(0,55÷1,7) ·10^{-6} m^3/sec, ω=30÷120 rad ·sec

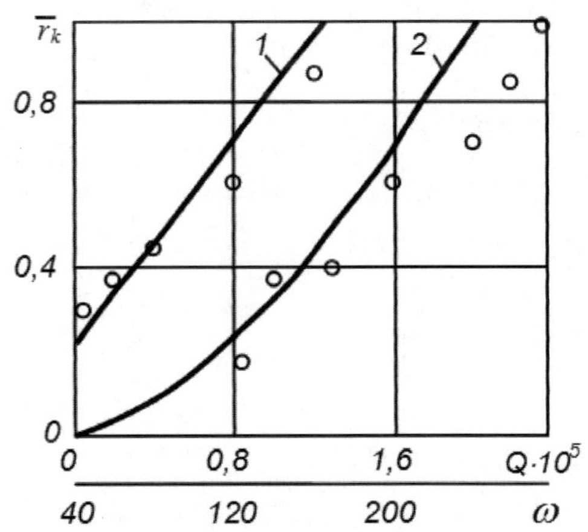

Fig.5. The dependence of the dimensionless critical radius of the flow rate and angular velocity ω: with n=0,77, K=1,57·10^{-1} Nsec^n m^2, ρ=1,017·10^3 kg/m^3, R=12,5·10^{-2} m, Θ_0=1,024 rad,

σ=53,3·10^{-3} N/m, $\alpha = \frac{\pi}{2}$ rad, $1-\bar{r}_K = f(\omega)$ with q=1·10^{-6} m^3/sec, $2-\bar{r}_K = f(q)$ with ω=60 rad ·sec

The rheological constants K and n were determined by the method of two capillaries of different lengths (excluding the initial section) on a constant pressure viscometer, ρ – densimeter, σ_3 – by the method of ring break, Θ_0 - largest lifting fluid when wetting the sample. Experiments were carried out on the setup described in [5]. In each experiment, a significant zone of deviations of the rupture radius was observed (~ 15%). For the experimental adopted their average value.

As can be seen from the graphs, there is a satisfactory agreement between the experimental and calculated data.

The study was carried out by a grant from the Russian Science Foundation (Project No. 18-79-10136).

IOP Conf. Series: Materials Science and Engineering **570** (2019) 012106 doi:10.1088/1757-899X/570/1/012106

References

[1] Razinov A And, Klinov A B, Dyakonov G S 2017 Kazan 860 p
[2] Zinnatullin NH, Musin DT, Gimranov FM, Sabirova NK 2004 Kazan 116 p
[3] Davydov AV, Bronkaya VV, Zinnatullin NX 2012 Herald of Kazan Technological University **15** **23** pp 145-47
[4] Zinnatullin NH, Antonov VV, Nafikov IM, Bulatov AA 1996 Engineering Physics Journal **69** 1 pp 112-17
[5] Zinnatullin N X, Nafikov I M 1984 News of higher educational institutions Ser. Chemistry and Chemical Technology **27** 6 pp 722-26
[6] Hartley D E, Murgatroya W 1964 Jnt. J. Heat Mass Transfer **7** 9 pp 1003-15
[7] Hobler T 1968 Chemia Stosowana **38** pp 265-74
[8] Mikielewicz J, Moszynski J R 1976 Jnt. J. Heat Mass Transfer **19** **7** 771-76
[9] Conlson H 1973 Chem. Jug. Tech. **45** **6** pp 362-68
[10] Ganiev BG, Bokov AE 1980 Physics Engineering Magazine 39 4 pp 581-91
[11] Charwat A F, Kelly R E, Gazley C 1972 J. Fluid Mech. **53** **2** pp 227-58
[12] Zinnatullin NH, Bulatov A.A., Nikolaev.SG 1994 Interuniversity collection: Mass transfer processes and apparatuses of chemical technology pp 97-101
[13] Zinnatullin NH, Vachagin KD, Tyabin NV 1965 Proceedings of the Kazan Institute of Chemical Technology 35 pp 146-53
[14] Zinnatullin NH, Vachagin KD, Tyabin NV 1968 Engineering Physics Journal 15 2 234-240
[15] Summ BA, Goryunov Yu V 1976 Moscow 231 p
[16] Zimon AD 1974 Moscow 195 p
[17] Levin VG 1959 Moscow 699 p

IOP Conf. Series: Materials Science and Engineering **570** (2019) 012107 doi:10.1088/1757-899X/570/1/012107

Application of neural networks in object recognition tasks for ADAS systems

R R Ziyatdinov, R A Biktimirov

Naberezhnye Chelny Institute KFU, Naberezhnye Chelny, Russia (Russia, 423810, Naberezhnye, Mira Ave., 68/19)

rust.kfu@gmail.com

Kazan Federal University, 18 Kremlyovskaya street, Kazan, 420008, Russian Federation

Abstract. Modern driver assistance systems (ADAS) require an environmental recognition function to inform the driver and for making management decisions. Neural networks are used to select and recognize objects in such systems. The paper presents the results of comparative analysis of various neural networks in object recognition problems. Experimental data showed that convolutional neural networks show the best results in recognition problems.

1. Introduction

Currently, the development of various driver assistance systems (ADAS systems) that improve the quality and safety of vehicle management is relevant. The standards define 6 levels of ADAS automation: from 0 (no automation) to 5 (full automation). Any of these levels involves the use of recognition systems of surrounding objects.

Various artificial neural networks are used in pattern recognition problems [1-5]. Most often, classical neural networks (network of radial basis functions, multilayer perceptron, etc.) are used. However, the analysis of experimental data obtained using such networks shows that the use of classical neural network architectures in recognition problems has little efficiency for the following reasons:

- recognized images are often large, which leads to an increase in the structure of the neural network;
- a large variety of parameters increases the size of the system, which results in an increased need for a larger training set and, therefore, an increase in the complexity of calculations and the time required for training the system;
- multiple neural networks are required to ensure efficient operation of the recognition system. This leads to an increase in the complexity of the task and the time of its execution;
- high sensitivity to various changes in the geometry of the input images, such as changes in the image angle, image scale, and other distortions [6].

In this regard, in addition to the multilayer perceptron, convolutional neural networks were used in the work. Such networks provide some resistance to small changes in scale, change of perspective, swerve, displacement and other distortions [7].

2. Comparative analysis

IOP Conf. Series: Materials Science and Engineering **570** (2019) 012107 doi:10.1088/1757-899X/570/1/012107

For comparative analysis we used the multilayer perceptron with elastic-propagation (RProp, 3 and 4 layers), multilayer perceptron with back-propagation (BackProp, 3 and 4 layers) and a convolutional neural network (CNN).

In the problem of image recognition (on the example of the base of handwritten numbers MNIST and the base of road signs GTCRB), the model YOLOv2 was used, which was designed to work with convolutional neural networks. It allows you to detect objects in real time, work with images of different sizes, offering a compromise between speed and accuracy.

the results of using convolutional neural network for recognition of handwritten digits: the ordinate axis indicates the number of correct recognition of 10,000 samples provided by the network, the axis of abscissa indicates iteration of training.

When using MNIST database, the highest accuracy for convolutional neural network was achieved at 1900 iterations with the number of correct recognitions at 9345. Accordingly, the recognition accuracy at this step is 93.5%. When using the road signs GTSRV recognition accuracy amounted to 98.6%.

The averaged test results of correctly recognized objects for various machine learning algorithms in the aggregate for two tests (the MNIST and GTSRB bases) are shown in Figure 2.

For the case of a multilayer perceptron, the most accurate was an architecture with 4 layers (2 hidden layers) with back propagation of error (MLP BackProp). However, the accuracy of this algorithm is extremely low (21.7%), which is insufficient for practical use in recognition systems.
For a convolutional neural network (CNN), the averaged result was 96%, which is the maximum in this study.

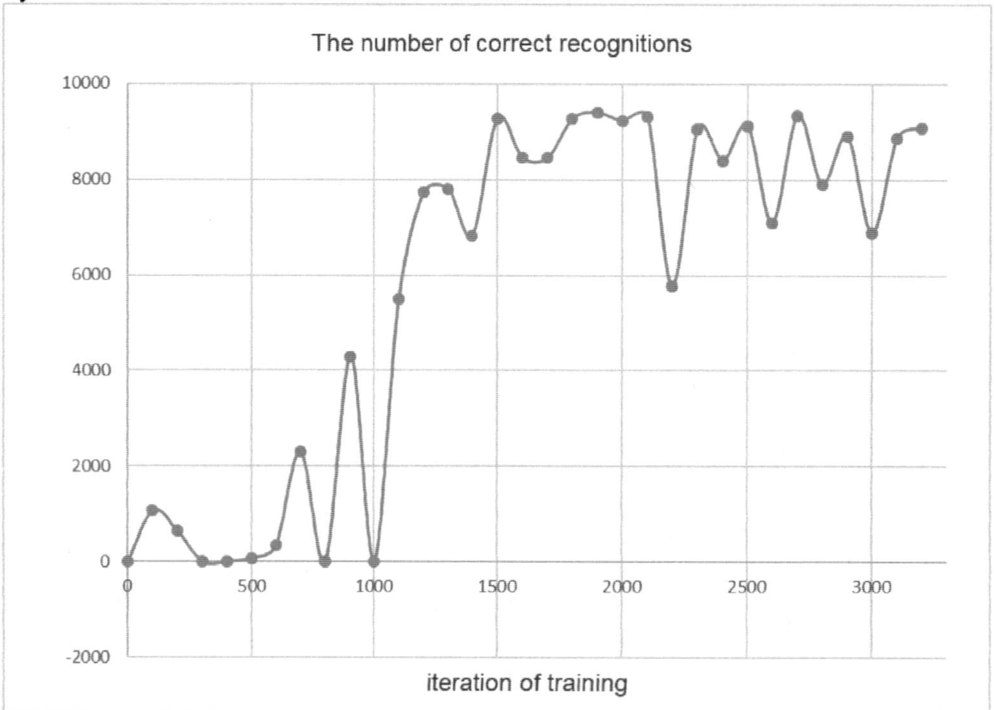

Figure 1 - The number of correct recognitions depending on the iteration of training of the convolutional neural network

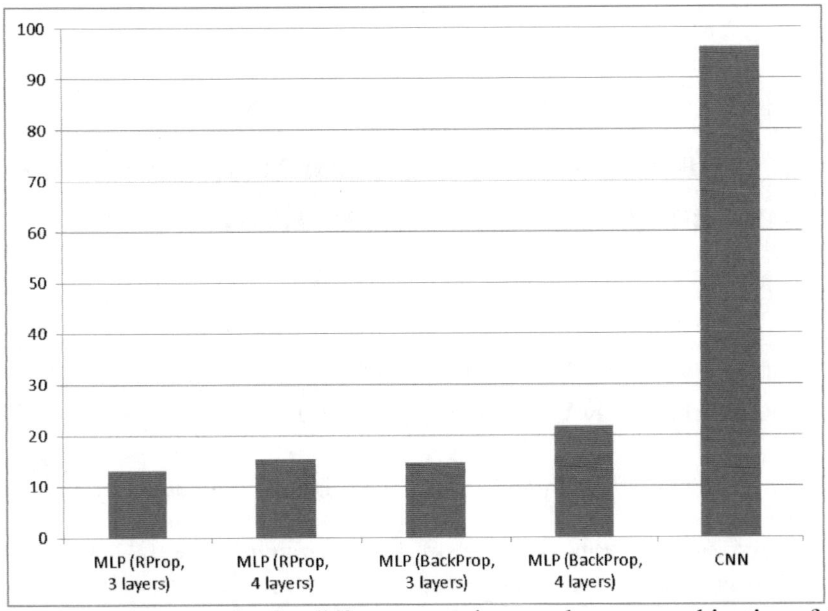

Figure 2 - Averaged test results for different neural networks on a combination of two tests

3. Conclusions

From the combination of two tests for the classical neural network, the most accurate was an architecture with 4 layers with back propagation of error. However, its accuracy was not sufficient for practical use. For a convolutional neural network, the average result was approximately 96%. Thus, for the practical implementation of the object recognition system in ADAS systems, convolutional neural networks are more applicable.

References

[1] Ziyatdinov R R, Shigabiev R R, Talipov D N 2017 *Automated road marking recognition system* Proceedings: IOP Conference Series: Materials Science and Engineering **012070**

[2] Ziyatdinov R R, Biktimirov R A 2017 *Automated road sign recognition system for ADAS systems* Proceedings: Innovative machine-building technologies, equipment and materials - 2017 ("IMTEM-2017") pp 153-58

[3] Shigabiev R R, Ziatdinov R R 2016 *The System of the lane-keeping assist with the use of neural network* Proceedings:: Information technologies. Automation. Updating and solution of problems of training of highly qualified personnel (ITAP-2016) pp 88-92

[4] Ziyatdinov R R, Biktimirov R A and Klochkova K V 2017 *The comparative analysis of classification of algorithms in the problems of pattern recognition* Journal of Fundamental and Applied Sciences V 9 pp 1790-97

[5] Ziyatdinov R R, Mukhametzyanov V R and Nabiullina G I 2015 *Application of neural networks for the diagnosis of depth sucker rod pumps* International Journal of Applied Engineering Research V 10 Is 24 pp 45022-26

[6] Yu N, Notkin B S, Sedov V A 2009 *Neuroiterative algorithm of tomographic reconstruction of the distributed physical fields in the fibreoptic measuring systems* Computer optics V 33 No 4 pp 446–55

[7] Biktimirov R A 2018 *Development of an intelligent pedestrian recognition system for ADAS systems: graduation thesis* Naberezhnye Chelny Institute of Kazan Federal University, Naberezhnye Chelny

IOP Conf. Series: Materials Science and Engineering **570** (2019) 012108 doi:10.1088/1757-899X/570/1/012108

Performance-based seismic evaluation methods for the estimation of inelastic deformation demands

M A Zubritskiy[1], O Y Ushakov[2], L S Sabitov[3,4]

[1] LLC «UralConceptProject», 22 Roza Luxemburg street, Ekaterinburg, 620075, Russia
[2] Ural State University, 19 Mira street, Ekaterinburg, 620002, Russia
[3] Kazan State Power Engineering University, 51 Krasnoselskaya street, Kazan, 420066, Russian Federation
[4] Kazan Federal University, 18 Kremlyovskaya street, Kazan, 420008, Russian Federation

zubrickiy_maksim@mail.ru

Abstract. The article presents experimental research and comparison of two performance-based seismic evaluation methods: nonlinear static procedure (NSP) / pushover analysis (PoA) and nonlinear time history analysis. It is established that nonlinear static procedure is good enough for prediction of plastic hinges localization, but difference between methods results is more than 15 % with a lack of seismic resistance.

1 Introduction

Performance-based method is focused on performance required in use for the business processes and the needs of the users, and then on the evaluations and verification of building assets result. It is the modern approach to earthquake resistant design.

Time history analysis is a rigorous numerical method by integrating differential equation of motion directly. Full time history give the response of a structure over time during and after the dynamic loads application represented by accelerogram (Fig. 1).

Nonlinear static analysis using pushover procedures are becoming increasingly common in engineering practice for building structures seismic evaluation. Seismic demands are computed by nonlinear static analysis of the structure, which is subjected to monotonically increasing lateral forces with an invariant height-wise distribution until a target displacement is reached. The purpose of pushover analysis is to estimate the expected performance of a structural system by evaluating its strength and deformation demands under seismic loads by means of a static nonlinear analysis, and comparing these demands to available capacities at the targeted performance levels. The result of the structure analysis is a Base Shear-Displacement diagram, the global force-displacement curve or capacity curve of the structure (Fig.2). This capacity curve provides valuable information about the response of the structure because it approximates how it will behave after exceeding its elastic limit. The capacity curve has several characteristic points which are represented in Fig 3.

IOP Conf. Series: Materials Science and Engineering **570** (2019) 012108 doi:10.1088/1757-899X/570/1/012108

Comparing two methods for seismic analysis its clear seems the time-history analisys is relatively more time consuming and costly.

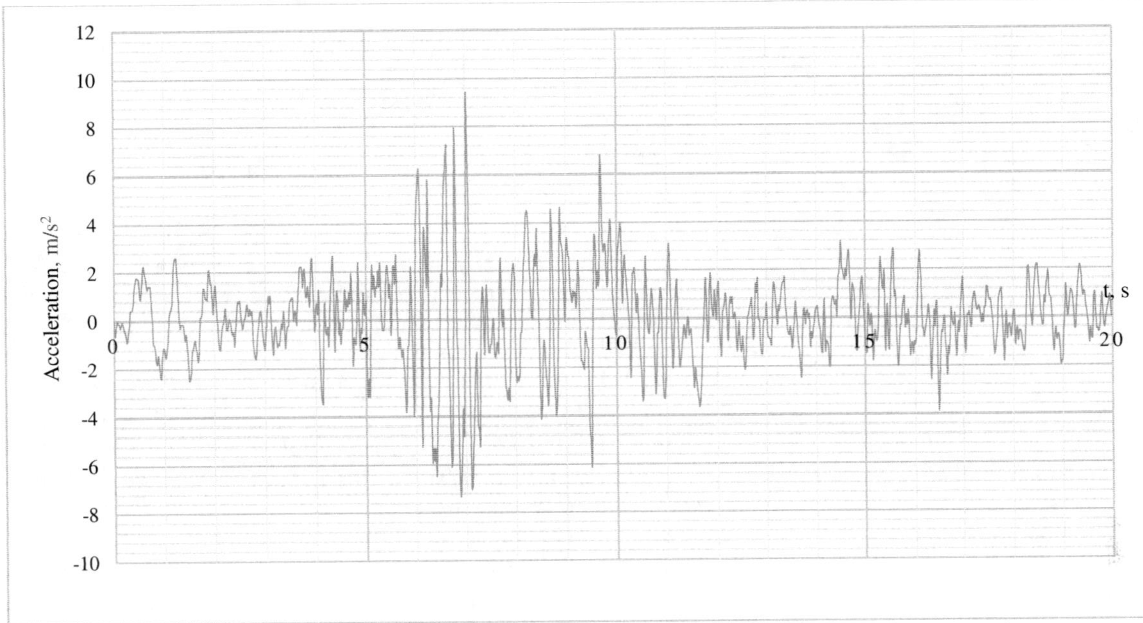

Figure 1. Accelerogramm radial component of the 1978 Tabas, Iran, Earthquake

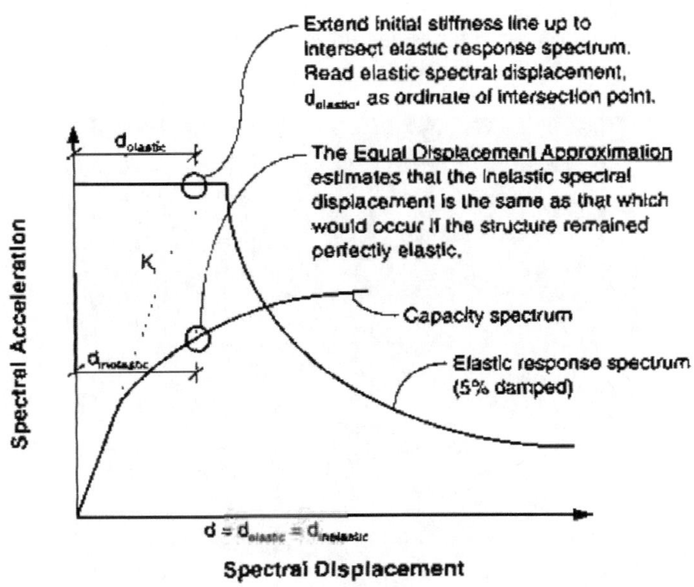

Figure 2. Targert displacement estimation using nonlinear static method (ATC-40)

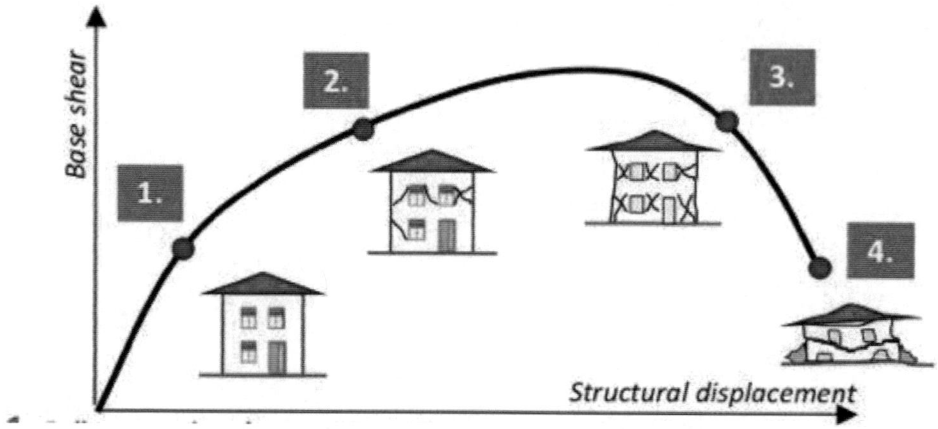

Figure 3. Capacity curve characteristic points:
1. Fully operational: Continuous service. Negligible structural and nonstructural damage. 2. Operational: Most operations and functions can resume immediately. Structure safe for occupancy. Essential operations protected, non-essential operations disrupted. Repair required to restore some non-essential services. Damage is light. 3. Life Safety: Damage is moderate, but structure remains stabl. Selected building systems, features, or contents may be protected from damage. Life safety is generally protected. Building may be evacuated following earthquake. Repair possible, but may be economically impractical. 4. Near Collapse: Damage severe, but structural collapse prevented. Nonstructural elements may fall. Repair generally not possible

2 Problem statement

Three masses 9-meters high column was selected as representative case study to carry out the performance-based seismic methods evaluation.

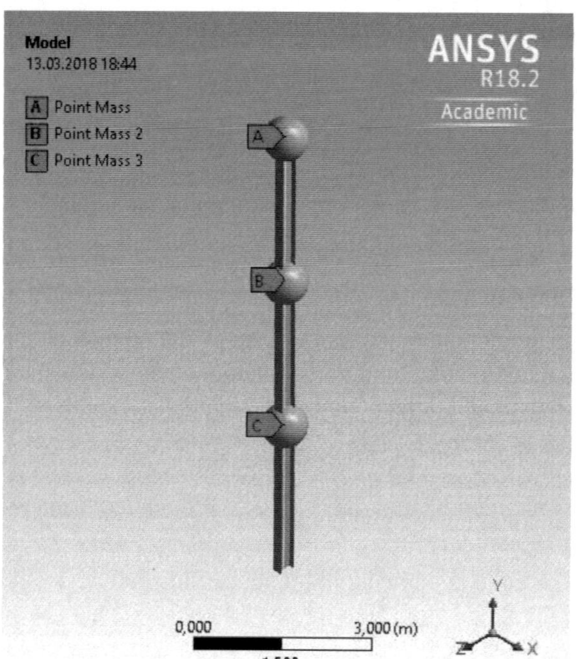

Figure 4. Dynamic model general view

IOP Conf. Series: Materials Science and Engineering **570** (2019) 012108 doi:10.1088/1757-899X/570/1/012108

As a construction material structural steel was chosen. Stress-strain diagram is shown on Fig. 5. To describe the non-linear behavior of the system elements the model of Bilinear Kinematic Hardening has been adopted. The diagrams of steel deformation under tension and compression are the same. The yield surface is described by the Von-Mises criterion and is a cylinder whose axis coincides with the axis of hydrostatic compression in the axes of the main stresses (Fig. 6). Damping parameters were calculated based on 1st and 3rd natural vibration frequency. Dynamic model characteristics are shown in Table 1.

The seismic excitation used for nonlinear time history and pushover evaluations is defined by a set of three strong ground motions:

1. Iran, 1978 г. (Erthq. 1);
2. El Centro, USA (California), 1979 г. (Erthq. 2);
3. Duzce, Turkey, 1999 г. (Erthq. 3).

Accelerogram records were taken from [14].

Table 1. Dynamic model characteristics

№	Nomination	Value		
		Erthq. 1	Erthq. 2	Erthq. 3
1	Cross-section, *mm*	*I-beam 300(h)x200(b)x15(b_f)x8(b_w)*		
2	Height, *mm*	*9000*		
3	Young modulus, *Pa*	$2e^{11}$		
4	Yield point, *MPa*	*270*		
5	Tangential modulus, *MPa*	$5.361e^{3}$		
6	Masses $m_a = m_b = m_c$, *kg*	*3000*	*10000*	*7000*
7	1st natural vibration frequency f_1, *Hz*	*0.853*	*0.45094*	*0.5637*
8	2nd natural vibration frequency f_1, *Hz*	*5.397*	*2.8677*	*3.5813*
9	3rd natural vibration frequency f_1, *Hz*	*13.799*	*7.3161*	*9.1411*
10	Damping parameter α_R, s^{-1}	*0.241*	*0.507*	*0.328794*
11	Damping parameter β_R, *s*	*0.00052*	*0.00234*	*0.001616*

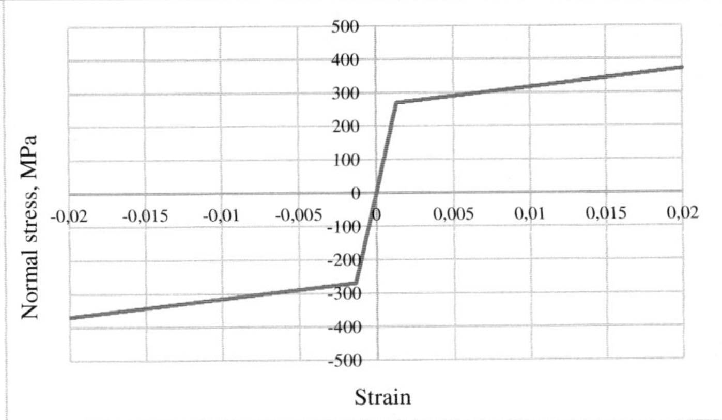

Figure 5. Stress-strain diagram of steel

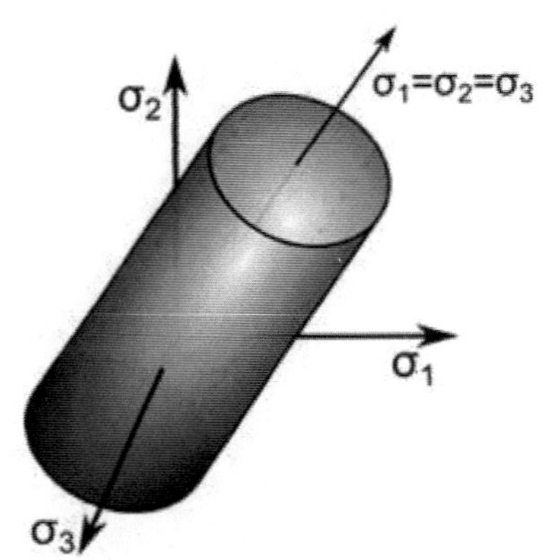

Figure 6. Von-Mises yield surface in the axes of the main stresses

IOP Conf. Series: Materials Science and Engineering **570** (2019) 012108 doi:10.1088/1757-899X/570/1/012108

3 Results

Seismic response were estimated by following parametrs set :
- Nodes horizontal displacements;
- Maximum bending moment ;
- Maximum shear forces.

Figures 7-9 show the top point horizontal displacement, bending moments and shear forces near the anchorage obtained from nonlinear time history analyses of Iran eathquaqe ground motions, respectively.

For the nonlinear static procedure «Pushover» module of software package Lira 10.6 were used. The characteristic point was obtained according to grapho-analytical method described in [3].

Numerical simulation results are summarized in Table 2.

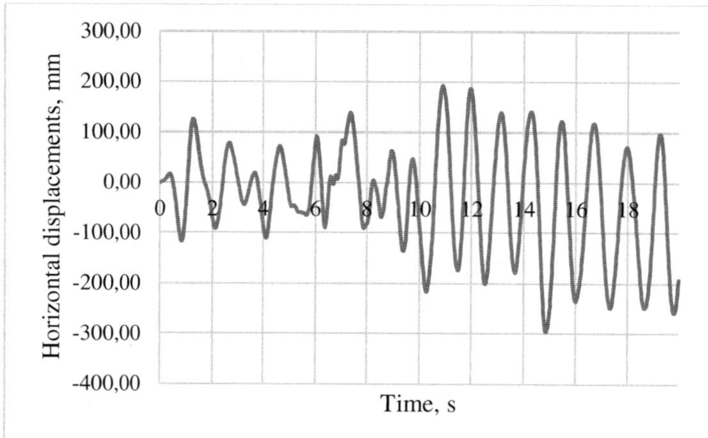

Figure 7. Horizontal displacements of the top point under seismic impact Erthq.1 [Δ_{max} = -295.70 mm at t = 14.88 s]

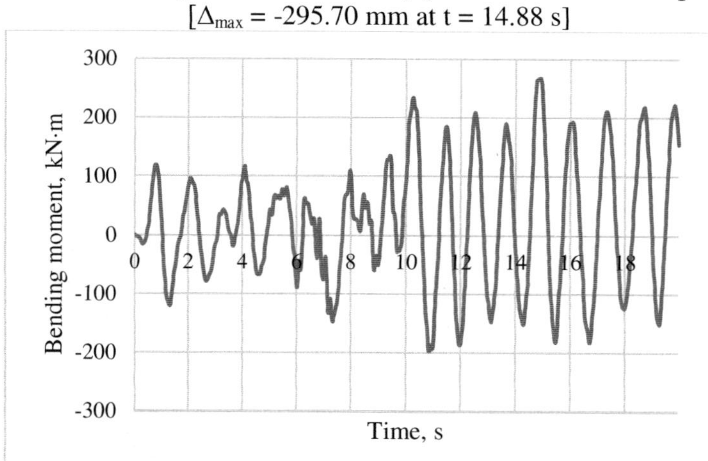

Figure 8. Bending moment near the anchorage under seismic impact Erthq.1 [M_{max} = 267.38 kN·m at t = 14.91 s]

ISTC-IETEM IOP Publishing
IOP Conf. Series: Materials Science and Engineering **570** (2019) 012108 doi:10.1088/1757-899X/570/1/012108

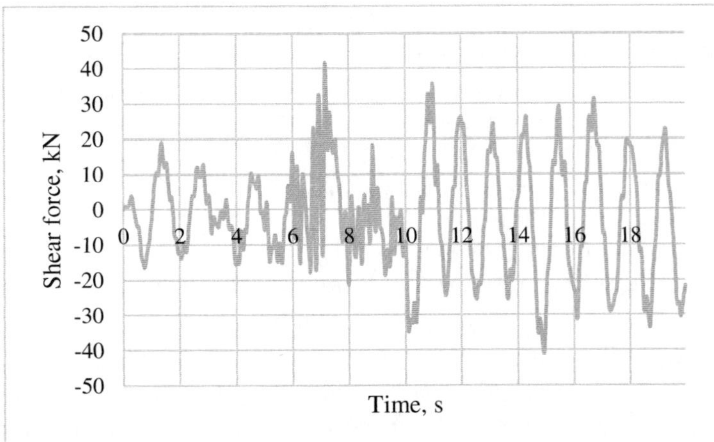

Figure 9. Shear force near the anchorage under seismic impact Erthq.1
[$Q_{max,1}$ = -41.19 kN at t = 14.92 s / $Q_{max,2}$ = -41.64 kN at t = 7.17 s]

Table 2. Analysis result comparison

№	Nomination	Value					
		Erthq.1		Erthq.2		Erthq.3	
1	Top point maximum horizontal deformation, *mm*	*-295.70* *-204.34*	*-30.9%*	*-282.15* *-274.21*	*-2.81%*	*-268.25* *-226.54*	*-15.55%*
2	Middle point maximum horizontal deformation, *mm*	*-163.80* *-108.23*	*-33.9%*	*-151.84* *-144.61*	*-4.76%*	*-140.87* *-119.72*	*-15.01%*
3	Lower point maximum horizontal deformation, *mm*	*-55.10* *-31.86*	*-42.2%*	*-46.79* *-44.38*	*-5.15%*	*-46.55* *-35.23*	*-24.32%*
4	Maximum bending moment near the anchorage, *kN·m*	*267.38* *235.24*	*-12.0%*	*270.3* *275.63*	*1.93%*	*271.54* *244.07*	*-10.12%*
5	Maximum shaer force near the anchorage, *кН*	*41.19* *31.43*	*-23.7%*	*43.73* *36.786*	*-15.8%*	*43.68* *32.58*	*-25.41%*
1. Numerator – nonlinear time history analysis; , denominator – nonlinear static procedure 2. Accuracy with a sign «-» means seismic resistance lack							

Conclusions

We can draw the following conclusions based on the completed research:

- Maximum horizontal displacements values differ by more than 10%. The difference between the horizontal displacements of the system nodes with decreasing height increases. The maximum difference of results is about 42%.
- The plastic hinges formation and structural failure are identical for different methods of seismic estimation.
- Seismic resistance lack in the nonlinear static procedure estimation is caused by the underestimation of higher vibration modes.
- Traditional Pushover analysis performed on the 1st vibration mode can only be used for low-rise buildings and low structures.
- To assess the high-rise buildings and structures seismic resistance modification of the non-linear static method are needed in terms of the distribution of inertial forces of the system taking into account the required number of higher vibration modes.

IOP Conf. Series: Materials Science and Engineering **570** (2019) 012108 doi:10.1088/1757-899X/570/1/012108

– New method development are needed to combine the inertial forces or the analysis results for each of the natural vibration forms for NSP, constructing a capacity curve and method for characteristic point searching. The obtained by new proposed method results_ should be verified_by the results obtained by the nonlinear time history analysis.

References

[1] SP 14.13330.2014. *Seismic building design code*

[2] STO NRU MGSU 2015. *Seismic resistance of buildings and structures*

[3] *Eurocode 8: Seismic Design of Buildings*. EUR 25204 EN – 2012

[4] Birbraer A N 1988 *Calculation of structures for seismic resistance* SPb

[5] O V Mkrtychev 2010 *Safety of buildings and structures during seismic and emergency impacts*

[6] O V Mkrtychev, G A Ginchvelashvili 2012 *Problems of accounting for nonlinearities in the theory of seismic resistance (hypotheses and delusions)*

[7] Kurzanov A M 2009 *The current state of rationing the calculation of structures for seismic load // Industrial and civil engineering* No 11 pp 52-53

[8] Polyakov S 1983 Earthquake-resistant structures of buildings: textbook for universities M 1983

[9] M F Barstein, N M Borodachev, L H Blumin and others 1981 *Dynamic calculation of structures for special dynamic effects*

[10] Kalkan E and Kunnath S K 2004 *Method of modal combinations for pushover analysis of buildings* Proc. of the 13th World Conference on Earthquake Engineering Vancouver, BC, Canada.

[11] *Applied Technology Council (ATC-40)* 1996 Seismic evaluation and retrofit of concrete buildings, Redwood, CA.

[12] FEMA-356 2000 *Prestandard and Commentary for the seismic rehabilitation of buildings, American Society of Civil Engineers (ASCE)* Reston VA

[13] Chopra A K, Geol R K 1999 *Capacity-demand-diagram methods based on inelastic design spectrum* Earthquake Spectra pp 637-56

[14] Chopra A 1995 *Dynamics of Structures: Theory and Applications to Earthquake Engineering* Prentice Hall

[15] Chopra A K, Goel R K 2000 *Evaluation of NSP to estimate seismic deformation: SDF systems* Journal of Structural Engineering 126(4) pp 482-90

[16] PEER Ground Motion Database // https://ngawest2.berkeley.edu/

IOP Conf. Series: Materials Science and Engineering **570** (2019) 012109 doi:10.1088/1757-899X/570/1/012109

Account for the contribution of higher modes under system seismic resistance estimation by nonlinear static method

M A Zubritskiy[1], O Y Ushakov[2],L S Sabitov[3,4]

[1] LLC «UralConceptProject», 22 Roza Luxemburg street, Ekaterinburg, 620075, Russia
[2]Ural State University, 19 Mira street, Ekaterinburg, 620002, Russia
[3] Kazan State Power Engineering University, 51 Krasnoselskaya street, Kazan, 420066, Russian Federation
[4]Kazan Federal University, 18 Kremlyovskaya street, Kazan, 420008, Russian Federation

zubrickiy_maksim@mail.ru

Abstract. The article presents higher modes accounting method under estimation of system seismic resistance by nonlinear static method. As part of the study, in order to verify the proposed method for finding the inertial forces modified system a complex of dynamic and static calculations was performed.It is established that proposed inertial forces modified system can significantly reduce system seismic resistance lack.

1Introduction

To find the most unfavorable system response by a nonlinear static method, it is necessary to perform at least two calculations with different inertial horizontal forces distribution along the system height:

- The pattern, based on lateral forces corresponding to the system natural vibrations main form. Seismic evaluation of the system with a given force distribution can be estimated only if the modal mass of the considered form is at least 75%;
- The pattern, based on inertial forces superposition of several vibration forms.In this case, the total modal mass of the considered vibration forms should not be less than 90%;
- The "uniform" pattern, based on lateral forces that are proportional to mass regardless of elevation (uniform response acceleration);
- The "modal" pattern, proportional to lateral forces consistent with the lateral force distribution determined in elastic analysis.

Further characteristic point search on the capacity curve is a laborious process required cumbersome graphs plotting and non-trivial calculations. In modern software package, a non-linear static method is implemented only based on only one inertial forces vibration mode. In addition, the system seismic response for a given impact is significantly less than the dynamic method results.

2 Higher vibrations forms accounting method

To determine the response of the system, taking into account the influence of higher vibration modes, we consider the following method with modified system of inertial forces.

IOP Conf. Series: Materials Science and Engineering **570** (2019) 012109 doi:10.1088/1757-899X/570/1/012109

The inertial forces modified system is the system based on forces superposition the SRSS-method, in which the top displacement of considered model will correspond to the total displacements obtained by response spectrum analysis.

According to [5] to destroy a material, no matter what the load is applied, it is necessary to expend the same amount of energy. Thus, the linear system deformation energy with ainertial forces modified system is identical to the system deformation energy, taking into account plastic deformations.

The seismic resistance evaluation next stage is to plot the capacity curve "Force at the base V_b - Displacement of the system top Δ" based on nonlinear static calculation for the system with one degree of freedom under the modified inertial forces system .

3 Problem statement and method testing

Three masses 9-meters high column was selected as representative case study to carry out the performance-based seismic methods evaluation.

As a construction material structural steel was chosen. Stress-strain diagram is shown on Fig. 2. To describe the non-linear behavior of the system elements the model of Bilinear Kinematic Hardening has been adopted. The diagrams of steel deformation under tension and compression are the same. The yield surface is described by the Von-Mises criterion and is a cylinder whose axis coincides with the axis of hydrostatic compression in the axes of the main stresses (Fig.3). Damping parameters were calculated based on 1st and 3rd natural vibration frequency. Dynamic model characteristics are shown in Table 1.

The seismic excitation used for nonlinear time history and pushover evaluations is defined by a set of three strong ground motions:

1. Iran, 1978 г. (Erthq. 1);
2. El Centro, USA (California), 1979 г. (Erthq. 2);
3. Duzce, Turkey, 1999 г. (Erthq. 3).

Accelerogram records were taken from [14].

Table 1. Dynamic model characteristics

№	Nomination	Value		
		Erthq. 1	Erthq. 2	Erthq. 3
1	Cross-section,*mm*		*I-beam 300(h)x200(b)x15(b_f)x8(b_w)*	
2	Height, *mm*		*9000*	
3	Youngmodulus, *Pa*		$2e^{11}$	
4	Yieldpoint, *MPa*		*270*	
5	Tangential modulus, *MPa*		$5.361e^{3}$	
6	Masses $m_a = m_b = m_c$, *kg*	*3000*	*10000*	*7000*
7	1st natural vibration frequencyf_1, *Hz*	*0.853*	*0.45094*	*0.5637*
8	2nd natural vibration frequencyf_1, *Hz*	*5.397*	*2.8677*	*3.5813*
9	3rd natural vibration frequencyf_1, *Hz*	*13.799*	*7.3161*	*9.1411*

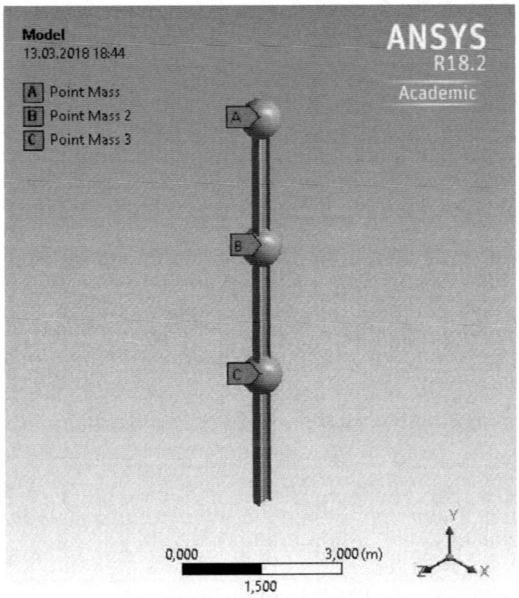

Figure 1. Dynamic model general view

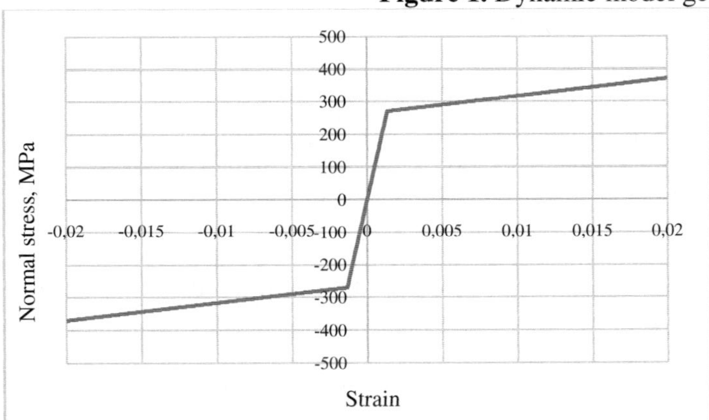

Figure 2. Stress-strain diagram of steel

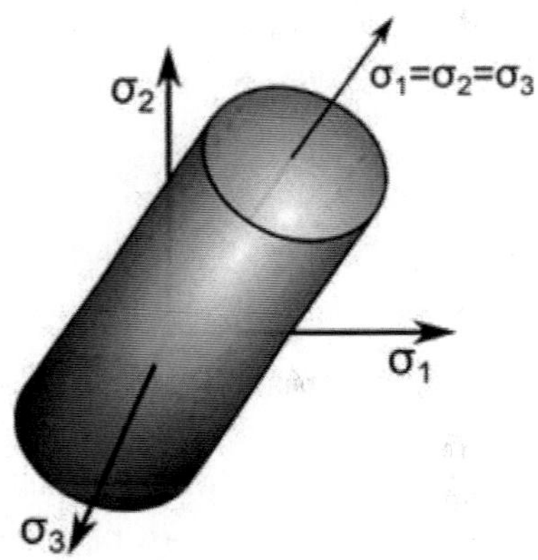

Figure 3. Von-Mises yield surface in the axes of the main stresses

To calculate the forces system for a given system during the seismic evaluation by the multimodal nonlinear static method, the initial data are the inertial forces of the first three vibration modes.

The inertial forces distribution and the forces resulting system for seismic impact Iran, 1978 (Erthq1) are obtained byresponse spectrum analysis.

To calculate the inertial force reduction coefficient, a linear static analysis was performed. The resulting coefficient value and the system's energy consumption is presented in Table 2.

The next step in seismic evaluation is to perform a non-linear static calculation under the action of a inertial forces modified system. Then the capacity curve is plotted in the coordinates of the "shearing force at the base — the top displacement of the system".

The characteristic point search is iterative: it is necessary to find such a point on the capacity curve so that the figure formed under the graph area corresponds to the target system energy consumption.

The results obtained for seismic impacts Erthq1..3 are summarized in Table 2.

Table 2. The results obtained by the multimodal nonlinear static method

IOP Conf. Series: Materials Science and Engineering **570** (2019) 012109 doi:10.1088/1757-899X/570/1/012109

№	Nomination	Value		
		Erthq.1	Erthq.2	Erthq.3
1	Inertial force at the upper node, *kg*	*4585.5*	*3492.1*	*3755.7*
2	Inertial force at the middle node, *kg*	*5995.0*	*4310.1*	*4529.5*
3	Inertial force at the lower node, *kg*	*5281.7*	*4112.7*	*3913.2*
4	Upper node maximum horizontal displacement obtained by response spectrum analysis, *mm*	*271.11*	*264.35*	*267.79*
5	Upper node maximum horizontal displacement obtained by static analysis with inertial forces modified system, *mm*	*717.78*	*536.55*	*565.96*
6	Reduction factorα	*0.3777*	*0.49268*	*0.47316*
7	Maximum system termination lateral force under the inertial forces reduced system action, *kN*	*58.754*	*57.57*	*56.602*
8	Potential strain energy / Energy consumption, *J*	*796.41*	*760.93*	*757.87*
9	Upper node maximum horizontal displacement obtained by multimodal pushover analysis, *mm*	*282.0*	*270.85*	*274.0*
10	Middle node maximum horizontal displacement obtained by multimodal pushover analysis, *mm*	*159.66*	*153.08*	*154.2*
11	Lower node maximum horizontal displacement obtained by multimodal pushover analysis, *mm*	*54.69*	*51.75*	*51.95*
12	Maximum bending moment near the anchorage obtained by multimodal pushover analysis, *kN·m*	*287.56*	*284.76*	*284.78*
13	Maximum shaer force near the anchorage obtained by multimodal pushover analysis, *kN*	*49.0*	*48.73*	*47.77*

4 Results

To estimate the responses error obtained by the multimodal nonlinear static method, it is necessary to compare the results with the responses obtained by the time history the direct dynamic method using the acceleration records Erthq1-3 [17].

IOP Conf. Series: Materials Science and Engineering **570** (2019) 012109 doi:10.1088/1757-899X/570/1/012109

Table3. Multimodal nonlinear static method error estimation under the seismic impact Iran, 1978 г. (Erthq1)

Parameter		Time history dynamic method	Multimodal nonlinear static method	Error, %
Horizontal displacement, *mm*	Upper node	*-295.70*	*-282.0*	*-4.63*
	Middle node	*-163.80*	*-159.66*	*-2.52*
	Lower node	*-55.10*	*-54.69*	*-0.74*
Maximum bending moment near the anchorage, *kN·m*		*267.38*	*287.56*	*+7.02*
Maximum shaer force near the anchorage, *kN*		*41.19*	*49.0*	*+15.9*

Table 4. Multimodal nonlinear static method error estimation under the seismic impact El Centro, USA (California) (Erthq2)

Parameter		Time history dynamic method	Multimodal nonlinear static method	Error, %
Horizontal displacement, *mm*	Upper node	*-282.15*	*-270.85*	*-4.00*
	Middle node	*-151.65*	*-153.08*	*+0.64*
	Lower node	*-46.79*	*-51.75*	*+9.58*
Maximum bending moment near the anchorage, *kN·m*		*270.3*	*284.76*	*+5.08*
Maximum shaer force near the anchorage, *kN*		*43.73*	*48.73*	*+10.3*

Table 5. Multimodal nonlinear static method error estimation under the seismic impactKobe, Japan, 1995 г. (Erthq3)

Parameter		Time history dynamic method	Multimodal nonlinear static method	Error, %
Horizontal displacement, *mm*	Upper node	*-268.25*	*-274.0*	*+2.10*
	Middle node	*-140.87*	*-154.2*	*+8.64*
	Lower node	*-46.55*	*-51.95*	*+10.39*
Maximum bending moment near the anchorage, *kN·m*		*271.54*	*284.78*	*+4.65*
Maximum shaer force near the anchorage, *kN*		*43.68*	*47.77*	*+8.56*

IOP Conf. Series: Materials Science and Engineering **570** (2019) 012109 doi:10.1088/1757-899X/570/1/012109

To assess the quality of the obtained data, the statistical processing was performed:

- The horizontal displacements average error is 2.16%; standard deviation - 5.59%;
- The bending moment average error is 5.58%; standard deviation -1.03%;
- The shaer force average error is 11.58%; standard deviation -3.13%.

Conclusions

Multimodal nonlinear static method was proposed to take into account the higher vibration modes influence under seismic evaluation.

According to the calculation results, it can be concluded that the proposed methodology calculation is useful.The results difference obtained by the time history dinamic methodcompared to results based on the multimodal nonlinear static method does not exceed 16%.

The developed method allows to evaluate the high-rise structures seismic resistance, as well as buildings and structures irregular in height. Comparing two methods for seismic analysis its clear seems the time-history analisys is relatively more time consuming and costly than multimodal nonlinear static method.

References

[1] *Seismic building design code* 2014SP 14.13330.2014.

[2] *STO NRU MGSU* 2015 Seismic resistance of buildings and structures

[3] *Eurocode 8: Seismic Design of Buildings* EUR 25204 EN – 2012

[4] Birbraer AN 1988*Calculation of structures for seismic resistance* SPb

[5] MkrtychevOV 2010*Safety of buildings and structures during seismic and emergency impacts*

[6] MkrtychevOV, GinchvelashviliGA 2012*Problems of accounting for nonlinearities in the theory of seismic resistance (hypotheses and delusions)*

[7] Kurzanov A M 2009*The current state of rationing the calculation of structures for seismic load*Industrial and civil engineeringNo 11 pp 52-53

[8] Polyakov S V 1983*Earthquake-resistant structures of buildings: textbook for universities*[9] BarsteinM F, BorodachevN M, BluminL H and others1981*Dynamic calculation of structures for special dynamic effects* M.

[10] Kalkan E and Kunnath S K 2004 *Method of modal combinations for pushover analysis of buildings* Proc. of the 13th World Conference on Earthquake Engineering, August 1-6, 2004, Vancouver, BC, Canada

[11] *Applied Technology Council (ATC-40)*1996 Seismic evaluation and retrofit of concrete buildings, Redwood, CA

[12] *FEMA-356*2000Prestandard and Commentary for the seismic rehabilitation of buildings, American Society of Civil Engineers (ASCE), Reston, VA.

[13] Chopra A K,Geol R K 1999*Capacity-demand-diagram methods based on inelastic design spectrum* Earthquake Spectra pp 637-56

[14] Chopra A 1995 *Dynamics of Structures: Theory and Applications to Earthquake Engineering* Prentice Hall

IOP Conf. Series: Materials Science and Engineering **570** (2019) 012109 doi:10.1088/1757-899X/570/1/012109

[15] Chopra A K, Goel RK 2000 *Evaluation of NSP to estimate seismic deformation: SDF systems*Journal of Structural Engineering 126(4) pp 482-90
[16] *PEER Ground Motion Database*https://ngawest2.berkeley.edu/
[17] ZubritskiyMA, UshakovOY, Sabitov LS 2019 *Performance-based seismic evaluation methods for the estimation of inelastic deformation demands*

IOP Conf. Series: Materials Science and Engineering **570** (2019) 012110 doi:10.1088/1757-899X/570/1/012110

Technological complexes of laser welding and hardening tool

V V Zvezdin [1], R M Hisamutdinov [1], R R Saubanov [1], R R Rakhimov [1] and S S Sirkin[2]

[1] Kazan (Volga) Federal University (KFU), Naberezhnye Chelny, Russia
[2] Kazan National Research Technical University named after A. N. Tupolev - KAI, Leninogorsk, Russia

irmaris@yandex.ru, rafisih88@mail.ru

Abstract. The control system of the automated laser technological complex for the technology of surfacing and hardening the cutting edge of the teeth of the mill with the positioning of the focus of laser radiation on the tooth is considered. Automating the process of laser hardening and surfacing of parts in mechanical engineering makes it possible to increase the efficiency of the technological process of tool restoration, with its specified quality indicators.

1. Introduction

Analysis of the development of modern engineering on the principles of industry 4.0 and robotics in the automotive industry [1] shows the widespread use of automated laser technology complexes (ALTC) surfacing, hardening, welding, cutting, etc. with the required quality of processing, as one of the effective methods to improve production efficiency by improving the productivity of technological processes (TP) and optimizing production resources [1-4].

An important factor in reducing the cost of production is a qualitative improvement in the technical level of production of mastered products in engineering [2-5]. With the advent of fiber lasers there was a sharp increase in demand for high-tech industrial equipment. [1-3].

2. Theoretical and experimental studies

A question of the process of automation welding and hardening the cutting edges of the teeth of the cutter facing tool is very important for production. Studies have shown the instability of the parameters characterizing the quality of TPs. Which are characterized by surface roughness, uniformity of chemical composition and microstructure in depth. A significant role in this is played by the stability and parameter values of laser technological complexes (LTC).

According to the results of experimental and theoretical studies on laser cladding and quenching, the main factors affecting the quality of the TPs are the stability of the set value of the power density, the quality and precision positioning of the focus laser radiation (LR) the cutting edge of the cutter teeth [1-2, 4].

In the course of research, a block diagram of ALTC was developed that meets the requirements for technological complexes (Fig. 1).

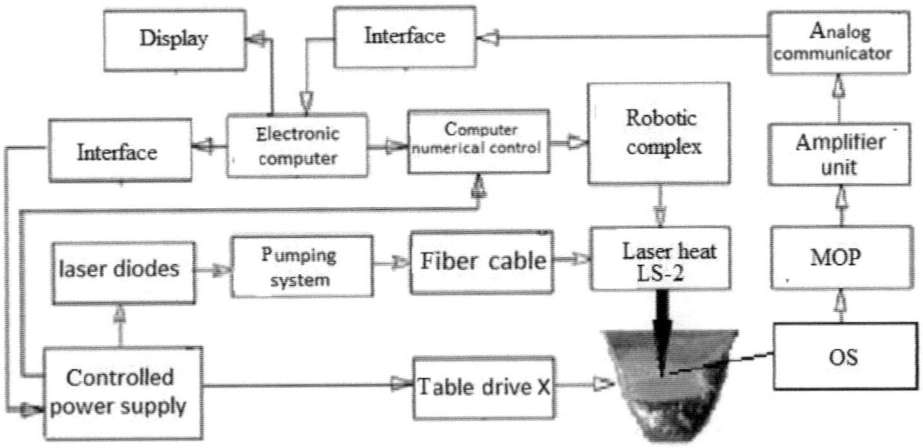

Figure 1. ALTC Structural Diagram:
when OS – Optical system, MOP – matrix photodetector

LR is fed by the optical head of the fiber ytterbium laser LS-2 to the top of the tooth surface, fixed on a robotic complex,and the focus of LR moves along the y, Z coordinates of the tooth. Moving the cutter on X is carried out by the drive of the desktop. Precision positioning of the focus on the cutting edge of the cutter teeth is provided by the elements of technical vision (ETV). It is an optical system, a matrix photodetector, an amplifier unit, an analog switch that converts parallel code into a serial and a communication interface with the control computer. An error signal according to the ratios of the irradiated squares MOP is supplied by converters on the computer where the developed algorithm is evaluated and issued on a rack CNC signal, adjustment of focus or for Y,Z. a controlled power source (CPS) receives a signal of the displacement X. [6-8].

The calculation of the active resistance of the photosensitive layer of AF shows its dependence on the area of illumination:

$$R_{pl} = \frac{\rho_{н}\rho_{o}l^{2}}{dl[(x-\Delta x)\rho_{н} + \Delta x\rho_{o}]} = \frac{\rho_{н}\rho_{o}l^{2}}{d[(S_{п} - S_{н})\rho_{н} + S_{н}\rho_{o}]} \qquad (1)$$

when R_{pl} – resistance of the photosensitive layer, $\rho_{Н}$, $\rho_{О}$ – resistivity of the respectively unlit and illuminated areas of the photodetector (OP); l – length of the photosensitive layer; $S_{П}$, $S_{Н}$ – the area of the photosensitive layer, respectively, full and unlit surfaces of the OP.

Correction for the drive Y remote change in the position of the focus of LR, taking into account the displacement of the orientation angle between the optical axis and the tangent surface of the tooth at the point of interaction of LR with the metal is determined by the expression [6]:

$$\Delta\varphi = f(R_{OPп} - R_{OPн}) \qquad (2),$$

when $\Delta\varphi$ – is the increment of the angle of displacement between the optical axis OP and the tangent of the tooth surface at the point influence of LR to metal; $R_{OPп}$, $R_{OPн}$– resistance of the photosensitive layer respectively, the total and unlit surfaces OP.

Figure 2 shows the location of the sensitive areas of the MOP relative to the top of the cutter tooth. The photoresistive matrix sensitive in the near LR region was chosen as the MOP.

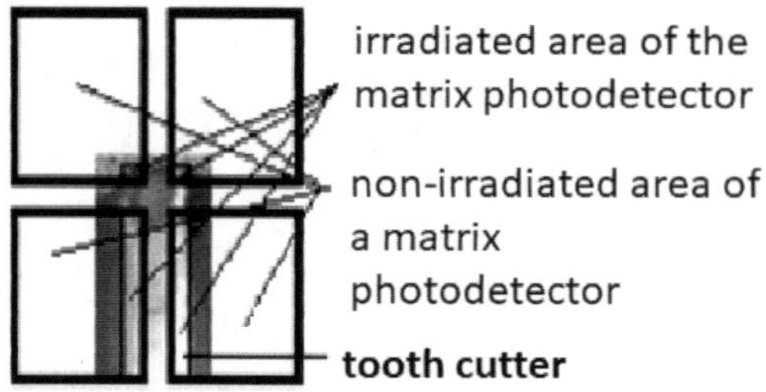

Figure 2. The location of the sensitive areas of the MOP relative to the top of the tooth cutter

Carried out laser cladding of powder brand BoroTec - Eutalloy® 10009 steel R18K5F2 using flux brand AN-43.

The technology of restoration of the cutting edge of the tool consists of a sequence of operations: cleaning the surface of the tooth of the cutter; surfacing powder; sharpening; laser hardening in the mode without melting and applying titanium nitride [2, 9]. Surfacing and hardening were carried out on a fiber ytterbium laser LS-2, with a wavelength of 1.06 μm, the top of the tooth surface was treated (Fig. 3.4).

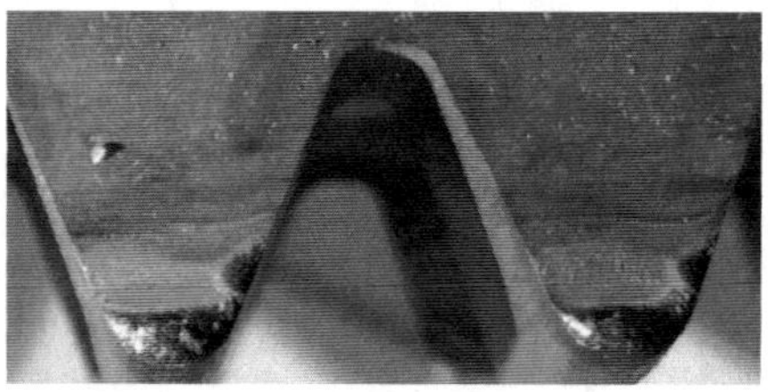

Figure 3. Picture of surfacing powder on milling cutter teeth

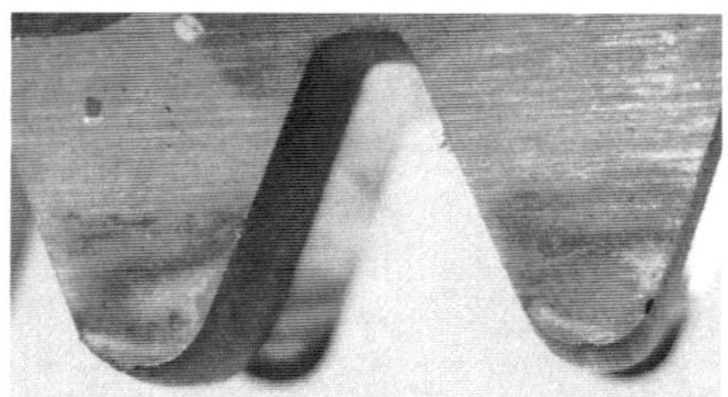

Figure 4. Picture of hardening milling cutter teeth

The microstructure of the hardened layer on the tops of the teeth LR a latent martensite and carbides. Carbide heterogeneity in the structure of the cutter satisfies GOST 19265-73. A graph of the dependence of microhardness on the depth of the layer shown in fig. 5.

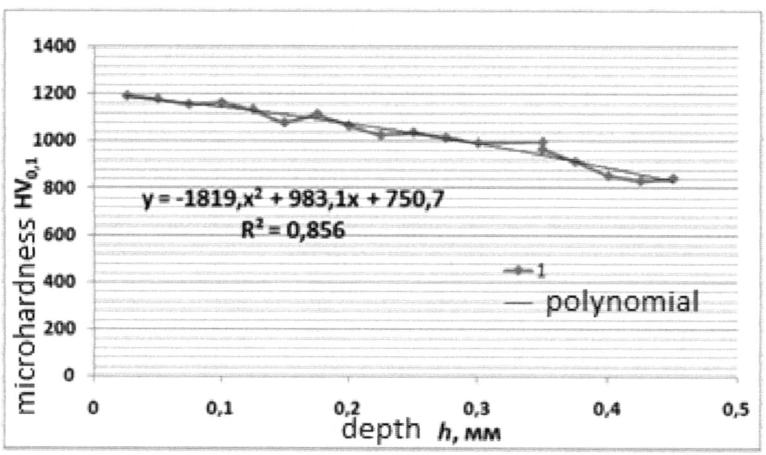

Figure 5. Microhardness versus hardened layer depth

The results of the chemical analysis of the metal of the sample correspond to the metal R18K5F2 GOST 19265-73.High hardness ($HV_{0,1}$ 1192) of the hardened layer increases the wear resistance of the cutter.

IOP Conf. Series: Materials Science and Engineering **570** (2019) 012110 doi:10.1088/1757-899X/570/1/012110

3. Conclusion

Tool edge wear is a consequence of changes in operating conditions, which is characterized by an increase in cutting and cutting forces at high temperature [1-4]. When implementing laser technology in the processing of tool steels, it is necessary to take into account of the quality LR [10]. The quality of LR defines parameters LTC and the quality of TP hardening and surfacing. The question of the impact of the distribution of intensity of energy in the spot, LR the quality of hardening and welding is still urgent.

References

[1] Zvezdin V V, Khisamutdinov R M, Rakhimov R R, Israfilov I H and Akhtiamov R F 2017 Method of control position of laser focus during surfacing teeth of cutters *IOP Conference Series: Materials Science and Engineering* Vol 240 Issue 1 012072

[2] Zvezdin V V, Khisamutdinov R M, Grechishnikov V A, Saubanov R R, Israfilov I Kh, Portnov S M and Yurasov S. Yu. 2018 Laser Machining of Tool Steels *Russian Engineering Research* Vol 38 No 12 pp 1038-41

[3] Zvezdin VV, Hisamutdinov, R M, Rakhimov R R, Israfilov I H and Saubanov R R 2018 Technology of overlay laser welding of durable powdery into blade edge of miller *IOP Conference Series: Materials Science and Engineering* Vol 412 Issue 1 **012083**

[4] Rakhimov R R, Saubanov R R and Israfilov I H 2017 Analysis of the impact of informative heat treatment parameters on the properties of hardening of the surface layers *Journal of Physics: Conference Series* Vol 789 Issue 1 **012040**

[5] Valiev R A, Bochkov V E, Bashkirov Sh Sh, Romanov E S and Chistjakov V A 1991 Mössbauer study of surface layers of high-speed steel after laser treatment *Hyperfine Interactions* 69(1-4) pp 589-92

[6] Zvezdin V V, Israfilov D I, Portnov S M, Saubanov R R, Rakhimov R R and Zvezdina N M 2015 Automatic control system of high-precision welding of preparations by the laser radiation at influence of the plasma torch *Proceedings of the higher educational institutions (Physics)* number 9-3 Vol 58 pp 51-54

[7] Gabdrakhmanov Az T, Israphilov I H, Galiakbarov A T, Samigullin A D and Gabdrakhmanov Al. T 2016 Improving the efficiency of plasma heat treatment of metals *Journal of Physics: Conference Series* Vol 669 **012014**

[8] Grigoryants A G, Perestoronin A V, Portnov S M, Zvezdin V V and Israfilov I H 2015 A system for automatic control of precision laser welding in engineering *Welding International* T 29 No 10 pp 801-804

[9] Saubanov R R, Zvezdin V V, Israfilov I H, Haybullin I I and Rakhimov R R 2014 Synthesis of oxidic powder in nonequilibrium low-temperature plasma with increase of indicators of quality of process *Journal of Physics: Conference Series* Volume 567, Issue 1 Article number 012034

[10] Rakhimov R R, Zvezdin V V, Israfilov I H, Nabiullina G I Saubanov R R 2014 Modeling of the laser heat treatment *News of the Tula State University* (Technical science) number 11-2 pp 476-84

IOP Conf. Series: Materials Science and Engineering **570** (2019) 012111 doi:10.1088/1757-899X/570/1/012111

The process control for improving the durability of the cutting edge of the teeth of the cutter at their recovery

V V Zvezdin, R M Hisamutdinov, R R Rakhimov, R R Saubanov and I H Israfilov

Kazan (Volga) Federal University (KFU), Naberezhnye Chelny, Russia
E-mail: irmaris@yandex.ru, rafisih88@mail.ru

Abstract. The features of the technology of laser deposition of wear-resistant powders on the cutting edge of the cutter tooth with control of the positioning of the focus of laser radiation are considered. The stabilization of the parameters of the laser technological complex and the positioning of the focus of the laser radiation makes it possible to increase the efficiency of the technological process of restoring the instrument, with its specified quality indicators.

1. Introduction

Analysis of the development of modern engineering shows the widespread introduction and use of automated laser processing complexes (ALPC) metal processing as one of the effective methods to improve production efficiency by increasing the productivity of technological processes (TP) [1-4].

The efficiency of production is characterized by increased productivity, the most complete use of production capacity, raw materials and material resources, achieving the best results at the lowest cost [1, 2]. The main role in the development of industry belongs to tool production, which determines the level of productive forces of the entire industry.

Modern technologies of development of tool production are: use of powder metallurgy; automation of technological processes, application of automatic lines by laser technological complexes; application of methods of thermal and thermochemical processing, wear-resistant coatings; expansion of the field of application of thermal processing methods. These directions, ultimately, contribute to the solution of the main tasks of increasing the efficiency of production and quality of products. One of the methods of reducing costs in the manufacture of products is the restoration of worn edges of tools by laser surfacing of wear-resistant powders.

2. Theoretical and experimental studies

The wear edge of the tool is the result of changes in the operating conditions that is characterized by an increased notching and cutting force when high temperatures [3, 4] (Fig. 1). When implementing laser technology in the processing of tool steels, it is necessary to take into account the quality of laser radiation (LR) [1-2, 5-7], the purity of the surface layer, the physical and chemical parameters of the applied surfacing powder, etc. The quality of the LR determines the parameters of laser technological complexes (LTC) and the quality of the TP surfacing. The question of the influence of the energy intensity distribution in the spot of LR the quality of the surfacing remains relevant.

Process control to improve the durability of the cutting edge of the teeth of the TP surfacing cutting edge of the teeth of the cutter is acutely in front of the tool production. Studies have shown the instability of the parameters characterizing the quality of the TP (Fig. 2). That characterized mainly by the mismatch of the center of the spot with the center of the top of the tooth of the cutter, as well as the surface roughness, heterogeneity of the chemical composition and microstructure in depth. A significant effect on TP deposition rate provide stability and the values of the parameters of LTS. It is necessary to consider the LTC as a set of interacting links of a complex system [5, 8-10]. It is an optical quantum generator, optical system (OS), LR interaction medium with metals, workpiece and the mechanism of its precision movement.

Figure 1. Fragment of router bit defective blade edge

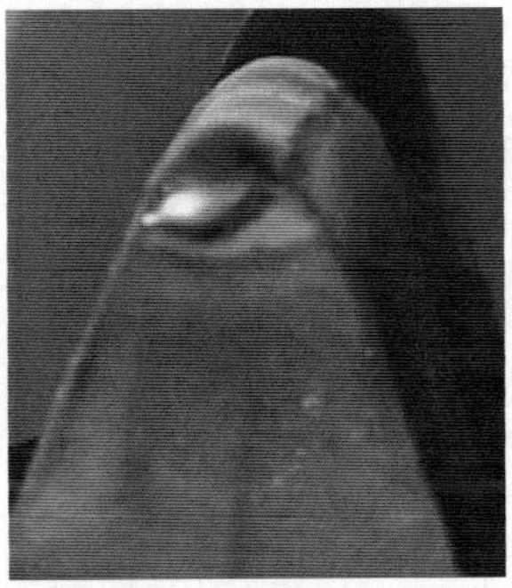

Figure 2. Fragment of router bit with laser cladding without positioning of focus of laser emission

According to the results of experimental and theoretical studies on laser cladding, the main factors affecting the quality of the TP are the stability of the set value of the power density, the quality and precision positioning of the focus LR the cutting edge of the cutter teeth [1, 3, 8, 11].

Figure 3 shows the location of the reflected radiation from the cutter tooth to the sensitive areas of the matrix photodetector (OP). Precise positioning is provided by the developed system of automated control of the position of the focus of LR when surfacing the teeth of the cutter [1]. Figure 4 shows a fragment of the tooth after laser welding with a precise positioning of the focus of LR.

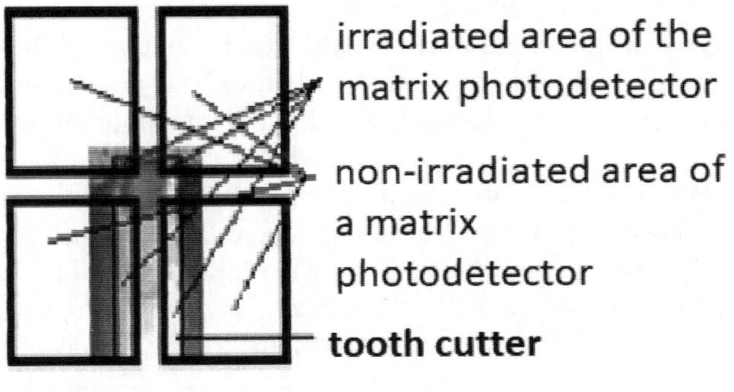

Figure 3. The location of the reflected radiation from the cutter tooth to the sensitive areas of the matrix photodetector (MOP)

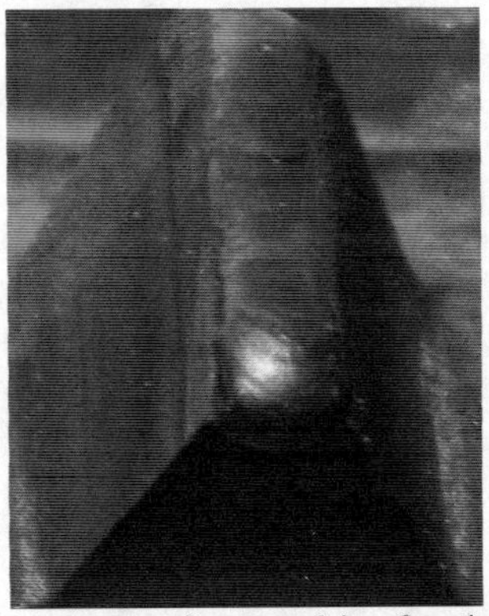

Figure 4. Fragment of router bit after laser cladding with precision positioning of focus of laser emission

For fast process of laser surfacing it is necessary to provide automated control of LTC with feedback on the parameters measured in real time of course of TP surfacing [8-11].

An experimental setup was developed, the block diagram of which is shown on the figure 5.

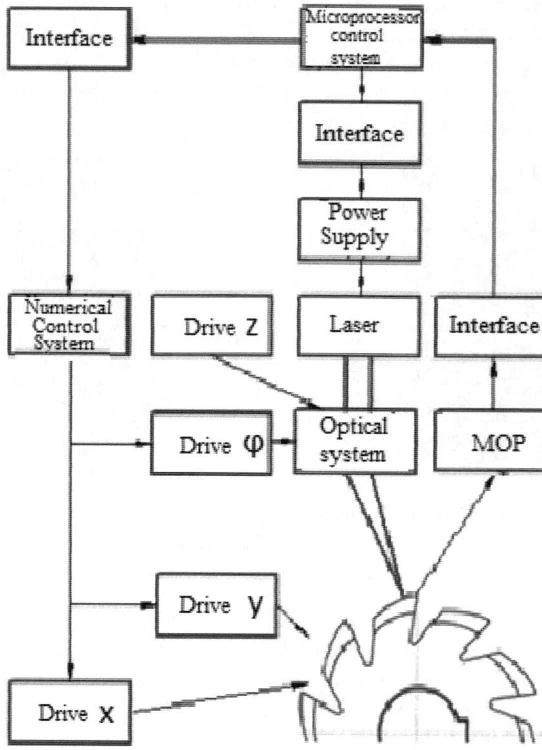

Figure 5. The structure of the ACS LTC

Control of the LTS parameters and their stabilization is carried out due to the negative feedback on the measured parameters from the interaction zone with the metal. This is achieved by analyzing the position of the focus of LR to the cutter tooth and the angle to the normal plane of the workpiece surface. The structure of the ACS LTC is a system that is characterized by a large number of feedbacks and is nonlinear. [1, 5, 10, 11].

The technology of restoring the cutting edge of the tool consists of a sequence of operations: cleaning the surface of the tooth cutter; surfacing powder; sharpening; application of titanium nitride [1]. Cladding and hardening of the vertices of the surface of the teeth powder brand BoroTec - Eutalloy® 10009 was carried out with a fiber ytterbium laser LS-2, with a wavelength of 1.06 μm (Fig. 4). Surface protection against oxidation is carried out with argon supplied through a special nozzle

3. Conclusion

Experimental studies of the impact LR on the steel cutter R18K5F2 show the possibility of optimizing the energy parameters of the LTC for surfacing to obtain the required quality indicators, which leads to increased durability of the cutting edge of the cutter teeth in their recovery. The developed ACS LTC meets the requirements for the accuracy of positioning relative to the edge of the cutter, provide stable quality indicators TP surfacing.

References

[1] Zvezdin V V, Khisamutdinov R M, Rakhimov R R, Israfilov I H and Akhtiamov R F 2017 Method of control position of laser focus during surfacing teeth of cuttersё *IOP Conference Series: Materials Science and Engineering* Volume 240, Issue 1 Article number 012072.

[2] Rakhimov R R, Saubanov R R and Israfilov I H 2017 Analysis of the impact of informative heat

treatment parameters on the properties of hardening of the surface layers *Journal of Physics: Conference Series* Vol 789 Is 1 **012040**

[3] Gabdrakhmanov Az T, Israphilov I H, Galiakbarov A T, Samigullin A D and Gabdrakhmanov Al.T 2016 Improving the efficiency of plasma heat treatment of metals *Journal of Physics: Conference Series* Volume 669 **012014**

[4] Valiev R A, Bochkov V E, Bashkirov Sh Sh, Romanov E S and Chistjakov V A 1991 Mössbauer study of surface layers of high-speed steel after laser treatment *Hyperfine Interactions* 69 (1-4) pp 589-92

[5] Grigoryants A G, Perestoronin A V, Portnov S M, Zvezdin V V and Israfilov I H 2015 A system for automatic control of precision laser welding in engineering *Welding International* T. 29, number 10 pp 801-4

[6] Zvezdin V V, Klochkova K V, Rakhimov R R, Saubanov R R, Israfilov I H and Pesoshin V A 2018 Technology ability of laser bonding of compacted graphite iron IOP Conference Series: Materials Science and Engineering Vol 412 Issue 1 **012084**

[7] Zvezdin V V, Hisamutdinov, R M, Rakhimov R R, Israfilov I H and Saubanov R R 2018 Technology of overlay laser welding of durable powdery into blade edge of miller *IOP Conference Series: Materials Science and Engineering* Vol 412 Issue 1 **012083**

[8] Bashmakov D A and Zvezdin V V 2018 Control of laser-field technological complexes for tool hardening *ARPN Journal of Engineering and Applied Sciences* 13(8) pp 2979-82

[9] Rakhimov R R, Zvezdin V V, Israfilov I H, Nabiullina G I Saubanov R R 2014 Modeling of the laser heat treatment *News of the Tula State University* (Technical science) number 11-2 pp 476-84

[10] Zvezdin V V, Israfilov D I, Portnov S M, Saubanov R R, Rakhimov R R and Zvezdina N M 2015 Automatic control system of high-precision welding of preparations by the laser radiation at influence of the plasma torch *Proceedings of the higher educational institutions (Physics)* Vol 58 ed Siberian Physico-Technical Institute (*Tomsk*) pp 51-54

[11] Zvezdin V V, Rakhimov R R, Saubanov R R, Israfilov I H and Akhtiamov R F 2017 Management of laser welding based on analysis informative signals *IOP Conference Series: Materials Science and Engineering* Vol 240 Is 1 **012073**

IOP Conf. Series: Materials Science and Engineering **570** (2019) 012112 doi:10.1088/1757-899X/570/1/012112

Physical and mechanical properties of fiber reinforced gypsum-based composite

R Kh Mukhametrakhimov[1], L V Lukmanova[1] and I R Gilmanshin[2,3]

[1]Kazan State University of Architecture and Engineering, Zelenaya str., 1, Kazan, 420043, Russia
[2]Kazan (Volga region) Federal University, Kremlevskaya str., 18, Kazan, 420008, Russia
[3]Kazan National Research Technical University named after A N Tupolev, Kazan, Russia

muhametrahimov@mail.ru

Abstract. Formulations of effective gypsum-fiber and gypsum-cement-fiber materials modified with active mineral additives and superplasticizer have been developed in the present experimental work. The influence of the type and length of reinforcing polypropylene fiber, the conditions of hardening and molding on the physical and mechanical properties of the specimens was studied.

1 Introduction

In addition to a lot of positive technical properties of gypsum, this mineral has considerable brittleness, which especially affect the use of thin-walled sheet materials and products (gypsum panel). Improvement of physical and mechanical properties, including strength, of products based on gypsum binder can be achieved by introducing into the composition of active mineral additives of natural and technogenic origin, chemical additives [1, 2]. The effectiveness of using active mineral additives such as thermally activated clay, ground limestone [3], ceramsite fume [4], metakaolin, white carbon, silica fume [5], bio-silica [6] in the gypsum binder is shown in different scientific works.

Artificial hardening of products with reinforcing materials (fibers) introduced into the composition of the molding material or being parts of the product is one of the promising area in solving this problem [7]. There is domestic and foreign experience in application of glass, polypropylene [8, 9], basalt [10] fibers, polyvinyl alcohol fibers [11] in the composition of the gypsum binder in combination with various mineral and chemical additives. Zone dispersed reinforcement with metal fibers, despite its effectiveness, does not provide the required corrosion resistance of thin-walled materials that are operated in humid environments [12]. However, there is a need for the batching of the composition depending on the chosen binder and additives. For the structural design of composite materials, in particular concrete, new methods are used, for example, diagram methods

[13], which make it possible to evaluate the stress-strain state of the calculated cross sections taking into account the physical nonlinearity of materials at all stages of loading - from the beginning of loading to failure [14].

Domestic industry has the technology of production of gypsum-fiber plates and panels at the present time. Vegetable fibers uniformly distributed in the gypsum-fiber are used as a reinforcement in these materials. The composition of gypsum-fiber mass consists of the following components: gypsum, fibrous substance, sizing additives. The technological process of production of gypsum-fiber panels consists of the following operations: preparing raw materials - gypsum and fibrous mass, mixing the mass, casting the product on the grid and then removing excess water from it, compacting gypsum-fiber panels under the press, bringing them to the desired thickness, drying, trimming and stacking panels in piles. Thus, it could be concluded that this technology has several significant drawbacks, namely:

- construction gypsum is unsuitable for the manufacture of products owing to its adhesion to the grid, seal closure – it makes it difficult to evacuate the mass;
- the need for adding sizing additives;
- complexity of the process;
- low water resistance of gypsum binder.

Furthermore, the use of vegetable fibers in the production of gypsum-fiber panels causes significant problems with biological corrosion and mold formation on the surface of products during their operation.

The aim of this work is to develop formulations of gypsum-fiber and gypsum-cement-fiber composites based on low-grade gypsum of the Republic of Tatarstan with high physical and mechanical properties with normal mixing density and natural hardening conditions.

It is necessary to determine the effect of the type, length and volume content of fibers in the gypsum paste on the strength characteristics of samples.

2 Materials and Methods

The following materials were used in the research process:

- binder: gypsum G6BII GOST 125-79 OOO Arakchinskiy gips
- active mineral additive: industrial waste of metallurgical industry - ferrosilicon ground to a specific surface 391 m^2/kg
- Portland cement PC 400 D20 produced Volsk cement plant
- superplasticizer Odolit-K (TU 5745-01-96326574-08)
- fibers: cellulose fibers (3-15 mm length), chopped basalt fibers (10-15 mm length), glass fibers (20 mm length), polypropylene fiber VSM-II (6, 12, 18, 24, 32 mm length).

The tests were carried out on standard beam samples 4x4x16 cm, from the molding mixture of normal thickness according to the method described in GOST 23789-79. The results of previous studies allowed to establish the optimal content of the superplasticizer Odolit-K, equal to 1% of the weight of the binder.

3 Results and Discussion

The results demonstrate high efficiency of cellulose fibers that increased the strength characteristics of plasticized high-strength gypsum G16 by 75-78%. However, similar tests on gypsum G6 did not have a similar effect; the increase in flexural strength was only 3-5%.

Test of samples containing basalt and glass fibers (dosages from 0.5 to 5%) showed no effectiveness due to the weak adhesion of the fibers to the gypsum stone based on low-grade gypsum.

Samples containing polypropylene fibers (6, 12, 18, 24 and 32 mm lenght) based on low- grade gypsum showed positive effect. The results of the study of the effect of polypropylene fiber on physical and mechanical properties of gypsum are presented at the Table 1.

IOP Conf. Series: Materials Science and Engineering **570** (2019) 012112 doi:10.1088/1757-899X/570/1/012112

Table 1. Test result.

Formulation number	Fiber brand	Fiber volume fraction, %	Water-gypsum ratio	Average density, g/cm^3	Flexural strength, MPa	Strength-density ratio	Relative flexural strength, %
1	VSM-II-6 (6 mm)	0	0.56	1.29	6.42	49.94	100.00
2		0.25	0.44	1.44	7.27	50.60	113.24
3		0.5	0.46	1.43	7.87	56.02	124.48
4		1	0.48	1.37	6.86	50.11	106.97
5		2	0.49	1.36	6.56	48.09	102.27
6		3	0.5	1.35	6.46	47.71	100.68
7	VSM-II-12 (12 mm)	0	0.56	1.29	6.42	49.94	100.00
8		0,25	0.44	1.46	6.94	47.44	108.10
9		0.5	0.46	1.44	7.96	55.36	124.13
10		1	0.48	1.38	8.05	58.46	125.50
11		2	0,49	1.36	7.32	53.79	114.06
12		3	0.5	1.36	7.15	52.48	111.38
13	VSM-II-18 (18 mm)	0	0.56	1.29	6.42	49.94	100.00
14		0.25	0.44	1.46	7.76	53.32	120.96
15		0.5	0.46	1.42	8.13	57.31	126.75
16		1	0.48	1.40	8.20	58.53	127.79
17		2	0.49	1.37	7.89	57.79	123.00
18		3	0.5	1.34	7.61	56.94	118.55
19	VSM-II-24 (24 mm)	0	0.56	1.29	6.42	49.94	100.00
20		0.25	0.44	1.48	7.61	51.46	118.62
21		0.5	0.46	1.42	6.27	42.12	94.39
22		1	0.48	1.39	5.56	39.44	86.69
23		2	0.49	1.36	4.75	34.89	74.06
24		3	0.5	1.37	4.29	31.39	66.93
25	VSM-II-32 (32 mm)	0	0.56	1.29	6.42	49.94	100.00
26		0.25	0.44	1.45	6.96	47.85	108.45
27		0.5	0.46	1.42	6.05	44.59	98.89
28		1	0.48	1.41	5.53	42.41	92.93
29		2	0.49	1.35	4.94	36.57	77.03
30		3	0.5	–	–	–	–

As shown in the Table 1, polypropylene fiber VSM-II-18 (18 mm) volume fraction in an amount of 0.5-1% demonstrates the maximum physical and mechanical properties. Further increase of the length and fiber volume fraction leads to their clumping and decrease in strength.

Studies of shrinkage deformation performed according to GOST 24544-81 showed that the samples reinforced with polypropylene fibers in the amount of 0.5-1% do not shrink, while the shrinkage of the control samples was 0.5-0.6% .

In order to increase the water resistance, Portland cement was introduced in the amount of 15, 20, 25% and ferrosilicon - 5% by weight of the binder. Test result is presented at the Table 2.

Table 2. Test result.

Gypsum content, %	Cement content, %	Water-gypsum ratio	Active mineral additive content, %	Density, g/cm^3	Flexural strength, MPa	Compressive strength, MPa		Softening coefficient
						dry samples	water-saturated samples	
85	15	0.56	5	1.70	3,.33	185.2	145.8	0.79
80	20	0.53	5	1.75	3.90	220.9	195.1	0.88
75	25	0.5	5	1.73	4.25	213.1	191.6	0.90

As can be seen from the Table 2, the increase of the cement content in the gypsum cement binder indicates an improvement in the mechanical performance and of the water resistance of samples at the age of seven days aged in wet conditions.

The optimal content of Portland cement in the gypsum cement binder modified with polypropylene fibers and the superplasticizer Odolit-K has also been determined. In order to avoid clumping and uniform distribution of fibers in the volume of the molding material, two-stage dispersion was applied. The results of experimental studies gypsum-fiber samples are shown at the Table 3.

Table 3. Test result.

Fiber content VSM-II-18, %	Water-gypsum ratio	Gypsum content, %	Cement content, %	Active mineral additive content, %	Density, g/cm^3	Flexural strength, MPa	Compressive strength, MPa
0.7	0.43	85	15	5	1.79	5.31	334,. 8
0.7	0.4	80	20	5	1.80	5.58	318.27
0.7	0.37	75	25	5	1.80	4.45	243.55

As shown in the Table 3, the most optimal value of the cement content in gypsum-fiber-fiber products is 20%.

Table 4 shows the results of tests of gypsum-cement-fiber samples at the age of seven days contained gypsum – 80%; (cement + active mineral additive) – 20%; fiber (18, 24, 32 mm length) – 0.7% by volume of the mixture; Odolit-K – 1% by weight of the dry binder.

Table 4. Test result.

Superplasticizer content,%	Water-gypsum ratio	Fiber length, mm	Density, g/cm^3	Flexural strength, MPa	Compressive strength, MPa
1	0.4	18	1.65	5.31	344.62
1	0.4	24	1.69	5.72	321.29
1	0.4	32	1.48	5.21	256.57

As can be seen from the test results presented at the Table 4, the flexural strength varies slightly while fiber length is increasing from 18 to 24 mm, therefore, gypsum-cement fiber samples reinforced with fiber length 18 mm are more suitable for manufacturing of gypsum-cement-fiber materials. Thus, the two-stage dispersion will be avoided and thereby the production technology of finished products will be facilitated.

Mechanisms of hydration processes of gypsum and cement binder differ significantly, that is why it is necessary to determine which of the components in gypsum-cement-fiber samples has a larger impact on the physical and mechanical properties. Steaming of samples was used in the present study. It was based on the assumptions that Portland cement can most actively interact with active mineral additive at higher temperatures and in an environment of saturated water vapor, which, in principle, can positively affect the strength characteristics of samples at a later time. To solve this problem, four series of samples of the same composition were made, three of which were steamed for 12 hours at a temperature of 80 °C, 90 °C and 100 °C, and then dried at 60 °C to constant weight. The fourth series were kept under natural conditions for three hours, and then also dried to constant weight for 24 hours at a temperature of 60 °C. The composition of the samples: gypsum binder G-6 – 80% by weight; Portland cement – 19%; active mineral additive – 1%. The test results of these samples are presented at the Table 5.

Table 5. Test result.

Curing conditions	Water-gypsum ratio	Density, g/cm^3	Flexural strength, MPa	Flexural strength, %
Steaming 80 °C + drying 60 °C	0.53	1.41	8.93	144.6
Steaming 90 °C + drying 60 °C	0.53	1.39	8.87	143.6
Steaming 100 °C + drying 60 °C	0.53	1.34	8.5	137.5
Hardening time+ drying 60 °C	0.53	1.32	6.18	100

As shown in Table 5, flexural strength of gypsum-cement-pozzolanic specimens subjected to drying after aging under natural conditions is substantially lower than the strength of specimens subjected to steaming. Obviously, it is caused by removal of moisture in the drying process, which violates the optimal conditions for the hydration of Portland cement in the gypsum-pozzolanic binder.

A difference is observed when steaming of the samples is occurred. As stated above, it is caused, by active interaction of Portland cement with active mineral additive at higher temperatures in an environment of saturated water vapor.

Thus, the test results showed that the optimal regime of heat treatment of a gypsum-cement-pozzolanic binder is steaming at 80 °C for 12 hours, followed by drying at 60 ^{0}C to constant weight.

It should be noted, steaming of the samples of all three compositions leads to an increase in the density of samples. This is due to the more complete involvement of Portland cement and active mineral additive in the process of formation of the paste matrix structure by increasing the volume of calcium hydrosilicates. The process of increasing the density of samples is accompanied by an increase in their strength characteristics and water resistance.

Conclusions

1. Formulations of gypsum-fiber compositions with high physical and mechanical properties under normal mixing density and natural hardening conditions based on low-grade gypsum of the Republic of Tatarstan were obtained. Thus, high flexural strength of the samples (8.13–8.20 MPa) and the absence of shrinkable deformations are achieved with a volume content of polypropylene fiber VSM-II-18 (18 mm) in an amount of 0.5–1%. Further increase the length and volume content of the fibers leads to their clumping and decrease the strength of the fibers.
2. The results demonstrate the steaming gypsum-cement fiber samples at a temperature of 80 °C for 12 hours, followed by drying at a temperature of 60 °C to constant weight leads to an increase flexural strength by 44.6% compared to samples subjected to drying after aging in natural conditions, due to the active interaction of Portland cement with active mineral additive at higher temperatures in an

environment of saturated water vapor. Thus, the developed gypsum-cement-fiber composite material has potential to obtain qualitatively new structural materials for the production of a wide range of building products and materials.

Acknowledgements

The authors were supported by grant offered by the President of the Russian Federation for state support of young Russian researchers-candidates of sciences (grant number MK-2509.2018.8).

References

[1] Muhametrahimov R Kh, Galautdinov A R and Lukmanova L V 2016 *Izvestiya KGASU* **5 (38)** 382-87
[2] Khaliullin M I, Nuriev M I, Rakhimov R Z and Gaifullin A R 2015 *Vestnik Kazanskogo tekhnologicheskogo universiteta* **18** 119-22
[3] Khaliullin M I, Rakhimov R Z and Faizrakhmanov I I 2017 *ZKG: ZEMENT - KALK - GIPS INTERNATIONAL. Bauverlag BV GmbH. Gutersloh* **5** 58-63
[4] Khaliullin M I and Nuriev M I 2016 *Izvestiya KGASU* **3 (37)** 225-30
[5] Morozova N N, Kuznetcov G V, Maysuradze N V, Akhtariyev R R, Abdrashitova L R and Nizamutdinova E R 2018 *Stroitelnyye materialy* **8** 26-30
[6] Izotov V S, Muhametrahimov R Kh and Galautdinov A R 2015 *Stroitelnyye materialy* **5** 20-23
[7] Vorobyev V A and Komar A G 1971 *Stroitelnyye materialy* (Moscow: Stroyizdat) p 496
[8] Hoang Nguyena, Paivo Kinnunena, Valter Carvellib, Mohammad Mastalia, Mirja Illikainen 2019 *Composites Part B: Engineering* **158** 328-38
[9] Hoang Nguyena, Paivo Kinnunena, Valter Carvellib, Mohammad Mastalia and Mirja Illikainen 2019 *Composites Part B* **158** 328-38
[10] Petropavlovskaya V B, Novichenkova T B, Buryanov A F, Solovyev V N and. Petropavlovskiy K S 2017 *Vestnik MGSU* **12 (111)** 1392-98
[11] Izotov V S, Muhametrahimov R Kh and Sabitov L S 2010 *Stroitelstvo i arkhitektura* **1 (17)** 119-25
[12] Cong Zhu Jianxin Zhang Jiahui Peng Wenxiang Cao Jiangsen Liu 2018 *Construction and Building Materials* **163** 695-05
[13] Radaikin O V, Sabitov L S, Kashapov N F and Gilmanshin I R 2018 *IOP Conference Series: Materials Science and Engineering* 012057
[14] Radaikin O V, Sabitov L S, Kashapov N F and Gilmanshin I R 2018 *IOP Conference Series: Materials Science and Engineering* 012058

IOP Conf. Series: Materials Science and Engineering **570** (2019) 012113 doi:10.1088/1757-899X/570/1/012113

Investigation of mechanical properties of fiber-cement board reinforced with cellulosic fibers

L V Lukmanova[1], R Kh Mukhametrakhimov[1] and I R Gilmanshin[2,3]

[1]Kazan State University of Architecture and Engineering, Zelenaya str., 1, Kazan, 420043, Russia
[2]Kazan (Volga region) Federal University, Kremlevskaya str., 18, Kazan, 420008, Russia
[3]Kazan National Research Technical University named after A N Tupolev, Kazan, Russia

muhametrahimov@mail.ru

Abstract. The article presents experimental research on the development of the optimal composition of cement-fibrous composite material for the manufacture of autoclaved fiber-cement boards with high physical, mechanical and economic parameters. The effect of the cement-sand ratio of the matrix, the specific surface area of the sand and the cellulosic fibers content on the physical and technical properties of the material is shown. It is established that the optimum cement-sand ratio of the matrix associated with a specific surface area of quartz sand 310 m^2/kg is 1:2. The optimal content of cellulosic fibers according to the criterion of increasing the strength is 5-7.5%.

1 Introduction

Hinged ventilated facades find their application for finishing the facades of newly erected buildings, as well as for the insulation of reconstructed buildings. Currently, various materials are used as a protective and decorative screen, but the questions associated with the development of new environmentally friendly and low cost materials that combine high decorative and performance qualities, are remain relevant. Fiber-cement boards attract particular attention in this regard [1, 2-4].

Fiber-cement boards are deservedly demanded by architects, designers and builders. They are used for finishing the facades of the accommodation building, office buildings, educational institutions, banks, metro stations and train stations, tunnels, bridges, racks [5].

These boards are an artificial composite stone building material resulting from solidification of a mixture consisting of cement, fibers (≈5-20% by weight of cement) and water. Nonlinear deformation models that have a high degree of reliability are used to assess the strength and deformability of composite materials, in particular concrete [6, 7].

The most common type of fiber cement is asbestos cement products. They have low cost and compete with other roofing and facade materials. However, asbestos fibers are carcinogenic, therefore, in many countries around the world, especially where natural reserves of asbestos are lacked, research related to the partial or complete replacement of asbestos with other types of organic or inorganic fibers remains relevant.

In this regard, the use of environmentally friendly raw materials of low cost required technical properties should be taken into account in the attempt to develop new cement-fiber composite materials.

The most significant effect can be achieved with the introduction of new asbestos-free cement-fiber materials on existing traditional production lines for the production of asbestos cement sheets with maximum use of existing equipment. There are studies of fiber-cement materials based on alkali-resistant glass grades (S-15GT) and basalt fibers [8], according to which the complete exclusion of asbestos from the composition and their replacing with glass or basalt fiber with respect to existing round-grid machines is practically impossible. It is caused by low cement-holding capacity of such fibers. Zone dispersed reinforcement with metal fibers, despite its effectiveness, does not provide the required corrosion resistance of thin-walled materials that are operated in humid environments [9].

The experience of using carbon fibers in fiber-cement boards is known, but high cost does not allow them to be fully utilized [10]. The use of basalt and glass fibers in cement systems is limited due to their low resistance to cement hydration products. However, currently this problem is being solved by modifying the surface of such fibers [11, 12].

The most preferred type of fibers, allowing to achieve high technical results, are cellulosic fibers [13]. A number of authors have shown the efficiency of using cellulosic fibers from bamboo [14, 15], rice straw [16]. Good quality results were obtained by VNII-project through creating the cement-fiber products with partial cement (up to 50%) replacing asbestos with cellulose fibers [17]. Cellulose was served as the fibrous material: HC-2 unbleached sulphate and bleached sulphite. Cellulose fiber-based sheets have passed a 10-year life cycle test. Strength and aesthetic properties of sheets have not deteriorated. Factory production of fiber-cement boards reinforced with cellulosic fibers in Russia is established at OAO LATO. Products fully meet the performance requirements [18].

According to the classification of dispersion-reinforced concrete [19], reinforcing fibers were proposed to be divided into two types: low-modulus and high-modulus, which are divided into metal, mineral and organic. Specifications, some of them are presented at the Table 1.

Table 1. Technical characteristics of fibers.

Fiber	Density, g/cm3	Tensile strength, MPa	Modulus of elasticity, MPa	Elongation at rupture, %
Polypropylene	0.9	400-770	3500-8000	10-25
Polyethylene	0.95	700	1400-4200	10
Nylon	1.1	770-840	4200	16-20
Acrylic	1.1	210-420	2100	25-45
Polyether	1.4	730-780	8400	11-13
Asbestos	2.6	910-3100	68000	0.6
Cellulose	1.2	300-500	10000	0.5-4
Glass	2.6	1050-3850	70000-80000	1.5-3.5
Steel Fiber	7.8	800-3150	200000	3-4

IOP Conf. Series: Materials Science and Engineering **570** (2019) 012113 doi:10.1088/1757-899X/570/1/012113

Carbonic	2.0	2000	245000	1
Carboniferous	1.63	7800	380000	2.1
Nylon	0.9	720	1900	24
Viscose	1.2	660	5600	14
Basalt	2.6	1600-3600	80000-110000	1.4-3.6
Wollastonite	2.9	200-400	10000	0.2-0.5
Kevlar	1.45	3600	150000	2.1-4
Polyacrylonitrile	1.2	900	20000	10

Cellulosic fibers according to this classification can be attributed to high modulus, organic, since the elongation at rupture is 0.5-4%, tensile strength 300-500 MPa, average fiber diameter 10-35 microns, length 200-1000 microns and more. Such characteristics of cellulosic fibers provide high impact strength and flexural strength of a composite material.

Cellulosic fiber has the following advantages: easy fluff, sufficient mechanical strength and flexibility, they are not carcinogenic. High adsorption capacity allows it to precipitate and firmly hold cement grains on its surface. Reinforcing ability determines the high mechanical bending and tensile strength of the hardened material. High alkali resistance of such fibers is also advantage.

This paper presents the results of the development of cement-fiber composite material based on cellulosic fibers for the manufacture of pressed autoclaved fiber-cement boards, which, depending on the functional purpose, can be used for cladding the outer walls of buildings, manufacturing suspended ceilings, sandwich panels, interior partitions, window sills slabs, as removable and fixed formwork, etc.

The purpose of the research was to determine the optimal composition of the cement-fibrous composite material for the manufacture of fiber-cement boards. The influence of the cement-sand matrix ratio (C/S), the specific surface area of the sand (SSA) and the content of cellulosic fibers on the physical and technical properties of the material were studied.

2 Materials and Methods

The composites of different composition were produced in thin plates, measuring 220x100x7 mm. The following materials were used: Portland cement CEM II / A-K (SH-P) 32.5H provided by Ulyanovsk Cement Plant, quartz sand from the Kama deposit SSA = 192, 210, 310, 410 m^2/kg, pulp of coniferous sulphate unbleached Solombalsky pulp and paper plant (GOST 11 208- 82).

Preparation of fiber-cement mixture was carried out in the following order: disintegration of cellulose fibers according to GOST 14363.4-89; after the dissolution was completed, the fiber was filtered through a filter screen; mixing fiber with cement and sand to form fiber-cement mass (FCM) with a concentration in water of 20-25%. After that, FCM was filtered.

Then FCM specimens were put in the mold and pressed according to the following regime: 10 min – pressure rise up to 600 N/cm^2, at the speed 10 kgf/s, 10 min – curing under pressure.

After curing for 1 hour, the obtained specimens were exposed to steam treatment in the steam chamber (according to the regime: 2 hours - temperature rise to 65 °C, isothermal curing for 4 hours, 2 hours - cooling), and then autoclaved treatment (according to 3 h - rise to a temperature of 175-180 °C, 6 h - isothermal curing, 3 h - cooling).

The tests were carried out according to GOST 8747-88. According to the results of experimental studies, a mathematical dependence of the flexural strength of the fiber-cement boards on the cement-sand ratio of the matrix was obtained, which is expressed by a second-degree polynomial:

$$Rg = -4.05x2 + 15.25x-1 \qquad (1)$$

under which maximum values are achieved at C/S = 1/1.6-1/2.1.

Hardening of autoclave materials is carried out as a result of chemical reactions between the components of the binder and water under conditions of high pressure and temperature. The main chemical process in autoclave curing is the interaction between calcium hydroxide, silica and water, accompanied by the formation of calcium hydrosilicates, which bond unreacted grains into artificial conglomerates. The reaction rate and strength of conglomerates increase as dispersion of raw materials increase. In this regard, determination of the optimal value of the fineness of the quartz sand used in the manufacture of fiber-cement board samples is of scientific interest. For this purpose, the tests were carried out on tiles samples with different specific surface area of the sand and cement-sand ratio 1: 1, 1: 1.5, 1: 2, 1: 2.5, 1: 3.

3 Results and Discussion

Dependence of the flexural strength of fiber-cement board samples on SSA of the quartz sand is shown in Figure 1.

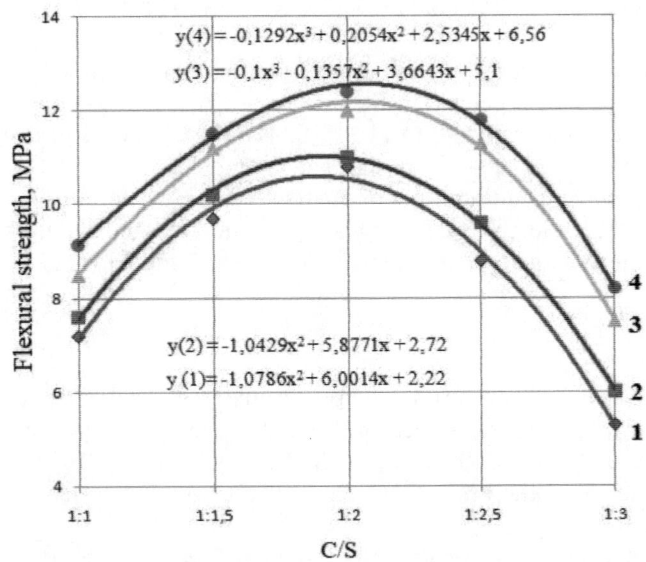

Figure 1 – The dependence of the flexural strength of the fiber-cement board samples on the C/S of the matrix with a constant fiber content of 5%, different dispersity of quartz sand (SSA): 1 – 192 m^2/kg, 2-210 m^2/kg, 3-310 m^2/kg, 4-410 m^2/kg.

As can be seen from Figure 1, the maximum parameters are achieved when C/S=1:2 and SSA is not less than 310 m^2/kg.

The effect of cellulosic fiber content at various values of the fineness of quartz sand on the flexural strength of the fiber-cement board samples are shown in Figure 2.

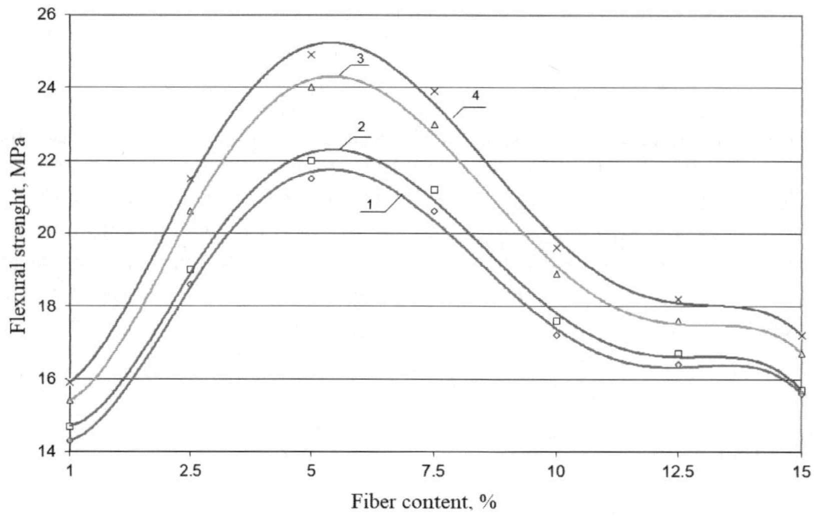

Figure 2 – The dependence of the flexural strength of fiber cement boards from fiber content: 1 – SSA=190 m²/kg, 2 – SSA =210 m²/kg, 3 – SSA =310 m²/kg, 4 – SSA =410 m²/kg.

Experimental studies have shown that maximum flexural strengths are achieved when fiber-cement board samples reinforced with cellulosic fibers of 5-7.5% with C/S= 1:2. While specific surface area of quartz sand increases from 190 to 210, 310, 410 m^2/kg, the flexural strength increases by 3, 12 and 16%, respectively.

The best indicator of the specific surface area of quartz sand is a 310 m^2/kg. As follows from Figure 2, a further increase of the specific surface area of the quartz sand slightly increases the flexural strength, but significantly increases the energy costs due to grinding time.

Conclusions

The optimal composition of cement-fibrous composite material for the manufacture of fiber-cement autoclaved boards with high physical, mechanical and economic parameters was obtained.

The obtained boards have a flexural strength 24.2 MPa, they reinforced with cellulosic fibers 5-7.5% with C/S= 1:2. In this case, the optimal indicator of the specific surface area of quartz sand is a value equal to 310 m_2/kg.

The obtained fiber-cement boards are weatherproof and frost-resistant, do not ignite and do not spread fire, have a low coefficient of thermal expansion, shock-resistant, moisture-proof, durable, environmentally safe, resistant to aggressive environment. They are much lighter than porcelain tiles; therefore, they can be successfully used for cladding of buildings with a limited load on the foundation.

Acknowledgements

The authors were supported by grant offered by the President of the Russian Federation for state support of young Russian researchers-candidates of sciences (grant number MK-2509.2018.8).

References

[1] Muhametrahimov R Kh, Galautdinov A R, Lukmanova L V, Khuzin A F, Lamberov A A and Egorova S R 2017 *MATEC Web of Conferences* 04005

[2] Muhametrahimov R Kh and Izotov V S 2014 *Stroitelnyye materialy* **1-2** 116
[3] Muhametrahimov R Kh 2016 *Izvestiya KGASU* **2** **(36)** 241-46
[4] Muhametrahimov R Kh, Galautdinov A R and Lukmanova L V 2018 *Izvestiya KGASU* **3** **(45)** 210-19
[5] Lunin E M and Baranov I M 2004 *Stroitelnyye materialy* **7** 16-17
[6] Radaikin O V, Sabitov L S, Kashapov N F and Gilmanshin I R 2018 *IOP Conference Series: Materials Science and Engineering* 012057
[7] Radaikin O V, Sabitov L S, Kashapov N F and Gilmanshin I R 2018 *IOP Conference Series: Materials Science and Engineering* 012058
[8] Volochek I Z, Khaldey T V, Ivanova V V and Chumadurova L I 1992 *Stroitelnyye materialy* **8** 25-26
[9] Izotov V S, Muhametrahimov R Kh and Sabitov L S 2010 *Stroitelstvo i arkhitektura* **1** **(17)** 119-25
[10] Solovyeva T A, Akchurin T K and Pushkarskaya O Y 2013 *Vestnik Volgogradskogo gosudarstvennogo arkhitekturno-stroitelnogo universiteta* **30** **(49)** 197-01
[11] Tonoli G H D, Rodrigues Filho U P, Savastano Jr H, Bras J, Belgacem M N and Rocco Lahr F A 2009 *Composites Part A: Applied Science and* Manufacturing **40** **(12)** 2046-53
[12] Starovoytova I A, Zykova Y S, Suleymanov A M, Semenov A N and Mishurova M V 2016 *Izvestiya KGASU* **3** **(37)** 217-24
[13] Ardanuy M, Claramunt J, Romildo Dias Toledo Filho 2015 *Construction and Building Materials* **79** 115-28
[14] Viviane da Costa Correia, Sergio Francisco Santos, Gonzalo Mármol, Antonio Aprigio da Silva Curvelo and Holmer Savastano Jr 2014 *Construction and Building Materials* **72** 65-71
[15] Fernando Pacheco-Torgal and Said Jalali 2011 *Construction and Building* Materials **25** 575-81
[16] Xiaoli Xie, Zuowan Zhou, Man Jiang, Xiaoling Xu, Zeyong Wang and David Hui 2015 *Composites Part B* **78** 153-61
[17] Grigoryeva L S, Rabey M B, Suleyman O V and Fisher I M 1992 *Stroitelnyye materialy* **10** 25-26
[18] OAO Latonit 1997 *Luchshiye fasady* **14** 60-61
[19] Rabinovich F N 2004 *Kompozity na osnove dispersno armirovannykh betonov. Voprosy teorii i proyektirovaniya, tekhnologiya, konstruktsii* (Moscow: ASB) p 560

Institute of Physics
Dirac House, Temple Back
Bristol BS1 6BE UK

ISSN: 1757-8981
ISBN 978-1-5108-9447-1

9 781510 894471

CURRAN ASSOCIATES INC.
proceedings.com